Nuclear Engineering

Mathematical Modeling and Simulation

Nuclear Engineering
Mathematical Modeling and Simulation

Zafar Ullah Koreshi
Air University, Islamabad,
Pakistan

ELSEVIER
ACADEMIC PRESS
An imprint of Elsevier

Academic Press is an imprint of Elsevier
125 London Wall, London EC2Y 5AS, United Kingdom
525 B Street, Suite 1650, San Diego, CA 92101, United States
50 Hampshire Street, 5th Floor, Cambridge, MA 02139, United States
The Boulevard, Langford Lane, Kidlington, Oxford OX5 1GB, United Kingdom

Notices
Knowledge and best practice in this field are constantly changing. As new research and experience broaden our understanding, changes in research methods, professional practices, or medical treatment may become necessary.

Practitioners and researchers must always rely on their own experience and knowledge in evaluating and using any information, methods, compounds, or experiments described herein. In using such information or methods they should be mindful of their own safety and the safety of others, including parties for whom they have a professional responsibility.

To the fullest extent of the law, neither the Publisher nor the authors, contributors, or editors, assume any liability for any injury and/or damage to persons or property as a matter of products liability, negligence or otherwise, or from any use or operation of any methods, products, instructions, or ideas contained in the material herein.

ISBN: 978-0-323-90618-0

For Information on all Academic Press publications
visit our website at https://www.elsevier.com/books-and-journals

Publisher: Charlotte Cockle
Editorial Project Manager: Sara Valentino
Production Project Manager: Niranjan Bhaskaran
Cover Designer: Christian J. Bilbow

Typeset by MPS Limited, Chennai, India

Dedication

To

My Mother

Mahbano Begum

My Father

Ambassador Dr. Samiullah Mujahid Koreshi (PhD)

My Wife

Nasia

My Sons

Mekael, Saif, and Zain

Contents

About the author

Zafar Ullah Koreshi [BSc (Hons) Nuclear Engineering, Queen Mary College, University of London (UK); MS, Nuclear Engineering, University of Wisconsin, Madison (USA), PhD Nuclear Engineering, University of Cambridge] is professor at Air University, having contributed as Dean Faculty of Engineering and is currently the Dean of Graduate Studies at Air University. His experience includes positions at the Pakistan Atomic Energy Commission, Dr. AQ Khan Research Laboratories, National University of Sciences and Technology, and at Air University, Islamabad. He has published in the *Annals of Nuclear Energy*, *Progress in Nuclear Energy*, *Nuclear Technology and Radiation Protection*, *ASME Journal of Nuclear Engineering and Radiation Science*, and in several other leading international journals. Dr. Koreshi has been Track Chair at the International Conference on Nuclear Engineering held in the United States, China, and Japan and has presented his research at the American Nuclear Society Annual Meetings. He is a member of the American Society of Mechanical Engineers (ASME), Professional Engineer Pakistan Engineering Council, and a life member of the Pakistan Nuclear Society. He has also received commendations as a reviewer for prestigious journals. Prof. Zafar Koreshi is also an associate editor of the *ASME Journal of Nuclear Engineering and Radiation Science*.

Foreword

This book has evolved from over four decades of involvement in nuclear engineering, industry, and academia beginning as an undergraduate at Queen Mary College, University of London, then at the University of Wisconsin, Madison (USA), followed by an engineering career, then at the University of Cambridge for my PhD. The last 35 years of my career have been at the Pakistan Atomic Energy Commission, at the National University of Sciences and Technology and the last two decades as professor at Air University.

Even though my teaching and administrative positions kept me busy, I always stayed close to my first academic passion: nuclear engineering, for which I was lucky to have learnt from professors Mike Williams and Charlie Maynard, Dr. Jeffery Lewins, and many other learned professors. To them, I owe all my knowledge and confidence developed as a student. Professor Williams took us students to our first conference in London when we had not the slightest clue as to what a conference was! From Charlie Maynard I learned not only from his lectures but from his personal values. From Jeffery Lewins, at Cambridge, I received encouragement and exposure to some of the finest opportunities to listen to great minds. When Jeffery Lewins was the president of the Institution of Nuclear Engineers, I met Dr. Otto Frisch and briefly spoke to him. He was one of the first three credited with the discovery and first experiments of nuclear fission! And of course I must mention my friendship with Prof. Abdus Salam, an accomplished Pakistani scientist and Nobel Laureate. In congratulating him at his office at Imperial College soon after his Nobel Prize was announced, I was elated and had many opportunities to accompany him on conferences and to receive his guidance and advice.

In writing this book, I feel that I have completed one duty—transferring whatever knowledge I gained in this field to the scientific community and especially to the young students studying nuclear engineering in universities all over the world.

This book is my way of presenting things to the body of knowledge that encompasses nuclear engineering. In it is the right proportion, I feel, of physics, mathematics, nuclear technology, diffusion theory, transport theory, Monte Carlo simulation, optimization, and applications and exercises that are essential to an academic nuclear engineer who will complete a PhD and continue in the profession.

The first two chapters of this book are the foundations, physics, and science of the atom (as known to an engineer); I think it is an appropriate amount of information to comprehend the alpha, beta, gamma, x- and neutron radiations that are modeled in nuclear engineering. A difficult concept, Doppler broadening, is explained in a relatively easy way! Chapter three is an overview of nuclear technology deliberating on nuclear reactors, their types and generations, the renaissance reactors, and their applications in marine propulsion, in rockets, and towards the exploration of deep space. This year three missions went to Mars, and we should all be getting some more pictures and information on what the red planet is like. This was a one-way trip so people could not be sent; in the future we can hope that nuclear rockets, with far more power than today's chemical fuels, will take and bring people to and from Mars. So, Chapter 3 gives a picture on the big systems developed in the nuclear world. I have put in nuclear fusion, both magnetic and inertial, as ITER could give humanity some hope and the next DEMO reactor could produce electricity from the technology of our Sun and from fuel made out of water!

Chapter 4 is on mathematical foundations, both deterministic and stochastic; again, in the right proportion! With this background, Chapter 5 shifts gears into modeling neutrons with simple diffusion theory which, in spite of its simplicity, is quite useful in the design of nuclear reactors and systems. This is followed by neutron transport, which is a step higher than diffusion, as it incorporates a full six-dimensional phase space formulation in Boltzmann's transport theory. This is where an advanced undergraduate student begins to take things a bit more seriously. And then, in Chapter 7, we step into the world of Monte Carlo (MC) simulation, where the power of *big computing* comes to our help in simulating processes to design nuclear reactors. Chapter 8 is a brief description of some well-known and extensively used computer codes in nuclear engineering.

With the foundations more-or-less clearly defined and understood, Chapter 9 deliberates on the mathematics of optimization and optimality as nuclear systems have not only to be designed but designed in the *best* way.

The next three chapters, Chapters 10–12, focus on MC simulation and compare MC with diffusion and transport to give the feel of a *unified* way of looking at the modeling of neutron transport and generalizing it to radiation transport. Chapter 13 reviews current research in the applications of optimization methods in nuclear engineering. Finally, Chapter 14 gives a little insight into MC simulations in medical physics; this is an area where humanity has benefitted so much from medical radio-isotopes to cure cancer and to undergo medical treatments and diagnostics.

In this journey, I have to acknowledge the support and encouragement of friends and colleagues, most notably Prof. Anil Prinja (QMC), Drs. Afzaal Malik, Tasneem Shah, Syed Arif Ahmad, and my colleagues at Air University and all the other organizations I have had the privilege to work. From my students, too, I have learned a great deal; from Dr Hamda Khan and Engr. Umair Aziz, and PG students and undergrads in all these years.

Even after all the effort I have put into this book, I must say, a lot remains and I hope this is just the first edition to be improved into newer editions.

I give this book to the next generations; this is my best effort and I hope that you will do much better. The caravan of knowledge moves slow but yes it moves.

In the end, the wealth of humans is the knowledge and values that we have and that has all come from our families, schools, and universities that we must value, love, and sustain. The custodians of that knowledge are the professors who have shaped this world and will take it to places that we cannot even imagine.

This book is addressed to the scientific community in radiation transport; as a text book, it would be suitable for a graduate level course spread over two semesters. This would prepare a PhD scholar to be "fully-equipped" to undertake research in *big simulations* for design optimization of present-day systems as well as for new systems, such as micro-nuclear power reactors for space exploration and possibly power generation on a planet such as Mars.

It is necessary to understand that while a great amount of subject matter is covered in this book, several gaps and shortcomings may appear. It is, in that sense, just a small contribution to the vast body of knowledge which has existed in this area from which several generations have benefitted.

I would like to thank Elsevier Publishers, particularly Ms. Maria Convey, Acquisitions Editor, Ms. Sara Valentino, Editorial Manager, Ms. Madeline Jones and Mr Niranjan Bhaskaran, Elsevier, for their continued support and encouragement.

Finally, my thanks to my mother and father for giving me the wealth of values, support, and opportunities, to my wife Nasia and to my sons Mekael, Saif, and Zain for their constant support and encouragement.

Zafar Koreshi

Air University

Islamabad, Pakistan

Chapter 1

The atom and nuclear radiation

This chapter comprises what can be considered as the essentials of nuclear engineering as a discipline that owes much of its present knowledge to theoretical science, discoveries, inventions, and technology.

Nuclear systems are based on the science of atomic and subatomic particles, namely neutrons, protons, electrons and the interaction of nuclear radiation (gamma and beta) and atomic radiation (X-rays) with matter.

These particles and radiations were discovered over many decades. Pioneering milestones include the discovery of X-rays by Wilhelm Conrad Röntgen at the University of Würzburg in 1895, spontaneous radioactivity by Antoine Henri Becquerel (University of Paris) in 1896, radioactivity by Pierre Curie and Marie Curie leading to the discovery of polonium and radium in 1898 (University of Paris). In 1897, Sir Joseph John Thomson, at the University of Cambridge discovered the electron. In 1919, one of his students Ernest Rutherford discovered the proton (Blackett, 1924; Rutherford, 1919) demonstrating transmutation: ${}^{14}_{7}N + {}^{4}_{2}He \rightarrow {}^{17}_{8}O + {}^{1}_{1}H$ and worked on radioactivity and named alpha and beta radiation. Another breakthrough was the reaction proposed as: ${}^{9}_{4}Be + {}^{4}_{2}He \rightarrow {}^{13}_{6}C +$ radiation, by Bothe and Becker and by Mme. Curie but with brilliant arguments, theory and experimentation shown to be a neutron by Chadwick (1932). In this paper Chadwick writes "If we suppose that the radiation is not a quantum radiation, but consists of particles of mass very nearly equal to that of the proton, all the difficulties connected with the collisions disappear, both with regard to their frequency and to the energy transfer to different masses." Chadwick goes on to say, "We may then proceed to build up nuclei out of a-particles, neutrons and protons, and we are able to avoid the presence of uncombined electrons in a nucleus" and concluding, among other properties of the proposed neutron that "...the mass of the neutron is about 1.006, just a little less than the sum of the masses of a proton and an electron." Thus Chadwick's correct interpretation of these experiments gave us the neutron. Clearly, Chadwick had no knowledge of quarks and thought that the neutron was an elementary particle. This chapter is on these particles and the nuclear and atomic radiations as well as on the atomic models for which quantum physics emerged as a field which explained phenomena based on particle-wave duality at very small dimensions.

In quantum physics, the "creators" of the field include Max Planck for the concept of the quanta, de Broglie for giving the de Broglie wavelength of matter waves, Bohr for the structure of the atom, Compton for demonstrating that electromagnetic radiation behaves as particles during their interaction with electrons (Compton scattering), Heisenberg for the uncertainty principle, Schrodinger for the wave function in the Schrodinger equation, Dirac for several developments in the field and significantly for quantum electrodynamics, and Pauli for the Pauli exclusion principle. Toward the photoelectric effect, Einstein's contribution demonstrated the absorption of an incident photon leading to the ejection of an orbital electron as one based on discrete values of energy rather than merely increasing the intensity of a lower frequency photon.

The knowledge on science is far from complete; the challenges that remain and are being pursued are the understanding of elementary particles and the search for a grand unification theory.

1.1 The atom

The Bohr-Rutherford model, proposed in 1913, resembles our solar system with planets going around the sun. This model was useful in explaining several observations but had serious deficiencies; one of them was that the release of electromagnetic radiation by the orbiting electrons would lead to a spiraling-in movement and an ultimate collapse of the atom. Thus improvements had to be made and three postulates were proposed regarding electron orbits, discretized angular momentum, and the ability of electrons to move from one orbit to another by discretized energy exchange.

With all the model contributions and research into new directions for the search for the smallest "indivisible" particles we come to the present knowledge of the atom in which we understand that all atomic and subatomic particles are built from elementary particles which are quarks and leptons.

Nuclear Engineering. DOI: https://doi.org/10.1016/B978-0-323-90618-0.00001-6

Thus an atom has a dense nucleus at its center, as shown in Fig. 1.1, consisting of neutrons with rest-mass m_n and protons with rest-mass m_p surrounded by a "cloud" of electrons with rest-mass m_e (Table 1.A1, Fundamental Physical Constants). An atom has an atomic number Z which is the number of protons or electrons, and a mass number A which is the number of protons and neutrons; thus $A = N + Z$. It is written as ZAX. The masses of atomic and subatomic particles are so small that they are expressed in terms of the atomic mass unit 1 u is defined as 1/12th the mass of an unbound carbon C^{12} atom (1 amu $= 1.66054 \times 10^{-27}$ kg). Neutrons and protons are made from quarks, which are indivisible elementary matter particles assigned properties of up, charm, top, down, strange and bottom. A neutron, as shown in Fig. 1.1, is two "up" quarks and 1 "down" quark while a proton is two "down" quarks and one "up" quark. The electron is itself an elementary particle in the family of leptons which consists of six particles. Of the four forces of nature: electromagnetic, strong, weak and gravitation, the nucleus is held together by the strong force carried by gluons. A more detailed insight into elementary particles would fall beyond the scope of this book and the reader can consult books and papers in particle physics (Kibble, 2015; Peskin, 2019). The neutron is stable as a bound particle within the nucleus but unstable as a free neutron decaying to a proton, electron and antineutrino with a mean lifetime of 885.7 ± 0.8 s.

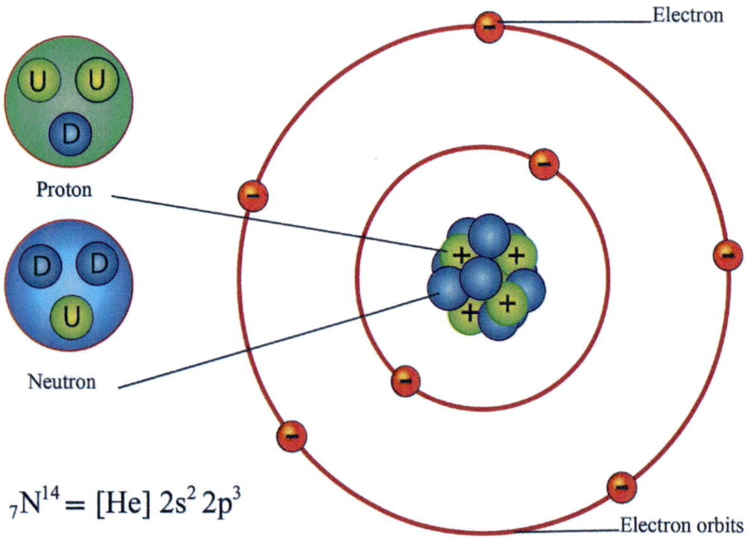

FIGURE 1.1 Bohr's model of an atom.

TABLE 1.A1 Fundamental physical constants.

Avogadro's number	N_{av}	6.022141×10^{-23}
Planck constant	h	$6.62609004 \times 10^{-34}$ m^2/kg/s
Electron charge	e	$1.602176565 \times 10^{-19}$C
Atomic mass unit	amu	1.66054×10^{-27} kg
Mass of neutron	m_n	1.67493×10^{-27} kg
		1.0086649156 amu
		939.550 MeV
Mass of proton	m_p	1.67262×10^{-27} kg
		1.0072764663 amu
Mass of electron	m_e	9.10938×10^{-31} kg
		5.485797×10^{-4} kg
Bohr radius	r_B	5.291772×10^{-11} m
Classical electron radius	r_e	2.817940×10^{-15}
Rydberg constant	R	10973731.568/m

Source: From https://physics.nist.gov/cgi-bin/cuu/Category?view = html&Atomic + and + nuclear.x = 85&Atomic + and + nuclear.y = 11

A measure of the size of the atom was obtained by Bohr from the radius of an electron orbit in the hydrogen atom from (1) the discretized angular momentum, and (2) the coulomb and centripetal forces

$$L = m_e v r = n\frac{h}{2\pi}$$

$$k\frac{Zq_e^2}{r_n} = \frac{m_e v_e^2}{r_n}$$

for which the radius of the nth shell is

$$r_n = \frac{n^2}{Z} r_B$$

where the Bohr radius is

$$r_B = \frac{h^2}{4\pi^2 m_e k q_e^2} = 0.52918 \quad 10^{-10} \quad \text{m} = 0.52918\,\text{Å}$$

in units of an Angstrom (1 Å = 10^{-10} m) used for an atomic dimension. Similarly the unit of fermi (1 fm = 10^{-15} m) is used for expressing the size of a nucleus. The radius of a nucleus, defined as the distance from its center to where its density falls to a half of its value from the center, is $R = R_0 A^{1/3}$ where $R_0 = 1.2$ fm.

For a hydrogen atom ($Z = 1$, $n = 1$) the Bohr radius is thus an estimate of the atomic radius. The energy of the shells from similar arguments is

$$E_n = -\frac{Z^2 e^4 m_0}{2\hbar^2 n^2}$$

The dual idea of particle and wave, and the nature of light is "connected" through the de Broglie wavelength

$$\lambda = \frac{h}{p}$$

where h is Planck's constant and p is the momentum given by $p = mv$.

Exercise 1:

1. From the definition of Avogadro's number, estimate the diameter of a carbon atom (density of carbon $\rho = 2.26$ g/ cm^3, A = 12 g (g-atom)$^{-1}$ and "packing fraction ε" ($0 < \varepsilon < 1$).
2. Calculate the radius of a carbon nucleus and compare the dimensions of the nucleus and the atom. Can the nucleus be considered to be a "point" nucleus as Rutherford did in his famous gold foil experiment?

In the Bohr model, there are four quantum numbers assigned to each electron: the principal quantum number n, the azimuthal quantum number $l = 0, 1, 2, \cdots n-1$, the magnetic quantum number $m_l = -l, -l+1 \cdots 0, l-1, l$ thus a total of $2l + 1$ states, and the spin quantum number $m_s = -1, /2, 1, /2$. These numbers refer to the size of the orbit ($n = 1$ is the innermost orbit, the next is $n = 2$ and so on) and its shape ($l = 0, 1, 2, 3, \ldots$ called the *s, p, d, f*,... type have spherical, dumbbell or complex shapes and are related to the energy of the electron.)

At the level of an atom, the equivalent Newton's second law of motion is the Schrodinger wave equation which gives the wave function $\Psi(x, t)$ for an electron when the quantum numbers i.e. energy descriptors are specified. The Schrodinger wave equation for nonrelativistic particle is

$$i\hbar\frac{\partial}{\partial t}[\Psi(x,t)] = \left[-\frac{\hbar^2}{2m}\frac{\partial^2}{\partial x^2} + V(x,t)\right]\Psi(x,t)$$

with classical and quantum "connections"

$$E \leftrightarrow i\hbar\frac{\partial}{\partial t}, P \leftrightarrow -i\hbar\nabla$$

The Schrodinger equation can be solved for a hydrogen atom, to show that the solution is a function of the radial position r, orthogonal and azimuthal angles θ, φ and the three quantum numbers n, l, m. The closest shell, called the K

shell, has $n = 1$ shell, for which the maximum number of electrons permissible, $2n^2$, is two electrons. Generally, the filling order for shells is 1s, 2s, 2p, 3s, 3p, 4s, 3d, 4p, 5s, ... The Pauli exclusion principle states that electrons that occupy the same orbital must have different spins, that is, $(2l+1)$ electrons of one spin can fill an orbit while $(2l+1)$ electrons of another, so that $2(2l+1)$ electrons can fill an orbit; thus for $l = 1$, 6 electrons can fill an orbit for $n = 2$. The following cases illustrate the filling order: hydrogen (Z = 1) = $1s^1$, helium (Z = 2) = $1s^2$, lithium (Z = 3) = $1s^2\ 2s^1$ = [He] $2s^1$, and calcium (Z = 20) = $2s^2\ 2p^6\ 3s^2\ 3p^6\ 4s^2$. Uranium is an example where the total angular momentum filling-in rule gives for (Z = 92)$1s\ ^2$ $(2s^2\ 2p^6)$ $(3s^2\ 3p^6\ 4s^2\ 3d^{10})$ $(4p^6\ 5s^2$ 4d 5p 6s 4f 5d 6p $5p^6\ 4f^{14}\ 5d^{10}\ 6s^2\ 6p^6\ 5f^3\ 6d^1\ 7s^2)$ = 2,8,18,32,21,9,2 electrons.

Electrons can travel fast, close to the speed of light, for which their kinetic energy (KE) is related to their momentum through the relativistic relation

$$\mathrm{KE_r} = \sqrt{(mc^2)^2 + p^2c^2} - mc^2$$

for which the wavelength is

$$\lambda = \frac{2\pi\hbar c}{pc} = \frac{2\pi\hbar c}{\sqrt{\mathrm{KE_r}^2 + 2\mathrm{KE_r}(mc^2)}}$$

For a hydrogen atom, the MATLAB program below computes the energies and velocities of orbital electrons in the ground state ($n = 1$) and two excited states ($n = 2, 3$).

MATLAB program 1: .

```
Z=1; h=6.62609004e-34; % Planck const m^2 kg/s
m=9.109e-31; % kg electron mass
e=1.6e-19;% charge on an electron C
k=9e9; % conversion Nm^2/C^2
c=3e8; % speed of light m/s
hbar=h/(2*pi);
R=Z^2*k^2*e^4*m/(2*hbar^2);
n=1:3; % three shells
EJ=-R./n.^2; EMeV=(1e19/1.6)*EJ; % energy in J and MeV of each shell
vel=sqrt(-2*EJ/m); velOverc=vel/c% vel of electrons in shell
lambdaAng=h./((1e-10)*m*vel);% lamda in Angstrom
Results
R= 2.1740e-18 J EJ = 1.0e-17 *[-0.2174    -0.0543    -0.0242] J
vel = 1.0e+06 *[ 2.1848    1.0924    0.7283] m/s
lambdaAng = [ 3.3295    6.6590    9.9886] Å
velOverc =      [0.0073    0.0036    0.0024]
```

The energies of the electrons in the first three states are −13.6, −3.39, and −1.51 eV, respectively. The lowest (ground) state has the largest negative value while the highest excited state would have the lowest value and a free electron would have a zero binding energy (BE). In these three states, the electrons have speeds less than 1% of the speed of light and thus can be considered as nonrelativistic.

Example 1: Use both nonrelativistic and relativistic expressions to calculate the KE of an electron ($m_e = 9.11 \times 10^{-31}$ kg) with speed $0.01c$ and $0.9c$. The classical $\mathrm{KE_c}$ and relativistic $\mathrm{KE_r}$ KE expressions are

$$\mathrm{KE_c} = \frac{1}{2}mv^2$$

and

$$\mathrm{KE_r} = (\gamma - 1)mc^2$$

where

$$\gamma = \frac{1}{\sqrt{1-(v/c)^2}}$$

v	γ	$KE_c(J)$	$KE_r(J)$	KE_c/KE_r
$0.01c$	1.0001	4.0995×10^{-18}	4.0998×10^{-18}	0.9999
$0.9c$	2.2942	3.3206×10^{-14}	1.0611×10^{-13}	0.3129

Clearly, the relativistic kinetic energy KE_r becomes larger than the classical kinetic energy KE_c and in the limit $v \to c, KE_r \to \infty$.

Nuclei, like atoms, have excited states at discrete levels and decay by emitting alpha, beta or gamma radiation. The emitted radiation subsequently interacts with matter in a number of ways described below. Just as atoms take part in atomic reactions (where the electrons participate), nuclei take part in nuclear reactions which will be discussed in detail in Chapter 2. In nuclear engineering, two energy-producing reactions are fission and fusion. In nuclear fission energy is released by the breaking-up of heavy nuclei such as uranium-235 and plutonium-239 into lighter nuclei. In nuclear fusion, the reaction of the sun and the stars, energy is released by the "fusing" or joining of light nuclei such as isotopes of hydrogen, into heavier nuclei such as helium. Fission is the basis of the present nuclear fission power reactors while nuclear fusion has yet to be achieved. The biggest fusion experiment is the 35-nation International Tokamak Experimental Reactor (ITER) located in France, which is expected to begin operation in 2025.

1.1.1 Nuclear stability

Inside the nucleus, the two competing forces that determine stability are the attractive nuclear force between neutrons and protons and the repulsive electric force between protons. For Z > ~20, $N/P > 1$ as seen in Fig. 1.2 and thus the nuclei are "above" the straight line $N = Z$. Neutron-rich nuclides get rid of their excess neutrons by emitting beta particles and at the very high end, for high A, nuclei emit heavier alpha particles. Nuclei with "magic numbers" for Z or (A-Z): 2, 8, 20, 28, 50, 82, 126 are strongly bound, more than their neighbors. Doubly magic numbers for $A = 4$, 16, 40,. . ., are bound even stronger than those with magic numbers.

In nature 176 of the 286 (over 60%) primordial stable nuclides have even mass number; 146 stable nuclides are even-even Z, A others are mixed while very few (~3%) are odd-odd.

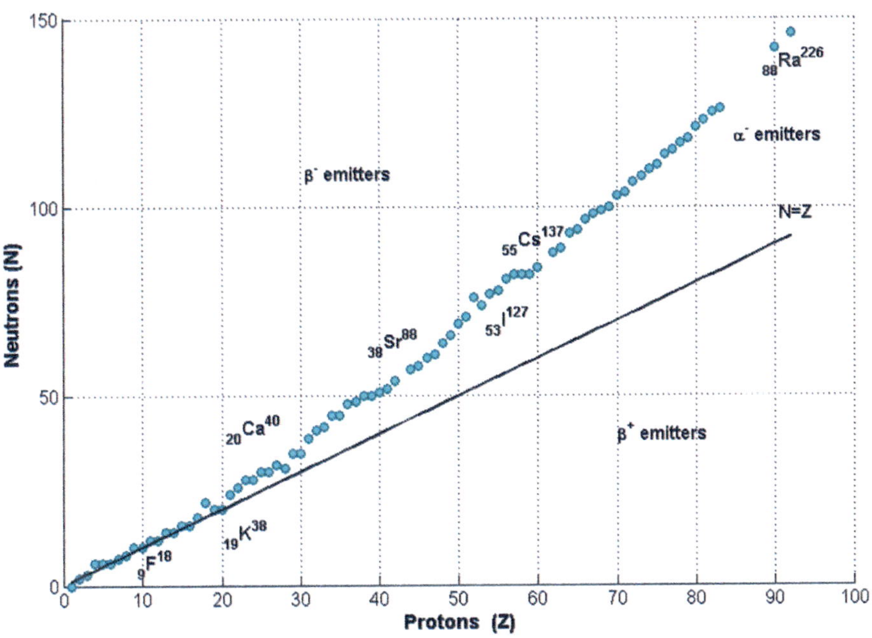

FIGURE 1.2 Nuclear stability (Z-N).

1.1.2 Binding energy

The nucleus is held together with a (negative) BE in which both pair up with opposite spins. This BE is the energy equivalent of the difference, called the "mass defect Δm" between the mass-energy of constituent nucleons $(Zm_p + Nm_n)c^2$and the nuclear mass-energy $M_n c^2$; thus by Einstein's equivalence $BE = \Delta mc^2$.

For nuclides with A > 20, the BE can be calculated by semiempirical "liquid drop model"

$$BE(A, Z) = a_V A - a_S A^{\frac{2}{3}} - a_C \frac{Z^2}{A^{1/3}} - a_A \frac{(Z-N)^2}{A} + \frac{(-1)^Z + (-1)^N}{2} \frac{a_P}{A^{1/2}}$$

where $a_V = 15.56\text{MeV}$, $a_S = 17.23\text{MeV}$, $a_C = 0.697\text{MeV}$, $a_A = 23.285\text{MeV}$, $a_P = 12.0\ 0$. The first term, proportional to the mass number A, represents the volume terms with a nucleon modeled as interacting only with its neighbors rather than with all other nucleons. The following three negative terms reduce the BE. The surface reduction is proportional to the surface area $\sim r^2$, or $A^{2/3}$. The next reduction term is due to coulombic repulsion between protons. The fourth term is a reduction due to the Pauli exclusion principle which allows two protons or two neutrons of opposite spin in each energy level. The last "pairing" term represents an experimentally observed phenomenon inversely proportional to the square root of the atomic mass number. Binding energies re plotted in Fig. 1.3 which shows low values at the left end where nuclear fusion, the reaction of the sun and the stars takes place. Both these reactions make the system "move" toward an increase in the BE per nucleon and are accompanied by the release of energy. This is followed by an increase up to 8.7542 MeV/nucleon for chromium, manganese, iron (8.8 MeV), cobalt, nickel, copper and zinc, then gradually falling to 7.5855 MeV/nucleon for uranium-238.

A nuclear reaction has Q value, defined as the difference of the rest-mass energies of the reactants r_1, r_2 and products p_1, p_2 $Q = [m_{r1} + m_{r2} - (m_{p1} + m_{p2})]c^2 = T_{p1} + T_{p2} - (T_{r1} + T_{r2}) = [(BE_{p1} + BE_{p2}) - (BE_{r1} + BE_{r2})]$ KE T is exothermal or exoergic if Q > 0 and with endothermic or endoergic if Q < 0.

Exercise 2: From the BE formula above, explain why the elements with odd Z are expected to be less abundant in nature?

1.2 Radioactive decay

When an unstable radionuclide spontaneously emits energy, it is said to be radioactive to an extent measured by the number of its disintegrations per second with 1 Becquerel (Bq) representing 1 disintegration per second. A commonly used unit is the curie with 1 curie (Ci) denoting 37 billion Bq.

Decay data of over 3000 radionuclides are maintained by Brookhaven National Nuclear Data Services and published monthly in *Nuclear Data Sheets* (Elsevier) which are produced mainly from the Evaluated Nuclear Structure Data File (ENSDF) maintained by the US National Nuclear Data File Center http://www.nndc.bnl.gov.

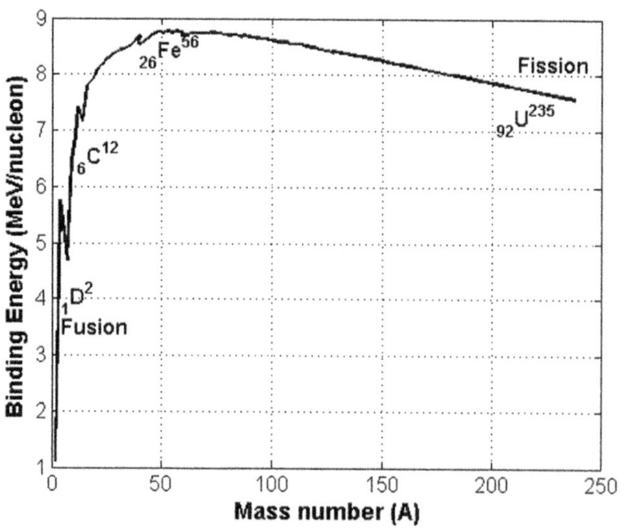

FIGURE 1.3 Binding energy per nucleon from the liquid drop model.

Radioactivity arises in nuclear reactors due to fission reactions resulting in fission fragments such as cesium and iodine. The number of such radioactive atoms can become significant, for example, about 6 atoms of cesium-137 are produced per 100 fission events. Considering that one fission reaction produces 3.20×10^{-11} J and that a 3000 MW (th) reactor would be undergoing $\sim 10^{20}$ fissions per second, the production of cesium-137 would be $\sim 6 \times 10^{18}$ atoms per second, or $\sim 2 \times 10^{26}$ atoms per full operating year. Later in this chapter, we will see how to estimate this quantity of atoms into mass from the Avogadro number.

Radioactive decay results in a "decay chain" where further nuclides are produced. Consider the decay of iodine-135 atoms due to its decay constant λ_I

$$\frac{d}{dt}I(t) = -\lambda_I I(t)$$

At any instant, the number of iodine-135 atoms is then

$$I(t) = I(0)e^{-\lambda_I t}$$

By the definition of half-life $t_{1/2}$ is the time for half the atoms from the beginning to have remained; thus $I(t_{1/2}) = I(0)/2$ so that

$$\frac{1}{2} = e^{-\lambda_I t_{1/2}}$$

yielding a relationship between the decay constant and half-life:

$$\lambda t_{1/2} = \ln 2 = 0.693$$

The activity of a radionuclide (disintegrations per second) is then $A = \lambda N$ Bq.

Iodine-131, for example, decays to xenon-131 through two reactions, namely

$$^{131}_{53}I \overset{8.02\mathrm{d}}{\rightarrow} {}^{131}_{54}\mathrm{Xe}* + \beta + \overline{\nu}_e$$

which is the predominant reaction, and

$$^{131}_{54}\mathrm{Xe}* \overset{11.9\mathrm{d}}{\rightarrow} {}^{131}_{54}Xe(\text{stable}) + \gamma(0.39\% \text{ of reactions in } ^{131}_{53}\mathrm{I} \text{ decay})$$

In a radioactive chain, there is the progression of a nuclide, such as tellurium-135, shown below. This results in the production of iodine-135, which successively decays to xenon-135, cesium-135, and barium-135.

$$^{135}_{52}Te \rightarrow {}^{135}_{53}I \rightarrow {}^{135}_{54}Xe \rightarrow {}^{135}_{55}Cs \rightarrow {}^{135}_{56}Ba$$

Just as the iodine-135 decay rate equation above, we can write down the equations for the number of atoms of each of the radioisotopes in the tellurium decay chain to obtain their concentrations as a function of time.

We shall see the important consequences of this decay chain in Chapter 3 in the context of nuclear reactor control. For now, we note that iodine has a "source" term (production by tellurium) and a "loss" term (decay to xenon-135). This permits us to write an ordinary differential equation for the number of atoms of iodine-135 $I(t)$ in terms of the number of atoms of tellurium-135 $T(t)$as

$$\frac{d}{dt}I(t) = \lambda_T T(t) - \lambda_I I(t)$$

Similarly, for xenon and cesium,

$$\frac{d}{dt}X(t) = \lambda_I I(t) - \lambda_X X(t)$$

$$\frac{d}{dt}C(t) = \lambda_X X(t) - \lambda_C C(t)$$

The above first-order ordinary differential equations, also known as the Bateman equations, can be expressed in matrix form and solved as an initial value problem

$$\frac{d}{dt}\boldsymbol{B}(t) = \boldsymbol{\lambda}\boldsymbol{B}(t)$$

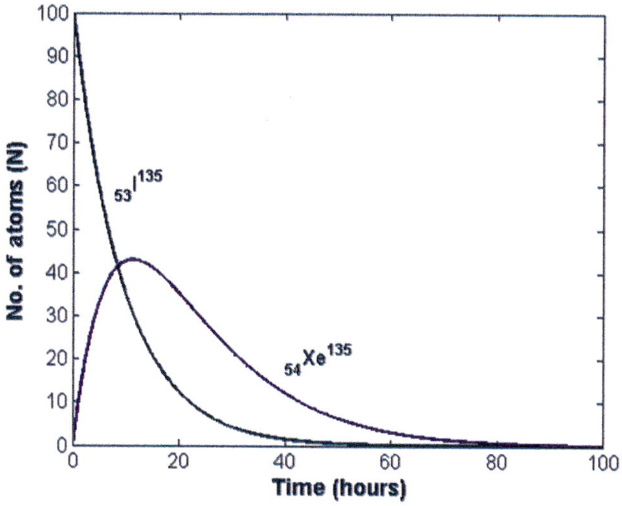

FIGURE 1.4 Radioactive decay of iodine-135 to xenon-135.

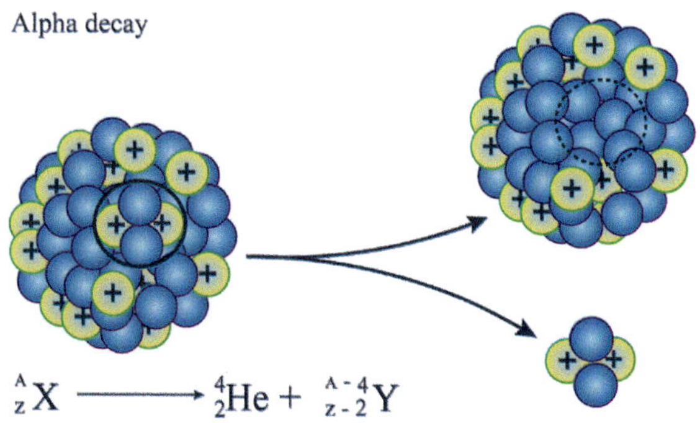

FIGURE 1.5 The process of spontaneous alpha decay.

The solution for $I(t)$ and $X(t)$ ignoring the initial concentration of $T(t)$ as well as the third equation for $C(t)$, an exact solution can be obtained by standard methods.

With initial conditions $I(0) = 100$ atoms and $X(0) = 0$, the solution, obtained with MATLAB (Annex 7) is shown in Fig. 1.4. Iodine-135 with a half-life of 6.6 h and xenon-135 with 9.2 h reaches a maximum xenon strength of 43.0370 atoms at 11.2 h. These decays and buildups are calculated in the event of a nuclear accident to determine the dynamics of radiation effects.

1.2.1 Alpha decay

The radioactive decay of heavy unstable nuclides shown in Fig. 1.2, with alpha particles, as shown in Fig. 1.5, results in the emission of a helium-4 nucleus and hence is associated with reduction of atomic number by 2 and of atomic mass by 2. Early work on radiation by Becquerel, Rutherford and Curie showed that radiation is emitted by atoms which change their form. The discovery of polonium and radium in 1898 from its presence in uranium ores, and the subsequent isolation of radium in metallic form by Marie and Pierre Curie in 1911 led to several experiments. Compounds of radium showed a faint blue light in the dark which was caused by the excitation, and subsequent de-excitation, of electrons in elements of the compounds caused by the spontaneous emission of radiation from radium. It was then discovered that the element has several isotopes which undergo alpha decay; for example, the isotopes Ra^{223}, Ra^{224} and Ra^{226} have a half-life 11.43 d, 3.66 d, and 1600 y each. The radium decay chain of Ra^{226} is

$$Ra^{226} \xrightarrow{\alpha, 1600\ y} Rn^{222} \xrightarrow{\alpha, 3.82d} Po^{218} \xrightarrow{\alpha, 3.05m} Pb^{214} \xrightarrow{\beta, 26.8\ m} Bi^{214} \xrightarrow{\beta, 19.7\ m} Po^{214} \xrightarrow{\alpha, 0.16\ ms} Pb^{210} \ldots$$
$$\xrightarrow{\beta, 22y} Bi^{210} \xrightarrow{\beta, 5d} Po^{210} \xrightarrow{\alpha, 138d} Pb^{206}(\text{stable})$$

decaying ultimately to Pb-206 which is stable. All radioactive chains end with a stable product. In this chain, radon-222 is an odorless radioactive decay present in earth as background radiation due to its transient production from uranium and thorium.

Other powerful alpha emitters include Polonium-210 emitting a 5.305 MeV alpha particle with the emission of one milligram equal to that from 5 g of radium and Plutonium-238 emitting a 5.456 MeV alpha. The alpha decay reactions of uranium-238 and plutonium-238 are

$${}^{238}_{92}\text{U} \rightarrow {}^{234}_{90}\text{Th} + {}^{4}_{2}\text{He}, Q_{\alpha} = 4.268\ \text{MeV}, T_{1/2} = 4.51 \times 10^{9}\ \text{y}$$

$${}^{238}_{94}\text{Pu} \rightarrow {}^{234}_{92}\text{U} + {}^{4}_{2}\text{He}, Q_{\alpha} = 5.593\ \text{MeV}, T_{1/2} = 87.7\ \text{y}$$

The energy of alpha particles ranges from ~ 1 to $\sim$ 11 MeV, although most alpha emitters produce alpha particles in the range 4–5 MeV. The number of alpha emissions is a function of the activity of an alpha emitter for which some commonly used sources are: Am-241, Pu-238, Pu-239, Po-210, Ra-226; and the portable sources, for example, Am-Be, Pu-Be, Ra-Be of 10–50 mCi. Alpha particles led to several new developments including an understanding of the structure of an atom in Rutherford's gold foil experiment and the discovery of the neutron by Chadwick by the alpha bombardment of beryllium in the reaction $49Be(\alpha, n)612C$.

1.2.2 Beta decay

In beta decay, a neutron-rich nuclide decays by the conversion of a neutron into a proton and an electron accompanied by the emission of an antineutrino. The antineutrino is an antiparticle (same mass but opposite electric charge and magnetic moment) of the neutrino. It is a lepton like an electron which is an elementary particle of spin ½. Unlike an electron the antineutrino has no electric charge. Antineutrinos cannot be easily detected as they pass the earth without any interactions. The beta decay of carbon-14 is

$${}^{14}_{6}C^{6} \rightarrow {}^{14}_{7}N^{7} + -10\beta + \nu, T_{1/2} = 5730\text{y}$$

Exercise 3: How is the beta decay of carbon-14 used in "carbon-dating" to estimate the age of fossils?

For excess of neutrons (beta β^{-} decay)

$$n \rightarrow p + \beta^{-} + \overline{\nu}$$

while excess of protons (positron β^{+} decay) leads to the emission of a positron, which is an antiparticle of an electron (Fig. 1.6)

$$p \rightarrow n + \beta^{+} + \nu$$

Reactions follow conservation laws for charge and lepton number conservation. The radioactive decays of iodine-131 and cesium-137, both powerful beta emitters with significant half-life and associated with cancer resulting from nuclear tests and explosions are shown in Figs. 1.7 and 1.8, respectively. Some positron emitting radionuclides with half-lives and mean energies are 611C(20.38 min, 0.385MeV), 713N(9.96 min, 0.491MeV), 917F, (64.5 s, 0.739MeV, 918F, (109.77 min, 0.2498MeV) and 1938K(7.636 min, 2.323MeV).

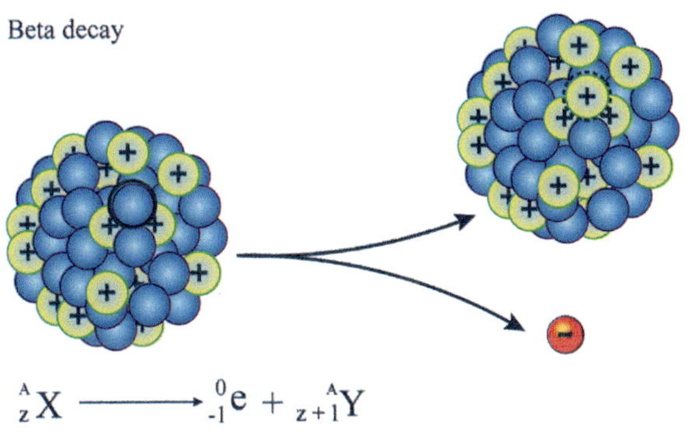

FIGURE 1.6 The process of spontaneous beta decay.

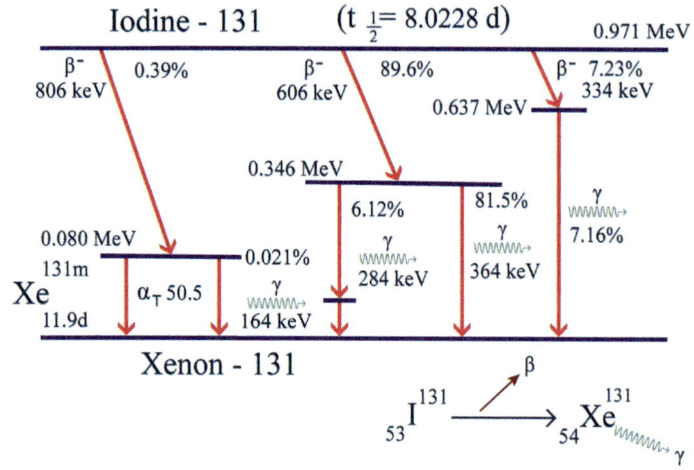

FIGURE 1.7 Radioactive decay of iodine-131.

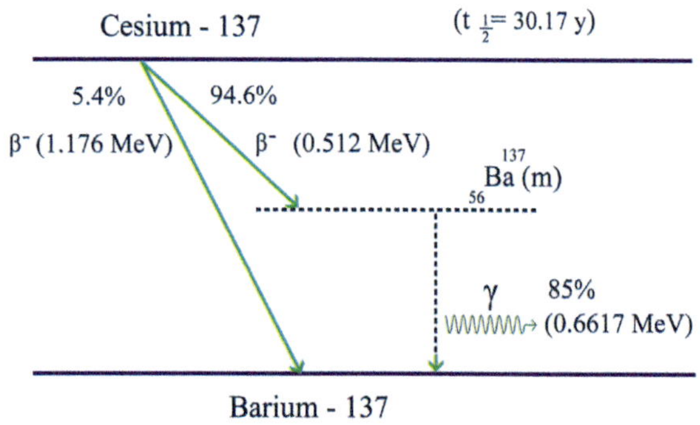

FIGURE 1.8 Radioactive decay of cesium-137.

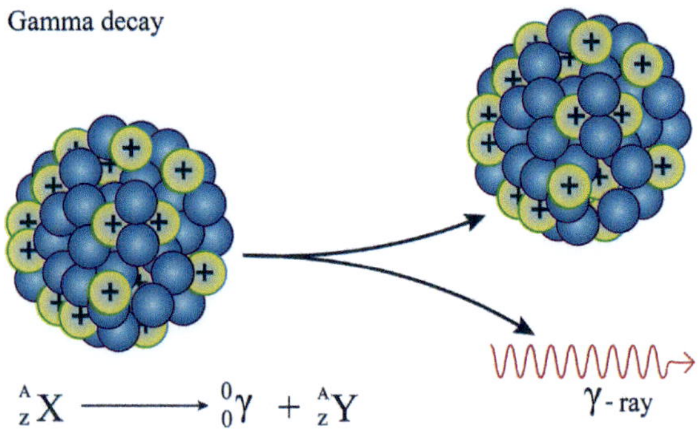

FIGURE 1.9 The process of spontaneous gamma decay.

1.2.3 Gamma decay

Gamma decay takes place as a result of the relaxation process of a nucleus in its excited state just as X-rays are emitted in the relaxation process of electrons. With discrete energy levels for nucleons, determined by quantum physics, each excited state is characterized by quantum numbers. The nuclear Shell Model describes the motion of each nucleon based on the forces it experiences from neighboring nucleons. Since the nucleus is tightly bound, the energy of gamma rays are much higher than that of X-rays, typically MeV compared with keV. A comprehensive list of intense ($>$ 1 keV) gamma rays, based on the ENSDF Brookhaven National Laboratory data, is given in Narita and Kitao (1992) (Fig. 1.9)

One of the strongest gamma rays produced has an energy of 7.11515 MeV (4.9% intensity) from the beta decay of 716*N* (Bielajew, 1990). One method of determining uranium enrichment is from the alpha decay of uranium-235 to thorium-231 which produces 4.3×10^4 gamma rays, of 185.7 keV, per gram uranium-235 per second.

It is important to note that energetic gamma rays can produce neutrons as in the photo-neutron (γ, n) reactions $49\text{Be}(\gamma, n)48Be$ and $12H(\gamma, n)11H$.

1.2.4 Radioactive nuclides in nuclear technologies

Nuclear power reactors: In nuclear power reactors, the main source of nuclides is the nuclear fission reaction which produces fission products as will be discussed in detail in the next chapter devoted to neutrons. These fission fragments give rise to the radiations discussed here. Some fission fragments are a problem in the operations of nuclear reactor; these are called poisons such as $53131Xe$, which can cause a reactor to shut down for some time. This results in a loss of power which is serious for reactors on ground but more so for propulsion reactors in submarines. In some cases, "poisons" such as burnable boron solutions are used to control the reactor from staying within permissible power bounds.

Radiation in nuclear weapons and accidents: In the early days of nuclear technology, weapons testing led to the release of radioactive nuclides in the atmosphere. In the 1945 nuclear explosions over Hiroshima and Nagasaki, casualties occurred due to the immediate blast and long-term radiation mainly from iodine and cesium isotopes. The iodine isotope $53131I$ is harmful and had been found over in the Chernobyl accident, one of the four nuclear reactors of the former Soviet RBMK-1000 Light Water Graphite (LWGR) were destroyed resulting in the release of $\sim$ 5% of radioactive core (Anspaugh, Catlin, & Goldman, 1988; World Nuclear Association, 2012) in the environment causing 28 fatalities, within weeks, due to acute radiation syndrome. In the Fukushima disaster, caused by the Tohoku 9.0 magnitude earthquake and tsunami, the radiation released was $\sim 18\ 10^{15}$ Bq of radioactive Cs-137 into the Pacific during the accident (Langlois, 2013). Water and soil samples nearby indicated the presence of strontium, iodine and cesium. The Japanese Red Cross reported that as of 2011, the casualties consisted of almost 16,000 people confirmed dead, almost 4000 missing and that approximately 90% of the deaths were due to drowning (INPO, 2011).

Exercise 4: Estimate the fraction of cesium and iodine concentrations from the Hiroshima and Nagasaki blasts remaining from 1945.

Space technologies: The exploration of space, starting from 1957 with increased interest in the 1960s followed by a long gap, has revived this year (2020) with three missions to Mars launched in July by China (Mallapaty, 2020), UAE (Gibney, 2020), and the United States (Witze, 2020). Traveling at about 20,000 kilometers per hour, they are expected to complete the 500 million kilometer journey to Mars in about seven months. In the US mission, the Perseverance Rover uses radioactive plutonium-238 for power. As described earlier, plutonium-238 is an alpha emitter but also emits beta and gamma radiations as well as neutrons from spontaneous fissions (Matlack & Metz, 1967). Thus for future missions, possibly with personnel in returnable rockets, radiation shielding would be required. Plutonium-238 generating 0.55 W/g (th) with a half-life of 88 years and strontium-90 giving 0.93 W/g (th) with a half- life of 28 years as well as others such as polonium-210, cesium-144 and curium-242 generate 141, 25, and 120 W/g (th) with lower half-lives of 0.378, 0.781, and 0.445 years, respectively, can all be used. In the Mars 2020 Perseverance Rover, plutonium-238 has been used in the form of plutonium oxide as in several previous US space missions. The radioactive decay heat in space systems can be converted into electrical energy by several methods such as the conventional ones used in power plants based on the Rankine or Brayton cycles, or by conversion to mechanical energy in a Stirling engine or by directly converting to electrical in a thermoelectric generator as in the Perseverance Rover. Such systems have a long history especially for space systems (Bennet, 1989; Demuth, 2003; Determan et al., 2011; El-Genk, 2008; Miskolczy & Lieb, 1990). NASA's latest multimission radioisotope thermoelectric generator (MMRTG) is based on earlier SNAP (Small Nuclear Auxiliary Power) systems uses General Purpose Heat Source (GPHS) Modules with a thermoelectric generator and radiator funs, to produce electrical power without moving parts and to use heat for spacecraft operation and temperature control.

Medical therapy: Over 200 radioisotopes are used in medicine such as technetium ^{99m}Tc, iodine ^{125}I, palladium ^{103}Pd, iridium ^{192}Ir, cesium ^{137}Cs, and cobalt ^{60}Co, for diagnostics and cancer therapy (Harkness-Brennan, 2018). They are used both for the imaging of organs, such as the thyroid, bones, heart, liver, and for the treatment of cancer of lungs, breast, colon and rectum, prostate, stomach, and liver, etc. The photons, protons, and positrons from these radiations are also used in tomography such as in Positron Emission Tomography (PET), in medical imaging procedures (Gray; Ulaner, 2018) using radio-tracers to determine changes such as in blood flow or in the condition of the heart muscle. In the COVID-19 pandemic beginning from 2019, PET has also been used (Loforte, Gliozzi, Suarez, & Pacini, 2020) for the detection of potential COVID-19 respiratory syndrome.

A relatively new therapy, boron neutron capture therapy (BNCT), based on the short range of alpha particles is being developed for the treatment of brain tumors. Improvements in cancer therapy, such as dose enhancement by radiation by nanoparticles and better computational methods for its estimation, are the focus areas of research. The reaction

$$^{10}_{5}B + ^{1}_{0}n \rightarrow ^{7}_{3}Li + ^{4}_{2}He$$

is used in boron neutron capture therapy (Kumada, 2014; Nedunchezhian, Aswath, & Thiruppathy, 2016; Suzuki, 2020) with the product ions producing very high linear energy transfer due to high stopping power and high ionization with ≈ 150 keV/μm from the alpha particle and ≈ 175 keV/μm from the lithium ion in pathways <10 μm which is comparable to the diameter of one cell. One of the obstacles in the development of BNCT has been the thermal neutron flux required for the reaction; for this the use of a tandem particle accelerator has been achieved with the required values of current and energy for the nuclear reaction to take place with increased quality (Dymova, Taskaev, Richter, & Kuligina, 2020).

Exercise 5: In what medical applications would alpha radiation be preferred over beta radiation?

Nondestructive testing: In industry, due to the penetrating power of gamma radiation with an ability to go through commonly used dense materials such as concrete and steels, nondestructive testing (NDT) is an established discipline with widespread applications covering the testing of roads, bridges, pipes and structural materials. The use of radiation for elemental identification and assaying, such as determining the purity of gold, is also carried out accurately with the use of X-ray fluorescence (XRF) using the absorption edges of the photoelectric cross-section.

1.3 Interaction of radiation with matter

As highlighted above, radiation is an area of major concern in nuclear engineering and thus its interaction with matter is central to calculations of activity of radionuclides as they are formed and as they decay in various radioactive chains.

Radiation, both ionizing and nonionizing, interacts in several ways. Some significant mechanisms are discussed below (Fig. 1.10).

1.3.1 Interaction of alpha rays with matter

As described in Section 1.2.1, alpha particles are energetic ionizing radiation consisting of a helium nucleus with an electrical charge of two units.

As alpha particles move in matter, they "see" atoms on the scale of a few Angstroms with "lots" of electrons and a tiny point nucleus on the scale of a few fermis. It is thus more probable that they interact with the electrons. In their movement, they constitute a charged particle with charge +2e moving in the vicinity of several electrons of charge −e moving in orbits of the target medium. The force between the alpha and an electron, in Newtons, is given by the coulomb field with the force

$$F = \frac{k(2e)(e)}{r^2}$$

where k is the conversion constant $N/m^2/C$ and r is the distance between them. In their interaction, three events are thus likely: removal of electrons from their orbits (ionization), excitation of the orbital electrons, or transfer of energy to the

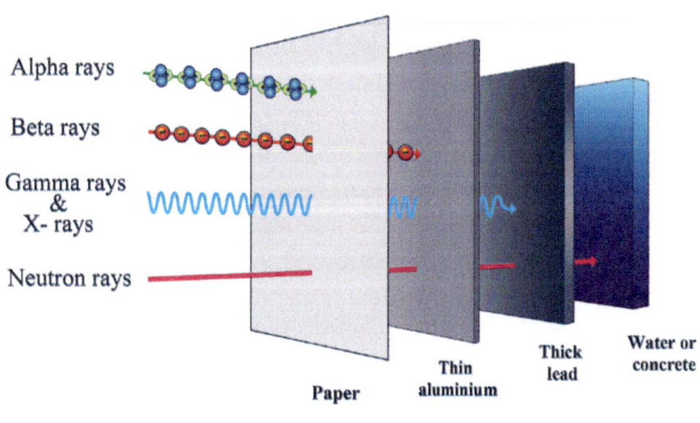

FIGURE 1.10 Interaction of radiation with matter.

"point" nuclei. Thus alpha particles can be understood to lose their energy via ionization, excitation and nuclear collisions.

Since alpha particles are heavy (compared with electrons), they move in a relatively straight line between collisions and undergo a series of "small-angle" deflections while the electrons receive energy from them and undergo deflection themselves. The collision can be modeled with the usual kinematics governed by conservation of momentum and energy. They lose energy by electronic excitation and ionization creating ion pairs. Due to their charge, they can be deflected by electric and magnetic field.

The stopping power $S(E)$ which is defined as the energy loss by collisions traveled, in units of energy/cm, is due to energy lost to electrons S_e and nuclei S_n both. Thus

$$-\frac{dE}{dx} = S_e + S_n$$

Clearly, the stopping powers would be expected to be functions of the alpha particle's mass, energy and electrical charge, and the atomic number of the target material as well as its density (which would determine the electron density).

An elementary two-body interaction with coulomb potential gives the electronic energy loss as

$$-\frac{dE}{dx} = \frac{4\pi N k^2 e^4}{m_e v_\alpha^2} \ln \frac{\gamma^2 m_e v_\alpha^3 f(Z)}{k e^2}$$

With the maximum energy transfer

$$W_{\max} = 2 m_e \gamma^2 v_\alpha^2$$

where

$$\gamma^2 = \frac{1}{1-\beta^2}$$

A quantum-mechanical relativistic expression is given by Bethe-Bloch as

$$S(E) = -\frac{dE}{dx} = \frac{4\pi N z^2}{\beta^2}\left(\frac{e^2}{4\pi\varepsilon_0}\right)^2 \left\{\ln\left(\frac{W_{max}}{I}\right) - \beta^2 - \frac{\delta}{2} - \frac{C}{Z}\right\}$$

in terms of the projectile atomic number z and speed v, and the density of electrons N and charge of electrons e and the ionization potential I of the slowing-down medium. The two correction terms (1) the "density effect" accounts for the shielding of distant electrons by the polarization of electrons caused by the electric field of the high-energy projectile, and (2) the "shell correction" (given the symbol "C") due to effects at low energy of tightly bound electrons.

The nonrelativistic Bethe formula

$$\frac{dE}{dx} = -\frac{4\pi N z^2}{m_e v^2}\left(\frac{e^2}{4\pi\varepsilon_0}\right)^2 \ln\left(\frac{2 m_e v^2}{I}\right)$$

The dependence of stopping power is $1/v^2$ so that low speed gives high energy loss (Table 1.1).

The collisions stopping power is the energy lost per cm due to coulomb collisions with the target atoms and electrons which result in ionization and excitation, while the nuclear stopping power is the loss transferred to recoiling atoms in an elastic collision. It is worth noting that the difference between the CSDA range and the projected range reduces for high energy

The range, defined as

$$R = \int_0^{E_0} \frac{dE}{S(E)}$$

can be interpreted as the mean range if at each point in the ion's track, the stopping power is the total of the collision and nuclear terms.

Alpha particles, which are "heavy" compared with electrons by a factor of 7273, lose their energy very quickly and thus travel small distances in typically straight lines before inducing ionization.

TABLE 1.1 Stopping power and range of alpha particles.

Material	Energy (MeV)	Stopping power (MeV $\frac{cm^2}{g}$)		CSDA Range	
		electronic	nuclear	($\frac{g}{cm^2}$)	(cm)
Dry air ($\rho = 1.225 \times 10^{-3} \frac{g}{cm^3}$)	1×10^{-3}	8.750×10^1	1.340×10^2	5.377×10^{-6}	4.389×10^{-3}
	1×10^{-1}	1.018×10^3	1.368×10^1	1.665×10^{-4}	0.1359
	1.00	1.922×10^3	2.103	6.698×10^{-4}	0.54722
	10	4.634×10^2	2.817×10^{-1}	1.309×10^{-2}	10.686
Water ($\rho = 1 \frac{g}{cm^3}$)	1×10^{-3}	9.891×10^1	2.282×10^2	3.273×10^{-6}	3.273×10^{-6}
	1×10^{-1}	1.131×10^3	1.921×10^1	1.425×10^{-4}	1.425×10^{-4}
	1.00	2.190×10^3	2.898	5.931×10^{-4}	5.931×10^{-4}
	10	5.340×10^2	3.844×10^{-1}	1.130×10^{-2}	1.130×10^{-2}

Source: ASTAR data from NIST.

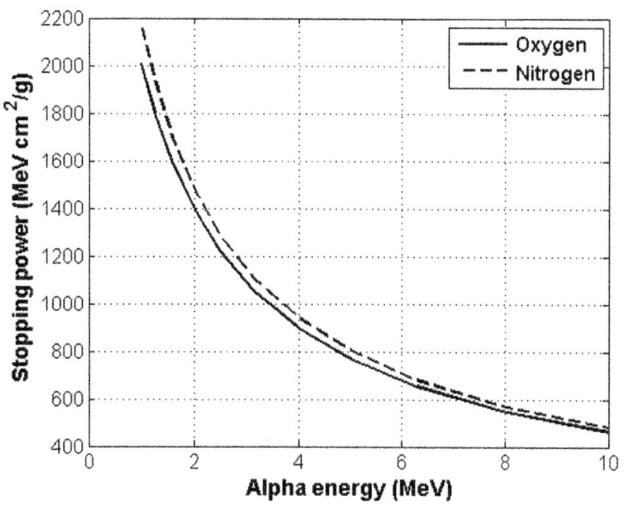

FIGURE 1.11 Stopping power for alpha particles in oxygen and nitrogen.

Example 2: Calculate the stopping power for alpha particles of KE 1−10 MeV in oxygen and nitrogen. From these, obtain the stopping power in air.

The MATLAB program implementing the Beth-Bloch formula (without the density factor and shell correction) is listed below, for the results shown in Fig. 1.11. Data used for mean ionization energy and density from NIST is included in the listing.

MATLAB Program 2: .

The values for ASTAR from NIST are reproduced below, showing excellent agreement. The stopping power for air can be obtained by incorporating the composition of air; thus (Table 1.2)

$$\frac{dE}{dx}|_{\text{air}} = 0.79\frac{dE}{dx}|_N + 0.21\frac{dE}{dx}|_O$$

Energy loss is a statistical quantity resulting in a distribution of path lengths giving a mean value higher than the CSDA value and a straggling is thus found. This becomes important in experiments where "projected ranges" of actual penetration of particles are measured and theoretically correlated with the CSDA range. The energy straggling is incorporated in simulation codes, as will be mentioned in Section 1.6 by, for example, a Gaussian distribution function. The mean energy loss in ionization from the Rutherford scattering cross-section is

```
% AlphaParticleBetheBlochFormula
AvNo=6.022e23 ; % avogadro number
mec2=0.510998; % MeV
K=4*pi*AvNo*re^2 *mec2;
fprintf(gid,'\n K=%12.4e ',K)

KE_low=1e-3;KE_hi=1e6; Steps=1000;
Del=(log10(KE_hi)-log10(KE_low))/Steps;
zP=2; AP=4; % alpha  % amu  1 amu = 931.5 MeV

for imat=1:2
    fprintf(gid,'\n\n Mat change')
   if (imat==1)
% Oxygen
ZT=8; rhoT=1.33151e-3; AT=16;% u
IT = 95e-6; % MeV
   end
  if (imat==2)
% Nitrogen
ZT=7; rhoT=1.16528e-3; AT=14;% u
IT = 82e-6; % MeV
   end

j=0;
for i=0:0.1:1
    j=j+1;
KE = 10^i; % Proj KE in MeV

beta=sqrt (1- (931.5/(931.5+KE/AP))^2);
gamma=sqrt(1/(1-beta^2));
Wmax=2*mec2*(gamma*beta)^2; % MeV
F=4*pi*AvNo*re^2*mec2*rhoT*ZT*zP^2/(AT*beta^2);
G=log(Wmax/IT)-beta^2;
H=F*G;
xx(j)=KE;
if(imat==1)
yyOx(j)=H/rhoT;% MeV cm^2 per g
end
if(imat==2)
yyN(j)=H/rhoT;% MeV cm^2 per g
end
end
end
figure(1)
set(gca,'FontSize',12)
plot(xx,yyOx,'k-','LineWidth',2);
hold on
plot(xx,yyN,'k--','LineWidth',2);
xlabel('\bf Alpha Energy (MeV)','fontsize',14)
ylabel('\bf Stopping Power (MeV cm^2/g)','fontsize',14)
%text(6,2100,'\bf Oxygen','fontsize',12);
legend('\bf Oxygen','\bf Nitrogen')
grid on
```

TABLE 1.2 Stopping Power from ASTAR NIST for alpha particles in oxygen and nitrogen.

Stopping power ($\text{MeV}\frac{\text{cm}^2}{\text{g}}$)					
Energy (MeV)	2	4	6	8	10
Nitrogen	1.406×10^3	8.985×10^2	6.785×10^2	5.523×10^2	4.691×10^2
Oxygen	1.329×10^3	8.602×10^2	6.519×10^2	5.313×10^2	4.520×10^2

$$\langle \overline{\Delta E} \rangle = \frac{2\pi z^2 e^4 NZx}{m_e v^2}$$

for which the fluctuations over a large number of collisions can be modeled with a Gaussian; for thin targets, that is, few collisions, the Vavilov and Landau distributions can be used (Peralta & Louro, 2014). The Gaussian probability distribution function (PDF) for the fraction of particles $f(E) = N(E)/N$ having energy in dE between E and $E + dE$ is

$$f(E) = \frac{1}{\sqrt{\pi}\sigma_{\text{str}}} \exp\left(-\frac{\left(E - \langle \overline{\Delta E} \rangle\right)^2}{\sigma_{\text{str}}^2} \right)$$

in terms of the mean energy $\overline{E}$ and the standard deviation σ_{str}which is given by

$$\sigma_{\text{str}}^2 = \langle \overline{\Delta E} \rangle E_{\max}\left(1 - \frac{1}{2}\beta^2\right)$$

The methods and implementation of sampling methods for PDF's will be discussed in Chapter 4.

1.3.2 Interaction of beta radiation with matter

Compared with alpha particles discussed in the previous section, beta particles have less mass and at energy 1 MeV they are at almost twice their rest-mass energy; they move 'swiftly' in comparison and thus have a longer range. At low energy, typically less than few MeV, the energy loss by electrons (and positrons) is mainly by excitation and ionization with little contribution from Møller (and Bhabha scattering). At high energy, typically greater than 10 MeV, the losses are mainly from bremsstrahlung, continuous electromagnetic braking radiation in the energy range of X-rays. The "critical energy" E_c defined as the energy at which ionization and bremsstrahlung losses are equal, decreases from about 300 MeV for hydrogen to $\sim$ 20 MeV for iron varying as

$$E_c = \frac{610\text{MeV}}{Z + 1.24}$$

The collision mechanism of electrons with electrons of the same mass implies that large energy transfer is possible. At high energies, the fraction of beta energy converted into photons is directly proportional to the atomic number of the target and energy of the electrons: $f_\beta^{Br} = 3.5 \times 10^{-4} ZE$.

The stopping power for electrons is the sum of the electronic ionization S_e and radiative loss S_r due to bremsstrahlung (Berger, Coursey, Zucker,& Chang, 2005; Taylor et al., 1970)

$$-\frac{dE}{dx} = S_e + S_r$$

The electronic part is similar to the ion-electron energy loss in the previous section. For electrons

$$S_e(E) = -\frac{dE}{dx} = \frac{2\pi r_e^2 m_e c^2 N z^2 Z}{\beta^2} \left\{ \begin{array}{c} \ln\left(\dfrac{m_e v^2 E}{2I^2(1-\beta^2)}\right) - \ln 2\left(\beta^2 - 1 + 2\sqrt{1-\beta^2}\right) \\ + (1-\beta^2) + \dfrac{1}{8}\left(1 - \sqrt{1-\beta^2}\right) \end{array} \right\}$$

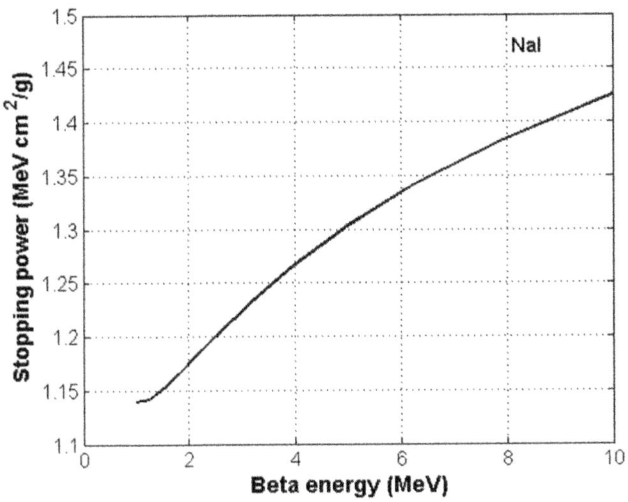

FIGURE 1.12 Stopping power for beta particles in sodium iodide (NaI).

TABLE 1.3 Stopping Power from ESTAR NIST for beta particles in sodium iodide (NaI).

Stopping power (MeV $\frac{cm^2}{g}$)					
Energy (MeV)	2	4	6	8	10
Collision	1.192	1.263	1.311	1.347	1.374
Radiative	0.1369	0.2857	0.4464	0.6145	0.7876

and the radiative loss is

$$S_r(E) = -\frac{dE}{dx}|_r = \frac{(z+1)z\rho_N E}{137m_e^2c^4}\left[4\ln\left(\frac{2E}{m_ec^2}\right) - \frac{4}{3}\right]$$

The stopping power $S_e(E)$ for beta particles in sodium iodide is calculated from the above expression with the MATLAB program listed below. For the KE of beta particles in the range 1–10 MeV, the stopping power is plotted as a function of energy in Fig. 1.12, which shows good comparison with the ESTAR stopping powers from NIST. The NIST values are plotted from 1×10^{-2}MeV to 1000 MeV for which $S_e(E)$ decreases from 11.16 MeV cm^2/g to 1.173 MeV cm^2/g at 1.5 MeV then increases slowly to 1.777 MeV cm^2/g at 1000 MeV. In contrast, the radiative stopping power $S_r(E)$ increases from 1.518×10^{-2}MeV cm^2/g at 1×10^{-2}MeV, equaling $S_e(E)$ at ~ 17.5 MeV, then increasing to 103.8 MeV cm^2/g at 1000 MeV.

The ESTAR NIST results are listed in Table 1.3 with the slow increase in $S_e(E)$ which is still higher than the radiative energy loss.

MATLAB program 3: .

Stopping power for beta particles in NaI

The range R(mg/cm^2) for beta particles of maximum energy E is given by the empirical relations

$$R = 412\ E^{1.265-0.0954\ln E}, 0.01\ MeV \le E \le 2.5\ MeV$$

$$R = 530\ E - 106,\ E > 2.5\ MeV$$

plotted for air, polyethylene, aluminum, iron, and lead in Fig. 1.13.

Thus energetic beta particles can travel more than a meter in air and are easily shielded by a few centimeters of lead. The effect of bremsstrahlung, for beta particles is shown in Fig. 1.14.

Stopping power for beta particles in NaI

```
% Electron Stopping Power.m
AvNo=6.022e23 ; % avogadro number
re=2.81794*1e-15; % m
mec2=0.510998; % MeV
ZT=45.798;AT=111;% NaI target At Wt
IT=452;% eV (for NaI)
rhoT=3.667;% g/cm^3
N=rhoT*AvNo/AT; % /cm^3

j=0;
for i=0:0.1:1
    j=j+1;
KE = 10^i; % Proj KE in MeV
%E=2.280;% MeV ke of projectile beta
gamma =(KE+mec2)/mec2;
beta=sqrt(1-1/gamma^2);

bb1 =2*pi*re^2*1*(mec2/beta^2)*N*ZT*1e6;% m^2 MeV m^(-3)=MeV/m
bb2 = log((beta^2*KE*mec2)/(2*(IT*1e-6)^2*(1-beta^2)));
bb3 = -(beta^2-1+2*sqrt(1-beta^2)*log(2))+(1-beta^2) +(1/8)*(1-sqrt(1-
beta^2));
bb = bb1*(bb2+bb3)
xx(j)=KE;
yy(j)=bb*1e-2/rhoT;% MeV cm^2 per g
fprintf(gid,'\n %4.0f %12.3e %8.4f %8.4f  %12.4e',j,xx(j),beta,gamma,yy(j))
end

figure(1)
set(gca,'FontSize',12)
plot(xx,yy,'k-','LineWidth',2);
xlabel('\bf Beta Energy (MeV)','fontsize',14)
ylabel('\bf Stopping Power(MeV cm^2/g)','fontsize',14)
text(8,1.47,'\bf NaI','fontsize',12);
grid on
```

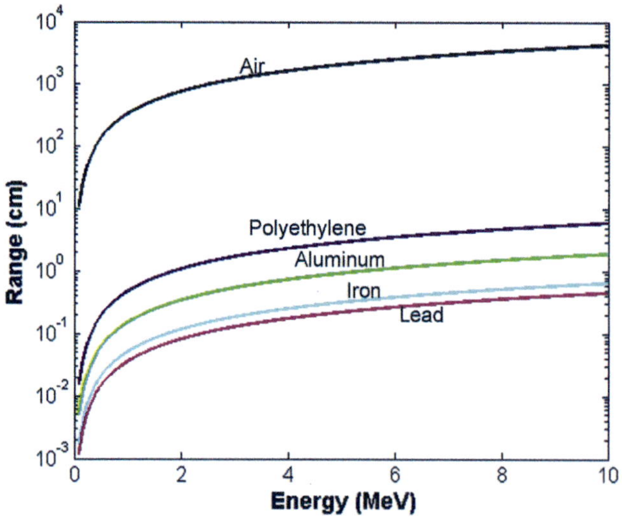

FIGURE 1.13 The range of beta particles as a function of energy.

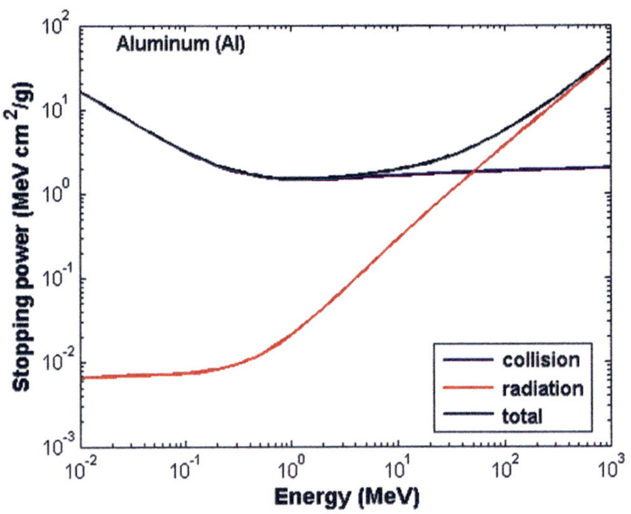

FIGURE 1.14 Stopping power of electrons in aluminum from ESTAR.

TABLE 1.4 Stopping power and range of beta particles.

Material	Energy (MeV)	Stopping power (MeV $\frac{cm^2}{g}$)		CSDA range	
		collision	radiative	$(\frac{g}{cm^2})$	(cm)
Dry air $(\rho = 1.225 \times 10^{-3} \frac{g}{cm^3})$	1×10^{-1}	3.633	4.222×10^{-3}	1.623×10^{-2}	13.249
	1.00	1.661	1.271×10^{-2}	4.912×10^{-1}	400.98
	10	1.979	1.795×10^{-1}	5.192	4238.4

Source: ESTAR data from NIST.

In order to compare the range of both alpha and beta in air, ASTAR (Table 1.1) and ESTAR (Table 1.4) values are compared; the range is far greater for beta than for alpha by factors of ~100, ~800, and ~400.

The specific ionization of beta particles (MeV/cm)

$$E_{n+1} - E_n = -\int_{s_n}^{s_{n+1}} \frac{dE}{ds} ds$$

$$\frac{dE}{dx} = \frac{2\pi q^4 NZ\left(3 \times 10^9\right)^4}{E_m \beta^2 \left(1.6 \times 10^{-6}\right)^2} \left\{ \ln \frac{E_m E_k \beta^2}{I^2(1-\beta^2)} - \beta^2 \right\}$$

can be used to estimate the number of ion pairs created by a beta particle as it slows down in an absorber in which the mean energy required to create an ion pair is W electron volts. The straggling of electrons, that is, the random differences between electrons is found from a PDF since the energy loss in a step is not the same for every step.

The collisional stopping power of water (liquid) produced by ESTAR (Berger et al., 2005) for electrons gradually decreases with energy, till about 1 MeV, to a constant value of ~2.0 MeV cm^2/g while the radiative loss becomes comparable at ~90 MeV then gradually rises to ~ 2.7 MeV cm^2/g. The total stopping power thus decreases with energy until the radiative term increase and then rises.

1.3.3 Interaction of gamma radiation with matter

Electromagnetic gamma rays, as discussed in Section 1.2.3, have high energy (~several MeV) and are thus very strong in terms of penetration in matter. In fact, one of the major problems in nuclear systems is gamma radiation for which detailed transport simulations are carried out.

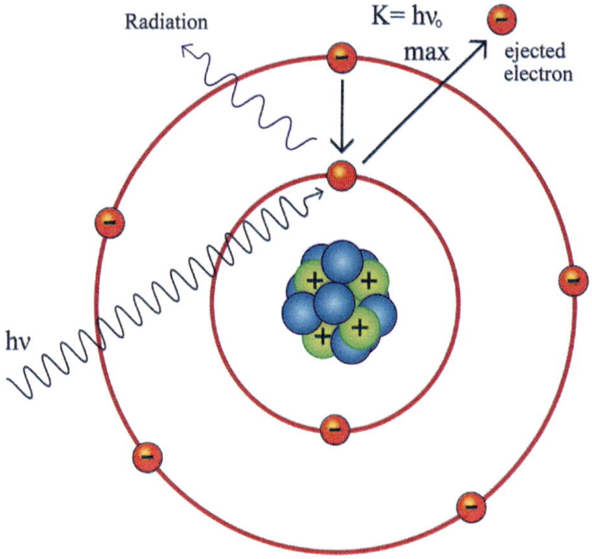

FIGURE 1.15 The photoelectric effect.

Gamma rays interact with matter in many different ways of which the three most significant are the photoelectric effect, Compton scattering and pair production. Some other forms of interaction are photoneutron reactions, quasi-deuteron disintegration and pion production at very high energies ($\sim>$ 1 GeV). The three main interactions combine to give the overall parameter which is used to calculate the overall attenuation of gamma transport in matter.

In the *photoelectric effect*, shown in Fig. 1.15, a photon of frequency ν and energy $E = h\nu$ interacts with an orbital electron of an atom, gets absorbed by the electron, and if its energy is greater than the BE of the electron, then the electron gets ejected with a maximum KE $K_{\max} = h(\nu - \nu_0)$ where $h\nu_0$ is the BE or the "work function" such that the threshold frequency of the emissive surface is ν_0. The photoelectric effect was explained by Einstein on the basis of quantum physics in which Planck had earlier stated that the transfer of energy takes place in integral amounts of quanta which depends on the frequency and not on the intensity of light. This was a revolutionary difference from classical physics in which light was considered a wave. Thus in the photoelectric effect, it is the frequency of the incident photon rather than its intensity which results in the ejection of an electron.

$$E_e = E_\gamma - E_B$$

When the host material is a high Z shield such as iron, lead, or tungsten, the number of electrons produced by the photoelectric effect is large. The X-rays produced from the photoelectric absorption are called fluorescent radiation and are used for element characterization. After the photoelectric effect has knocked out an electron, an outer electron can fill the vacancy to produce the fluorescent emission. Subsequently the excess energy can eject another outer shell electron producing an Auger electron. Similar events can be triggered in subshells. XRF is used as a nondestructive technique (NDT) for multielemental identification such as determining the purity of gold (*Handbook of Practical X-Ray Fluorescence Analysis*, 2006).

For the photoelectric effect to take place, the frequency of an incident photon must be higher than one of the frequencies correspond to the binding energies of the "saw-tooth" absorption edges for *K, L*,.,. shell electrons shown for lead in Figs. 1.16 and 1.17. Usually, the most important photoelectric effect is for the *K* shell electrons, for which the approximate formula is $E_K = (Z-1)^2 13.5eV \sim 88.573$ keV (the square of the effective nuclear charge times the Bohr energy orbit energy for the first orbit in hydrogen.

The probability of a photoelectric interaction σ_{PE} depends on the energy of the incident photon, the atomic number of the element and the BE of the electrons it is proportional to Z^n/E^3, $n = 4$–4.8.

The shells and associated energies of elements from hydrogen to lawrencium are *K, L*, M_1 to M_7, N_1 to N_7, O_1 to O_5, P_1 to P_5, Q_1. BE in subshells of a free atom (eV) are listed in Table 1.5.

High *Z* materials would therefore have more photoelectric absorption and would thus be good gamma shields. Thus one of the best shields is depleted uranium with a density of 19.1 g/cm^3, about five times better than lead with a density of 11.35 g/cm^3.

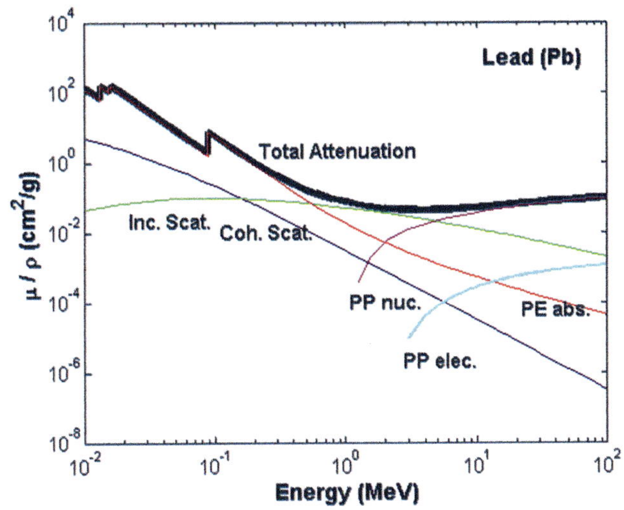

FIGURE 1.16 The mass attenuation coefficient for the photoelectric effect.

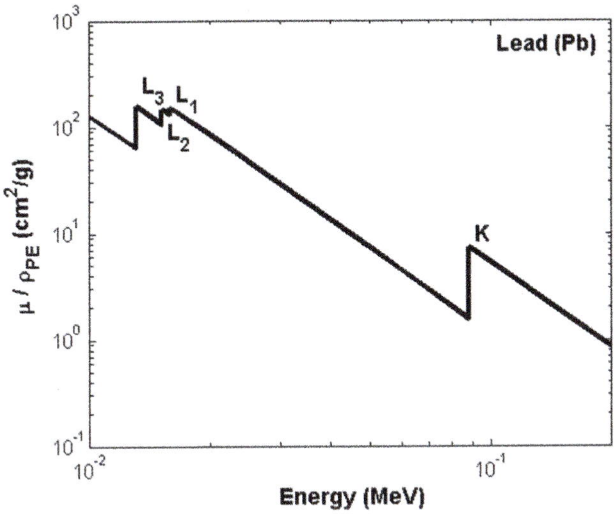

FIGURE 1.17 The photoelectric absorption edges for lead (Pb).

TABLE 1.5 Shell energies (eV).

Z	Element	Electronic configuration	*K*	*L1*	*L2*	*L3*	*M1*	…	*Q1*
1	H	$1s^1$	13.60						
2	He	$1s^2$	24.59						
3	Li	[He] $2s^1$	58	5.392					
4	Be	[He] $2s^2$	115	9.322					
5	B	[He] $2s^2 2p^1$	192	12.93	8.298				
⋮	⋮	⋮	⋮	⋮	⋮	⋮	⋮	⋮	⋮
20	Ca	$1s^2 2s^2 2p^6 3s^2 3p^6 4s^2$	4041	441	353	349	46	…N2(6.113)	
82	Pb	[Xe] $4f^{14}$ $5d^{10}$ $6s^2$ $6p^2$	88000	15860	15200	1304			
92	U	[Rn] $5f^3 6d^1 7s^2$	115611	21762	20953	17171	5553	…	6
98	Cf	[Rn] $5f^{10} 7s^2$	134967	26008	25103	19907	6733	…N3 (1273)	
⋮	⋮	⋮	⋮	⋮	⋮	⋮	⋮	⋮	⋮
103	Lw[a]	[Rn] $5f^{14} 7s^2 7p^1$	153040		30091	29101	22356	…	7

[a]*Lawrencium symbol changed by IUPAC to Lr.*

For a photon of high energy, in comparison with the K shell BE, the atomic absorption is σ_K $\sigma_K = 4\sqrt{2}\phi_0 \frac{Z^5}{137^4}\left(\frac{m_e c^2}{h\nu}\right)^{7/2}$
where

$$\phi_0 = \frac{8\pi}{3}\left(\frac{e^2}{m_e c^2}\right) = 0.6651 \times 10^{-24}\text{cm}^2$$

and scattering is considered as that of scattering of light from free electrons given by the Thompson cross-section ϕ_0 in the nonrelativistic case $h\nu \ll m_e c^2$ but modified to the Klein-Nishina cross-section for relativistic energies.

At higher energies than that compared with K shell binding energies, typically $m_e c^2 = 0.511 MeV$ the photoelectric effect becomes less important and scattering of photons, called *Compton scattering*, is considered as taking place with free electrons. This is reasonable since the K shell edge goes to ~0.11 MeV for the heaviest nuclei. Coherent scattering is classical scattering or Thomson scattering with the cross-section $\phi_0$for low-energy (in comparison with the ionization energy) photons passing in the vicinity of an outer electron of an atom. This causes the electron to vibrate with the frequency of the photon and to emit another photon of the same frequency and energy, but different direction, of the incoming photon which no longer exists. Coherent scattering is similar to elastic scattering in the sense that energy is not lost in the interaction. In contrast to coherent scattering, incoherent scattering changes the frequency and energy as if the electrons were randomly fluctuating.

The kinematics of coherent scattering was first shown by Compton as a two-body collision (Fig. 1.18) where the massless photon could be treated as a particle. In Compton scattering, at intermediate energies (few hundred keV to low MeV) and predominantly at low-Z material, a photon transfers some of its energy to an electron and continues to undergo multiple scattering. A photon of energy $E = h\nu$ will have momentum $p = E/c = h\nu/c$ with wavelength $\lambda = c/\nu$.

In Compton scattering, a collision results in scattering and recoil which changes the frequency of the photon. Consider the collision of an energetic (massless) photon with an electron of rest-mass m_0 with rest energy $E_0 = m_0 c^2$ in which the photon is scattered to the right upwards an angle θ and the electron recoils to the right downwards an angle φ.

The pre- and postcollision two-body parameters are shown below.

Collision parameters

"Particle"	Energy		Momentum	
	Before collision	After collision	Before collision	After collision
Photon	$E = h\nu$	$E' = h\nu'$	$p = h/\lambda$	$p' = h/\lambda'$
Electron	$E_0 = m_0 c^2$	$E_e = mc^2$	$p_e = 0$	$p_e' = \frac{1}{c}\sqrt{E_e^2 - E_0^2}$

With conservation of momentum and energy the equations are:

$$\lambda' - \lambda = \frac{hc}{E_0}(1 - \cos\theta) = \lambda_c(1 - \cos\theta)$$

$$h\nu + E_0 = h\nu' + E_e$$

where the Compton wavelength $\lambda_c = 2.426 \times 10^{-10}$cm

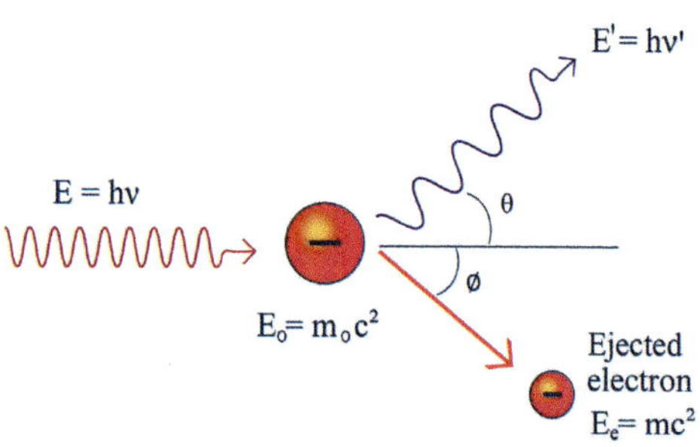

FIGURE 1.18 Compton scattering for a photon.

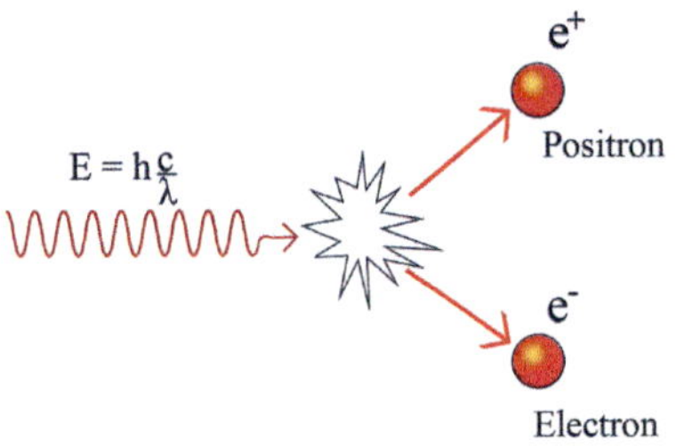

FIGURE 1.19 Pair production for a photon.

$$E_e = \sqrt{(p'_e c)^2 + E_0^2}$$

The energy of the photon after collision E' is thus found to be related to its energy before collision E and it scattering angle θ by

$$\frac{1}{E'} - \frac{1}{E} = \frac{1}{E_0}(1 - \cos\theta)$$

As seen, $\lambda' = \lambda$ for forward scattering $\theta = 0$ and the maximum change is for backscattering $\theta = \pi^c$ for which the electron gets maximum energy. The probability of Compton scattering is given by the Klein-Nishina (K-N) differential cross-section, that is, the probability that the photon gets scattered into solid angle $d\Omega = 2\pi d\theta$, so that the total cross-section is $\sigma = \int \frac{d\sigma}{d\Omega} d\Omega$, and

$$\frac{d\sigma}{d\Omega} = \frac{1}{2} r_e^2 \left(\frac{\lambda}{\lambda'}\right)^2 \left(\frac{\lambda}{\lambda'} + \frac{\lambda'}{\lambda} - \sin^2\theta\right)$$

with units of barns per steradian. At low incident energy, in comparison with the electron energy $m_e c^2$ when the scattering is close to elastic, the final and initial wavelength are the same, while for high-energy scattering is inelastic. In the Thomson regime, the *K-N* cross-section reduces to the Thomson "unpolarized" cross-section

$$\frac{d\sigma}{d\Omega}|_T = \frac{1}{2} r_e^2 (2 - \sin^2\theta)$$

The total cross-section for an electron is

$$\sigma_C = \frac{3}{8\epsilon}\phi_0 \left\{ \left[1 - \frac{2(\epsilon + 1)}{\epsilon}\right] \ln(2\epsilon + 1) + \frac{1}{2} + \frac{4}{\epsilon} - \frac{1}{2(2\epsilon + 1)^2} \right\}, \epsilon = \frac{h\nu}{m_e c^2}$$

For an atom, the Compton scattering cross-section is thus $N\sigma_C$. The K-N cross-section reduces to the value from the classical Thomson as forward scattering of the photon increases with increasing energy.

At yet higher energies, *pair production* is the dominant interaction mechanism for gamma resulting in energy deposition and the subsequent creation of a beta particle and a positron as shown in Fig. 1.19. This ejected positron, in turn, when it has a minimum energy 1.022 MeV, annihilates with an electron to produce two 0.511 MeV gammas.

The cross-section for pair production for sufficiently high energy $h\nu \gg 137 m_e c^2 Z^{-1/3}$ can be written in the asymptotic form

$$\sigma_P = Z^2 r_e^2 \left(\frac{28}{9} \ln 183\, Z^{-1/3} - \frac{2}{27}\right)$$

The total cross-section (barns) for gamma interaction with matter is thus

$$\sigma_t^\gamma = \sigma_{PE} + N\sigma_C + \sigma_P$$

With nuclei, the photons can be energetic enough to be absorbed and eject protons, (γ, p) reactions, or neutrons in what are called the photoneutron (γ, n) reactions. Gammas can also elastically or inelastically scatter for neutrons. However neutron interactions will be discussed in a separate chapter. An excellent reference for photon cross-sections is the photon cross-sections, Attenuation Coefficients, and Energy Absorption Coefficients from 10 keV to 100 GeV (Hubbell, 1969).

With the cross-section defined above, the liner attenuation coefficient with units of cm^{-1} together with the mass attenuation coefficient $\tilde{\mu}$ with units of cm^2/g can be defined as

$$\mu = N\sigma_t^{\gamma}, \tilde{\mu} = \frac{\mu}{\rho}$$

The intensity of photons can then be found from the source intensity I_o across a shield of thickness x as

$$I(x) = I_o e^{-\mu x}$$

The above expression would hold under ideal conditions, that is, a point anisotropic source of intensity I_o incident on a thin shield. For a point isotropic source, the intensity would be given as

$$I(R) = I_o \frac{e^{-\mu R}}{4\pi R^2}$$

to include both geometrical and material attenuation. For a broad beam or thick shield, the effects of scattering of radiation inside a shield is better represented by including a "buildup" factor B, a dimensionless number greater than one representing the effect of secondary radiation over that of primary radiation in which the gamma radiation passes straight through without appreciable scattering. The buildup factors for exposure rate $\dot{X}$ in place of intensity I above are given in the Radiological Health Handbook (US Dept of Health Education and Welfare, 1970). For a point isotropic source in water, the buildup factors, tabulated for values in the range $E = 0.255 - 10.0\text{MeV}$ show that at 0.255 MeV, B increases from 3.09 for $\mu x = 1$, to 982 for $\mu x = 20$, thus the effect becomes more significant as μx increases. For fixed thickness, $\mu x = 1$ with increase in energy from 0.255 to 10 MeV, B decreases from 3.09 to 1.33, while at $\mu x = 10$, for the same energy increase, B decreases from 982 to 5.98. With the same trends, the maximum buildup factors are 141 for aluminum, and 55.6 for iron. These trends however diverge for tin as for $\mu x = 20$, B increases from 18.8 at 1 to 33.4 at 10 MeV. Similar differences are reported for tungsten and lead, while for uranium the maximum buildup factor is 28.5 for 10 MeV and $\mu x = 20$.

For a mixture or compound

$$\tilde{\mu} = \sum_i w_i(\mu/\rho)_i$$

The coefficients for some materials are given in Table 1.6.

For gamma interactions with materials, the thickness of a shield is expressed as a Half Value Layer (HVL) as one which would reduce the intensity by a half of its original intensity, or a Tenth Value Layer (TVL) which would reduce the intensity to one-tenth. Thus for a 1 MeV gamma, the HVL of lead is 0.76 cm while that of iron is 1.52 cm.

Exercise 6: Compare the severity of radiation from alpha particles, beta particles, X-rays, and gamma rays in the context of shielding personnel in the vicinity of a space station with a nuclear-powered radioisotope generation system.

1.4 Sources and effects of radiation

We are all exposed to natural radiation originating from cosmic radiation and decay chains of which the three main ones are from thorium-238 called the thorium series, uranium-238 called the uranium or radium series, and uranium-235 called the actinium series. These produce radon, an inert noble gas, which among other isotopes such as potassium-39 and potassium-40 goes into the lungs and causes further damage by alpha decay with a half-life of 3.8 days.

TABLE 1.6 Mass attenuation coefficients $\tilde{\mu}$ for 1 MeV photons.

Material	Air	Water	Concrete	Iron	Lead	Glass, lead[a]	Polyethylene
$\tilde{\mu}(\text{cm}^2/\text{g})$	0.06358	0.07072	0.06495	0.05995	0.07102	0.06914	0.07262

[a]*Glass, lead has* <*Z/A*> *= 0.42101, I = 526.4 eV, ρ = 6.220 g/cm³, composition (weight fraction) oxygen: 0.156453, silicon: 0.080866, titanium: 0.008092, arsenic: 0.002651, lead: 0.751938.*

Source: From https://physics.nist.gov/PhysRefData/XrayMassCoef/ElemTab/z82.html.

In addition to α, β and γ radiations described above, neutron radiation is important in all fission processes in the production of energy as well as in nuclear systems used for various applications in industry. Neutrons can be produced from fission in nuclear reactors, spontaneous fission sources, accelerator-based neutron generators and from other neutron producing reactions such as (α, n) and (γ, n) reactions.

A neutron producing nuclear fission reaction of uranium-235 is

$$01n + 92235U \rightarrow 54140Xe + 3894Sr + 201n + 200\text{MeV}$$

which is the basis for thermal energy in nuclear reactors.

Portable neutron sources include americium-beryllium (Am-Be) and californium Cf^{252} with properties listed in Table 1.7 and neutron spectra shown in Fig. 1.17.

Californium-252 has a shorter half-life than Am-Be but is a more intense source; a small 10 μg pellet of Cf-252 emits 2.31 10^7 n/s with an average energy less than that from Am-Be. Both are intense sources with diverse and wide-ranging applications.

The energy spectrum of Cf-252 shown in Fig. 1.20 is of the form $f(E) = Ce^{-E/a}\sinh\sqrt{bE}$ $(a = 1.025,\ b = 2.926)$ with a probable energy of 0.7 MeV and average energy 2.1 MeV, while Am-Be produces relatively more high-energy neutrons.

The Am-Be source is an example of the production of neutrons from an alpha emitter (Am) in the presence of beryllium from an (α, n) reaction such as

$$He_2^4 + Be_4^9 \rightarrow C^{13*} \rightarrow \begin{cases} B_6^{12} + n_0^1 + 5.6\text{MeV} \\ C_6^{12} + n_0^1 + \gamma 4.4\text{MeV} + 1.2\text{MeV} \end{cases}$$

The most intense source of neutrons are fission reactors, with a fast neutron ($\sim$2 MeV) yield $\sim 10^{12}$ n/MW though small accelerator-based neutron generators, such as the van de Graff and Cockroft-Walton generators were used in the early days with accelerated charged particles striking low-Z targets. Accelerator-based sources were developed mainly

TABLE 1.7 Am-Be and Cf-252 neutron sources.

Source	Am-Be	^{252}Cf
Type	α-emitter	Spontaneous
Half-life (years)	432.2	2.645
Specific activity (Ci/g)	–	532
Neutron yield (n s^{-1} Ci^{-1})	2–2.4 × 10^6	4.4 × 10^9
Average energy (MeV)	~ 4.2 MeV	~ 2
Maximum energy (MeV)	~ 11 MeV	~ 10

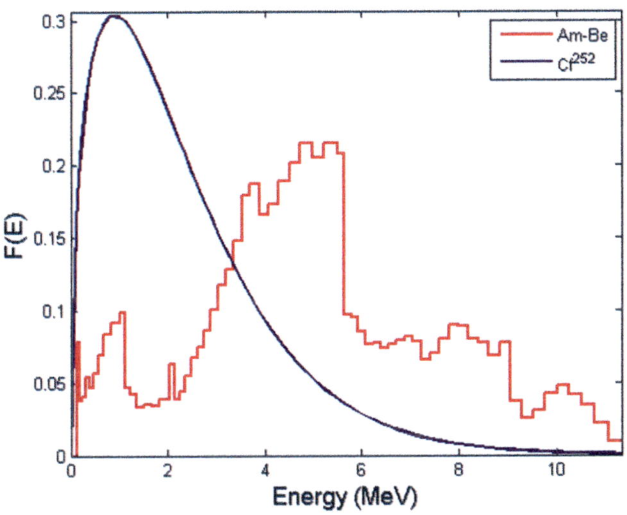

FIGURE 1.20 Neutron energy spectra of Am-Be and Cf^{252}sources.

for materials research (Granada, Santisteban, Dawidowski, & Mayer, 2012; Verbeke, Leung, & Vujic, 2000) and are now being improved for specialized uses such as cancer therapy (Dymova et al., 2020; Letourneau et al., 2017; Marchix et al., 2018). The latest technology for small intense portable neutron generators is based on the D-D and D-T fusion reactions

1.4.1 Radiation dose

The first units used in radiation were the Becquerel (Bq = 1 disintegration per second) and the curie (3.7×10^{10} disintegrations per second). The amount of any radioisotope that emits 1 Ci depends on its half-life $t_{1/2}$ from which its decay constant λ is found; the number of atoms required to produce 1 Ci are then

$$N_{Ci} = \frac{3.7 \times 10^{10}}{\lambda}$$

from which the mass of that radioisotope required for an activity of 1 Ci can be found

$$M_{Ci} = \frac{N_{Ci}}{N_{Av}}$$

The natural background radiation source from cosmic and decay chains is about 2 mrem/day, or about 0.730 rems (7.30 mSv) in one year (Cacuci, 2010). As mentioned in the previous section, the natural background radiation exists mainly due to decay chains from thorium and uranium; thus the human body contains radioactivity from isotopes such as radon and potassium and has an activity of about 8000 Bq (a few micro Curies). For a neutron source, the half-life of Cf-252, given in Table 1.7, is 2.645 years so that the amount of Cf-252 producing 1 Ci would be

$$M_{Ci} = \frac{3.7 \times 10^{10} A t_{\frac{1}{2}}}{0.693\, N_{Av}} = \frac{3.7 \times 10^{10} \times 2.645 \times 365 \times 24 \times 3600 \times 252.08163}{0.693 \times 6.022 \times 10^{23}} = 1.9\text{mg}$$

Thus 1.9 milligram of Cf -252 gives 1 Ci which is a specific activity of ~526 Ci/g.

The neutron dose from Cf^{252} is 2.2–2.3 $\times$ 10^3 rem m^2/g/h (22–23 Sv m^2/g/h) so that a 25 μg source emitting 5.75 $\times$ 10^7 n/s would give a neutron dose at 1 m in air of 55–57.5 mrem/h or 2.66 $\times$ 10^{-15} Sv/s per source neutron.

The dose from 1 μg of ^{252}Cf at 1 m in air is 0.0221 mSv/h (2.21 mrem/h) from fast neutrons and 0.0019 mSv/h from gamma rays. Thus it is necessary to shield the neutrons as well as any secondary radiation produced from it; only then it can be safely used.

Neutron dose	0.59–0.73 μSv/h at 1 m/GBq	22–23 Sv m^2/g/h
Gamma dose	0.68 μSv/h at 1 m/GBq	1.6 Sv m^2/g/h

The gamma dose rate is quoted at $R = 100$ cm in air, of 1.42368E-20 (0.1346) Sv/source. Assuming the source strength of 1 g Cf^{252} to be 2.4 $\times$ 10^{12} n/s the gamma dose is 0.123 mSv m^2/g/h from secondary photons produced by neutron interactions. With a simulation in the photon-only mode, assuming an average gamma energy of 0.8 MeV, the gamma dose is 3.40568 $\times$ 10^{-17} (0.0001) Sv/source gamma which is 1.6249 Sv m^2/g/h.

This has been estimated using an average neutron emission of 3.768 neutrons/fission and 2.314 $\times$ 10^{12} n/g.s i.e. 0.6141 $\times$ 10^{12} fissions/g.s and a gamma emission yield of 8.30 $\pm$ 0.08/fission with an average energy of 0.80 $\pm$ 0.01 MeV, Thus the gamma emission is estimated as 5.09703 $\times$ 10^{12} gamma/g.s = 1.8349 $\times$ 10^{16} gamma/g.h = 0.6249 Sv/g.h.

Exercise 7: What would be the effect of radon gas Rn-222 on the lungs when it undergoes alpha decay?

The units used for quantifying the effects of nuclear radiation are the absorbed dose (D), the equivalent dose (H) and the effective dose (E) defined as follows:

1.4.2 Absorbed dose

The SI unit of absorbed radiation dose is a *Gray* (Gy) (Commission, 2007; IAEA, 2014; *Radiation Effects and Sources*, 2016) defined as an absorbed energy dose of 1 J/kg. The absorbed dose (D) is the specific energy absorbed by a material. Traditionally, it has been expressed as a rad which is defined as 100 erg/g. Thus

$$1\ \text{Gy} = 1\ \text{J/kg} = 10^7\ \text{erg}/10^3\ \text{g} = 100\ \text{rad}.$$

Historically, the unit of *roentgen* (R) has been used to represent the exposure (E) of *radiation* that results in the generation of one electrostatic unit of charge in 1 cm^3 of air at STP. It is thus better to use the roentgen for electromagnetic radiation.

Thus

$$1\text{elementary charge(charge on an electron)} = 4.8 \times 10^{-10}\text{esu}$$

$$1\text{esu} = \frac{1}{4.8 \times 10^{-10}}\text{elementary charges}$$

One pair requires 32.5 eV

$$1\text{esu} = \frac{32.5}{4.8 \times 10^{-10}}eV = 6.76 \times 10^{10}\ eV$$

and 1 $eV = 1.6 \times 10^{-19}\ J$, therefore

$$1\text{esu} = \frac{32.5}{4.8 \times 10^{-10}}\text{eV} = 1.0816 \times 10^{-8}\text{J}$$

Thus 1.0816×10^{-8} J is deposited in 1 cm^3 dry air at STP and with air density $\rho = 0.00129\ \text{g/cm}^3$ this is 0.0084J/kg or 84erg/g, since 1 J = 10^7 ergs.

Thus

$$1\text{R} = 1\frac{\text{esu}}{\text{cm}^3} = 0.0084\frac{\text{J}}{\text{kg}} = 84\frac{\text{ergs}}{\text{g}} = 0.258\frac{\text{C}}{\text{g}}$$

1 R = 2.58 × 10^{-4} C/kg. The exposure of 1 C/kg is equivalent to 33.8 J/kg, or 33.8 Gy.

Example 3: If dry air, with an ionization energy (I) of 32.5 ev/ion pair (~34 J/C), is subjected to an exposure of 1 R, the absorbed dose is D = E I = 2.58 × 10^{-4} × 32.5 = 8.4 × 10^{-3} J/kg = 8.4 mGy.

From the roentgen, the CGS units of rem (roentgen equivalent man) is defined as a measure of the stochastic effect of low level ionizing radiation on the body (1 roentgen = 0.96 rem).

1.4.3 Equivalent dose

Since we would like to quantify the effect of radiation, two more factors must be considered viz (1) the type of radiation, and (2) the response of a tissue of a living organism to that radiation. For these effects, the quantities defined are the equivalent dose (H) and the effective dose (E).

Since the absorbed dose represents the amount of energy absorbed, another unit is required to represent the "effect" of the dose. The SI unit of the "effect" called the "equivalent dose" is a *Sievert* (1 Sv = 100 rems) defined as the absorbed dose in grays multiplied by a "quality factor" Q which is a measure of the effect of a radiation. For X-rays, beta and gamma radiation, Q ~ 1 while for neutrons an average value of Q ~ 10 can be used.

A chest X-ray which gives a dose of 0.1 mSv (0.01 rem = 10 mrem) which is the equivalent of about one week of natural background radiation. A procedure such as a CT scan of the abdomen and pelvis is about 20 mSv and a PET/CT scan is ~ 25 mSv which are very high doses equivalent to over 200 chest X-rays.

The equivalent dose depends on the rate of energy deposition of a radiation as it collides with the host material in its interactions. For a tissue T subjected to a radiation R with absorbed dose $D_{T,R}$, the equivalent dose to that tissue (H_T) is obtained by the cumulative effect of all the radiations on the tissue:

$$H_T = \sum_R W_R D_{T,R},$$

where W_R is the radiation weighting factor, formerly called the quality factor Q, resulting from the linear energy transfer of a particular radiation. For X-rays, γ-rays, and generally for β-rays, $W_R \sim 1$, while for α-particles, heavy recoil nuclei and neutrons, $W_R$varies between 1 and 20 depending on the energy of the radiation.

Example 4: Two radiations *viz* an X-ray, and energetic neutrons, are incident on tissues of lung and skin with the data provided below. Estimate the equivalent dose (H).

Radiation	W_R	$D_{lungs,R}$ (rads)	H_{lungs} (rems)	$D_{skin,R}$ (rads)	H_{skin} (rems)
X-rays	1	1	1	2	2
Neutrons	10	0.2	2	0.2	2

From the above, the equivalent dose to the lungs is $H_{lungs} = 1 \times 1 + 10 \times 0.2 = 3$ rems (0.03 Sv). Similarly for skin, $H_{skin} = 1 \times 2 + 10 \times 0.2 = 4$ rems (0.04 Sv).

From a medical procedure in a hospital, a single X-ray of the chest, abdomen and pelvis would give an equivalent dose of 10, 60 and 70 mrem while a full-body CT scan would give 1000–2000 mrem ($>1-2$ years of background radiation).

1.4.4 Effective dose

To represent the tissue-specific response, the stochastic health risk is quantified by a weighting factor W_T representing the relative radio-sensitivity of a tissue. The overall effective dose E is thus expressed as

$$E = \sum_T W_T H_T,$$

The weighting factors W_T, (Lamarsh and Baratta, p. 510) for some organs are listed below (Table 1.8).

The units of effective dose are also rems (*Sv* in SI).

Thus the overall effective dose is

$$E = \sum_T W_T H_T = W_{lungs} \cdot H_{lungs} + W_{skin} \cdot H_{skin} = 0.12(3) + 0.01(4) = 0.40\text{rems}.$$

All the above quantities can be expressed as rates $\dot{D}, \dot{H}$ and $\dot{E}$ with units of rem/h, mrem/s, Gy/hr, Gy/s, mSv/s, etc.

1.4.5 Radiation safety limits

Radiation safety limits are prescribed by the International Commission on Radiological Protection (ICRP) and the US Nuclear Regulatory Commission (USNRC) for the general public as well as professionals working with or exposed to any form of radiation. According to US, Article 20.1201 Occupational annual dose limits for adults, the total effective dose equivalent limit is set at 5 rems (0.05 Sv). The USNRC Code for Federal Regulations 10 CFR Part 19 requires that" all individuals who, in the course of their employment, are likely to receive a dose of more than 100 mrem in a year, must receive adequate training to protect themselves against radiation."

No serious radiation effects have been normally seen for doses less than about 5 rems (0.05 Sv). Beyond this limits, changes in blood chemistry have been observed. At 70 rems, vomiting takes place followed by hair loss, diarrhea, and

TABLE 1.8 Weighting factor W_T for organs.

Tissue	ICRP 26[a]	ICRP 60[a]
Gonads	0.20	–
Breast	0.15	0.05
Red bone marrow	0.12	0.12
Lung	0.12	0.12
Thyroid	0.03	0.05
Bone surfaces	0.03	0.01
Skin	–	0.01
Other organs	0.06	0.05

[a]*ICRP publications 26, 60.*

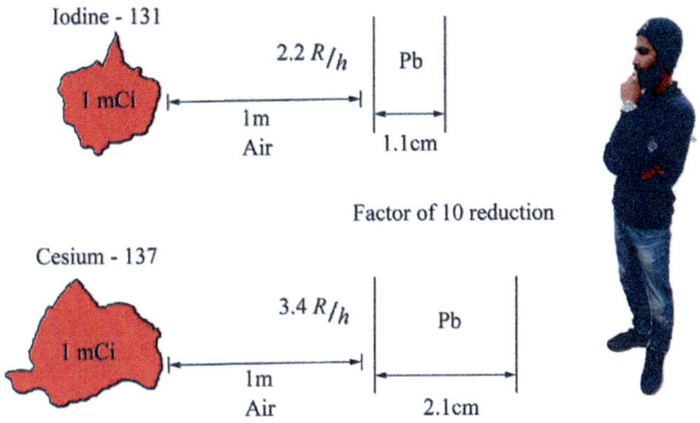

FIGURE 1.21 The radiation dose in air from 1 mCi of iodine-131 and cesium-137.

hemorrhage at about 100 rems. Possible death within two months can occur at 400 rems (4 Sv). Persons who receive doses exceeding 5000 rems (50 Sv) die "within few hours of exposure" probably by the failure of the central nervous system.

As Fig. 1.21 shows, the dose from 1 mCi of iodine-131 and cesium-137 one meter away in air is 2.2 R/h and 3.4 R/h respectively. Assuming a quality factor of 1, this still gives high dose rates which can be reduced to one-tenth of their values with thin lead shields (1–2 cm thick).

With the above estimates, we can now get an idea of the magnitudes of radiation at Fukushima where the dose rates were as high as 1.2 rems/h which is at the site boundary; compare this with the maximum (occupational) permissible dose rate of 5 rems/y (0.05 Sv/y); it comes to the permissible dose of over 2000 years! Even thirty-seven miles (60 km) away from the site, the dose rate was measured to be 0.8 mrem/h; much higher than the maximum permissible.

1.4.6 Radiation detection

Radiation detectors, both passive and active, are used extensively for routine monitoring of personnel in radiation environments in nuclear reactors and medical facilities. They are small passive dosimeters, lightweight and wearable on the chest such as the common thermoluminescent dosimeter (TLD), a detector capable of measuring doses from 0.01 mGy to 10 Gy, used for personal dose monitoring typically over a period of a few months. TLDs produce luminescence (visible light) by the thermal effect of ionizing radiation on a chemical compound such as calcium sulfate and lithium fluoride. Other detectors are also used on the body for displaying radiation exposure or as a ring when radiation sources are being handled. Several other uses include space monitoring to detect radiation, or further to identify radiation sources from the measurements of the energy spectra. The basic purpose of these dosimeters is to detect and measure radiation to ensure that, in accordance with regulations (10 CFR Part 20, and ICRP) whole body dose limits of 5 rems/y, or 50 mSv, as the total effective dose equivalent (TEDE) are not exceeded.

The modes of operation of detectors are based on the ionization caused by charged-particle alpha and beta radiation, as well as by the photoelectric, Compton and pair production interactions of gamma rays. In its simplest form, a gas-filled cylindrical ionization chamber with a central positively charged anode and a negatively charge wall cathode, both with a potential, attract electrons and positive charges created by the energy deposited by the incident radiation. As the potential is increased, the mobility of charges is also increased to give a flow of measurable current until the signal is proportional to the intensity of the incident radiation. Proportionality is achieved when each electron, from the ion pair created by the incident radiation, in turn produces its own secondary ionization and builds an avalanche with a high multiplication. In this mode of operation, with an applied voltage typically of a few thousand volts, the ionization chamber is called a proportional counter. At a higher voltage, the electric field surrounding the anode becomes weaker due to a discharge caused by several avalanches spreading into the volume of the gas which reduces further secondary ionization. Proportional counters operate in a voltage range which inhibits this discharge. Beyond this voltage, a positive space charge builds around the anode and the discharge terminates.

Proportional counters are better for alpha particle detection and measurements due to their short range, as described in Section 1.3.1; they are also used for low-energy electrons produced by X-rays and for beta particles. In the Geiger-Mueller mode, each pulse is counted and an audible signal indicates the presence of radiation; the GM tube is thus a robust pulse counter.

Compact solid state detectors are used for measuring incident radiation the production of electron-hole pairs in a semiconductor p-n junction due to the small band gap (in silicon 1.14 eV and in germanium 0.67 eV). These gaps, being much less than the ~ 32 eV energy required to create an ion pair in a gas detector, gives an edge to semiconductor detectors which thus have better efficiency and resolution. Another very important application of germanium-based semiconductor detectors is in nondestructive gamma ray spectroscopy which permits accurate measurement of excited levels of nuclei (see Section 1.3.3). One important application of gamma ray spectroscopy is the determination of uranium enrichment by measuring the 185.7 keV gamma rays emitted in the decay of uranium-235 to thorium-231.

The property of certain materials to "scintillate," that is, to emit visible or ultraviolet light by energy deposition of incident radiation is used for the detection of gamma rays. Though the resolution is lower than that for solid state detectors, the long range of gamma rays makes scintillation detectors attractive.

Some commonly used inorganic scintillators with high photoelectric cross-sections, and hence high efficiency, are sodium iodide and cesium iodide activated with thallium NaI(TI), CsI(TI), and silver activated zinc-sulfide. Organic materials, such as plastics, are also used; however, due to their low atomic number, they are better suited to the detection of radioisotopes with low-energy beta emissions (Section 1.4.2).

Cherenkov detectors are used for the detection of very high velocity ($\beta > 1/n$) subatomic particles by photomultiplier tubes with light sensitive "photocathodes." For electrons, this velocity is $\sim$ 1 MeV which is easily reached by electrons, while for protons and neutrons it is in the GeV range. The speed of light in a material of refractive index n can be less than the speed of light in vacuum c since $n = c/v$; thus for vacuum, for example, $n = 1$ and for water $n = 1.33$; thus light moves at $\sim 75\%$ of the speed of light in vacuum; thus a 500 MeV muon ($v \sim 0.98c$) would be traveling faster than light and would produce a radiation such as the bluish white tinge of light observed by Cherenkov in the water pool of a nuclear reactor.

For more detailed information on the energy spectrum of a radiation source, or of several sources, a multichannel analyzers (MCA) is used to obtain "pulse height" tallies. The principle used in a MCA is that incident radiation which produces charge Q would produced a voltage V = Q/C which could be "discriminated" by its' electronics to permit a measurement of the energy spectrum of the incident radiation. As will be illustrated in a simulation exercise (Section 1.6), pulse height tallies are now available in simulation codes so that the energy spectrum can be identified for various sources.

In nuclear forensics, and for security applications, activation techniques such as prompt neutron activation, are also used for the detection and identification of nuclear material and explosives.

Neutrons are detected by the secondary ionization they produce during their interaction with matter, for example, in their passage in air or a gas, they can create ionization which can be detected in the form of a voltage (Knoll & Kraner, 1981). In a commonly used gas detector containing boron trifluoride gas (BF_3), the detector consists of a cylinder (of aluminum, brass or copper) filled with BF_3 gas at a pressure of 0.5–1 atmospheres.

The nuclear reaction with cross-section shown in Fig. 1.22 that takes place in the gas is

$$510B + 01n \rightarrow 37Li + 24He.$$

The products $37Li$ and α ($24He$) travel in opposite directions after the collision creating ion pairs in the BF_3 gas. When the lithium is left in the ground state (about 6% of the time) the particles have 2.792 MeV to create ion pairs. In the other case where lithium is left in the excited state, the KE available is 2.310 MeV and hence a smaller resulting signal. In the ground state of lithium, the energy carried by the α particle is 1.78 MeV while the remainder1.012 MeV is carried by the lithium.

In naturally occurring boron, the B-10 percentage is 20% while B-11 is 80% and the boron (n, α) cross-section is attractive for the isotope B-10. Thus it is necessary to enrich the gas in B-10. Thus in practice the enrichments for B-10 is increased to 96%. In the detection system, the cylinder is the detector (cathode) while a single thin wire running down the axis of the tube is the anode.

1.5 Atomic densities of elements and mixtures

Several elements and mixtures are used in standard nuclear engineering for various materials and processes used in the front-end fuel cycle, in nuclear systems and in the water and reprocessing "back-end" fuel cycle. The first step in the modeling process is thus preparing the atomic and molecular data. For criticality calculations, the atomic densities of the following are calculated: uranium-235, uranium-238, plutonium, natural boron, water, boron carbide, uranium dioxide fuel, UO_2F_2 in solution with water and several other commonly used materials (Harmon et al., 1994). An excellent sourcebook for modeling and simulation data is the Compendium of Material Composition Data for Radiation Transport Modeling (McConn et al., 2011) which contains data for 372 materials from A-150 tissue equivalent plastic to zirconium hydride including explosives and medical materials.

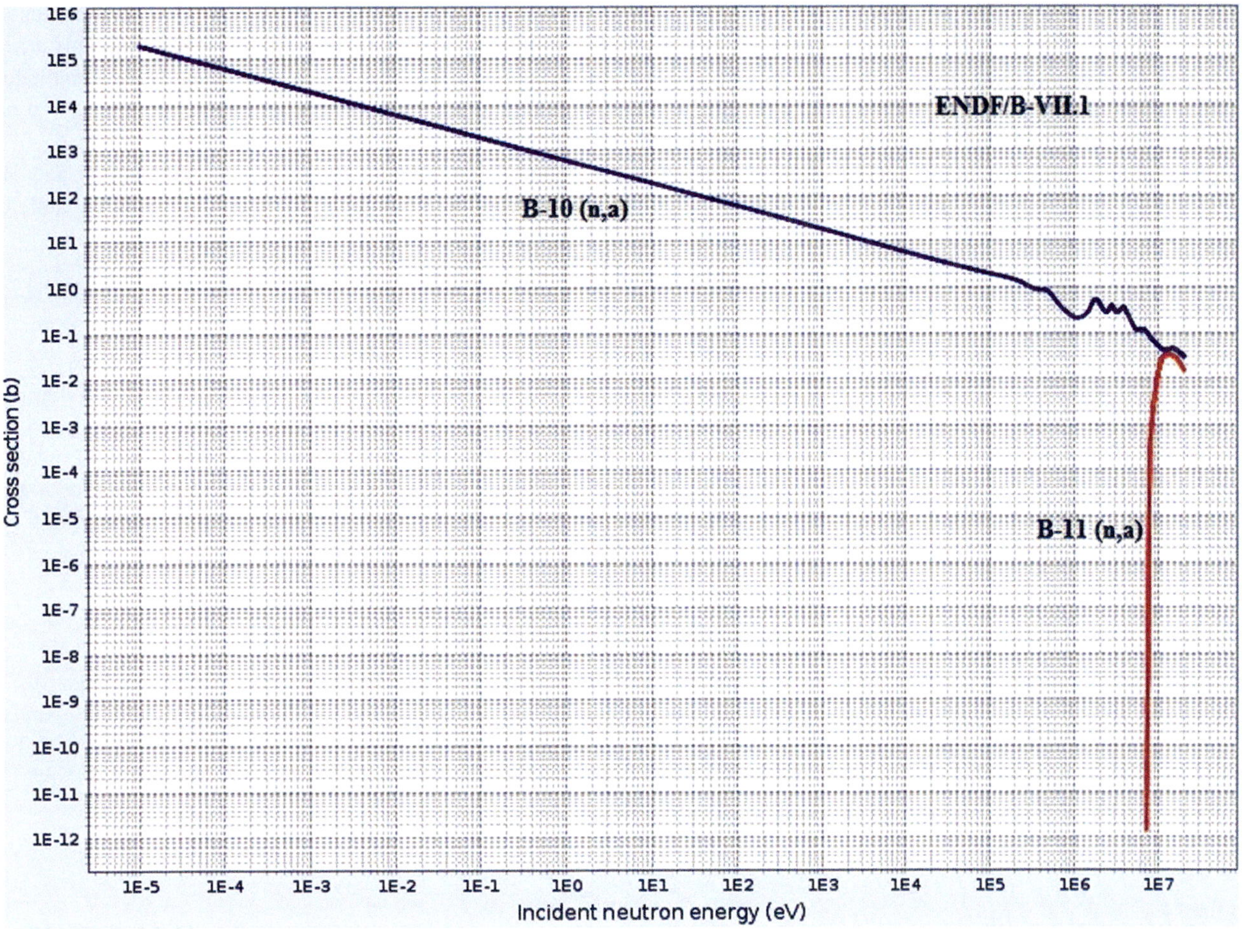

FIGURE 1.22 B(n,a) cross-section.

The atomic number density N is given by the expression $N = \rho N_{av}/M$, where ρ is the gram-density (g/cm^3), N_{av} is Avogadro's Number (6.023 $\times$ 10^{23} atoms $\cdot$ gm-atom^{-1} for an element, or molecules $\cdot$ g/mol for a molecule). This definition of Avogadro's number is crucial to the understanding of number density. As an example, consider the number density of water molecules, of hydrogen atoms and oxygen atoms in such molecules. The number of water molecules can be found as

$$N_{H_2O} = \frac{\rho_{H_2O} N_{av}}{M_{H_2O}} = \frac{1 \times 6.023\ 10^{23}}{18} = 0.3346 \times 10^{23} \text{ molecules cm}^{-3}$$

From the above, we can find the number of hydrogen and oxygen atoms: $N_H = 2N_{H_2O}$, and $N_O = N_{H_2O}$ with units of atoms/cm^3.

Example 5: Calculate the number density of an element given its density and molecular weight.

For a single element, we know that Avogadro's number of atoms N_{av}, or one gram-atom, would weigh its atomic weight A, so one gram would have N_{av}/A atoms, and for a density ρ g/cm^3, there would be

$$N = \frac{\rho N_{av}}{A} \text{atomscm}^{-3}.$$

Calculate the number density of pure U^{238} with $\rho = 19.1$ g/cm^3 and $A = 238.0508\text{g} \cdot (\text{g.atom})^{-1}$

$$N = \frac{19.1X0.602310^{24}}{238.0508} \frac{\text{gcm}^{-3} \cdot \text{atoms}(\text{g} \cdot \text{atom})^{-1}}{\text{g}(\text{g} \cdot \text{atom})^{-1}}$$

which gives $N = 0.0483310^{24}$ atomscm^{-3}.

Example 6: Calculate the atomic fractions and atomic weight of an element given weight fractions of its constituent elements.

Derivation from first principles: Consider two elements i and j, of atomic weight A_i g · (g.atom)$^{-1}$ and A_j g · (g.atom)$^{-1}$with weight fractions $\varepsilon = w_i/w$, and$1 - \varepsilon = w_j/w$, where $w = w_i + w_j$, respectively. Then, for N_i and N_j atoms/cm^3, since Avogadro's number of any substance k weighs A_k grams, we can write

$$N_i = \frac{\rho_i N_{av}}{A_i}, \text{and } N_j = \frac{\rho_j N_{av}}{A_j}, \text{so that} \frac{N_i}{N_j} = \frac{\rho_i A_j}{\rho_j A_i} = \frac{w_i A_j}{w_j A_i} = \frac{\varepsilon A_j}{(1-\varepsilon)A_i}$$

and

$$\varepsilon = \frac{A_i N_i}{A_i N_i + A_j N_j}.$$

The above can be readily used to express the atomic fractions α_k of each element in terms of the "enrichment" ε. Thus in this two-component mixture

$$\frac{N_i}{N_j} = \frac{\alpha_i}{\alpha_j} = \frac{\varepsilon A_j}{(1-\varepsilon)A_i}$$

from which

$$\alpha_i = \frac{\varepsilon A_j}{(1-\varepsilon)A_i + \varepsilon A_j}, \text{with} \alpha_j = 1 - \alpha_i.$$

From the above, we can write the mass of 1 g.atom, or its atomic weight $\overline{A}$ as

$$\overline{A} = \alpha_i A_i + \alpha_j A_j$$

so that

$$\frac{1}{\overline{A}} = \frac{\varepsilon}{A_i} + \frac{1-\varepsilon}{A_j}.$$

A relation between atomic fraction and weight fraction can be readily obtained as

$$\frac{\alpha_i}{w_i} = \frac{\overline{A}}{A_i}.$$

Application: Consider U^{235} and U^{238} mixed with weight fractions $\varepsilon = w_5 = 0.03$ and $1 - \varepsilon = w_8 = 0.97$, respectively. The average atomic weight of the mixture $\overline{A}$ and the atomic fractions α_5, α_8 are required to be determined.

The average atomic weight is found as

$$\frac{1}{\overline{A}} = \frac{\varepsilon}{A_i} + \frac{1-\varepsilon}{A_j} = \frac{0.03}{235.04} + \frac{0.97}{238.05}, \overline{A} = 237.9586,$$

and the atomic fractions are

$$\frac{\alpha_5}{w_5} = \frac{\overline{A}}{A_5} = \frac{237.9586}{235.04}, \alpha_5 = 0.0304.$$

Similarly

$$\frac{\alpha_8}{w_8} = \frac{\overline{A}}{A_8} = \frac{237.9586}{238.05}, \alpha_5 = 0.9696.$$

Example 7: Calculate the density of a mixture prepared from two elements of given densities and weight fractions.

Derivation from first principles: Consider a substance made by mixing two elements A and B, of density ρ_A g/cm^3 and ρ_B g/cm^3 with weight fractions w_A, andw_B, respectively. The density of the mixture ρ_{mix} is found from the 'first principles' approach as follows.

Consider a volume of 1 cm^3, in which the volume fractions of elements A and B are V_A and V_B respectively. Then

$$V_A + V_B = 1.$$

Thus for x grams of the mixture, which in this case is also ρ_{mix}, the individual amounts are ρ_A V_A and $\rho_B V_B$ respectively, and so

$$\rho_A V_A + \rho_B V_B = x.$$

Thus the volume fractions can be found, from which the density is

$$\frac{1}{\rho_{mix}} = \frac{w_A}{\rho_A} + \frac{w_B}{\rho_B}.$$

The above can be generalized, for a $n-$ component mixture, to

$$\frac{1}{\rho_{mix}} = \sum_{i=1}^{n} \frac{w_i}{\rho_i}.$$

Example 8: Calculate the number density of each element in a molecular substance of given density and weight fraction.

Calculate the atomic densities of U^{235}, U^{238}, and O_2 in U (4 wt. %) O_2 fuel pellets. Assume that the density of UO_2 is 10.9 g/cm^3.

Calculate the average atomic weight of U:

$$\frac{1}{\overline{A}_U} = \frac{\varepsilon}{A_i} + \frac{1-\varepsilon}{A_j} = \frac{0.04}{235.04} + \frac{0.96}{238.05}, \overline{A}_U = 237.9281.$$

The atomic fractions are found to be $\alpha_5 = 0.0405, \alpha_8 = 0.9595$.

The molecular weight of UO_2 can now be determined, since one molecule of UO_2 has one atom of U and one molecule (2 atoms) of oxygen.

$$\overline{A}_{UO_2} = \overline{A}_U + 2\overline{A}_O = 237.9281 + 2(16) = 269.9269.$$

Calculate the molecular density of UO_2:

$$N_{UO_2} = \rho_{UO_2} N_{av} / \overline{A}_{UO_2} = 0.02432\ 10^{24} \text{molecules}/\text{cm}^3.$$

The atomic density of U and O_2 can now be determined as

$$N_U = N_{UO_2} = 0.02432\ 10^{24} \text{atoms}/\text{cm}^3$$

and

$$N_O = 2N_{UO_2} = 0.04864\ 10^{24} \text{atoms}/\text{cm}^3.$$

From the atomic fractions of U^{235} and U^{238}, calculate the atomic densities of U^{235} and U^{238}

The atomic fractions are $\alpha_5 = 0.04049$ and $\alpha_8 = 0.95951$, and from Eq.(-) the individual atomic densities of U^{235} and U^{238} are found to be

$$N_5 = 0.00098\ 10^{24} \text{atoms}/\text{cm}^3$$

and

$$N_8 = 0.02334\ 10^{24}\text{atoms/cm}^3.$$

Preparing such "mixture" cross-sections is a crucial "preprocessing" exercise for retrieving elemental cross-section data from a data library and multiplying by the constituent number densities.

1.6 Mathematical modeling and simulation

The interaction mechanisms of charged particles and photons discussed in the previous sections are the basis of mathematical modeling and computer simulation using the relevant databases and computer programming. Their applications are diverse such as in shielding, radiation therapy, plasma physics, and astrophysics. For mathematical modeling, the most accurate representation of particle transport is by the Boltzmann Transport Equation (BTE) which describes the transport of particles in phase space but suffers due to its complexity as an integro-differential equation making analytical solutions possible only for regular geometries restricted to simple scattering laws. In a deterministic approach, the linear BTE has an asymptotic limit expressed as the Fokker-Planck equation (BT-FPE) (POMRANING, 1992) equation with roots in quantum and statistical mechanics as an evolution of the probability distribution of a system with drift and diffusion. In another formulation, the simplification of the "in-scattering" integration term resulted in a form of the Fokker-Planck equation convenient for modeling electron, ion and photon transport.

The BTE is a central focus area, along with its stochastic counterpart: the Monte Carlo method which is an outcome of the integral form of the BTE. These aspects form the core of this book in chapters developed to their development and applications.

As will be highlighted, the integro-differential BTE was, in the pre-supercomputing era, the main model for neutron transport in nuclear engineering, and thus a lot of effort has gone into analytical solutions for idealized models, and the development of powerful theoretical and numerical models in 3D codes capable of solving realistic problems with discretization in a finite number of energy groups, angle discretization as in the discrete ordinates (S_N) and spherical harmonics (P_N) methods, and space discretization with finite-element, finite volume, nodal and several advancements of numerical methods.

In simulations, it is possible to model realistic configurations and use elaborate theoretical models with extensive databases to obtain reliable answers. Real-world systems and engineering designs are not restricted to regular geometry and hence the first step is to model the geometry. This is fortunately possible by methods, such as combinatorial geometry, which will also be described in detail. The second step is the use of elaborate physics models with the support of voluminous data compiled from sophisticated models, empirical formulas and experiments. The power of simulation is thus strong enough to undertake realistic design analysis in nuclear engineering, and it is precisely the strength of such mathematical modeling enforced and advanced by simulation with modern high performance computing that is the essence of education today.

For simulation, efficient numerical methods are continuously being developed for deterministic as well as stochastic methods. These will be covered in detail in later chapters, especially the Monte Carlo method which gains its widespread acceptance from powerful methods as well as from the features of high performance computing such as processing speed and vector and parallel processing.

Computer codes provided by contributing research and academic institutions are maintained at organizations including the International Atomic Energy Agency (IAEA), the OECD Nuclear Energy Agency (NEA), and Radiation Safety Information Computational Center (RSICC) at Oak Ridge National Laboratory. Extensive databanks are available from several sources such as the National Institute for Science and Technology (NIST), the Korean Atomic Energy Research Institute (KAERI), and the National Nuclear Data Center (Bielajew, 1990).

Of the several available particle transport codes, some are SRIM (Ziegler, Ziegler, & Biersack, 2010), electron gamma shower EGS (Kawrakow & Rogers, 2003), the Monte Carlo particle simulation code GEANT (Research, 2020), AlfaMC (Peralta & Louro, 2014), MCNPX (Waters et al., 2007), the electron and photon coupled Monte Carlo code PENELOPE (Salvat et al., 2011), the multiparticle transport code FLUKA (Battistoni et al., 2016). For electron transport ETRAN (Muraz et al., 2020; Seltzer, 1991) with the stopping power databases from ESTAR, ASTAR and PSTAR for electrons, alpha particles and protons (Berger et al., 2005) are extensively used.

Both deterministic (based on the BTE) and stochastic (based on MC) methods are extensively used for charged-particle simulations and have their advantages and disadvantages. Decisive factors are clearly, their ability to model realistic design problems and the underlying physics, and to give reliable results with computational efficiency (Adams & Martin, 1992; Patel, Warsa, & Prinja, 2020). In the former, this means acceleration schemes while in the latter it means a careful selection of the simulation parameters which include several artifacts such as biasing, variance reduction, and importance sampling.

Exercise 8: What is the commonality between the Boltzmann Transport Equation, the Neutron Transport Equation, and the Fokker-Planck equation?

1.6.1 Alpha particle transport simulation

The Monte Carlo simulation of alpha particles is based on simulating events in the transport of a particle from its "birth" as a source particle to its "death" when it has lost all of its energy during interactions or has escaped from the system. One birth-to-death sequence is called a random walk or a history. The word "random" is used because laws of probability apply at each "event" in a history so the next one could consist of a different sequence. Many such histories are followed to accumulate the tallies of interest. Each event in the random walk is modeled using the Bethe-Bloch stopping power formula described in Section 1.3.1. The flowchart, in a very simplified form, is shown in Fig. 1.23. Simulation parameters, such as the number of histories to be simulated, the number of materials to be used, and cut-off parameters are specified. The tallies (e.g., energy deposited, transmitted energy, energy straggling, the CSDA range and the projected range) are initialized. The energy loss and step size (% of energy allowed for each step) is specified so that the step size is assigned.

A new particle with given parameters is started, and "located" in a volume and given a step length in its initial direction. At the end of the step, its position and energy are updated and a check is carried out to determine if the new position is in the same volume or whether it is a boundary crossing or system escape. The counters are updated accordingly.

For charged-particle transport, the interaction physics is based on the coulomb electric force in collisions between ions and electrons or ions and nuclei. One method used is based on the Continuous Slowing-Down Approximation (CSDA) in which it is assumed that at every point in the trajectory, the energy loss by the alpha particles is gradual modeled by the unrestricted (i.e. not cut-off energy to specify catastrophic events) stopping power.

For large energy losses in a target, the "thick target model" is used in which a Gaussian straggling model is used to estimate the energy loss. Conversely, for intermediate and low energy losses the Vavilov and Landau distributions have been used (Peralta & Louro, 2014; Ziegler, Ziegler, & Biersack, 2010). Stopping powers are used from various models or from databases such as ASTAR (Berger et al., 2005) for both the nuclear and electron contributions.

Postcollision angles are determined from multiple scattering distributions. The change of angle in the scattering process is by a series of many small angles to constitute "multiple scattering"; this is modeled by a Gaussian distribution. In the case of large angle scattering, where the alpha particle would lose a large part of its energy the distribution of angles would have a Rutherford scattering behavior with larger tails than a Gaussian. In the AlphaMC code, the standard deviation of the deflection angles varies as $\theta_0 = \frac{13.6\text{MeV}}{pc\beta} Z_\alpha \sqrt{\frac{s}{X_0}}$ so that the true path length is then $t = s + \frac{K}{4} s^2$ where $K = \frac{\left[13.6\text{MeV}\left(\frac{Z_\alpha}{pc\beta}\right)\right]^2}{X_0} \sim \frac{{\theta_0}^2}{s}$; this correction is important as alpha particle energy decreases and standard deviation θ_0 increases as well as for high Z_α.

The history is continued until the particle comes to a stop or escapes from the system. Many more such histories are simulated to obtain reliable statistical results with standard deviation and variance.

As the alpha particles slow down, their stopping power increases due to the Bragg peak followed by a sudden end of the trajectory.

The range of alpha particles is one of the several quantities estimated from such simulations. The range calculated from CSDA is based on the evaluation of the integral $R_{\text{CSDA}} = \int_0^{\text{KE}_\alpha} dE \frac{dE}{dx}^{-1}$; projected range is the mean value of the maximum penetration depth computed from the distribution function of particles.

1.6.2 Interaction of electrons with matter

Following Section 1.3.2, the simulation of electron transport (Olbrant & Frank, 2010) is based on the interaction of beta radiation incorporated through a realistic simulation of elastic and inelastic processes using forward-direction scattering and transport coefficients extracted from the ICRU 77 database (Berger et al., 2007). This reduction introduces the stopping dower into the BT-FPE. One application carried out with this formulation for light propagation in biological

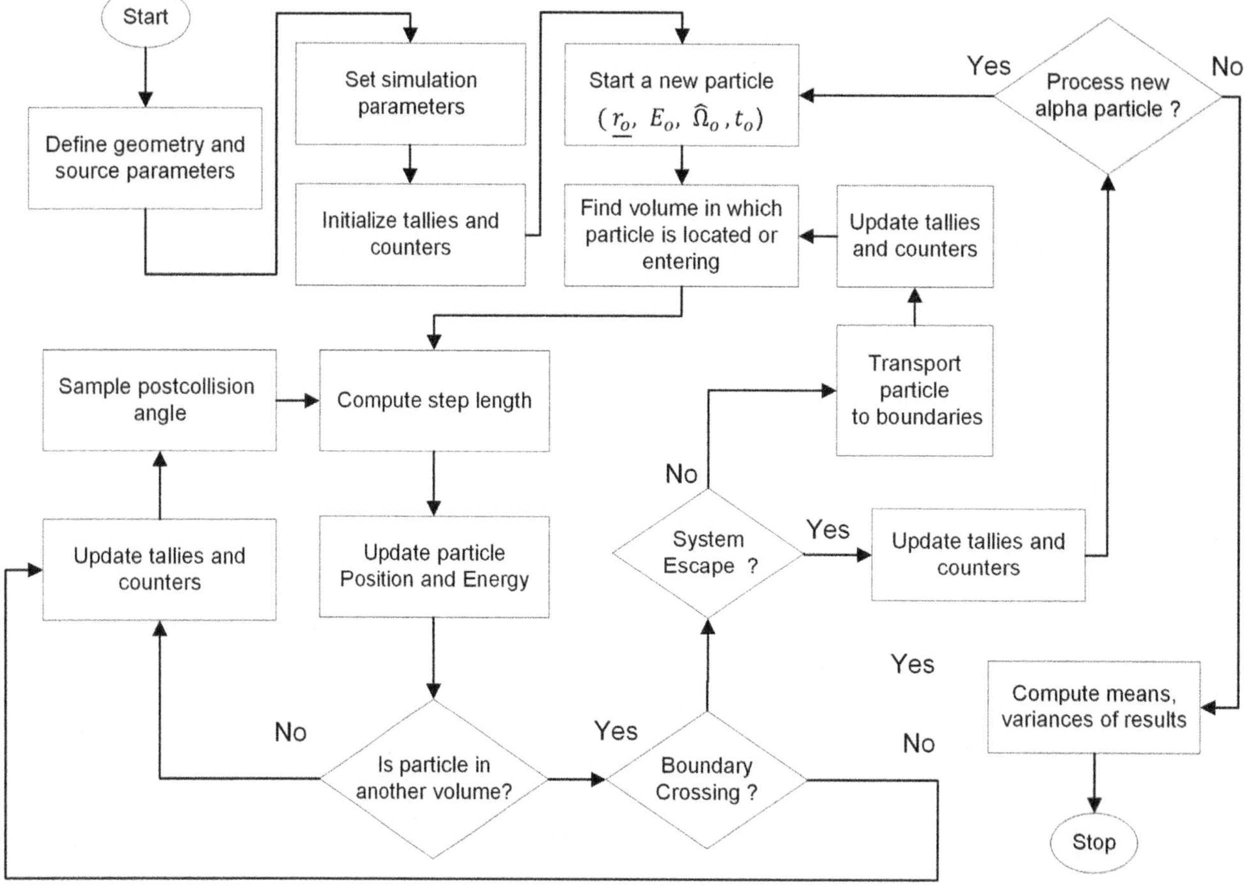

FIGURE 1.23 Flowchart of alpha particle transport simulation.

tissues calculates reflectance and transmittance in the liver tissue (González-Rodríguez & Kim, 2008; Olbrant & Frank, 2010) and compares results with the transport equation discrete ordinates method taken as a benchmark. In another application, the absorbed dose $D(\bar{r})$, from 5- and 10-MeV electron beams in a semi-infinite water phantom, is calculated via the stopping power and the angular flux from the transport equation. Good agreement is found with the standard physics MC code GEANT and with PENELOPE. The compared results are for cancer therapy applications of an electron beam of 10-MeV on muscle (0–1.5 cm) followed by bone (1.5–3 cm) and then lungs (3–9 cm).

In the stochastic approach, the histories of a large number of electrons are followed and tallies gathered during the simulation to obtain estimates based on means, variances and the laws of probability.

Electron transport is much more "complicated" than that for alpha particles due to the large number of collisions dominated by the long-range coulomb force with a large number of continuous deflections. Due to the large number of small-angle collisions, the simulation of electron transport can become computationally laborious if each interaction is processed. Thus a "condensed history" approach is used in which a single event is taken to represent several, possibly hundreds of thousands, small-angle collisions.

The classic paper on electron transport by Berger (1963), using Monte Carlo simulation written in FORTRAN and implemented on an IBM704 computer, led to the development of the ETRAN code (Seltzer, 1991). In this paper, Berger gives two classes of simulation strategies varying essentially in terms of their selection of slowing-down energy and multiple scattering models. A good understanding of the simulation process, based on a random walk, can be obtained from Berger's "Class I" model based on the simplest approach as shown in Fig. 1.24. The process consists of the following steps:

1. At the start of the simulation, the history counter is set to $n = 0$ for a maximum number of histories N to be simulated, and the slab thickness, material and the source parameters $\bar{r}_0, E_0, \hat{\Omega}_0, t_0$ are given. Tallies and counters, such

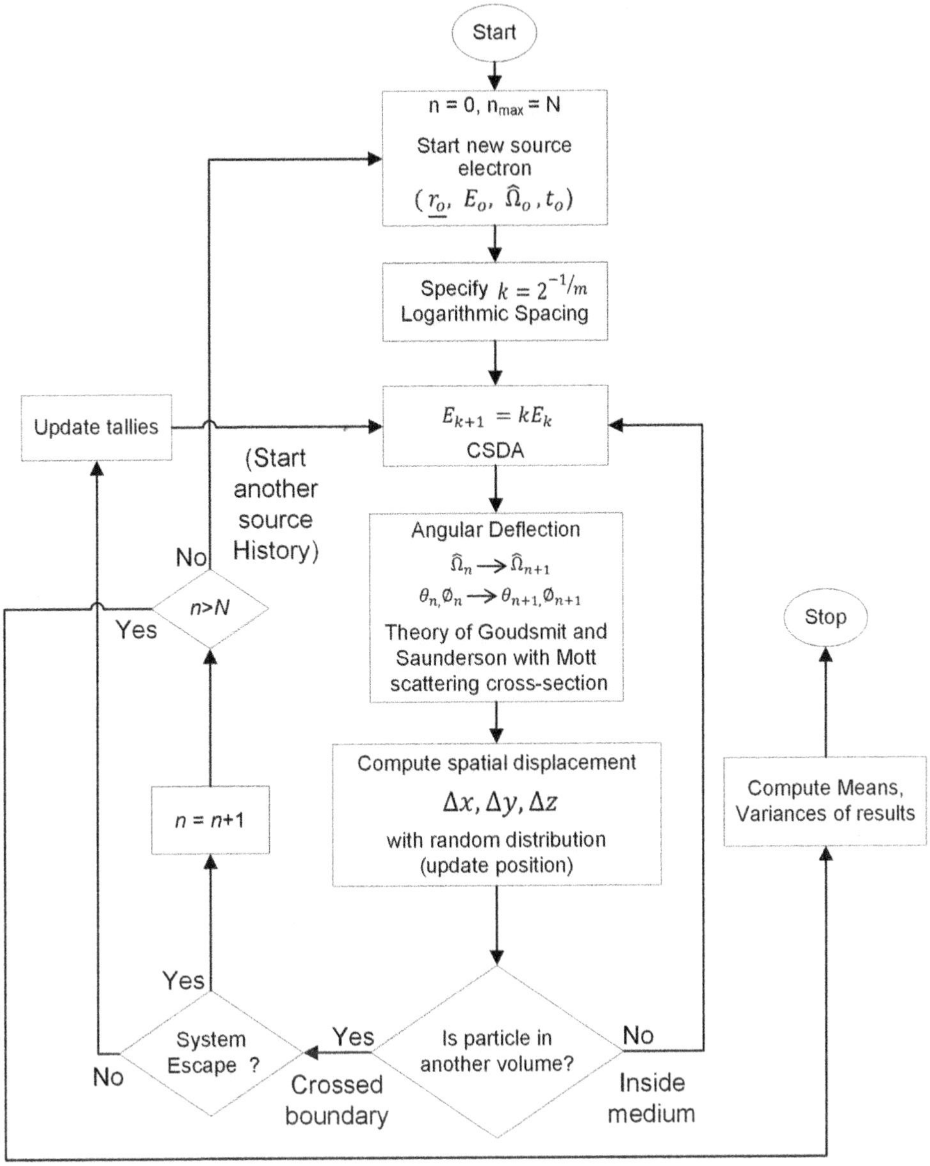

FIGURE 1.24 Flowchart of electron transport simulation.

as the number of steps in the random walk, absorptions, transmissions and reflections are initialized. Berger's program does not carry out variance reduction schemes which are an important feature of modern codes.

2. The energy loss index $k = 2^{-\frac{1}{m}}$, for $E_{n+1} = kE_n$ in a condensed step is specified for the number of steps in a logarithmic spacing in which the electron loses half its energy; for example, $m = 4$ would consist of four steps in which the electron energy would be $E_0, E_1 = 2^{-\frac{1}{4}}E_0, E_2 = 2^{-\frac{1}{2}}E_0, E_3 = 2^{-\frac{3}{4}}E_0$ and $E_4 = 2^{-1}E_0$.
3. Angular deflections are computed considering kinematics with mean square deflections in multiple scattering calculated from the Goudsmit-Saunderson model in conjunction with the Mott scattering cross-section; accounting for deflection of electrons. The spatial displacement of the electron is computed by accounting for straggling through random variables.
4. The energy loss of the electron in the medium is computed once a collision has been processed.
5. Checks are made for boundary crossing to update counters and begin a new history until the maximum specified histories are processed.
6. The tallies are estimated from arithmetic means, if variances are not calculated. Most codes now compute variances and relative standard errors to give figures of merit to establish convergence.

The flowchart of a class II algorithm, used in EGS4, is different from the class I described above (Bielajew and Rogers) as it samples the distance to an interaction rather than specifying the step size. It also calculates the energy loss rather than sampling it from a distribution. Such differences have an effect on the results and thus simulation, especially Monte Carlo simulation, is considered both an art and a science where parameters are selected based on some intuitive sense of what is "more important" and otherwise. These aspects of simulation, such as importance sampling and variance reduction, will be covered later in this book. A similar selection of parameters is required in MCNP (Hughes, 1996) and its upgrade (Grady Hughes, 2014) which does not use restricted stopping powers and builds distribution tables such as energy-dependent histograms for angular bins from Goudsmit-Saunderson for substep lengths and scattering angles, as well as for bremsstrahlung production probabilities. These parameters are selected in MCNP on the PHYS:E card where default, or user-specified values are used for the simulation. The values include the upper limit for electron energy, the choice of whether photons will produce electrons, whether electrons will produce photons, the use of full bremsstrahlung or simple angular distribution approximations, the use of sampled or expected straggling, and the number of electron-induced X-rays, the number of knock-on electrons, the number of photon-induced secondary electrons and the option of producing bremsstrahlung at each substep.

The simulation used in MCNP for positrons is identical to that for electrons distinguishing the particles for purpose of tallying only. For electron transport in aluminum, Berger sets $m = 16$, that is, 16 steps for the electron to lose half its energy. This covers the range of slowing-down from 2.0 to 0.03125 MeV in 97 steps as shown in Fig. 1.25 below. The step size is

$$\Delta_s = \frac{(k-1)E_s}{\left|\frac{dE(s')}{ds'}\right|}$$

From Fig. 1.24 it is seen that that in the range 0.01−2 MeV the stopping power decreases; thus while slowing-down, the energy E_s and stopping power increases thus $\Delta_s$decreases. In Berger's simulation, the step size is $\Delta_s = 0.057476\ \text{g/cm}^2$ at 2 MeV for the first energy interval in steps 1−16, while in the last interval (steps 81−96), $\Delta_s = 0.000184\ \text{g/cm}^2$. In all intervals, the deflection angle increases from 11.4 degrees in the first interval to 16.5 degrees in the fourth interval and then decreases to 14.7 degrees.

For multiple scattering, Goudsmit and Saunderson (1940a, 1940b) derived the exact angular distribution for the scattering angle as a Legendre series. Here

$$A_{GS}(\omega)\sin\omega\, d\omega = \sum_{i=0}^{\infty}\left(l + \frac{1}{2}\right) exp\left\{-\int_0^s G_l(s')ds'\right\} P_l(\cos\omega)\sin\omega d\omega$$

where

$$G_l(s) = 2\pi N \int_0^{\pi} \sigma(\theta, s)\{1 - P_l(\cos\theta\}\sin\theta d\theta$$

The GS distribution peaks at small values of $\omega \sim 10^\circ - 25^\circ$ as shown for a calculation for a pencil beam of 20 MeV electrons passing through a 0.25 cm thick slab of water (Rogers & Bielajew, 1990).

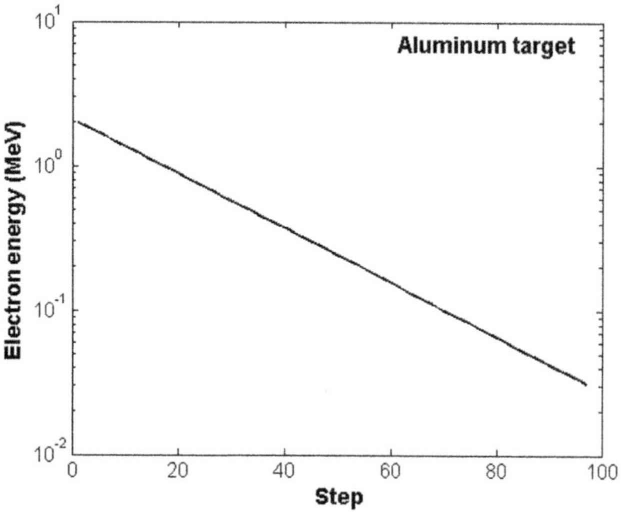

FIGURE 1.25 Electron energy versus steps in simulation in aluminum from 2.0 to 0.03125 MeV for $m = 16$.

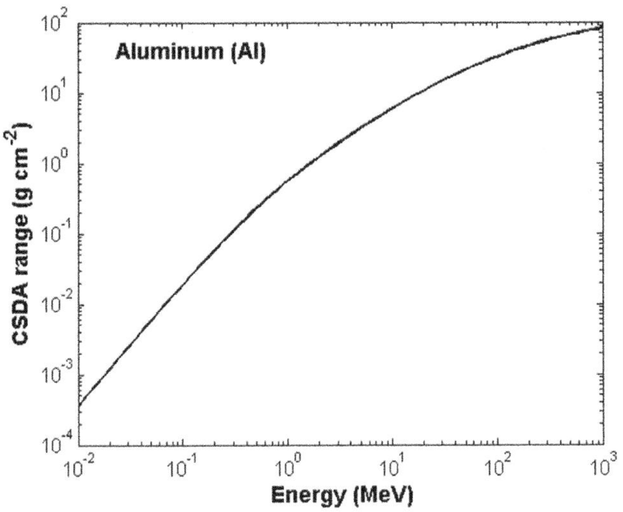

FIGURE 1.26 Continuous Slowing Down Approximation range in aluminum from ESTAR.

The problem considered by Berger was a one-dimensional slab with a given source in phase space $P(\bar{r}_0, E_0, \hat{\Omega}_0, t_0)$. The simulation was carried out to estimate the transmission, reflection, and absorption of electrons in a slab of aluminum for which the range is shown in Fig. 1.26.

The effect of keeping the smallest step size for all energy values would thus be seen to be computationally inefficient.

In other strategies for simulation, path length is modeled with log spacing so that energy loss is constant per step or a mixed-log model depending on where the electron is (near a boundary or inside the medium), or fixed path length making it shorter to keep angular deflections small.

Similarly, energy loss models can be continuous slowing-down or fluctuation models. Angular deflections are computed considering kinematics with mean square deflections in multiple scattering as in GEANT4 for a detailed description of multiple scattering collisions calculated from single-scattering Rutherford scattering law, Gaussian distribution or Molière distribution (considering the occasional large angle event which are not incorporated in the Gaussian model). In GEANT4, the Lewis theory computes moments of the spatial distribution as well as described by Urban (2005, 2006) with results given for distributions of transmission, energy deposition and spectra of MeV electrons in aluminum layers.

In a comparison between the Molière and Goudsmit-Saunderson model for the Rutherford and Mott models for electrons and positrons, Berger has shown that the angular distribution for both electrons and positrons, slowing-down in aluminum from 1 to 0.9576 MeV, has ~40% each for small-angle scatterings (~41% for 0–15 degrees and ~42% for 15–30 degrees) and the rest as follows: ~12%, 3%, ~1%, ~0.36%, 0.27%, 0.08% and 0.02% for each successive 15 degree interval to 180 degrees. Thus the chance for a "catastrophic" event in which a large deflection occurs with a strong influence on the subsequent history has a probability of occurrence for a large number of collisions in a step. Alternately, catastrophic collisions can be explicitly incorporated by specifying a cut-off energy loss beyond which such an event occurs.

In more detailed models, such as in MCNP (Werner, 2017; X-5 Monte Carlo Team, 2008) simulations, production of secondary electrons are also considered (electron-induced X-rays, electrons produced from electron-impact ionization called "knock-on" electrons, and bremsstrahlung photons) particles shown in MCNP5 print Table 86 below.

In MCNP, the algorithms used are the Goudsmit-Saunderson theory for angular deflections, and the Landau theory for modeling energy loss fluctuations. Straggling is accurately modeled when path length steps are less than 0.5 g/cm^2 (McLellan, Med Phys, 1994).

There is considerable variation in commonly used codes. In the EGS4 code (Bielajew and Rogers) electron and positron models include standard models for low-energy Møller and Bhabha scattering as part of restricted (implying less than a specified cut-off) collision stopping power, atomic excitation, soft bremsstrahlung and elastic particle multiple scattering, as well as provisions for modeling "catastrophic" phenomena including large energy loss scatterings. The parameters, such as cut-off thresholds, selected in a model have a bearing on the results. Clearly low thresholds will use more computer time but would give better results which may not always be needed beyond some level of accuracy. In energy deposition computations, both thresholds of secondary particle creation and bremsstrahlung production would have an effect on the results and a user would need to model with carefully selected parameters considering whether the nature of the problem corresponds to a thin or thick target.

Comparisons have been made between MCNP, GEANT, and PENELOPE (Archambault & Mainegra-Hing, 2015) to highlight the effect of algorithms in the simulations. For beam energies of 0.5, 1.0, and 5.0 MeV on a water-filled sphere, the estimated energy depositions showed good agreement for the single-scattering calculations. However, differences of up to 5% have been found between EGSnrc and PENELOPE for some problems such as the "in air" case for decreasing radius at 0.5 MeV.

1.6.3 Interaction of gamma radiation with matter

In photon transport simulation shown in Fig. 1.27, a Monte Carlo simulation would begin, as for alpha and electron transport, with a description of the geometry, materials, and photon source. Then the simulation parameters such as number of photon histories (random walks) to simulate, "weight cutoffs," the interaction physics treatment.

In a code like MCNP, the user has the option to use "simple photon transport," where, for example, for high-energy photons, coherent (Thomson) scattering and fluorescent photons production is ignored, or the "detailed physics transport" in which coherent scattering and fluorescent photons are transported and form factors $F(Z,E,\theta)$ are used to account for binding effects. The Thomson 'unpolarized' differential cross-section (Section 1.3.3) is then written as

$$\frac{d\sigma(Z,E,\theta)}{d\Omega}|_T = \frac{1}{2}r_e^2\left(2-\sin^2\theta\right)F(Z,E,\theta)$$

For which data is available for several material. The simple model reduces simulation effort and makes assumptions that the photoelectric effect is an absorption process and its simulation continues or stops according to statistical criteria. For Compton scattering, with appropriate probability, the collision is simulated and the photon transport continues while the recoil energy is deposited at the collision site and can be used to generate a recoil electron for further transport.

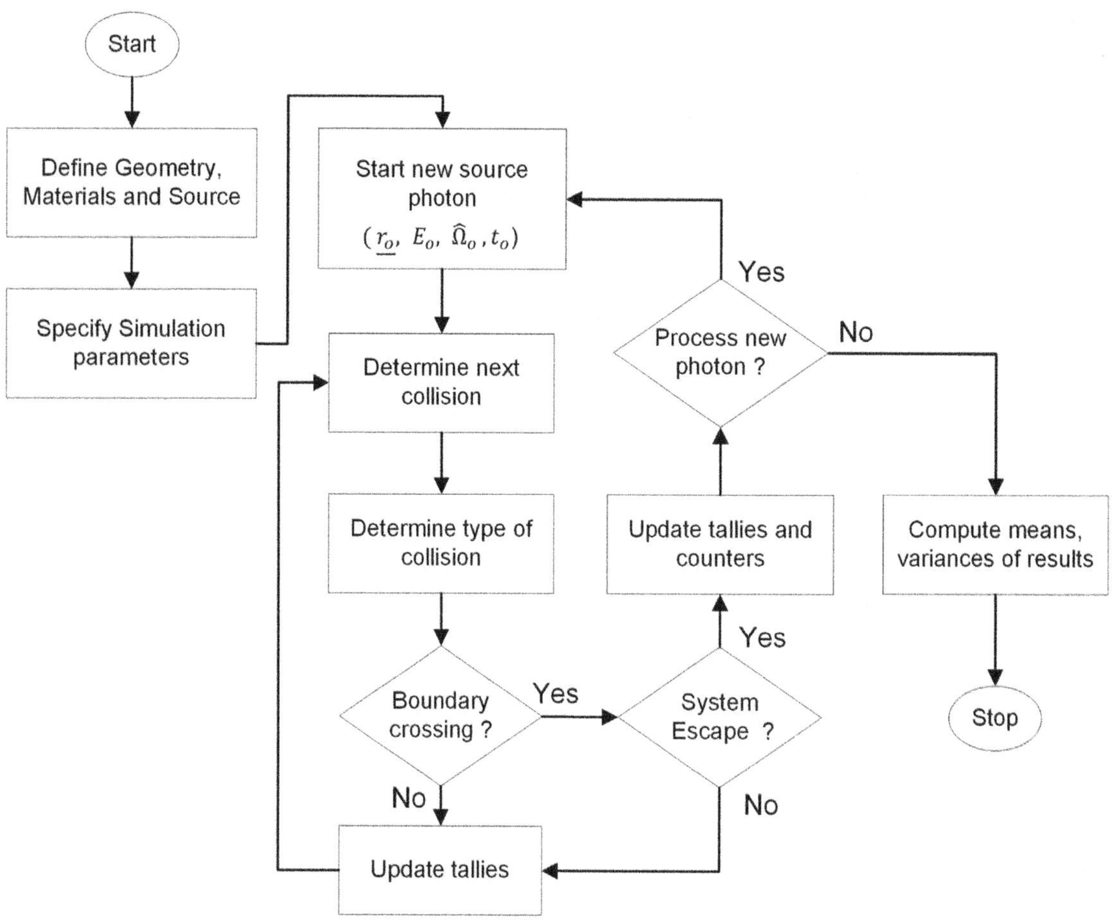

FIGURE 1.27 Flowchart of photon transport simulation.

A typical photonuclear data library would contain the following interaction data for photons.

The simulation can be specified to run single mode or mixed-mode (photon-electron). Electron interaction data for Z = 1 to 94 contain stopping power parameters, bremsstrahlung data as well as data for K-edge energies, Auger electrons.

Exercise 9: In the transport simulations for alpha particles, beta particles, and photons, what are common features in Monte Carlo simulations that may contribute to better results and what would be their effect on their computation time?

1.6.4 Radiation dose from Calfornium-252 gamma source in water

Many practical problems require mixed-mode simulations where, for example, a photon transport simulation would include the transport of electrons produced which could subsequently result in the production of X-rays. In such mixed-mode "P-E" simulation, the electrons created by photons are banked and stored for later transport. When the electron mode is not initiated, then a thick target bremsstrahlung (TTB) model is used in which electrons are generated but locally stopped. When electrons are not transported, then their energy is assumed to be locally deposited.

This mini-simulation is a step toward understanding the elaborate physics that is incorporated into simulation tools.

Consider a point source emitting S gammas per second incident from the left in a straight line (anisotropic) striking a rectangular material labeled a detector "D" as shown in Fig. 1.28. We carry out a simulation to estimate how many gammas reach across the transverse water layers.

The gamma source, Cf^{252} placed at the center, has an energy spectrum shown in Fig. 1.29. This simulation is carried out to tally the currents and dose inside the transverse water shells of radii 0.2, 0.4, 0.6, 0.8, 1.0, 1.2, 1.4, 1.6, 1.8 and 2.0 cm and length 1 cm.

The MCNP program Book1_06 is attached in Annex 1 and the MATLAB program from which these figures were produced, is listed in Annex 3.

This simulation uses the photo-atomic data for hydrogen and oxygen from the Evaluated Nuclear Data File ENDF/B-VI and electron data from the file el03. Each region is assigned equal statistical importance for photons and electrons. In the simulation, the maximum photon energy is 100 MeV.

A typical code output describing the libraries used would read: The mcplib04 library used is 1000.04p 1898 ENDF/B-VI Release 8 with photo-atomic data for hydrogen (1-H mat 100) and oxygen 8000.04p 3272 ENDF/B-VI Release 8 photo-atomic data for 8-O mat 800 02/07/03 indicating the release version and the element ID.

For the electron step size, MCNP uses the value $k = 2^{-1/8}$ to compute steps on a grid from a high value, in this case E = 100 MeV down to a given low value, 1.079 keV, which at an average loss of 8.3% per step requires 133 steps. At these steps, the collision and radiation stopping powers are determined from the formulas described in Section 1.3 (Bethe, 1930). These major steps are broken down into m substeps where the value m ranges from 2, for $Z < 6$ to 15 for $Z > 91$. In this simulation, $m = 3$. The range table gives stopping power (collision, radiation, total), the range from CSDA, the value of "drange" which, divided by the density gives the step size in cm. Five lines of the 133-line table 85 of MCNP, steps 1, 2, 10 and 100 give the following values for the energy, stopping powers, CSDA range and "drange."

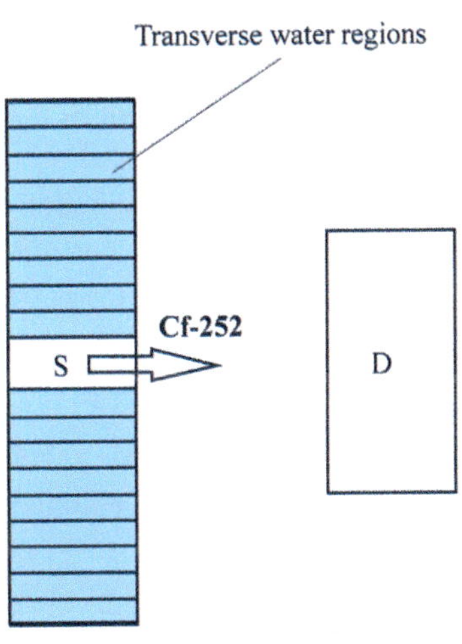

FIGURE 1.28 Source and detector configuration.

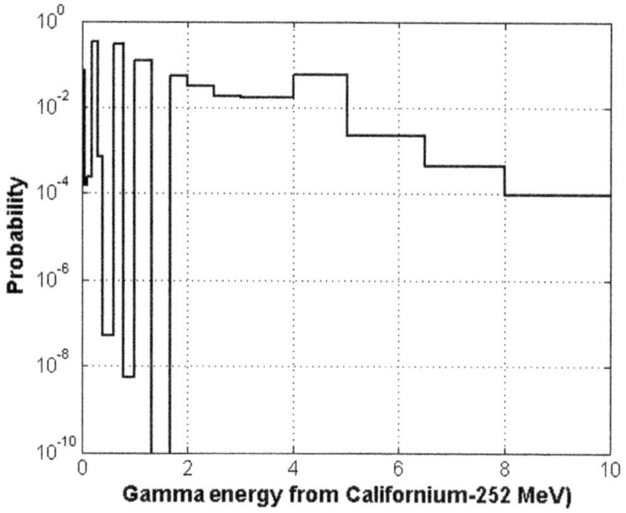

FIGURE 1.29 Probability distribution function of cobalt-60 gamma source.

TABLE 1.9 MCNP simulation steps for electron transport.

Step	Energy (MeV)	Stopping Power (MeV cm^2/g)		Range (MeV cm^2/g)	drange (MeV cm^2/g)
133	1.079×10^{-3}	1.140×10^{-2}	2.888×10^{-3}	4.873×10^{-6}	7.647×10^{-7}
100	1.8826×10^{-2}	13.8	3.938×10^{-3}	7.70×10^{-4}	1.095×10^{-4}
9	50	2.140	1.145	19.83	1.286
2	91.7	2.195	2.220	30.74	1.765
1	100	2.203	2.438	32.58	1.833

Table 1.9 shows that the energy from step 1 to step 2, and all successive steps, decreases by 8.3% due to the selection of the value $2^{1/8}$. Similarly, half the energy is lost in eight collisions from step 1 to step 9. The next two columns give the collision and radiative stopping powers, while the fourth column gives the CSDA range and the last column gives the "drange" value giving a step size of 1.833 cm in the first step, which reduces to 7.647×10^{-7} at the 133rd step where the electron energy is 1.079 keV.

As shown in the flowchart for electron transport, the condensed random walk is sampled using interpolated data from the grid. Each path length s (of length "drange") in the first major step consists of s/m minor steps for which the Goudsmit-Saunderson model is used for angular deflection. At each collision point, the probability of secondary particle (fluorescent radiation or knock-on electrons) production is estimated. Bremsstrahlung photons are produced according to a Poisson distribution with energy sampled from distribution tables. The electron energy is subsequently reduced to account for this production. Similarly, the ionization from K shell impact and Auger electrons is also included in the cross-section library.

The electron secondary production table is generated for the same energy grid as above, with values for bremsstrahlung, X-rays and knock-on electron cross-sections. For energy straggling, the term responsible for the difference of the actual path from a straight path, sampling is done from Landau's distribution function which has been further extended by Blunck and Leisegang to estimate the variance of the straggling Gaussian (Hughes, 1996).

The simulation is started by sampling photons from the specified source. In this case, all source photons are generated at the origin with direction cosine (1,0,0); the energy is sampled from the PDF (Fig. 1.28).

For 1 million particles simulated, the photon summary shows that the average source energy is 0.97764 MeV, for which the production is mainly from bremsstrahlung (722 tracks, energy 2.2502 10^{-5} MeV) and from photon annihilation; thus for every track started 1.0008 tracks are simulated. The loss terms are mainly from escape (986466, average energy 0.96453 MeV), capture (14224, 2.6118 10^{-4} MeV) and pair production (98, 4.0133 10^{-4} MeV). On the average there were only 67256 ($\sim$6.7%) photon collisions.

Similarly, the electron summary table shows creation of electrons from pair production (196, 3.0117 10^{-4} MeV), from Compton recoil (47805, 1.2507 10^{-2} MeV), photoelectric effect (14224, 2.6118 10^{-4} MeV) and from knock-on events (623758, 2.4104 10^{-3} MeV). Thus total production was 685983 electrons with an energy of 1.5480 10^{-2} MeV. The total electron energy would thus be obtained by multiplying by the source term; for example if there are 10^6 photons per second emitted from the source, the electron energy produced would be 1.5480 10^4 MeV. The electrons were "lost" in the simulation by escape (4642, 4.8246 10^{-3} MeV) and by energy cut-off (681341, 6.6631 10^{-4} MeV).

The above summary tables should be carefully analyzed to understand the essential phenomena taking place in the problem of interest.

From the overview of the accounting described in the summaries, the activity table gives estimates of the average track, that is, movement of photons in each transverse water region. From this table, we learn that there were 9554 collisions with hydrogen atoms and 52381 collisions with oxygen atoms in the innermost water region (cell 1). We also see that in the transverse direction, the number of collisions reduces drastically from innermost regions in the transverse direction due to the "current" or boundary conditions reducing from 3.15% (0.0057) to 0.63% (0.0126) with the standard relative error increasing about three times, as seen in Fig. 1.30.

In later chapters, we will learn how we can reduce the errors in estimates to get better more reliable results. Similarly, the photon flux across the right surface of the inner source cylinder region is 7.95881 10^{-2} photons/cm^2 (0.0005) with a surface area 12.5664 cm^2. The volume averaged photon flux in the transverse regions is shown in Fig. 1.31 to decrease from 3.9512 (0.0001) photons/cm^2-s to 5.992 10^{-4} (0.0126) photons/cm^2-s. In the detector region, the flux is 3.0178 10^{-1} (0.0002) photons/cm^2-s.

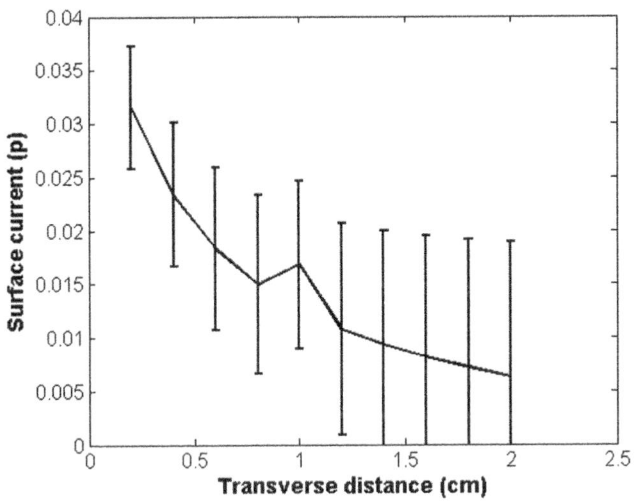

FIGURE 1.30 Surface current of photons across transverse surfaces.

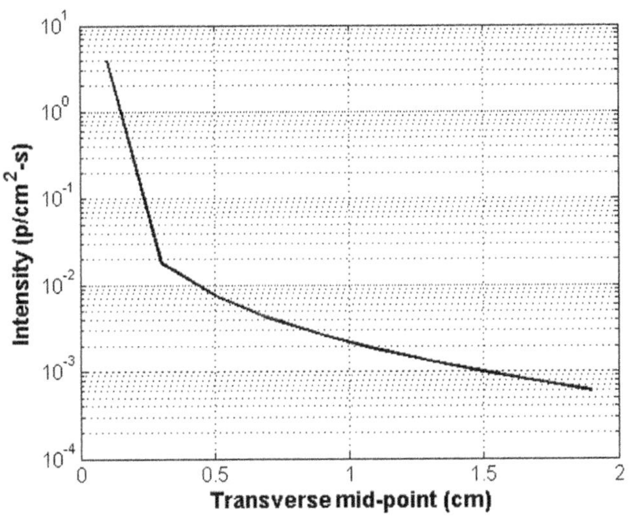

FIGURE 1.31 Intensity of photons across transverse surfaces.

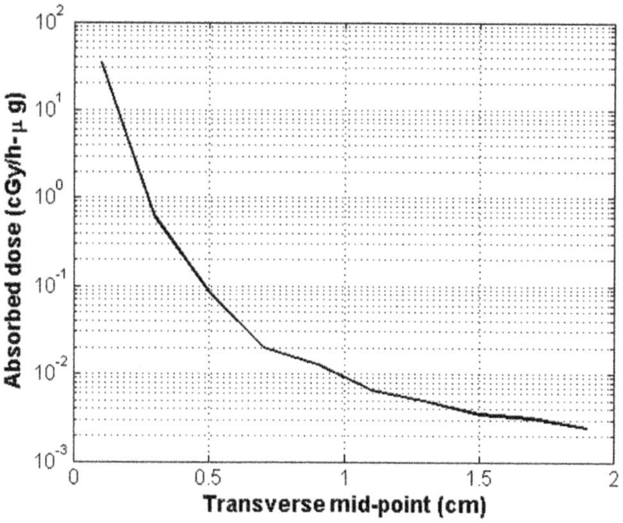

FIGURE 1.32 Energy deposition in transverse regions.

TABLE 1.10 Radiation dose rate (rem/h) in water regions.

Data set	Inner region	Outer region
H*10	6.15572 10^{-3} (0.0008)	2.68817 10^{-7} (0.0132)
ICRP-21	5.95151 10^{-3} (0.0009)	2.21992 10^{-7} (0.0134)
ANSI-1977	6.78671 10^{-3} (0.0008)	3.31944 10^{-7} (0.0131)

Exercise 10: How would you interpret the reduction in the photon surface current versus distance in the context of interaction of gamma rays with matter (Section 1.3.1)?

Another important quantity is the photon energy deposition shown in Fig. 1.32. This follows the same patterns as in the previous tallies, that is, a decrease from 7.62 10^{-3} (0.0072) MeV to 1.02176 10^{-5} (0.0926) photons/cm^2-s.

The dose estimates using three different data sets (Monte Carlo Team, 2005), H*10, ICRP-21and ANSI-1977, give the radiation in rem/h shown in Table 1.10 for the inner and outer regions. The reduction by four orders of magnitude is an important estimate of the amount of attenuation of a californium source due to water.

This simulation took 0.64 min on an Intel(R) Core(TM) i7–2620M CPU@ 2.70 GHz 32-bit operating system.

Capabilities developed

A basic understanding of

1. an atom
2. stability
3. binding energy
4. radioactivity
5. radiation and interaction mechanisms of alpha and beta particles and gamma rays
6. radiation in nuclear systems
7. mixed-mode simulation
8. skills for computing atomic densities for commonly used materials

Nomenclature

English

c speed of light

f_{β}^{Br} fraction of incident beta energy converted by Bremsstrahlung into photon energy

h Planck's constant
$\hbar$ reduced Planck's constant $\hbar = h/2\pi$
k conversion factor
l azimuthal quantum number
m_l magnetic quantum number
m_s spin quantum number
n principal quantum number
n number of shell
n neutron
$\mathbf{p}$ momentum
p proton
q_e charge on an electron
r_B Bohr radius
r_e electron radius
r_n radius of the nth shell
$t_{1/2}$ half-life
$\mathbf{u}$ atomic mass unit
v speed of electron
w_i weight fraction of the ith component in a mixture

English capital

A relative atomic mass
B binding energy as a function of A and Z
C coulomb
D absorbed dose
E exposure
E equivalent dose
E_k kinetic energy of beta particle
E_n energy of the nth shell
E_m electron mass energy equivalent (0.511 MeV)
F form factor in Compton scattering
H equivalent dose
I intensity
I mean ionization and excitation potential of absorber atoms (MeV)
$I(t)$ atoms of iodine-135 at time t
KE_c kinetic energy (classical)
KE_r kinetic energy (relativistic)
L angular momentum
N number of absorber atoms/cm^3
N_{av} Avogadro's number
NZ number of absorber electrons/cm^3
Q quality factor
R range
R roentgen
S target width
Sv Sievert
$T(t)$ atoms of tellurium-135 at time t
V volume
$X(t)$ atoms of xenon-135 at time t
X_0 material radiation length
Z atomic number of absorber

Greek lower

α	alpha particle
α_i	atomic fraction of *i*th constituent in mixture
β	beta particle
β	$\beta v/c$ speed of beta particle relative to speed of light *c*
β^+	positron (positively charged beta particle)
γ	gamma ray
γ	$\frac{1}{\sqrt{1-(v/c)^2}}$
ε	packing fraction
ε_0	permittivity of free space
θ	orthogonal angle
θ_0	standard deviation of deflection in alpha transport Gaussian model
λ	decay constant
λ	wavelength
μ	linear attenuation coefficient
μ	linear attenuation coefficient
ν	neutrino
$\overline{\nu}$	antineutrino
ρ	density
σ	cross-section
ϕ_0	Thomson cross-section
φ	azimuthal angle

Greek capital

Ω	solid angle
Ψ	wave function

Abbreviations

A	angstrom
Bq	Becquerel
Ci	curie
eV	electron volt
fm	Fermi
Gy	gray
ICRP	International Commission on Radiological Protection
MeV	million electron volt
Rad	radiation absorbed dose
Rem	Röntgen equivalent man
TLD	thermoluminescent dosimeter
USNRC	US Nuclear Regulatory Commission

Problems

1. Given atomic fractions: U^{234} (0.0057%), U^{235}(0.72%), and U^{238}(99.27%), find the average atomic weight and the corresponding weight percentages.
2. Given that $Q = [(BE_{p1} + BE_{p2}) - (BE_{r1} + BE_{r2})]$, if $Q > 0$ and the reaction is exothermic, what is the implied by the statement that "the BE of the products is more that the binding energies of the reactants"?
3. In the tellurium-135 decay chain (Section 1.2), with a half-life of tellurium of 11 s, $\lambda_I = 2.874 \times 10^{-5}$/s, $\lambda_X = 2.027 \times 10^{-5}$/s, assume initial conditions and solve the rate equations for iodine-135 and xenon-135.
4. To propose a power source for the Mars rover, compare three alpha emitters, radium, polonium-210 (140 W/g, gamma dose 0.012 Gy/h), and plutonium-238. From their specific heat, estimate the temperatures they will attain for a capsule of half a gram.
5. Consider an alpha particle with charge *ze* with KE moving 'head on' toward a nucleus with charge *Ze* and assume both to be point particles having an elastic collision. As the alpha particle approaches the heavier nucleus it comes to a point beyond which it can not continue so it "stops and turns back," that is, scattering angle 180 degrees, write

TABLE 1.P1 Data for Godiva, Jezebel and Jezebel 23.

Model	Godiva	Jezebel	Jezebel23
Radius (cm)	8.7407	6.3849	5.9838
Density (g/cm^3)	18.74	15.61	18.424
Composition (atoms/barn-cm)	U^{235} 4.4994e−2 U^{238} 2.4984e−3 U^{234} 4.9184e−4	Pu^{239} 3.7047e−2 Pu^{240} 1.7512e−3 Pu^{241} 1.1674e−4 Ga^{69} 8.26605e−4 Ga^{71} 5.48595e−4	U^{233} 4.6712e−2 U^{234} 5.9026e−4 U^{238} 2.8561e−4 U^{235} 1.4281e−5
Mass (g)	52419.98	17019.77	16534.98

the conservation of energy for kinetic and potential and show that the closest distance between the two is $d = \frac{2Ze^2}{4\pi\varepsilon_0(KE)} = \frac{197.32Z}{137KE}$ fm where the KE is in MeV.

6. Estimate the wavelength of the bremsstrahlung photons from a beta particle emitted from P^{32} with a maximum energy of 1.71 MeV if all its energy is lost in a single collision.
7. Give reasons to support the choice of a low-Z shield for beta particles even when the range is shorter than in a high Z absorber.
8. Given that the mean energy required to create an ion pair in air is 33.7 eV, estimate the number of ion pairs created by a beta particle of KE 2 MeV. For the mean ionization and excitation potential, use the approximate formulas $I \sim 19.0\text{eV}$ for $Z = 1$, $I \sim 11.2 + 11.7Z$ eV for $2 \leq Z \leq 13$, and $I \sim 52.8 + 8.71Z\ eV$ for $Z > 13$,.
9. Calculate atomic densities for the following:
   a. Natural uranium with $\rho = 19.1\ g/cm^3$ and atomic fractions U^{238} 0.992745 U^{235} 0.007200.
   b. Bare Pu^{239} metal delta phase 100% Pu239 with $\rho = 15.8\ g/cm^3$.
   c. Given the following data for the fast critical assemblies Godiva, Jezebel and Jezebel23 (Cullen et al., 2007) determine the weight fractions of each of the materials listed.
10. Calculate atomic densities of the fuels
    a. UO_2 of density 10.5 $g/cm^3$with a U^{235} enrichment of 17%.
    b. PuO_2 of density 11.46 g/cm^3, with weight fractions of O_8^{16} 0.118055 and Pu_{94}^{239} 0.881945.
    c. U-10 wt% Zr alloy powder of density 15.48 $g/cm^3$with uranium consisting of 17wt.% U^{235} and the rest U^{238} (Table 1.P1).
11. Calculate atomic densities of the reflectors
    a. Beryllium metal density 1.85 g/cm^3
    b. Beryllium oxide with a density of 3.01 g/cm^3, and weight fractions $_4Be^9$ 0.360320, $_8O^{16}$ 0.639680
12. Find the atomic densities in boron carbide given its density $\rho = 2.52\ g/cm^3$ (weight fraction $B_5^{10} = 0.782610$, $B_5^{11} = 0.217390$) Atomic Weight = 55.24, Answer: $N(B_4C) = 0.0277$, $N(B_{nat}) = 0.1108$, $N(B_5^{10}) = 0.02205$, $N(B_5^{11}) = 0.08875$, $N(C) = 0.0277$
13. Find the atomic densities in a solution of UO_2F_2 with a uranium enrichment of 5%, density of U^{235} of 0.04 g/cc and a given ratio of hydrogen to fissile atoms (H/X) of 500.U(4.89)O_2F_2 solution $N_5 = 1.0889 \times 10^{-4}$, $N_8 = 2.0909 \times 10^{-3}$, $N_F = 4.3996 \times 10^{-3}$, $N_H = 5.7058 \times 10^{-2}$, $No = 3.2929 \times 10^{-2}$, $N_t = 9.6586 \times 10^{-2}$ (atomic densities are in units of 10^{24} atoms/cm^3
14. Calculate the atomic densities in the following structural materials
    a. stainless steel consisting of Fe with 18% chromium by weight, 8% nickel and 0.08% carbon, find the atomic densities
    b. aluminum (atomic weight 26.9815, density $\rho = 2.7\ g/cm^3$)

References

Adams, M., & Martin, W. (1992). Diffusion synthetic acceleration of discontinuous finite element transport iterations. *Nuclear Science and Engineering*. Available from https://doi.org/10.13182/NSE92-A23930.

Anspaugh, L., Catlin, R., & Goldman, M. (1988). The global impact of the chernobyl reactor accident. *Science (New York, N.Y.)*, *242*(4885), 1513–1519. Available from https://doi.org/10.1126/science.3201240.

Archambault, J. P., & Mainegra-Hing, E. (2015). Comparison between EGSnrc, Geant4, MCNP5 and penelope for mono-energetic electron beams. *Physics in Medicine and Biology*. Available from https://doi.org/10.1088/0031-9155/60/13/4951.

Battistoni, G., Bauer, J., Boehlen, T. T., Cerutti, F., Chin, M. P. W., Dos Santos Augusto, R., Ferrari, A., et al. (2016). The FLUKA Code: An accurate simulation tool for particle therapy. *Frontiers in Oncology*. Available from https://doi.org/10.3389/fonc.2016.00116.

Bennet, G.L. (1989). A look at the Soviet space nuclear power program." In *IECEC-89*. Crystal City, Washington D.C. <https://fas.org/nuke/space/sovspace.pdf>.

Berger, M.J. (1963). Monte Carlo calculation of the penetration and diffusion of fast charged particles. Methods in Computational Physics.

Berger, M.J., J.S. Coursey, M.A. Zucker, & J. Chang. (2005). *ESTAR, PSTAR, and ASTAR: Computer programs for calculating stopping-power and range tables for electrons, protons, and helium ions (Version 1.2.3)*. National Institute of Standards and Technology.

Berger, M. J., Jablonski, A., Bronic, I. K., Powell, C. J., Salvat, F., & Sanche, L. (2007). Elastic scattering of electrons and positrons. *Journal of the International Commission on Radiation Units and Measurements*, *7*(1), https://doi.org/10.1093%2Fjicru_ndm001.

Bethe, H. A. (1930). Zur Theorie Des Durchgangs Schneller Korpuskularstrahlen Durch Materie. *Annalen der Physik*, *5*, 325.

Bielajew, A.F. & D.W.O. Rogers. *Lecture notes: Electron Monte Carlo simulation*. Ottawa, Canada. <http://rchttp://www.kek.jp/research/egs/docs/pdf/nrc-pirs0394.pdf>.

Blackett, P. M. S. (1924). The ejection of protons from nitrogen nuclei, photographed by the Wilson method. *Proceedings of the Royal Society*, 349–360.

Bielajew & Rogers (1990). Brookhaven National Brookhaven National Laboratory. (accessed 10 Jan 2021 Elsevier accessed 8 January 2021 <https://www.nndc.bnl.gov/nudat2/reCenter.jsp?z = 7&n = 9>.)

Cacuci, D.G. (2010). Handbook of nuclear engineering, Volume 2. *Engineering*.

Chadwick, J. (1932). The existence of a neutron. Proceedings of the *royal society*. <https://doi.org/10.1098/rspa.1932.0112>.

Commission, E. (2007). ICRP: ICRP Publication 103. ICRP, 2007. *The 2007 recommendations of the international commission on radiological protection*. ICRP Publication 103. Ann. ICRP 37 (2–4).

Demuth, S. F. (2003). SP100 space reactor design. *Progress in Nuclear Energy*. Available from https://doi.org/10.1016/S0149-1970(03)90003-5.

Determan, W.R., G.A. Johnson, A.J. Zillmer, T.E. Dix, C.Y. Lu, M. Tosca, & L.A. Trager. (2011). SNAP-10A Safety Test Program. In *Nuclear and emerging technologies for space, NETS-2011*.

Dymova, M. A., Taskaev, S. Y., Richter, V. A., & Kuligina, E. V. (2020). Boron neutron capture therapy: Current status and future perspectives. *Cancer Communications*. Available from https://doi.org/10.1002/cac2.12089.

El-Genk, M. S. (2008). Space nuclear reactor power system concepts with static and dynamic energy conversion. *Energy Conversion and Management*. Available from https://doi.org/10.1016/j.enconman.2007.10.014.

Elsevier. *Nuclear data sheets*. <https://www.sciencedirect.com/journal/nuclear-data-sheets>.

Gibney, E. (2020). Arab world's first Mars probe takes to the skies. *Nature*. Available from https://doi.org/10.1038/d41586-020-02170-2.

González-Rodríguez, P., & Kim, A. D. (2008). Light propagation in tissues with forward-peaked and large-angle scattering. *Applied Optics*. Available from https://doi.org/10.1364/AO.47.002599.

Goudsmit, S., & Saunderson, J. L. (1940a). Multiple scattering of electrons. *Physical Review*. Available from https://doi.org/10.1103/PhysRev.57.24.

Goudsmit, S., & Saunderson, J. L. (1940b). Multiple scattering of electrons. II. *Physical Review*. Available from https://doi.org/10.1103/PhysRev.58.36.

Grady Hughes, H. (2014). *Enhanced electron-photon transport in MCNP6*. In. doi:10.1051/snamc/201403105.

Granada, J. R., Santisteban, J. R., Dawidowski, J., & Mayer, R. E. (2012). Neutron transmission: A powerful technique for small accelerator-based neutron sources, In*Physics Procedia*. Available from https://doi.org/10.1016/j.phpro.2012.03.015.

Gray, S. (Ed.). *Handbook of positron emission tomography illustrated edition*. Hayle Medical.

Handbook of Practical X-Ray Fluorescence Analysis. (2006). *Handbook of practical X-Ray fluorescence analysis*. doi:10.1007/978-3-540-36722-2.

Harkness-Brennan, L. (2018). *An Introduction to the physics of nuclear medicine*. doi:10.1088/978-1-6432-7034-0.

Harmon, C.D., II, R.D. Busch, J.F. Briesmeister, & R.A. Forster. (1994). *Criticality calculations with MCNP: A primer*.

Hubbell, J.H. (1969). Photon cross sections, attenuation coefficients, and energy absorption coefficients from 10 KeV to 100 GeV. *National Standard Reference Data System*.

Hughes, H.G. (1996). Treating electron transport in MCNP. In *Proceedings of the training course on the use of MCNP in radiation protection and dosimetry held in Bologna-Italy, May 13–16, 1996*. Los Alamos National Laboratory. <https://mcnp.lanl.gov/pdf_files/la-ur-96-4583.pdf>.

IAEA. (2014). Radiation protection and safety of radiation sources: International basic safety standards general safety. *IAEA Safety Standards*.

INPO. (2011). *Special report on the nuclear accident at the Fukushima Daiichi Nuclear Power Station*. INPO 11-005.

Kawrakow, I. & D.W.O. Rogers. (2003). The EGSnrc code system: Monte Carlo simulation of electron and photon transport. System.

Kibble, T. W. B. (2015). The standard model of particle physics. *European Review*. Available from https://doi.org/10.1017/S1062798714000520.

Knoll, G. F., & Kraner, H. W. (1981). Radiation detection and measurement. *Proceedings of the IEEE*. Available from https://doi.org/10.1109/PROC.1981.12016.

Kumada, H. (2014). Neutron sources. *Comprehensive biomedical physics*. Available from https://doi.org/10.1016/B978-0-444-53632-7.00613-4.

Langlois, L. (2013). IAEA action plan on nuclear safety. *Energy Strategy Reviews*, *1*(4), 302–306. Available from https://doi.org/10.1016/j.esr.2012.11.008.

Letourneau, A., Marchix, A., Tran, N. H., Chauvin, N., Menelle, A., Ott, F., & Schwindling, J. (2017). Development of compact accelerator neutron source. *EPJ web of conferences*. Available from https://doi.org/10.1051/epjconf/201714603018.

Loforte, A., Gliozzi, G., Suarez, S. M., & Pacini, D. (2020). Contributory role of positron emission tomography in a left ventricular assist device recipient at the time of COVID-19 pandemic. *ASAIO Journal*. Available from https://doi.org/10.1097/MAT.0000000000001176.

Mallapaty, S. (2020). China's successful launch of mars mission seals global era in deep-space exploration. *Nature*. Available from https://doi.org/10.1038/d41586-020-02187-7.

Marchix, A., Letourneau, A., Tran, H. N., Chauvin, N., Menelle, A., Ott, F., & Schwindling, J. (2018). Saclay compact accelerator-driven neutron sources (SCANS), In*Journal of Physics: Conference Series*. Available from https://doi.org/10.1088/1742-6596/1046/1/012009.

Matlack, G.M. & C.s.F. Metz. (1967). *Radiation characteristics of Plutonium-238*. New Mexico. <https://www.osti.gov/servlets/purl/4495451>.

McConn, R. J., Jr, Gesh, C. J., Pagh, R. T., Rucker, R. A., & Williams, R. G., III (2011). *Compendium of material composition data for radiation transport modeling*. Pacific Northwest National Laboratory.

Miskolczy, G. & D.P. Lieb. (1990). Radioisotope thermionic converters for space applications. In *Proceedings of the intersociety energy conversion engineering conference*. doi:10.1109/iecec.1990.716874.

Monte Carlo Team. (2005). MCNP - A general Monte Carlo N-particle transport code, Version 5. *Los Alamos Nuclear Laboratory*. doi:LA-UR-03−1987.

Muraz, J.-F., Santos, D., Ghetta, V., Giraud, J., Marpaud, J., Hervé, M., ... Forlino, M. (2020). Development of a regenerated beryllium target and a thermal test facility for compact accelerator-based neutron sources. *EPJ web of conferences*. Available from https://doi.org/10.1051/epjconf/202023103003.

Narita, T. & K. Kitao. (1992). *List of strong gamma-rays emitted from radionuclides*.

Nedunchezhian, K., Aswath, N., Thiruppathy, M., & Thirugnanamurthy, S. (2016). Boron neutron capture therapy - A literature review. *Journal of Clinical and Diagnostic Research*. Available from https://doi.org/10.7860/JCDR/2016/19890.9024.

Olbrant, E., & Frank, M. (2010). Generalized Fokker-Planck Theory for electron and photon transport in biological tissues: Application to radiotherapy. *Computational and Mathematical Methods in Medicine*. Available from https://doi.org/10.1080/1748670X.2010.491828.

Patel, J. K., Warsa, J. S., & Prinja, A. K. (2020). Accelerating the solution of the SN equations with highly anisotropic scattering using the Fokker-Planck approximation. *Annals of Nuclear Energy*. Available from https://doi.org/10.1016/j.anucene.2020.107665.

Peralta, L., & Louro, A. (2014). AlfaMC: A fast alpha particle transport Monte Carlo code. *Nuclear Instruments and Methods in Physics Research, Section A: Accelerators, Spectrometers, Detectors and Associated Equipment*. Available from https://doi.org/10.1016/j.nima.2013.11.026.

Peskin, M. E. (2019). Concepts of elementary particle physics. *Concepts of Elementary Particle Physics*. Available from https://doi.org/10.1093/oso/9780198812180.001.0001.

Pomraning, G. C. (1992). The Fokker-Planck operator As an asymptotic limit. *Mathematical Models and Methods in Applied Sciences*. Available from https://doi.org/10.1142/s021820259200003x.

Radiation Effects and Sources. (2016). *Radiation effects and sources*. doi:10.18356/b1749f17-en.

Research, CERN: European Council for nuclear. (2020). *GEANT4 physics reference manual*.

Rogers, D. W. O., & Bielajew, A. F. (1990). Monte Carlo techniques of electron and photon transport for radiation dosimetry. *The Dosimetry of Ionizing Radiation*. Available from https://doi.org/10.1016/b978-0-12-400403-0.50009-9.

Rutherford, E. (1919). LIV. Collision of α particles with light atoms. IV. An anomalous effect in nitrogen. *The London, Edinburgh, and Dublin Philosophical Magazine and Journal of Science*. Available from https://doi.org/10.1080/14786440608635919.

F. Salvat, J.M. Fernández-Varea, J. Sempau. (2011). PENELOPE-2011, a code system for Monte Carlo simulation of electron and photon transport. *Simulation*.

Seltzer, S. M. (1991). Electron-photon Monte Carlo calculations: The ETRAN code. *International Journal of Radiation Applications and Instrumentation. Part*. Available from https://doi.org/10.1016/0883-2889(91)90050-B.

Suzuki, M. (2020). Boron Neutron Capture Therapy (BNCT): A unique role in radiotherapy with a view to entering the accelerator-based BNCT era. *International Journal of Clinical Oncology*. Available from https://doi.org/10.1007/s10147-019-01480-4.

Taylor, L.S. et al. (1970). *ICRU Report 16: Linear energy tranfer*. Washington D.C.

Ulaner, G. A. (2018). *Fundamentals of oncologic PET/CT* (1st ed.). Elsevier.

Urban, L. (2005). A model for multiple scattering in Geant4. In *Proceedings of the Monte Carlo topical meeting*.

Urban, L. (2006). A multiple scattering model in Geant4. In *Proceedings of the CERNOPEN-2006-077*.

US Dept of Health Education and Welfare (1970). In John C Villforth and George R Shultz (Ed.). *Radiological health handbook*. Maryland.

Verbeke, J. M., Leung, K. N., & Vujic, J. (2000). Development of a sealed-accelerator-tube neutron generator. *Applied Radiation and Isotopes*. Available from https://doi.org/10.1016/S0969-8043(00)00262-1.

Waters, L.S., G.W. McKinney, J.W. Durkee, M.L. Fensin, J.S. Hendricks, M.R. James, ... & D.B. Pelowitz (2007). The MCNPX Monte Carlo radiation transport code. In *AIP conference proceedings*. doi:10.1063/1.2720459.

Werner, C.J. (ed.). (2017). *MCNP (R) user's manual code Version 6.2*.

Witze, A. (2020). NASA has launched the most ambitious Mars Rover ever built: Here's what happens next. *Nature*. Available from https://doi.org/10.1038/d41586-020-02257-w.

World Nuclear Association. (2012). Chernobyl accident 1986. *World Nuclear Association* 1986 (April 1986): 1−14. <http://www.world-nuclear.org/info/chernobyl/inf07.html>.

X-5 Monte Carlo Team. (2008). MCNP - A general Monte Carlo N-particle transport code, Version 5 volume I: Overview and theory. *Los Alamos Nuclear Laboratory*. doi:LA-UR-03−1987.

Ziegler, J. F., Ziegler, M. D., & Biersack, J. P. (2010). SRIM - The stopping and range of ions in matter. *Nuclear Instruments and Methods in Physics Research, Section B: Beam Interactions with Materials and Atoms*. Available from https://doi.org/10.1016/j.nimb.2010.02.091.

Chapter 2

Interactions of neutrons with matter

The interaction of a neutron with matter is influenced by the presence of atoms it encounters during its transport; the electrons it comes across have a relatively smaller mass and are unable to cause any significant deflection while interactions with nuclei are orders of magnitude less due to the smaller chance of "coming in the way." Similarly, the probability of neutron-neutron collisions is also small. This leaves neutron-nuclei encounters which form the basis of neutron interactions in matter. The single quantity that determines the probability of interaction is a microscopic cross-section which is like the circular cross-section of a sphere. The higher the cross-section, the higher will be the probability of interaction and vice versa. Further, neutrons carry no electrical charge and do not therefore cause direct ionization. Neutron detectors, discussed in Chapter 1, thus use indirect ionization in a gas to induce a measurable current. The neutron, as a free neutron, is unstable and decays with the formation of a proton and a beta particle ${}_0^1n \rightarrow {}_1^1H + {}_{-1}^{0}e$.

The basic concept of neutron interaction in matter is the matter-wave duality, in which the analysis considers a neutron as a wave function. This analysis will be briefly reviewed in this chapter. It is based on quantum physics, and draws several concepts from optics, and one model is therefore called the *optical model* of neutron interaction.

2.1 Kinetic theory

We start our description by considering the kinetic theory of gases, for which Boltzmann formulated the Boltzmann transport equation (BTE) described in Chapter 1 in the context of the Fokker-Planck equation for charged particles and the photon transport equation for photons (gamma rays and X-rays). The BTE is applicable to neutron transport as well and provides a 'complete' model of neutron transport in terms of a statistical *distribution function* and uses the microscopic cross sections from their underlying quantum descriptions. The works of Boltzmann and Max Planck are both early 20th century at which time kinetic theory of gases was refined as a basis for modeling transport phenomena and quantum physics was born. There was no knowledge then of elementary particle physics or that a neutron was composed of quarks. Thus present-day knowledge on neutron interaction with matter rests on quantum physics, which is the basis of all data used in later chapters of this book for carrying out neutron and photon transport in nuclear systems. This data is compiled in elaborate form in large 'libraries' that is used for simulating phenomena and obtaining engineering design information.

For the distribution function, a kinetic theory description of a gas, containing n_0 molecules, says that all molecules do not have the same energy but a distribution $n(E)$, modeled as a Maxwellian distribution

$$n(E) = \frac{2\pi n_0 E^{1/2}}{(\pi kT)^{3/2}} e^{-E/kT} \tag{2.1}$$

with units of molecules per unit energy.

In "velocity space," the distribution function $n(v)$, with units for the speed of molecules v is obtained from the above as $E = 1/2mv^2$, so that $n(v)dv = n(E)dE$ gives

$$dn(v) = \frac{4\pi n_0 v^2}{(2\pi kT/m)^{3/2}} e^{-mv^2/kT} \quad dv \tag{2.2}$$

Thus what we understand as macroscopic quantities, such as temperature and pressure, represent the average of a statistically large sample of gas molecules each moving about randomly with their own temperature.

Nuclear Engineering. DOI: https://doi.org/10.1016/B978-0-323-90618-0.00002-8

The most probable energy E_p and most probable speed v_p are obtained by setting the first derivative for $n(E)$ in Eq. (2.1) and for $n(v)$ in Eq. (2.2) to zero

$$\frac{dn(v)}{dv}|_{v_p} = 0$$

is

$$v_p = \sqrt{\frac{2kT}{m}}$$

Thus the most probable energy is $E_p = \frac{1}{2}kT$. The average speed $\overline{v}$ and average energy $\overline{E}$ can be found as

$$\overline{v} = \frac{\int_0^\infty vn(v)dv}{\int_0^\infty n(v)dv} = \sqrt{\frac{8kT}{\pi m}} = \frac{2}{\sqrt{\pi}}v_p$$

$$\overline{E} = \frac{\int_0^\infty En(E)dE}{\int_0^\infty n(E)dE} = \frac{3}{2}kT$$

The distributions $n(v)$ and $n(E)$ are plotted in Fig. 2.1 with the transformations

$$x = \frac{v}{\sqrt{\frac{2kT}{m}}},\ n(v) = \frac{4n_0}{\pi^{1/2}\sqrt{\frac{2kT}{m}}}x^2e^{-x^2}$$

and for $n(E)$

$$x = \frac{E}{kT},\ n(E) = \frac{2n_0}{\pi^{1/2}kT}x^{1/2}e^{-x}$$

so that the vertical axis is in units of n_0/kT for $n(E)$, and in units of $n_0\sqrt{m/(2kT)}$ for $n(v)$ so that $v_p = 1$ in Fig. 2.1. At room temperature, $T = 300\ K$, the most probable speed of a neutron is thus

$$v_p = \sqrt{\frac{2kT}{m}} = \sqrt{\frac{2 \times 1.38066 \times 10^{-23} \times 300}{1.67493 \times 10^{-27}}} = 2239\ \text{m/s}$$

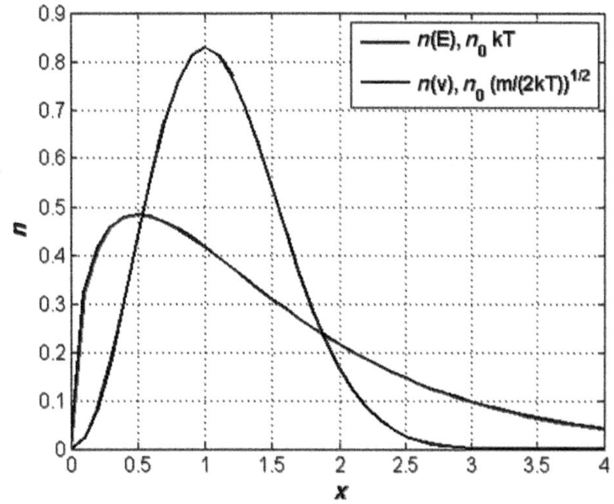

FIGURE 2.1 Maxwellian distribution.

for which $E_0 = 0.025$ eV. Neutrons in thermal equilibrium with their surroundings at room temperature with a speed of ~ 2200 m/s and energy 0.025 eV are referred to as *thermal neutrons*. Neutrons with higher speeds, with energies on the order of several MeV, are referred to as *fast neutrons*.

Exercise 2.1: *Thermal flux* show that the function $\phi(E) = nv$ can be written as $\phi(E) = \phi_{th}M(E,T)$ where

$$M(E,T) = \frac{E}{(kT)^2}e^{-E/(kT)}$$

where $\phi_{\text{th}} = \sqrt{\frac{8kT}{\pi m}}n_0 \equiv \overline{v}_M n_0$ and $\int_0^\infty M(E,T) = 1$. Compare $\overline{v}_M$ with v_p found above.

Matlab Example 2.1: *Distribution function* plot the distribution functions $n(v)$ and $n(E)$

```
k=1.38066e-23;% J/K
i=0;
for x=0:0.1:4
 % for E, x=E/kT    for v, x=v/sqrt(2kT/m)
 i=i+1;
nE(i)=(2/pi^0.5)*sqrt(x)*exp(-x);
nv(i)=(4/pi^0.5)*x^2*exp(-x^2);
xx(i)=x;
end
```

It will be seen in subsequent chapters that the distribution function requires knowledge of the phase space comprising seven variables (three variable of space, two of angle, one of energy and one of time) for a complete description of transport. Clearly this is beyond the scope of deterministic treatment and thus moments of the distribution function are often used. This is left as an exercise for the reader to appreciate the physical significance of moments which, in principle, can be used to reconstruct a distribution function.

Exercise 2.2: *Moments of distribution function* find expressions for the zeroth, first, and second moments of the distribution function $n(E)$ and describe their physical significance.

In nuclear engineering, the fundamental quantity of interest is the neutron flux $\phi\left(\overline{r}, E, \hat{\Omega}, t\right) = n\left(\overline{r}, E, \hat{\Omega}, t\right)v$ with units of neutrons per cm^2 per second per unit energy interval per steradian. With some simplifications, the integrated or "averaged" flux can also be defined as $\phi = nv$ where n is the number of neutrons per unit volume, in a domain, and v is the neutron speed. Neutron flux has units of neutrons (cm/s) cm^3 and is thus a 'distance traveled' in a volume, or a track length (neutrons $\cdot$ cm^2/s). The neutron current is the vector form of the flux, given by $\overline{J} = \hat{N} \bullet \overline{v}n$ where $\hat{N}$ is a unit normal vector on area dA so that the net number of neutrons leaving dA is $\overline{J}dA$ neutrons per second. Note that $\overline{v} = \hat{\Omega}v$ and $\hat{N} \bullet \hat{\Omega} > 0$ implies that neutrons are leaving the surface and conversely $\hat{N} \bullet \hat{\Omega} < 0$ implies neutrons entering a surface.

2.2 Types of neutron interactions

We now return to the particle-wave duality by which neutron interactions with matter are described. A neutron has a de Broglie wavelength (Chapter 1) which varies from $\sim 10^{-14}$ m for a fast neutron to 10^{-10} m for a thermal neutron causing thermal neutrons to have a larger cross-section. At the quantum level, the discrete angular momentum $L = l\hbar$ can have values for waves $s, p, d, \ldots$ for which $l = 0, 1, 2, \ldots$ while at the classical level, the angular momentum of the incident neutron is the product of its linear momentum and its impact parameter. Thus if an elastic collision is to take place, it can happen at the lowest level of angular momentum for which the impact parameter must be of the order of a fermion (nuclear size) which corresponds to the lowest wave scattering, that is, s-wave scattering ($kR \ll 1$). As shown in Fig. 2.2, for a neutron scattering off uranium 238, the angular momentum quantum number varies as $l \sim 0.0019E^{1/2}$ (Ein eV) so that $l = 1$ at $E \sim 277$keV. Thus s-wave scattering in uranium 238 can be assumed to hold for $E \lesssim 277$keV, while p-wave scattering will become significant at $E > 277$keV. For neutrons colliding with light nuclei, this 'boundary' will increase to the right i.e. s-wave scattering can be assumed to hold till higher energies as depicted in Fig. 2.3.

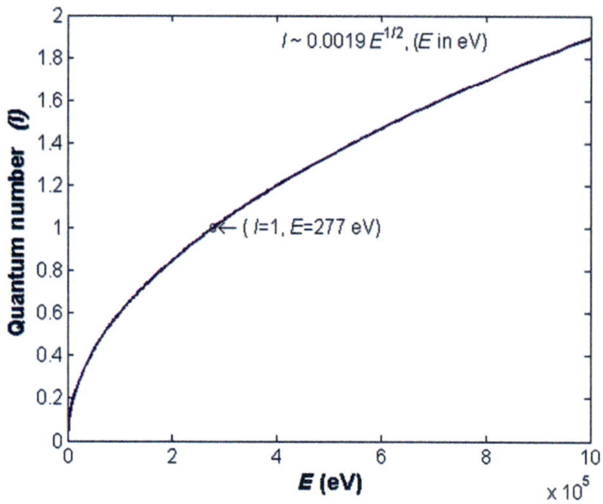

FIGURE 2.2 Angular quantum number versus kinetic energy of neutron.

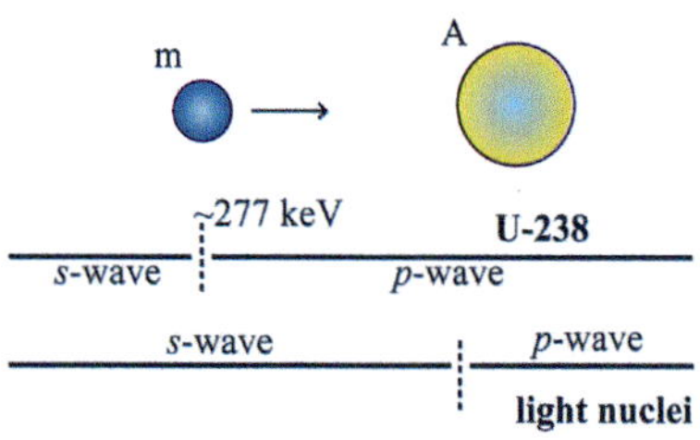

FIGURE 2.3 *s*-wave ($l = 0$) and *p*-wave ($l = 1$) scattering for neutron scattering.

Matlab Example 2.2: *Angular quantum number* plot the angular quantum number l versus energy E for elastic scattering.

```
m=1;A=238;mu=m*A/(m+A);c=2.99e8;
h=4.1413e-15;% eV-s
m=1.6749e-27;% kg
mE=m*c^2*1e19/1.6; %eV
Radn=1.2e-15;% m
RadU238=Radn*A^(1/3);
KK=(2*pi/(h*c))*sqrt(2*mE)*(Radn+RadU238)
E=0:10:100;
l=KK*sqrt(E); % E in eV
```

The de Broglie wavelength is used to represent the neutron in the optical model (Hodgson, 1994; Lamarsh, 1966; Reuss, 2008) as an incident plane wave with amplitude $\psi \sim e^{ikx}$ and wave number $k = 2\pi/\lambda$ which is related to the neutron energy as $k \sim \sqrt{E}$. The incident wave is represented as a superposition of an infinite number of partial waves characterized by the angular momentum and divided into cylindrical zones of radii which are multiples of h. It can then be shown that elastic scattering is an *s*-wave phenomenon as presented with simplified reasoning above. These models were presented at a time when the neutron was considered to be a particle and a wave but its composition as quarks was unknown.

For the moment, theory presents neutron interactions as either potential scattering, in which there is no physical collision but a scattering such that the neutron and nucleus both undergo deflections, or an actual collision where the neutron enters the nucleus, stays in the compound nucleus for some time and then exits from the nucleus.

Elastic scattering, shown in Fig. 2.4, is thus considered to be an interaction, far from a resonance, where no compound nucleus is formed but a two-body collision appears to take place. It can be modeled as potential scattering in terms of the radius of the neutron ($R_n = 1.2 \times 10^{-15}$ m), and the radius of the target nucleus ($R_A = R_n A^{1/3}$m) when it

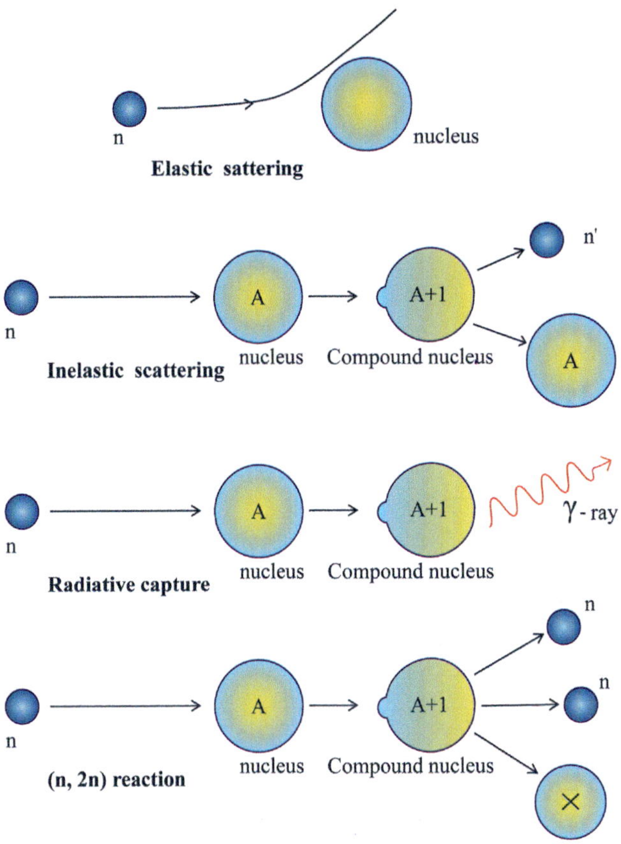

FIGURE 2.4 Selected neutron interactions.

takes the value $\sigma_s = 4\pi R_A^2$. In the presence of a resonance, the elastic scattering cross-section has been modeled in terms of a single resonance and potential scattering as well an "interference" term. All neutron interactions, such as inelastic scattering, radiative capture and neutron multiplication $(n, 2n)$ shown in Fig. 2.4, other than elastic scattering away from a resonance, undergo compound nucleus formation.

In an inelastic reaction an incident neutron enters the target nucleus adding its binding and kinetic energy to the compound nucleus, resides for some time in the compound nucleus, and then exits with some energy retained in the compound nucleus. The emitted particle can be a neutron itself or some other particle exiting from the nucleus. While a neutron is in the compound nucleus, it exchanges energy with other nucleons; the outcome depends on whether or not it imparts energy sufficient for a nucleon to be ejected. An 'ejection' takes place on the order of $\sim 10^{-17}$s, which is instantaneous for all practical purposes, but large compared with the nuclear time, of the order of $\sim 10^{-21}$s. When a neutron emerges from such an inelastic reaction, we describe it as a (n, n') reaction. In case of high incident energy, above a threshold, a neutron can result in multiplication $(n, 2n)$ or $(n, 3n)$ reactions.

A possible reaction is a direct interaction causing charged particles to be expelled from a nucleus as in the case of such as (n, α) and (n, p) reactions.

In neutron interactions where a neutron enters the target nucleus and has insufficient energy to cause ejection of a nucleon, it gets absorbed while the relaxation energy can be emitted from an isomer in the form of a gamma ray. When the incident energy corresponds to one of the excited states of a nucleus, 'resonance' can take place where the cross-section becomes very large. These states are called 'metastable' states if they exist for $\sim 10^{-17}$s.

2.2.1 Neutron scattering in the lab and center of mass systems

Neutron scattering with target nuclei is indeed a complicated phenomenon both theoretically and experimentally.

Scattering experiments are conducted in the laboratory (L) system with an incident neutron of velocity v_L incident on a stationary nucleus, while another system, called the center of mass (CM) system is used for mathematical simplicity, as shown in Fig. 2.5.

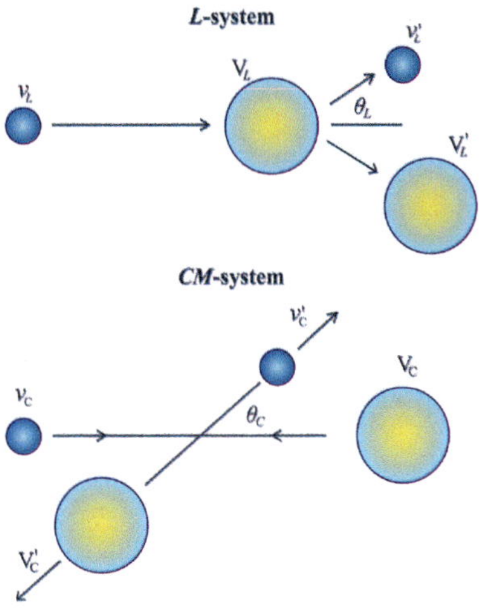

FIGURE 2.5 Elastic scattering on a neutron and nucleus.

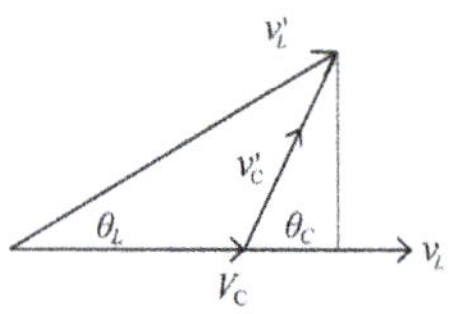

FIGURE 2.6 Scattering angle in Lab and center of mass systems.

		Laboratory		Center of mass	
		Neutron	Nucleus	Neutron	Nucleus
Precollision	Speed	v_L	V_L	v_C	V_C
	Energy	E_L	E_{AL}	–	–
Post collision	Speed	v_L'	V_L'	v_C'	V_C'
	Energy	E_L'	E_{AL}'	–	–
	Scattering angle	θ_L	–	θ_C	–

Some simplicities of the CM system are as follows:

1. Velocities before and after collision: $v_C = v_C'$, $V_C = V_C'$
2. Momentum before collision and after collision is zero.

The *CM* system is useful for another property of scattering called isotropic scattering. The two systems are connected, as shown in Fig. 2.6; the velocity of the center of mass V_{CM} observable in the *L*-system. The incident neutron has direction $\hat{\Omega}$ in the L-system and scatters into the $\hat{\Omega}'$ direction with $\hat{\Omega} \bullet \hat{\Omega}' = \cos\theta_L$; notice that the azimuthal angle φ is not shown in Fig. 2.6 as it is considered unchanged.

Applying the conservation of momentum

$$(A+1)V_{CM} = v_L + AV_L \tag{2.3}$$

When $V_L = 0$, the velocities of the neutron and nucleus in the CM system are

$$v_C = \frac{A}{A+1} v_L,$$

and

$$V_C = -\frac{1}{A+1} v_L.$$

Energy conservation, assuming a target nucleus initially at rest, gives

$$E_L = E_L' + E_{AL}'$$

which can be shown with the conservation of horizontal and vertical momentum, to yield a relationship between the final energy of a neutron E'_{nL} and its initial energy E_{nL}:

$$E'_L = \frac{E_L}{(A+1)^2}\left[\cos\theta_L + \sqrt{A^2 - \sin^2\theta_L}\right]^2.$$

The above can be written as

$$E'_L = \frac{1}{2}E_{nL}[(1+\alpha) + (1-\alpha)\cos\theta_C] \tag{2.4}$$

where

$$\alpha = \left(\frac{A-1}{A+1}\right)^2$$

We can see from the above that if $\theta_C = 0$, $E'_L = E_L = E'_{L\max}$ (forward scattering) and when $\theta_C = \pi$, $E'_L = E'_{L\min} = \alpha E_L$ (back-scattering).

Thus for hydrogen, $A = 1$, all the energy can be lost by a neutron in a single collision, while for a heavy nucleus such as U^{238}, $\alpha = 0.9833$ so that a neutron can lose just a small amount of energy since its minimum energy after collision is $0.9833E_L$; a 1 MeV neutron striking a U^{238} nucleus will thus have an energy between 0.9833 and 1 MeV so that the maximum energy transferred to the U^{238} nucleus can be 0.0167 MeV. The energy (in the Lab system) after a collision of a neutron with nuclei of hydrogen, carbon, iron and U238 is shown in the Fig. 2.7 below. For hydrogen, it is seen that on average 50% of the neutron's energy is lost in a collision.

From the horizontal and vertical components:

$$v'_L\cos\theta_L = V_{CM} + v'_C\cos\theta_C \tag{2.5}$$

and

$$v'_L\sin\theta_L = v'_C\sin\theta_C \tag{2.6}$$

yielding the relationship:

$$\tan\theta_L = \frac{\sin\theta_C}{1/A + \cos\theta_C}. \tag{2.7}$$

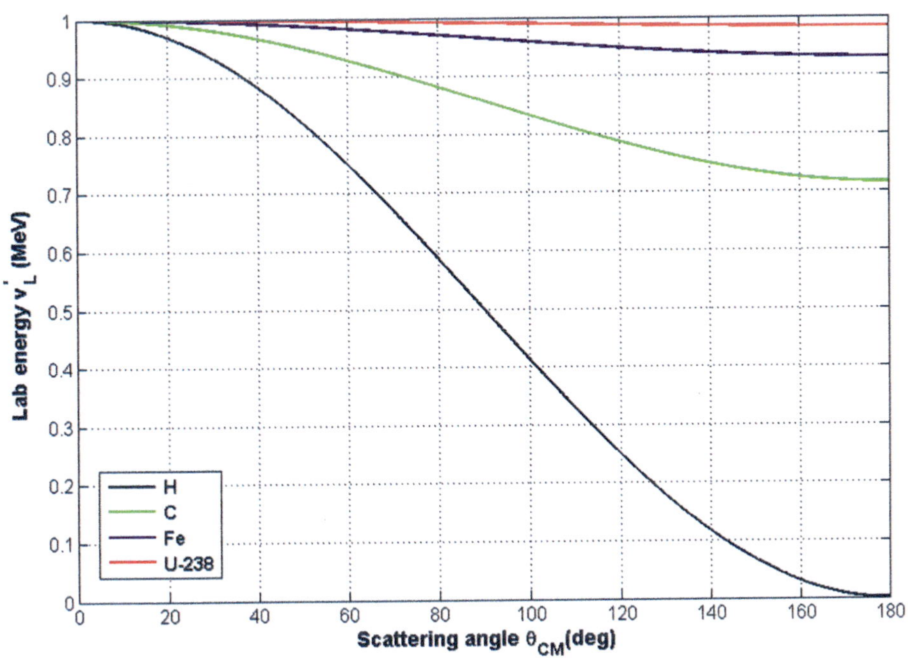

FIGURE 2.7 Neutron energy in Lab system after a collision (initial energy is 1 MeV).

Thus for heavy nuclei such as U^{238} $(A \gg 1)$ $\theta_L \approx \theta_C$ since the center of mass velocity V_{CM} is much less than the incident neutron velocity in the laboratory system. This is a significant statement in the context of expressing isotropic scattering in the C-system, which will differ from results of the L-system for light target nuclei.

Exercise 2.3: *Scattering angles:* from the scattering angles in the Lab and CM systems [Eq. (2.7)] find a relationship between the solid angles $d\Omega = \sin\theta \, d\theta \, d\varphi$ in both systems.

The scattering cross-section has a smooth variation at low energies, like the absorption cross-section, but extends to somewhat higher energies (typically a few MeV), then exhibits broad resonances, and then smoothens off.

At low energies, neutrons exhibit s-wave scattering for light as well as heavy nuclei, which is predominantly isotropic in the center of mass system. At higher energies and for larger nuclear radius, the scattering is p-wave scattering which is forward-biased rather than isotropic. Thus a neutron scattering off a U^{238} nucleus is bound to be forward-biased *i.e.*, favoring forward scattering rather than higher angles of back-scattering. Light nuclei, with a high threshold for inelastic reactions, are more likely to undergo elastic collisions at low energies while heavy nuclei will mainly undergo inelastic scattering.

Exercise 2.4: *Solid angle:* from elementary considerations of solid geometry, consider a volume element in a solid sphere of radius R and show that $dV = \int_0^R r^2 \int_0^{2\pi} d\varphi \int_0^{\pi} \sin\theta \, d\theta$. Calculate the solid angle subtended by an electron on a nucleus.

2.3 The microscopic cross-section

The microscopic cross-section, with units of area, represents the measure of an interaction at the nuclear level. It can be expressed as a probability of an interaction of a certain type. It is visualized as the cross-sectional area (cm^2) of a sphere which a neutron can "see" as it moves in matter. A small cross-section indicates a small probability of that particular interaction. Since atoms and neutrons are very small, these cross sections are expressed in units of $10^{-24}\ cm^2$ which was code-named a "barn" during the Manhattan Project days.

Each interaction, such as an elastic scattering, inelastic scattering, absorption, fission, radiative capture etc. has an associated cross-section. The sum of all cross sections is called the 'total' cross-section σ_t

$$\sigma_t = \sigma_s + \sigma_a + \sigma_f + \sigma_{n,\gamma} + \sigma_{n,2n} + \ldots$$

The intensity of neutrons reduces as they interact with matter, as described in Section 1.3.3 for photons, since the reduction dI is proportional to the intensity I, the atomic density N, the rate of interaction $N\sigma_t$ and the volume of material $dV = Adx$, where A and dx are the cross-section area and thickness of the slab respectively as shown in Fig. 2.8.

The intensity of neutrons can then be found from the source intensity I_o across a shield of thickness x from elementary considerations. The intensity across a thin strip dx is

$$I(x + dx) = I(x)(1 - N\sigma_t dx)$$

so that the change in intensity across the strip is $dI(x)$ given as

$$dI(x) = I(x + dx) - I(x) = -I(x)N\sigma_t dx$$

from which

$$\frac{dI(x)}{I(x)} = -N\sigma_t dx$$

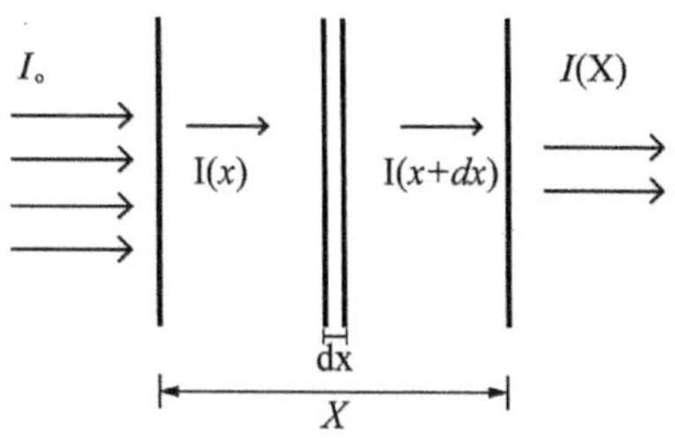

FIGURE 2.8 Intensity reduction.

Finally, integration over the width of the slab gives

$$\int_0^X \frac{dI(x)}{I(x)} = -\int_0^X N\sigma_t\,dx$$

$$\ln\frac{I(X)}{I(0)} = -N\sigma_t X$$

$$I(X) = I_o e^{-N\sigma_t X} \tag{2.8}$$

The above expression would hold under ideal conditions, that is, a point anisotropic source of intensity I_o incident on a thin shield. For a point isotropic source, the intensity would be given as

$$I(R) = I_o \frac{e^{-N\sigma_t R}}{4\pi R^2} \tag{2.9}$$

to include both geometrical and material attenuation.

Exercise 2.5: *Intensity*: given an incident intensity of 10^8 neutrons/cm^2/s on a composite slab consisting of a water region 5 cm thick followed by an iron slab 3 cm thick, find the transmitted intensity.

Material	At. Weight*	Density (g/cm^3)	σ_a (b)	σ_s (b)
Iron (Fe)	55.847	7.87	2.53	11
Water (H_2O)	18.0153	1.00	0.664	103

1. Based on C^{12} (12 amu)

To obtain a probability of scattering, a normalized probability distribution function can be written as

$$f(E)dE = \frac{1}{\sigma(E)}\frac{d\sigma}{dE}dE$$

so that this probability, from initial energy E to final energy E' (in the Lab system) is

$$P(E \to E')dE' = \frac{\sigma_s(\theta)d\Omega(\theta,\varphi)}{\sigma_s} \tag{2.10}$$

Now, since $d\Omega(\theta) = \sin\theta_C d\theta_C d\varphi = d\mu d\varphi$ and with no preferential azimuthal angle, then

$$P(E \to E')dE' = \frac{2\pi\sigma_s(\theta)d\mu}{\sigma_s}$$

which, in the case of isotropic scattering in the center of mass system reduces to

$$P(E \to E')dE' = \frac{d\mu}{2} = -\frac{1}{2}\sin\theta_C\, d\theta_C \tag{2.11}$$

From Eq. (2.4)

$$\frac{dE_L'}{d\theta_C} = \frac{E_L}{2}(1-\alpha)\sin\theta_C$$

giving

$$P(E \to E') = \frac{1}{E(1-\alpha)} \tag{2.12}$$

This is a significant result in the context of *s*-wave scattering discussed in the previous section for which scattering is isotropic in the CM system as is the case for light nuclei such as carbon 12 for low-energy neutrons. When $kR < 1$, for light nuclei the scattering is always isotropic in the CM system while for $kR \gg 1$, due to *p*-wave interference, the scattering is anisotropic (in the forward direction).

Exercise 2.6: *Probability distribution*: Consider the forward and backward scattering cases for the differential scattering cross-section by assuming linear forms and sketch the probability $P(E \rightarrow E')$ vs the final energy E' which should be constant for the isotropic case given above.

Exercise 2.7: *Moments of scattering cross-section*: Expand the angular scattering cross-section

$$\Sigma_s\left(\bar{r}, E' \rightarrow E, \hat{\Omega}' \rightarrow \hat{\Omega}\right) = \sum_{l=0}^{M} \frac{2l+1}{4\pi} \Sigma_{sl}(\bar{r}, E' \rightarrow E) P_l\left(\hat{\Omega}' \bullet \hat{\Omega}\right)$$

to $M = 1$ and explain what advantage there would be in using this expansion.

To further understand the process of slowing down, consider an infinite medium of hydrogen with a uniform distributed source emitting S neutrons/cm^3/s with energy E_o. Assuming no absorption in hydrogen, we can write the probability of scattering [Eq. (2.12)] into any energy E in dE as

$$P(E_o \rightarrow E)dE = \frac{dE}{E_o}$$

Thus a neutron colliding with a hydrogen atom can lose all its energy as was shown in Fig. 2.7.

Exercise 2.8: *Final energy after scattering*: Calculate the probability that a neutron at an initial energy $E = 1\text{MeV}$ scattering in hydrogen will have its final energy in the range 0.5 MeV−1 MeV.

The resonance region has irregular peaks which can be described by models such as the Breit-Wigner model which takes into account quantum-mechanical phenomenon and compound nucleus formation.

Fig. 2.9 shows the total ENDF/B-VIII.0 cross sections of selected light nuclei lithium-7, beryllium-9 and carbon-12, in the energy range $1 \times 10^{-5} - 1 \times 10^{-8}$ eV plotted from http://atom.kaeri.re.kr/. In a low-energy range, 1×10^{-5} eV-1×10^{-3} eV, the cross sections behave as

$$\sigma_t = A + \frac{B}{\sqrt{E}} \tag{2.13}$$

(A, B are constants). It is readily seen on a linear-linear scale that the carbon-12 cross-section, for example falls off rapidly from ~25 b at 1×10^{-4} eV to ~9 b at 1×10^{-3} eV and then remains constant till ~0.1 MeV. This low-energy cross-section dependence is usually referred to as the $1/v$-dependence and holds for light nuclei. At very low temperature, below the thermal neutron energy ~0.025 eV, the cross sections show temperature-dependent variations due to crystalline effects.

In the energy range from ~0.1 to ~ 10 MeV, resonances of varying widths and intensities are observed extending to about 20 MeV for carbon. At lower energies, the resonances are identifiable and are called *resolved resonances* while at higher energies they become unresolved. For selected heavy nuclei, U-233, U-235, and Pu-239, the ENDF/B-VIII.0 total fission cross sections are shown in Fig. 2.10 with the same three distinct regions for light nuclei but with the resonance region having moved toward lower energies (few electron volts) and overlapping into 'unresolved' resonances.

Figs. 2.9 and 2.10 plotted from the *point data* evaluations available in ENDF/B files (Brown et al., 2018) are *continuous energy* evaluations. From the figures, three energy groups, can be identified *viz* thermal, intermediate and fast, typically in the ranges $0 \lesssim 0.1$ eV, 1 eV—few kilo electron volts, and few keV—MeV. In practice, these point cross sections are processed and *group averaged cross sections* are prepared for use in nuclear design calculations. The nuclear data processing system NJOY (Macfarlane, Muir, Boicourt, & Kahler, 2012) is used to prepare multigroup cross-section data in general form or for specific codes and applications. For commonly used transport codes such as ANISN (Engle, 1967), WIMS-D (Deen & Woodruff, 1995) and MCNP (Pelowitz et al., 2013), NJOY uses the modules DTFR, WIMSR and ACER.

Some nuclear reactions, such as nuclear fission and multiplication take place when an incident neutron has energy in excess of some *threshold energy* and are usually high energy reactions. One such reaction is the multiplication $(n, 2n)$ reaction for which cross sections are shown in Fig. 2.11 for Be-9, U-238, and Pu-239.

Cross sections of selected elements are shown in Table 2.1 to indicate their probabilities of interactions. Thus light nuclei such as carbon are predominantly scattering materials at thermal energy while plutonium-239 has very high fission cross-section, and indium has very high absorption cross-section. These values have engineering design

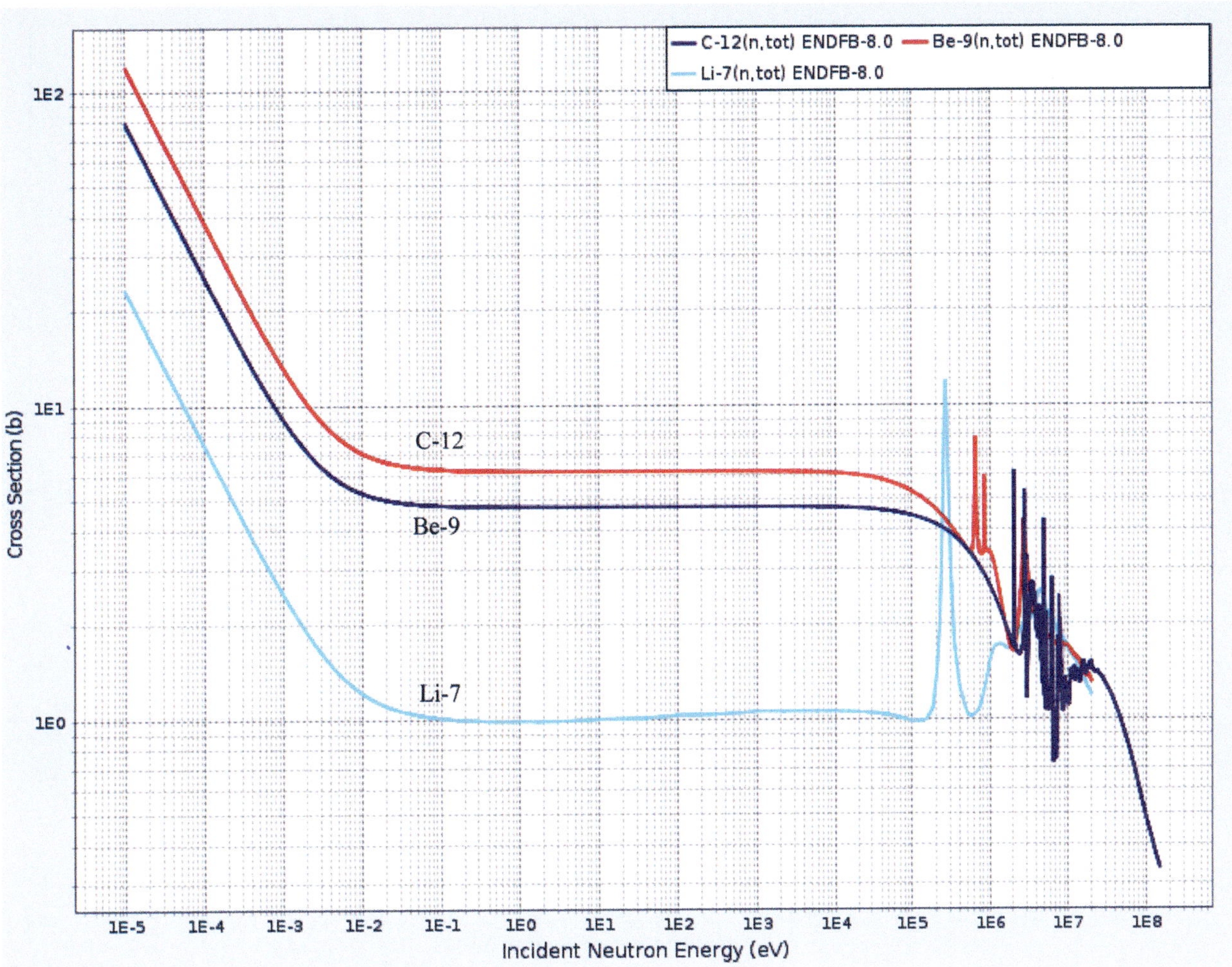

FIGURE 2.9 Total microscopic cross sections of Li-7, Be-9 and C-12.

applications. Similarly, boron-10 and xenon-135 have (thermal) absorption cross sections 3840 b and 2.65 Mb; thus boron-10 is used as an absorber rod in nuclear reactors while xenon-135 is highly undesirable as it absorbs valuable thermal neutrons which could otherwise have continued to produce fissions in the uranium fuel of a power reactor.

When microscopic cross sections are averaged over the energy spectrum in a nuclear reactor, they become application-specific, and are more efficiently used than point cross sections, for carrying out large simulations. Plutonium 239 cross sections are given in Table 2.2 averaged over a Maxwellian spectrum

$$<\sigma_M(T)> = \frac{2}{\sqrt{\pi}} \frac{\int_{E_L}^{E_U} \sigma(E,T)M(E,T)dE}{\int_{E_L}^{E_U} M(E,T)dE} \tag{2.14}$$

for the energy range $E_L = 10^{-5}$eV, $E_U = 10$eV, a $1/E$ spectrum over ($E_L = 10^{-5}$eV, $E_U = 10$keV)

$$<\sigma_{ri}(T)> = \int_{E_L}^{E_U} \sigma(E,T)\frac{1}{E}dE \tag{2.15}$$

and a fission $\chi(E)$ spectrum

$$<\sigma_f(T)> = \frac{\int_{E_L}^{E_U} \sigma(E,T)\chi(E)dE}{\int_{E_L}^{E_U} \chi(E)dE} \tag{2.16}$$

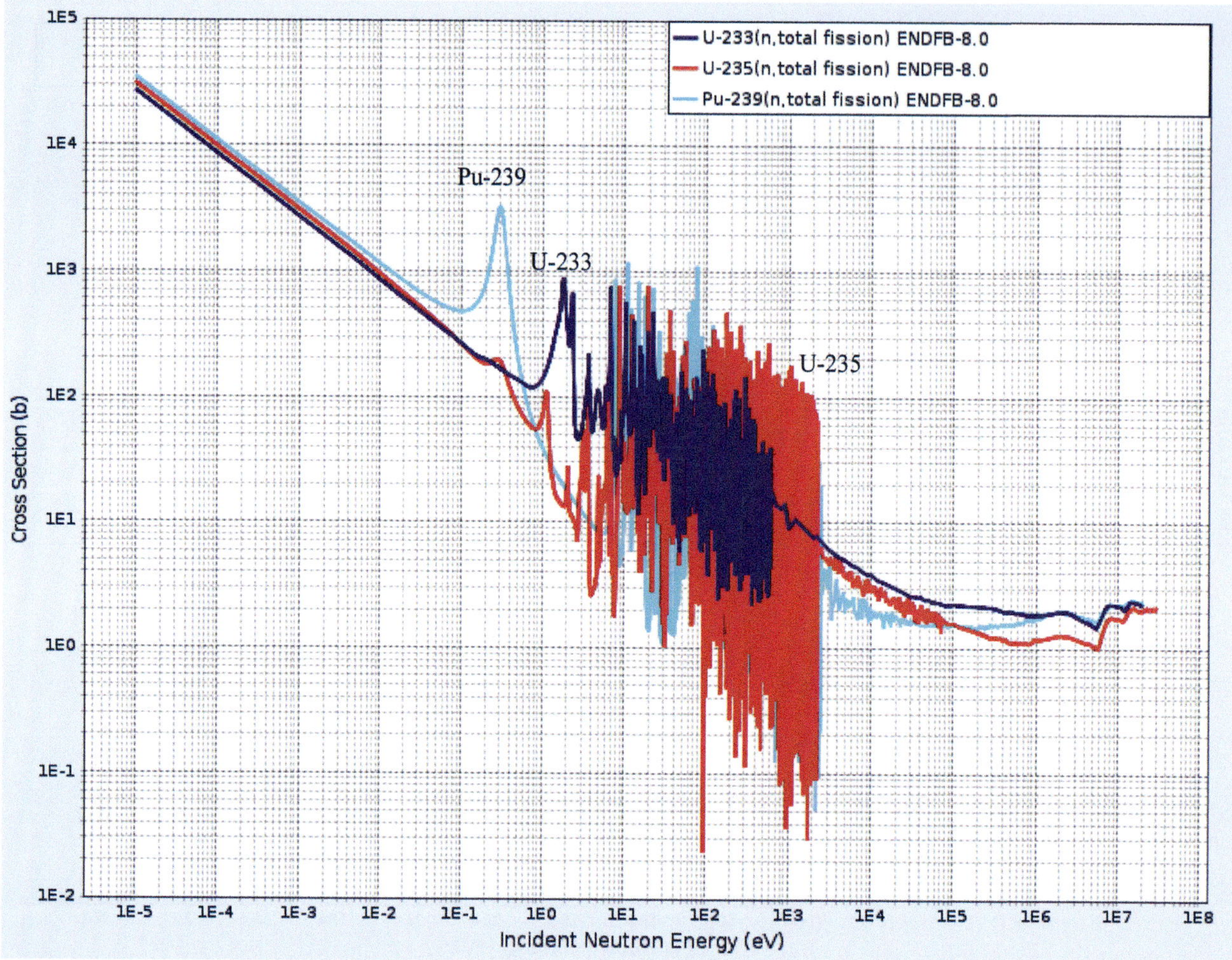

FIGURE 2.10 Total microscopic fission cross sections of U-233, U-235 and Pu-239.

where

$$\chi(E) = \sqrt{\frac{4}{\pi a^3 b}}\exp\left(-\frac{ab}{4}-\frac{E}{a}\right)\sinh\left[(bE)^{\frac{1}{2}}\right]$$

$$(E_L = 10^{-5}\text{eV},\ E_U = 20\text{MeV},\ a = 0.988\text{MeV},\ b = 2.249\text{MeV}^{-1}).$$

The Westcott g-factor is the ratio of the Maxwellian averaged cross-section to the 2200 m/s averaged cross-section and has the value ~ 1 for $1/v$ nuclides

$$g(T) = \frac{<\sigma_M(T)>}{\sigma(0.0253\text{eV}, T)}$$

Calculations of neutron thermal cross sections, Westcott factors, resonance integral averaged cross sections for 843 ENDF materials from major nuclear libraries are given at various temperatures (Pritychenko & Mughabghab, 2012).

Exercise 2.9: *Generate group cross sections*: Write a program to generate thermal group cross sections for Pu-239 by curve-fitting the data in Fig. 2.10 (in the energy range 0–0.1 eV) and weighting them with a Maxwellian distribution [Eq. (2.14)].

MT numbers are listed in the ENDF-6 Formats Manual (Trkov, Herman, & Brown, 2018). Nuclear cross sections from the National Nuclear Data Center, Brookhaven National Laboratory released as ENDF/B-VIII.0 (Brown et al., 2018) are the latest version with IAEA standards and benefit from experimental data and improvements in theory in simulation.

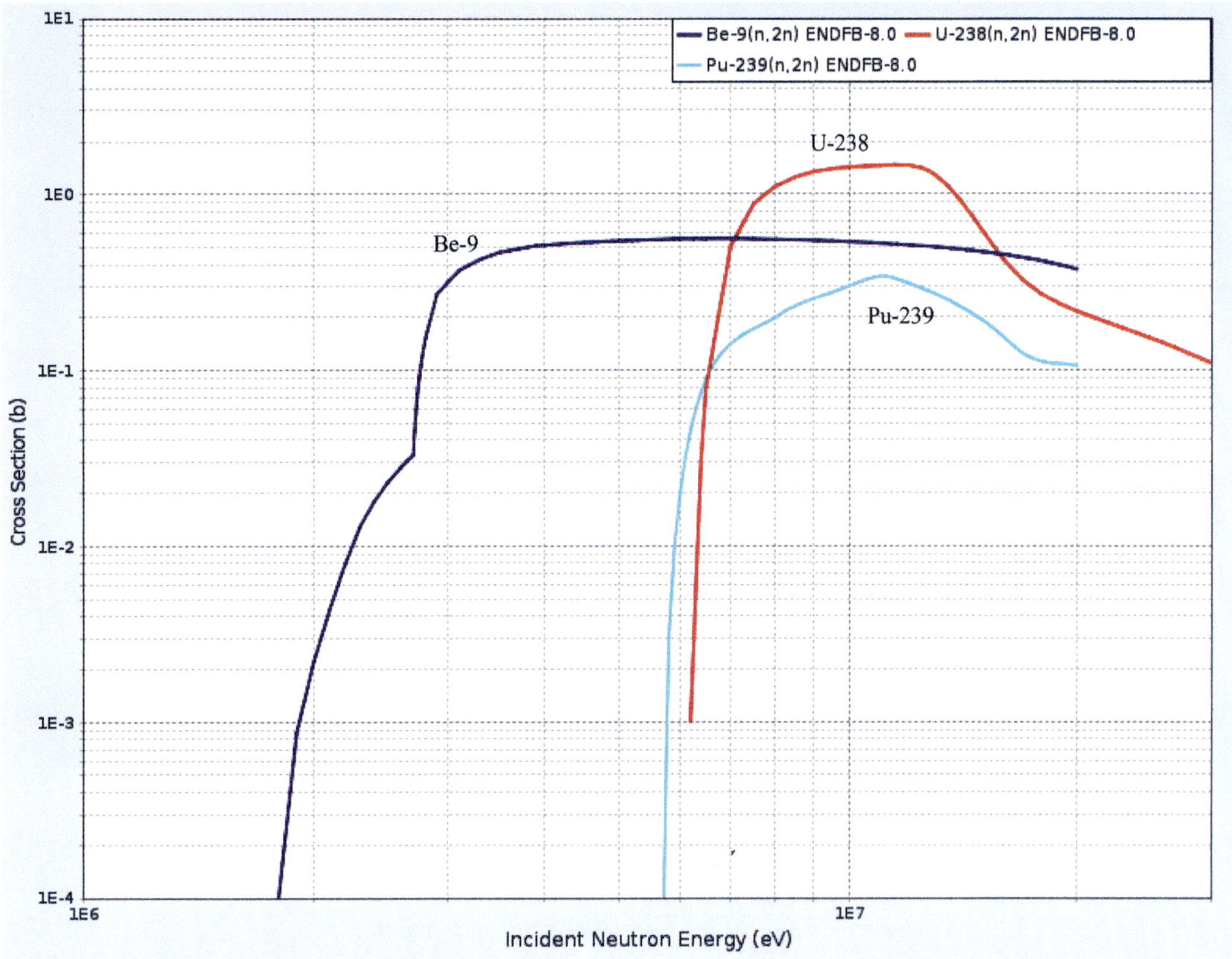

FIGURE 2.11 (n,2n) reactions Be-9, U-238, and Pu-239.

TABLE 2.1 Thermal neutron absorption cross sections of selected elements.

Element	σ_a (b)	σ_s (b)
$^{12}_{6}C$	0.0034	4.8
$^{238}_{92}U$	7.6	8.3
Mn	13.3	2.3
In	197	2.2
$^{239}_{94}Pu$	1022	9.6

2.4 The macroscopic cross-section

The macroscopic cross-section Σ is an interaction rate per unit distance traveled by the neutron obtained by multiplying the microscopic cross-section σ (a measure of the probability of interaction expressed as an area) by the atomic density N. From Section 1.5, for the data given in Table 2.3, with $\rho = 19.1 \mathrm{gcm}^{-3}$, the thermal macroscopic absorption cross-section Σ_a of natural uranium is $\Sigma_a = \sum_{i=1}^{3} N_i \sigma_{a,i}$, where $N_i = \alpha_i \overline{A}$. The atomic weight of natural uranium with the above

TABLE 2.2 Averaged cross sections of Pu-239.

MT[a]	Reaction	0.025 eV	Maxwellian average $<\sigma_M(T)>$	g-factor $g(T)$	Resonance integral $<\sigma_{ri}(T)>$	14-MeV	Fission spectrum average $<\sigma_f(T)>$
1	(n,total)	1.028 kb	1.110 kb	1.080	–	5.933 b	7.835 b
2	(n,elastic)	8.813 b	9.728 b	1.104	–	2.944 b	4.396 b
4	(n,inelastic)		E-thr = 7.894 keV			416.4 mb	1.581 b
16	(n,2n)		E-thr = 7.894 keV			229.6 mb	3.325 mb
17	(n,3n)		E-thr = 7.894 keV			2.318 mb	988.9 nb
18	(n,fission)	747.4 b	790.5 b	1.058	301 b	2.334 b	1.802 b
102	(n,gamma)	271.5 b	310.1 b	1.142	179.7 b	847.4 μb	52.61 mb

[a]*MT numbers from ENDF-6 Table formats.*
Source: https://wwwndc.jaea.go.jp/cgi-bin/Tab80http://WWW.cgi?/data/JENDL/JENDL-4-prc/intern/Pu239.intern.

TABLE 2.3 Thermal absorption cross-section of some uranium isotopes.

Nuclide	Atomic abundance (%)	Atomic weight (u)	σ_a(b)
U-234	0.0057	234.04	103.47
U-235	0.72	235.04	680.8
U-238	99.27	238.05	2.73

specified atomic fractions is $\overline{A} = 238.0179$. This gives an atomic density

$$N_U = \frac{19.1 \times 0.6023\ 10^{24}}{238.0179} = 0.0483\ 10^{23} \quad \text{atoms} \bullet \text{cm}^{-3}.$$

from which the individual atomic densities $N_{234}, N_{235}, N_{238}$ can be calculated. The thermal macroscopic cross-section is $\Sigma_a = 0.3678/\text{cm}$.

Exercise 2.10: *Weight fractions of isotopes*: From the atomic abundance, find the weight fractions of the uranium isotopes in Table 2.3.

The mean free path is the free-flight distance traveled between interactions and is the inverse of the macroscopic cross-section; thus

$$\lambda = \frac{1}{\Sigma}$$

Exercise 2.11: *Half value layer thickness*: For neutrons from Cf-252 calculate the Half Value Layer (HVL) in polyethylene, water, concrete and lead. From the HVL for gamma rays (Section 1.3) repeat your calculation to find the HVL considering both neutron and gamma emissions from Cf-252.

Exercise 2.12: *Average number of collisions*: From the definition of the mean free path, estimate the average number of collisions a neutron of 1 MeV and a gamma of 2 MeV would have in a 5-cm thick slab of iron. Justify any assumptions you make.

2.5 Flux measurement

In a nuclear reactor or any nuclear system with a source of neutrons, the energy spectrum is typically spread over several orders of magnitude. Thus a wide range of neutrons are available for flux measurements based on the exposure or *fluence* (IAEA, 1970) defined as the time integral of the neutron flux density with units of neutrons/cm^2.

TABLE 2.4 Transmutations of Mn-55, Ag-109, In-115, Au-197, and Dy-164 by neutrons.

Parent nuclide			Reaction	Radioactive nuclide		
Identity	% abundance	σ_{act} (b)[a]		Identity	$T_{1/2}$	Radiations, MeV[b]
Mn 55	100	13.4 ± 0.3	(n, γ)	Mn 55	2.6 h	$\beta^-(2.81, 1.04, 1.0.65), \gamma(0.82, 1.77, 2.06)$
Ag 109	48.65	2.8 ± 0.5	(n, γ)	Ag110m	270 days	$\beta^-(I.T), \gamma$(complex)
Ag 109	48.65	110 ± 20	(n, γ)	Ag110	24.5 s	$\beta^-(2.24, 2.82), \gamma(0.66, 0.9)$
In 115	95.8	145 ± 15	(n, γ)	In 116 m	54 min	$\beta^-(1.0, 0.87, 0.60), \gamma(0.14 - 2.19)$
In 115	95.8	52 ± 6	(n, γ)	In 116	13 s	$\beta^-(2.95), \gamma$(none)
Au 197	100	96 ± 10	(n, γ)	Au 198	2.7 days	$\beta^-(0.96, 0.29, 1.37), \gamma(0.41, 0.68, 1.09)$
Dy 164	28.2	<1000	(n, γ)	Dy 165	2.4 hr	$\beta^-(1.25, 0.88, 0.42), \gamma(0.1 - 0.76)$

[a]*for neutrons of 2200 m/s.*
[b]*beta energies are maximum energies with average energy about 40% of maximum.*
Source: Nuclear Engineering Handbook, 7-7, (1958), D. J. Hughes et al., "Neutron Cross Sections," BNL-325, McGraw Hill Book Company, Inc., New York 1955; J. M. Hollander, I. Perlman and G. T. Seaborg, Table of Isotopes, *Revs. Med. Phys.* 25(2): 469–651 (1953). *Natl. Bur. Standards Circ.* 499.

Thermal flux is measured by the activation of thin foils in a nuclear reactor for sufficient time to cause transmutation with beta emission, in most cases, which can subsequently be counted. For a flux ϕ_0 for neutrons at a most probable speed of 2200 m/s corresponding to a temperature of $T_0 = 293.6°K$, the integral value thermal flux for a reactor temperature T can be found as

$$\phi_{th} = \phi_0 \sqrt{\frac{4T}{T_0}}$$

Foils of manganese, silver, indium, gold, and dysprosium with isotopes and activation cross sections listed in Table 2.4, are used.

The slowed down (epithermal) and thermal neutron components of the spectrum are separated using cadmium covers which discriminate between the two ranges, like a high-pass filter, due to the very high thermal absorption cross-section of cadmium. Two measurements are carried out; in the first, a bare foil detector is used where the total activity is measured and in the second, the foil is covered by cadmium so that it is subject to only epicadmium neutrons which pass through the cadmium cover.

Fig. 2.12 shows the activation cross sections for Ag-109, Cd-113, In-115, and Au-197 with capture resonances, particularly of indium at 1.41 eV and gold at 4.9 eV, and high absorption cross-section of cadmium below ~0.4 eV. Cobalt-59 has also been used as a detector; its main advantages are that it exists in the form of this isotope with 100% abundance, it is easy to make wires or thin disks and it can be used at high temperatures.

Fast neutron fluence is measured from threshold reactions such as (n, α) and (n, p) reactions using threshold activation detectors (Hyvönen-Dabek & Nikkinen-Vilkki, 1980) which have a good response to neutrons of a typical fission spectrum. Commonly used detectors for dosimetry in materials use Fe^{54}(n,p) and Ni^{58}(n,p) monitors while for fission monitors for very long irradiations, Ti^{46}(n,p), Sc^{46} and Cu^{63}(n, α) are used. Other methods include BF_3 counters, ionization chambers, U^{235} fission chambers, and semiconductor neutron spectrometers and silicon detectors (Shafronovskaia, Zamyatin, & Cheremukhin, 2014; Zamyatin, Cheremukhin, & Shafronovskaya, 2017).

A thin foil of material such as Ag 109, In 115, and Au 197 is exposed to the neutron flux, which undergoes transmutation to produce the radioactive nuclide. The amount of induced radioactivity will be proportional to the incident intensity, exposure time and activation cross-section. Across the foil, the intensity will be given by Eq. (2.8) and the number of nuclides $P(t)$ can be found from the first-order rate equation (Section 1.2),

$$\frac{dP(t)}{dt} = \Sigma_{act}\phi - \lambda P \tag{2.17}$$

In the production term, $\Sigma_{act}\phi$, the neutron flux has been assumed to be time-dependent but space-independent due to the assumption of a thin foil. Eq. (2.17) gives

$$P(t) = \frac{\Sigma_{act}\phi}{\lambda}\left(1 - e^{-\lambda t}\right)$$

When the foil has been exposed for T seconds, the number of atoms per unit volume of the foil, of radionuclide produced is $\Sigma_{act}\phi T$. The activity of the foil $A = \lambda P$ is subsequently measured in, for example, an ionization chamber

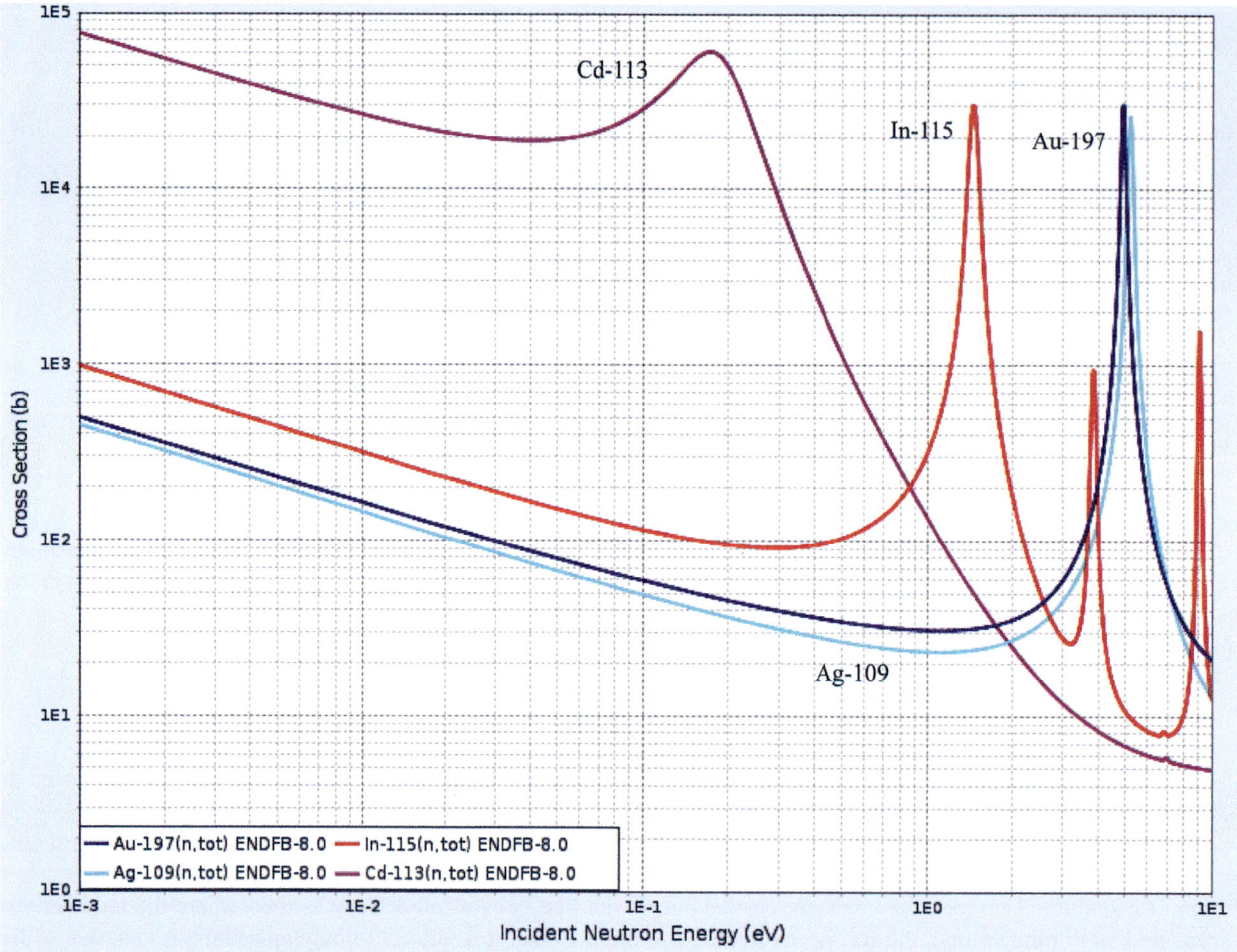

FIGURE 2.12 Cross sections of Ag-109, Cd-113, In-115, Au-197.

Exercise 2.13: *Foil activation*: A thin gold foil of mass 0.4 grams is placed in a nuclear reactor for one hour; subsequent measurements for the activity give 10^4 disintegrations per second. Estimate the thermal flux.

2.6 Reaction rates

The purpose of calculating the neutron flux in a system is to obtain reaction rates R_x where

$$\mathscr{R}_x = \Sigma_x \phi V. \tag{2.18}$$

all volumetric reaction rates and surface leakages are obtained from the flux.

In a more general form

$$R_x = \int dx \quad N(x) \int dE \quad \sigma_x(E)\phi(x,E) \tag{2.19}$$

The neutron flux can be used to estimate the reaction rate of a particular reaction within a region of interest. As an example, the number of fission reactions taking place per second $\mathscr{R}_f$ in a volume V is given by

$$R_x = N\sigma_x\phi$$

In the Evaluated Nuclear Data File ENDF/B library, the ENDF/B-6 Formats Manual lists the MT numbers, from 1 to 999, of available reaction cross-section data. These are given for the incident neutron energy in the Laboratory system. Each set has one (or more) Q value(s) depending on the possible number of outcomes of a reaction.

TABLE 2.5 Commonly used reaction cross-section MT numbers.

MT	Reaction	Description
1	(n,total)	Total cross-section
2	(z,z0)	Elastic scattering cross-section for incident particle z = all particles (n, p, e...)
3	(z,nonelas.)	Nonelastic neutron cross-section.
4	(z,n)	Production of one neutron in the exit channel. Sum of the MT = 50-91
16	(z,2n)	Production of two neutrons and a residual1. Sum of MT = 875-891, if they are present.
19-21	(n,f), (n,nf), (n,2nf)	First-, second-, and third-chance neutron-induced fission cross-section
27	(n,abs)	Absorption; sum of MT = 18 and MT = 102 through MT = 117
102	(z,γ)	Radiative capture
105	(z,t)	Production of a triton, plus a residual. Sum of MT = 700-749, if they are present.

Source: ENDF/B-6 Formats Manual.

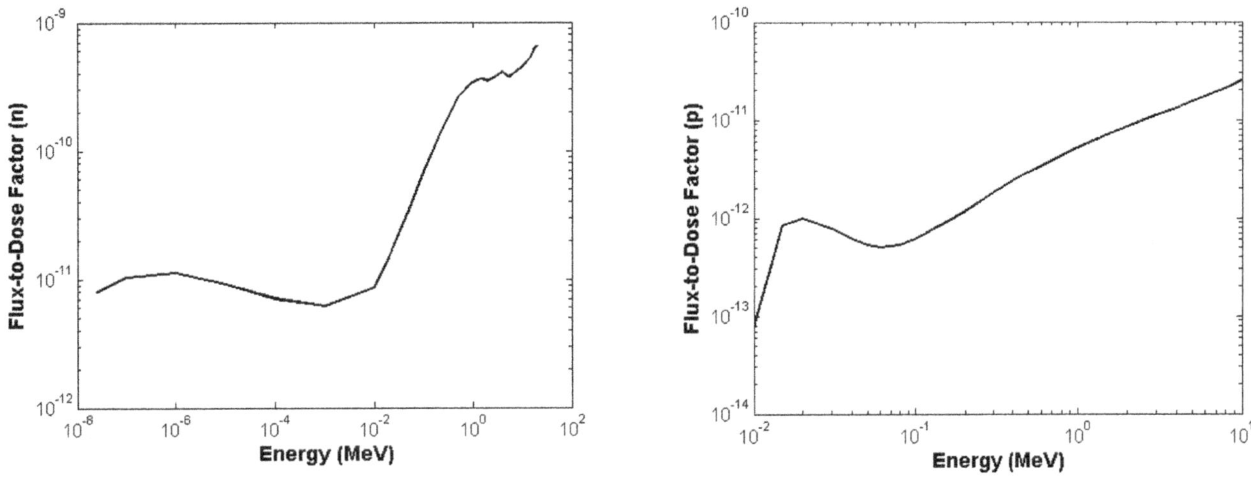

FIGURE 2.13 Flux-to-dose conversion factors for neutrons and gammas.

In the following chapters in this book, simulations will be based on computing the neutron (or photon) flux in a system and then using the reaction cross-section data, with the appropriate MT number, to compute reaction rates.

Some commonly used MT numbers are listed in Table 2.5.

Several other forms of data are also assigned MT numbers such as MT = 500 for total charged particle stopping power (Section 1.3), 502 for total photon interaction, 522 for photoelectric absorption, 527 for electro-atomic bremsstrahlung. Note that cross sections for charged particles and photons are also included.

Exercise 2.14: *Reaction MT numbers:* Which reaction rate MT numbers would be used to calculate the energy released from fission reactions in UO_2 fuel in the thermal, epithermal and fast groups?

Radiation dose can also be calculated from the neutron and photon flux-to-dose conversion factors shown in Fig. 2.13; the H*10 dose due to neutrons and photons a distance 1 m away from a point isotropic source, are used to estimate the radiation dose.

2.7 Neutron slowing down, diffusion and thermalization

In the transport process, a neutron at a source energy E_0, slows down to some energy, E, as it loses energy, or gains lethargy $u = \ln E_0/E$, in collisions by elastic and inelastic scattering. Since the threshold of inelastic reactions for low mass nuclei is very high (~MeV) the energy loss by inelastic scattering is mainly from heavy nuclei. The fractional energy loss of neutrons with light nuclei can be very high while only a small fraction of its energy is lost in collisions with heavy nuclei such as uranium.

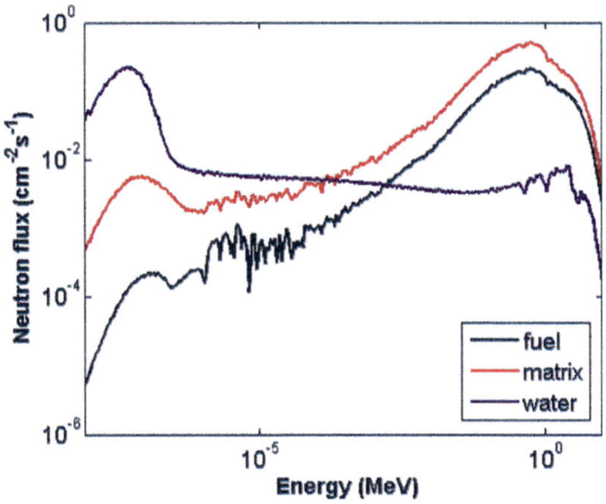

FIGURE 2.14 Typical neutron flux spectra in a micronuclear reactor.

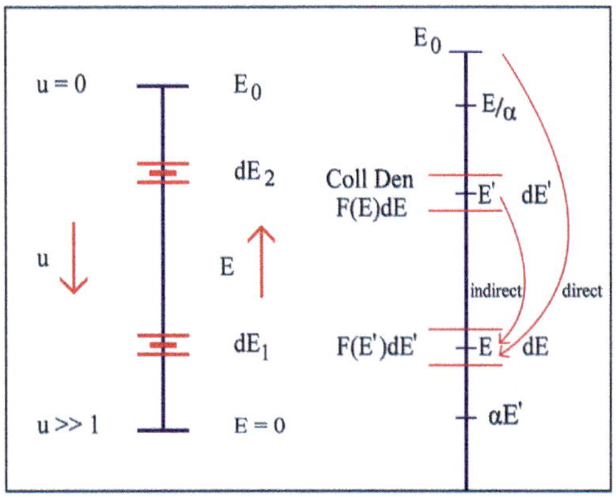

FIGURE 2.15 Energy and lethargy scattering diagrams.

The neutron spectra in a micronuclear reactor (Aziz, Koreshi, Sheikh, & Khan, 2020) with a predominantly fast spectrum in the matrix and a thermal spectrum in the water reflector is shown in Fig. 2.14. In fast reactors, for example, the spectrum is "hard," that is, predominantly high energy ($\sim$MeV) while in thermal reactors, it is "soft," that is, shifted toward lower energies. Since the fission cross-section is low at high energies and *vice versa*, fast reactors will have a higher flux to maintain reactor power. Typical orders of magnitude of neutron flux in thermal and fast reactors are $\sim 10^{12}$ n/cm^2/s and $\sim 10^{15}$ n/cm^2/s respectively.

Ignoring space dependence, the behavior of the energy-dependent flux $\phi(E)$ in an infinite medium can be found from elementary considerations. As shown in Fig. 2.15 the slowing down of a neutron from source (initial) energy E_0 in the case of hydrogen medium (left fig) can take place to final energy $E = 0$ as discussed in Section2.3. Note that the source energy corresponds to zero lethargy since $u = \ln E_0/E$ since a source neutron has high energy and hence is the opposite of a lethargic neutron. The energy range $0 < E < E_0$ thus corresponds to a lethargy range $\infty > u > 0$. The figure on the right in Fig. 2.15 shows neutron collisions in a medium for which $A > 1$. In this case, a source neutron can have energy in the range $\alpha E_0 \le E \le E_0$ while a neutron at E' can have energy E'' in the range $\alpha E' \le E'' \le E'$.

Defining collision density $F(E) = \Sigma_t(E)\phi(E)dE$ as the total number of interactions at E in dE, the two contributions will consist of a 'direct' contribution from the source and an 'indirect' contribution from collisions taking place at some higher energy which lead to scattering into dE. Using Eq. (2.12), these two terms (for $\alpha = 0$) are

1. source neutrons falling into $dE = SdE/E_o$
2. $\int_{E'=E}^{E'=E_o} \Sigma_t(E')\phi(E')dE'P(E' \to E)dE = dE \int_{E'=E}^{E'=E_o} \frac{\Sigma_t(E')\phi(E')}{E'} dE'$

The balance equation for the collision density in dE is

$$F(E)dE = \frac{SdE}{E_o} + dE \int_{E'=E}^{E'=E_o} \frac{F(E')}{E'} dE' \tag{2.20}$$

The above integral equation can easily be solved by differentiating to yield

$$\frac{dF}{dE} = -\frac{F}{E}$$

from which the neutron flux is

$$\phi(E) = \frac{S}{\Sigma_t E} \sim \frac{1}{E} \tag{2.21}$$

Another quantity of interest is the slowing down density $q(E)$ which is defined as the average number of neutrons at some energy $E' > E$ falling below energy. The collision density is thus a reaction rate while the slowing down density is analogous to the current $\overline{J}$. It can be shown by similar reasoning that for the case of an infinite hydrogen medium with a uniform source and in the absence of absorption, $q(E) = S$.

For neutron moderation for A > 1, the entire energy domain is not available for a neutron after collision, thus the *Placzek discontinuities* shown in Fig. 2.16 are found for the collision density based on whether a neutron has had one or more collisions while reaching the energy interval of interest. On the horizontal axis is the collision density for hydrogen ($A = 1, \alpha = 0$) while the infinitely heavy medium ($A = \infty, \alpha = 1$) has very little (or zero) energy transfer in a collision hence it can have an infinite number of collisions in the first interval. For all other media, the wiggles or transients die out as shown and the collision density assumes a constant value (in lethargy).

When discontinuities appear only near the source, then in the asymptotic energy region, the solution can be shown to be

$$F = \frac{S}{\xi E} \tag{2.22}$$

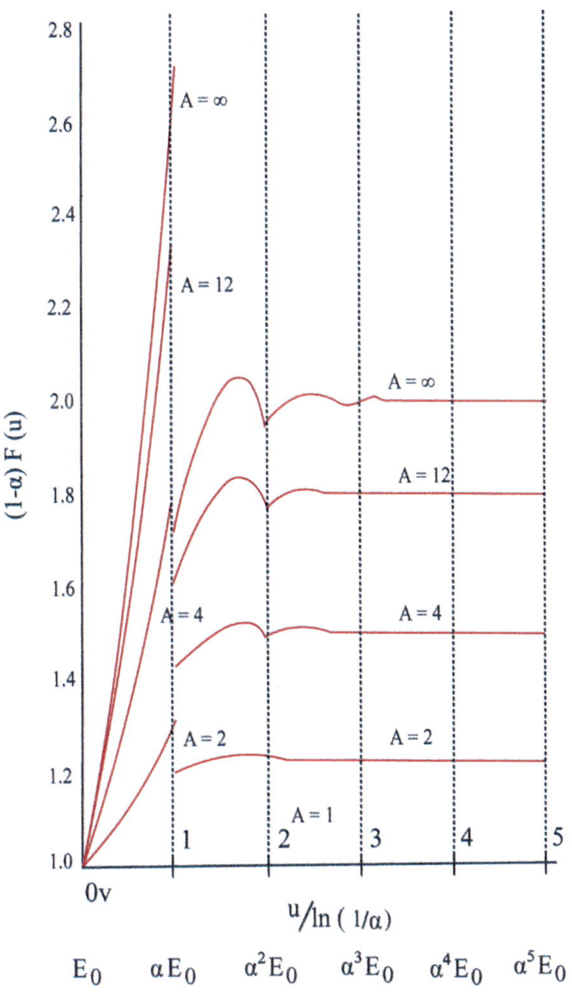

FIGURE 2.16 Placzek wiggles for neutron scattering.

and subsequently

$$\phi(E) = \frac{S}{\xi \Sigma_t E} \sim \frac{1}{E} \tag{2.23}$$

where ξ, or $\overline{\Delta u}$, is the average change in lethargy per collision can be obtained using the probability for a neutron going from $E \rightarrow E'$

$$P(E \rightarrow E') = \frac{1}{E(1-\alpha)}$$

so that

$$\overline{\Delta u} = \int_{\alpha E}^{E} dE' \ln\left(\frac{E}{E'}\right) P(E \rightarrow E') \tag{2.24}$$

where the integral has been carried out over final energy for the range permissible. This gives

$$\overline{\Delta u} = \xi = 1 - \frac{(A-1)^2}{2A} \ln\left(\frac{A+1}{A-1}\right) \cong \frac{2}{A+2/3}.$$

Thus for neutrons colliding in hydrogen $\xi = 1$ i.e. the change in lethargy is maximum while for A >> 1, $\xi \ll 1$. The number of collisions needed for a neutron to slow down from 1 MeV to 1 eV assuming isotropic scattering in the CM system is shown in Table 2.6.

Exercise 2.15: *Collision density*: Write an expression for the collision density for neutrons reaching the energy interval *dE* (Fig. 2.15) after one collision $F_1(E)$.

The smallest number of collisions is for hydrogen (13.6) while the largest is for uranium (1649); these values have are important for the selection of materials for achieving a neutron spectrum of interest.

Matlab Example 2.3: *Number of collisions*: Calculate the number of collisions in hydrogen, carbon, iron and uranium-238 for a fission neutron at 1 MeV to slow down to energy 1 eV.

```
A=12; Alpha = ((A-1)/ (A+1))^2
xi=1-((A-1)^2/(2*A))*log((A+1)/(A-1))
E1=1e6; E2=1;n=(log(E1/E2))/xi
```

Exercise 2.16: *Lethargy*: Calculate the average gain in lethargy per collision $\overline{\xi}$ for light water and heavy water.

Extending the above analysis for hydrogen, it can be shown that for A >> 1

$$q(E) = \int_{E'=E}^{E'=E/\alpha} F(E') dE' \frac{E - \alpha E'}{(1-\alpha)E'} = C\xi \tag{2.25}$$

TABLE 2.6 Number of collisions needed to slow down from 1 MeV to 1 eV.

Nucleus	Mass no	α	ξ	n
Hydrogen	1	0	1.0000	13.8
Deuterium	2	0.111	0.7253	19
Beryllium	9	0.640	0.2066	66.9
Carbon	12	0.716	0.1578	87.6
Oxygen	16	0.779	0.120	115.2
Sodium	23	0.840	0.0845	163.5
Iron	56	0.931	0.0353	391.5
Uranium	238	0.983	0.0084	1649

where for no absorption in which case all neutrons eventually slow down past energy E, $C = S/\xi$, so that again $q(E) = S$. Extensions to the case of nonmonoenergetic source show the same result for hydrogen as obtained for the monoenergetic source considered above. For mixtures, again a similar dependence is found except that properties are averaged over the elements of the mixture.

The above results can be used to estimate a moderating time t_m defined as the time taken by a neutron to slow down from some high energy, (e.g., fission energy) to some cutoff energy E_m. If the cutoff is set at 1 eV, then it is above the thermal range and generally 'beyond' the resonances. The energy group lower than E_m, is the 'thermal' group; we now have two distinct regions for slowing down and diffusion. In the first region, the neutrons will slow down or undergo *moderation*; in the second, from 1 eV down to some equilibrium energy, neutrons will *diffuse* for a further diffusion time t_d until they are captured or reach equilibrium.

The moderating time is easily found from Fermi's "age" theory which assumes an "average" behavior for the slowing down region and assumes that Fick's law is valid for all energies, which is clearly a poor assumption especially for hydrogenous media. The average number of collisions $\bar{n}$ is found using $\overline{\Delta u}$ [Eq. (2.24)] giving

$$\bar{n} = \frac{du}{\overline{\Delta u}} = \frac{dE}{E\bar{\xi}}$$

In Fermi's continuous slowing down model, an average behavior is assumed for each neutron undergoing collisions. A continuity equation for the neutrons crossing "boundaries" of a lethargy interval is written, using Fick's Law which introduces neutron flux ϕ in terms of the slowing down density $q(u)$. Using the relationship

$$q(r,u) = \int_{u-\xi}^{u} F(r,u')du'$$

the "Fermi age equation" is written as

$$\nabla^2 q(r,\tau) = \frac{\partial q(r,\tau)}{\partial \tau} \tag{2.26}$$

where the neutron age τ (with units of area) is defined as

$$\tau = \int_0^u \frac{D(u)}{\xi\Sigma_s(u)} du \tag{2.27}$$

The interpretation of the age is possible by expressing the lethargy increase as $du = \xi\Sigma_s dx$, i.e. the increase in lethargy per collision multiplied by the number of scattering collisions. Thus

$$\tau = \int_0^u \frac{D(u)}{\xi\Sigma_s(u)} \xi\Sigma_s(u)dx = \overline{D}l(u)$$

where the distance $l(u)$ is a measure of the movement which results in a lethargy increase from 0 to u. Eq. (2.26) can be solved to obtain the slowing down density q with boundary conditions for the source $q(r,0) = S(r)$, flux and current continuity at the interface conditions requiring, and at the surface where q vanishes, The flux continuity is

$$\phi(r,u) = \frac{q(r,u)}{\xi\Sigma_s(u)}$$

while the current continuity is to preserve

$$J = -D\nabla\phi(r,u) = -\frac{D}{\xi\Sigma_s(u)}\nabla\phi(r,u)$$

For a plane source in an infinite slab, the solution to the Fermi age equation is very illustrative for the physical understanding of neutron age. The slowing down density is

$$q(x,\tau) = \frac{S}{\sqrt{4\pi\tau}} e^{-x^2/4\tau} \tag{2.28}$$

which gives a picture of neutrons moving outwards from a source in the form of a Gaussian distribution as they settle into the host medium. In 3D, the number of neutrons that started at $r = 0$ and have slowed down to within a shell dr at r is

$$q(r,\tau) = 4\pi r^2 dr \frac{S}{(4\pi\tau)^{3/2}} e^{-r^2/4\tau} \tag{2.29}$$

The probability of finding a neutron in a spherical shell is thus

$$p(r) = 4\pi r^2 \frac{1}{(4\pi\tau)^{3/2}} e^{-r^2/4\tau} \tag{2.30}$$

Since

$$\int_0^\infty r^2 e^{-r^2/4\tau} dr = \frac{4\tau\sqrt{4\pi\tau}}{4}$$

the total probability is

$$\int_0^\infty drp(r) = \frac{4\pi}{(4\pi\tau)^{3/2}} \int_0^\infty drr^2 e^{-r^2/4\tau} = \frac{4\pi}{(4\pi\tau)^{3/2}} \frac{4\tau\sqrt{4\pi\tau}}{4} = 1$$

With this probability, the variance or spread will be a measure of the age of neutrons in terms of the physical spread. The second moment (measure of variance) is then

$$\int_0^\infty r^2 p(r)dr = \overline{r^2} = 6\tau$$

The physical significance of "age" is that

$$\tau = \frac{1}{6}\overline{r^2}$$

The neutron age, shown in Fig. 2.17 is thus a measure of how far a neutron travels, or what its age becomes, in going from a high energy to some low energy.

The measured age of fission neutrons slowing down to the indium resonance energy at 1.45 eV is given in Table 2.7. Neutrons in water moderate in much shorter a distance than they do in graphite implying that neutrons reach further into a nonhydrogenous material than into a hydrogenous material.

The moderation time for slowing down from E_0 to E is obtained by integration:

$$t_m = \frac{2}{\overline{\Delta u}\Sigma_s}\left(\frac{1}{v_m} - \frac{1}{v_0}\right) \sim \frac{2}{\overline{\Delta u}\Sigma_s v_m}, \ (v_0 \gg v_m) \tag{2.31}$$

The moderating time for water, heavy water and beryllium is $\sim 1.0\,\mu s$, $\sim 8.1\,\mu s$, and $\sim 9.3\,\mu s$ respectively. Estimates for the diffusion time, obtained from the speed and mean free path as: $t_d = v_{th}/\lambda_a$, give $\sim 100.0\,\mu s$, $\sim 1.5 \times 10^5\,\mu s$, $\sim 4.3 \times 10^3\,\mu s$ and $\sim 23\,\mu s$ for water, heavy water, beryllium and graphite. The diffusion length L, of neutrons is defined as

$$L^2 = \frac{D}{\Sigma_a}, \text{ where } D = \frac{1}{3\Sigma_{tr}}$$

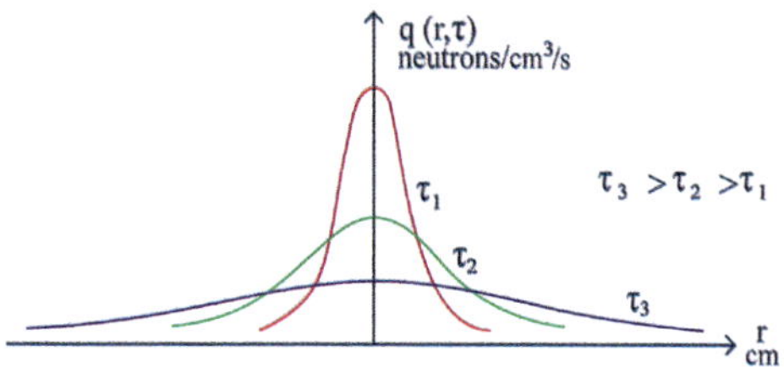

FIGURE 2.17 Neutron slowing down density.

TABLE 2.7 Age-to-thermal (fission to 1.45 eV) of selected materials.

Moderator	H_2O	D_2O	Be	BeO	Graphite
τ_{In} (cm^2)	~ 26	111	85	80	311

Source: Lamarsh, J. R. Introduction to Nuclear Engineering, p.203.

TABLE 2.8 Diffusion length of selected materials.

Moderator	Diffusion length L (cm)
H_2O	2.85
D_2O	116
C	54

Source: Murray R. L., Nuclear Energy, Elsevier, 6th (Ed.) p.55.

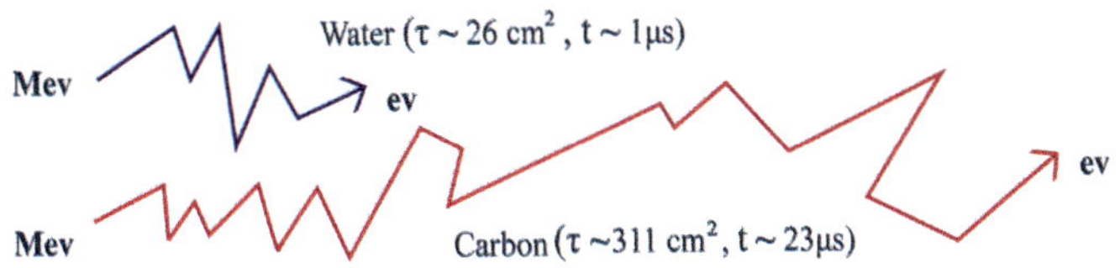

FIGURE 2.18 Neutron collision paths in water and carbon.

in terms of the transport cross-section $\Sigma_{tr} = \Sigma_s(1 - \overline{\mu}_0)$ for average cosine of the scattering angle $\overline{\mu}_0 = 2/(3A)$. Table 2.8 lists the values of some diffusion length in some moderating materials.

A schematic of the moderation and diffusion of neutrons in water and graphite is shown in Fig. 2.18.

Total cross-section <65>, <<35>> carbon and water, good picture <<p.38>>, theory <<39>>.

When neutrons reach thermal energies, they are in equilibrium with the host nuclei which have similar energies, and hence neutrons can receive energy from the nuclei, or up-scatter in contrast to their slowing which was down-scattering only. The transport and energy exchange of thermal neutrons takes into account the binding of nuclei in the host medium. It is found that the cross sections of water are substantially different from the cross-section obtained by the sum $\sigma_{\text{water}} = 2\sigma_{\text{H}} + \sigma_{\text{O}}$. The distribution is

$$F_a = \frac{2\pi nN}{(\pi kT)^{3/2}}\left(\frac{2}{\mu}\right)^{1/2}\int_0^{\infty}\sigma_s(E)e^{-E/kT}dE$$

$$F_a = \int_0^{\infty}\sum\nolimits_a(E)\phi_{\mu}(E)dE \tag{2.32}$$

where $\sum_a(E) = N\sigma_a(E)$ and $\phi_\mu(E)$ is given by

$$\phi_\mu(E) = \frac{2\pi n}{(\pi kT)^{3/2}}\left(\frac{2}{\mu}\right)^{1/2}Ee^{-E/kT} \tag{2.33}$$

The thermal corrections for thermal flux at temperature $T$and cross sections with 2200 m/s neutrons is:

$$\frac{\nu_T}{\nu_o} = \left(\frac{T}{T_o}\right)^{1/2}$$

$$\phi_T = \frac{2}{\sqrt{\pi}}\left(\frac{T}{T_o}\right)^{1/2}\phi_o, \quad \text{and} \quad \overline{\Sigma} = \frac{\sqrt{\pi}}{2}g(T)\left(\frac{T_o}{T}\right)^{1/2}\Sigma(E_o) \tag{2.34}$$

where $g(T)$ is the non-$1/v$ Westcott factor shown in Table 2.9 for U-235, U-238, and Pu-239.

Exercise 2.17: *Gold foil activation*: Repeat the gold foil calculation of Exercise 2.13 with the thermal cross-section assuming the temperature in the reactor was 100°C.

TABLE 2.9 Non-*1/v* factors.

T, °C	U^{235}		U^{238}	Pu^{239}	
	g_a	g_f	g_a	g_a	g_f
20	0.9780	0.9759	1.0017	1.0723	1.0487
100	0.9610	0.9581	1.0031	1.1611	1.1150
200	0.9457	0.9411	1.0049	1.3388	1.2528
400	0.9294	0.9208	1.0085	1.8905	1.6904
600	0.9229	0.9108	1.0122	2.5321	2.2037
800	0.9182	0.9036	1.0159	3.1006	2.6595
1000	0.9118	0.8956	1.0198	3.5353	3.0079

2.8 Resonance cross-section

Resonances for light and heavy nuclei cross sections were seen in Figs. 2.9 and 2.10 to occur as peaks where the cross sections jump to very high values. An understanding of the phenomenon of resonance is based on the quantum-mechanical structure of the nucleus where excited levels exist, and when an incident neutron adds energy to the nucleus which matches the level of an excited state, the cross-section shows a sudden jump. The terminology of resonance is similar to that in conventional mechanical and electrical systems where the amplitude of a response suddenly increases when an induced signal matches a natural frequency of the system.

The probability of an interaction in which a compound nucleus is formed is the product of the probability that a compound nucleus is formed, and subsequently that it relaxes in a particular emission such as a neutron emission or a gamma emission. A reaction $\alpha \rightarrow \beta$ where $\alpha = n + A$ and $\beta = b + B$ thus proceeds in two stages as $\alpha \rightarrow \mathrm{CN} \rightarrow \beta$.

This probability $P(t) = |\Psi(t)|^2$ is obtained from the wave function

$$\Psi(t) = \Psi(0)e^{-iE_0t/\hbar}e^{-t/(2\tau)}$$

To find the probability of finding a state at energy E, taking the Fourier transform

$$\Phi(E) = \frac{1}{\sqrt{2\pi}}\int_{-\infty}^{\infty} dt\, e^{iEt/\hbar}\Psi(t)$$

yields

$$\Phi(E) = \frac{i\hbar\Psi(0)}{\sqrt{2\pi}}\frac{1}{(E-E_0)^2 + \left(\frac{i\hbar}{2\tau}\right)}$$

Now

$$P(E) = |\Phi(E)|^2$$

which gives $P(E) = \frac{\hbar^2|\Psi(0)|^2}{2\pi}\frac{1}{(E-E_0)^2 + \left(\frac{\hbar}{2\tau}\right)^2}$

The width of a resonance, denoted by Γ (energy) is given by the Heisenberg uncertainty relation

$$\Gamma \cong \frac{\hbar}{\tau}$$

where τ is the lifetime. Thus

$$P(E) = \frac{N_0}{(E-E_0)^2 + \Gamma^2}$$

where the normalization constant is

$$N_0 = \frac{\hbar^2|\Psi(0)|^2}{2\pi}$$

The cross-section for compound nucleus formation is thus

$$\sigma_{CN}(E) \sim \frac{1}{(E-E_0)^2 + \Gamma^2/4} \tag{2.35}$$

$$\sigma_{CN}\left(E = E_0 \pm \frac{\Gamma}{2}\right) = \frac{1}{2}\sigma_{CN}(E = E_0)$$

With nuclei at rest ($T =$ 0°K), the ith cross-section is given by the Breit–Wigner formula

$$\sigma_i E = \pi \tilde{\lambda}_1^2 g \left(\frac{E_0}{E}\right)^{1/2} \frac{\Gamma_n \Gamma_i}{(E-E_0)^2 + \Gamma^2/4} \tag{2.36}$$

where $\Gamma = \Gamma_n + \Gamma_i$ and

$$\tilde{\lambda}_1 = \frac{\hbar}{\sqrt{2\mu E_0}}, g = \frac{(2J_C + 1)}{(2J_n + 1)(2J_A + 1)}$$

with

$$x = \frac{2}{\Gamma}(E - E_0), \sigma_1 = \frac{4\pi \tilde{\lambda}_1^2 g \Gamma_n}{\Gamma}, \sigma_i(x) = \frac{\sigma_1 \Gamma_i}{\Gamma}\frac{1}{1 + x^2}$$

For radiative capture $i = \gamma$, for fission $i = f$, ...

For the scattering cross-section, the Breit–Wigner formula has $i = n$ with the potential scattering $p = 4\pi R^2$ and an interference term

$$\sigma_s = \pi\gamma g \frac{\Gamma_n^2}{(E-E_0)^2 + \Gamma^2/4} + 2\sqrt{\pi g p}\frac{\Gamma_n(E - E_0)}{(E-E_0)^2 + \Gamma^2/4} + p \tag{2.37}$$

$$\sigma_s(x) = \frac{\sigma_1 \Gamma_n}{\Gamma}\frac{1}{1 + x^2} + \frac{2\sigma_1 R}{\tilde{\lambda}_1}\frac{x}{1 + x^2} + 4\pi R^2 \tag{2.38}$$

The total cross-section is

$$\sigma_t = \pi g \frac{\Gamma_n \Gamma}{(E-E_o)^2 + \Gamma^2/4} + 2\sqrt{\pi g p}\frac{\Gamma_n(E - E_0)}{(E-E_o)^2 + \Gamma^2/4} + p \tag{2.39}$$

At higher energies, for the above light nuclei $\gtrsim$ 20 MeV, the total cross-section falls off rapidly as $1/v$. This behavior for light nuclei is also observed for some magic-number heavy nuclei.

The total cross-section for uranium-238 shown in Fig. 2.19 has resonances in the intermediate region similar to those for U-233, U-235, and Pu-239 (Fig. 2.10). Comparing the resonance regions of Figs. 2.5 and 2.6, it is seen that the resonances of uranium-238 extend in the region from ~ 6 eV to ~ 20 keV are sharper and resolved at the lower limits but become unresolved at higher energies. The cross sections for heavy nuclei generally exhibit this trend.

The first few resonances of uranium-238 are at 6.67, 20.8, 36.7, and 66.03 eV with relative widths 27.5, 31.8, 63, and 49 meV, respectively. These correspond to excited states of the nucleus which occur at discrete levels similar to the excited states of atoms; the difference being that these nuclear excited states are separated at energies of MeV rather than in eV for atoms.

When $\Gamma \sim 27$meV, $\tau \sim 2.44 \times 10^{-14}$ s which is a very 'long' time and hence a gamma would be emitted as in the time in between, no nucleon had sufficient energy to be emitted. Sharp peaks are thus radiatively captured peaks while broad resonances represent resonance scattering.

Each excited state is called an isomer and with the ground state, it makes an "isomer pair." The return to the ground state is by gamma emission which could be "delayed" due to spin effects. This delay is orders of magnitude

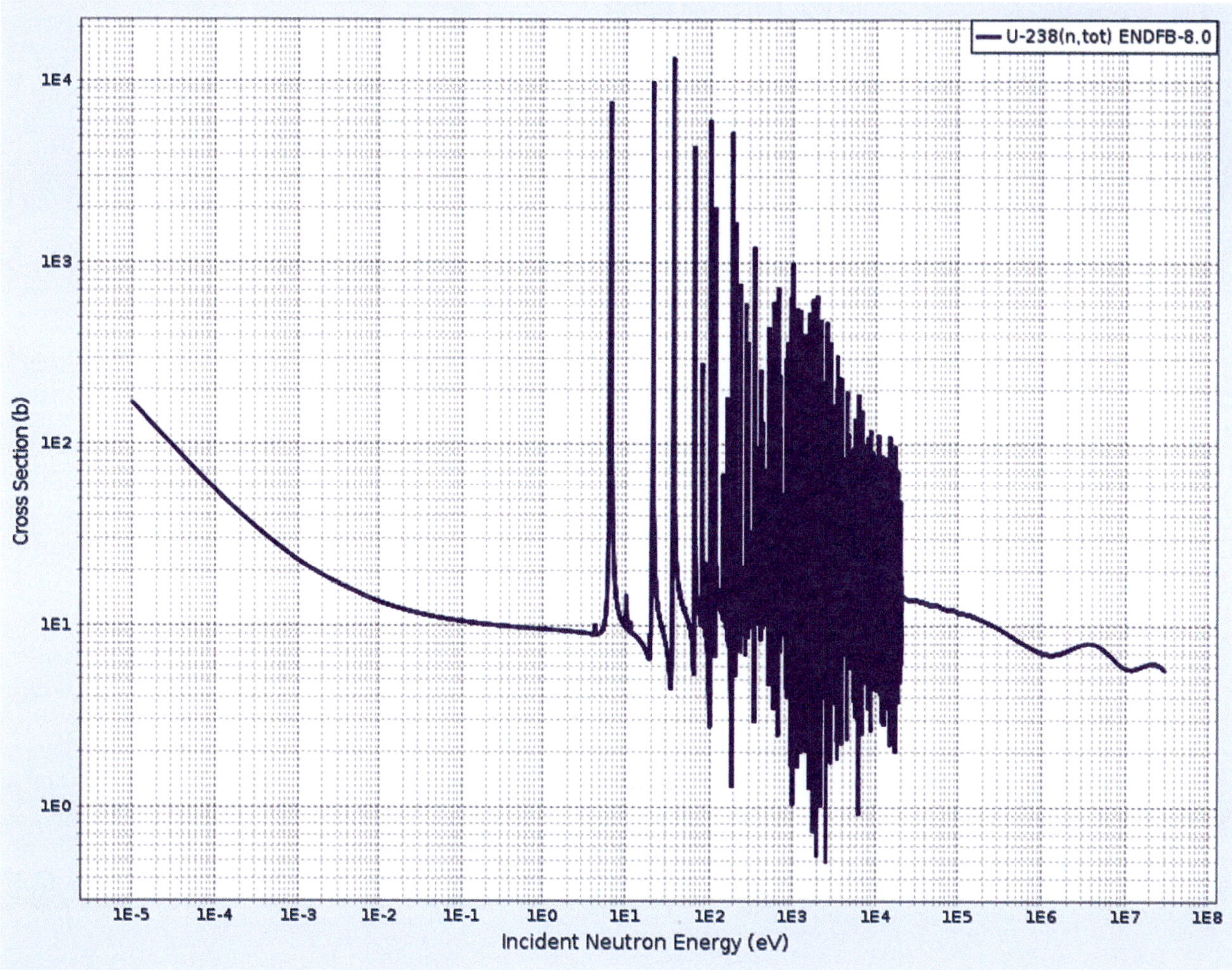

FIGURE 2.19 Total microscopic cross-section of U-238.

of compound nucleus formation which is $\sim 10^{-14}$ s (compared with $\sim 10^{-22}$ s for a new nuclear bond to form). The distance between the excited states decreases, going to higher energies, till the isomers overlap in the "continuum" domain. States which do not return to the ground state immediately and 'hold on' for a while are called metastable isomers (Fig. 2.20).

Fig. 2.21 shows the total and scattering cross sections of U-238 showing the relative magnitudes and asymmetry. The potential scattering cross-section is the value of $\sigma_s$far from the resonance [Eq. (2.38)].

A simple model to understand a resonance cross-section is the Breit–Wigner one-level model [Eq. (2.36)] demonstrated in Matlab example 2.4.

Matlab Example 2.4: *Breit-Wigner Formula*: Demonstration of Breit–Wigner single resonance formula.

```
m1=1;m2=238;mu=m1*m2/(m1+m2);h=6.626e-34;Gamman=1.493;Gamma=24.493;
g=1; amu=1.66e-27; lambda1B=(h/(2*pi))/sqrt(2*mu*amu*6.67*1.6e-19);
sigma1=4*pi*lambda1B^2*g*(1.493/24.493);
x=(2/Gamma)*(6.67-6.67); R=1.2e-15*(238)^(1/3);
sigmaS=sigma1*Gamman/Gamma +4*pi*R^2
SigmaAbs=pi*lambda1B^2*g*(Gamman*(Gamma-Gamman))*4/Gamma^2
SigmaT=pi*lambda1B^2*g*(Gamman*(Gamma))*4/Gamma^2+4*pi*R^2
```

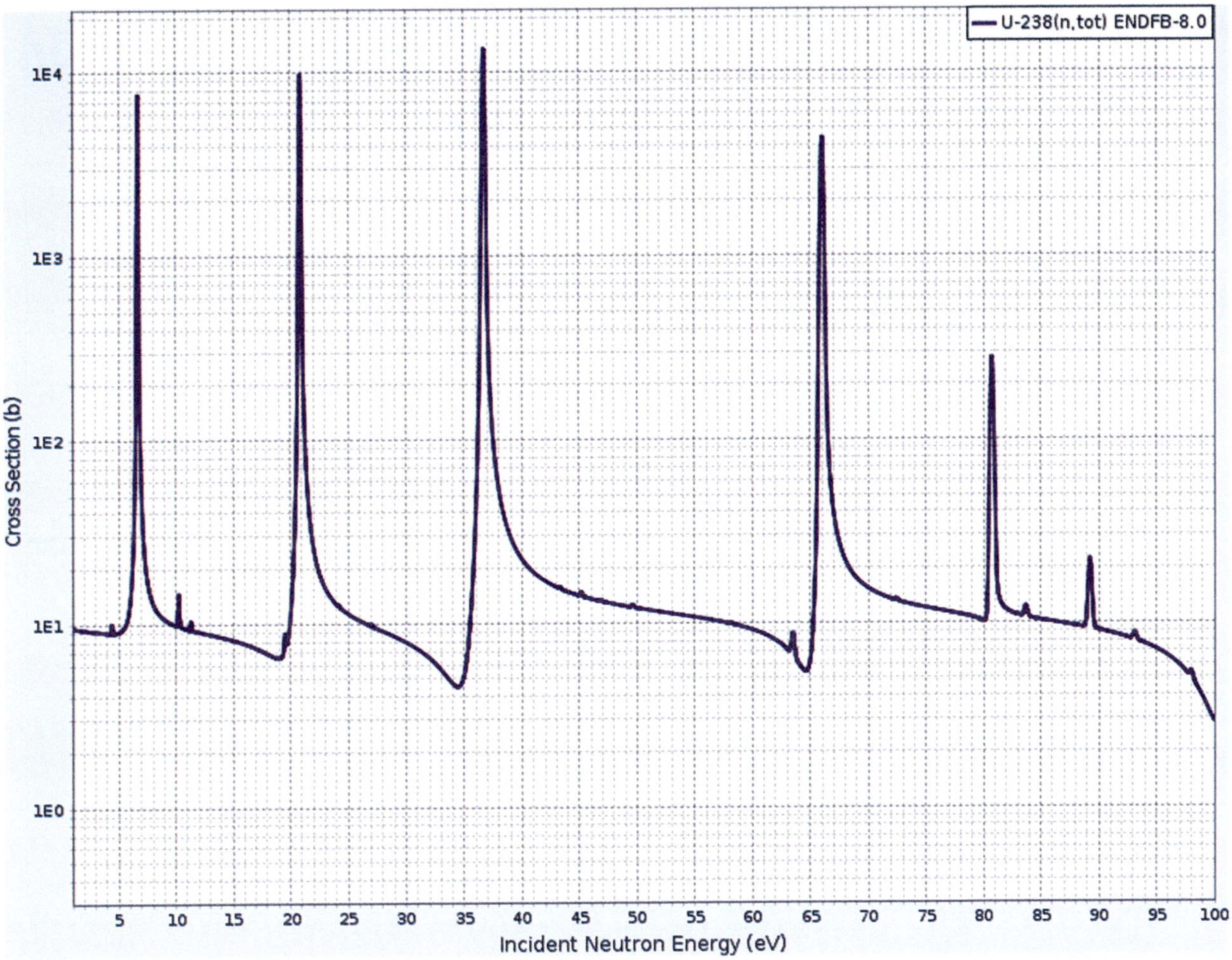

FIGURE 2.20 First few resonances of U-238.

The above discussion has assumed that nuclei are at rest, or at absolute zero K. when the cross-section depends on the energy of the neutron, the nuclide and quantum-mechanical phenomena. As temperature increases, the thermal vibration of nuclei increases and within the nucleus more states are available for capture by the neutron. When relative motion between neutrons and nuclei becomes important, the radiative capture cross-section is

$$\overline{\sigma}_\gamma(E,T) = \frac{\sigma_1 \Gamma_\gamma}{\Gamma}\left(\frac{E_1}{E}\right)^{\frac{1}{2}} \psi(\varsigma, x) \sim \frac{\sigma_1 \Gamma_\gamma}{\Gamma}\psi(\varsigma, x) \tag{2.40}$$

where

$$\psi(\varsigma, x) = \frac{\varsigma}{2\sqrt{\pi}} \int_{-\infty}^{\infty} \frac{\exp\left[-\frac{1}{4}\varsigma^2 (x-y)^2\right]}{1+y^2} dy$$

This effect of *Doppler broadening* is shown in Fig. 2.22 with the zero K value plotted from the Breit–Wigner formula (Matlab example 2.4).

From Fig. 2.22, plotted for Eq. (2.40) and demonstrated in Matlab Example 2.5, it is seen that the area under the curves remains the same as they reduce in value but increase in width with an increase in the probability of resonance capture.

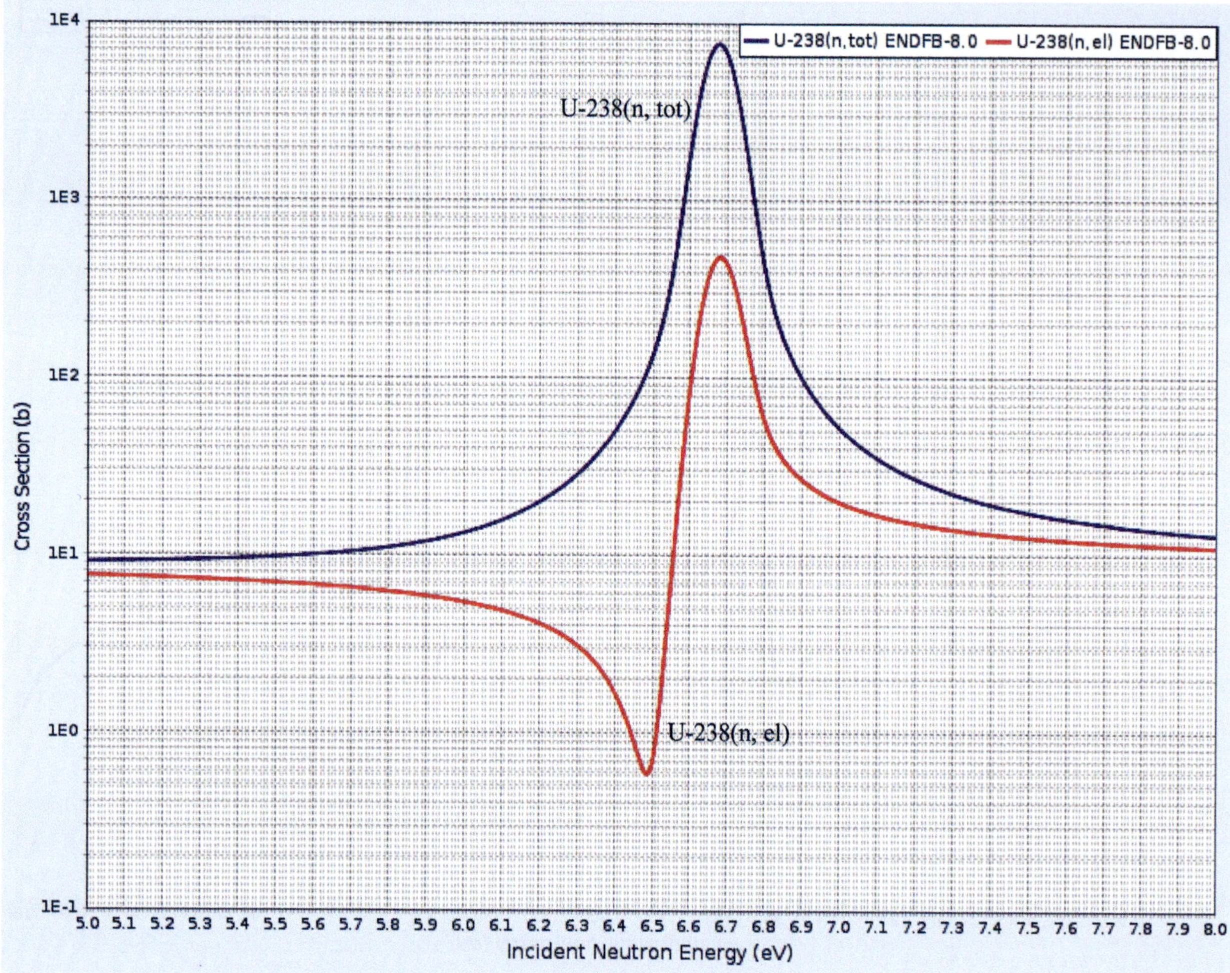

FIGURE 2.21 Total and scattering microscopic cross-section at first resonance of U-238.

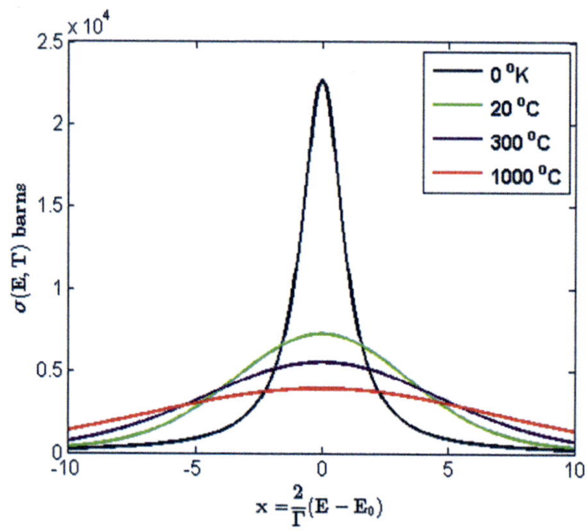

FIGURE 2.22 Doppler broadening due to temperature.

Matlab Example 2.5: *Doppler Broadening*: Doppler broadening due to temperature.

```
% DopplerTemp.m
k=8.617e-5; % eV/K
m1=1; m2=238;mu=m1*m2/(m1+m2); h=6.626e-34;
Gamman=1.493e-3; Gamma=24.493e-3; Gammag=Gamma-Gamman;
g=1; amu=1.66e-27; E1=6.67; % resonance energy
lambda1B =(h/(2*pi))/sqrt(2*mu*amu*E1*1.6e-19);
sigma1=4*pi*lambda1B^2*g*(Gamman/Gamma);

fun = @(y,zeta,x) (exp((-(zeta^2)/4)*(x-y).^2))./(1+y.^2);
T=293; GammaD=sqrt(4*E1*k*T/m2);% Doppler Width
Zeta = Gamma./GammaD;
j=0;
for jj = 1:0.05:11
j=j+1; x=jj-1;
q1 = integral (@(y) fun(y, zeta, x),-Inf, Inf);
psi=(zeta/(2*sqrt(pi)))*q1;
xx(j)=x;
sigma(j)=sigma1*Gammag*psi/Gamma *1e4*1e24;% in b
end
```

The resonance escape probability can be written as

$$p_i = 1 - \int_{E_{i-}}^{E_{i+}} \frac{\Sigma_a}{\Sigma_s + \Sigma_a} \frac{dE}{\bar{\xi} E} \tag{2.41}$$

but this expression cannot be solved analytically since the flux across a resonance is not known analytically. In the above, scattering can be considered to take place due to the moderator while absorption would be due to the presence of absorber (fuel). A dilution ratio representing the moderator to fuel ratio is used in the analysis, with a dilution cross-section σ_d defined as

$$\sigma_d = N_M \sigma_{sM} / N_{aA}$$

Thus approximate methods are used starting from a single resonance model with infinite and finite dilution with the narrow resonance (NR) and the narrow resonance infinite mass (NRIM) models with and without the effect of temperature. When resonances are separated and narrow, they can be treated separately, and the resonance escape probability can be expressed as the product of individual probabilities p_i as

$$\begin{gathered} p = \prod_{i=1}^{N} p_i \\ p = \exp\left[-\int_{E_{i-}}^{E_{i+}} \frac{\Sigma_a}{\Sigma_s + \Sigma_a} \frac{dE}{\bar{\xi} E} \right] \end{gathered} \tag{2.42}$$

with $\Sigma_a = N_A \sigma_{\gamma A}$ and $\Sigma_P = N_A \sigma_{PA} + \Sigma_{sM} =$ constant. The scattering cross-section is written as

$$\sigma_s(x) = \frac{\sigma_1 \Gamma_n}{\Gamma} \frac{1}{1+x^2} + \frac{2\sigma_1 R}{\bar{\lambda}_1} \frac{x}{1+x^2} + 4\pi R^2 = \sigma'_{sA} + \sigma_{PA}$$

with

$$\sigma_{\gamma A}(x) = \frac{\sigma_1 \Gamma_\gamma}{\Gamma} \frac{1}{1+x^2}$$

giving

$$p = \exp\left[-\int_{E_{i-}}^{E_{i+}} \frac{\Sigma_a}{N_M \sigma_{sM} + N_A(\sigma'_{sA} + \sigma_{PA}) + \Sigma_a} \frac{dE}{\bar{\xi} E} \right]$$

or in terms of the dilution cross-section $\sigma_d = \Sigma_p / N_A$

$$p = \exp\left[-\frac{N_A}{\bar{\xi} \Sigma_P} \int_{E_{i-}}^{E_{i+}} \frac{\sigma_{\gamma A}}{1 + (N_A/\Sigma_P)(\sigma'_{sA} + \sigma_{\gamma A})} \frac{dE}{E} \right] \tag{2.43}$$

The resonance escape probability is then

$$p = \exp\left[-\frac{N_A I}{\bar{\xi}\Sigma_P}\right] \tag{2.44}$$

with the effective resonance integral, for resonance in the range (E_{i-}, E_{i+}) defined as

$$I = \int_{E_{i-}}^{E_{i+}} \frac{\sigma_{\gamma A}}{1 + (N_A/\Sigma_P)(\sigma'_{sA} + \sigma_{\gamma A})} \frac{dE}{E} \tag{2.45}$$

In Eq. (2.44) the average lethargy gain averaged over the moderator and absorber is

$$\bar{\xi}(E) = \frac{\xi_M \Sigma_{sM} + \xi_A \Sigma_{pA}}{\Sigma_{sM} + \Sigma_{pA}} = \frac{\xi_M \Sigma_{sM} + \xi_A \Sigma_{pA}}{\Sigma_P}$$

Thus $\bar{\xi}(E)\Sigma_P = \xi_M \Sigma_{sM} + \xi_A \Sigma_{pA}$ and finally

$$p = \exp\left[-\frac{N_A I}{\xi_M \Sigma_{sM} + \xi_A \Sigma_{pA}}\right] \tag{2.46}$$

Eq. (2.46) is used for calculating the resonance escape probability in reactors.

Exercise 2.18: *Resonance escape probability*: Given a graphite moderator-uranium (U238) absorber system with the data: $N_M = 7.96 \times 10^{22}$ cm^{-3}, $N_M = 3.98 \times 10^{20}$ cm^{-3}, $\sigma_{SM} = 4.8$ b, $\sigma_{\gamma A} = 7.64$ b, $I = 73$ b, calculate the resonance escape probability.

2.9 Nuclear fission

Nuclear fission, first observed by Otto Hahn and Fritz Strassmann in December 1938, named and conceptualized by Lisa Meitner and Otto Frisch as depicted in Fig. 2.23, was a revolutionary breakthrough that led to the production of sustainable energy from a nuclear reaction. This discovery laid the basis for the Manhattan Project in which nuclear weapons were designed, tested and used at the end of the second world-war in 1945. Subsequently, nuclear fission was developed for several useful applications including naval propulsion, electricity generation from nuclear reactors, and the production of radioisotopes for medicine and agriculture.

2.9.1 The fission process

The nuclear fission reaction takes place when a neutron incident on a nucleus gets absorbed in it leading to the breakup of the nucleus into two fission fragments accompanied with the release of two or more energetic neutrons, radiation and a burst of energy. Several combinations of fission fragments can be emitted; one possible reaction shown in Fig. 2.23 is

$${}_0^1n + {}_{92}^{235}U \rightarrow {}_{54}^{140}Xe + {}_{38}^{94}Sr + 2{}_0^1n + 200\,\text{MeV}$$

with a neutron striking a ${}_{92}U^{235}$ atom resulting in a compound nucleus ${}_{92}U^{236}$ which fissions almost spontaneously with energy released of the order of 200 MeV (3.2 10^{-11} J). The mean lifetime of the compound nucleus formed was initially

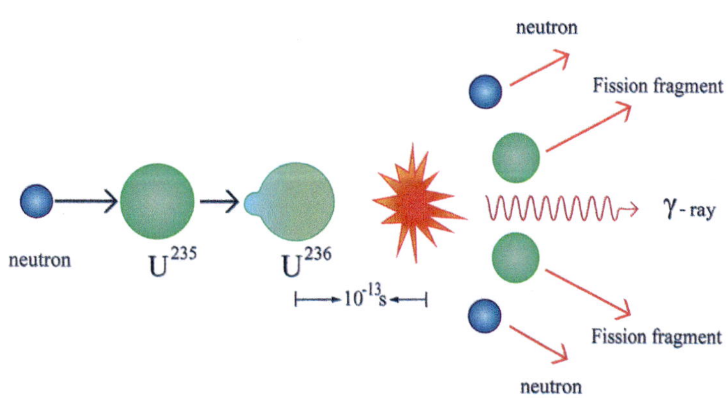

FIGURE 2.23 Nuclear fission.

TABLE 2.10 Spontaneous fission rates.

Nuclide	Half-life (years)	Spontaneous fission rate (n/g-s)
Uranium 235	7.04×10^8	7.43×10^{-4}
Uranium 238	4.50×10^9	1.80×10^{-2}
Plutonium 238	87.70	2.64×10^3
Plutonium 239	2.40×10^4	2.30×10^{-2}
Plutonium 240	7.00×10^3	8.43×10^2
Plutonium 242	7.04×10^8	1.80×10^3
Curium 242	7.04×10^8	1.80×10^7
Curium 244	7.04×10^8	1.17×10^7

estimated (Feather, 1939) to be of the order of $\sim 10^{-13}$ s. By $\sim 10^{-12}$ s, the fission fragments have lost their energy and come to rest with the subsequent release of radiation.

This is indeed a very small amount of energy compared with that involved in applying a 10 N force and moving an object by 1 m (10 J). The power produced by a "small" nuclear reactor capable of supplying electricity to a small city is $\sim$300 MW which means energy of $\sim 9.46\ 10^{15}\ J$ in one year. To produce this much energy, we would need to fission $\sim 2.96\ 10^{26}$ atoms, or about 490 gram-atoms ($\sim$ 115 kg) of U^{235}. But that would require a 100% enrichment of uranium and the fission of every single atom in the material which is not possible because natural uranium has only 0.71% of the isotope U-235 and the absorption cross-section of U-238 is much lower than that of U-235 (Table 2.3). Due to this and other reasons, a nuclear reactor producing 300 MW of power would require about 30,000 kg of uranium fuel enriched to only 3% of the isotope U^{235}. With this much fuel, it would only be a cylinder of height and diameter 2 m which means that very high energy density is achievable, and that nuclear fission is a great source of energy.

During fission. there is also a probability that three fragments (two main and a light particle) and neutrons are produced. This emission occurs in less than one in hundred fissions. The process is sometimes called ternary fission as a third fragment is produced. In Table 2.5, cross-section MT numbers in ENDF/B files are given for first-, second- and third-chance fissions for reactions (n,f), (n,nf), and $(n,2nf)$ reactions, respectively.

While fission can be *induced* as discussed above, it also takes place on its own; such *spontaneous fission* is due to a probabilistic arrangement where tunneling leads to the occurrence of fission. The nuclear fission yields listed in ENDF/B-VIII.0 library released in 2018 by the National Nuclear Data Center Brookhaven National Laboratory have sub-libraries No.5 for spontaneous fission yields with nine elements (uranium-238, curium-244,246,248, californium-250,252, einsteinium-253 and fermium-243,256) and No. 11 for induced fission yields with 31 elements (from thorium-227 to fermium-255). Table 2.10 lists the spontaneous fission rates of some uranium, plutonium and curium isotopes. The high spontaneous fission of Pu-238, together with its half-life of $\sim$ 88 years, makes it an attractive radioisotope power source (Section1.0)

The reason that spontaneous fission takes place is the instability caused by a high *N/P* ratio (stability curve Section 1.1); this ratio remains 1.0 till calcium 40 and then increases to $\sim$1.5 causing an excess of neutrons for heavy nuclides which is "corrected" by alpha decay. In fission, neutrons are released but the ratio remains high so that beta decay takes place in which N is reduced while mass number remains the same charge increases (Fig. 2.24).

The *N/P* ratio is then reduced by successive isobaric (same number of A) beta decays β^- with, in most cases, increasing half-lives until a stable configuration is achieved such as in the Krypton 93 decay to stable niobium 93 (Hanson et al., 2016) (Fig. 2.25).

2.9.2 Critical energy

Fission takes place with the compound nucleus formation as an intermediate stage in which the binding energy as well as the kinetic energy of the absorbed neutron is added to the target nucleus. This energy is then shared with the nucleons while the compound nucleus readjusts to a new form depending on the energy sharing process. To break the nucleus, energy must be sufficient enough to give it potential energy larger than it had in its original form; the threshold energy is the height of the coulomb barrier

$$E_C = \frac{Z_1 Z_2 k e^2}{(R_1 + R_2)}$$

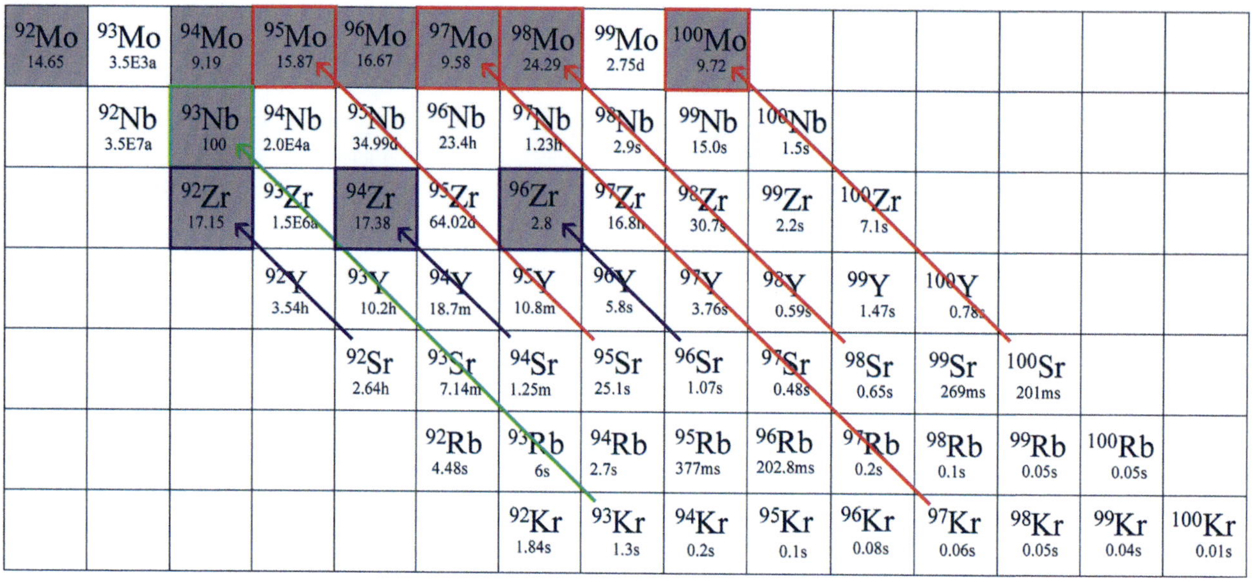

FIGURE 2.24 An isobaric decay chain.

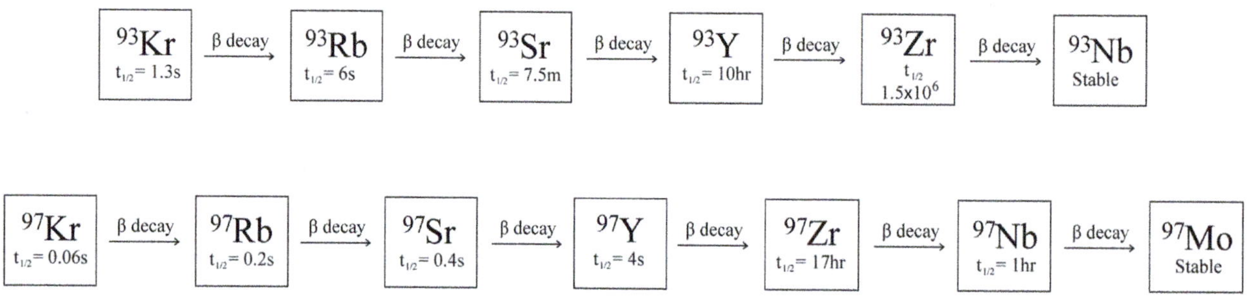

FIGURE 2.25 Isobaric decay chains in uranium and plutonium fission.

The difference between this energy and the Q value of fission is then a minimum requirement called the *critical energy* for fission to take place.

The Q value of a nuclear reaction was defined in Section1.1 as the difference of the rest mass energies of the reactants r_1, r_2 and products p_1, p_2:

$$Q = \left[m_{r1} + m_{r2} - \left(m_{p1} + m_{p2}\right)\right]c^2$$

with $Q > 0$ for endothermic or endoergic and $Q < 0$. For fission, the Q value varies depending on the fission fragments; it is thus measured from the energy deposited over all fissions. For the fission reaction in Example 2.1, the Q value is $\sim$182 MeV.

Example 2.1: *Q value*: calculate the Q value of the fission reaction $^{235}_{92}U + ^{1}_{0}n \rightarrow ^{144}_{56}Ba + ^{90}_{36}Kr + ^{1}_{0}n$

Reactants		Products	
Reactant	Mass (a.m.u)	Product	Mass (a.m.u)
$^{235}_{92}U$	235.044	$^{144}_{56}Ba$	143.92(11.5 s)
$^{1}_{0}n$	1.0082	$^{90}_{36}Kr$	89.92 (32.32 s)
		$2 \times ^{1}_{0}n$	2.0164
Total	236.0522		235.8564
$Q = 0.1958 \times 931.5 = 182.39$ MeV			

Source: Isotope mass was obtained from the Commission on Isotopic Abundances and Atomic Weights. http://www.ciaaw.org, 92235U https://ciaaw.org/uranium.htm, 56144Ba https://pubchem.ncbi.nlm.nih.gov/element/Bariumsection = Atomic-Mass-Half-Life-and-Decay.

Critical energies of some fissile nuclides is shown in Table 2.11.

Matlab Example 2.6: Critical energy.

```
Z=92; Z1=Z/2;Z2=Z/2;A=238;A1=A/2;A2=A/2;
h=6.62609004e-34; % Planck const m^2 kg/s
m=9.109e-31; % kg electron mass
Re=2.8179e-15;% m radius of electron
e=1.6e-19;% charge on an electron C
k=9e9;% conversion Nm^2/C^2
c=3e8; % speed of light m/s
hbar=h/(2*pi);
CoulEn=Z1*Z2*k*e^2/((Re/2)*(A1^(1/3)+A2^(1/3)));%J
CoulEnMeV=CoulEn*1e13/1.6;% MeV
Q=212;% MeV for fission
CritValue=CoulEnMeV-Q
```

The energy released in fission of U235 is shown in Table 2.12. With the exception of the energy associated with neutrinos, all energy is recoverable. Most of the fission energy, over 80%, is carried by fission fragments; this is released within the fuel as will be discussed in detail in Chapter 3. Charged particles such as beta rays, as discussed in Chapter 1, will also have a short range. Gamma rays from fission, as well as those produced from radiative capture, would travel a few centimeters into the surroundings of the fuel, while neutrons would be expected to go little further.

From elementary considerations, the energy produced in fission can be estimated from the binding energy curve. For U238 the binding energy is 7.5 MeV/nucleon; if the nucleus breaks into two nuclei of mass 238/2 = 119 each, for which the binding energy per nucleon is 8.4 MeV/nucleon then the gain is 0.9 MeV/nucleon so that the energy released would be 238*0.9 ~ 214 MeV. These numbers compare well with Table 2.12 where a detailed breakdown is given.

TABLE 2.11 Critical energy.

Nuclide	Critical energy (MeV)
Th 232	5.9
U 233	5.5
U 235	5.75
U 238	5.85
Pu 239	5.5

Source: Lamarsh, Introduction to Nuclear Reactor Theory, p.85.

TABLE 2.12 Energy released in a nuclear fission reaction.

Product	MeV
Fission fragment (kinetic energy)	166–168
Fission neutrons (kinetic energy)	5
Prompt gamma rays	7
Fission product gamma rays	7
Fission product beta particles	5–8
Neutrinos	10–12
Total	*200–207*

2.9.3 Fission yield

The fission yield is defined in terms of the *independent fission yield* which is the number of atoms of a specific nuclide produced "immediately" in 100 fission reactions, the *cumulative fission yield* which is the total number of atoms of a nuclide produced in 100 fission reactions, and the *chain yield*, which is the total chain yield, after the emission of prompt neutrons, for fission is cumulative yield of the last chain member. There are at least 500 nuclides which can be produced in fission directly or by beta decay of precursors depending on the incident energy and the mass and charge distributions. Several models have been proposed to estimate the fission yield. One such model is based on the partition of the nuclear yield on isomeric states (Madland & England, 1977; Okumura, Kawano, Jaffke, Talou, & Chiba, 2018) and the spin distribution of the fragments.

We cannot say with certainty what the fission fragments will be as there is a probability of emission of fragment fragments with conservation of atomic number and atomic mass (before and after reaction). For U^{235}, the distribution of fission fragments is shown in Fig. 2.26. This shows that there is a high probability of getting two fission fragments of mass numbers 95 and 140. Some nuclides close to A = 140 are Tellurium 135, Iodine 135, neon 135, Cesium 135 and Barium 135.

A possible nuclear fission reaction is ${}_0^1n + {}_{92}^{235}U \to {}_{54}^{140}Xe + {}_{38}^{94}Sr + 2{}_0^1n + 200\,\text{MeV}$, with both fission fragments xenon and samarium being unstable like other fission fragments and undergo beta decay.

2.9.4 Number of neutrons emitted in fission

The number of neutrons emerging from a fission reaction, ν, varies between zero and seven; for U^{235} the average number of neutrons is $\overline{\nu} \sim 2.5$. As a function of energy,

$$\nu(E) = \nu_0 + aE$$

where $\nu_0 = 2.43$, $a = 0.065/\text{MeV}$ in the range 0–1 MeV, and $\nu_0 = 2.45$, $a = 0.150/\text{MeV}$ for $E > 1$ MeV. In U 233, $\nu_0 = 2.48$, $a = 0.075/\text{MeV}$ in the range $0 - 1$ MeV, and $\nu_0 = 2.41$, $a = 0.136/\text{MeV}$ for $E > 1$ MeV. In Pu 239, $\nu_0 = 2.87$, $a = 0.148/\text{MeV}$ in the range $0 - 1$ MeV, and $\nu_0 = 2.91$, $a = 0.133/\text{MeV}$ for $E > 1$ MeV. These *prompt neutrons* are ejected by evaporation from the fission fragments, such as ${}_{36}^{87}Kr$, instantaneously, i.e. within $\sim 10^{-14}$ to 10^{-12} s of a fission reaction. Some fission fragments, such as ${}_{35}^{87}Br$ (decay time55s) emit neutrons with a little delay which can be from a fraction of a second to a minute as shown in Table 2.13 for thermal fission in uranium 235. These *delayed neutrons*, though less than one percent of the total number of neutrons emitted in fission, play a beneficial role

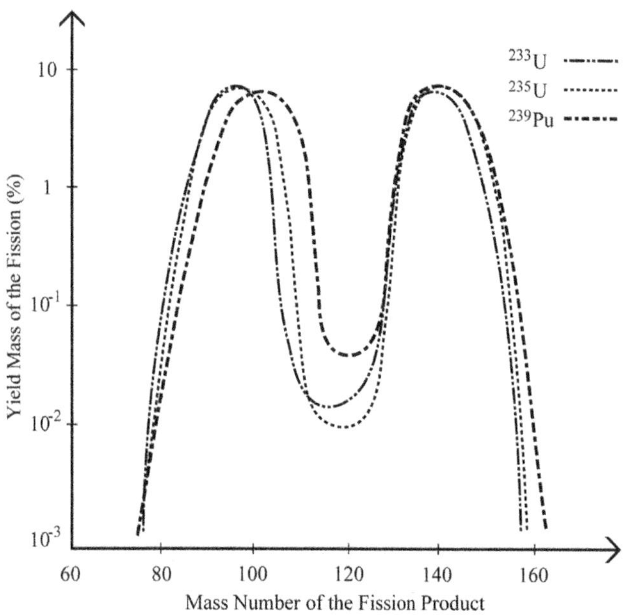

FIGURE 2.26 Fission fragment yield for U233, U235 and Pu239.

TABLE 2.13 Prompt and delayed neutrons for thermal fission in U235.

Group i	Half-Life (s)	Decay constant (λ_i, s)	Energy (keV)	Yield, neutrons per fission	Fractions (β)
1	55.72	0.0124	250	0.00052	0.000215
2	22.72	0.0305	560	0.00346	0.001424
3	6.22	0.111	405	0.00310	0.001274
4	2.30	0.301	450	0.00624	0.002568
5	0.610	1.14		0.00182	0.000748
6	0.230	3.01		0.00066	0.000273
				Total yield:	0.0158
				Total Delayed fraction(β):	0.0065

Source: Lamarsh and Baratta, Introduction to Nuclear Engineering, p88. Based in part on G. R. Keepin, Physics of Nuclear Kinetics, Reading, Mass: Addison-Wesley 1965.

in the control of nuclear reactors. The delayed neutron yield was measured by Keepin in 1957 (Keepin, Wimett, & Zeigler, 1957) and measured by Krick and Evans (Evans, 1974) for fast fission of U-233, U-235, U-238, Pu-239, Pu-240 and Th-232 averaged over 0.1-1.8 MeV. For U-235, the measured value by Krick and Evans is 0.0163 $\pm$0.0013.

Fission is also accompanied with prompt gamma ray, within the first 69 ns of thermal neutron fission from U-235, with an energy distribution $n_\gamma(E)$ neutrons emitted per MeV per fission (Peele & Maienschein, 1970).

$$n_\gamma(E) = \begin{cases} 6.6 & 0.1 < E < 0.6\text{MeV} \\ 20.2e^{-1.78E} & 0.6 < E < 1.5\text{MeV} \\ 7.2e^{-1.09E} & 1.5 < E < 10.5\text{MeV} \end{cases}$$

where E is in MeV.

2.9.5 Fissile and fertile materials

Nuclides, such as uranium-233, uranium-235, and plutonium-239, which undergo fission directly are called *fissile nuclides* while nuclides that need to be transmuted (change of the nucleus) to become fissile nuclides are called *fertile nuclides*. As uranium resources are finite, and plutonium does not exist in nature, the potential role of fertile nuclides has remained an important concern.

Two important "fertile" nuclides found in nature are thorium-232 and uranium-238, which can be induced to undergo transmutation by two-stage nuclear reactions

$$^{232}_{90}\text{Th} + ^{1}_{0}n \rightarrow ^{233}_{90}\text{Th} + \gamma$$

$$^{233}_{90}\text{Th} \xrightarrow{\beta^-(22\text{min})} ^{233}_{91}\text{Pa} \xrightarrow{\beta^-(27\text{day})} ^{233}_{92}\text{U}$$

and

$$^{238}_{92}\text{U} + ^{1}_{0}n \rightarrow ^{239}_{92}\text{U} + \gamma$$

$$^{239}_{90}\text{U} \xrightarrow{\beta^-(23.5\text{min})} ^{239}_{93}\text{Np} \xrightarrow{\beta^-(2.3\text{day})} ^{239}_{94}\text{Pu}$$

The thermal and fast cross sections for some fissile and fertile nuclides are given in Tables 2.14 and 2.15, respectively (Nuclear Energy Agency, 2015). These tables use a thermal spectrum for an averaged Pressurized Water Reactor (PWR) and a fast spectrum for a reference fission spectrum. The data of Tables 2.14 and 2.15 for U-233, U-235 and Pu-239 shown in Fig. 2.10 show very high fission cross sections for these fissile nuclides with U-233 and Pu-239

TABLE 2.14 Averaged cross sections for fissile/fertile nuclides (thermal spectrum).

Thermal spectrum	Th^{232}	U^{233}	U^{235}	U^{238}	Pu^{238}	Pu^{239}	Pu^{240}	Pu^{241}
Neutron capture (n,g) (b)	5.010	11.270	14.380	14.898	37.052	83.656	479.200	51.600
Fission (n,f) (b)	0.009	79.710	53.310	0.040	2.767	142.644	0.370	139.300

Source: NEA, OECD, 2015.

TABLE 2.15 Averaged cross sections for fissile/fertile nuclides (fast spectrum).

Fast spectrum	Th^{232}	U^{233}	U^{235}	U^{238}	Pu^{238}	Pu^{239}	Pu^{240}	Pu^{241}
Neutron capture (n,g) (b)	0.093	0.068	0.085	0.067	0.096	0.057	0.099	0.117
Fission (n,f) (b)	0.078	1.893	1.229	0.321	1.997	1.793	1.376	1.649

Source: NEA, OECD, 2015.

formed from Th-232 and U-238, respectively; in fact, both U-233 and Pu-239 have higher fission cross sections, than for U-235, for both thermal and fast spectra.

The data of these tables is valid for a fresh core as it does not consider the changes in the spectra due to changes in the fuel composition as the fuel is consumed.

2.9.6 The fission spectrum

Neutrons emerging from a fission reaction can have energies ranging from 0 to 10 MeV; a typical average value is 2 MeV. The number of neutrons emerging from fission with energies in the range E to $E+dE$ is $\nu(E)\chi(E)dE$, where $\chi(E)$ is the fission spectrum. The following empirical expression, shown is the Watt fission spectrum

$$\chi(E) = \left(\frac{\pi a^3 b}{4}\right)^{1/2} e^{ab/4} e^{-E/a} \sinh\left[(bE)^{\frac{1}{2}}\right] \tag{2.47}$$

where the constants a, b are listed in Table 2.16 for thermal and fast fissions at 1 and 14 MeV in thorium-232, uranium-233, uranium-235, and plutonium-239.

The shapes of the fission spectra for the nuclides is shown in Fig. 2.27.

Some other fits to the experimental fission energy spectrum are the Maxwell spectrum

$$\chi(E)dE = \frac{2\pi}{(\pi T)^{3/2}}\sqrt{E}\ \exp\left(-\frac{E}{T}\right)dE$$

($T = 1.33$ MeV for uranium) for which the average energy is: $E_{av} = 3/2T \sim 2\text{MeV}$, and the Cranberg spectrum

$$\chi(E)dE = \frac{2\exp(-AB/4)}{\sqrt{\pi A^3 B}}\exp\left(-\frac{E}{A}\right)\sinh\sqrt{BE}\,dE$$

(A = 0.965 MeV, B = 2.29 MeV^{-1} for uranium). For some other fission spectra used for modeling and simulation, the reader is referred to the MCNP Manual (Werner, 2017).

Exercise 2.19: *Watt Spectrum*: Find the most probable and average energy of neutrons from the Watt fission spectrum.

TABLE 2.16 Constants *a* and *b* for the Watt fission spectrum.

Nuclide energy	Th-232			U-233			U-235			Pu-239		
	th	1 MeV	14 MeV	th	1 MeV	14 MeV	th	1 MeV	14 MeV	th	1 MeV	14 MeV
a MeV	1.0888	1.1096	1.17	0.977	0.977	1.0036	0.988	0.988	1.028	0.966	0.966	1.055
b MeV^{--1}	1.6871	1.6316	1.461	2.546	2.546	2.6377	2.249	2.249	2.084	2.842	2.842	2.383

Source: ENDF/B.

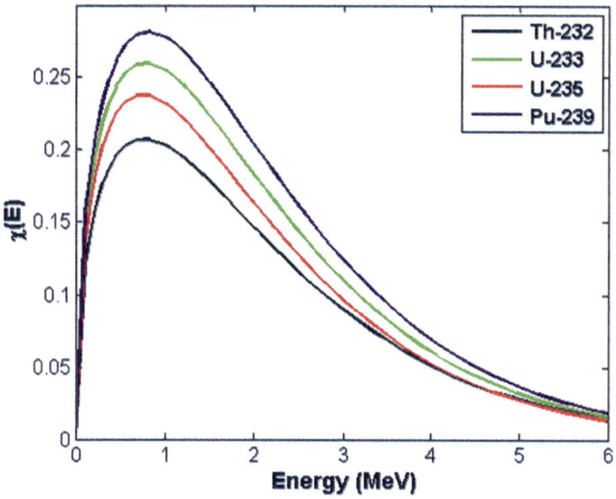

FIGURE 2.27 Watt fission energy spectrum.

2.10 Criticality

When fission takes place, the number of neutrons can rapidly multiply leading to the release of a large explosive energy. It thus needs to be controlled so that the population remains steady as in the case of a nuclear power reactor.

Three such regions can be identified in terms of the system multiplication $k^{(g)}$ in the gth generation defined as the neutron population in a generation divided by the neutron population in a previous generation

$$k^{(g)} = \frac{n^{(g)}}{n^{(g-1)}} \tag{2.48}$$

The regions are

$$k^{(g)} \begin{cases} <1 & \text{subcritical} \\ =1 & \text{critical} \\ >1 & \text{supercritical} \end{cases}$$

In a nuclear reactor, the objective is to produce steady electrical power so that the objective is to keep the system critical. It is important to understand that nuclear criticality can be achieved at different power levels, for example, a 300 MW reactor, a 600 MW reactor and a 1200 MW reactor are all kept critical. When an experiment in a lab is being carried out, it is usually subcritical for safety purposes while when a system is designed to release an explosive burst of energy, such as a nuclear weapon, then it is designed to become supercritical.

Nuclear fission can result in the uncontrolled multiplication of neutrons as fission proceeds if neutrons are not captured or do not escape from the system. If we assume that a fission reaction is accompanied with the release of two neutrons, then N fissions will result in 2^N neutrons. Thus for $N = 10$ fission reactions, there will be $2^{10} = 1024$ neutrons and ten "generations" later, this number will have grown to $2^{20} = 1{,}048{,}576$, and subsequently by the 30th generation, there will be $2^{30} = 1.0737\ 10^9$ neutrons. Such a runaway situation is called super-criticality as it is accompanied by the uncontrolled release of energy as in a nuclear bomb or a nuclear accident.

Exercise 2.20: *Generations in fission*: Estimate the number of generations needed for the release of energy equivalent to the energy from 1 ton of TNT explosive.

Neutrons 'born' as fission neutrons at energies of the order of 1 MeV, with energy distribution as modeled by the Watt spectrum (Fig. 2.27) slow down by collisions, as they travel in a medium, as discussed in Section 2.7, until they achieve equilibrium with their surroundings at 'thermal' energies. To understand criticality, we consider a simplified picture shown in Fig. 2.28.

For S "fast" neutrons injected into a system, $P_F S$ neutrons leak out of the system boundaries; of the remaining $(1-P_F)S$, the fast fission multiplication increases the number to $\varepsilon(1-P_F)S$. These fast neutrons, while colliding with host nuclei, lose their energy gradually and reach the "resonance" region where $\varepsilon p(1-P_F)P_T S$ escape the resonances,

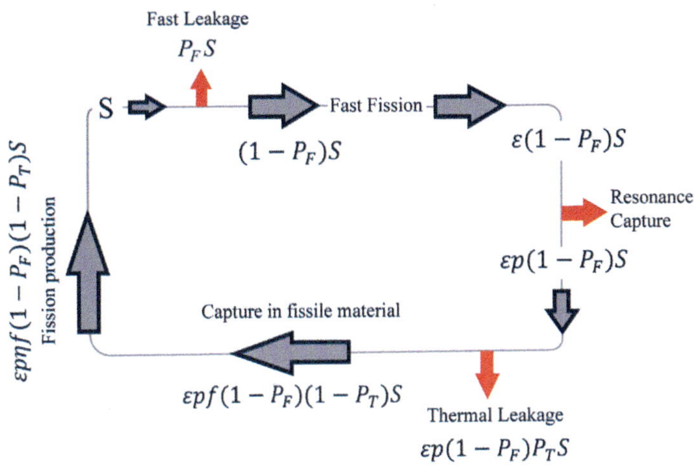

FIGURE 2.28 The processes contributing to the system multiplication.

while $\varepsilon(1-p)(1-P_F)S$ are "captured," and continue to slow down to thermal energies. At this "lower-energy," some neutrons are captured in moderator and structural material while $\varepsilon pf(1-P_F)(1-P_T)S$ are absorbed in fissile material and thus $\varepsilon p\eta f(1-P_F)(1-P_T)S$ neutrons are produced by fission. In summary, the cycle began with S neutrons and ended with $\varepsilon p\eta f(1-P_F)(1-P_T)S$ neutrons. This picture gives us an estimate of the 'multiplication' of the system, viewed as the number of neutrons emerging in the "next generation" divided by the corresponding number in the previous generation. This is called the effective multiplication of the system and is written from Eq. (2.48) as

$$k_{\text{eff}} = \frac{\varepsilon p\eta f(1-P_F)(1-P_T)S}{S} \tag{2.49}$$

The above can be simplified for an "infinite" system from which there is no leakage. In that case, $k_{\text{eff}} = k_\infty\, P_{NL}$, where k_∞ is the infinite multiplication factor

$$k_\infty = \varepsilon p\eta f \tag{2.50}$$

and P_{NL} is the nonleakage probability due to the finite size of the system expressed as the product nonleakage probabilities $P_{NL} = (1-P_F)(1-P_T)$. For a finite system, the effective multiplication factor is

$$k_{\text{eff}} = k_\infty \quad P_{NL,F} \quad P_{NL,T} \tag{2.51}$$

From the cycle in Fig. 2.28, the quantities are defined as follows:

1. epsilon ε, the fast fission factor is a measure of the number of fissions caused by "fast" neutrons; thus we would need to specify the cutoff energy E_{th} above which neutrons are classified as fast neutrons. For neutrons in the energy range $(0, E_0)$ where E_0 is the maximum energy of neutrons, with the fast group (E_{th}, E_0) and the thermal group $(0, E_{th})$, ε is defined as

$$\varepsilon = \frac{\text{no.of fission reactions}}{\text{no.of fission reactions by thermal neutrons}} = \frac{\int dV \int_0^{E_0} \Sigma_f(r,R)\phi(r,E)dE}{\int dV \int_0^{E_{th}} \Sigma_f(r,R)\phi(r,E)dE} = 1 + \frac{\int dV \int_{E_{th}}^{E_0} \Sigma_f(r,R)\phi(r,E)dE}{\int dV \int_0^{E_{th}} \Sigma_f(r,R)\phi(r,E)dE} \tag{2.52}$$

2. the resonance escape probability p [Section 2.8, Eq. (2.46)] is defined as the probability of escaping resonance capture while slowing down from fast to thermal energies. In terms of the absorbing (fuel) atoms and the scattering (moderator) atoms, and the effective resonance integral I,

$$p = \exp\left[-\frac{N_A I}{\xi_M \Sigma_{sM} + \xi_A \Sigma_{pA}}\right]$$

3. *eta* η is defined as

$$\eta = \frac{\text{no.of fission neutrons produced}}{\text{no of thermal neutrons aborbed in fuel}} = \frac{\int \upsilon\Sigma_f \phi_{\text{th}} dV}{\int \Sigma_{aF}\phi_{\text{th}} dV} = \frac{\upsilon\Sigma_f}{\Sigma_{aF}} \tag{2.53}$$

TABLE 2.17 One-Group cross sections for Uranium-235.

U-235	υ	σ_f (b)	σ_c (b)	σ_s (b)	σ_t (b)	$\frac{\upsilon\sigma_f}{\sigma_a}$
(A)	2.50	1.3	–	4.0	5.3	2.25
(B)	$\upsilon\sigma_f = 5.297$		$\sigma_a = 2.844$	$\sigma_{tr} = 8.246$	–	1.8625

Source: (A) Bell and Glasstone, Nuclear Reactor Theory: p.126; (B) Reactor Physics constants ANL-5800.

4. and the thermal utilization f as

$$f = \frac{\text{thermal neutrons absorbed in fuel}}{\text{total no of thermal neutrons absorbed}} = \frac{\int_0^V \Sigma_{aF}\phi_{\text{th}} dV}{\int_0^V (\Sigma_{aF} + \Sigma_{a,\text{other}})\phi_{\text{th}} dV} \tag{2.54}$$

where $\Sigma_{a,\text{other}} = \Sigma_{a,\text{M}} + \Sigma_{a,\text{structural materials}} + \ldots$ is the thermal macroscopic thermal cross-section for absorption in fuel, moderator with absorption cross-section $\Sigma_{a,\text{M}}$ and structural material etc. If absorption is considered to take place only in fuel and moderator, then for a homogeneous mixture of fuel and moderator (so that they occupy the same volume)

$$f = \frac{\Sigma_{aF}}{\Sigma_{aF} + \Sigma_{aM}} \tag{2.55}$$

In a most elementary analysis, the criticality for an 'infinite' system thus depends on four factors *viz* ε, p, η *and* f.

Example 2.2: *Six-factor formula*: One-group criticality of a bare sphere of uranium-235 using the six-factor formula, one-group diffusion theory, one-speed neutron transport and Monte Carlo simulation.

For a one-group calculation, the data of Table 2.17 is used.

2.10.1 Diffusion theory

Based on the four-factor formula [Eq. (2.50)], k_∞ can be calculated from the data. A sphere of U-235 will have a fast spectrum since there is no light-weight material; the fast fission the neutron spectrum will be fast. The fission factor, resonance escape and thermal fuel utilization,$\varepsilon \sim 1, p \sim 1, f \sim 1$, so that $k_\infty = \eta = 2.25$. This says that an infinitely large system will have no leakage and its multiplication will be very high; in fact supercritical. The nonleakage probabilities for fast and thermal neutrons will be derived in Chapter 4 (diffusion theory) for one- and two-group equations. For now we accept the results to demonstrate calculation of k_{eff}; the critical equation. The fast and thermal nonleakage probabilities are

$$P_{NL,F} = \frac{1}{1 + B^2\tau} \tag{2.56}$$

and

$$P_{NL,T} = \frac{1}{1 + L^2B^2} \tag{2.57}$$

respectively, so that

$$k_{\text{eff}} = 1 = \frac{k_\infty}{(1 + B^2\tau)(1 + L^2B^2)} \tag{2.58}$$

in terms of the buckling B, that is, the curvature of the neutron flux, given by

$$B_m^2 = \frac{k_\infty - 1}{L^2} = B_g^2 = \left(\frac{\pi}{R_c}\right)^2 \tag{2.59}$$

for a sphere with critical radius R_c. Criticality requires that the material buckling B_m and geometrical buckling B_m are equal. The critical radius is then found as

$$R_c = \pi\sqrt{\frac{L^2}{k_\infty - 1}} \tag{2.60}$$

In Eq. (2.55), the diffusion length is interpreted as the fast diffusion length L_f since there is no thermalization of neutrons.

$$L^2 = \frac{D}{\Sigma_a} = \frac{1}{3\Sigma_{tr}\Sigma_a}$$

The results are: $D = 0.8413\,\text{cm}$, $L^2 = 6.1568\,\text{cm}^2$, critical radius $R_c = 8.3935\,\text{cm}$, and the critical mass $M_c = \frac{4}{3}\pi R_c{}^3 = 46.443\,\text{kg}$. This is to be compared with the critical radius and mass of a pure U-235 bare sphere (critical radius 8.46 cm and critical mass 48 kg) given by de Volpi (DeVolpi, 1982) listed in Table 2.20.

Diffusion theory is an elementary model and useful for obtaining approximate solutions; a better formalism is provided by transport theory which is covered in some detail in Chapter 6. Here, we use some results from the *one-speed transport equation* which has a diffusion approximation little better than the one-group diffusion model.

2.10.2 Transport theory

The one-speed transport equation (Bell & Glasstone, 1979) reduced from the continuous energy neutron transport equation has been solved analytically for several idealized geometries and scattering models. Exact form solutions for criticality in slabs and spheres (Bell & Glasstone, 1979; Kaper, Lindeman, & Leaf, 1974; Öztürk, 2012; Rawat & Mohankumar, 2011) have been obtained which are both elegant and useful but of course need validation with elaborate computer simulations.

The transport theory asymptotic flux in a sphere of radius R is

$$\phi_{as}(r) = \frac{A}{r}\sin\frac{r}{|\nu_0|} \tag{2.61}$$

where ν_0 is a solution of the transcendental equation

$$1 = c\nu_0 \tanh^{-1}\frac{1}{\nu_0} = \frac{c\nu_0}{2}\ln\frac{\nu_0 + 1}{\nu_0 - 1} \tag{2.62}$$

For the mean number of secondaries c, that is, the number of neutrons emerging from an interaction

$$c = \frac{\overline{\nu}\sigma_f + \sigma_s + \sigma_{n,2n} + \cdots}{\sigma_t} \tag{2.63}$$

The sphere is approximately critical when the asymptotic flux is zero at the extrapolated radius $\tilde{R}$; thus $\phi_{as}(\tilde{R}) = 0$ at $\tilde{R} = R_c + x_0 = \pi|\nu_0(c)|$.

The solutions of Eq. (2.62) are real for $c < 1$ and complex for $c > 1$; an approximation for ν_0 is

$$\frac{1}{\nu_0{}^2} = \frac{3(1-c)}{c}\left[1 - \frac{9}{5}\frac{1-c}{c} - \cdots\right] \tag{2.64}$$

To find the critical radius ν_0 is determined form the transcendental equation [Eq. (2.62)]

$$R_c = \pi|\nu_0| - x_0 \tag{2.65}$$

and the extrapolation distance x_0 can be calculated from the Mark P_1 boundary condition

$$x_0 = \frac{1}{\sqrt{3}}\left[1 - \frac{1}{3}(c-1) + \frac{1}{5}(c-1)^2 + \cdots\right] \tag{2.66}$$

From the data of Table 2.17,

In the spherical harmonics P_1 "diffusion approximation," the neutron flux is

$$\phi(x) = \frac{1}{2}\sqrt{\frac{3}{1-c}}\ e^{-}|x|/L \tag{2.67}$$

where the diffusion length is

$$L = \frac{1}{\sqrt{3(1-c)}} \tag{2.68}$$

Criticality verification benchmark sets for one-group U-235 and Pu-239 bare spheres are listed in Table 2.17 (Sood, Arthur Forster, & Kent Parsons, 2003); these give the value $k = 1$ for the critical radii R_c listed.

2.10.3 Monte Carlo simulation

For criticality calculations, as for most other nuclear design calculations, Monte Carlo (MC) simulation is one of the most elaborate methodologies due to its capability of handling realistic 3D geometries and interaction models. Details of the Monte Carlo method will be covered in Chapter 8; here only the results are presented and discussed in the context of this example.

The objectives of the MC simulation are i- to estimate k_{eff}, ii- to obtain the values of ε, p, η and f in the four-factor formula, and iii- the leakage of neutrons.

The composition of the Godiva sphere is: (wt.%) U-235 93.71, U-238 5.27, U-234 1.02 with a density 18.74 g/cm^3 and radius 8.7407 cm. The Godiva assembly was one of the first critical experiments for the experimental validation of calculations as well as for kinetics experiments; it has continued since the 1950s with the present Godiva IV assembly for research on nuclear as well as high energy lasers (Thompson et al., 2020). Several hand calculation methods have also been developed for critical assemblies as well as for reactor configurations (Bowen & Busch, 2005; Caplin, Duluc, & Richet, 2015); these are usually based on one- and two-group diffusion theory as well as interacting array models and validating with codes such as MCNP (Brown, 2009; Harmon, Busch, Briesmeister, & Forster, 1994).

In MC simulation, you only get the answer you ask for; in deterministic codes, the solution i.e. neutron flux and reactions rates are given in the problem domain specified in the input files. Thus the tallies and group boundaries need to be specified. For this, we need to have a basic understanding of the physics and engineering of the problem. That was the purpose of the preceding sections of this chapter.

So, now we take a look at Fig. 2.29, in the context of the underlying theory covered in Section 2.8 and Figs. 2.19–2.21, to 'see' the resonance structures of both U-235 and U-238. The impressions we get are: i- both isotopes have similar resonance structures, ii- there are three distinct energy regions, iii- resonances are spread over the energy range, few eV–1 keV for U-235 and few eV to ~ 2 keV for U-238, and iv- U-238 has higher resonance cross-section values at lower energies. Fig. 2.29 shows resolved resonances at lower energies and unresolved resonances at higher energies.

In order to select bin boundaries for energy groups, another detailed picture of the resolved resonances is required; for this we take a look at Figs. 2.30 and 2.31.

From Fig. 2.31, we are in a position to select bin boundaries to tally absorptions in the low-energy resonance regions.

While Fig. 2.29 suggests group boundaries: $0 - 1\text{eV}$, $1\text{eV} - 2\text{keV}$, $2\text{keV} - 14\text{MeV}$, Fig. 2.31 suggests "finer" boundaries $1, 1.6, 2.5, 4, 5.2, 5.7, 6.6, 7, 7.5$ and 10 eV and 2keV, 1MeV, and 14MeV. This process tells that MC simulation is both an art and a science.

The simulation was carried out for 5000 neutrons per batch, 200 batches, and 20 "skip" cycles; all these terms will be covered in detail in this book. The interactions of one million neutrons in the uranium sphere were simulated with a computational effort of 1.95 min on an Intel (R) Core i7-2620M CPU @ 2.70 GHz processor with a 32-bit Operating System and 8.00 GB RAM.

The energy-dependent neutron flux obtained from simulation is shown in Fig. 2.32. It appears to be a "hard" spectrum, essentially all above 1 keV which is understandable as there is no light material present in Godiva.

From Fig. 2.32, it can be expected that resonances will play no part in the neutron population as neutrons will most probably not reach the resonance energies during their slowing down process. The energy group boundaries for neutron flux: 300 equal-lethargy bins from 0-10 MeV. This would not have been expected in the presence of water as in a Light Water Reactor.

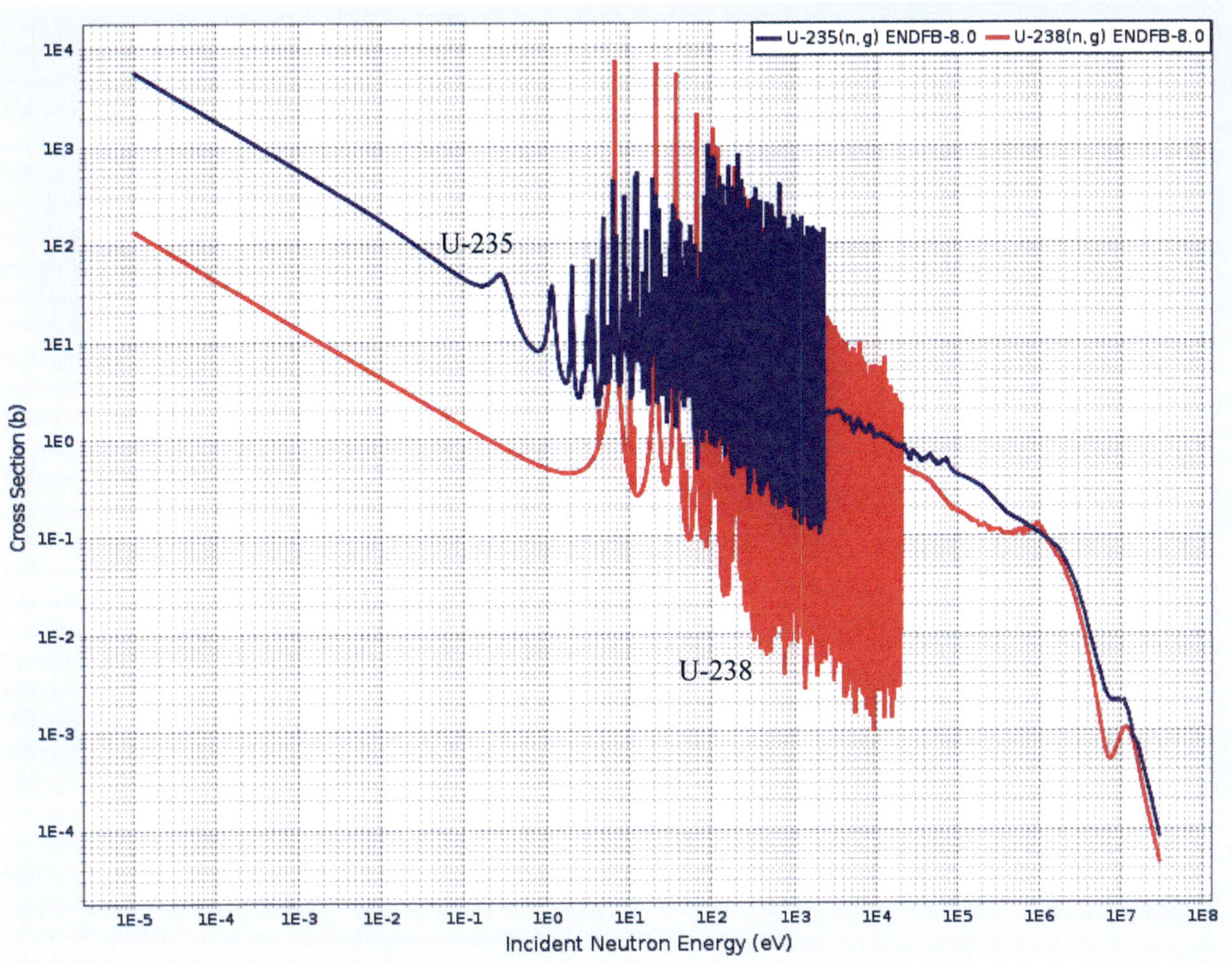

FIGURE 2.29 Radiative capture cross sections of U-235 and U-238.

The average flux for one source neutron is $\overline{\phi} = 2.4213 \times 10^{-3}$ (0.0006) neutrons per cm^2 per second. The tallies, their bin structures and results from simulation are given in Table 2.18.

Reactions tallied: $\Sigma_t\overline{\phi}V$, $\overline{\upsilon}\Sigma_f\overline{\phi}V$, $\Sigma_f\overline{\phi}V$, $\Sigma_{(n,2n)}\overline{\phi}V$, $\Sigma_{(n,3n)}\overline{\phi}V$, $\Sigma_{(n,fx)}\overline{\phi}V$, $\Sigma_{(n,f)}\overline{\phi}V$ in energy groups $0-1$eV, 1 eV $-$ 2 keV, 2 keV $-$ 14 MeV;

Results showed no radiative captures $\Sigma_{(n,\gamma)}\overline{\phi}V$ in groups with bin boundaries at 1, 1.6 2.5, 4, 5.2 5.7 6.6, 7, 7.5 and 10 eV since there were no neutrons at these energies. The radiative captures in the range 10 eV$-$2 keV were found to be 3.58282×10^{-5} (0.1048) and in the range 1 MeV$-$14 MeV there were 8.68463×10^{-3} (0.0014); the total captures being $4.48660\ 10^{-3}$ (0.0014). Thus in the four-factor formula, it is justifiable to set $\varepsilon = p = 1$. Similarly there is no moderator so that fuel utilization $f = 1$ and $k_{eff} = \eta P_{NL,f}$. The system multiplication based on the average of three estimators is $k_{eff} = 0.99739$. From the tallies, $\eta \sim 2.3319$ and $P_{NL,f} \sim 0.4263$ giving $k_{eff} \sim 0.9941$.

Exercise 2.21: *Godiva*: From the results of the above Monte Carlo simulation, write a conservation equation balancing the total number of interactions with all other interactions. What information can you obtain from this balance equation?

Several benchmarks are given for criticality validation calculation (Sood et al., 2003) with one-group data for U-235 and Pu-239 given in Table 2.19. Notice that these values give a critical radius for a U-235 sphere which is different from that of Godiva (Table 2.22). The same is true for Pu-239 from the values in Tables 2.21 and 2.22.

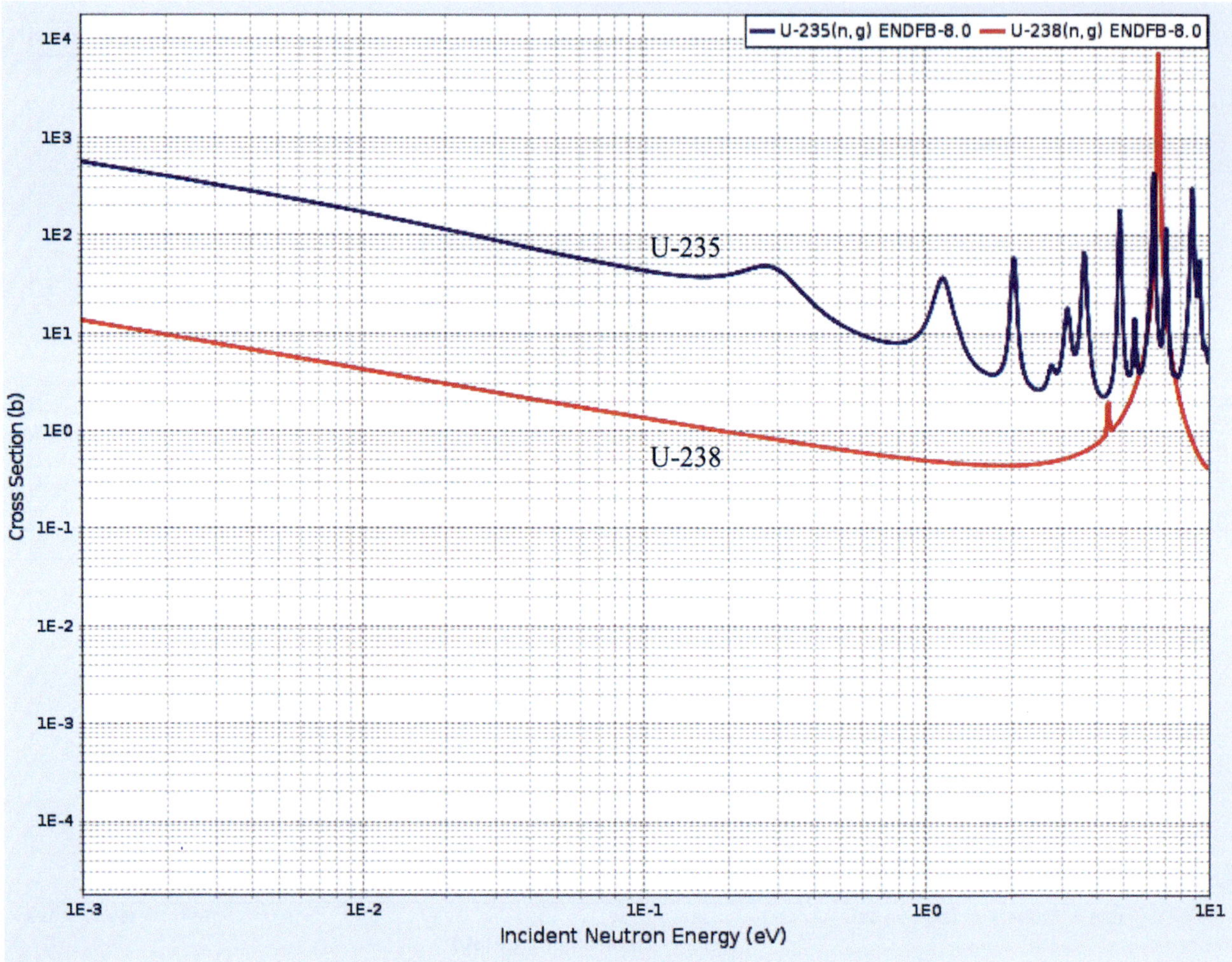

FIGURE 2.30 Low-energy range radiative capture cross sections of U-235 and U-238.

It is important to understand that criticality depends on a number of factors such as the enrichment of fissile materials, their geometrical shapes, and the environment which they are placed in. The first step in a nuclear system such as a nuclear reactor is to determine its critical composition and size. The data that must be used for such calculations is strongly dependent on the neutron spectrum. Only a basic understanding of the system can be obtained from generalized data such as that shown in Table 2.20, for example, the difference in the fission cross-section for three situations viz thermal energy σ(fs,th), a reactor spectrum σ(fs,rs) and a fast spectrum σ(fs,f). Some basic questions that can be answered by looking at the data in Table 2.20 are:

1. Would it more hazardous to sit in the vicinity of a U-235 or a U-238 sphere in terms of neutron emissions from spontaneous fission?
2. Which isotope of plutonium would be most suitable for a thermoelectric generator?
3. What would be the effect of water placed in the vicinity of a Godiva sphere?

In the early days of nuclear technology, the most important challenge was to determine the critical masses of fissile nuclides and experimentally verified for bare and reflected spheres of uranium and plutonium listed in Table 2.21. It is important to note that criticality can be achieved by significant reductions of fissile material when the core is surrounded with a "reflector" such as U-238 though with a penalty of increased mass.

In Table 2.21, the critical mass for U-235 is 48 kg for a pure U-235 sphere; this is different from the Godiva bare sphere (Rowlands et al., 1999) which has a weight composition given in Table 2.22.

A study of neutronic systems can be carried out with the use of diffusion theory, transport theory or Monte Carlo simulation which are based on group averaged cross-section data and spectrum calculations described in this chapter.

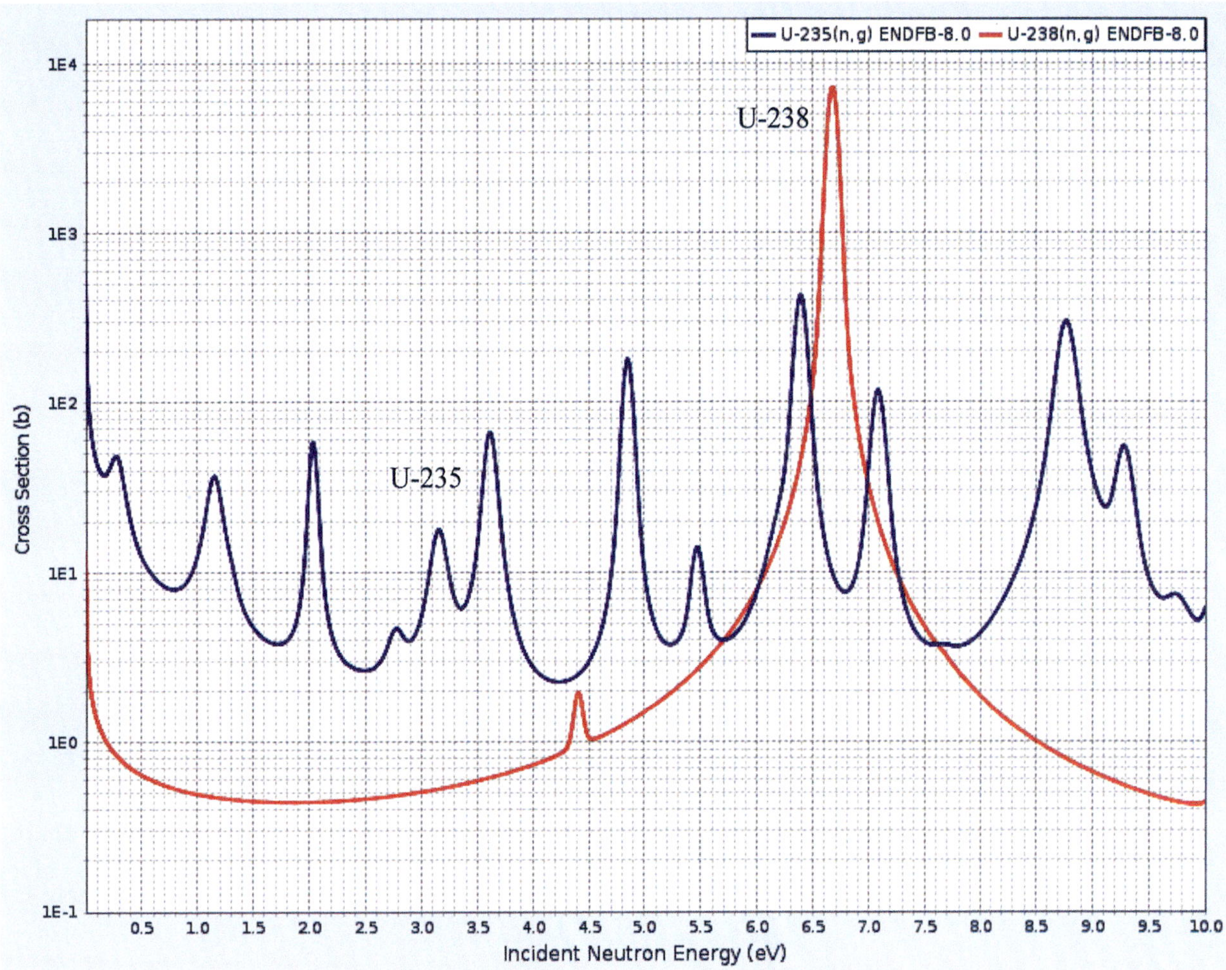

FIGURE 2.31 Low-energy resolved resonances of U-235 and U-238.

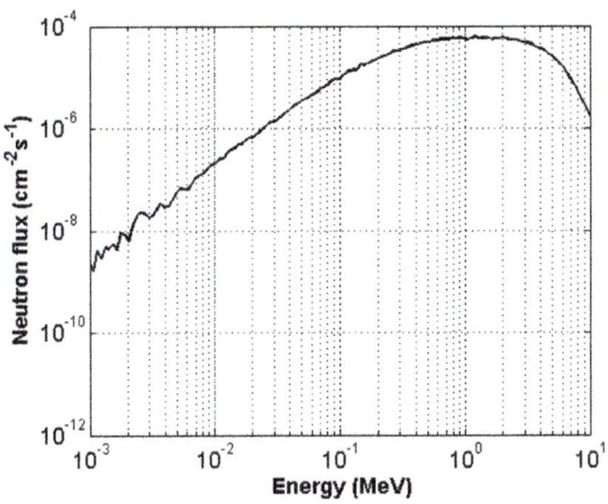

FIGURE 2.32 Energy-dependent neutron flux in Godiva.

TABLE 2.18 Monte Carlo simulation results for Godiva.

Tally	Quantity estimated	Reaction MT number	Energy Bins			
			0 – 1eV	1eV – 2keV	2keV – 14MeV	Total
Neutron current	J	–	–	–		0.5737 (0.0006)
Total interactions	$\Sigma_t\overline{\phi}V$	1	–	2.3219×10^{-4} (0.1002)	2.6531 (0.0007)[a]	2.6534 (0.0007)
Fission reactions	$\Sigma_f\overline{\phi}V$	−6	–	7.1939×10^{-5} (0.1023)	3.8385×10^{-1} (0.0006)	3.8392×10^{-1} (0.0006)
Neutrons produced from fission	$\overline{\upsilon}\Sigma_f\overline{\phi}V$	−6 −7	–	1.7509×10^{-4} (0.1023)	9.9697×10^{-1} (0.0006)	9.9715×10^{-1} (0.0006)
Multiplication (n,2n)	$\Sigma_{(n,2n)}\overline{\phi}V$	16	–	–	–	2.6531×10^{-3} (0.0076)
Multiplication (n,2n)	$\Sigma_{(n,3n)}\overline{\phi}V$	17	–	–	–	4.5917×10^{-6} (0.1428)
Total fission[b]	$\Sigma_{(n,fx)}\overline{\phi}V$	18	–	–	–	3.7727×10^{-1} (0.0006)
fission	$\Sigma_{(n,f)}\overline{\phi}V$	19	–	–	–	6.5429×10^{-3} (0.0010)

[a]*relative standard error; read as 2.6531(1 ± 0.0007).*
[b]*sum of (n,f),(n,n′f), (n,2nf),(n,3nf).*

TABLE 2.19 One-Group isotropic cross sections and critical radius of bare spheres.

Material	υ	Σ_f (cm)	Σ_c (cm)	Σ_s (cm)	Σ_t (cm)	$\frac{\upsilon\Sigma_f}{\Sigma_a}$ (cm)	R_c (cm)
U-235 (a)	2.70	0.065280	0.013056	0.248064	0.32640	2.25	7.428998
Pu-239 (a)	2.84	0.081600	0.019584	0.225216	0.32640	2.612903	6.082547

Source: Sood, A., Forster, A., and Parsons, D. K. Analytical Benchmark Test Set for Criticality Code Verification, Progress in Nuclear Energy, Vol. 42, No. 1, pp. 55-106, 2003.

TABLE 2.20 Some useful actinide physical properties.

Nuclide	α (th) b/b	α (f) b/b	ν (th) n/fs	ν (f) n/fs	σ (fs, th) b	σ (fs, rs) b	σ (fs, f) b	σ (c, th) b	σ (c, rs) b	τ (total) y	N n/s-kg	Q W/kg
Th-232				2.10	0		0.08	7		1.4E10	0.1	
U-233	0.09		2.45	2.61	528		1.90	50		1.6E5		
U-234				2.60	0	3.9	1.50		281	2.5E5	1.7	
U-235	0.17	0.27	2.40	2.56	580	409	1.30	100	105	7.1E8	0.2	
U-236				2.50		3.2	0.80	6	74	2.4E7	1.7	
U-236				2.30							3.7E7	2E4
Np-237		7		2.85	0	2.5	1.4	170	359	2.2E6	~1	
U-238				2.50	0	0.6	0.3	3	8	4.9E9	17	
Pu-238				(3)		19	2.5	500	366	8.8E1	2.6E6	567
Pu-239	0.36	0.22	2.86	3.08	742	1060	1.8	243	433	2.4E4	30	1.9
Pu-240		1.6	2.8	3.10	0	2.5	1.4	300	897	6.6E3	1E6	7.1
Pu-241	0.38	0.13	2.96	3.0	1010	1110	1.8	420	334	1.4E1		3.4

th = thermal, f = fast, rs = reactor spectrum, fs = fission, c = capture.
α = capture-to-fission ratio, ν = neutron yield, σ = cross-section (in barns).
τ = half-life, Q = specific power (heat), N = spontaneous fission neutron rate.
E = exponent (to power of ten).
Neutron emission rate from (α, n) reaction in ^{238}Pu oxide is 1.4E7 n/s-kg.
Some infinity-dilute resonance integers (0.6eV-10 MeV):
α(^{239}Pu) = 0.6; σ(c) (^{240}Pu) = 8500 b; σ(c) (^{242}Pu) = 1120b.
Source: DeVolpi, A., Denaturing Fissile Materials, Progress in Nuclear Energy, Vol. 10, No. 2, pp. 161-220: from Table 2.

TABLE 2.21 Critical solid spherical systems: bare and reflected with 10 cm U.

Material	Density (g/cc)	Configuration	Core		Ref.	Total mass (kg)
			Radius (cm)	Core mass (kg)	Ref. mass (kg)	
U^{233}	18.9	Bare	5.87	16	–	16
		Ref.	4.2	5.7	221	227
U^{235}	18.9	Bare	8.46	48	–	48
		Ref.	5.8	15.7	300	316
Pu^{239} α	15.9	Bare	5.49	11	–	11
		Ref.	3.8	4.5	204	209
Pu^{239} δ	19.5	Bare	5.68	15	–	15
		Ref.	4.7	7.0	243	250

Source: DeVolpi, A., Denaturing Fissile Materials, Progress in Nuclear Energy, Vol. 10, No. 2, pp. 161-220: Table 3.

TABLE 2.22 Bare critical assemblies.

Model	Godiva	Jezebel	Jezebel23
Radius (cm)	8.7407	6.3849	5.9838
Density (g/cm^3)	18.74	15.61	18.424
Composition (atoms/barn-cm)	U^{235} 4.4994e-2 U^{238} 2.4984e-3 U^{234} 4.9184e-4	Pu^{239} 3.7047e-2 Pu^{240} 1.7512e-3 Pu^{241} 1.1674e-4 Ga^{69} 8.26605e-4 Ga^{71} 5.48595e-4	U^{233} 4.6712e-2 U^{234} 5.9026e-4 U^{238} 2.8561e-4 U^{235} 1.4281e-5
Mass (g)	52419.98	17019.77	16534.98

Problems

2.1. Describe the Boltzmann equation for the kinetic theory of gases and the significance of its moments.

2.2. From the size of aU-238 nucleus, calculate the potential scattering cross-section and compare with the figures for U-238 cross sections in this chapter.

2.3. How would you find the average lethargy gain for a mixture consisting of four elements with given atomic number densities $N_i, i = 1, 2, 3, 4$?

2.4. Write a formula for the intensity reduction of a spherical multilayered shield of neutrons.

2.5. What is the significance of the cadmium ratio in the measurement of neutron flux?

2.6. How can the neutron age used in a "modified" one-group formula to compute neutron leakage in a finite system?

2.7. On an energy-flux figure, for a narrow resonance sketch the flux and the absorption cross-section and describe the phenomenon known as self-shielding. Show that the lethargy dependent flux decreases inversely with the total cross-section.

2.8. For the first resonance of U-238 at 6.67 eV calculate the total cross-section given in Eq. (2.39) ignoring the interference term.

2.9. Write a formula for estimating the critical energy of fission in Pu-239.

2.10. Calculate the value of c for the Godiva one-speed data given in Table 2.17 and solve the transcendental equation [Eq. (2.62)].

2.11. Calculate the extrapolation distance x_0 from the Mark boundary condition [Eq. (2.63)] and estimate the critical radius for Godiva.

2.12. Given a homogenous mixture of a fissile material and water, outline the steps for calculating the critical dimensions of slab system. The given data is i-concentration of the fissile material c_F grams per liter, and ii- the non-1/v g_i, factors and cross sections $\sigma_i$for both materials, $i = abs, f$. For a slab system, the buckling is $B = \pi / \tilde{a}$, where the physical slab thickness is $a = \tilde{a} - 2d$, $\tilde{a}$ is the extrapolated thickness and $d = 0.71\lambda_{tr}$. Justify any other assumptions you make.

2.13. In the above problem (criticality for a homogeneous system) consider the effect of slowing down (Section 2.8) including both neutron age τ and the diffusion length L to calculate the nonleakage probabilities in the six-factor formula.

2.14. In the above problem (criticality for a homogeneous system), outline the procedure if you are required to find the concentration of the fissile material to make a slab of given dimensions critical.

Nomenclature

English lower case

c	mean number of secondary neutrons per collision
f	thermal utilization factor
k_{eff}	effective multiplication
k_∞	infinite system multiplication
m	mass of neutron
$n(E)$	number of molecules with energy E per unit interval dE

n_0	total number of molecules
t_m	moderating time
v_c	velocity of neutron in center of mass system
V_c	velocity of nucleus in center of mass system
v_l	velocity of neutron in laboratory system
$\bar{v}_M$	average Maxwellian neutron velocity
v_p	most probable velocity
v_r	relative speed of neutron and nucleus in the laboratory system
x_0	extrapolation distance

English upper case

A	relative mass of nucleus (relative to neutron mass)
B	buckling
D	diffusion coefficient
E_0	resonance center of mass energy
E_C	critical energy
E_c	kinetic energy of neutron and nucleus center of mass energy
E_{AL}	kinetic energy of target nucleus before collision (Lab system)
E'_{AL}	kinetic energy of target nucleus after collision (Lab system)
E_L	kinetic energy of neutron before collision (Lab system)
E'_L	kinetic energy of neutron after collision (Lab system)
E_p	most probable energy
F	collision density
I	effective resonance integral
J	neutron current
L	diffusion length
P	potential scattering
P	resonance escape probability
$P_{F,T}$	leakage probability (F = fast, T = thermal)
$P_{NL,F,T}$	nonleakage probability (F = fast, T = thermal)
$\mathscr{R}_x$	reaction rate type x
R_c	critical radius
S	source
T	temperature
V	volume
V_{CM}	velocity of center of mass
V_l	velocity of nucleus in the laboratory system
J	spin

Greek lower case

μ	reduced mass $m_1m_2/(m_1+m_2)$
α	capture-to-fission ratio
α	$\left(\frac{A-1}{A+1}\right)^2$
ε	
η	number of fission neutrons emerging per thermal neutron absorbed
$\bar{\xi}$	average gain in lethargy per collision
λ	mean free path
$\tilde{\lambda}_1$	reduced wavelength
ν_0	transport relaxation length
ν	neutrons emerging from fission reaction
χ	fission distribution function
θ_L	scattering angle in the lab system
θ_{CM}	scattering angle in the CM system
τ	neutron age

τ half-life
τ lifetime of compound nucleus
ϕ_{as} asymptotic flux
$\overline{\phi}$ average flux
φ azimuthal angle

Greek upper case

Γ measure of the width of a resonance (energy)
Γ_i/Γ probability of *i*th particle emission from a resonance
Γ_γ/Γ probability of gamma emission from a resonance
Σ_a macroscopic transport cross-section
Σ_f macroscopic fission cross-section
$\Sigma_{(n,\gamma)}$ macroscopic radiative capture cross-section
Σ_{tr} macroscopic transport cross-section
$\Sigma_{(n,2n)}$ macroscopic $(n, 2n)$cross-section
$\Sigma_{(n,3n)}$ macroscopic $(n, 3n)$cross-section
$\Sigma_{(n,fx)}$ sum of $(n,f), (n,n'f), (n,2nf), (n,3nf)$ cross sections
$\Sigma_{(n,f)}$ macroscopic fission cross-section
Φ probability
Ψ wave function

Abbreviations

L Laboratory system
CM center of mass system
CN compound nucleus
MT material reaction number

References

Aziz, U., Koreshi, Z. U., Sheikh, S. R., & Khan, H. (2020). A study of criticality and thermal loading in a conceptual micronuclear heat pipe reactor for space applications. *Nuclear Technology and Radiation Protection, 35*. Available from https://doi.org/10.2298/NTRP2003208A.

Bell, G. I., & Glasstone, S. (1979). *Nuclear reactor theory*. New York: Robert E. Krieger Publishing Co., Inc.

Bowen, D G. & Busch, R D. (2005). Hand calculational methods for criticality safety - A Primer. In *Proceedings of the NCSD topical meeting (American Nuclear Society Nuclear Criticality Safety Division)*.

Brown, D. A., Chadwick, M. B., Capote, R., Kahler, A. C., Trkov, A., Herman, M. W., Sonzogni, A. A., et al. (2018). ENDF/B-VIII.0: The 8th major release of the nuclear reaction data library with CIELO-project cross sections, new standards and thermal scattering data. *Nuclear Data Sheets*. Available from https://doi.org/10.1016/j.nds.2018.02.001.

Brown, F B. (2009). A review of best practices for Monte Carlo criticality calculations. In *Proceedings of the American Nuclear Society - nuclear criticality safety division topical meeting on realism, robustness and the nuclear renaissance*.

Caplin, G., Duluc, M., & Richet, Y. (2015). On the benefit of using both hand calculation methods and computer experiments algorithms to solve complex nuclear criticality safety problems. In *Proceedings of the international conference on nuclear criticality safety*.

Deen, J. R., & Woodruff, W. L. (1995). *WIMS-D4M user manual Rev.0*. Illinois: Champaign-Urbana.

DeVolpi, A. (1982). Denaturing fissile materials. *Progress in Nuclear Energy*. Available from https://doi.org/10.1016/0149-1970(82)90022-1.

Engle, W. W. Jr. (1967). *A users manual for ANISN: A one dimensional discrete ordinates transport code with anisotropic scattering*. Tennessee.

Evans, A. E. (1974). *Delayed neutrons from fission: Present status of measurements of yields, group half-lives and abundances and spectra*. Los Alamos, New Mexico, USA. <https://inis.iaea.org/collection/NCLCollectionStore/_Public/06/197/6197971.pdf>.

Feather, N. (1939). The time involved in the process of nuclear fission [1]. *Nature*. Available from https://doi.org/10.1038/143597a0.

Hanson, S K., Pollington, A, D., Waidmann, C. R., Kinman, W. S., Wende, A. M., Miller, J. L., Selby, H. D. et al, (2016). Measurements of extinct fission products in nuclear bomb debris: Determination of the yield of the trinity nuclear test 70 y later. In *Proceedings of the National Academy of Sciences of the United States of America*. Available from https://doi.org/10.1073/pnas.1602792113.

Harmon, C. D., II, Busch, R. D., Briesmeister, J. F., & Forster, R. A. (1994). *Criticality calculations with MCNP: A primer*.

Hodgson, P. E. (1994). *The nucleon optical model*. <https://doi.org/10.1142/9789812815538_0003>.

Hyvönen-Dabek, M., & Nikkinen-Vilkki, P. (1980). Determination of neutron generator fast neutron spectra using threshold detectors. *Nuclear Instruments and Methods*. Available from https://doi.org/10.1016/0029-554X(80)90824-1.

IAEA. 1970. *Neutron fluence measurements, Technical Reports Series No. 107*. Vienna, Austria.

Kaper, H. G., Lindeman, A. J., & Leaf, G. K. (1974). Benchmark values for the slab and sphere criticality problem in one-group neutron transport theory. *Nuclear Science and Engineering*. Available from https://doi.org/10.13182/nse54-94.

Keepin, G. R., Wimett, T. F., & Zeigler, R. K. (1957). Delayed neutrons from fissionable isotopes of uranium, plutonium, and thorium. *Physical Review*. Available from https://doi.org/10.1103/PhysRev.107.1044.

Lamarsh, J. R. (1966). *Introduction to nuclear reactor theory*. Addison-Wesley Publishing Company.

Macfarlane, R. E., Muir, D. W., Boicourt, R. M., & Kahler, A. C. (2012). *The NJOY nuclear data processing system, Version 2012*. LA-UR-12-27079, Los Alamos National Laboratory.

Madland, D. G., & England, T. (1977). Influence of isomeric states on independent fission product yields. *Nuclear Science and Engineering*. Available from https://doi.org/10.13182/NSE77-A14501.

Nuclear Energy Agency, OECD. 2015. *Introduction of thorium in the nuclear fuel cycle short- to long-term considerations*.

Okumura, S., Kawano, T., Jaffke, P., Talou, P., & Chiba, S. (2018). 235U(n, f) independent fission product yield and isomeric ratio calculated with the statistical Hauser–Feshbach Theory. *Journal of Nuclear Science and Technology*. Available from https://doi.org/10.1080/00223131.2018.1467288.

Öztürk, H. (2012). The criticality calculations for one-speed neutrons in a reflected slab with anisotropic scattering using the modified UN Method. *Kerntechnik*. Available from https://doi.org/10.3139/124.110272.

Peele, R. W., & Maienschein, F. C. (1970). The absolute spectrum of photons emitted in coincidence with thermal-neutron fission of Uranium-235. Oak RIdge, TN.

Pelowitz, D. B., Goorley, J. T., James, M. R., Booth, T. E., Brown, F. B., Bull, J. S., Cox, L. J., et al. (2013). MCNP6 User's Manual. *La-Cp-13-00634*.

Pritychenko, B., & Mughabghab, S. F. (2012). Neutron thermal cross sections, Westcott factors, resonance integrals, Maxwellian averaged cross sections and astrophysical reaction rates calculated from the ENDF/B-VII.1, JEFF-3.1.2, JENDL-4.0, ROSFOND-2010, CENDL-3.1 and EAF-2010 evaluated data Librari. *Nuclear Data Sheets*. Available from https://doi.org/10.1016/j.nds.2012.11.007.

Rawat, A., & Mohankumar, N. (2011). Benchmark results for the critical slab and sphere problem in one-speed neutron transport theory. *Annals of Nuclear Energy*. Available from https://doi.org/10.1016/j.anucene.2011.02.011.

Reuss, P. (2008). *Neutron physics*. Cedex, France: EDP Sciences.

Rowlands, J, Bosq, J. C., Brinckman, J. J., Fort, E., Giffard, F., Jacqmin, R., Rimpaul, G., et al. (1999). *Intercomparison of calculations for Godiva and Jezebel*.

Shafronovskaia, A. I., Zamyatin, N. I., & Cheremukhin, A. E. (2014). Method of measuring fast neutron fluence using the planar silicon detectors. In *Proceedings of the twenty-fourth Russian particle accelerator conference, RuPAC 2014*.

Sood, A., Arthur Forster, R., & Kent Parsons, D. (2003). Analytical benchmark test set for criticality code verification. *Progress in Nuclear Energy*. Available from https://doi.org/10.1016/S0149-1970(02)00098-7.

Thompson, N., Hutchinson, J., Bahran, R., Hayes, D., Myers, W., Arthur, J., Bounds, J., et al. (2020). National Criticality Experiments Research Center (NCERC) - Capabilities and recent measurements. *EPJ Web of Conferences*. Available from https://doi.org/10.1051/epjconf/202023918003.

Trkov, A., Herman, M., & Brown, D. A. (2018). *ENDF-6 formats manual, data formats and procedures for the evaluated nuclear data files ENDF/B-VI, ENDF/B-VII and ENDF/B-VIII*. Upton, NY 11973-5000. <http://www.nndc.bnl.gov>.

Werner, C. J. (ed.). (2017). *MCNP (R) User's manual code Version 6.2*.

Zamyatin, N. I., Cheremukhin, A. E., & Shafronovskaya, A. I. (2017). Measuring fluence of fast neutrons with planar silicon detectors. *Physics of Particles and Nuclei Letters*. Available from https://doi.org/10.1134/S1547477117050156.

Chapter 3

Nuclear reactors and systems

A nuclear reactor is based on the utilization of energy from a nuclear reaction such as fission or fusion. In a nuclear power reactor, the thermal energy from nuclear reactions is converted into mechanical energy, which can subsequently be converted into electrical energy which we know as electricity that powers our homes, cities, and industries. Energy is what makes our 21 century world so different from earlier centuries. It is in our interest to sustain the production of energy and develop better, cleaner, and more sustainable forms of energy that will make it possible for human beings to live high-quality lives.

There has been renewed worldwide interest in nuclear energy due to its success, resilience and innovations leading to competitive high-density power systems compatible with sustainable development goals.

Nuclear reactors have been operating since the first power reactor went critical on December 1, 1942 producing only 60 W just enough to light a bulb. Since then, the nuclear industry has undergone great advancements; the first generation reactors have mostly retired and advanced third generation reactors are being commissioned and several are connected to the grid. The contribution to global electricity now exceeds 11% while the nuclear industry faces competition from renewable solar and wind energy technologies.

In the last 80 years of nuclear power production, the nuclear industry has seen a very high safety record with three exceptions namely Three Mile Island (USA, 1979) Chernobyl (Ukraine, 1986) and Fukushima (Japan, 2011). In the Three Mile Island (TMI) accident, partial core melt resulted due to a loss of primary coolant; the reactor shut down within a second, there was no injury or loss of life by significant radioactive release into the environment. Chernobyl was a major accident due to both design and operational failures; the radiation release was significant and the accident caused a number of deaths and thyroid cancers. The Fukushima accident was also a major accident resulting from a large earthquake followed by a Tsunami causing damage to the cores of reactors and large-scale evacuations. With the knowledge that followed from detailed analyses, nuclear reactor design and practices have improved to reduce the probability of such occurrences significantly.

This chapter presents a review of the current status of nuclear reactors leading to the next generation, Gen-IV, designs aimed for the coming decades, followed by a birds'-eye view of the modeling and simulation requirements of the nuclear industry. The technology of nuclear reactors is based on particles and radiation (electrons, protons, neutrons, X-rays and gamma-rays) and nuclear reactions discussed in the preceding chapters. Here, the overall systems will be discussed and, in their context, engineering requirements and challenges to be addressed by scientists, engineers and designers will be highlighted.

3.1 Status of nuclear power

Nuclear power contributes to over 11% of the world's electricity with generation units established in the 1950s. This section reviews the generations of nuclear power reactors from the early 1950s to the next generation reactors.

3.1.1 Generations of nuclear power

Nuclear reactors are classified in several ways and can now, with hindsight, be categorized in generations with the first generation beginning in the early 1950s to the current beginning of what will soon be called the fourth generation.

According to the neutron spectrum, coolant and moderator a reactor is classified as a *thermal reactor* for a thermal (low energy) spectrum and conversely as a *fast reactor* when it has a fast (relatively high energy) spectrum, with little neutron moderation. Going by the coolant and moderator, a water cooled and moderated reactor is called a light water reactor (LWR), which can be subclassified as a pressurized water reactor (PWR) or a boiling water reactor (BWR), a

Nuclear Engineering. DOI: https://doi.org/10.1016/B978-0-323-90618-0.00003-X

heavy water (D_2O) cooled and moderated reactor is called a pressurized heavy water reactor (PHWR), when gas is used as coolant, it is called a gas cooled reactor (GCR).

Reactors are also classified according to their capacity as shown in Table 3.1.

In earlier generations, several variants were tested and developed with materials and moderators such as beryllium and beryllium oxide to verify their compatibility with water, sodium and lead-bismuth alloys. The supercritical water reactor, operating beyond the critical point of water, was tested at high temperatures to enhance the thermodynamic efficiency. Organic coolants were tested for their moderating properties and corrosion reduction. The behavior of materials under intense radiation was a deciding factor in their selection. The latest developments in reactor design, *Generation-IV systems*, aim for compact architecture, the use of liquid coolants (instead of water) for higher operating temperatures instead of water, enhanced safety systems and new proliferation-resistant fuel technologies. Small and modular reactors further ease the factory manufacturing process and since they require less water than larger reactors, so they can be located in off-grid locations with limited water supply.

The nuclear submarine program of the Navy supported and benefited commercial land-based PWR development; the same is true for space nuclear systems that benefited the development of compact systems.

The first power-producing nuclear reactor at the Calder Hall plant, in the United Kingdom, started generating electricity in 1951. These reactors, for example, Calder Hall (GCR), Dresden-1 (BWR), Shippingport (PWR), Fermi-I (SFR), Kola-I (PWR) and Kola-II Russian Vodo-Vodyanoi Energetichesky PWR (VVER) reactors were part of the first generation reactors (IAEA, 2020).

Generation I (1950s–1970s) refers to these earliest nuclear reactors. These reactors were the products of military- and submarine-inspired efforts taking place with huge financial inputs, infrastructure developments in the post-Second Word War era. This knowledge and experience, with massive financial expenditures, found useful application in the construction and subsequent grid connections to the grid. Between 1954 and 1960, there were fifteen (15) operational reactors with capacity 1087 MW(e).

Generation II (1970s–1990s) reactors, at higher capacity, were being deployed on a large scale with the objectives of improving the performance, increasing capacity and fuel burnups with better safety and increased lifetime from 30–40 year to 50–60 years. The number of reactors increased to 84 units with a capacity of 16,656 MW(e). Canada had developed the PHWR from its Chalk River experience and was looking forward to its first exports. The UK had improved the gas-cooled reactors, and the Soviet Union had upgraded the RBMK (Reactor Bolshoy Moshchnosty Kanalny) graphite moderated reactors. In the next decade 161 reactors were added to the grid taking the number of NPPs to 245 with a capacity exceeding 133 GW(e). The US had 47 commercial reactors, USSR had 10, followed by Canada (4), India (3), Argentina (3) and Japan (2), while China, Germany, and South Korea were the countries that had none. The size of reactors had increased to ~3000 MW(t)/1000 MW(e), with a few ~3500 MW(t)/1200 MW(e) in the US, 1500 MW(t)/655 MW(t) in the UK and going up to ~3200 MW(t)/925 MW(e) in Russia. The United States had 31 PWRs and 16 BWRs, while the UK had GCRs, and Russia had four types namely the large LWGR (RBMK-1000) 3200 MW(t) to the small LWGR *Bilibino* 62 MW(t)/11 MW(e), and intermediate PWRs (VVER) ~1375 MW(t)/440 MW(e) reactors. The emphasis in these systems was to go toward large designs for favorable economies of scale. These were large-scale power plants, for example, Bruce (CANDU), Kalanin (PWR), Kursh 1–4 (LWGR), Palo Verde (PWR), Grand Gulf (BWR) and Fukushima II (BWR).

In the 1980s, the Republic of Korea and Germany added seven (7) and six (6) reactors, respectively, to their commercial grids while China still had none.

The average capacity of nuclear reactors had increased from ~200 MW(e) in 1970 to ~540 MW(e) by 1980 and to ~760 MW(e) by 1990. Thus, the fivefold increase in the number of commercial NPPs had undergone a capacity increase of about four times. These numbers indicate the sharp rise of the nuclear industry and the operational confidence that led to the capacity increase.

TABLE 3.1 Nuclear reactor classifications by size.

Reactor size	Standard terminology	Power (MWe)	Typical applications
Large	NPP	>1000	Baseload power
Medium	–	700–1000	Baseload power
Small	SMR	100–700	Flexible grid power
Very small	vSMR	<100	Remote locations
Micro	MNR	<10	Propulsion
Battery	Nuclear battery	<1	

The two events that affected the awareness levels in the nuclear industry in this time period were TMI and Chernobyl.

In 1975, the "Reactor Safety Study" report WASH-1400, known as the Rasmussen Report (Bartel, 2016) was the "first full-scope use of Probabilistic Risk Assessment PRA techniques" that contributed to the awareness of quantifying the risk of what can go wrong, its likelihood, and consequences. Published 4 years before TMI, the WASH-1400 Report could correctly identify that a small scale Loss of Coolant Accident (LOCA) could initiate core damage which is a design basis accident that occurred at TMI.

Generation III (1990 onwards) nuclear reactors were considered to have begun, with the purpose of design improvements producing simpler and rugged designs which would lead to standardization and more efficient licensing and construction processes. Technological improvements included enhanced passive safety features, higher operating lifetimes of up to 60 years, higher burnup for better fuel utilization and reduced waste, reduced core melt and damage probability, and reduced power generation cost. These reactors required certifications from the European Utility Requirements (EUR) in Europe and by EPRI (Electric Power Research Institute)/URD (Utility Requirements Document) in the United States.

In Europe, the European Pressurized Reactor, also called the Evolutionary Power Reactor (EPR) was the first to obain the EUR license; the first EPR unit became operational in CHina (Taishan 1) in 2018. Anotehr large Generation III Advanced Power Reactor APR1400, Section 3.3, is a 1400 MW(e) large PWR designed with safety barriers and protection on "defense in depth" implemented to five levels of protection. These include successive barriers to prevent the release of radioactive material to the environment and an integral design (steam generators inside the Reactor Pressure Vessel) to prevent a LOCA.

wGeneration III + had realized innovative and evolutionary designs in which safety features were enhanced, radiation exposures reduced and the maintenance and operations made simpler. In 1994, China had its first connection to the grid; a 300 MW(e) PWR *Qinshan*-I. Globally, between 1990 and 2000, fifty-two (52) new reactors were connected to the grid. The number of reactors increased in number to 416 by 1990 with a capacity of ~318 GW(e).

The Gen-III +, Advanced Passive Reactor AP1000, Section 3.3, is an upgrade of the Gen-III AP600 at 1100 MW (e) is also a relatively large reactor. Its distinguishing features include a relatively large pressurizer connected to one of the hot leg pipes variable fuel enrichment (2.3% − 4.8%), fuel burnable absorbers for better lifespan, high average burnup of 60 GWd/t, and thermodynamic efficiency ~34%. A modular design is applied to the containment and for the fabrication of equipment such as the reactor internals and the condenser. The advancements to the AP600 include low maintenance and high availability. The AP1000 has good safety margin for shutdown and core damage frequency $\sim 5 \times 10^{-7}$/reactor year (RY). It meets very high standards of reliability and cost affordability and satisfies sustainable development goals, mitigating climate change and reducing carbon emissions associated with fossil fuel thermal plants. These reactors included the CANDU 6, AP600, the Advanced Power Reactor APR1400, and the advanced boiling water reactor (ABWR).

The Fukushima earthquake in March 2011, led the nuclear industry to take a fresh look at the nuclear industry. The evolution of nuclear power reactors and designs (IAEA, 2020) from the early prototypes of the first generation reactors of the 1950s thus led to large-scale deployment with second generation designs in the 1970s.

In the period 1990–2000, there were 52 new reactors connected on their grids. This was the period of the beginning of evolutionary *Generation III+* designs, such as the Advanced Passive AP1000 Reactor. In the next two decades, from 2000 to 2010 there were 32 new reactors and from 2010 to end of December 2019, there were 58 new reactors connected to the grid.

The most notable programs were those of China with forty-eight (48) new reactors, India with sixteen (16), Japan twenty-six (26) until 2010, South Korea twenty-one (21) and Russia seventeen (17). Thus, from 2000 until the beginning of 2020, out of the 90 new reactors, 71 were in Asia and twelve (12) in Russia.

As of the end of 2019, there were 448 operational nuclear reactors in the world. Most of the reactors in the world are PWRs (68%), followed by BWRs (14%), PHWRs (11%), GCRs (3%), LWGR (3%), and Fast Breeder Reactors (FBRs) (1%). There are 305 operating PWRs in the world (IAEA, 2020) out of which 42 produce less than 600 MWe and 263 produce greater than or equal to 600 MWe. Out of 54 new constructions, 44 are PWRs. By far, the PWR program is the most successful in the world contributing to over 60% of the total nuclear power reactors.

The number of nuclear reactors by region indicates the rising trend of nuclear energy in Asia, the stagnation in the United States, and the decline in Europe and the United Kingdom. Asia (Far East) is leading with both operational and under-construction reactors counted, followed by Northern America, Western Europe, Central and Easter Europe, then Middle East and South Asia and Africa.

In China, nuclear power development is taking place at a phenomenal level; with its 49 reactors and a ~48 GW(e) capacity, 16 reactors are under construction. This will modernize China's power infrastructure and reduce its heavy dependence on coal which presently contributes to over 65% of the electricity generation. The carbon dioxide emissions of China at 28% of the world's total places it at first position and hence its' move to nuclear power is bound to in line with sustainable development goals. The new builds are mainly large reactors such as AP1000, *Hualong* One (based on the GenII + CPR1000 and AP1000 technologies) and the CAP1400 whose extensions will be 1400 and 1700 MW reactors with China holding intellectual property rights. With reactors under construction, as well as 43 planned reactors, China with 107 nuclear power plants (NPPs) is set to become the world's number one nuclear power program with a capacity exceeding 100 GW(e).

It is important to consider the number of permanent shutdowns, and its long-term effect on nuclear power at the global level. As of December 2019, there have been 186 permanent shutdowns with Europe seeing 62, USA 37, UK 30, Japan 27 and Canada 6. By February 2020, the number of shutdowns had increased in Europe with closures of 30 reactors in Germany, six in Sweden, four each in Italy, Ukraine and Bulgaria, and three each in Spain and Slovakia. Many of these have been due to the expiration of design life. The number of shutdowns peaked in 2011 with the Fukushima disaster in March 2011 leading to the closure of four units.

Another peak in shutdowns was in 2019 mostly due to the aging of reactors and expiration of licenses. The average age of nuclear reactors in the world is about 20 years while the US reactors on the average have an age of about 40 years.

3.1.2 Reactors shut down

The design life of Generation-1 nuclear reactors was generally about 30–40 years; several reactors were shut down before the expiry of this design life due to economics of their electricity generation or other operational problems. In spite of this design life span, the operational life span, as in improved reactors, can be much longer as there are no strong technical limits to prevent extending reactor operations by another 20 years, and possibly up to 80 years. In the United States, for example, with 96 operational reactors, as of January 1, 2021, providing about 20% of the electricity, it is likely that most will get life extensions to go up to possibly 80 years.

Of the shut down reactors across the world, 57 are PWRs, 50 are BWRs, and 38 are graphite-moderated GCRs; this probably marks the end of the UK-type GCRs and a decline in BWRs although four are under construction and nine are planned. The PWRs, however, are on the ascendance, with 44 under construction and 57 reactors planned.

The UK program of early Magnox GCRs, the first in the world to have generated electricity, had to shut down reducing its fleet to AGRs and a PWR With these shutdowns, The UK is considering the Chinese Hualong Design; similar to the two 1100 MWe units which China has constructed in Pakistan out of which one was connected to the grid in March 2021.

With its quick progress, China has also entered into foreign agreements most notably with Pakistan and fourteen other countries including Romania, Argentina, the UK, Iran and Turkey. In Pakistan, 300 MW PWRs are in operation at *Chashma* as the C-1 to C5 series and in December 2020, the fuel loading of the Karachi-2 (K2) *Hualong* 1100 MW was followed by criticality in March 2021 and grid connection the same month. Out of 8 nuclear reactors in Pakistan, seven PWRs (~3700 MW) have been set up by China.

3.1.3 The future of the nuclear power industry

The future of the nuclear industry is thus of vital concern to several countries. As the largest operator of nuclear reactors, the United States currently has 96 operational reactors at 58 NPPs; these have an average age of ~40 years which approaches the limit of the design lifespan. These reactors provide ~20% of the total electricity generation of the US.

After the Fukushima earthquake disaster of March 11, 2011, the nuclear energy prospects have suffered a setback. In earlier days, Japan had 61 NPPs, out of which 27 were shut down and one was under construction. Out of the remaining operational 33 NPPs, 16 were PWRs and 17 were BWRs. Currently 31 NPPs have a capacity of 31.7 GW(e).

Japan's energy situation is heavily dependent on oil imports and hence it has been looking toward a reduced dependency on fossil fuels, as well as on nuclear energy after Fukushima; leaving renewables as the preferred source. In spite of this, the share of nuclear is planned to be ~20% by 2030.

Since its inception in the 1950s, the nuclear industry has had an excellent safety record; it has been a reliable clean energy baseload power source contributing to ~16% of the electricity at one time and reducing now to ~ 11%. From the three accidents that have taken place, TMI, Chernobyl and Fukushima, considerable knowledge has been gained and incorporated into new designs especially the Gen-IV designs.

Based on the statistics and trends discussed above, there are conflicting signals on the growth of the nuclear industry.

The key advantages of nuclear energy are:

1. clean and low-carbon energy generation; if col or gas would be burnt instead of nuclear, hundreds of millions of tons of carbon dioxide woud be released in to the atmosphere, worsening the environmental problems,
2. high-density energy,
3. safe track record, and
4. reliability.

The nucler industry faces challenges due to

1. high capital cost and
2. radioactive high-level waste management.

Some sustainable growth strategies for the nuclear industry are

1. technology improvements and enhanced safety in large power reactors,
2. development of small modular reactors (SMRs) to become commercially viable by the 2030s, and
3. hydrogen production from nuclear reactors.

3.2 Nuclear reactor systems

Nuclear reactors are huge and capital intensive power infrastructures integrating technologies that are both conventional and hi-tech. Each NPP as a system consists of several subsystems. The International Atomic Energy Agency has classified the major systems as

1. The primary system
2. Balance of plant
3. Spent fuel storage
4. Nonelectrical systems

Essentially, a nuclear reactor is a system that produces thermal energy from fissions in the fuel elements placed in its core. The thermal energy can be converted into any other form of energy from which "useful" work can be extracted. In the form of a mathematical statement, this is the First Law of Thermodynamics which states that *energy can neither be created nor destroyed but it can be converted from one form to another.* In the PWR core, the heat produced Q raises the enthalpy H of the water entering at some temperature and pressure T_i, P_i, and the heat is used to produce work W in the turbine from which electricity is generated. After the steam has been used, it loses its energy, flows through a condenser and is pumped back into the steam generator. The efficiency η of a reactor is a measure of the ratio of net work to heat generated. Reactors are designed in a way that maximizes their efficiency; for this it should produce heat at the highest temperature possible and reject heat at the lowest temperature possible.

The subsystems (IAEA, 2007) of each system are classified as follows:

The reactor vessel (RV) which includes the reactor core and the control systems, the coolant system, the pressurizer and the steam generator with safety and protection systems.

In the reactor core, the components include the fuel assembly with fuel rods and cladding surrounded by the moderator and coolant. The engineering design of the reactor core covers the areas of neutronics, radiation transport, thermal hydraulics, structural mechanics and control systems. Detailed calculations are used to obtain design and operating parameters for the amount and placement of fissile material in the core, the flow of coolant, and the control of reactivity. The system pressure, in a PWR for example, is maintained by a pressurizer which ensures the state of the coolant and maintains the high pressures required.

At the heart of the control system is the control rod mechanism and the calculation of operational load following movements as well as maintaining reactivity margins for safe shutdown in the event of any departure from the prescribed set points. In addition to control rods, nuclear reactors use burnable absorbers such as boron, gadolinium, europium in the form of compounds. Boric acid, for example, is dissolved in the coolant to reduce the reactivity in case it increases. Another operational feature, to "flatten the flux" is to provide the fuel rods with a burnable poison at the beginning of a cycle to enhance the performance of the reactor.

A steam generator is a major component where heat transfer from the nuclear system takes place to the working fluid such as water which gets converted to hot "superheated" steam which is intended to drive a turbine. Here, the shell and tube exchange process is dependent on the flow rate and thermal capacity of the fluid as well as the physical condition of the system. The design and operation parameters of a steam generator are based on thermodynamics and structural analysis.

Safety systems are based on several independent and interdependent subsystems to ensure operational functionality in case of any minor or major departures from set points. These utilize containment pressure suppression systems, core spray pumps, flowrate regulators, and other components, which includes the turbine and condenser.

In the "secondary" or conventional part of a nuclear plant the major components are turbine(s), generator(s), condenser, feed-water system, emergency power supply systems and fire protection systems.

For spent fuel storage, the reactor has subsystems comprising spent fuel pool and an interim storage facility.

The major areas are discussed in the context of specific reactor designs in the following subsections.

3.2.1 Pressurized water reactor

The PWR, illustrated in Fig. 3.1, is classified as a light water (thermal) reactor since it uses light (ordinary) water as the moderator and coolant. As discussed in Chapter 2 (interaction of neutrons with matter) fission neutrons colliding with a hydrogenous light material are able to slow down and thermalize in a small number of collisions. This gives a "thermal" flux for which the fission cross-section is higher than that for fast neutrons.

The major components in the primary side of a PWR are the pressure vessel, the core, the pressurizer, the steam generator, the coolant pump and the control system shown in Fig. 3.1 with the residual heat removal system in a separate building. The secondary side is the "balance of plant" which is common with conventional power plants.

These are common to all PWRs though the number of loops may vary depending on the size of the plant. A "four-loop" Westinghouse PWR core, shown in Fig. 3.2, has four steam generators and a pressurizer within the pressure vessel. The terminology refers to the number of steam generators and primary circuit coolant pumps. The internals of the core are shown in detail in Fig. 3.2B.

The fuel rods and control rods are vertically placed within the reactor while "cold" coolant fluid enters from the top of the core on one side flows downwards then upwards and exits as a hot fluid from the top on the other side. In the steam generator, heat is transferred to the secondary side fluid (water in a PWR) from which superheated steam flows to the turbine where electricity is generated. The secondary loop is closed through the condenser and return pump.

Some representative figures for the Westinghouse Advanced Passive reactors AP600 and AP1000, and the KEPCO Advanced Power Reactor APR1400 are shown in Table 3.2. AP1000 and APR1400 are typical of the "big" league of reactors, with the European Pressurized Reactor (EPR) 4590 MW(t)/1660 MW(e) *Taishan* units 1 and 2 as mentioned in the previous section.

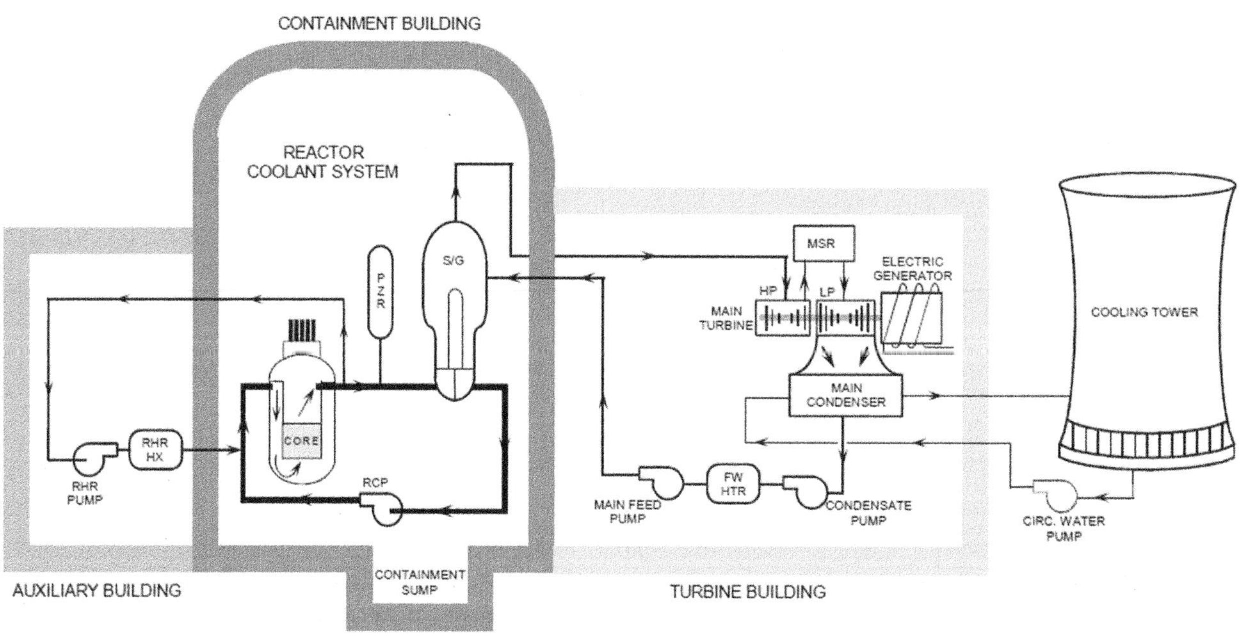

FIGURE 3.1 Schematic diagram of a pressurized water reactor.

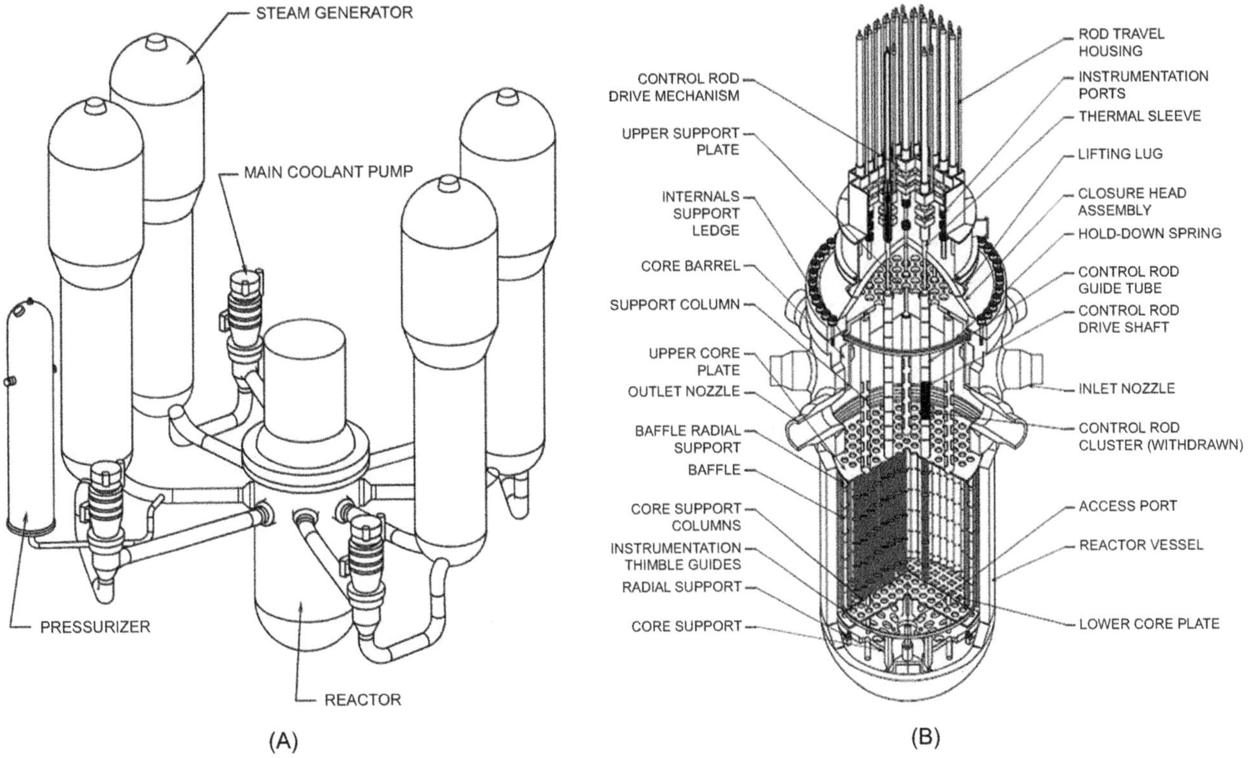

FIGURE 3.2 (A) A four-loop PWR, (B) PWR core. *https://www.nrc.gov/reading-rm/basic-ref/students/for-educators/04.pdf. Courtesy of Westinghouse Electric Company (c) 2021. All rights reserved.*

TABLE 3.2 Overall characteristics of AP-PWR and APR1400 nuclear reactors.

Parameter	AP600	AP1000[a]	APR1400[b]
Thermal power (MW(t))	1940	3400	3983
Net electrical power (MW(e))	600	1100	1400
Efficiency (%)	31	32	35
Plant life span (y)	60	60	60
Average plant factor (%)	93	93	>90
Equivalent core diameter (m)	2.921	3.04	3.63
Active core height (m)	3.658	4.267	3.81

[a] *AP1000 https://aris.iaea.org/PDF/AP1000.pdf.*
[b] *APR1400 Advanced Power Reactor https://aris.iaea.org/PDF/APR1400_2020May.pdf https://aris.iaea.org/sites/core.html.*
Source: AP1000 iaea.org https://www.osti.gov/etdeweb/servlets/purl/20256613; AP600 https://aris.iaea.org/PDF/AP-600.pdf.

As Table 3.2 shows, the reactors are "high power density," for example, 3983 MW(t) in a volume of $\sim$ 40 m^3 is $\sim$100 MW(t)/m^3. With such high power densities, PWRs require very high levels of safety and control. The efficiencies of these reactors is still a maximum of 35% due to thermodynamic limits of the Rankine cycle. These efficiencies are optimized by minimizing the energy losses in the system. The lifetimes of the PWRs given in Table 3.2 are $\sim$60 years although there is no technical limit other than the usual wear and tear of components. These plants have to be safe to avert any unplanned departure from design conditions. Some safety goals of the AP1000 are: core damage frequency $< 5.09 \times 10^{-7}$ per RY, occupational radiation exposure $<$ 0.7 person-Sv/RY, and operator action time of half an hour.

A design comparison of a typical 1000 MW(e) PWR with a small 300 MW(e) reactor is given in Table 3.3 illustrating significant differences in size and fuel and coolant flow.

TABLE 3.3 Physical design parameters of AP1000 and PWR 300 MW(e).

Core	AP1000[a]	CHANSNUPP[b] (300 MW)
Active height (m)	4.267	2.9
Equivalent diameter (m)	3.04	2.486
Fuel material	Sintered UO_2	UO_2
Fuel enrichment (%)	2.35/3.40/4.45	2.4/2.67/3.0
Fuel inventory (t)	96.176	35.917
No. of assemblies	157	121
Rods per assembly	264	204
Assembly pitch (cm)	–	20.03
Rod pitch (cm)	1.26	1.33
Geometry (no. of rods)	17 × 17	15 × 15
Pin outer dia. (mm)	9.5	10
Pellet height (mm)	9.83	10
Pin height (mm)	–	3210
Clad/thickness (mm)	0.5715	0.70
Clad material	ZIRLO	Zircalloy
Moderator	H_2O (liquid)	H_2O (liquid)
Moderator inventory (MT)	–	57
Coolant	H_2O (liquid)	H_2O (liquid)
Primary pumps	4	2
No. of loops	2	2

[a]*https://www.westinghousenuclear.com/new-plants/ap1000-pwr.*
[b]*Mian, Z. and Nayyar, A. H. (1999). Pakistan's Chashma Nuclear Power Plant, PU/CEES Report. No. 321, Center for Energy and Environmental Studies, Princeton Environmental Institute, Princeton University, Princeton, NJ 08544–5263, Dec. 1999, p. 62.*
Source: AP1000 fuel data from https://www.nrc.gov/docs/ML0715/ML071580895.pdf.

TABLE 3.4 Process variables of AP1000 and CHASNUPP.

Core	AP1000*	CHANSNUPP** (300 MW)
Burnup (MWD/t)	60,000	30,000
Max. centerline temperature (°C)	2593	1806
Max. clad temperature (°C)	–	345.7
Coolant	Water	Water
Coolant flow rate (tph)	14300 kg/s, (51480 tph)	24,000
Coolant inlet temperature (°C)	279.4	288.5
Coolant outlet temperature (°C)	324.7	315.5
Pressure (kg/cm^2)	155.13	155
Geometry	17 × 17	15 × 15
Moderator temperature at full load (°C)	303	302

**https://www.nrc.gov/docs/ML0715/ML071580895.pdf.*
***Mian, Z., and Nayyar, A. H., Pakistan's Chashma Nuclear Power Plant, PU/CEES Report. No. 321, Center for Energy and Environemntal Studies, Princeton Environemental Institute, Princeton University, Princeton, NJ 08544-5263, Dec. 1999, p. 62.*

The scale-up in process variables of both systems is given in Table 3.4 showing a double discharge burnup, higher operating centerline temperature and higher coolant flow rate.

The control systems of both reactors use Ag-In-Cd control rods and borated dilution

3.2.2 Boiling water reactor

In a BWR (Fig. 3.3), heat is carried by the coolant (water) flowing upwards through the core; this steam-water mixture then passes through separation systems at the top of the core, and hot dry steam leaves the core. The hot dry steam goes

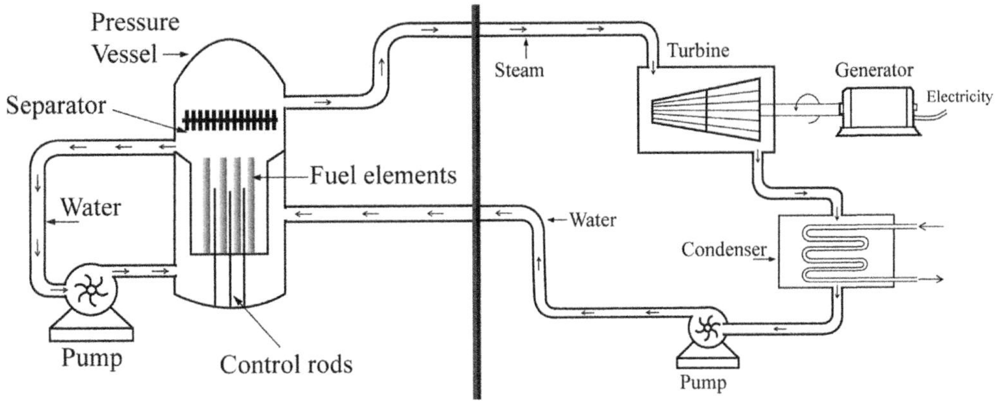

FIGURE 3.3 Schematic design of a Boiling Water Reactor.

TABLE 3.5 Physical design parameters of BWR (Fukushima).

Main specifications	Electric output (MW)		1100
	Start of construction		May-73
	Start of commercial operation		Oct-79
	Reactor type		BWR5
	Containment type		Mark II
	Main contractor		GE/Toshiba
Nuclear reactor	Heat output (MW)		3293
	Number of fuel assemblies		764
	Full length of fuel assemblies (in.)		176
	Number of control rods		185
	Reactor pressure vessel (RPV)	Inner diameter (in.)	252
		Height (in.)	906
		Total weight (short ton)	827
		Design pressure (psi)	1249.9
		Design temperature (°F)	576
	Primary containment vessel (PCV)	Height (ft)	157.5
		Diameter of cylindrical portion (ft)	32.8 (top)
		Diameter of spherical portion (ft)	82.0 (bottom)
		Suppression pool water amount (kgal)	845.4
		Design pressure (psig)	40.6
		Design temperature (°F)	340 (DW) 221 (SC)
Steam turbine	Number of revolutions (rpm)		1500
	Steam temperature (°F)		540
	Steam pressure (psig)		950
Fuel	Type		Uranium dioxide (Unit 3 contains MOX)
	Uranium (ton)		132

Source: https://www.tepco.co.jp/en/nu/fukushima-np/outline_f1/index-e.html.

to the turbine where thermal energy is converted into mechanical energy and eventually to the generator where mechanical energy is converted into electrical energy.

On exit from the turbine, the "low quality" steam and water is condensed into water and pumped back into the core.

The physical design parameters of Unit 6 of the Fukushima BWRs are shown in Table 3.5. The units 1–4 of the Fukushima Daiichi NPP were abolished on April 19, 2012, and units 5, 6 were abolished on January 31, 2014 in accordance with the Electric Utility Law. The Fukushima Daiichi BWRs reactors are GE designs of the 1960s known as Mark I containment. They came into commercial operation from 1971 onwards; the earlier units were 460 and 780 MW (e) while unit 6 given in the Table was 1100 MW(e) During normal operation they all had a core outlet temperature of

TABLE 3.6 The advanced boiling water reactor (ABWR).

Parameter	ABWR (GEH[a], HGNE[b], and Toshiba)	ABWR-II GE-Hitachi
Thermal power (MWt)	3926	4960
Net electrical power (MWe)	1350	1638
Efficiency (%)	34	33
Plant life span (y)	60	60
Average plant factor (%)	90	93
Equivalent core diameter (m)	5.163	5.41
Active Core height (m)	3.81	3.71

[a]*General Electric Hitachi.*
[b]*Hitachi General Electric Nuclear Energy.*
Source: ABWR https://aris.iaea.org/PDF/ABWR(Hitachi-GE)_2020.pdf, ABWR-II https://aris.iaea.org/PDF/ABWR-II.pdf, Core data https://aris.iaea.org/sites/core.html.

286°C under a pressure of 6930 kPa and with 115–130 kPa pressure in dry containment. The operating pressure was about half that in a PWR. A drawback of the safety system was that the reactors had analog instrumentation in a ground floor building instead of digital controls on an elevated floor; thus, information could not be accessed remotely.

For improved designs, the 1350 MW(e) advanced ABWR, with overall parameters listed in Table 3.6, has an improved plant life of 60 years, plant availability of 87%, longer refueling period of 24 month intervals, and reduced probability of core damage of 10^{-5} per RY. The enhanced safety systems include an emergency core cooling system designed to ensure emergency cooling in the event of a design basis LOCA, so that the cladding temperature does not reach the limit of 2200°F (1204°C) in accordance with USNRC 10CFR50.46. Concerns of radiation have been addressed by reducing both the radiation exposure (limited to 100 man-rem/year) and radiation waste.

As of December 2019, there were 65 operational BWRs while 50 were shut down, four were under construction, and nine are planned. Most of the operational units were of capacity greater than 600 MW(e).

In Japan, 16 BWRs (out of which 10 were in Fukushima) were permanently shut down. In the US, 13 BWRs, mostly connected to the grid in the 1960s, (Big Rock Point US-155, Bonus US-014, Dresden-1 US-10, Elk River US-11, GE Vellecitos US-018, Humboldt Bay US-133, Lacrosse US-409, Millstone-1 US-245, Oyster Creek US-219, Pathfinder US-130, Pilgrim-1 US-293, Shoreham US-322, Vermont Yankee US-271) were permanently shut down. In Europe 19 BWRs were permanently shut down: 10 in Germany, 2 in Italy, 4 in Sweden, and 1 each in Switzerland, Spain and Netherlands.

Based on current information (IAEA, 2020), the only countries planning BWRS are Japan (six ABWRs, 3926 MW(t) each) and USA (1 ESBWR 4500 MW(t) and 2 ABWR 3926 MW(t)) while two ABWRs are under construction in Japan.

3.2.3 Pressurized heavy water reactor

The design of the CANDU PHWR can be traced back to the Manhattan Project when the Chalk River Laboratory was established in 1944 for supplying spent fuel from the plutonium production reactor NRX for the US weapons program. The NRX suffered a partial meltdown and was replaced by the higher flux NRU to increase production.

The experience from NRX and NRU led to the CANDU design starting with a small 20 MW(e) reactor, called the Nuclear Power Demonstration (NPD). Subsequently, the 200 MW(e) Douglas Point reactors were built and commissioned in 1966. The reactor power was increased to 500 MW by increasing the diameter of the calandria from ~ 8 to ~ 10 cm with a corresponding increase in the number of fuel elements per fuel bundle from 19 to 28.

The PHWR uses natural uranium fuel rods, heavy water cooled and moderated "horizontal" reactor with several simple and safe features. The overall plant schematic is shown in Fig. 3.4.

The fuel is bundled in a pressure tube typically 8–10 cm in diameter and placed horizontally in pressure tubes within a calandria (a horizontal cylindrical tube). Several calandria tubes are inserted into the horizontal moderator tank. Coolant flows though horizontal channels between fuel rods as shown in Fig. 3.4. Overall physical design parameters of CANDU variants are listed in Table 3.7 from the relatively low power Indian PHWR to the Enhanced CANDU 6 design.

Another variant of the PHWR design is the 300 MW(e) thorium-fueled Indian advanced heavy water reactor (AHWR) designed and developed by the Bhabha Atomic Research Center (BARC). This uses a vertical pressure tube with boiling light water coolant and heavy water moderator reactor The design incorporates passive safety systems working on natural

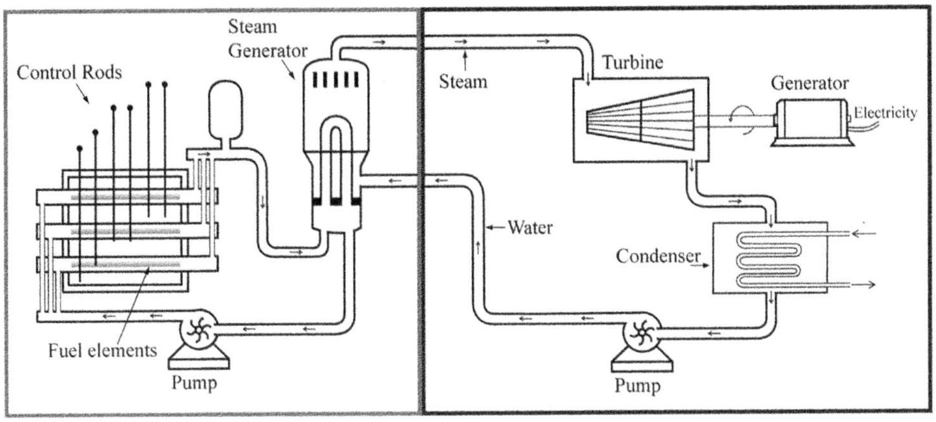

FIGURE 3.4 Schematic diagram of a pressurized heavy water reactor.

TABLE 3.7 Physical design parameters of PHWR designs.

Parameter	IPHWR-220	APHWR (BARC)	Enhanced CANDU 6
Thermal power (MWth)	754.50	920	2084
Net electrical power (MW(e))	235.81	284	740
Efficiency (%)	26.5	–	35.5
Plant life span (y)	40	100	60
Average plant factor (%)	>90	90	90
Equivalent core diameter (m)	4.51	6.9	7.595
Active Core height (m)	4.95	3.5	6.28

Source: IPHWR-220: https://aris.iaea.org/PDF/IPHWR-220.pdf, EC6: https://aris.iaea.org/PDF/EC6.pdf, APHWR:Bhabha Atomic Research Center http://www.barc.gov.in/reactor/ahwr.pdf.

laws such as force of gravity and heat absorption in a gravity-driven water pool, independent shutdown systems, passive poison injection into the moderator for a wired-system failure. Due to boiling in the coolant, a steam generator is replaced by a simpler steam drum; similarly a number of major components such as primary coolant pumps are not required. All reactivity temperature coefficients (fuel, channel, void and moderator) are negative. Minor actinide production, including plutonium production, is reduced due to the thorium fuel making the fuel proliferation-resistant.

As of December 2019, there are 49 PHWRs operational in world; with 19 in Canada, 18 in India, 3 in South Korea, 2 in China, 2 in Romania, 3 in Argentina, and 1 in Pakistan, and 1 is under construction.

3.2.4 Gas cooled reactor

Gas cooled Magnox (magnesium-aluminum alloy cladded fuel rods) reactors, as depicted in Fig. 3.5 below, were constructed in the 1950s for both commercial and military use. The first Magnox reactors at *Calder Hall*, commissioned in August 1956, and decommissioned in 2003, were also the first nuclear reactors to produce electricity on a commercial basis, though their main purpose was plutonium production. Each of the four 268 MW(t)/60 MW€ reactors used natural uranium fuel, graphite moderator, and carbon dioxide gas as the coolant.

The first generation Calder Hall reactors were decommissioned and succeeded by the second generation *Windscale Advanced Gas Cooled Reactor*. As pioneer of the civil nuclear program, the UK had 26 Magnox units, with grid connections between 1962 and 1971, including Berkely, Calder Hall, Chapelcross, Dungeness, Hinkley Point, Hunterston, Oldbury, Sizewell, Trawsfynydd and Wylfa. These have all been decommissioned by Magnox Ltd. on behalf of the UK government.

The Calder Hall reactors had natural uranium metal fuels, in the form of one cast bar, with element length 1.016 m and 6 elements in a channel. The reactor had an active core height of 6.4 m and an active core diameter of 9.45 m. The last Magnox reactor, *Wylfa*, produced power about seven times more thermal power, but with about 10 times more electrical output than that from Calder Hall. Thus efficiency had been significantly increased mainly by increasing both the mass of uranium in the core and the total gas flow.

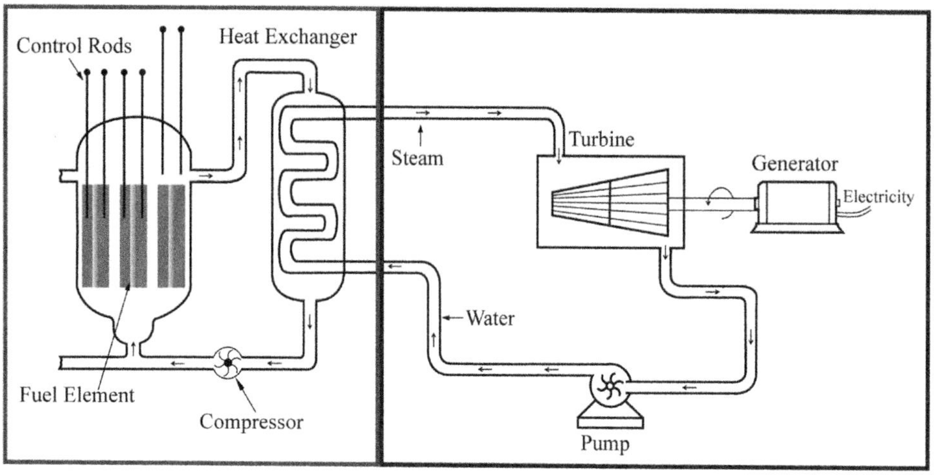

FIGURE 3.5 Schematic diagram of a gas cooled reactor.

TABLE 3.8 Physical design parameters of the *Torness* GCR.

Parameter	Torness AGR
Thermal power (MW(t))	2 × 1623
Net electrical power (MW(e))	2 × 660
Efficiency (%)	40.7
Plant life span (y)	Grid connection 1989; lifespan increased to 2030[a]
Average plant factor (%)	
Equivalent core diameter (m)	9.5
Active core height (m)	8.3

[a]*https://www.world-nuclear-news.org/C-EDF-Energy-extends-lives-of-UK-AGR-plants-1602164.html.*
Source: Nonbel, E. (1996). Description of the advanced gas cooled type of reactor (AGR), Riso National Laboratory, Roskilde, Denmark, Nov. 1996. https://www.osti.gov/etdeweb/servlets/purl/448871.

Currently there are 14 operational reactors, all in the UK (IAEA) at 6 sites and 7 stations. Table 3.8 shows the overall data for Torness GCR which is based, as the previous Magnox reactors, on graphite moderator and CO_2 coolant but with a change of fuel from metal natural uranium to 2.2% enriched uranium in UO_2 fuel. These changes have reduced the physical size of the plant, raised outlet operating pressure and gas flow rate as well as the gas temperature, resulting in 40.7% efficiency.

The AGRs need to improve in their design to be competitive with LWRs particularly PWRs to which they have lost the market, even though GCRs were the first nuclear reactors to produce commercial power.

3.2.5 Fast breeder reactor

The FBR program began in the 1940s with the objective of using high enriched uranium to produce small-size reactors with liquid metal coolants. This would increase the overall plant efficiency as well as the neutron economy by "breeding" plutonium, which does not occur in nature, from a fast neutron spectrum. These efforts led to the experimental breeder reactor (EBR-I) which in 1951 was the first reactor to demonstrate the production of electricity. The EBR-I, followed by EBR-II and the French Phoenix FBR were all shut down after just a few years of operation due to several technical difficulties.

At present the only two fast reactors in operation, BN-600 and BN-800, are in Beloyarsk in Russia (IAEA, 2017). As shown in Fig. 3.6, a liquid metal cooled fast reactor has a primary circuit in which high temperature liquid metal, such as sodium, flows the core and exchanges heat in an intermediate heat exchanger with a coolant, such as water. From there on, the conventional part of the power conversion unit is the same as discussed in the previous sections for the PWR, BWR and GCR.

The BN-600 reactor with overall parameters and design evolutions shown in Table 3.9, has been in operation in Beloyarsk since the last 30 years. The design improvements were made due to lessons learned from operational experiences

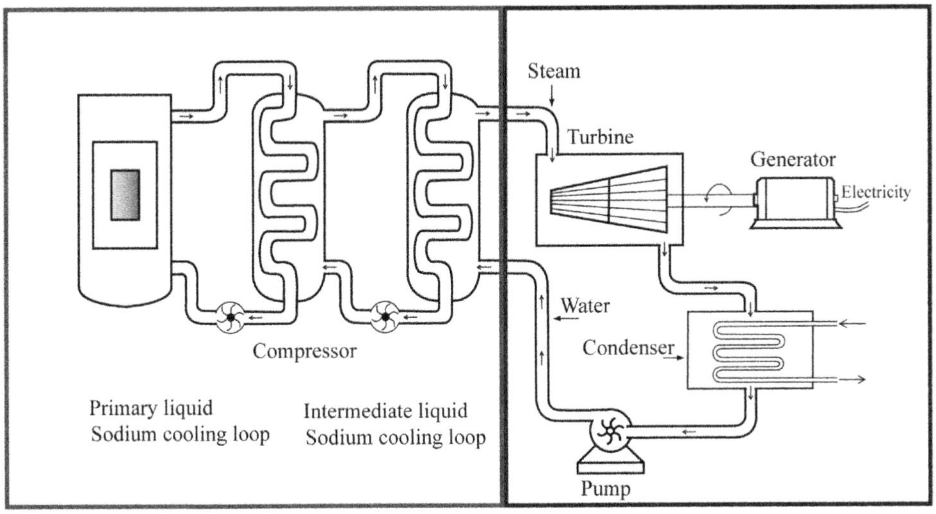

FIGURE 3.6 Schematic diagram of a Liquid Cooled Fast Reactor.

TABLE 3.9 Characteristics of the BN-600 Sodium cooled Fast Breeder Reactor (FBR).

Parameter	Reactor Type I	Type M	Type M-1
Thermal power (MWth)	1470	1470	1470
Net electrical power (MWe)	600	600	600
Efficiency (%)			
Plant life span (y)			
Average plant factor (%)	–	–	–
Equivalent core diameter (m)	2.058	2.058	2.058
Active core height (m)	0.750	1.000	1.030
Fuel inventory in core (kg)	8260	11630	12090
Average fuel burnup (MWd/kg U)	42.5	44.5	60.0
Fuel enrichment (U-235)			
Low enrichment zone	21	17	17
Medium enrichment zone	–	21	21
High enrichment zone	33	26	26

LEZ, low enrichment zone; *MEZ*, medium enrichment zone; *HEZ*, high enrichment zone.
Fast Reactors and Related Fuel Cycles: Next Generation Nuclear Systems for Sustainable Development FR17, Proceedings of an International Conference Yekaterinburg, Russian Federation, 26–29 June 2017. https://www-pub.iaea.org/MTCD/Publications/PDF/STIPUB1836web.pdf.
Source: IAEA Tecdoc 1569, p. 123.

the power of the plant was kept the same while increasing the active core height from 750 mm to 1030 mm, "flattening" the flux by incorporating three enrichment zones, and increasing the fuel inventory of the core from ~8.3 tons to ~12 tons; all these modifications improved fuel utilization enhancing the burnup by almost 50%, to 60 MWd/kg U.

A power upgrade to BN-600, the sodium cooled BN-800 reactor was commissioned for commercial operation in 2016 at the Beloyarsk NPP. In addition to these two operational reactors, the BN-1200 design is under development for technical and economic feasibility.

The directions for possible improvements in fast reactor design are under investigation; several material selection decisions are yet to be taken such as the selection of sodium, lead, gas, molten salt, supercritical water cooled systems, as well as hybrids, such as accelerator-driven systems. As work continues toward innovative Gen-IV designs for the next decades, safety and sustainability are being enhanced and being incorporated in new designs.

Out of 54 new constructions, one 470 MW FBR was under construction as of December 2019 in India and one 1500 MW(t)/600 MW(e) CFR-600 sodium cooled fast reactor (SFR) is under construction in Fujian province, China.

In spite of the major advantages of FBRs including their efficiency, compact design, and plutonium breeding, their safety and proliferation resistance are issues that must be addressed for better public acceptance.

3.3 Marine propulsion reactors

3.3.1 Introduction

The use of nuclear reactors extend far beyond production of electricity for cities; they are used for numerous other applications such as marine propulsion for submarines and icebreakers, radioisotope production for medicine and agriculture, testing of materials, desalination and plutonium production. This section presents an overview of the applications of nuclear reactors for marine propulsion that includes the propulsion systems of merchant ships, icebreakers, and the nuclear navies with submarines, aircraft carriers and cruisers (Fig. 3.7).

Nuclear reactors for marine propulsion are attractive, in comparison with fossil fuels since they do not need air to operate and hence submarines can stay under water for long periods, they have high power density, long operational periods, and low operational costs.

3.3.2 US nuclear submarine program

Lobner (2018) presents the timeline for development of marine nuclear power in the US from Enrico Fermi's briefing to the Navy Department in 1939, followed by the first controlled critical pile (CP-1) established under Fermi's leadership on 2nd December 1942, to Manhattan Project, Alvin Weinberg's first description of the PWR, the Gunn-Abelson 1946 nuclear plant design concept for a submarine, involvement of Argonne National Lab and Westinghouse to the beginning of the construction work on the Submarine Thermal Reactor (STR) Mark 1 Prototype in 1950. The contractors included Bechtel (B), Combustion Engineering (C), General Electric (G), and Westinghouse (W). Their codes appear in the naming of ship power systems; for example, A1W represent aircraft carrier generation 1 built by Westinghouse. Other first alphabetic codes are C for cruiser, D for destroyer and S for submarine. It was on the 1st of August 1946 that the US Atomic Energy Commission was established. Two reactor designs actively pursued in the 1950s were based on the liquid metal cooled fast reactor and the STR series.

After the second world war, the US Navy was quick to replace diesel and battery systems on their fleet with nuclear power systems, starting from the nuclear-powered submarine *USS Nautilus* in 1955, powered by the STR Mark II (S2W) reactor. The three big projects at that time were *USS Nautilus* in January 1955 STR Mark II Thermal PWR, *USS Seawolf* in 1955 SSN-575 Liquid Metal Cooled Reactor S2G and *USS Skipjack* in May 1956, SSN-585 Advanced Submarine Fleet Reactor (ASFR). The poor performance of the liquid metal reactor S2G on *USS Seawolf* led to its removal and replacement by the Nautilus PWR. This was probably the turning point that gave the field to PWRs which Westinghouse spearheaded for commercial reactor designs.

In 1958, keels were laid for the first US civilian nuclear-powered merchant ships *NS Savannah*, the SSN *USS Scorpion*, and the *USS George Washington*.

Work on a GCR program began in the US in 1956. By 1959, the liquid metal design for submarine reactors had been discarded in favor of PWRs.

By 1962, the US Navy had rapidly built up a fleet of 26 nuclear submarines which included the aircraft carrier USS Enterprise powered with eight PWRs and the cruiser USS Long Beach powered with two PWRs. At that time, 30 more nuclear submarines were under construction in a large nuclear submarine industrial infrastructure.

FIGURE 3.7 USS Gerald R Ford aircraft carrier. https://upload.wikimedia.org/wikipedia/commons/b/b2/USS_Gerald_R._Ford_%28CVN-78%29_underway_on_8_April_2017.JPG (in public domain, permitted use).

3.3.3 Former Soviet/Russian nuclear submarine program

In Russia, work on a submarine using a nuclear propulsion reactor was initiated in 1952 (Reistad & Olgaard, 2006). Within 4 years after *USS Nautilus*, the first Soviet nuclear submarine *Leninsky Komsomol* was commissioned in June 1958 and became operational in 1959 powered with two water cooled reactors 1.4 kW each in addition to two diesel generators and two auxiliary electric motors. Around the same time, the first nuclear-powered civilian surface ship and icebreaker *Lenin*, was constructed and commissioned in 1959 powered with three PWR nuclear reactors.

3.3.4 Submarine programs: UK, France, China, India and Pakistan

The S5W (Submarine 5th Generation Westinghouse) nuclear reactor propulsion system was provided to the UK, under agreement, to power the UK's first submarine HMS Dreadnought commissioned in 1960. S5W provided 15,000 shaft horsepower (shp) equivalent to 11 MW for the submarine to achieve speeds of 20 knots (37 km/h) on surface and 28 knots (52 km/h) submerged. The S5W Westinghouse design enabled Rolls Royce to develop its first PWR in 1965 which powered the UK's second nuclear submarine.

Apart from the UK which had nuclear submarines together with the US in the 1950s, France, China and India had developed significant programs independently.

The first French nuclear submarine *Le Redoutable* (S611), commissioned in December 1971 was also powered by a PWR.

The first Chinese nuclear submarine was commissioned in 1974; at present four ballistic missile submarines (SSBNs) with two additional *Jin*-class SSBNs added in 2020, and six attack submarines (SSNs) form the nuclear part of the total submarine fleet estimated at 60 (NTI: https://www.nti.org/analysis/articles/china-submarine-capabilities/)

The first Indian nuclear submarine *INS Arihant* (S2 for Strategic Strike) SSBN was commissioned in August 2016 and is powered by an 83 MW(e) PWR with 40% enriched uranium. This was followed by the *INS Arighat* ballistic missile submarine (S3) which will be part of a fleet of four nuclear-powered submarines to be armed with ballistic missiles carrying nuclear warheads (Mian, Ramana, & Nayyar, 2019). In addition to these nuclear submarines, India has 15 diesel-electric attack submarines.

The submarine fleet of Pakistan has five diesel-electric attack submarines and three mini submarines; collaborative efforts with China are underway for the construction of eight Type 39 Yuan-Class attack submarines (Mian et al., 2019; Pakistan Navy, 2021) including two Agosta-70 commissioned in 1979–80 and three Agosta-90B, all diesel-electric vessels purchased from France and subsequently assembled and made entirely in Pakistan.

3.3.5 Modern-day submarines

Today, forty countries have attack submarines, but only have five nuclear-weapon states – China, France, Russia, the United Kingdom, United States – have nuclear-powered attack submarines while India has two SLBMs.

For naval submarines, they offer the added advantage of enhanced capability of stealth movement and quiet propulsion under water enabling them to undertake combat and intelligence-gathering activities. However, nuclear ships are expensive; for example, the USS Gerald R. Ford aircraft carrier, length 1,092–1,106 ft and height nearly 250 ft and displacement ~ 100,000 tons, with two nuclear reactors on board, would cost about $12.9 billion which is roughly the cost of four 1000 MW nuclear power reactors. The total naval reactors budget request in 2018 was in excess of US $1.47 billion in three heads namely naval reactors operations and infrastructure (US$ 467 million), naval reactors development (US$ 473 million), and S8G prototype refueling (US$ 190 M). The weapons carried on submarines include Tomahawk missiles, Trident missiles (SSBN), antiship cruise missiles, nuclear-tipped missiles, cruise missiles, submarine launched ballistic missiles (SLBMs), intercontinental ballistic missiles (ICBMs), and multiple independently targetable reentry vehicles (MIRVs).

Today, the US has 83 nuclear submarines including 10 aircraft carriers and 1 research vessel, with over 50 attack submarines (SSNs), 14 strategic ballistic missile submarines (SSBNs), 11 nuclear aircraft carriers (CVNs) including 10 Nimitz-class CVNs and 1 Ford-class CVN.

The ship service life is 30–33 years and reactors are mainly S5W, S6W, S9G (Submarine platform 9th Generation design by General Electric). The size of the subs varies from the smallest Virginia-class with a length of 377 ft, Beam 34 ft to the largest Ohio class with a length of 560 ft, beam 33 ft. The subs travel at speeds from 20 + knots for the large subs and from 25 + knots for the smaller subs.

Most marine reactors are PWRs while some are liquid metal cooled reactors; in both cases, the steam produced from the reactor heat is used either directly to power the propulsion system or indirectly by producing electricity from turbo-electric generators.

The former Soviet Union produced 246 nuclear submarines, each with two PWRs, with most decommissioned in the 1990s, leaving Russia with 58. Typical mission times for first generation submarines were 60 days with underwater (at depths of 200–300 m) speeds of 20–25 knots. At the height of the cold war in the early 1980s, both the US and the then USSR were actively pursuing their programs with the total estimated to be over 460 nuclear-powered submarines. Between 1955 and 1995, three generations and classes of nuclear submarines have been classified (Reistad & Olgaard, 2006) for attack, cruise missile and ballistic missile classes including 56, 102 and 78 submarines built with a total of 246. A few others were built between 1974 and 1996 bringing the total built to 264. Russia has continued with the development of nuclear-powered icebreakers, for maintaining the Northern Sea Route in the frozen Arctic coast of the North Pole region which is mostly free of ice only 2 months of the year and connects the Atlantic and Pacific Oceans. From the first icebreaker *Lenin*, several are operational with *5 let Pobedy* being the world's largest icebreaker powered by two OK-900A (171 MW(t) each) nuclear reactors. The design parameters of *Lenin's* first generation LEU OK-150 nuclear reactor and the HEU KLT-40 reactor are given in Table 3.10.

In today's scenario, nuclear ships including submarines, aircraft carriers and cruisers armed with hundred-kiloton or more nuclear warheads and ballistic missiles as well as stealthy intelligence-gathering mobile platforms are part of military strategy for global dominance. All of this has been made possible by fast moving nuclear-powered ships capable of maneuvering battle conditions to defend themselves as well as to operate in the attack mode. In future, this list is bound to expand, with Canada, Pakistan, Australia, Iran, Israel, and Brazil taking interest in the development of nuclear submarines.

3.3.6 Technical features

There are major differences in the requirements of land-based and water-based nuclear systems such as the availability of space, longer refueling period, power maneuverability, physical movement, vibrations, and corrosion. Space requirements necessitate compactness of nuclear systems which translates immediately into the need for higher fuel enrichment. Fortunately, the power requirements are much less than that for typical land-based 1000 MW power plants.

Some distinguishing features of submarine nuclear reactors (Ragheb, 2011) are

1. The use of HEU, as high as 93%–96% enriched weapon grade uranium, in PWRs.
2. The use of high burnup fuels such as U-Zr, U-Al and ceramics.
3. The requirement of "lifetime cores" which means no refueling for ~50 years.
4. The requirement of overcoming dead time from xenon 54135*Xe* (half life 9.2 hours) poisoning.
5. Internal radiation (neutron and gamma) shielding.
6. Operating temperatures ~600–700 $°F$ (maximum fuel temperatures) in earlier U-Zr fuel elements, increasing to ~1700$°F$ in UO_2 fuel clad in stainless steel with beryllium moderator and reflector to give an intermediate neutron spectrum and hence a higher thermodynamic efficiency.
7. Noise reduction from coolant pumps and speed reduction gearboxes.
8. Power levels typically 10–200 MW(e) compared with 300–1200 MW(e) for commercial plants.
9. Containment vessel design to withstand pressures (~13 atmospheres for earlier designs) that could be attained in accident conditions.

Some variations in operating parameters are also mentioned in the literature; these could reflect changes in operating conditions. Similarly, design improvements were made from OK-150 due to the LOCA; these included moving the inlet and outlet tubes of the reactor tank to the top of the tank to avoid operator error resulting in loss of coolant from drainage, as had occurred in the *Lenin* reactor.

From Table 3.10, the total number of fuel rods is 7,704 with 189 fuel elements containing 36 fuel rods each and 30 elements containing 30 rods each. The diameter of each rod was 6.1 mm. With the burnup of 1 kg equivalent to ~1000 MWd, OK-150 achieved ~20% full burnup. The control rods were inserted vertically, and flux flattening was achieved by boron-10 burnable poison in the shroud tubes of the central rods.

For the present KLT-40 design, similar to OK-900, the burnable poison used is natural gadolinium with fuel elements. Shielding for most reactors is steel-water layers on the sides (radially) and concrete on top.

In Russian submarine reactors, the first generation water cooled and water-moderated reactors had enrichments of ~20% power limited to 70 MW(t) with separated primary and secondary system. These plants all had U-Al alloy fuels

TABLE 3.10 Nuclear submarine reactors OK-150 and KLT-40.

	OK-150, first core load	KLT-40
General		
Time period	1950s	Present
Reactor power (MW(t))	90[a]	135[c]
Burnup (MWd)	18,000–20,000[b]	62,000–68,000
Pressure vessel diameter/height (m)	2/5	
Core		
Core height (m)	1.58	1.0
Core diameter (m)	1.0	1.21
Fuel		
U-enrichment (%)	5	90
Mass of U235 in core (kg)	85	150.7
Fuel material	UO_2	U-Zr-alloy
Number of fuel elements	219	241
Fuel element lattice pitch (mm)	64	72
Fuel element lattice type	Triangular	Triangular
Shroud, outer diameter (mm)	54	60
Shroud material	Zr-alloy?	Zr-alloy?
Number of fuel pins per element	36	53
Fuel pin diameter (mm)	6.1	5.8
Fuel pin lattice pitch (mm)	Not available	7.2
Cladding thickness (mm)	0.75	Not available
Cladding material	Zr-alloy or SS	Zr-alloy
Helium gas gap fuel and cladding (mm)	0.05	–
Fuel pellet diameter (mm)	4.5	Not given
Coolant		
Loops	2	–
Steam generator per loop	1	–
Coolant pump per loop	2	
Pressurizers (total number); in Russian, called *volume compensators*	4	–
Inlet temperature (°C)	248	278
Outlet temperature (°C)	325	318
Pressure (bar)	200	130
Coolant flow rate (m^3/h)	1000	
Steam		
Output (t/h)	360	
Pressure (bar)	29	40
Temperature (°C)	310	290

[a]*Modified to 159 MWth for OK-900 in 1967–70, then to 171 MWt for nuclear-powered icebreakers* Arktika, Sibir, Rossia, Sovetskiy, Soyus, *and* Yamal.
[b]*Burnup was increased to 29,000–38,000 MWd for OK-900 lin 1967–70, later increased to 88,000–96,000 MWtd.*
[c]*In Sevmorput; for icebreakers Taimyr and Vaigatch, 171 MWt.*
Source: Reistad, Nuclear Plants Russia, p. 18.

with enrichment varying from 5.45 to 21% with 30–50 kg U-235 in the core, except for K-140 which had a higher power (90 MW(t)), 21% enrichment and 116.3 kg U-235. With fuel densities of the order of 10 g/cm^3, the core height suggested is ~ 1 m. Improvements for second generation systems, such as the reorganization of reactor systems and reducing the number of reactors from two to one as well as the use of lighter material made the reactors more compact with more assemblies at ~20% enrichment or less assemblies with ~40% enrichment; shaft power and mission time were increased. Third generation systems requiring more power contain more fissile material than earlier designs. These reactors used heavy biological (steel-water-concrete) shielding amounting to ~50% of the plant weight. For control systems, europium rods, in the form of Eu_2O_3 were used.

3.3.7 HEU/LEU submarine reactors

The use of HEU makes it possible to meet the space constraints of submarines as well as to overcome, to a significant extent, the "dead time" caused by the buildup of xenon poison.

The present life of US submarine HEU cores is 33 years for the Virginia-class attack submarines being extended to 42 years for the cores of the Columbia-class ballistic missile submarines. All UK submarine reactors are HEU fueled PWRs based on earlier Westinghouse designs with typical power levels of 15,000 shaft horsepower (78 MWth/11 MW (e)) for the PWR-1 design, and 27,500 shaft horsepower (145 MW(t)/20.5 MW) for the PWR-2 design with improvements planned for PWR-3. (Lobner, 60 years of marine power https://www.lynceans.org/wp-content/uploads/2015/09/Part-4_UK-France-Others-60-yrs-of-marine-nuc-power.pdf)

Commercial power reactors use low enriched uranium (LEU) fuel; for example, LWRs typically use 3%–5% enriched fuel while fast reactors use 13%–19.5% enriched fuel. Enrichment below 20% U235 concentration in uranium, is called LEU while above 20%, called Highly Enriched Uranium (HEU) is prohibited for commercial nuclear reactors as well as research reactors which were earlier based mainly on HEU. Another classification for uranium enrichment is WG-HEU (Weapons Grade Highly Enriched Uranium) which is above 85% enrichment.

The role of PWRs for nuclear propulsion seems to be as large as that for commercial power; it may further strengthen SMR technologies and extend into higher power densities for micronuclear reactors (MNRs) making it possible to enable return journeys to Mars. However, as more countries aspire for nuclear submarines, the use of weapons grade uranium will increase as it is not prohibited by international law. At present HEU is used by submarines of the USA, Russia, and the United Kingdom (Philippe & von Hippel, 2016). The use of HEU by the US nuclear submarines is ~2.5 tons which is ~60% of global navel HEU and sufficient for manufacturing 100 nuclear weapons every year. This may be slow due to the long experience particularly of the US Navy, with over 60 years of operational experience, 30 different designs and a good safety record, the conversion of HEU cores to LEU cores would be inevitable just as in the case of research reactors. The technical difficulties that would need to be surmounted would be the decrease in power of LEU cores of similar size to the HEU cores; this would indeed require design changes in the space allocations for submarine reactors.

As the lifespan of older submarines has come to an end, several have been decommissioned and dismantled in elaborate and integrated programs. Out of ~465 nuclear submarines built in the world, over 300 have most likely been decommissioned and largely dismantled. It has been reported that in the dismantling effort, in most cases "20–45 percent uranium-235 has been used as fuel...," that all but one US submarine has one reactor while most Russian submarines had two reactors, that the "cores of these reactors typically hold between 200 and 300 fuel assemblies each containing up to a few tons of fuel rods," and that the thermal power varies from 10 MW in older submarines to 200 MW in newer classes of submarines (Kopte, 1997). There have been serious issues and concerns in the dismantling process due to the radiation levels of the spent fuel, and other liquid and solid radioactive waste, and the safety of procedures as well as the waste disposal procedures, and long-term ecological effects, in land sites or in deep oceans.

3.4 Plutonium production reactors

The term "production" for a reactor represents the production of plutonium; since plutonium does not occur in nature (except for in trace amounts in naturally occuring uranium ores), it must be produced. Plutonium is more fissile than uranium-235 (Chapter 1) resulting in smaller-sized weapons. In nuclear power reactors, plutonium is produced during normal operation from the transmutation of uranium-238. Typically, 5 kg of Pu-239 and 2 kg are produced ton of fuel in a 1000 MW(e) PWR over a period of about 3 years. It is separated from the spent fuel in a reactor by reprocessing methods. Some of its main isotopes are Pu-238, Pu-239 (fissile), Pu-240 (fertile), Pu-241 (fissile), and Pu-242.

The purpose of production reactors has been to produce weapons grade plutonium (Pu-239 with less than 8% Pu-240).

As discussed in the previous section, the NRX reactor with natural uranium and heavy water, and Magnox reactors with natural uranium metal, graphite, and CO_2 were production reactors. Similarly, breeder reactors, fast or thermal, have been utilized for breeding plutonium in a fast reactor such as the Liquid Metal Fast Breeder Reactor (LMFBR), the gas cooled fast reactor (GCFR) and the Lead Cooled Fast Reactor (LFR).

The present reactor grade plutonium (55%–70% Pu-239, 30%–35% nonfissile Pu, and more than ~19% Pu-240) production from power reactors in the world is about 70 tons annually. It is estimated that there are hundreds of tons of weapons grade plutonium in the world. This is clearly a tremendous radiation and proliferation hazard and a major area of concern for the public acceptance of the nuclear industry. Similarly, thermal breeders can be used with thorium 232 to breed the fissile fuel uranium 233.

The EBR-I built at Idaho (USA) became the first plant to produce electricity on the December 20, 1951 and to demonstrate in 1953 that a reactor could breed fissile fuel. The EBR-I reactor operated for 12 years but was shut down in 1963. EBR-I was followed by the SFR EBR-II which achieved criticality in 1965, produced 62.5 MWt/20 MWe with 65% enriched uranium fuel and demonstrated several safety systems and procedures for a HEU reactor. EBR-II was shut down after about 30 years operations in 1994. A similar breeder reactor Phoenix was built and commissioned in France in December 1973 and operated at 590 MW(t)/233 MW(e) demonstrating a breeder ratio 1.16. The plant was shut down in 2009. The reasons for closing down breeders were most probably technical difficulties such as their shutdown for refueling resulting in low capacity factors, and safety and proliferation concerns. Two power reactors, similar to breeders operating in Russia are BN-600 and BN-800 which are discussed in the next section.

In thermal reactors, plutonium is produced as a by-product from the radiative capture of uranium-238 in the fuel. It is separated by reprocessing spent fuel and is then fed again into reactors as a *mixed oxide fuel* (MOX). The amount of plutonium produced in a 1000 MW(e) LWR with ~25 tons spent fuel per fuel is ~290 kg, of which more than half is the fissile plutonium 239. This fraction is higher for CANDU and Magnox reactors; for the former with low burnup, it can be ~80%.

The early-day production reactors in the US have been Hanford and Savannah River estimated to have produced 103.4 tons of plutonium in the 50 years from 1944 to 1994. In Russia, according to IAEA (IAEA TECDOC-1591) estimates the spent fuel reprocessed till 2002 exceeded 300,000 metric tons. Today's stockpiles of plutonium are "sufficient for tens of thousands of nuclear weapons" https://vcdnp.org/plutonium-stockpiles-causes-and-solutions/ and efforts are underway in several countries to limit their production and seek sustainable solutions.

3.5 Small modular reactors

As described in Section 3.1, a SMR is defined by the IAEA as a NPP with a capacity in the range 100–700 MW(e) (Bari, 2014; IAEA, 2018).

As of December 2019 there were 97 small and medium operational reactors with gross capacity less than 700 MW out of 443 operational reactors worldwide (IAEA, 2020, p. 65). The rationale to develop SMRs includes specialized applications, such as powering of production facilities, desalination, hydrogen production, district heating, reaching off-grid remote locations, and providing incremental demands to large nuclear reactors.

The technology of SMRs is mostly well-known and derived from the earlier designs of the 1970s and 1980s. Out of the 45 SMR designs being pursued, most are in the conceptual or detailed design phase and none of the innovative SMR designs are commercially available. Some designs (e.g., NuScale 50 MW/module; mPower 125 MW; Westinghouse 200 MW; PRISM 311 MW; SMART 100 MW; 4S 10 MW; HTR-10 10 MW; HTR-PM 250 MW) are moving toward licensing. The USNRC issued a Standard Design Approval to NuScale in September 2020 https://www.nuscalepower.com/about-us/faq#ES1 making it possible to move ahead with the development of small plants such as the 77 MW(e) NuScale Power Module which can be integrated into a 12 module plant with an output of 924 MW(e).

3.5.1 Design features of SMRs

The design features of SMRs, some of which are listed for land-based reactors in Table 3.11 and for marine-based and other designs in Table 3.12, are based on innovative designs with scaling-down of large NPPs essentially integrating all primary components inside a RV resulting in a small size, in addition to passive safety systems. The water cooled SMRs are limited to outlet core temperatures of the 300°C range as compared to the high temperature reactors with enhanced efficiencies in Table 3.12.

The Westinghouse lead fast reactor (LFR), listed in Table 3.12 and shown in Fig. 3.8, with a power capacity of >450 MW(e), is classified as a medium-output modular, passively-safe plant uses MOX fuel. It is cooled by lead which has a high boiling point of 174°C. It has a fast spectrum with high burnup core and a high operating temperature. It has innovative design features that include the use of advanced materials to enable high temperature operation leading to good thermodynamic efficiency; of the order of 40%–50% at temperatures in the range 600°C–700°C.

The safety features of the WLFR include DC power capacity for over 72 hours and no events for which immediate operator action is needed. The integral design eliminates concerns of LOCAs. Lead has better performance than that observed with sodium. Reactor shutdown is achievable with automatically-actuated passive cooling systems, without the need for instrumentation and control signals or moving parts and, with lessons learned from Fukushima, passive heat removal of decay heat is possible by radiation transfer between the RV and the guard vessel (GV) as well as by the use of naturally–circulating air. Another advantage is the use of smaller size turbomachinery as no condenser is

TABLE 3.11 Overall parameters of some water cooled SMRs.

No.	SMR		Power MW t/ MWe	Design	En. %	Fuel	Core In /Out °C	P (MPa)	RPV H/D (m)	Burnup GWD/t	Life (y)	Design	Primary Circ.
1	CAREM	A	100/~30	Int. PWR	3.1	UO_2	284/326	12.25	11/3.2	24	40	UCP	NC/PSC
2	ACP100	C	385/125	-do-	<4.95	-do-	286/319.5	15	10/3.5	<52	60	BDF	FC
3	CAP200	C	660/>200		4.2	-do-	289/313	15.5	8.845/3.280	37	60	CDF	FC
4	DHR	C	400/none	Pool type	<5.0	-do-	68/98			30	60	BDF	FC
5	IRIS	I	1000/335	Int. PWR	4.95	-do-	92/330	15.5	21.3/6.2	65 (max)	60	BDF	FC
6	DMS	J	840/ 300	BWR	< 5	-do-	186/287	7.17	15/4.8	< 60	60	BDF	NC
7	IMR	J	1000/350	Int. PWR	4.8	-do-	329/345	15.51	17/6	> 40	60	CDF	NC
8	SMART	K	330/100	Int. PWR	<5	-do-	296/323	15	15.5/6.5	<60	60	LC	FC
9	ELENA	R	3.3/0.068	PWR	15.2	-do-	311/328	19.6	3.7/1.25	57600/273900	-	CDF	NC
10	KARAT-45	R	180/45–50	BWR	4.5	-do-	180/286	7	11.5/3.10	45.9	80	CDF	NC
11	KARAT-100	R	360/100	BWR	4	-do-	104/286	7	13.25/4	45.9	80	CDF	NC
12	RITM-200	R	175/50	Int. PWR	<20	-do-	277/313	15.7	8.5/3.3	-	60	UD	FC
13	RUTA-70	R	70/NA	Pool type	3	-do-	75/102	atm	17.25/3.2	25–30	60	CDF	FC
14	UNITHERM	R	30/6.6	PWR	19.75	-do-	249/330	16.5	9.8/2.9	1.15	25	CDF	NC
15	VK-300	R	750/250	SP BWR	4	-do-	190/285	6.9	13.1/4.535	41.4	60	DD	NC
16	UK SMR	UK	1276/443	PWR	<4.95	-do-	296/327	15.5	11.3/4.5	55–60	60	MC	FC
17	mPower	U	575/195	Int. PWR	<5	-do-	290.5/318.9	14.8	27.4/4.15	<40	60	UD	FC
18	NuScale	U	160/50	Int. PWR	<4.95	-do-	258/314	12.8	17.8/3.0	>30	60	URR	NC
19	SMR-160	U	525/160	PWR	4.95	-do-	209/321	15.5	15/3	45	80	PD	NC
20	WSMR	U	800/>225	Int. PWR	<5	-do-	294/324	15.5	28/3.7	>62	30	CDF	FC

A, Argentina; *C*, China, IIC-Iris International Consortium; *J*, Japan; *K*, Republic of Korea; *RF*, Russian Federation; *U*, America.
Design status: *UCP*, under construction as prototype; *BDF*, basic design finished; *LC*, license certified; *UD*, under development; *DD*, detailed design; *MC*, mature concept; *URR*, under regulatory review; *PD*, preliminary design in progress; Primary circulation: *NC PSC*, natural circulation, pressure suppression containment; *FC*, forced circulation.
Source: Advances in Small Modular Reactor Technology Developments, A Supplement to: IAEA Advanced Reactors Information System (ARIS), 2018 Edition.

TABLE 3.12 SMRSs: marine-based LWRs, HTGCR, Molten Salt, eVinci.

No.	SMR		Power MW Th/e	Design	En. % Fuel	Core In/ Out, °C	P (MPa)	RPV H/D (m)	B GWD/t	Life (y)	Design	Primary Circ.
1	ACPR50S	C	200/50	Loop type PWR	<5 UO_2	299.3/321.8	15.5	7.2/2.2	<52	40	CDF	FC
2	KLT-40S	R	150/35	PWR	18.6 -do-	280/316	12.7	4.8/2	45.4	40	UT	FC
3	RITM-200M	R	175/50	Int. PWR	<20 -do-	277/313	15.7	9.2/3.5	-	60	UD	FC
4	VBER-300	R	917/325	Int. PWR	4.95 -do-	292/328	16.3	9.3/3.9	50	60	LS	FC
Gas cooled												
1	HTR-PM	C	2x250/210	GCR	8.5 Sph.	250/750	7	25/5.7 (inner)	90	40	UC	FC
2	GTHTR300	J	<600/100∼300	Prismatic HTGR	14 UO2 TRISO	587−633/ 850−950	7	23/8	120	60	BDF	FC
3	GT-MHR	R	600/288	Mod. He	LEU or WPu Coated particle fuel	490/850	7.2	29/8.2	100−720	60	PDF	FC
4	PBMR-400	S	400/165	high temperature GCR	9.6% Pebble bed with coated particle fuel	500/900	9	30/6.2	92	40	PD	FC
Liquid Metal Cooled												
1	4S Toshiba	J	30/10	(pool type)	< 20 (U-Zr alloy)	355/510	atm	24/3.5	34	60	DD	FC
2	BREST-OD-300	R	700/300		∼13.5 U-Pu /N	420/535	Low pressure	17.5/26	61.45	30	DDS	FC
3	LFR-TL-X	N	15/5; 30/10; 60/20	(pool type)	19.75 LEU, cylindrical cassette	360/420	atm	3.5/2	40	30	CD	FC
4	WLFR[a]	U	950/>450	Pool type	≤19.75 Oxide, with provision for transition to UN	420/600	Nearly atm	8.0/7.5	≥100	60−100	CD	FC
Molten Salt												
1	IMSR[b]	Ca	400/190	MSR	<5 Molten salt fuel	625−660/ 670−700	< 0.4	7.0/3.6	-	60	CDF	FC
2	CMSR[c]	D	250/100−115	MSR	(U 1.1%, Pu 69% fissile) Sodium-actinide fluoride	600/700 or 700/900	1	2.5/2.1	250	60	CD	FC
3	CAWB[d]	D	50/20	MSR	_ LiF-ThF4	_	_	_	_	3−5	CD	FC
4	ThorCon[e]	I	557/250	Th. MSR	19.7 LiF-ThF4	565/704	1.2	12/8	220	80	CBD	FC
1	eVinci	U	0.6−4−40/0.2−15	Monolithic core HP	19.5 UO2 or UN	NA	NA	NA	<10	10	TRL 5	HP

UT, under test; *LS*, licensing stage; *PDF*, preliminary design completed; *S*, South Africa; *DDS*, detailed design startup in early 2023; *P*, heat pipe; *D*, Denmark; *Ca*, Canada.
[a]*Westinghouse Lead Fast Reactor.*
[b]*Integral Salt Molten Reactor, Canada.*
[c]*Compact Molten Salt Reactor, Seaborg Technologies, Canada.*
[d]*Copenhagen Atomics Waste Burner, Denmark.*
[e]*ThorCon International, Indonesia, 557 mWth/250 MWe (per module).*
Source: Advances in Small Modular Reactor Technology Developments, A Supplement to: IAEA Advanced Reactors Information System (ARIS), 2018 Edition.

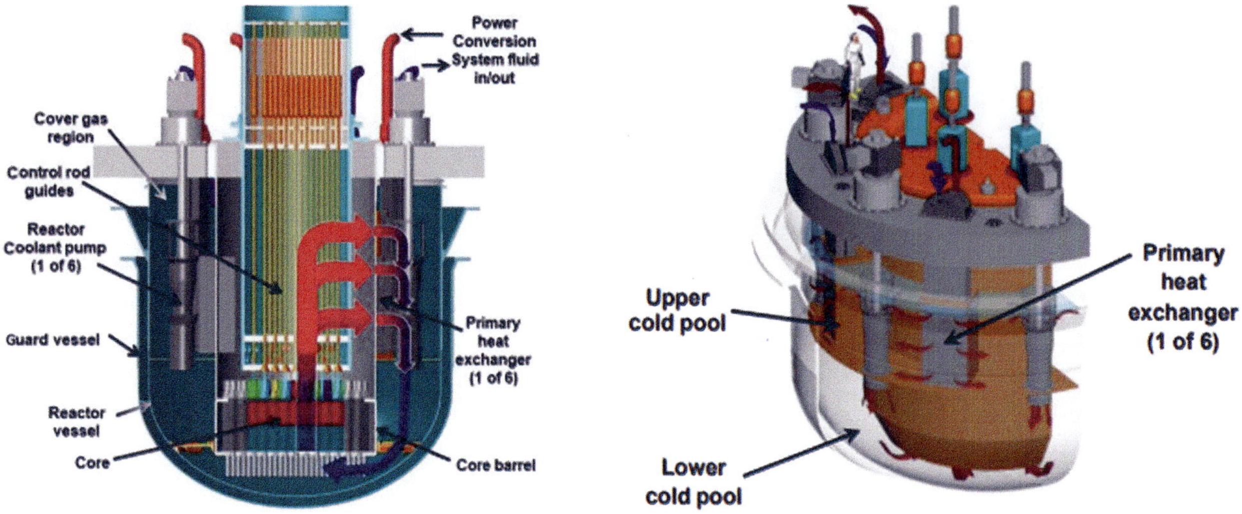

FIGURE 3.8 Westinghouse liquid cooled fast reactor (WLFR). *Courtesy of Westinghouse Electric Company (c) 2021. All rights reserved.*

TABLE 3.13 Physical design parameters of LFR and Evinci Westinghouse designs.

Core	WLFR[a]	Evinci[b]
Reactor type	Lead cooled fast reactor	Monolith core with heat pipes
Power (MW(th)/MW (e))	950/>460	0.6–4–40/0.2–15
RPV height/diameter (m)	Approx. 8.0/7.5	N/A
In-house plant consumption	~15 MWe	N/A
Fuel	Oxide (UO_2 or MOX)	UO_2 or UN
Coolant	Pb	Heat pipes/metal hydride moderator
Core discharge burnup	≥100 MWd/kgHM	<10 GWd/ton for a 1.2 MWth reactor
Fuel material	UO2, MOX then UN	UO2, UN
Fuel enrichment	≤19.75%	19.5
Fuel cycle (months)	24	120
Main reactivity control	Rods	Ex-core control drums
Approach to engineering safety	Passive, inherent IAEA passive safety category B goal	Passive, inherent IAEA passive safety category B goal
Design life (years)	60 for components, 100 for structures	10
Plant footprint (m^2)	N/A	<1500
Distinguishing features	Low pressure reactor vessel; high BP coolant, no moving parts or actuation for decay heat removal; high efficiency	Transportable reactor (vSMR) operating semiautonomously, no moving parts or actuation for decay heat removal; high efficiency
Design status	Conceptual design	TRL-5 (Technology Readiness Level 5 with USNRC and CNSC) Canadian Nuclear Safety Commission

[a]*https://www.westinghousenuclear.com/new-plants/lead-cooled-fast-reactor.*
[b]*https://www.westinghousenuclear.com/new-plants/evinci-micro-reactor.*

required with supercritical CO_2 in the power conversion system. The Westinghouse LFR is anticipated to start construction of full-scale prototype by ~2030 and convert to a first of a kind (FOAK) commercial unit by about 2035.

Another new very small, portable and advanced Westinghouse design *eVinci* (Table 3.13) is modular, transportable, factory-built, fueled and assembled configuration with installation time of less than 30 days. With minimal moving parts and high reliability, it has the advantages of a small footprint and a near-zero Emergency Planning Zone (EPZ). Some overall design parameters of the WLFR and eVinci designs are compared in Table 3.13.

A step forward for eVinci was its selection, on the March 10, 2021 by the Department of Energy (US DOE) with a DOE funding of ~US$ 12.9 million out of a total value of ~US$ 28.6 million, to "prepare for the Nuclear Demonstration Unit (NDU) through design, analysis, testing and licensing to manufacture, site and test the NDU by 2022" Source: SMR-Book, IAEA (Status of eVinci on March 10, 2021, https://www.energy.gov/ne/articles/us-department-energy-further-advances-nuclear-energy-technology-through-industry-awards).

The SMRs at various stages are: one mature concept, six basic design, four preliminary design, 24 conceptual design, five under development, four detailed design, one final design, one certified design, one licensing stage, three under construction (CAREM, KLT-40S, HTR-PM). Both NuScale and eVinci are set to progress for the commercial markets.

3.5.2 Very small modular reactor

At the low power level, a very small modular reactor (vSMR) such as the Toshiba 4S (Super Safe Small and Simple) 10 MW(e) reactor (Table 3.13) with active core height: 2.5 m, and diameter 1.16 m has a small core size with. 1.69 t fissile U-235 inventory, out of 9.23 t U, has high burnup (average 34 GWd/t) and low source term for radiation release in case of an accident.

Both the 4S designs (30 and 50 MW) are small and address Gen-IV requirements including nonproliferation, essentially due to the absence of blanket breeders typically incorporated in fast reactors, and passive safety in LEU reactor systems. One of the most attractive features of the 4S reactor is that refueling is not required in the entire 30-year operation.

Key design and technology issues of SMRs are based on their inherent features which include:

- Small factory-built systems leading to reduction in on-site work
- A single (integrated) pressurized vessel containing steam generator so that a large break LOCA is eliminated from DBAs
- A small core will have less risk due to less accident radioactive release
- Passive safety systems, for example, natural gravity-driven circulation in LWR SMRs

The above factors lead to reductions in the EPZ from 5 to 25 km, suggested by IAEA for 100–1000 MWth reactors, to possibly 1000 ft as listed by Babcock and Wilcox for its mPower reactor. SMRs can thus be shifted closer to the location requiring power in contrast to present-day NPPs. Thus resilient and isolated micro grids of high reliability can be set up.

As mentioned in Section 3.4, one issue of concern for SMRs is the amount of plutonium production which over the 30 year operation period of 4S could be as high as 11 kg after 2 years, 82 kg after 15 years and 159 kg after 30 years. Unfortunately, this is a negative aspect of a SMR of the 4S type which means that proliferation resistance is an issue that would require careful security standards and practices.

3.5.3 Generation-IV reactors

With increased global competitiveness from conventional fossil-based and renewable energy systems, the nuclear industry finds itself in a position in which it has to offer compact, reliable, flexible and safe systems addressing additional public concerns on the long-term disposal of nuclear waste and proliferation of sensitive nuclear material. A nuclear renaissance depends on the ability of the nuclear industry to meet the challenges of free markets. As discussed in Section 3.1.3, the present developments indicate growth in Asia and a down-swing in Europe and the United States.

To address these challenges, new reactor designs have been considered as *Generation-IV* reactors for the coming years. These are "new" in the sense of having improved designs and technologies on old systems.

Essentially, the capability of nuclear reactors to respond to natural and operational events, potentially leading to accident conditions, has been increased. The lessons learned from over 60 years of nuclear power generation with three accidents (TMI, Chernobyl, Fukushima) have contributed to design changes. These have led to enhanced performance objectives with improved safety, passive and self-actuated shutdown systems, better decay heat remove, and severe accident prevention.

The overall objectives translate into a shift from water systems to nonwater systems with higher operational temperatures and higher power densities.

The Generation-IV International Forum (GIF), created in January 2000 to study potential systems, consists of fourteen partners (Argentina, Australia, Canada, People's Republic of China, Euratom, France, Japan, Republic of Korea, Russian Federation, Republic of South Africa, Switzerland, United Kingdom, the United States and Brazil as a nonactive member). This Forum has selected six reactor technologies "to support the next generation of innovative nuclear systems." The systems are

1. Gas Cooled Fast Reactor (GCFR)
2. Lead Cooled Fast Reactor (LFR)
3. Molten Salt Reactor (MSR)
4. Supercritical Water Cooled Reactor (SCWR)
5. Sodium Cooled Fast Reactor (SFR)
6. Very High Temperature Reactor (VHTR)

These systems, as stated above, were selected on the basis of their enhanced safety, improved economics and new products such as hydrogen for transport applications, reduced nuclear waste and increased proliferation resistance. While Gen-III + systems were evolutionary, Gen-IV systems are revolutionary in terms of their energy conversion systems as well as the nuclear fuel cycle technologies.

Some highlights from the *Technology Roadmap Update for Generation-IV Nuclear Energy Systems*, (Nuclear Energy Agency [NEA], 2014), are:

1. the importance given to fast reactors is mainly due to their breeding capability reducing the magnitude of waste disposal,
2. the development of advanced separation technologies will avoid separation of sensitive fissile material for any undesirable diversion,
3. significant resources dedicated to the development of the SFR and VHTR, are due to the considerable historical effort associated with these technologies,
4. materials testing and corrosion, and lead chemistry, core instrumentation, fuel development (MOX for first core, then minor actinide MA-bearing fuels); and possibly nitride fuel for lead cooled reactors will be carried out with the Russian reactor BREST-OD-300 as the reference design, and that,
5. efforts will be dedicated toward improving safety and operation with advanced instrumentation and control systems, prevention and mitigation of sodium fires, and the overall prevention and mitigation of severe accidents with potential of large energy releases.

For the 2400 MW(t) Gas Cooled Fast Reactor, the 75 MW(t) two-loop He/water ALLEGRO experimental demonstrator in Czech Republic is a test-bed to demonstrate the viability of the thermal tolerance of (U-Pu)C fuel, the functionality of the He/gas heat exchangers and of safety systems. Based on utilizing the advantages of both fast and high temperature reactors, ALLEGRO also tests the utilization of process gas for hydrogen production.

Similarly, the Pb-Bi-cooled Fast Reactor (LFR) SVBR-100 test-bed BREST-300 in Russia will carry out tests and study functionality of components.

The MSR is an old concept from the 1950s to 1970s; at that time they were mainly thermal-neutron-spectrum graphite-moderated reactors but since 2005 the liquid-fueled MSR research and development efforts have focused on fast spectrum designs.

For the SCWR two baseline concepts are the pressure-vessel-based and pressure-tube-based LWR designs operating at high temperature and high pressure above the thermodynamic critical point of water (374°C, 22.1 MPa). The neutron spectrum depends on the core design.

For the SFR, there is an extensive body of knowledge operational reactors operational, for example, in China, India, Japan and Russia. The SFR uses liquid sodium as the reactor coolant, at low pressure and high outlet temperature 500–550°C. Much of the basic technology for the SFR has been established in former fast reactor programs and was further confirmed by the EBR-I and EBR-II in the US, from *Phénix* in France, the BN-600 reactor in Russia and the *Monju* fast reactor in Japan.

The VHTR is a graphite-moderated, helium-cooled thermal reactor with high core exit temperatures, in the range 700°C–950°C. In the Japanese HTTR, the reactor core is a prismatic-block type such as the Japanese HTTR, while the Chinese HTR-10 is of a pebble bed type. For these reactors, TRISO fuel is used at high operating temperature (up to 1250°C) giving high burnups (up to 200 GWd/tHM).

Generally SMRs and Gen-IV reactors can be classified based on their core exit temperature; a temperature of ~300°C is a water cooled SMR, ~500–600°C is a liquid cooled fast reactor, ~600–700°C is a MSR, and greater than ~700°C is a gas cooled fast reactor. As a rule of thumb, the higher the exit temperature, the higher will be the thermal efficiency of the reactor power plant.

3.5.4 Radiation source term

Although the nuclear industry has generally had a good record in terms of radiation release to the environment, with the exceptions of TMI, Chernobyl and Fukushima, the dose in the vicinity of a reactor is a public concern. A person living within 80 km from a NPP would have an annual dose of about 400 mrem (4 mSv), out of which half would be from radon.

In case of an accident at a plant with a release of radiation in the environment, the "source term" estimates for a power plant would be mainly strontium, iodine and xenon. For a 300 MW(e), the inventory of fission products is given in Table 3.14.

Source: CHASNUPP (p. 67 Zia Mian).

In the case of a core melt, for rupture of cladding, the assumed release fraction from fuel is generally assumed to be 1% for xenon, krypton, cesium and iodine. Gen-IV reactors such as the Lead Cooled Fast Reactor lead will have better shielding to retain iodine and cesium at temperatures up to 600°C, thereby reducing the source term in case of release of volatile fission products from the fuel.

TABLE 3.14 Source term for a 300 MW(e) PWR.

Fission product	Inventory (Ci/MW(e))
Kr-85	560
Kr-85m	24,000
Kr-87	47,000
Kr-88	68,000
Sr-89	94,000
Sr-90	3700
Sr-91	110,000
Y-91	120,000
Mo-99	160,000
Ru-103	110,000
Ru-106	25,000
Te-129m	5300
Te-131m	13,000
Te-132	120,000
Sb-127	6100
Sb-129	33,000
1−131	85,000
1−132	120,000
1−133	170,000
1−134	190,000
1−135	150,000
Xe-131m	1000
Xe-133	170,000
Xe-133m	6000
Xe-135	34,000
Xe-138	170,000
Cs-134	7500
Cs-136	3000
Cs-137	4700
Ba-140	160,000
La-140	160,000
Ce-144	85,000
Np-239	1.64 × 106

Source: CHASNUPP (p. 67 Zia Mian).

3.6 Nuclear fusion

Our planet earth revolves around the sun, at the center of our solar system, which is our source of light, heat and gravitation. The sun is a massive star by our standards, but otherwise of moderate size in comparison with other stars, about 50 times larger than the earth consisting of over 90% (atomic) hydrogen and about 9% helium, with a core temperature understood to be over 15 million °C and a surface temperature of ~5500°C. At its core, it undergoes nuclear fusion reactions where hydrogen is converted into helium. Clearly, when all the hydrogen has been consumed in such reactions, the life of the sun will have come to an end and it would no longer be capable of sustaining its role in the solar system.

The big question for humans is whether nuclear fusion, of the same type happening in the sun and stars, can be replicated on earth in our scientific laboratories. Can such high temperatures be maintained; do we have materials strong enough to withstand such excessive heat loads, and can this energy be harnessed to produce electricity? If all these questions can be addressed, then we have a virtually inexhaustible source of energy since hydrogen and its isotopes occur in the vast oceans of earth.

First, we need to understand how the sun undergoes nuclear fusion reactions. Before it became a star, the sun was a collection of hydrogen and helium gases which gained density and formed gravity. The gravity pulled nuclei into a core and became heavy enough to form into a star. As the nuclei came close together, they fused and reactions were initiated, that continue today.

On earth, we first need the fusion fuels: hydrogen and its isotopes deuterium and tritium. Hydrogen is available in (light) water in the form of H_2O while deuterium is available in the ratio 1:8000 in sea water. Tritium does not occur in nature and has to be produced from nuclear reactions such as the bombardment of lithium by neutrons.

3.6.1 The fusion reaction

From the mass defect and binding energy it is evident that energy can be produced from nuclear reactions in two cases (Section 1.3); for heavy nuclei breaking into smaller nuclei, called *nuclear fission*, and for light nuclei combining or fusing into heavier nuclei, called *nuclear fusion*.

Fig. 3.9 illustrates the nuclear fusion reaction of two isotopes of hydrogen namely deuterium and tritium. The fusion reaction results in the production of alpha particles (helium) and a neutron

$${}_1^2H + {}_1^3H \rightarrow {}_2^4He + {}_0^1n \quad (17.6\text{MeV})$$

The energy released in this reaction, 17.6 MeV; the alpha particle carries 3.6 MeV which is deposited "locally" due to the relatively short range of an alpha particle in matter (Section 1.1) while the neutron carries 14 MeV which can escape from the system as neutrons have a relatively larger range (Section 2.1). The physics of nuclear fusion is reviewed in Annex.

The figure of merit is the *Lawson* criterion, which for DT fusion is

$$n\tau_E \geq 1.5 \times 10^{20} \quad \text{m}^{-3}\text{s}$$

The goal of achieving nuclear fusion by plasma confinement is to achieve breakeven by meeting this criterion.

Of the three confinement schemes namely gravitational confinement, magnetic confinement and inertial confinement, the latter two have been investigated in scientific laboratories. Gravitational confinement is an appropriate description of nuclear fusion that takes place in the sun where very large confinement times are evident. Here, on earth it is only possible to try magnetic confinement fusion (MCF) and inertial confinement fusion (ICF) for which, so far, it has not been possible to achieve even the short confinement times which would enable nuclear fusion reactors to operate in the way that nuclear fission reactors can for years.

To further appreciate the technological challenges of advanced nuclear systems, and before that, the modeling and simulation efforts necessary, both MCF and ICF are briefly reviewed.

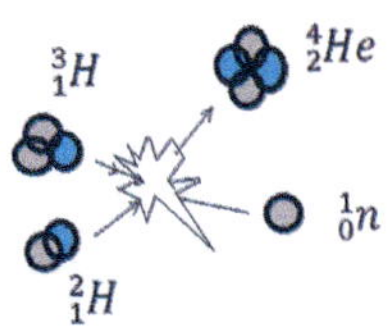

FIGURE 3.9 DT nuclear fusion reaction.

3.6.2 Magnetic confinement fusion

When a plasma, that is, free ions and electrons, is in a magnetic field, the charged particles experience a Lorentz force, which causes them to gyrate about a magnetic field line. The force is proportional to the magnetic field, and the radius of gyration is directly proportional to the perpendicular velocity (perpendicular to the magnetic field) and inversely proportional to the applied field. A plasma could be pinched by the application of a strong electric field too. Thus a strong field leads to a gyration of very small radius and a charged particle is confined about the field. Both electrons and ions move in circular orbits about a field line. Due to their longitudinal speed they drift away from their orbit. One of the first attempts of magnetic confinement were based on the pinch and mirror concepts; thus in a magnetic mirror, in which both ends had strong field lines would bounce the plasma back and forth between the two ends. The concept of mirrors was one of the earliest for confinement schemes from the 1960s; the Mirror Test Facility at Lawrence Livermore Laboratory was designed in the 1970s but research work by then had shifted to several other devices such as the stellerator and the Soviet Tokamak design. By the 1980s, the Joint European Torus (JET) at Culham, England, and the Tokamak Fusion Test Reactor at Princeton as well as other tokamaks around the world were operated to demonstrate scientific feasibility.

Currently the Tokamak project is the leading effort with work underway on the International Thermonuclear Experimental Reactor (ITER), illustrated in Fig. 3.10, an international project with 35 states, to demonstrate scientific and technological feasibility of fusion.

The basic principle of plasma confinement is the use of magnetic coils placed around a vacuum vessel in the shape of a toroidal chamber. These produce toroidal and poloidal fields which close upon themselves. In the vacuum of the chamber, the plasma is injected in the form of gaseous fuel. With the induced current, the gas is ionized into the plasma state. The ions and electrons are confined along magnetic field lines and energized to fusion temperatures of up to 300 million °C.

The first plasma is scheduled to be introduced in ITER when its construction phase will be over by December 2025.

$$36Li + 01n \rightarrow 24He + 13T + 4.8MeV$$

$$37Li + 01n \rightarrow 24He + 13T + 01n - 2.46MeV$$

A schematic of a nuclear fusion power reactor is shown in Fig. 3.11; the toroidal chamber where the hot plasma undergoes fusion temperatures at few hundreds of million °C is the source of thermal energy. The core is surrounded by a lithium blanket where tritium breeding is achieved by the 14 MeV neutrons produced from the DT reaction. The

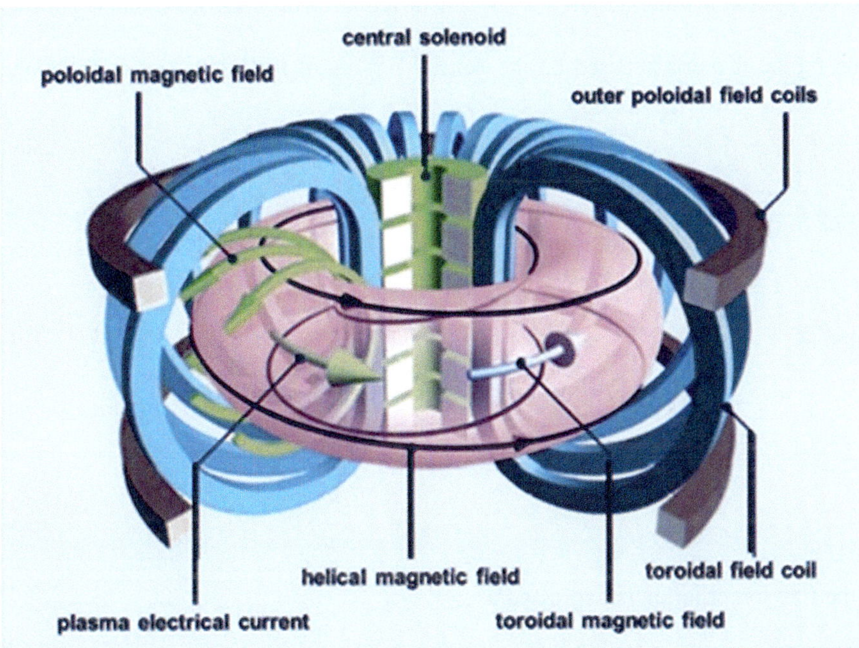

FIGURE 3.10 The ITER Tokamak for magnetic fusion. *Courtesy ITER @iter.org https://www.iter.org/newsline/-/3037.*

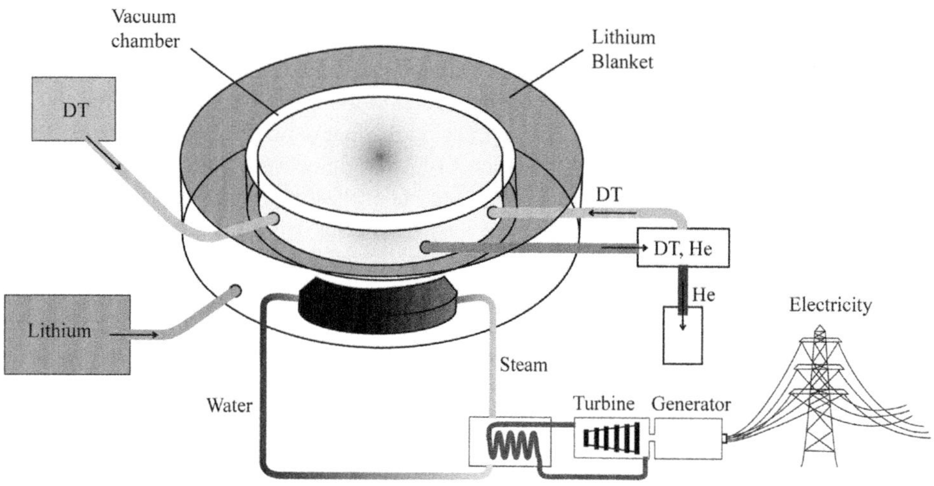

FIGURE 3.11 Schematic of a nuclear fusion power plant.

thermal energy is carried out of the reactor into a conventional power conversion system very similar to that described in previous sections for nuclear fission reactors.

The fusion reactor, though a power plant, has a virtually inexhaustible fuel supply which comes from water in the form of deuterium; tritium is bred within the reactor. There is no fissile fuel or radiation hazard with the exception of some neutron and gamma radiation, but no safety, waste disposal, or proliferation issues. It is thus the ultimate energy source which would most probably not find any competitor.

3.6.3 Inertial confinement fusion

Fusion can also take place when a deuterium and tritium pellet is heated to very high temperatures, as illustrated in Fig. 3.12 for direct- and indirect-drive mechanisms, and held together merely by its inertia for sufficient time to achieve Lawson's criterion $n\tau_E \geq 1.5 \times 10^{20}\ \ \mathrm{m}^{-3}\mathrm{s}$. In this approach, called *inertial confinement fusion*, the high temperatures are obtained by intense heating of a tiny pellet to burn its surface, causing ablation, or evaporation, of a thin layer which expands and, by newton's third law of motion, causes an inward moving shock wave. Thus the outwards explosion results in an inwards implosion which, if carefully shaped to be symmetric, results in the compression of a spherical target and subsequent heating and confinement to enable fusion.

The dynamics of ICF, in the context of external heating, requires a slightly different form of the Lawson criterion to model the energy and its transport.

Since ICF involves compression, the figure of merit is based on the density $\rho = mn$ of the imploding material and the speed of the imploding material

$$v = \sqrt{\frac{kT}{m_i}}$$

The time scale would then be, in terms of the pellet radius R

$$\tau_E \sim \frac{R}{v}$$

and the Lawson criterion could be written as

$$\rho R \sim C\frac{T^{3/2}}{\langle \sigma v \rangle}$$

where

$$C \equiv \frac{12\ \ k^{3/2} m_i^{\ 1/2}}{E_c}$$

With the appropriate "best values" the Lawson criterion is expressed as

$$\rho r > 3\mathrm{g/cm}^2$$

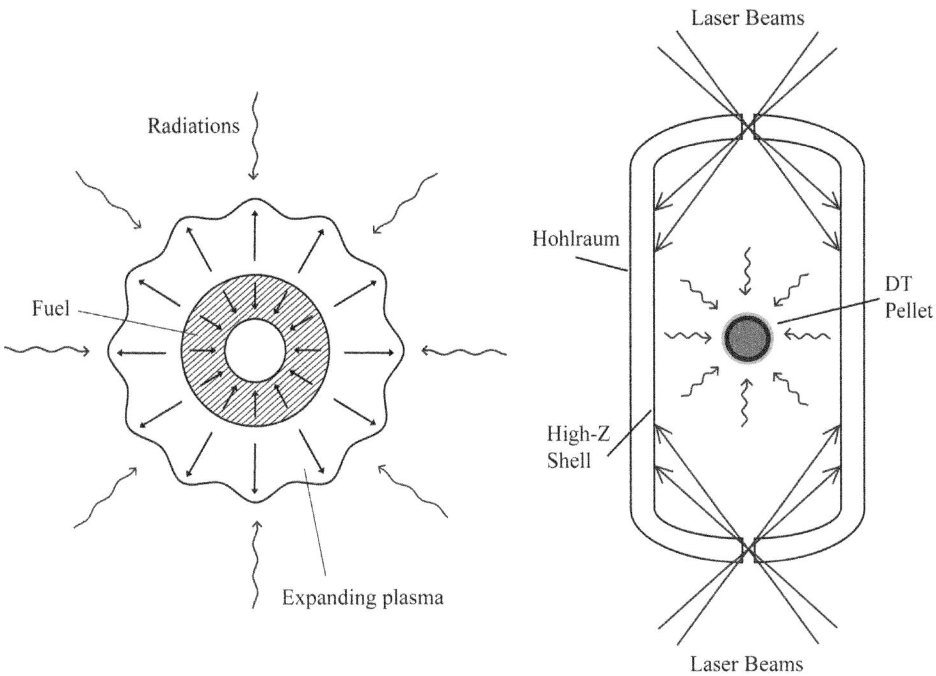

FIGURE 3.12 Laser compression of a pellet by direct- and indirect-drive.

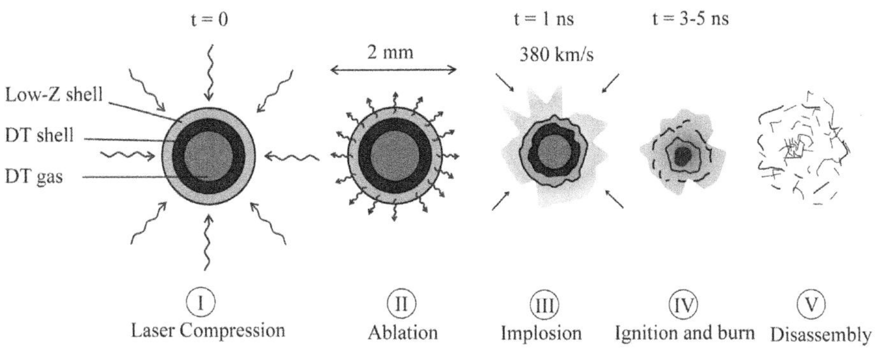

FIGURE 3.13 Stages of implosion in a direct-drive laser ICF.

The confinement properties depend on the areal density of the compressed core, $\rho R = \int_0^R \rho(r)dr$

$$\rho R\left(\frac{T}{4.7}\right)^{2.2} > 1$$

Research into ICF has been going on for decades with significant investments in facilities for delivering large amounts of laser energy ($\sim$MJ) into tiny DT pellets ($\sim$1 mm diameter) over very short periods of time driving imploding DT shells up to over 360 km/h compressing it to initiate ignition in a "hot spot" at the center as illustrated in Fig. 3.13.

The idea is to hold this hot spot for sufficient time to cause a thermonuclear burn wave in its vicinity which is surrounded with cold and dense material and still collapsing. As the hot plasm burns, its pressure and hence its expansive force begins to increase; it competes with the imploding surrounding until the pressure is so much that the entire assembly explodes; whatever fusion energy is produced is that between the ignition of the hot spot and disassembly. These types of experiments comprise direct-drive ignition and have the best chance for succeeding for spherical pellets. There are several other "requirements" to ensure success, such as slow isentropic compression and perfect implosion symmetry.

The gain G is expressed as

$$G = \frac{E_{\text{out}}}{E_{\text{in}}}$$

where fusion output E_{out} is achieved typically from a high-powered laser providing the input energy E_{in}. The gain value of the JET has been $\sim$0.67 while ITER aims for a value of 10 producing 500 MW(e) for an input of 50 MW(e).

Research efforts in the US (Betti & Hurricane, 2016), Europe (Tikhonchuk, 2020), Japan (Horioka, 2018), China (He, 2016; He & Zhang, 2007) and several other countries, are focused on achieving larger values for the Lawson criterion.

The best-performing indirect-drive implosions with a 1.9 MJ laser have "accelerated the DT mass to about 360–380 km/s, reaching a fuel kinetic energy of about 12 kJ and producing about 26 kJ of fusion energy" (Betti & Hurricane, 2016).

Some of today's challenges in ICF are the development of advanced schemes to reduce losses and mitigate instabilities, and improve implosion technologies.

3.7 Space propulsion

Space travel must have fascinated humans from times immemorial; history records the proposal for rockets over 100 years ago by the Russian theoretician Konstantin Ziolkovsky. The then Soviet work, finding its way to Europe, culminated in Germany in the 1930s. There, under the leadership of Werner von Braun, the German rocket program produced the first rocket to fly into space, the V-2. Since the 1960s spacecraft with nuclear heaters and power systems have been operating

Space propulsion can be achieved by a number of routes including chemical combustion, nuclear thermal energy to a propellant and direct nuclear fission and nuclear fusion energy. Liquid and solid propellants based on combustion are most common for missile and space systems from the first ballistic missiles to NASA's 2020 Mission to Mars. Before the present mission, the most notable has been the Apollo landing of three US astronauts on the Moon in 1969. These were followed by the Viking, Pioneer, Voyager, Galileo, Ulysses, and Cassini missions. The first use of Pu-238 as a source for radioisotope power was for Pioneer 10.

3.7.1 Conventional rocket designs

Ballistic missiles and rockets today are based on the German V-2 *Vergeltungswaffe*-2 rocket, technical name A-4 shown in Fig. 3.14 designed by the group led by Wehrner von Braun and used in large quantities in the second world war. On its first test flight on October 3, 1942, the V-2 flew 80 km high, and as such, was the first rocket ever to reach such a height. The V-2 was a short-range ballistic, surface-to-surface, missile about 300 km, liquid-propelled missile that carried a 1000 kg chemical warhead and was used in Europe against Allied targets. It was fired at an angle to reach its target as control systems were not very advanced in those days. Still, it had gyroscopes to determine angles and an accelerometer for engine cutoff. It could reach altitudes of about 80 km during its powered flight; if fired vertically it could go as high as about 200 km and achieve a maximum speed of 2880 km/h compared with 1100 km and 24,100 km/h for the most advanced long-range intercontinental ballistic missiles (ICBM), *Minuteman* III, developed in 1964–70 for the US land-based nuclear strategic force. The submarine launched ballistic missile (SLBM) *Trident* D-5, with two other SLBMs *Polaris* (IRBM) and *Poseidon* (IRBM) is also an intercontinental submarine launched ballistic missile (SLBM), deployed in 1990 with a unit cost of US$ 29.1 million, launched from SSBN submarines of the types discussed in the previous section.

The overall design parameters of the V-2 rocket, up-scaled to the present Minuteman III and Trident D5 missile designs is shown in Table 3.15. With the same basic design concept, there are differences in the size, the propellant and casing materials, and in the thrust. Minuteman III has three rocket motors of length and diameter 18.6/5.5, 9.1/4.3 and 5.4/4.3 ft with thrusts 200,400, 60,700, and 34,500 lb_f, respectively.

The Trident II D5, first deployed in 1990 is part of missile inventory on the US Ohio class and British Vanguard class submarines also has a three-stage rocket motor system.

3.7.2 Space exploration

The V2 design was based on ethanol and liquid oxygen with ethanol having a higher heat of vaporization ($\sim$846 kJ/kg) than kerosene ($\sim$251 kJ/kg) which means that ethanol would take longer to vaporize keeping the engine from melting. These technologies were improved to design better performing systems.

From missiles to rockets, the thrust and size changes, for example, The Saturn V rocket that carried Apollo 11, the first human mission to the Moon, generated 34.5 million Newtons (about 7.7 million lbf) thrust at launch https://solarsystem.nasa.gov/news/382/10-things-rockets-we-love-saturn-v/ which is about twenty times the thrust of Minuteman III.

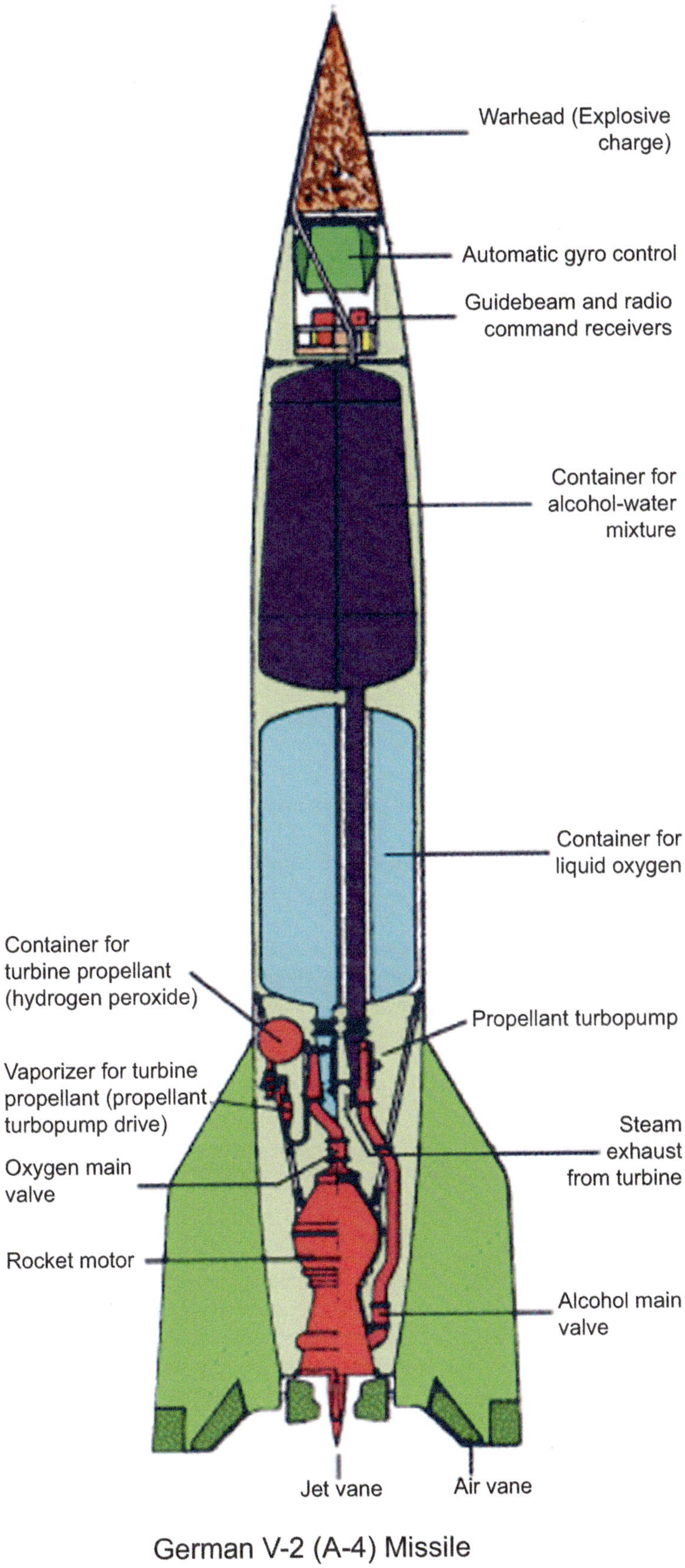

FIGURE 3.14 The V-2 rocket. *Courtesy NASA grc.nasa.gov 2021.*

TABLE 3.15 Physical design of V-2, Minuteman III and Trident D5 missiles.

Design parameter	V-2	Minuteman III	Trident D5
Classification	SRBM	ICBM	ICBM (SLBM)
Range (km)	320	13,000	12,000
Length (m)	14	18.2	13.42
Diameter (m)	1.65	1.85	2.11
Fuel	75% ethanol, 25% water	Three-stage Solid	Three-stage solid
Oxidizer	Liquid oxygen	Combined in solid	Combined in solid
Warhead	chemical	thermonuclear	Thermonuclear MIRV
Launch weight (kg)	~13,000	34,467	59.090
Specific impulse (s)	203–239	Varies by stage	
Thrust (kN)	264.90	935	
Control system	Gyroscope/accelerometer	INS/IGS	INS, SRS

Source: V-2: http://www.worldwar2facts.org/v2-rocket-facts.html http://www.astronautix.com/v/v-2.html ICBM: https://missilethreat.csis.org/country/united-states/.

The rocket, launched from Cape Kennedy on July 16, 1969 carrying three astronauts, Neil Armstrong, Michael Collins, and Edwin Aldrin, landed on the Moon from where on July 20, 1969, Armstrong voice was heard with the historical sentence "... on small step for a man, one giant leap for mankind." The specific impulse of Saturn V with Liquid Oxygen (LOX)/Rocket Propellant (RP) powered with Rocketdyne engines was 265 seconds for the first stage and 424 seconds for the second and third stages with LOX/LH_2.

Fifty years later, NASA's mission to Mars with the Perseverance Rover was launched on a two-stage Atlas V-541 rocket from Cape Canaveral, Florida, on July 30, 2020 landed on February 18, 2021. Weighing 531,000 kg fully fueled with the spacecraft on top, the 58 m high rocket fueled with type RP-1 kerosene fuel/ liquid oxygen, in the first stage, and liquid hydrogen/liquid oxygen in the second stage Centaur RL-10 engine, can provide a thrust of 3.8 million Newtons. The thrust-to-weight ratio of the system at ~0.8 is better than that for previous space rockets.

The NASA Mars Mission 2020 to the *Red Planet* along the trajectory shown in Fig. 3.20 took 6 months and 19 days to travel a distance of about 470 million km with an average speed of 98,000 km/h (~27 km/s)

The RD-180 engine, used in the Atlas V rocket of the Mars 2020 mission is a high performance engine LOX/kerosene engine built and marketed 50–50 by NPO Energomash (Russia) and Pratt and Whitney (USA). Some technical features of the RD-180 engine, used in Atlas III and Atlas V launchers, are its specific impulse of 337.8 seconds in vacuum, thrust of 0.9 million lb_f, and a chamber pressure of 257 atmospheres. The launch vehicle has stage 1 and 2 Atlas V and Centaur engines with four solid rocket boosters; their combined thrust is over 5 million lb_f. Before the 2020 Mars mission, Apollo missions 4, 6, 8 used 1.5 million lb_f thrust while Apollo missions 9–17 used 1.522 million lb_f thrusts.

These are the limits of chemical propulsion systems, and thus define the limits of deep space travel with severe constraints of delivering payloads with reusable rockets.

3.7.3 Nuclear rocket designs for deep space exploration

To improve the capability of rockets to perform better than the chemical propulsion systems described above, there has been an awareness for decades that nuclear propulsion may hold the key to the future of deep space exploration.

The work on nuclear rockets began in the 1950s with the Project Rover and the Nuclear Engine for Rocket Vehicle Application (NERVA) shown in Fig. 3.15; design improvements and testing continued with development of state of the art systems. Nuclear thermal propulsion (NTP) engines were demonstrated from the 1950s to 1970s during Project Rover and later by the NERVA (Nikitaev & Thomas, 2020). Fuel is uranium nitride (UN) (supplies the heat) and propellant (that carries the heat) is hydrogen. Recently, efforts in deep space exploration have intensified with several missions to the Moon and the most recent (2020) missions to Mars (NASA, 2021).

The nuclear option for rockets is an attractive one due to its very high heating value compared with other fuels, as shown in Table 3.16.

The values in Table 3.16 show that the nuclear advantage is orders of magnitude higher in comparison to chemical fuels. Further, as concluded from an overview of Gen-IV reactors, high temperature systems would become available in the near future, with liquid metal and GCRs ideal for safe and efficient power. With these advantages, a nuclear rocket

TABLE 3.16 Heating values of some conventional fuels compared with uranium fuels.

Fuel	Heating value (MJ/kg)[a]
Hydrogen (H_2)	120–142
Petrol/gasoline	44–46
Crude oil	42–47
Hard black coal	25
Natural uranium in LWR	500 GJ/kg
Uranium enriched to 3.5% in LWR	3900 GJ/kg
Natural uranium in FNR	28,000 GJ/kg

[a]*A Giga-Joule is a large amount of energy; the explosive yield of one ton of chemical explosive trinitrotoluene is TNT 4.18 GJ.*
Source: World Nuclear Association https://www.world-nuclear.org/information-library/facts-and-figures/heat-values-of-various-fuels.aspx.

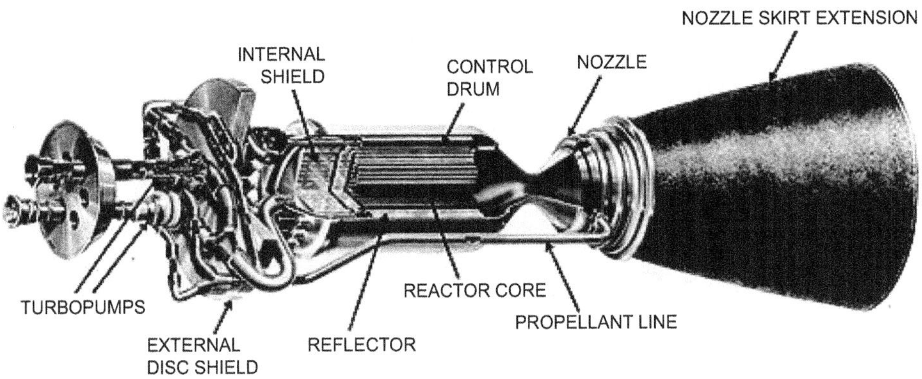

FIGURE 3.15 NERVA rocket engine. *Courtesy NASA grc.nasa.gov 2021.*

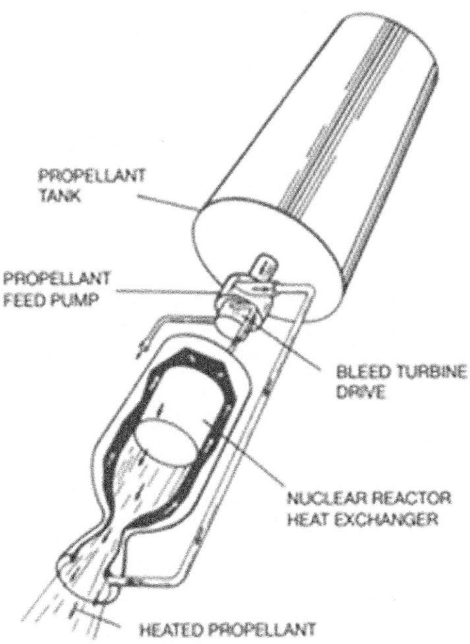

FIGURE 3.16 The concept of a nuclear propulsion rocket system.

would essentially be used a thermal energy source for heating a propellant, as shown in Fig. 3.16 giving it high exhaust velocity and thrust.

NTP offers an alternative to CPT since more thermal energy could be provided to propellants; this is likely to result in higher nozzle exit velocities and hence higher thrust. If realized, NTP could reduce the travel time for outer space exploration.

It is worth estimating the comparative advantage of NTP over CPT with regards to propulsion efficiency and specific impulse for deep space exploration and, specifically, the possibility of reducing flight time to Mars and then the possibility of a return flight as well. In the present NASA mission, the rocket is used for launching the spacecraft carrying the Perseverance Rover and descent systems.

The NTP engine, unlike chemical systems which depend on combustion to produce thrust, uses a propellant pumped into a nuclear reactor heated to high temperatures and with thermal energy converted to kinetic energy, the monopropellant exits the nozzle with very high velocity.

LEU NTP engines have been extensively studied by Aerojet Rocketdyne (AR) with the objective of providing a system to NASA for payload mass and high thrusts to enable crewed missions from the 2030s through the 2050s (Joyner et al., 2020). These include moving heavy cargo within 200 days to Mars.

LEU thermal spectrum systems based on ceramic-metallic (CERMET) fuel were moderated using materials in a structural assembly [tie-tubes (T/T)] similar to the NERVAs design. The LEU NTP approach has been shown to be technically feasible based on work from 2016 to 2019.

Significant conceptual design evaluations have been completed for the LEU NTP design approach at 25,000-lb$_f$ thrust and have shown high specific impulse and thrust-to weight ratios at or above 3:1. Analyses has been carried out with results showing that a NTP can be used for Mars architectures for crewed missions using a 25,000-lbf (111-kN) NTP engine system.

A nuclear thermal rocket does not require oxygen; it is thus a *monopropellant* system. It needs the highest specific impulse for high temperature, low molecular weight, and chamber pressure; thus a high temperature reactor and hydrogen fuel are the best combination. For hydrogen, $M = 2.016$ g/mol, while for on the average propellants, $M \sim 13.8$ g/mol with $T \sim 2800$K; for all other conditions fixed, the increase in specific impulse is more than twice, from $I_{sp} \sim 300$ to ~ 770 seconds. A nuclear rocket (Fig. 3.17) could give $I_{sp} \sim 900$ seconds with significant advantage in deep space exploration missions, such as reducing mission time and carrying crew and equipment. The power of a nuclear thermal rocket depends most significantly on the temperature achievable; the higher the better. Thus, the SMRs and Gen-IV technologies reviewed in the previous section are all candidates, provided space and weight constraints are satisfied. These are the same challenges characteristic of Gen-IV systems such as the technological development of advanced fuels and materials as well as the optimization of design and operational parameters.

Bimodal nuclear thermal rocket engines, working for both propulsion and power generation, have been shown (Clough) to reduce the weight of space vehicles to the Moon, Mars, and beyond. The NERVAs NERVA reactor can be used for both propulsion and power; a 316 MW(t) reactor that produces a thrust of 66.6 kN and specific impulse a specific of 917 seconds can be run at 73.8 kW(t) to produce the necessary 16.7 kW electric power with a Brayton cycle generator with a flow as illustrated in Fig. 3.16.

Four engines, using HEU and liquid hydrogen, have been simulated for the RL-10, RL-60 and CERMET engines. The results from these simulations are given in Table 3.17. In all cases, the specific impulse far exceeds the capabilities of chemical fuels while providing high thrust-to-weight ratio.

The improvements considered to the above HEU designs have included optimization of LEU core configurations (Joyner) to maximum specific impulse and minimum the core reactor mass.

Better fuel materials such as UN with a matrix of molybdenum and tungsten to raise the melting temperature led to an improvement in performance. Several other candidate materials, for example, zircaloy, zirconium hydride, and zirconium carbide have been considered for cladding uranium dioxide fuels to raise the operating conditions.

Similar to the issue of HEU in submarine reactors, space systems will require LEU fuels in NTP systems. It has been shown that modifications can be made to fast systems in the NERVA to achieve a thermal spectrum for which LEU-based nuclear thermal rockets can offers distinct advantages over existing HEU systems.

TABLE 3.17 Engine comparisons for NERVA and CERMET systems.

Description	Small NERVA	Small CERMET	Large NERVA	Large CERMET
Engine diameter (cm)	87.7	55	98.5	68
U-235 mass (kg)	27.5	177.3	36.8	376.7
Reactor mass (kg)	1435	1215	2645	2722
Engine total mass (kg)	1730	1450	3305	3327
Thrust (kN)	33.2	33.8	113.3	111.2
Specific impulse (s)	899.6	885.4	911.5	824.2
Thrust-to-weight	1.96	2.38	3.49	3.41

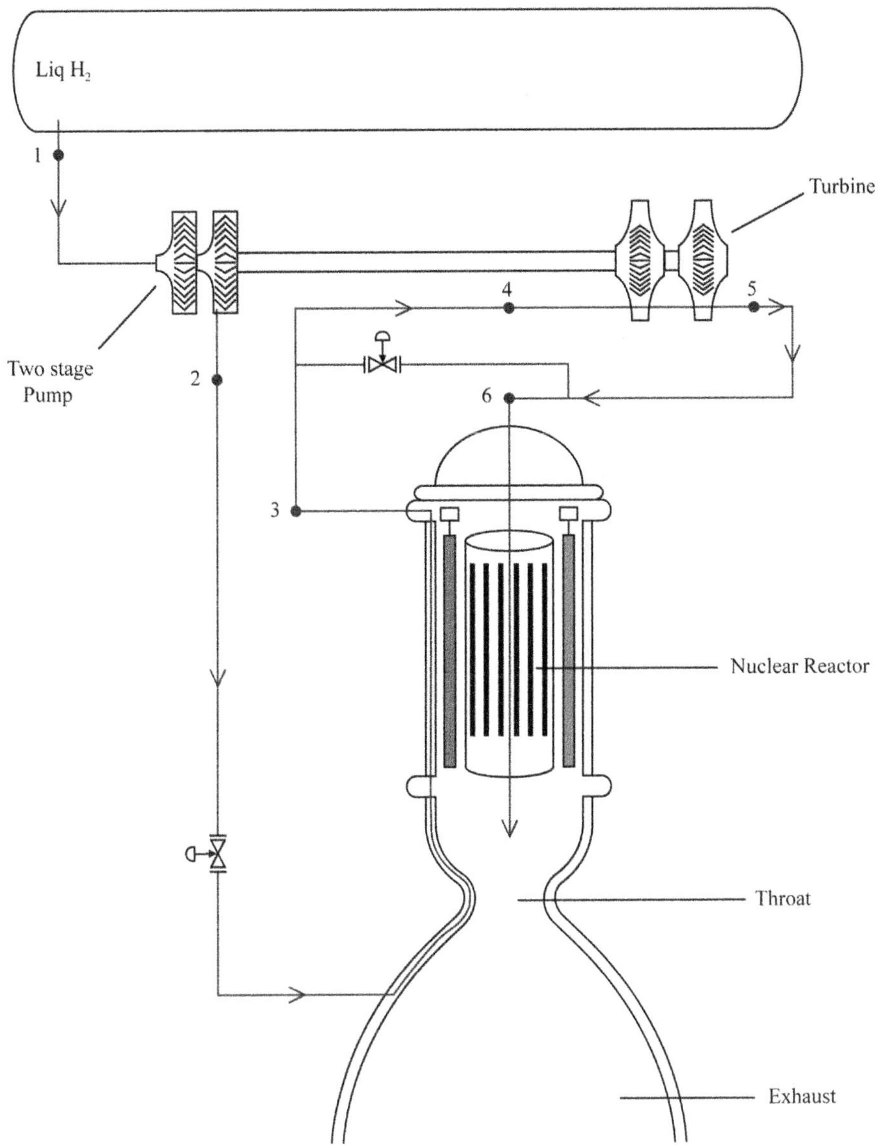

FIGURE 3.17 Engine flow in NERVA-type nuclear propulsion rocket engine.

TABLE 3.18 Design characteristics of some SNAP systems.

Characteristic	SNAP-2A	SNAP-8	SNAP-10A
Power (kW)	3	35	0.58
Design lifetime (a)	1	1	1
Reactor power (kW)	55	600	43
Reactor outlet (K)	920	975	833
Fuel and spectrum	U-ZrH thermal	U-ZrH thermal	U-ZrH thermal
Coolant	Na-K-78	Na-K-78	Na-K-78
Power conversion	Rankine (Hg)	Rankine (Hg)	Thermoelectric (Si-Ge)
Turbine inlet (K)	895	950	777 K/610 K
Unshielded weight (kg)	545	4545	295

Source: IAEA.

3.8 Nuclear power systems in space

The two kinds of nuclear power systems used in space have been the radioisotope thermal generators (RTGs) and small nuclear auxiliary power (SNAP) reactor systems. The new systems under development are more efficient and are classified as nuclear batteries or MNRs. These are briefly described in the following sections.

3.8.1 Radioisotope thermal generators

As described in Chapter 1, the alpha decay of heavy nuclides can be used as a thermal source of energy which can be directly converted to electrical energy in, for example, a thermoelectric generator. Such systems, called RTGs have been used since the 1960s for satellite and space probes. The most recent is a Pu-238 powered RTG to supply electrical power to the Perseverance Rover in the NASA Mission to Mars 2020. The thermal energy from the radioisotope Pu-238 is used to charge the batteries as well as to keep the various systems in the rover at their correct operating temperatures. The hot side from such a source and the cold side at space temperatures are both connected to a thermocouple which converts thermal energy into electricity. For the Perseverance Rover, about 110 W(e) is produced from a 2 ft by 2 ft, 45 kg power system containing 4.8 kg Pu-238 with an operational life of about 14 years.

3.8.2 Small nuclear auxiliary power systems

In addition to the RTGs used for heat and very small amounts of power, nuclear power reactors were also developed and used in space systems. The *Romashka* reactor developed and used by the former Soviet Union, in the 1950s and 1960s, was a nuclear reactor, with 49 kg HEU in a core of diameter 241 mm and height 351 mm, with an electrical output of 460–475 W produced from a thermoelectric generator weighing 635 kg. By 1970, the *Topaz* thermionic reactor converter

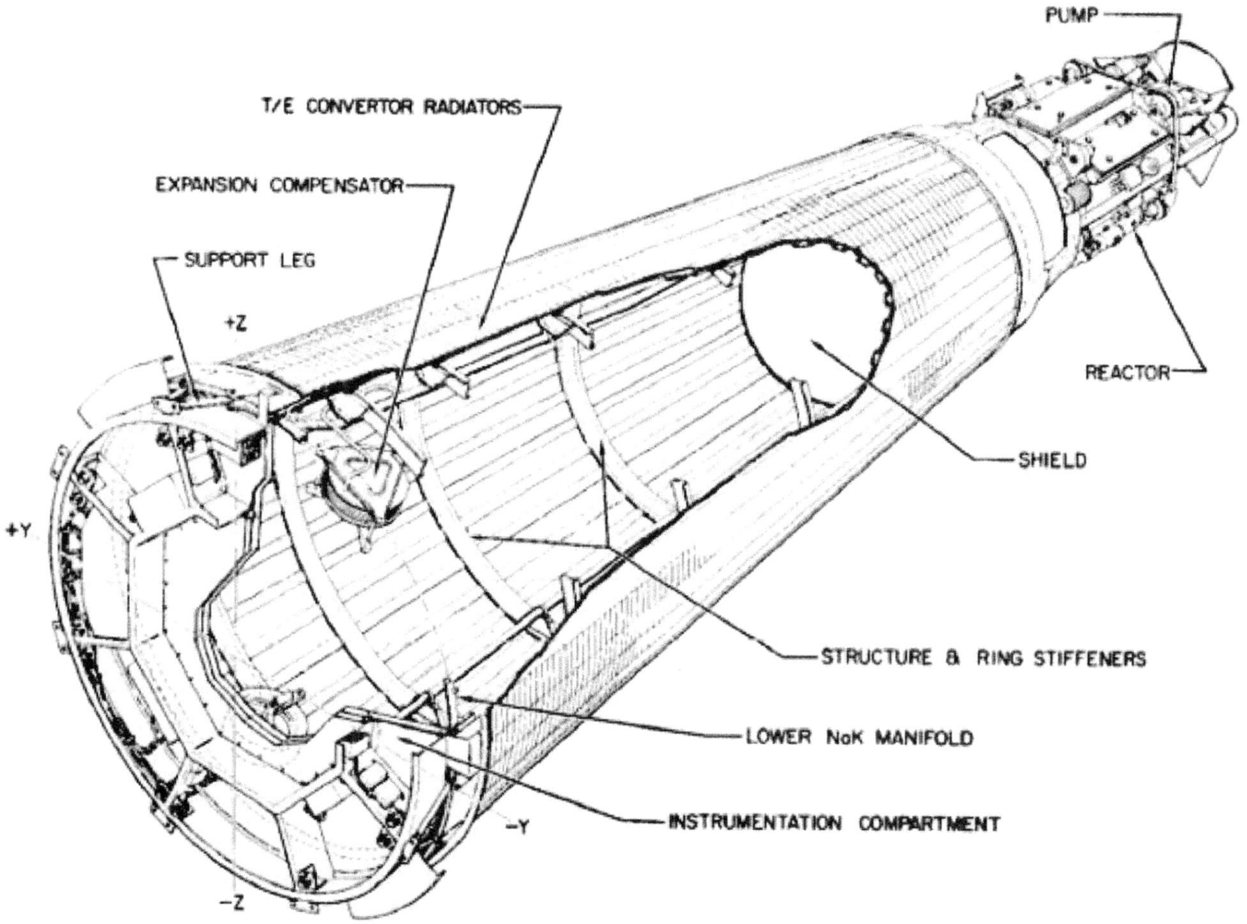

FIGURE 3.18 The SNAP-10A reactor. *Courtesy NASA grc.nasa.gov 2021.*

was developed based on the "boiling" of electrons, used as the working fluid, from a very hot static emitter surface typically at 1800K collected at a cooler surface typically at about 1000K. This electrical plant had electrons as the working fluid. From the 1970s to the late 1980s, the former Soviet Union operated a series of the Cosmos spacecraft.

SNAP power systems, such as the SNAP-10A shown in Fig. 3.17, based on the Soviet program, were developed in the 1960s.

Design characteristics of some SNAP (Fig. 3.18) models are listed in Table 3.18; these systems were intended for use in the kilowatt electrical range but were not successful.

Small power systems, in the range of 3–20 kW(e), have also been proposed for converting nuclear fission thermal energy using thermoelectric generators with heat pipes instead of the conventional coolant with pump flow. The Heat pipe Operated Exploration Reactor is based on this concept.

3.8.2.1 *Nuclear reactors for space*

With missions to the Moon and Mars (Fig. 3.19), several designs of MNRs for space have been investigated with renewed vigor. These represent improvements on the early designs of the 1950s and 1960s.

Very small, *micronuclear* (MNR) reactors have been designed for space and marine applications. One such design is shown in Fig. 3.18 with a solid matrix core comprising fuel and heat pipes surrounded by radial and axial reflector, absorber, and shielding region. The reactor core is embedded with cylindrical fuel rods of UN and heat pipes in a monolith, surrounded with radial and axial reflectors and radiation shields, with boron carbide (B_4C)-tipped control drums embedded within the radial reflector. Thermal energy produced from nuclear fission in the critical core is transported by heat pipes from the core to the TEG. The objective of this design is power conversion from nuclear thermal energy in the core to electrical energy in the TEG without the use of rotating machinery such as pumps and turbines. The shielding is designed so that permissible safety levels are maintained.

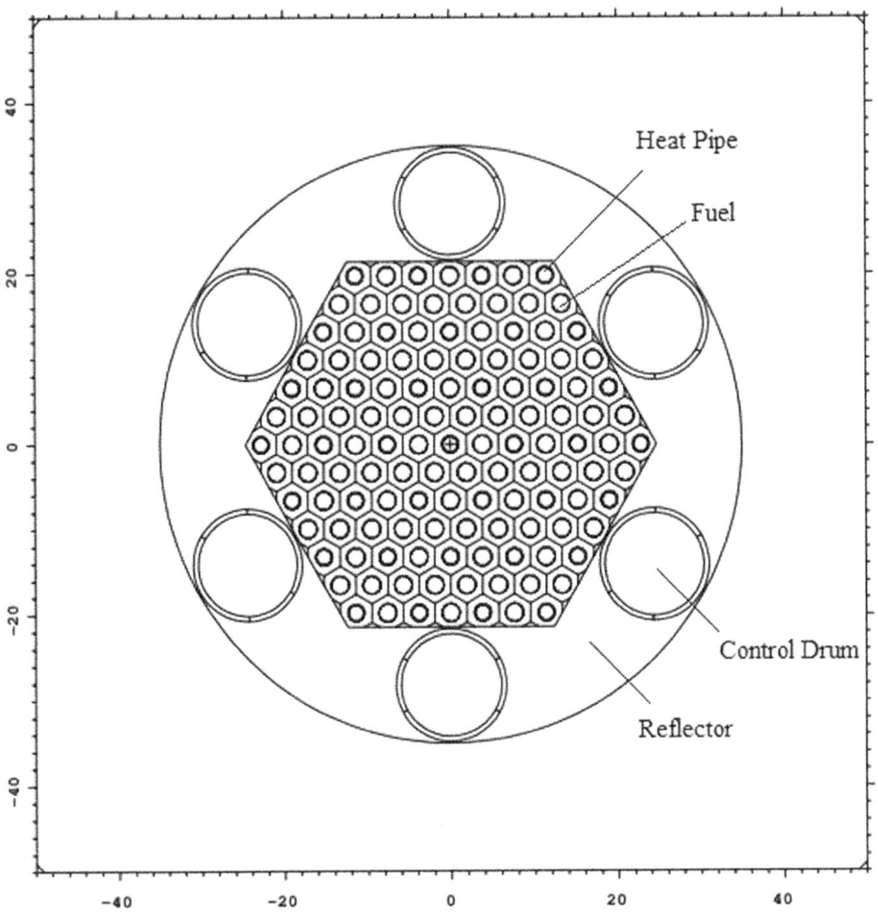

FIGURE 3.19 Core of a micronuclear heat pipe reactor.

The main features of the overall core design are listed in Table 3.19 and Table 3.20. The thermal power is 700 kW with a thermoelectric conversion system giving 35 kW(e) which gives an efficiency of only 5%.

The total uranium fuel is ~149 kg, out of which ~104 kg is U-235, with a compact core within a barrel of radius 35 cm and height 40 cm. The thermal output of this reactor can be as high as ~700 kW(t) with lithium in heat pipes; with

TABLE 3.19 Overall characteristics of a MNR.

Parameter	Material
Power kW(t)/kW(e)	700/35
Fuel/enrichment	UN (70%)
Heat pipe	Li (wick Mo-12Re)
Reflector/configuration	BeO/side and top
Control system	Control drums in reflector
Radiation shield/configuration	Water and tungsten/side, top, and bottom

TABLE 3.20 MNR design parameters.

Design parameter	Data
Fuel rod	
Diameter (cm)/Height (cm), Density (g/cm^3)	2.03/40, 13.6
Cross-section area of fuel rod (cm^2)	3.2365
Mass fuel rod (kg)	1.7607
Mass fraction U^{235}/ U^{238}	0.6608/0.2743
Mass fraction N	0.0560
Mass of U (kg)/Mass of U^{235} (kg)	1.6621/1.1635
Fuel and fissile material	
No. of fuel rods	90
Area of fuel rods (cm^2)	291.2893
Mass of fuel rods (kg)/Mass of fuel rods/length (kg/cm)	158.46/3.9615
Mass U (kg)/Mass of U^{235} (kg)	149.5890/104.7150
Heat pipes	
No. of heat pipes	37
RadiiVapor/Wick/Liq/Struc (cm)	0.90/1.00/1.02
Area of heat Pipe (cm^2)	3.2685
Total area of heat pipes (cm^2)	120.9345
Den of vapor (g/cm^3)	0.01
Den of wick, structure (g/cm^3)	12.0, 12.0
Mass (kg), Mass/length (kg/cm)	12.8538/0.321345
Matrix	
Material, density (g/cm^3)	Nb-1Zr, 6.55
Area of Hexagon (cm^2)	1585.1
Area (Hexagon-Fuel-HP) (cm^2)	1172.9
Mass (kg), Mass/length (kg/cm)	331.9531
Core	
Mass (kg)	503.2669
Mass/length (kg/cm)	12.5817
Axial shield	
Top and Bottom shield water	35 cm
Top and Bottom shield tungsten	40 cm

electrical output depending on the thermoelectric efficiency. With typically ~5%, the MNR is thus capable of providing ~35 kW(e). The shielding is the only major disadvantage in terms of weight when tungsten becomes necessary for gamma dose reduction. It is necessary to minimize the shield weight to suit space and permissible dose requirements. The consequences of heavy shield will be acceptable for on-ground specialized power applications but unfavorable for space missions where a compromise on the dose might be necessary for conforming to any propulsion constraints.

Further, the use of HEU with over 90% enriched fuel will not be able to satisfy the requirements of criticality safety in case of a launch abort; thus fuel enrichment in the range 55%–80% will be preferable.

In the kilopower range, 1–10 kW(e), reactors can be smaller and useful for space and specialized surface applications. Most recently, a 5 kW(t)/1 kW(e) prototype called KRUSTY (Kilopower Reactor Using Stirling Technology) was fueled with a solid core HEU (Poston, Gibson, Godfroy, & McClure, 2020) U-8Mo, 32.2 kg fuel (27.7 kg U235) was demonstrated at Los Alamos National Laboratory at fuel temperatures greater than 800°C. The heat removal system uses sodium as the fluid and nickel wick in heat pipes connected to a Stirling engine. Thus, there are no movable parts.

This compares with a larger 2400 kW(t)/120 kW(e) 70% enriched UN fueled reactor with a 310 kg core and a thermoelectric generator (Sun et al., 2018). An advantage of using high-density UN fuel is a reduction in the core size In another conceptual design by Wang et al. (Chenglong, Sun, Simiao, Wenxi, & Suixheng, 2020), a 500 kW(t), 25 kW(e) 65% enriched UN fueled HPR is given with a core weight also of 310 kg. Heat pipes of sodium, lithium and potassium have been studied for a number of designs (El-Genk & Tournier, 2011) and appear promising for near future space applications.

3.9 Conclusions

This chapter reviewed nuclear reactor systems and designs that have evolved over the last seven decades. Some, like MCF and ICF have yet to be demonstrated with technological and commercial feasibility.

The objective of this chapter was to develop a realization of the nature and scale of nuclear technologies.

For nuclear power reactors, the need to demonstrate economically competitive reactors with enhanced safety and proliferation-resistant features were identified as the most important goals of Generation-IV reactors.

Nuclear propulsion reactors, especially for submarine, require a transition from HEU to LEU designs which must be addressed but may not be achievable for several reasons.

For nuclear fusion reactors, ICF appears to be at its very early stages of development while MCF is anticipated to show technological promise with success in ITER by 2025.

Nuclear propulsion is an area that appears both necessary and promising due to the limits of chemical propulsion technologies that restrict deep space exploration and habitation of Mars.

In all these areas, there is a need for more R&D efforts, with more extensive modeling and computationally powerful simulation as well as validations with new technologies such as advancements in fuel and separation technologies.

Some identifiable challenges in modeling and simulation include

1. The development of advanced neutronic and thermal-hydraulic coupling models
2. A detailed safety analysis with coupled nuclear models
3. An improvement of computational tools
4. Validation of existing calculation tools and nuclear data libraries
5. Simulation of fuel burnup for developing high burnup minor actinide (neptunium, americium, curium) bearing fuels
6. Modeling the clad behavior at high temperature
7. Modeling the core-cooling capability
8. Modeling passive heat removal system
9. Modeling hydrogen production methods
10. Modeling the effect of high core outlet temperatures 700–950°C
11. Carrying out system integration studies
12. Carrying out a full design optimization

Problems

1. Why is it not necessary to use heavy water in a PWR?
2. What are the significant design differences between a BWR and a PWR?
3. What is the purpose of breeding in a FBR?

4. When did enriched uranium become available to NPP designers and why was the STR Mark II designed as a thermal reactor rather than a fast reactor?
5. If you had to choose from a PWR or a liquid metal cooled reactor (LMCR) for a submarine, with the objective of space constraints and enhanced power, which would you select and for what reasons? Consider the following factors: energy spectrum, size, pressure vessel, moderator, coolant, xenon poisoning, refueling, coolant melting point, coolant-water interaction.
6. Comment on the following Russian design for RM-1 and VM-40A reactors in terms of their advantages and disadvantages.

   RM-1
   Fuel:90% enriched U-Be_{13} dispersed in a beryllium matrix
   Amount of U-235 in core: 90 kg
   Fuel rod pellet diameter ~10 mm covered with 0.1 mm MG clad in SS with 0.5 mm thickness
   Number of fuel rods: ~3000
   Coolant: eutectic lead-bismuth alloy (44.5 wt.% Pb, 55.5 wt.% Bi), melting point ~125°C
   VM-40A
   Fuel:90% enriched U-Be_{13} dispersed in a beryllium matrix
   Amount of U-235 in core: 200 kg
   Coolant: eutectic lead-bismuth alloy (44.5 wt.% Pb, 55.5 wt.% Bi), melting point ~125°C
7. In view of the critical mass of Godiva (Section 3.2), comment on the amount of U-235 in a submarine reactor core.
8. How is high temperature achieved in Gen-IV reactors?
9. Neutronics of a submarine core: For a U-Zr fuel consisting of 15 wt.% Zr and 85 wt.% U enriched to 93% in U-235, calculate the fast and thermal cross-sections and comment on the size of a submarine reactor core using this fuel. How much would the size differ from that of a commercial PWR?
10. Estimate the natural uranium required for a submarine reactor 150 MW(t) operating at an average of 25% power for 1 year. Assume that 50% of all U235 atoms undergo fission and that 60% of the natural uranium goes into the submarine.

Nomenclature

English (lower case)

c_p	specific heat at constant pressure
c_v	specific heat at constant volume
e	internal energy
k	thermal conductivity
k	Boltzmann constant
m_i	mass of ion
p	pressure

English (upper case)

A^*	cross-section area at throat
A_e	cross-section area at exit (diverging part)
E_c	energy of charged particle
G	gain
$\tilde{H}$	extrapolated height
H	enthalpy
I_{sp}	specific impulse
M	molecular weight
P_e	exit pressure
P_e	inlet pressure
Q	heat
R	universal gas constant
T	absolute temperature
Q	heat
T	temperature

U	internal energy
W	work

Greek lower case

β	delayed neutron fraction
ϵ	expansion ratio
γ	ratio of specific heats c_p/c_v
η	efficiency
ρ	density
τ_E	energy confinement time

Abbreviations and acronyms

Ci	curie
kN	kilo newton
Sv	Sievert
Zr	zirconium
ABWR	advanced boiling water reactor
AP	advanced power
BARC	Bhabha Atomic Research Center
BWR	boiling water reactor
CANDU	Canadian Deuterium Uranium
CERMET	ceramic metal
CHASNUPP	Chashma nuclear power plant
D_2O	deuterium oxide
FBR	Fast Breeder Reactor
GCFR	gas cooled fast reactor
GCR	gas cooled reactor
GW(e)	gigawatt electric
HEU	highly enriched uranium
HTR	high temperature reactor
IAEA	international atomic energy agency
ICBM	intercontinental ballistic missile
ICF	inertial confinement fusion
INS	inertial navigation system
ITER	international thermonuclear experimental reactor
LEU	low enriched uranium
LOCA	loss of coolant accident
LFR	lead cooled fast reactor
LMFBR	liquid metal cooled Fast Breeder Reactor
LWGR	light water gas cooled reactor
MIRV	multiple independently targetable reentry vehicle
MCF	magnetic confinement fusion
MNR	micronuclear reactor
MOX	mixed oxide
MSR	molten salt reactor
MW(e)	megawatt electric
MW(t)	megawatt thermal
MWd	megawatt-day
NASA	National Aeronautics and Space Administration
NERVA	Nuclear Engine for Rocket Vehicle Application
NTP	nuclear thermal propulsion
NPP	nuclear power plant
PHWR	pressurized heavy water reactor

PWR	pressurized water reactor
RTG	radioisotope thermal generator
SCR	sodium cooled fast reactor
SCWR	supercritical water cooled reactor
SHP	shaft horsepower
SLBM	submarine launched ballistic missile
SMR	small modular reactor
SNAP	small nuclear auxiliary power
SSBN	submersible ship, ballistic, nuclear
SRS	stellar reference system
TNT	trinitrotoluene
USNRC	United States Nuclear Regulatory Commission
VHTR	very high temperature reactor
vSMR	very small modular reactor
VVER	water-water energetic reactor

References

Bari, R. A. (2014). Proliferation resistance and physical protection (PR&PP) in small modular reactors (SMRs). *Handbook of Small Modular Nuclear Reactors*, 219–236. Available from https://doi.org/10.1533/9780857098535.2.219.

Bartel, R. (2016). *WASH-1400 the reactor safety study*. NUREG/KM-0010. Washington, DC.

Betti, R., & Hurricane, O. A. (2016). Inertial-confinement fusion with lasers. *Nature Physics*. Available from https://doi.org/10.1038/NPHYS3736.

Chenglong, W., Sun, H., Simiao, T., Wenxi, T., & Suixheng, Q. (2020). Thermal-hydraulic analysis of a new conceptual heat-pipe cooled small nuclear reactor system. *Nuclear Engineering and Technology*, *52*, 19–26.

El-Genk, M. S., & Tournier, J.-M. P. (2011). Uses of liquid-metal and water heat pipes in space reactor power systems. *Frontiers in Heat Pipes*. Available from https://doi.org/10.5098/fhp.v2.1.3002.

He, X. T. (2016). The updated advancements of inertial confinement fusion program in China. *Journal of Physics: Conference Series*. Available from https://doi.org/10.1088/1742-6596/688/1/012029.

He, X. T., & Zhang, W. Y. (2007). Inertial fusion research in China. *European Physical Journal D*. Available from https://doi.org/10.1140/epjd/e2007-00005-1.

Horioka, K. (2018). Progress in particle-beam-driven inertial fusion research: Activities in Japan. *Matter and Radiation at Extremes*. Available from https://doi.org/10.1016/j.mre.2017.08.002.

Huba, J. D. (2011). NRL PLASMA FORMULARY Supported by the office of naval research. *Plasma Physics*.

IAEA. (2007). *Nuclear power plant design characteristics: Structure of nuclear power plant design characteristics in the IAEA power reactor information system (PRIS)*. Vienna: International Atomic Energy Agency Nuclear Po. Available from https://www-pub.iaea.org/mtcd/publications/pdf/te_1544_web.pdf.

IAEA. (2017). Fast reactors and related fuel cycles: Next generation nuclear systems for sustainable development FR17. In *Proceedings of an International Conference Yekaterinburg, Russian Federation, 26–29 June 2017*. Yekaterinburg, Russian Federation: International Atomic Energy Agency. Available from https://www-pub.iaea.org/MTCD/Publications/PDF/STIPUB1836web.pdf.

IAEA. (2018). *Advances in small modular reactor technology developments*. Vienna.

IAEA. (2020). *Nuclear power reactors in the world*. Vienna: International Atomic Eneergy Agency, Edited by Series No. 2. IAEA-RDS-2.

Joyner, C. R., Eades, M., Horton, J., Jennings, T., Kokan, T., Levack, D. J. H., ... Reynolds, C. B. (2020). LEU NTP engine system trades and mission options. *Nuclear Technology*. Available from https://doi.org/10.1080/00295450.2019.1706982.

Kopte, S. (1997). *Nuclear submarine decommissioning and related problems*. Bonn.

Lobner, P. (2018). *Marine nuclear power: 1939–2018 Part 2A: United States-Submarines*.

Mian, Z., Ramana, M. V., & Nayyar, A. H. (2019). Nuclear submarines in South Asia: New risks and dangers. *Journal for Peace and Nuclear Disarmament*. Available from https://doi.org/10.1080/25751654.2019.1621425.

NASA. (2021). MARS 2020 mission perseverance rover. https://mars.nasa.gov/mars2020/timeline/cruise/.

Nikitaev, D., & Thomas, L. D. (2020). Seeded hydrogen in nuclear thermal propulsion engines. *Journal of Spacecraft and Rockets*. Available from https://doi.org/10.2514/1.a34711.

Nuclear Energy Agency [NEA]. (2014). Technology roadmap update for generation IV nuclear energy systems. *Gen IV International Forum*.

Pakistan Navy. (2021). Pakistan submarine capabilities. *Nuclear Threat Initiative*. Available from https://www.nti.org/analysis/articles/pakistan-submarine-capabilities/.

Philippe, S., & von Hippel, F. (2016). The feasibility of ending HEU fuel use in the U.S. Navy. *Arms Control Today*, November.

Poston, D. I., Gibson, M. A., Godfroy, T., & McClure, P. R. (2020). KRUSTY reactor design. *Nuclear Technology*. Available from https://doi.org/10.1080/00295450.2020.1725382.

Ragheb, M. (2011). Nuclear naval propulsion. *Nuclear Power – Deployment, Operation and Sustainability*. Available from https://doi.org/10.5772/19007.

Reistad, O., & Olgaard, P. L. (2006). *Russian nuclear power plants for marine applications*. Roskilde, Denmark.

Sun, H., Wang, C., Ma, P., Liu, X., Tian, W., Qiu, S., & Su, G. (2018). Comnceptual design and analysis of a multipurpose micro nuclear reactor power source. *Annals of Nuclear Energy, 121*, 118–127.

Tikhonchuk, V. T. (2020). Progress and opportunities for inertial fusion energy in Europe. *Philosophical Transactions of the Royal Society A: Mathematical, Physical and Engineering Sciences*. Available from https://doi.org/10.1098/rsta.2020.0013.

ANNEX: the physics of nuclear fusion

For fusion to take place, the nuclei must be brought close to each other. As their energy is increased, the atoms are stripped of their electrons and ionize into the plasma state. To get ions closer to fuse, the coulomb barrier must overcome the repulsive forces. Once the barrier is crossed, the nuclear strong attraction forces enable fusion.

$$V(r) = \frac{Z_1 Z_2 e^2}{4\pi\varepsilon_0 r}$$

where

$$k = \frac{1}{4\pi\varepsilon_0} \sim 9 \quad \times 10^9 \text{Nm}^2/\text{C}^2$$

The potential energy is then

$$V(r) = \frac{Z_1 Z_2 e^2}{4\pi\varepsilon_0 r} = 9 \quad \times 10^9 \quad Z_1 Z_2 \frac{(1.6 \times 10^{-19})^2}{r} \bullet \frac{10^6}{10^{-15}} \text{MeV}(r \text{in fm})$$

Thus

$$V(r) = \frac{1.44 Z_1 Z_2}{r} \text{MeV}(r \text{in fm})$$

For D-D (with $Z_1 = Z_2 = 1$), and $r = 10$ fm, $V(r) = 0.144$MeV. Taking the radius of a nucleus as

$$r = 1.2 \quad A^{1/3} \text{fm}$$

the radius of a deuterium nucleus is 1.5119 fm; at twice this distance, that is, if two D nuclei are touching each other, $V = 0.4762$ MeV. This energy corresponds to a temperature found from $E = k_B T$ from which

$$T = \frac{E}{k_B} = \frac{0.4762 \quad \times 10^6}{8.617 \times 10^{-5}} \sim 5.5 \times 10^9 \text{K}$$

If the coulombic barrier is the only dominant barrier, then the thermal energy is calculated to be in excess of 10^9 K for D-D. This is the theoretical critical ignition temperature below which fusion should not be expected to be possible. However, fusion does take place at temperatures lower than this critical estimate. Two phenomena make fusion possible below this critical temperature. The first is the Maxwellian distribution of energies and the second is the tunneling of ions through the coulomb barrier; both these phenomena lower the critical temperature. If this were not the case, then fusion would not have been possible on the sun, as its core temperature is of the order of 1.5×10^7 K which is lower than the critical temperature for D-T fusion.

At a given temperature, or energy, all the ions do not have the same energy; in fact, most have a *probable* energy while a few have much lower and much higher energies. This is illustrated in Fig. 3.A1 for the distributions $n(E), n(v)$; for $n(E), x = E/kT$; for $n(v), x = v\sqrt{m/2kT}$.

The probability of quantum-mechanical tunneling, given by the Gamow factor, is $\sim e^{-E_G/\sqrt{E}}$ while the number of particles at high energies due to their Maxwellian distribution (Section 2.1) is $\sim e^{-E/kT}$; thus at high energies the number of ions tunneling through the barrier is $\sim e^{-\sqrt{E_G/E}} e^{-E/kT}$. This is a result obtained from the Schrodinger equation (Section 1.1). Here, the Gamow energy E_G is given by

$$E_G = 2 \quad m_r c^2 (\pi\alpha Z_1 Z_2)^2$$

where E_G, Gamow energy; m_r, reduced mass; α, fine structure constant $\alpha = e^2/(4\pi\varepsilon_0\hbar) \sim 137$

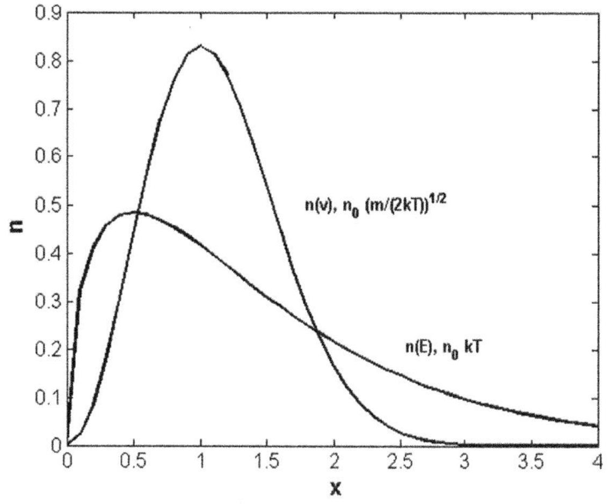

FIGURE 3.A1 Distribution functions $n(v)$ and $n(E)$.

The transmission probability has a maxima at

$$E_M = E_G\left(\frac{kT}{2E_G}\right)^{2/3}$$

Several other fusion reactions are also possible but the reactions between isotopes of hydrogen are more likely than the reactions between heavier nuclei such as helium due to the stronger coulomb barrier in case of the latter. The D-D fusion reactions possible are

$$^2_1H + ^2_1H \rightarrow \begin{cases} ^3_2He + ^1_0n \\ ^3_1H + ^1_1H \end{cases}$$

The fusion reaction rate for two species with n_1 and n_2 atoms cm^{-3} is given by

$$R = n_D n_T \langle\sigma v\rangle \ \text{cm}^{-3}\text{s}^{-1}$$

where $\langle\sigma v\rangle$, the Maxwellian-averaged cross-section is a function of the temperature T of the reaction. For DD and DT reactions,

$$\langle\sigma v\rangle_{DD} = 2.33 \times 10^{-14} T^{-2/3} \exp\left(-18.76 T^{-1/3}\right) \text{cm}^3\text{s}^{-1}$$

and

$$\langle\sigma v\rangle_{DT} = 3.68 \times 10^{-12} T^{-2/3} \exp\left(-19.94 T^{-1/3}\right) \text{cm}^3\text{s}^{-1}$$

where T is in keV (1 eV ~ 11,600K). The functions $\langle\sigma v\rangle_{DT}$ and $\langle\sigma v\rangle_{DT}$, have maximum values of 3.6 × 10−17 cm^3/s at 825.3 keV and 5×10^{-15} cm^3/s at 991 keV respectively, that is, ~ 100 million degrees.

The fusion cross-sections, from the Naval Research Lab's plasma physics formulary are

$$\sigma_F(E) = \frac{A_5 + \left((A_4 - A_3E)^2 + 1\right)^{-1} A_2}{E\left(\exp\left(A_1 E^{-\frac{1}{2}}\right) - 1\right)}$$

with the coefficients listed in Table 3.A1

The fusion cross-sections for DD and DT reactions are shown in Fig. 3.A2.

The DD cross-sections are orders of magnitude lower than the DT reaction and so the DT reaction is preferred.

The power density released in the form of charged particles (Huba, 2011) is

$$P_{DD} = 3.3 \times 10^{-13} n_D^2 \langle\sigma v\rangle_{DD} W/\text{cm}^3$$

and

$$P_{DT} = 5.6 \times 10^{-13} n_D n_T \langle\sigma v\rangle_{DT} W/\text{cm}^3$$

TABLE 3.A1 Coefficients of the fusion cross-sections.

Coefficient	$D(T,n)He^4$	$D(D,p)T^3$	$D(D,n)He^3$
A_1	45.95	46.097	47.88
A_2	50200	372	482
A_3	1.368×10^{-2}	4.36×10^{-4}	3.08×10^{-4}
A_4	1.076	1.22	1.177
A_5	409	0	0

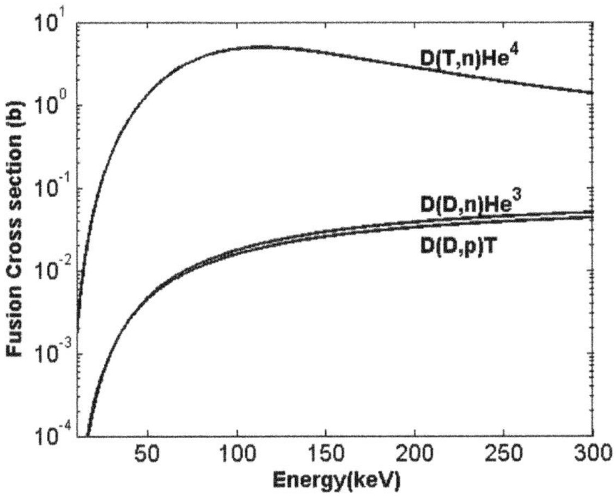

FIGURE 3.A2 D-D and D-T fusion cross-section.

Assuming $n_D = n_T = n/2$, the rate of reactions is

$$R = \frac{1}{4} n^2 \langle \sigma v \rangle$$

The energy of the hot plasma is

$$E = 3nk_B T$$

As a first estimate, equating the power produced by fusion, with the assumption that the neutron energy is lost and the charged particle energy E_c remains (due to the short range of ions) to the power lost during the energy confinement time, τ_E, gives an energy balance

$$\frac{1}{4} n^2 \langle \sigma v \rangle E_c \geq \frac{3nk_B T}{\tau_E}$$

which is rearranged to give

$$n\tau_E \geq \frac{12k_B}{E_c} \frac{T}{\langle \sigma v \rangle}$$

For D-T, the minimum value of $T/\langle \sigma v \rangle$ at 26 keV ($\sim$260 million K) and the values of k_B and E_c gives the *Lawson criterion* for DT fusion

$$n\tau_E \geq 1.5 \times 10^{20}\ \text{m}^{-3}\text{s}$$

The Lawson criterion is fundamental figure of merit to the realization of fusion as source of energy. In simple words, it says that the product of plasma density n and confinement τ_E must exceed the quantity $1.5 \times 10^{20}\ \text{m}^{-3}\text{s}$ which means the higher the density, the lower the confinement time and vice versa. To achieve breakeven through Lawson's criterion is the goal of all experiments.

Chapter 4

Mathematical foundations

This chapter reviews the mathematical foundations and knowledge required for understanding the formulation, performing a simulation, and coding for numerical implementation of problems in nuclear engineering. The material is presented in the following order: general mathematics covering ordinary and partial differential equations, mathematics specific to the neutron diffusion and transport equations followed by probability and statistics for Monte Carlo methods used in simulation. The objective is, again, to provide a comprehensive review of the mathematics applicable to nuclear engineering, usually spread over different subjects, in one resource available for study before or during the phase when problems are encountered.

Modeling in simulation in nuclear engineering covers several areas including neutron diffusion and transport, fluid dynamics, heat transfer, dynamics and control, and structural mechanics as depicted in Fig. 4.1.

The mathematical formulation of the governing equations is mostly in the form of differential equations, integral equations and integro-differential equations. Other than the deterministic formulations, simulation is carried out based on probability distributions and estimates in a stochastic formalism.

The objectives of mathematical modeling and simulation in nuclear engineering incude the design and performance anlayses of nuclear systems in many cases to compute the power produced in a nuclear reactor and to ensure it remains within safe critical bounds. For example, in a model of a reactor core shown in Fig. 4.2, the innovative nuclear reactor design has a core at the center, contains fuel rods and control rods placed vertically in assemblies, moderator and coolant and is surrounded by radiation shielding. The fluid removes the heat from the fissioning fuel rods and, as described in Section 4.3 for reactor designs, transfers its thermal energy to produce steam for a turbine and an electrical generator.

The mathematical models described in this chapter are intended to be applied for the computation of neutron and photon flux inside the core, carrying out a dynamic and safety analysis as well as to estimate the radiation leaving the

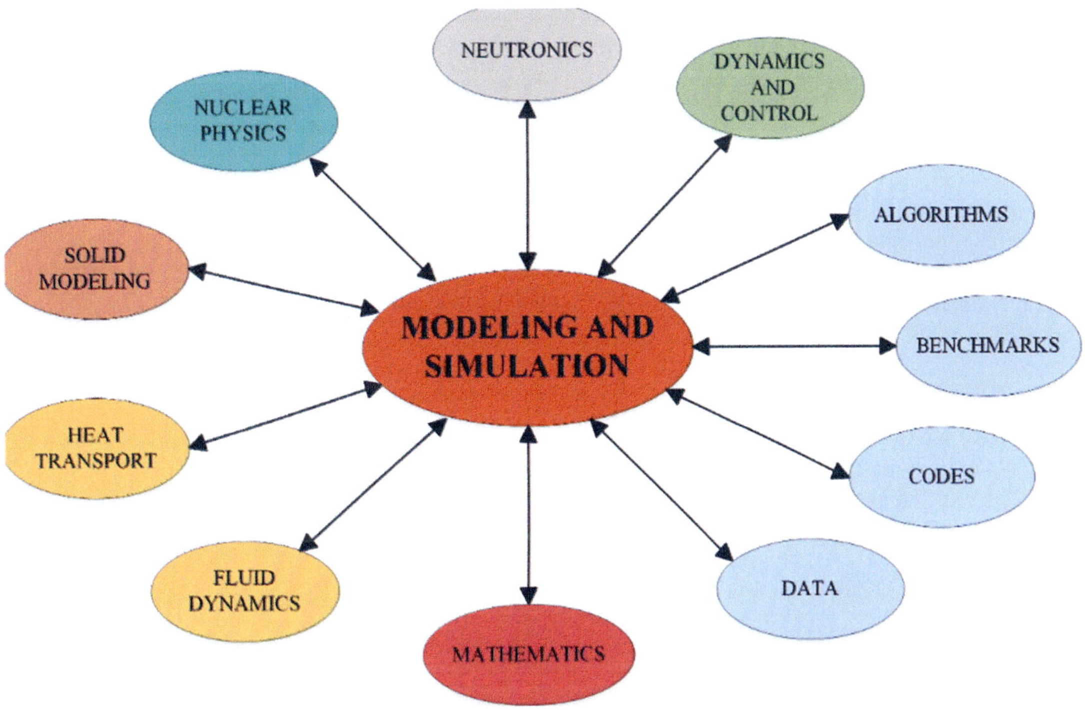

FIGURE 4.1 Modeling and simulation in nuclear engineering.

Nuclear Engineering. DOI: https://doi.org/10.1016/B978-0-323-90618-0.00004-1

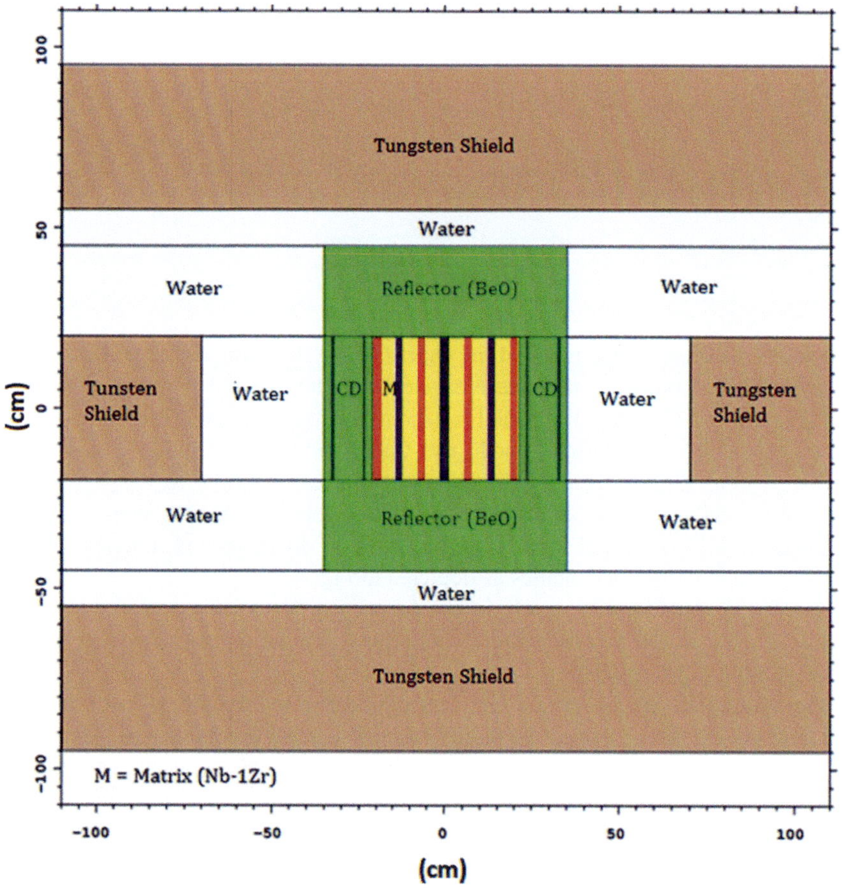

FIGURE 4.2 Model of a micronuclear reactor (side view).

core for which radiation shielding is designed as shown by the presence of water and tungsten shields in Fig. 4.2. The underlying physics of nuclear radiation and radiation transport was covered in Chapter 1 for charged particles and electromagnetic radiation and in Chapter 2 for neutron interactions with matter.

4.1 Ordinary differential equations

Ordinary differential equations (ODEs) are used to represent systems with one independent variable such as the spatial variable x, and can, in many cases, be solved exactly (Baker, 2016; Polyanin & Manzhirov, 2006; W. G. & D. Zwillinger, 1993). ODEs can be first-, second-, or higher order, where the order represents the power of the highest derivative of the dependent variable with respect to the independent variable.

A linear ODE is of the nth-order if it can be expressed as

$$y^{(n)} + f_1(x)y^{(n-1)} + \ldots + f_{n-1}(x)y' + f_n(x)y = R(x) \tag{4.1}$$

where $y^{(n)}$ is the nth derivative of the function y. If the RHS $= 0$, the ODE is called a homogeneous linear ODE. If any of the coefficients are a function of the dependent variable y, the ODE is nonlinear.

Thus, first-order differential equations are linear if they can be expressed as $y^{(1)} + f(x)y = R(x)$. They are easiest solved by the use of an integrating factor $\exp(\int f(x)dx)$; thus, the ODE $y^{(1)} + 2xy = x$ can be written as

$$\frac{d}{dx}ye^{x^2} = xe^{x^2}$$

and by integrating both sides

$$y(x)e^{x^2} = \int xe^{x^2} + C$$

giving the results

$$y(x) = \frac{1}{2} + Ce^{-x^2}.$$

Solving Nth-order ODEs is possible by reducing them to N first-order ODEs which, when linear, are easily cast into state-space equations amenable to standard matrix methods of linear algebra.

Examples of nonlinear first- and second-order ODEs are

$$y\frac{dy}{dx} - x = 0$$

and

$$a_1\frac{d^2y}{dx^2} + a_2\left(y^2 - 1\right)\frac{dy}{dx} + a_3y = 0.$$

If an ODE has coefficients that are functions of the independent variable x,

$$\frac{dy}{dx} + 2xy = x$$

it is called a variable coefficient linear ODE.

Equations of the form

$$\frac{dy}{dx} + p(x)y = g(x)$$

have an exact solution

$$y(x) = e^{-\int p(x)dx}\int g(x)e^{\int p(x)dx}dx.$$

In standard form, the Sturm–Liouville theory applies to second-order linear ODE boundary value problems (BVPs) that can be written as

$$\frac{d}{dx}\left[p(x)\frac{dy}{dx}\right] + q(x)y = -\lambda w(x)y(x) \tag{4.2}$$

where the functions $p(x), q(x), w(x)$ are the coefficient functions and λ, if it exists, is called the eigenvalue which is a scalar for which the equation is balanced and has eigenfunctions associated with the eigenvalue.

As an example, the steady-state neutron diffusion (Section 4.5) equation

$$\nabla \bullet D\nabla\phi - \Sigma_a\phi + S = 0 \tag{4.3}$$

and the steady-state heat conduction equation

$$\nabla.k\nabla T + S = 0 \tag{4.4}$$

are both second-order ODEs which can be expressed in the form of the Sturm–Liouville equation.

Let us therefore review a method of obtaining an exact solution of a simple form of such an equation: the linear nonhomogeneous form

$$a_1(x)\frac{d^2y}{dx^2} + a_2(x)\frac{dy}{dx} + a_3(x)y = F_{\text{ext}}$$

which arises in several areas of science and engineering such as the Schrodinger equation discussed in Chapter 1, forced vibrations in mechanical systems, including the Euler equation, the Bessel equation, the Airy equation. The Legendre equation. Eq. (4.2) also represents the 1-D motion in the mass-spring-damper problem, when ϕ represents the displacement $x(t)$, and x represents time t, where the restoring force of the spring is represented by Hooke's Law and an external time-dependent force F_{ext} is applied. The procedure of solving this equation is to seek a complementary solution $y_c(x)$ from the homogeneous equation and a particular solution $y(x)$ from the inhomogeneous term, and express the solution as

$$y(x) = y_c(x) + y_p(x).$$

Thus, in operator form

$$\left(a_1\hat{D}^2 + a_2\hat{D} + a_3\right)y_c(x) = 0$$

The operator $\hat{D}$ is treated as a scalar and roots are obtained for $a_1m^2 + a_2m + a_3 = 0$. For roots m_1, m_2, the solution is expressed as $y_c = \sum_{i=1}^{2} A_ie^{m_ix}$. For the particular solution

$$y_p(x) = \frac{1}{\left(a_1\hat{D}^2 + a_2\hat{D} + a_3\right)}F_{\text{ext}}.$$

The solution

$$y_p(x) = \sum_i A_ie^{m_ix}$$

where m_i are found from the particular solution and the coefficients A_i are determined from the given boundary conditions which can be Dirichlet (with the dependent variable specified on the boundaries), Neumann (with the normal derivative specified on the boundaries), or mixed boundary conditions.

Exercise 4.1: Find the exact solution of the equation.

$$D\frac{d^2\phi}{dx^2} - \Sigma_a\phi = 0$$

for a planar source at $x = 0$ emitting S neutrons $\text{cm}^{-2}\,\text{s}^{-1}$ with the boundary conditions: (1) Dirichlet *b. c*: finite flux $\phi(x)$, and (2) source Neumann condition: $\lim_{x\to 0} J(x) = S/2$; ($L^2 = D/\Sigma_a$). *Answer*: $\phi = \frac{SL}{2D}e^{-x/L}$.

Example 4.1: Find the exact solution of the second-order ODE.

$$D\frac{d^2\phi}{dx^2} - \Sigma_a\phi = -S$$

for a planar source at $x = 0$ emitting S neutrons $\text{cm}^{-2}\,\text{s}^{-1}$ in a finite slab of thickness $2d$. With the boundary conditions: (1) $\phi(a + d) = \phi(-a - d) = 0$, and (2) source condition: $\lim_{x\to 0} J(x) = S/2$; ($L^2 = D/\Sigma_a$). First obtain the complementary solution using the homogeneous equation

$$\left(\hat{D}^2 - 1/L^2\right)\phi_c(x) = 0.$$

The roots are $m_1 = 1/L$, $m_2 = -1/L$, giving the complementary solution

$$\phi_c(x) = C_1e^{x/L} + C_2e^{-x/L}$$

For the particular solution

$$\phi_p = \frac{1}{(\hat{D} - 1/L)(\hat{D} + 1/L)}(-S)$$

which can be written in partial fractions as

$$\phi_p = \frac{1}{2}\left[\frac{1}{(L\hat{D} - 1)} - \frac{1}{(L\hat{D} + 1)}\right](-SL^2)$$

and expressed as

$$\phi_p = \frac{1}{2}\left[\left(1 - L\hat{D}\right)^{-1} + \left(1 + L\hat{D}\right)^{-1}\right](SL^2)$$

Now expanding $\left(1 - L\hat{D}\right)^{-1} = 1 + L\hat{D} - \left(L\hat{D}\right)^2 + \cdots$, gives

$$\phi_p = SL^2$$

The solution is thus

$$\phi(x) = A_1 \cosh x/L + A_2 \sinh x/L + SL^2$$

and the coefficients A_1, A_2 can be found from the boundary conditions. Using the boundary conditions gives

$$\phi(x) = \frac{SL}{2D} \frac{\sinh\left[(a+d-|x|)/L\right]}{\cosh\left[\frac{a+d}{L}\right]}.$$

In case there are a number of sources, placed at different locations, the Green's function can be used for calculating the flux in the medium.

4.1.1 The Poisson equation: steady-state heat conduction in 1-D

The steady-state heat conduction equation (Eqs. 4.3 and 4.4) with Fick's law $\overline{q} = -k\overline{\nabla}T$, gives the steady-state heat conduction (as for neutron diffusion) equation with heat generation for constant k in regular geometry:

$$\text{rectangular } \frac{d^2T}{dx^2} + \frac{q'''}{k} = 0$$

$$\text{cylindrical } \frac{1}{r}\frac{d}{dr}\left(r\frac{dT}{dr}\right) + \frac{q'''}{k} = 0$$

$$\text{spherical } \frac{1}{r^2}\frac{d}{dr}\left(r^2\frac{dT}{dr}\right) + \frac{q'''}{k} = 0$$

of the form $\nabla^2\phi = S$, which is called the Poisson equation; in the homogeneous case ($S = 0$), the equation reduces to the Laplace equation. These equations can be solved by the standard procedure given above.

Exercise 4.2: For a fuel rod considered as a plate of thickness $2a$ and clad of thickness b with the boundary conditions:

$$T(0) = T_m, \left.\frac{dT}{dx}\right|_{x=0} = 0$$

obtain the temperature distribution, in the fuel and clad, and the heat flow using the AP1000 data from Chapter 3 (Table 3.3): fuel radius = 4.095 mm, gap thickness 0.082 mm, clad thickness 0.5715 mm.

The AP1000 fuel rods are of a cylindrical configuration with cross section. In this model, we can assume that the "half-thickness" of a fuel plate is a equal to the radius of the fuel (Fig. 4.3).

Assume that the thermal conductivity of UO_2 (IAEA, 2008) is given as

$$k_f(\tau) = \frac{100}{7.5408 + 17.692\tau + 3.6142\tau^2} + \frac{6400}{\tau^{5/2}}\exp\left(-\frac{16.35}{\tau}\right) \text{W/m K}$$

where

$$\tau = \frac{T}{1000}$$

and for the cladding, $k_c = 36\text{W/mK}$.

Recall that the steady-state heat equation

$$\nabla^2 T + \frac{q'''}{k} = 0$$

is valid for constant thermal conductivity while k_f is generally a function of temperature or space in the case of a composite wall. Thus to solve the equation, it will be easier to assume that k_f is constant. When this assumption is valid, the above ODE can be easily solved to yield the temperature distribution

$$T(x) = T_m - \frac{q'''}{2k_f}x^2$$

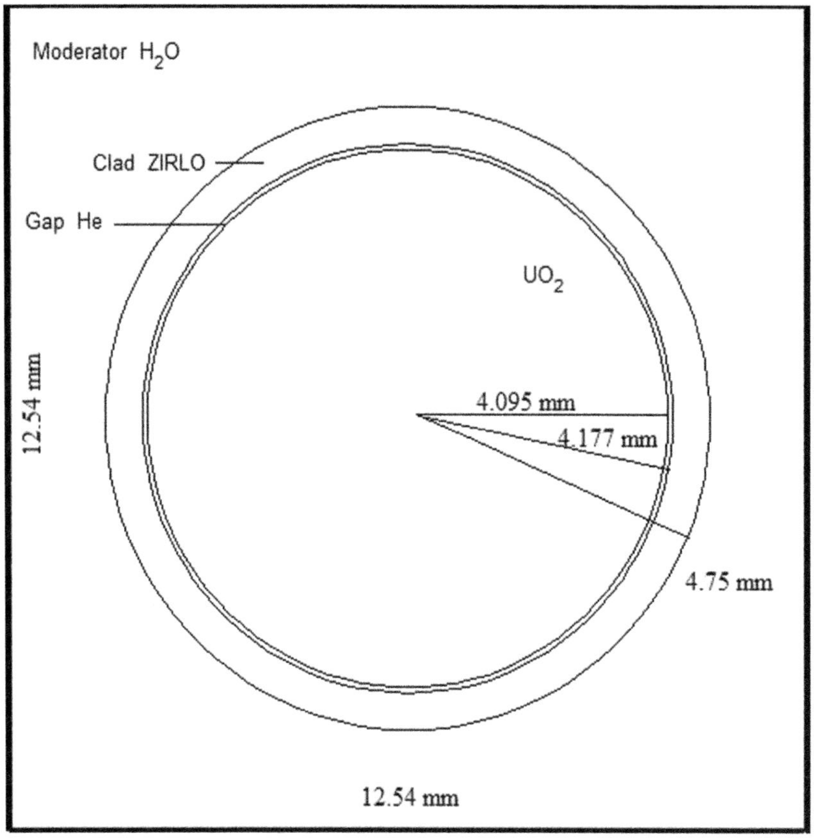

FIGURE 4.3 AP1000 fuel rod in a moderator cell.

and the heat flow is found by integrating from $\int_0^a q\,dx = -k\int_0^a dT(x)$. Also

$$q = \frac{T_m - T_s}{R_f}$$

where $T(a) = T_s$ and the "resistance" (analogous to Ohms Law $V = IR$) is

$$R_f = \frac{a}{2k_f A}.$$

In the cladding, the temperature is found by integrating $\nabla^2 T = 0$ to get

$$T(x) = C_1 x + C_2$$

in terms of the constants of integration C_1 and C_2.

With the outer surface temperature $T(a+b) = T_c$, the temperature in the cladding is

$$T(x) = T_s - \frac{x-a}{b}(T_s - T_c).$$

Thus the overall heat flow is

$$q = \frac{T_m - T_c}{R_f + R_c}$$

where the resistance of the cladding is

$$R_c = \frac{b}{k_c A}.$$

In the above exercise, the thermal conductivity is a function of the dependent variable $T(x)$; thus the governing equation in plane geometry is the nonlinear equation

$$k_f(T)\frac{d^2T}{dx^2} + q''' = 0.$$

In some cases, it is possible to obtain exact solutions of such nonlinear equations, but generally numerical methods are required as will be discussed in this chapter.

4.1.2 Coupled first-order ODEs: the point kinetics equations

The Point Kinetics Equations (PKE) in a lumped model of a nuclear reactor, for the neutron density $n(t)$ and the concentration of the ith nuclide $C_i(t)$ from Chapter 2, are

$$\frac{dn(t)}{dt} = \frac{\rho(t) - \tilde{\beta}}{\Lambda} n(t) + \sum_{i=1}^{6} \lambda_i C_i(t)$$

and

$$\frac{dC_i(t)}{dt} = \frac{\beta_i(t)}{\Lambda} n(t) - \lambda_i C_i(t), i = 1, \cdots 6$$

with initial conditions $n(0)$ and $C_i(0)$. These are two coupled first-order ODEs that can be solved by combining them into one second-order ODE, or by taking Laplace transforms

$$N(s) = \int_0^\infty n(t)e^{-st}dt$$

and

$$\Gamma_i(s) = \int_0^\infty C_i(t)e^{-st}dt$$

which yields

$$N(s) = \frac{\ell\left[n(0) + \sum_i \frac{\lambda_i C_i(0)}{s+\lambda_i}\right]}{\ell s + \beta - \rho_o - \sum_i \frac{\beta_i \lambda_i}{s+\lambda_i}}.$$

The solution, obtained by inversion, is

$$n(t) = \sum_{i=1}^{7} A_i e^{\omega_i t}$$

with each root ω_i satisfying the "inhour equation"

$$\rho_0 \quad = \beta + \ell\omega - \sum_i \frac{\beta_i \lambda_i}{\omega + \lambda_i}.$$

Since the total delayed fraction $\beta = \sum_i \beta_i$

$$\rho_0 \quad = \ell\omega + \sum_i \frac{\beta_i \omega}{\omega + \lambda_i}.$$

Thus the roots ω_i can be found explicitly or by solving the transcendental equation giving the solution.

Exercise 4.3: Solve the coupled equations PKE to show, for one group of delayed neutrons, that the solution is.

$$\begin{bmatrix} \hat{n}(t) \\ \hat{C}(t) \end{bmatrix} = a_0 \begin{bmatrix} 1 \\ \dfrac{\beta}{\Lambda(\lambda + \omega_0)} \end{bmatrix} e^{\omega_0 t} + a_1 \begin{bmatrix} 1 \\ \dfrac{\beta}{\Lambda(\lambda + \omega_1)} \end{bmatrix} e^{\omega_1 t}$$

where the roots are

$$\omega_{0,1} = \frac{1}{2}\left[\left(\frac{\rho-\beta}{\Lambda} - \lambda\right) \pm \sqrt{\left(\frac{\rho-\beta}{\Lambda} - \lambda\right)^2 + 4\lambda\frac{\rho}{\Lambda}}\right].$$

Evaluate the roots for $\rho = 0.2\beta$ given the thermal fission U235 data

$$\lambda = 0.0767\text{s}^{-1}, \beta = 0.00650, \Lambda = \frac{\beta}{\lambda} = 0.0847\text{s}$$

and plot the solutions. In general, there is one real root with the same sign of the applied reactivity while all other roots are negative and real.

4.2 Partial differential equations

Partial differential equations (PDEs) represent systems where the differential equation has two or more independent variables such as spatial dimensions x, y, z, orthogonal and azimuthal angles θ, φ, energy E, and time t. A linear second-order PDE

$$a(x,y)\frac{\partial^2\phi}{\partial y^2} + b(x,y)\frac{\partial^2\phi}{\partial x\partial y} + c(x,y)\frac{\partial^2\phi}{\partial x^2} + d(x,y)\frac{\partial\phi}{\partial y} + e(x,y)\frac{\partial\phi}{\partial x} + f(x,y)\phi = 0 \tag{4.5}$$

is classified as elliptic, parabolic, or hyperbolic if $b^2 - 4ac$ is less than, equal to, or greater than zero, respectively.

Using the classification criterion specified above, we can classify the PDE:

$$\frac{1}{v}\frac{\partial\phi(r,t)}{\partial t} = \text{D}\frac{\partial^2\phi}{\partial x^2} - \Sigma_a\phi(r,t) + S(r,t)$$

with $a = b = 0, c = D$; thus $b^2 - 4ac = 0$, as a parabolic equation.

This implies a certain kind of boundary conditions: a Dirichlet or Neumann boundary condition on an open surface to be specified for a stable solution. Similarly, elliptical PDEs require Dirichlet or Neumann boundary conditions on a closed surface surrounding the region of interest, while hyperbolic PDEs require Cauchy boundary conditions on an open surface.

The Handbook of Nonlinear Partial Differential Equations (Polyanin & Zaitsev, 2004) gives solutions for nonlinear parabolic, hyperbolic, and elliptic equations, nonlinear equations with exponential, hyperbolic, logarithmic, and trigonometric nonlinearities in one, two, or more space variables, as well as for higher order equations. The exact methods given for solving nonlinear PDEs include characteristic equations and canonical forms.

The solution procedure is usually to convert PDEs to ODEs and then solved to obtain exact analytical solutions where possible, or a system of algebraic equations $\bar{\bar{A}}\bar{x} = \bar{B}$ where $\bar{\bar{A}}$ is a matrix, $\bar{x}$ is the vector being solved, and $\bar{B}$ is a known "force" vector. The solution is thus obtained, from standard numerical techniques, such as Gaussian elimination, Gauss–Siedel iterative methods, LU decomposition methods, etc., by inversion of the matrix $\bar{\bar{A}}$ giving $\bar{x} = \bar{\bar{A}}^{-1}\bar{B}$.

4.2.1 Equations of fluid dynamics

Examples of PDEs used in thermal hydraulics are the continuity equation

$$\frac{\partial\rho}{\partial t} + \nabla\bullet(\rho\bar{u}) = 0,$$

the conservation of momentum equation

$$\frac{\partial\rho\bar{u}}{\partial t} + \nabla\bullet(\rho\bar{u}\bar{u}) = -\overline{\nabla}p + \overline{\nabla}\tau + \rho\bar{g},$$

and the conservation of energy equation

$$\frac{\partial\rho e}{\partial t} - \frac{\partial p}{\partial t} + \nabla\bullet(\rho\bar{u}e) = \nabla\bullet\left(k\overline{\nabla}T\right) - p\nabla\bullet(\bar{u}) + \Phi + S$$

To "close" the above system, an equation of state

$$\rho = \rho(P,T)$$

is used for solving for the six independent variables $\rho, \bar{u}, p, e$.

4.2.2 The 1-D time-dependent heat conduction

A 1-D time-dependent form of Eq. (4.4) for temperature, with constant thermal conductivity k, from an energy balance, is

$$\frac{\partial^2 T}{\partial x^2} + \frac{q'''}{k} = \frac{1}{\alpha}\frac{\partial T}{\partial t}$$

where $\alpha = k/\rho c$ is the thermal diffusivity of the material and can be solved with given boundary conditions for a 1-D slab shown in Fig. 4.4.

Assuming separability $T(x,t) = X(x)\Gamma(t)$ and setting equal to a separability constant $-\beta^2$

$$\frac{1}{X(x)}\frac{d^2X(x)}{dx^2} = \frac{1}{\alpha\Gamma(t)}\frac{d\Gamma(t)}{dt} = -\beta^2$$

the PDEs are converted to a set of ODEs

$$\frac{d\Gamma(t)}{dt} + \alpha\beta^2\Gamma(t) = 0$$

$$\frac{d^2X(x)}{dx^2} + \beta^2 X(x) = 0$$

with solutions $\Gamma(t) = e^{-\alpha\beta^2\tau}$ and $X(\beta_m x)$. This is an eigenvalue problem, as it satisfies a boundary condition such as $X'|_0 = 0$ and for $X(x) = \cos\beta x$, $X'|_0 = = -\beta\sin\beta x = 0$, so either $\beta = m\pi$, $m = 0, 1, 2, \ldots$. The solution is thus

$$T(x,t) = \sum_{m=1}^{\infty} c_m X(\beta_m x) e^{-\alpha\beta_m{}^2\tau}.$$

The spatial part of the solution is

$$F(x) = \sum_{m=1}^{\infty} c_m X(\beta_m x)$$

for which the coefficients are determined using orthogonality

$$\int_0^L X(\beta_m, x)X(\beta_n, x)dx = \begin{cases} 0 & \text{for } \quad m \neq n \\ N(\beta_m) & \text{for m} = n \end{cases}$$

where the normalization integral $N(\beta_m)$ is defined as

$$N(\beta_m) = \int_0^L [X(\beta_m, x)]^2 dx$$

and

$$c_m = \frac{1}{N(\beta_m)} = \int_0^L X(\beta_m, x)F(x)dx$$

to yield the time-dependent solution

$$T(x,t) = \sum_{m=1}^{\infty} e^{-\alpha\beta_m{}^2\tau} \frac{1}{N(\beta_m)} X(\beta_m, x) \int_0^L X(\beta_m, x')F(x')dx'.$$

Here F(x) is an arbitrary function and the X functions depend on the boundary conditions.

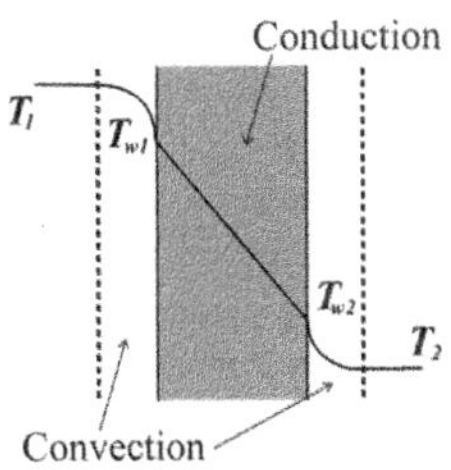

FIGURE 4.4 Boundary conditions on a 1-D slab.

4.2.3 Laplace equation: 2-D steady-state heat conduction

For 2-D steady-state conduction with constant k and no heat generation, that is, $q''' = 0$, Eq. (4.4) is the Laplace equation

$$\left(\frac{\partial^2}{\partial x^2} + \frac{\partial^2}{\partial y^2}\right)T(x, y) = 0$$

for a square plate of width W and height H. Given the Dirichlet boundary conditions $T(x, H) = T_u$ and $T(W, y) = T(0, y) = T(x, 0) = T_s$ the equation is solved by assuming separability $T(x, y) = X(x)Y(y)$; this gives

$$\frac{1}{X}\frac{d^2X}{dx^2} + \frac{1}{Y}\frac{d^2Y}{dy^2} + \lambda^2 = 0$$

with two second-order ODEs

$$\frac{1}{X}\frac{d^2X}{dx^2} = -\beta^2, \quad \frac{1}{Y}\frac{d^2Y}{dy^2} = -\gamma^2,$$

where $\beta^2 + \gamma^2 = \lambda^2$. The solutions of X and Y are trigonometric functions with four constants determined by the boundary conditions and from the orthogonality of Legendre polynomials. The exact solution is

$$T(x, y) = (T_u - T_s)\frac{2}{\pi}\sum_{n=1}^{\infty}\frac{(-1)^{n+1} + 1}{n}\sin\left(\frac{n\pi x}{W}\right)\frac{\sinh\left(\frac{n\pi y}{W}\right)}{\sinh\left(\frac{n\pi H}{W}\right)} + T_s.$$

The above solutions were considered for steady-state and constant thermal conductivity k. A numerical finite-difference solution for this 2D plate is given in Section 4.5.1. In general, if $k(T) = f(T)$, the equation on a square plate is an elliptic nonlinear Poisson equation.

$$k(T)\left(\frac{\partial^2}{\partial x^2} + \frac{\partial^2}{\partial y^2}\right)T(x, y) + q''' = 0.$$

Some homogeneous cases for which numerical methods are compared with exact solutions (Filipov & Faragó, 2018; Yeih, 2020) are for $k(T) = T$ on a circle of unit radius and on an annular region; $k(T) = e^T$ and on a square plate $0 \leq x \leq 1, 0 \leq y \leq 1$.

For $k(T) = e^T$, the exact solution is $T(x, y) = \log(xy + 5)$, shown in Fig. 4.5, which is compared with numerical solutions which are generally used for nonlinear PDEs.

Since it is not always possible to obtain exact solutions, numerical methods are used as discussed later in this chapter.

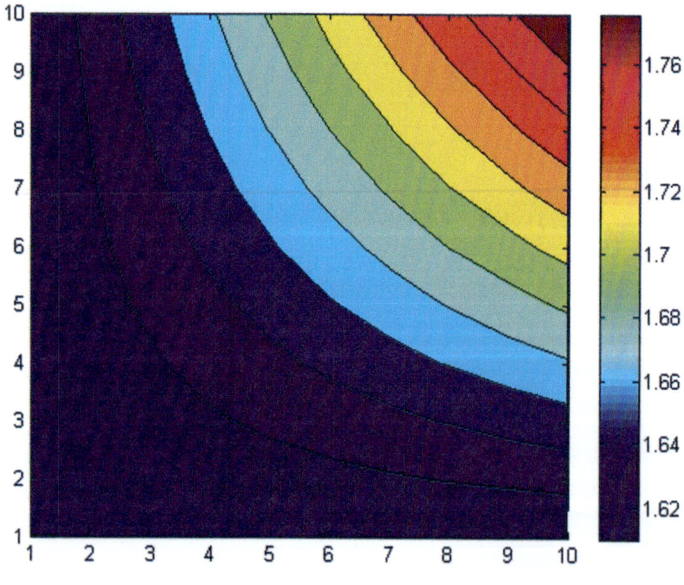

FIGURE 4.5 Temperature field $T(x,y)$ on a square plate ($k(T) = e^T$).

4.2.4 Heat conduction in 2-D and 3-D

In rectangular, cylindrical, and spherical coordinate systems, the governing heat conduction equation can be written as

$$\frac{1}{r^n}\frac{\partial}{\partial r}\left(r^n k\frac{\partial T}{\partial r}\right) + q''' = \rho c\frac{\partial T}{\partial t}$$

with $n = 0$ (slab), 1 (cylinder), an 2 (sphere).

Thus, in rectangular geometry,

$$k\left(\frac{\partial^2}{\partial x^2} + \frac{\partial^2}{\partial x^2}\right)T(x,y) + q''' = \rho c\frac{\partial T}{\partial t}$$

This equation is solved assuming separability, similar to the 1-D case. In the steady-state heat conduction equation, for 2-D or 3-D problems, the governing equation becomes a PDE with the appropriate forms of the Laplacian.

The Laplace equation is solved by assuming separability when the resulting equations can be written as ODEs of the form for harmonic motion. With the relevant boundary conditions, solutions can be expressed in terms of modes, or eigenfunctions, for eigenvalues. The solutions can be expressed in terms of an infinite series and have orthogonal properties.

In 3-D, the rectangular coordinate equation for thermal conductivity k which may be space- or temperature-dependent, the equations are

$$\frac{\partial}{\partial x}\left(k\frac{\partial T}{\partial x}\right) + \frac{\partial}{\partial y}\left(k\frac{\partial T}{\partial y}\right) + \frac{\partial}{\partial z}\left(k\frac{\partial T}{\partial z}\right) + q''' = \rho c\frac{\partial T}{\partial t}$$

For a cylinder $x = r\cos\phi, y = r\sin\phi$

$$\frac{1}{r}\frac{\partial}{\partial r}\left(k\frac{\partial T}{\partial r}\right) + \frac{1}{r^2}\frac{\partial T}{\partial \phi}\left(k\frac{\partial T}{\partial \phi}\right) + \frac{\partial}{\partial z}\left(k\frac{\partial T}{\partial z}\right) + q''' = \rho c\frac{\partial T}{\partial t}$$

For a sphere, $x = r\cos\phi\sin\theta, y = r\sin\phi\sin\theta$, and $z = \cos\theta$ (Fig. 4.6)

$$\frac{1}{r^2}\frac{\partial}{\partial r}\left(kr^2\frac{\partial T}{\partial r}\right) + \frac{1}{r^2\sin^2\theta}\frac{\partial}{\partial \phi}\left(k\frac{\partial T}{\partial \phi}\right) + \frac{1}{r^2\sin\theta}\frac{\partial}{\partial \theta}\left(k\sin\theta\frac{\partial T}{\partial \theta}\right) + q''' = \rho c\frac{\partial T}{\partial t}$$

Some boundary conditions and initial conditions need to be specified to solve the above: Neumann, Dirichlet, or mixed are given in Table 4.1.

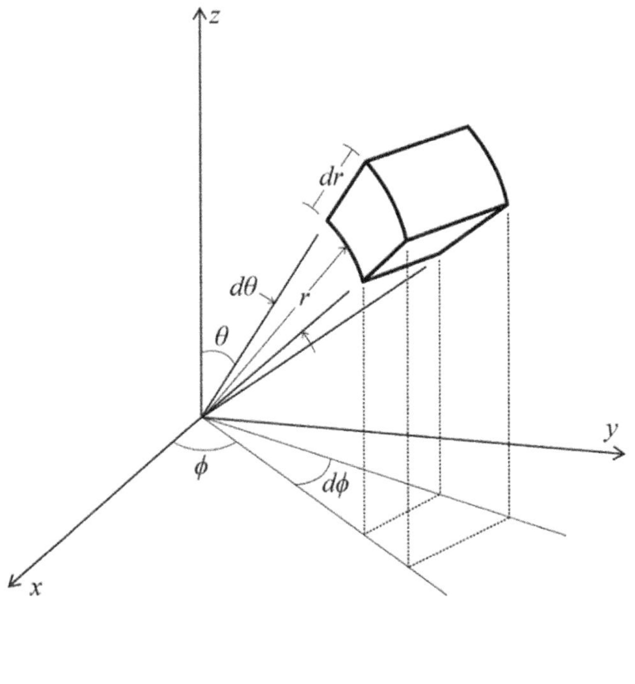

FIGURE 4.6 Solid angle.

TABLE 4.1 Dirichlet, Neuman and mixed boundary conditions.

Boundary type	Boundary condition
Temperatures at surfaces	$T(0,t) = T_1,\ T(L,t) = T_2$
Heat fluxes (W/m^2)	$q = -k\frac{\partial T}{\partial x}$
Insulated boundary	$\frac{\partial T(0,t)}{\partial x} = 0$
Reflecting boundary	$\frac{\partial T(L/2,t)}{\partial x} = 0$
Convection boundary condition	$-k\frac{\partial T(0,t)}{\partial x} = h_1[T_{\infty 1} - T(0,t)]$
Radiation boundary condition	$-k\frac{\partial T(0,t)}{\partial x} = \varepsilon_1\sigma\left[T^4_{\text{surr},1} - T^4(0,t)\right]$
Interface boundary conditions	$T_A(x_0,t) = T_B(x_0,t),\ -k_A\frac{\partial T_A(x_0,t)}{\partial x} = -k_B\frac{\partial T_B(x_0,t)}{\partial x}$

In the time-dependent case, assuming separability $T(r,t) = \psi(r)\Gamma(t)$ in the usual manner

$$\frac{1}{\psi(r)}\nabla^2\psi(r) = -\frac{1}{\alpha\Gamma(t)}\frac{d\Gamma(t)}{dt} = -\lambda^2$$

to get the Helmholtz equation

$$\nabla^2\psi(r) + \lambda^2\psi(r) = 0$$

Which can be reduced to ODEs in coordinate systems including the rectangular, cylindrical, and spherical systems.

Exercise 4.4: For 3-D Cartesian geometry, extend the analysis of Sections 4.4.2.3 and 4.2.3 to find the exact solution for the time-dependent temperature given the initial conditions.

In cylindrical geometry, with the same procedure followed above, for

$$\frac{1}{\psi}\left(\frac{\partial^2\psi}{\partial r^2} + \frac{1}{r}\frac{\partial\psi}{\partial r} + \frac{1}{r^2}\frac{\partial^2\psi}{\partial\phi^2} + \frac{\partial^2\psi}{\partial z^2}\right) = \frac{1}{\alpha\Gamma(t)}\frac{d\Gamma(t)}{dt} = -\lambda^2$$

and $(r,\phi,z) = R(r)\Phi(\phi)Z(z)$, the ODEs are

$$\frac{1}{Z}\frac{d^2Z}{dz^2} = -\eta^2,\quad \frac{1}{\Phi}\frac{d^2\Phi}{d\phi^2} = -v^2, \text{and} \frac{1}{R_v}\left(\frac{d^2R_v}{dr^2} + \frac{1}{r}\frac{dR_v}{dr}\right) - \frac{v^2}{r^2} = -\beta^2$$

The separated ODEs and their solutions are

$$\frac{d^2Z}{dz^2} + \eta^2 Z = 0 Z(\eta,z)\text{:}\sin\eta z \text{and} \cos\eta z$$

$$\frac{d^2\Phi}{d\phi^2} + v^2\Phi = 0 \Phi(v,\phi)\text{:}\sin v\phi \text{and} \cos v\phi$$

$$\frac{d^2R_v}{dr^2} + \frac{1}{r}\frac{dR_v}{dr} + \left(\beta^2 - \frac{v^2}{r^2}\right)R_v = 0 R_v(\beta,r)\text{:}J_v(\beta r)\text{and}Y_v(\beta r)$$

$$\beta^2 + \gamma^2 + \eta^2 = \lambda^2$$

$$\frac{d\Gamma}{dt} + \alpha\lambda^2\Gamma = 0\Gamma(t)\text{:}e^{-\alpha\lambda^2\tau}$$

Bessel functions are solutions of the νth order (ν is a nonnegative real number) second-order ODEs

$$\frac{d^2y}{dx^2} + \frac{1}{x}\frac{dy}{dx} + \left(1 - \frac{\nu^2}{x^2}\right)y = 0$$

There are functions of the first kind, which are regular at the origin (e.g., I_0, I_1, J_0, J_1), and functions of the second kind which are singular at the origin (e.g., Y_0, Y_1, K_0, K_1). The solutions which are linear combinations of $J_\nu(x)$ and

$Y_\nu(x)$. These are related as

$$Y_\upsilon(x) = \frac{\cos\pi\upsilon J_\upsilon(x) - J_{-\upsilon}(x)}{\sin\pi\upsilon}$$

for $J_{-\nu} = (-1)^\nu J_\nu$. When the differential equation is written in a modified form:

$$\frac{d^2y}{dx^2} + \frac{1}{x}\frac{dy}{dx} - \left(1 + \frac{\upsilon^2}{x^2}\right)y = 0$$

the solutions are a linear combination of $I_\nu(x)$ and $K_\nu(x)$: $K_\nu(x) = \frac{\pi}{2}\frac{I_{-\nu}(x) - I_\nu(x)}{\sin\pi\nu}$. Functions of the first and second kind are shown in Fig. 4.7.

Now, consider the 1-D cylindrical form of the neutron diffusion equation (Eq. 4.3)

$$D\left(\frac{d^2}{dr^2} + \frac{1}{r}\frac{d}{dr}\right)\phi(r) - \Sigma_a\phi(r) + S(r) = 0$$

For a "line source" where the source extends along a line on the *z*-axis at *r = 0*, the homogeneous form of the equation will apply to all *r* excluding the source. This will be the modified Bessel equation form given above with $\nu = 0$ and a

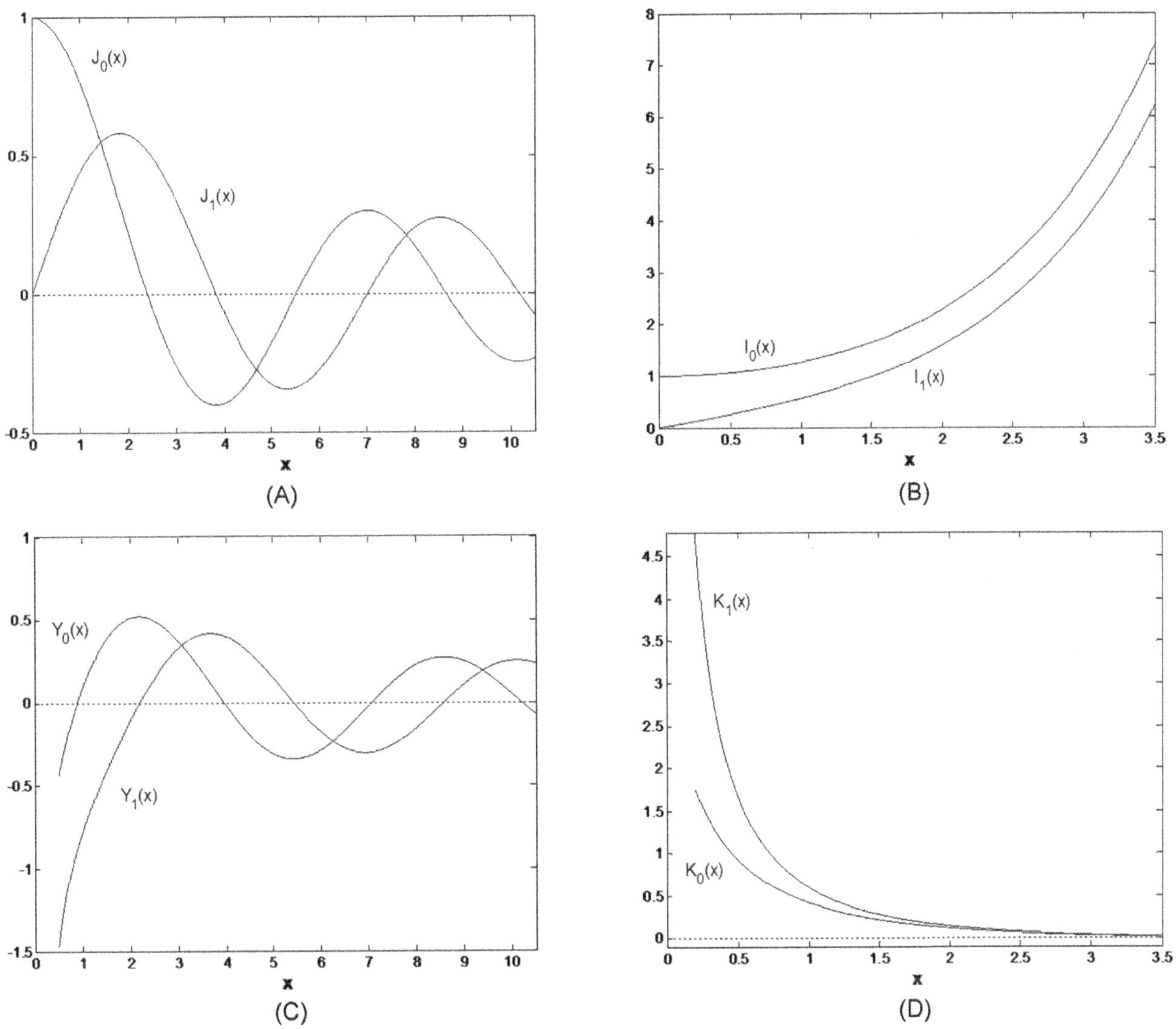

FIGURE 4.7 (A–D) Bessel functions of the first and second kind.

solution of the form

$$\phi(r) = A_1 I_0(r) + A_2 K_0(r)$$

where A_1 and A_2 are constants. Since we require a solution at $r \neq 0$, and the function $I_0(r)$ grows with increasing r while $K_0(r)$ remains bounded and decreases, the solution will be $\phi(r) = A_2 K_0(r)$. For the Bessel functions, the series forms are:

$$I_0(x) = 1 + \frac{x^2}{4} + \frac{x^4}{64} + \frac{x^6}{2304} + \ldots \text{ and } K_0(x) = -\left(\gamma + \ln\frac{x}{2}\right) I_0(x) + \frac{x^2}{4} + \frac{3x^4}{128} + \frac{11x^6}{13,824} + \ldots$$

where $\gamma = -0.577215665\ldots$ is Euler's constant.

4.2.4.1 Spherical geometry

In spherical coordinates, the temperature field, in general, is $T = T(r, \theta, \phi, t)$ given by the governing equation

$$\frac{\partial^2 T}{\partial x^2} + \frac{2}{r}\frac{\partial T}{\partial r} + \frac{1}{r^2 \sin\theta}\frac{\partial}{\partial \theta}\left(\sin\theta \frac{\partial T}{\partial \theta}\right) + \frac{1}{r^2 \sin^2\theta}\frac{\partial^2 T}{\partial \phi^2} = \frac{1}{\alpha}\frac{\partial T}{\partial t}$$

with $\mu = \cos\theta$

$$\frac{\partial^2 T}{\partial r^2} + \frac{2}{r}\frac{\partial T}{\partial r} + \frac{1}{r^2}\frac{\partial}{\partial \mu}\left[\left(1-\mu^2\right)\frac{\partial T}{\partial \mu}\right] + \frac{1}{r^2(1-\mu^2)}\frac{\partial^2 T}{\partial \phi^2} = \frac{1}{\alpha}\frac{\partial T}{\partial t}$$

defining a new variable $V = r^{1/2} T$ and simplifying for azimuthal symmetry, temperature is a function of r, μ, t:

$$\frac{\partial^2 V}{\partial r^2} + \frac{1}{r}\frac{\partial V}{\partial r} - \frac{1}{4}\frac{V}{r^2} + \frac{1}{r^2}\frac{\partial}{\partial \mu}\left[\left(1-\mu^2\right)\frac{\partial V}{\partial \mu}\right] = \frac{1}{\alpha}\frac{\partial V}{\partial t}$$

With separability

$$V(r, \mu, \phi, t) = \Gamma(t) R(r) M(\mu)$$

the ODEs are

$$\frac{d}{d\mu}\left[\left(1-\mu^2\right)\frac{\partial M}{\partial \mu}\right] + n(n+1)M = 0$$

$$\frac{d^2 R}{dx^2} + \frac{1}{2}\frac{dR}{dr} + \left[\lambda^2 - \left(n + \frac{1}{2}\right)^2 \frac{1}{r^2}\right] R = 0$$

with elementary solutions

$$P_n(\mu) \text{ and } Q_n(\mu)$$

$$J_{n+1/2}(\lambda r) \text{ and } Y_{n+1/2}(\lambda r)$$

where $P_n(\mu)$ and $Q_n(\mu)$ are Legendre functions of the first and second kinds, respectively.

$$\frac{d\Gamma}{dt} + \alpha\lambda^2 \Gamma = 0 \Gamma(t) : e^{-\alpha\lambda^2 \tau}$$

The above can also be written as, with $x = r\sin\theta\cos\varphi$, $y = r\sin\theta\sin\varphi$, $z = r\cos\theta$, and $f(r, \theta, \varnothing) = R(r)\Theta(\theta)\Phi(\varphi)$, the equations are

$$\frac{\partial^2 \Phi}{\partial \varphi^2} = -m^2 \Phi(\varphi)$$

$$\frac{1}{\sin\theta}\frac{\partial}{\partial \theta}\left(\sin\theta \frac{\partial \Theta}{\partial \theta}\right) - \frac{m^2}{\sin^2\theta}\Theta(\theta) = -l(l+1)\Theta(\theta)$$

$$\frac{1}{r^2}\frac{\partial}{\partial r}\left(r^2 \frac{\partial R}{\partial r}\right) - \frac{l(l+1)}{r^2} R = \lambda_\alpha R(r)$$

where $\Theta(\theta)\Phi(\varphi) = Y_{l,m}(\theta,\varphi)$. Here, l, m are integers, and $l \geq |m|$, or $-m \leq l \leq m$, and

$$Y_{l,m}(\theta,\varphi) = \sqrt{\frac{2l+1}{4\pi}\frac{(l-m)!}{(l+m)!}}P_l^m(\cos\theta)e^{im\varphi}$$

where $Y_{l,m}$ are spherical harmonics and P_l^m are associated Legendre functions. Note that

$$P_l^m(x) = (-1)^m\left(1-x^2\right)^{m/2}\frac{d^m}{dx^m}P_l(x)$$

For azimuthal symmetry, $m = 0$ (azimuthal symmetry) $P_l^m(x) = P_l(x)$. The orthogonality condition gives the orthogonality condition

$$\int_{-1}^{1} P_l^m(x)P_n^m(x)dx = \frac{2}{2n+1}\frac{(n+m)!}{(n-m)!}\delta_{ln}$$

The spherical harmonics functions are orthogonal and normalized; being complete, any function can be expressed in terms of spherical harmonics. The solution can thus be written in the form:

$$f_\alpha(r,\theta,\varnothing) = Y_{l,m}(\theta,\varnothing)\frac{1}{\sqrt{r}}J_{l+\frac{1}{2}}\left(j_{l+1/2,n}r/a\right)$$

where $j_{l+1/2,n}$ represents the nth zero of the Bessel function $J_{l+\frac{1}{2}}$. $\alpha = (n,l,m), n = 1,2,3,\cdots$.

Legendre's equation is

$$\frac{1}{\sin\theta}\frac{\partial}{\partial\theta}\left(\sin\theta\frac{\partial\Theta}{\partial\theta}\right) + l(l+1)\Theta(\theta) = 0,$$

which, for $x = \cos\theta$, can be expressed as an eigenvalue equation

$$\left(\frac{d}{dx}\left(1-x^2\right)\frac{d}{dx}\right)\Theta(x) = -l(l+1)\Theta(x)$$

for $-1 \leq x \leq 1$, has solutions $P_l(x)$, called Legendre polynomials, defined by Rodriguez' formula (Abramowitz & Stegun, 1964)

$$P_l(x) = \frac{1}{2^l l!}\frac{d^l}{dx^l}\left(x^2-1\right)^l.$$

These polynomials are complete and have the orthogonality condition:

$$\int_{-1}^{1} P_l(x)P_m(x)dx = \frac{2}{2l+1}\delta_{lm}$$

$$P_l(-x) = (-1)^l P_l(x).$$

Note that for the normalization $P_n(1) = 1$ for Legendre polynomials given in Table 4.2, it is easily seen for these cases. The recurrence relation (Abramowitz & Stegun, 1964)

$$(n+1)P_{n+1}(x) - (2n+1)xP_n(x) + nP_{n-1}(x) = 0$$

TABLE 4.2 Legendre polynomials.

l	$P_l(\mu)$	l	$P_l(\mu)$
0	1	1	μ
2	$\frac{1}{2}(3\mu^2-1)$	3	$\frac{1}{2}(5\mu^3-3\mu)$
4	$\frac{1}{8}(35\mu^4-30\mu^2+3)$	5	$\frac{1}{8}(63\mu^5-70\mu^3+15\mu)$
6	$\frac{1}{16}(231\mu^6-315\mu^4+105\mu^2-5)$	7	$\frac{1}{16}(429\mu^7-639\mu^5+315\mu^3-35\mu)$
8	$\frac{1}{128}(6435\mu^8-12012\mu^6+6930\mu^4-1260\mu^2+35)$		

can be used to obtain higher terms. Since the Legendre polynomials form a complete set, a function such as $g(z, \mu)$ can be expressed as

$$g(z, \mu) = \sum_{l=0}^{\infty} g_l(z) P_l(\mu)$$

$$g_l(z) = \frac{2l+1}{2} \int_{-1}^{1} g(z, \mu) P_l(\mu) d\mu$$

with moments $g_l(z)$. We can thus express $g(z, \mu)$ as: Isotropic: $g(z, \mu) = g_0(z)$; First-order angular: $g(z, \mu) = g_0(z) + \mu g_1(z)$; Second-order angular: $g(z, \mu) = g_0(z) + \mu g_1(z) + \frac{1}{2}\left(3\mu^2 - 1\right) g_2(z)$ and higher orders.

Exercise 4.5: Show the connection between the spherical geometry equation for 1-D spherical geometry.

$$\frac{\partial^2 T}{\partial r^2} + \frac{2}{r}\frac{\partial T}{\partial r} = \frac{1}{\alpha}\frac{\partial T}{\partial t}$$

and the one-dimensional rectangular geometry equation.

4.2.5 Flux formulation

In the flux formulation of the 1-D time-dependent equation (Ozisik, 1993)

$$\frac{\partial^2 T}{\partial x^2} = \frac{1}{\alpha}\frac{\partial T(x, t)}{\partial t}$$

can be written in terms of the heat flux

$$q(x, t) = -k\frac{\partial T(x, t)}{\partial x}$$

as

$$\frac{\partial^2 q}{\partial x^2} = \frac{1}{\alpha}\frac{\partial q(x, t)}{\partial t}$$

$(0 < x < \infty, t > 0)$ with boundary conditions

$q(x, t) = f_0 =$ constant at $x = 0, t > 0$

$q(x, t) = 0$ for $t = 0$With $Q(x, t) = q(x, t) - f_o$

$\frac{\partial^2 Q}{\partial x^2} = \frac{1}{\alpha}\frac{\partial Q(x,t)}{\partial t}$ in $0 < x < \infty, t > 0$

$Q(x, t) = -f_0 =$ constant at $x = 0, t > 0$

$Q(x, t) = 0$ for $t = 0$ and

$$Q(x, t) = -f_o \text{erf}\left(\frac{x}{\sqrt{4\alpha t}}\right)$$

$$q(x, t) = f_o + Q(x, t) = f_o\left[1 - \text{erf}\left(\frac{x}{\sqrt{4\alpha t}}\right)\right]$$

and

$$q(x, t) = f_o \text{erfc}\left(\frac{x}{\sqrt{4\alpha t}}\right)$$

From $q(x, t)$, the temperature distribution is found by integrating the above

$$T(x, t) = \frac{f_o}{k}\int_x^{\infty} \text{erfc}\left(\frac{y}{\sqrt{4\alpha t}}\right) dy$$

and

$$T(x,t) = \frac{2f_0}{k}\left[\left(\frac{\alpha t}{\pi}\right)^{1/2} e^{-x^2/4\alpha t} - \frac{x}{2}\operatorname{erfc}\frac{x}{\sqrt{4\alpha t}}\right]$$

Exercise 4.6: Compare the heat flux formulation with the Fermi Age formulation and comment on the similarities or differences.

4.3 Integral equations

Several problems in science and engineering, such as the Boltzmann transport equation, are formulated in terms of integral equations where the unknown appears in an integral (Kanwal, 2013; Polyanin & Manzhirov, 2008). In general, a differential equation can be converted into an integral equation. It is important to note that initial value problems as well as boundary value problems can be converted into equations.

Integral equations are classified according to the limits on the integral, the occurrence of the unknown function and the homogeneity of the equation. Some equations with their classification are listed in Table 4.3.

In the above equations, $K(x, y)$ is called the "kernel of the integral operator."

Exercise 4.7: Show that the IVP.

$$\frac{d^2y}{dx^2} - 5\frac{dy}{dx} + 6y = 0;\, y(0) = 0, y'(0) = -1$$

can be converted to a Volterra equation of the second kind

$$g(x) = (6x - 5) + \int_a^x (5 - 6x + 6t)g(t)dt$$

Exercise 4.8: Show that the BVP.

$$\frac{d^2y}{dx^2} + xy = 1, y(0) = 0, y(1) = 1$$

can be converted to a Fredholm integral equation of the second kind.

$$y(x) = \frac{1}{2}x(1 + x) + \int_0^1 K(x, t)y(t)dt$$

Obtain the kernel $K(x, t)$.

TABLE 4.3 Classification of integral equations.

Equation	Name	Type	
$u(x) = \int_a^b K(x,y)f(y)dy$	Fredholm	Homogeneous	I
$f(x) = \int_a^b K(x,y)f(y)dy$	Fredholm	Homogeneous	II
$f(x) = \int_a^b K(x,y)f(y)dy + g(x)$	Fredholm	Inhomogeneous	II
$u(x) = \int_a^x K(x,y)f(y)dy$	Volterra	Homogeneous	I
$f(x) = \int_x^b K(x,y)f(y)dy$	Volterra	Homogeneous	II
$f(x) = \int_a^x K(x,y)f(y)dy + g(x)$	Volterra	Inhomogeneous	II

The above equations are called nonsingular linear integral equations since the unknown function appears inside one or more integrals and linear operations are performed on it. Examples of nonlinear inegral equations are

$$y(x) = \int_a^b K(x,t)[y(t)]^2 dt$$

$$y(x) = \int_a^b \int_a^b K(x,s,t)y(s)y(t)dsdt, a \le s \le b$$

If the function to be determined in the above equations, $f(x)$, appears as $h(x)f(x)$ and $h(x)$ is neither zero or one, then the equation is called a linear integral equation of the third kind.

In this section, the integrals considered are nonsingular, linear, and either homogeneous or nonhomogeneous.

Some of the methods to solve integral equations are:

1. differentiating under the integral sign using Leibnitz's rule
2. using successive approximations
3. using a resolvant kernel
4. Laplace transform for an integral of the convolution type

The formula for Leinnitz's rule is

$$\frac{d}{dx}\int_{a(x)}^{b(x)} F(x,t)dt = \int_{a(x)}^{b(x)} \frac{\partial F}{\partial x} dt + F[x,b(x)]\frac{db}{dx} - F[x,a(x)]\frac{da}{dx}$$

A Laplace transform method is used for an integral equation of the convolution type

$$y(x) = f(x) + \int_a^b K(x-t)y(t)dt$$

$$y(x) = f(x) + \lambda \int_a^x K(x-t)y(t)dt$$

where the convolution theorem holds for transforms such as Laplace and Fourier transforms and algebraic properties apply, that is, additive, associative, distributive and multiplicative.

One way of solving integral equations is by integral transforms, for example, for Fredholm integrals, when $K(x,y) = K(x-y)$, the Fourier transform can be used while for Volterra equations, the Laplace transform can be used.

Example 4.2: Fredholm Equation by successive approximations.

$$f(x) = g(x) + \int_a^b k(x,y)f(y)dy$$

Write as a fixed-point equation to define a transformation operator

$$Tf = f$$

where the operator T is

$$Tf(x) = g(x) + \int_a^b k(x,y)f(y)dy$$

By the contraction Mapping Theorem, the integral equation has a unique solution in C([*a*, *b*]).

The successive approximations are

$$Tf_0 = g + Kf_0$$

$$T^2 f_0 = T(Tf_0) = T(g + Kf_0) = g + K(g + Kf_0) = g + Kg + K^2 f_0$$

$$T^3 f_0 = T(T^2 f_0) = g + Kg + K^2 g + K^3 f_0$$

$$\vdots$$

$$T^n f_0 = g + Kg + K^2 g^2 + \cdots + K^{n-1} g^{n-1} + K^n f_0$$

Since

$$\lim_{n\to\infty} T^n f_0 = 0$$

the function is obtained as

$$f = \lim_{n\to\infty} T^n f_0 = \sum_{n=0}^{\infty} K^n g.$$

As an example, consider the inhomogeneous Fredholm equation

$$f(x) = 1 + \int_0^1 xf(y)dy$$

Write the upper limt as α to demonstrate convergence; then

$$f(x) = 1 + \int_0^\alpha xf(y)dy$$

for $0 < \alpha < 1$.

With any starting approximation $f_0 \in c([0, 1])$ successive approximations

$$Tf\left(x\right) = 1 + \int_0^\alpha xf(y)dy$$

are obtained. Let $f_0(x) = 1$. Then

$$f_2(x) = Tf_1(x) = 1 + \int_0^\alpha xf_1(y)dy = 1 + \int_0^\alpha x(1+\alpha y)dy = 1 + x\left[\alpha + \frac{1}{2}\alpha^3\right]$$

$$f_3(x) = Tf_2(x) = 1 + \int_0^\alpha xf_2(y)dy = 1 + \int_0^\alpha x\left(1 + y\left(\alpha + \frac{1}{2}\alpha^3\right)\right)dy$$

$$= 1 + x\left[\alpha + \frac{1}{2}\alpha^3 + \frac{1}{2^2}\alpha^5\right]$$

Continuing, the nth approximation is

$$f_n(x) = 1 + x\left[\alpha + \frac{1}{2}\alpha^3 + \frac{1}{2^2}\alpha^5 + \ldots + \frac{1}{2^{n-1}}\alpha^{2^{n-1}+1}\right]$$

So that the solution may be reconstructed as

$$f(x) = \lim_{n\to\infty} f_n(x) = 1 + x\sum\nolimits_{n=0}^{\infty} \frac{1}{2^n}\alpha^{2n+1}$$

$$= 1 + x\alpha\sum\nolimits_{n=0}^{\infty}\left(\frac{\alpha^2}{2^n}\right)^n = 1 + x\alpha \bullet \frac{1}{1-\alpha^2/2}$$

which gives the solution

$$f(x) = 1 + \frac{2\alpha}{2-\alpha^2}x = 1 + 2x.$$

This satisfies the starting integral equation. In the solution, the denominator converges for $\alpha \in \left(-\sqrt{2},\sqrt{2}\right)$; the solution is not valid for and furthermore that the solution satisfies for any $\alpha = \pm\sqrt{2}$.

Example 4.3: Solve the Fredholm integral equation.

$$y(x) = e^x + \lambda\int_0^1 2e^x e^t y(t)dt$$

Let

$$C = \int_0^1 e^t y(t)dt$$

Then $y(t) = e^t(1 + 2C\lambda)$, and using this, the value of C can be found as $C = \frac{(e^2 - 1)}{2[1 + \lambda(e^2 - 1)]}$, where $\lambda \neq 1/(e^2 - 1)$

The solution is

$y(x) = \frac{e^x}{1 - \lambda(e^2 - 1)}$, where $\lambda \neq 1/(e^2 - 1)$

A resolvent kernel $R(x, t; \lambda)$ can be obtained for both nonhomogeneous Fredholm and Volterra integrals to remove the unknown function from under the integral sign; for a Fedholm integral equation

$$y(x) = f(x) + \lambda \int_a^b k(x, t)y(t)dt$$

the equation then becomes

$$y(x) = f(x) + \lambda \int_a^b R(x, t; \lambda)f(t)dt$$

which is easily solved since $f(x)$ is given.

Then this resolvent kernel satisfies the following relationship:

$$R(x, t; \lambda) = K(x, t) + \lambda \int_a^b k(x, s)R(s, t; \lambda)ds$$

giving

$$R(x, t; \lambda) = \sum_{m=1}^{\infty} \lambda^{m-1} K_m(x, t)$$

Exercise 4.9: Classification of an integral equation.

The integral equation for Chandrasekhar's H function (Chandrasekhar, 1960) appears in the exact solution of the neutron transport equation which will be used in Chapter 6. How would you classify this equation?

$$H(\mu) = 1 + \mu H(\mu) \int_0^1 \frac{\psi(\mu')}{\mu + \mu'} H(\mu')d\mu'$$

Exercise 4.10: Fredholm equation by successive approximations. Consider the integral equation

$$y(x) = \frac{3}{x}e^x - \frac{1}{2}xe^x - \frac{1}{2} + \frac{1}{2}\int_0^1 ty(t)dt.$$

Let

$$K_1(x, t) = K(x, t) = t$$

so that the kernel is

$$K_m(x, t) = \int_0^1 K(x, s)K_{m-1}(s, t)ds$$

show that

$$K_m(x, t) = \int_0^1 K(x, s)K_{m-1}(s, t)ds = \left(\frac{1}{2}\right)^{m-1} t$$

and the resolvant kernel is

$$R(x,t;\lambda)=\sum_{m=1}^{\infty}\lambda^{m-1}K_m(x,t)=\sum_{m=1}^{\infty}\left(\frac{1}{2}\right)^{m-1}\left(\frac{1}{2}\right)^{m-1}t$$

$$=\sum_{m=1}^{\infty}\left(\frac{1}{4}\right)^{m-1}t=t\left[1+\frac{1}{4}+\left(\frac{1}{4}\right)^2+\dots\right]=\frac{4t}{3}$$

from which

$$y(x)=f(x)+\lambda\int_0^1 R(x,t;\lambda)f(t)dt$$

and the solution is easily obtained as

$$y(x)=\frac{3}{x}e^x-\frac{1}{2}xe^x-\frac{e}{3}+1.$$

Exercise 4.11: Solve the Volterra equation $u(x)=\int_0^x e^{x-y}u(y)dy$, given the boundary condition $u(0)=0$ by differentiating with respect to x.

Exercise 4.12: Solving the Volterra equation using the convolution theorem.

Solve the Volterra equation $u(x)=f(x)+\int_0^x k(x-y)u(y)dy$ by taking the Laplace transform.

4.3.1 An important integral equation for neutron transport

In the previous section, the integral equation of a Volterra form was obtained from a PDE

$$\phi(x,\mu)=\frac{1}{\mu}\int_0^x e^{-\frac{x-y}{\mu}}f(x)dy+Ce^{-x/\mu} \tag{4.6}$$

When this equation is integrated over μ, the equation can be written in operator form as

$$f(x)=\hat{K}(x-y)f(x)+g(x) \tag{4.7}$$

where

$$\hat{K}(x-y)f(x)\equiv\int d\mu\frac{1}{\mu}\int_0^x e^{-\frac{x-y}{\mu}}f(x)dy$$

and

$$g(x)\equiv\int d\mu Ce^{-x/\mu}.$$

We will return to this form in the chapter on Transport Theory to obtain a numerical solution to an important benchmark in nuclear engineering.

4.3.2 Integral equations in neutron transport

Several "special" functions (Abramowitz & Stegun, 1964) are used to represent source distributions and neutron flux especially for the separation of variables to facilitate exact solution or numerical procedures.

The delta function $\delta(x)$ is visualized to be zero everywhere and have a large value at $x=0$. In engineering, especially in the context of a neutron source, it is used to represent a source at some spatial position, angle or time. As an example a source defined as

$$S(x,E,\Omega,t)=\delta(x-x_0)\delta(E-E_0)\delta(\Omega-\Omega_0)\delta(t-t_0)$$

represents a unit source located at position x_0 with energy E_0, emitting neutrons in the solid angle $d\Omega_0$ about Ω_0 at time t_0. The response to this unit point instantaneous source is the Green's function which can be used to find the flux distribution from a general source.

Among the several definitions of the δ function, consider the normal distribution with an infinitesimal variance σ^2:

$$\delta(x) = \lim_{\sigma \to 0} \frac{1}{\sigma\sqrt{2\pi}} e^{-\frac{x^2}{2\sigma^2}}$$

and a variation, used in complex integration, of the form:

$$\delta(x) = \frac{1}{\pi} \lim_{\varepsilon \to 0} \frac{\varepsilon}{x^2 + \varepsilon^2}$$

One of the properties of the δ functions is

$$\int_{-\infty}^{\infty} f(x)\delta(x-a)dx = f(a)$$

4.4 Integro-differential equations

An integro-dfifferential equation has the unknown function in the derivative term as well as under the integral sign. An example is the Volterra integro-differential equation

$$\frac{d^n\phi(x)}{dx^n} = f(x) + \lambda \int_0^x K(x,y)\phi(y)dy \tag{4.8}$$

From Table 4.3 this is classified as an Volterra equation of the second kind since the unknown function $\phi(x)$ is on both sides of the equation in the derivative and under the integral sign, and the presence of $f(x)$ makes it nonhomogeneous. Similar to the successive approximation methods described above, the Adomian decomposition method can be used to determine the solution from recurrence relations when an exact solution is not possible.

Consider the equation

$$\frac{d\phi(x)}{dx} = 1 - \int_0^x \phi(y)dy$$

with the initial condition $\phi(0) = 0$. The simplest method is to apply a Laplace transform

$$s\Phi(s) - \Phi(0) = \frac{1}{s} - \frac{\Phi(s)}{s}$$

Rearranging

$$\Phi(s) = \frac{1}{(s^2+1)}$$

giving the solution as $\phi(x) = \sin x$. There are several other methods such as the variational iteration method and the Laplace transform method as applied here as well for the convolution integral. Now consider the first-order PDE (Bell & Glasstone, 1952)

$$\mu \frac{\partial}{\partial x}\phi(x,\mu) + \phi(x,\mu) = f(x) + S(x,\mu)$$

where $f(x) = \frac{c}{2}\int_{-1}^{1} d\mu' \phi(x,\mu')$. A solution for this can be obtained by multiplying both sides by the integration factor $e^{-x/\mu}$ and integrating over x:

$$\phi(x,\mu) = \frac{1}{\mu}\int_0^x e^{-\frac{x-y}{\mu}} f(x)dy + Ce^{-x/\mu}$$

where C, a constant of integration, can be obtained from the given boundary condition.

To prepare for the mathematical formulation to come in the following chapters, consider an integro-differential equation of the form

$$\left(\mu\frac{\partial}{\partial x}+1\right)\phi(x,\mu)=\frac{c}{2}\int_{-1}^{1}d\mu'\phi(x,\mu')+S(x,\mu).$$

A general solution is obtained in the classical *albedo* problem for a semiinfinite half-space $x\in(0,\infty)$, shown in Fig. 4.6, for an incident source $\phi(0,\mu)=1$ at the free boundary $z=0$. For $c\leq 1$ (nonmultiplying medium) and without an extraneous source, that is, $S(x,\mu)=0$, this problem was considered originally for radiative transfer (Chandrasekhar, 1960) (Fig. 4.8).

Note that transport is along the s direction although the equation is written for the x variable, that is, it is a one-dimensional representation of two-dimensional transport. The connection between the variables is

$$s=\frac{x-x'}{\mu}$$

and, since the contribution to the flux $\varnothing(x,\mu)$ will come from all x', we need to connect the variable through the transformation

$$\frac{d}{ds}=\frac{\partial x'}{\partial s}\frac{\partial}{\partial x'}=-\mu\frac{\partial}{\partial x'}$$

The equation thus becomes

$$\left[\frac{d}{ds}-1\right]\phi(x',\mu)=-\frac{c}{2}\phi(x')$$

where

$$\phi(x')\equiv\int_{-1}^{1}d\mu'\quad\phi(x',\mu')$$

Multiplying by the integrating factor e^{-s} and integrating from 0 to s

$$\phi(x,\mu)=\phi(x',\mu)e^{-\frac{x-x'}{\mu}}+\frac{c}{2\mu}\int_{x'}^{x}\phi(t)e^{-\frac{x-t}{\mu}}dt$$

The neutron transport equation, describing the angular flux distribution in a nuclear reactor is an integro-differential equation in phase space

$$\begin{aligned}\Omega\bullet\nabla\phi(\vec{r},\Omega,E)+\Sigma_t(\vec{r},E,t)\phi(\vec{r},\Omega,E)=&\int_0^{\infty}dE'\quad\int d\Omega'\Sigma_s(\vec{r},E',\Omega'\bullet\Omega)\phi(\vec{r},\Omega',E')\\&+\frac{1}{4\pi}\chi(E)\int_0^{\infty}dE'\int d\Omega'\nu\Sigma_f(E')\phi(\vec{r},\Omega',E')+S(\vec{r},\Omega,E).\end{aligned}\tag{4.9}$$

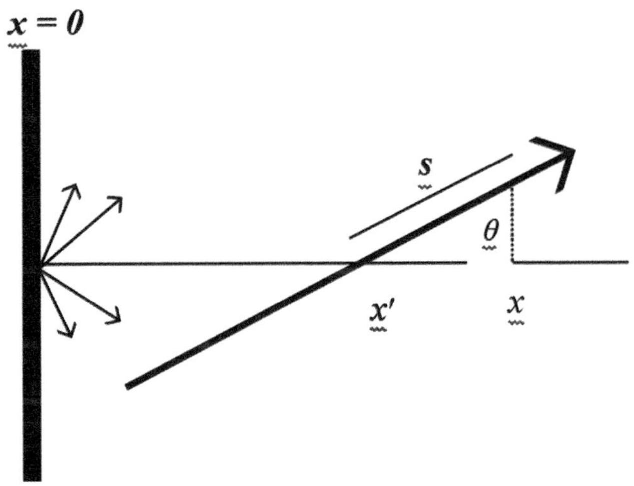

FIGURE 4.8 Neutron transport in 1-D.

During reactor operation, the composition $N_i(t)$ of isotopes at any time t is found by solving the integro-differential *burnup* equation coupled with the point kinetics equation described earlier in this chapter. This time-dependent integro-differential equation is

$$\begin{aligned}\frac{dN_i(t)}{dt} = \sum_j \left[\int_0^\infty \gamma_{ji}(E,t)\sigma_{f_j}(E,t)\phi(E,t)dE\right]N_j(t) + \sigma_{c_{i-1}}(E,t)\phi(E,t)N_{i-1}(t) + \lambda_{i'}N_{i'}(t)\\ -\left\{\left[\int_0^\infty \left(\sigma_{f_i}(E,t) + \sigma_{c_i}(E,t)\phi(E,t)dE\right)\right] + \lambda_i\right\}N_i(t).\end{aligned} \tag{4.10}$$

The composition of an ith nuclide changes with time due to creation in processes a to c and destruction to process d and e as follows:

a. fission of some isotope j which produces the ith by an amount γ_{ji},

$$\left[\int_0^\infty \gamma_{ji}(E,t)\sigma_{f_j}(E,t)\phi(E,t)dE\right]N_j(t)$$

b. radiative capture of an $(i-1)$th nuclide:

$$\sigma_{c_{i-1}}(E,t)\phi(E,t)N_{i-1}(t)$$

c. radiative decay of another nuclide $\lambda_{i'}N_{i'}(t)$

d. fission and capture in which some other nuclide is produced

$$\left[\int_0^\infty \left(\sigma_{f_i}(E,t) + \sigma_{c_i}(E,t)\phi(E,t)dE\right)\right]N_i(t)$$

e. its radioactive decay $\lambda_i N_i(t)$

These calculations are coupled with reactor kinetics on the short-term (~seconds and minutes) as described by the PKE as well as on the long-term (~days and months) with the burnup equations. Such coupled kinetic-dynamics, which incorporate dynamics with fuel burnup, as illustrated in Figs. 4.9 and 4.10, are used for reactor start-up simulations. The short-term changes in reactivity affect the reactor power during start-up which could take about 5 days to

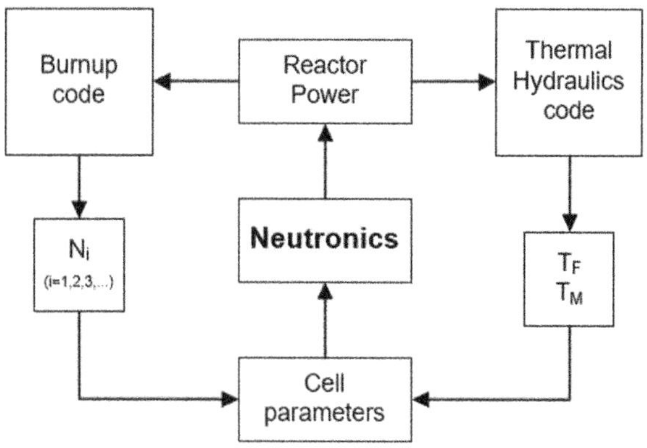

FIGURE 4.9 Coupling of neutronics with thermal hydraulics.

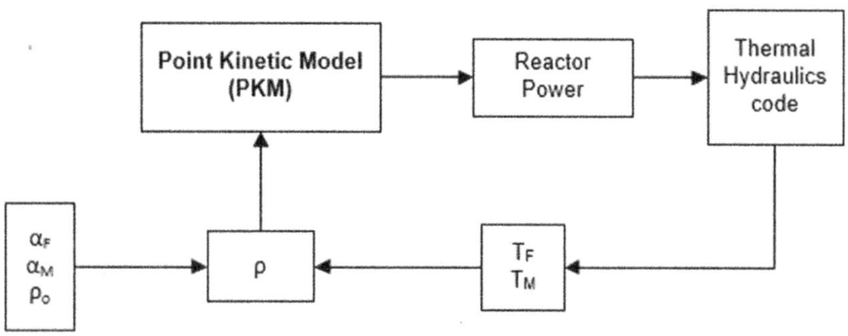

FIGURE 4.10 Coupling of neutronics with thermal hydraulics and point kinetics.

reach 100% full power. Consider a model (Johnson, Lucas, & Tsvetkov, 2010) based on the point kinetic equations, Bateman equations for nuclide transmutation, and a lumped thermal model computing the dynamics of the power and fuel temperature (Bell & Glasstone, 1979; Hetrik, 1993; Lamarsh & Baratta, 2001). This is a simplified *lumped* model which considers the reactor as a point, that is, spatial dependence is ignored.

The flow of the calculation procedure is as follows

1. Initial conditions specified include the fuel vector consisting of atomic densities of actinides $N_i(t)$, fuel T_F and moderator temperatures T_M, feedback coefficients α_F, α_M and source term $Q(t)$, and the input reactivity ρ_0; this gives the reactivity $\rho(t)$ in Eq. (4.11).
2. The neutron density $n(t)$ and precursor concentrations $C_i(t)$ are determined from Eqs. (4.12) and (4.13), giving the flux $\phi(t) = n(t)v$, and reactor power $P(t)$ (Eq. 4.14).
3. The new system multiplication k is found from Eq. (4.15), which gives the reactivity $\rho(t)$ and the change in reactivity $\Delta\rho(t)$ from Eq. (4.16).
4. The fuel temperature $T_F(t)$ is updated from the reactor power $P(t)$, using the heat balance Eq. (4.17); the temperature is then used to calculate the coefficient of reactivity $\alpha_F(t)$ from Eq. (4.18).
5. The fuel vector $\overline{N}$ is updated by solving the Bateman equations (Eq. 4.19) for all the actinides considered in the model [24 actinides in (Johnson et al., 2010)].
6. The generation time Λ is calculated from the time for absorption Λ_∞, for absorption or leakage Λ_0, based on the new fuel vector and system multiplication k, from Eq. (4.20).

$$\rho(t) = \rho_0 + \alpha_F(T_F(t) - T_{F0}) + \alpha_M(T_M(t) - T_{M0}) \tag{4.11}$$

$$\frac{d\phi(t)}{dt} = \frac{\rho(t) - \tilde{\beta}}{\Lambda}\phi(t) + \sum_{i=1}^{6}\lambda_i C_i(t) + Q(t) \tag{4.12}$$

where

$$\frac{dC_i(t)}{dt} = \frac{\beta_i(t)}{\Lambda}\phi(t) - \lambda_i C_i(t), \quad i = 1, \cdots 6 \tag{4.13}$$

and

$$P(t) = E_R \Sigma_f \phi(t) V \tag{4.14}$$

The system multiplication is

$$k_{\text{eff}}(t) = \frac{\upsilon\Sigma_f(t)}{\Sigma_a(t)}\frac{1}{1 + L^2B^2} \tag{4.15}$$

Where the numerator is calculated from the atomic densities $N_i(t)$ in the fuel vector $\overline{N}$

$$\upsilon\Sigma_f(t) = \sum_i \upsilon_i \sigma_{f,i} N_i(t)$$

The reactivity and the change in reactivity are calculated (in \$) from

$$\rho = \frac{k-1}{k}, \quad \Delta\rho(\$) = \frac{k' - k}{\beta k k'} \tag{4.16}$$

and

$$mc_p\frac{dT_F(t)}{dt} = P(t) - hA(T_F(t) - T_M(t)) \tag{4.17}$$

The temperature coefficients of reactivity are

$$\alpha_i = \frac{d\rho}{dT_i}, \quad i = F, M \tag{4.18}$$

The Bateman equations are

$$\frac{dN_Z^A}{dt} = \left(N_Z^{A-1}\sigma_c - N_Z^A\sigma_c\right)\phi(t) - \lambda N_Z^A + \lambda N_*^* \tag{4.19}$$

and

$$\Lambda_\infty = \frac{1}{v\Sigma_a}, \Lambda_0 = \frac{\Lambda_\infty}{1 + L^2B^2}, \Lambda = \frac{\Lambda_0}{k_{\text{eff}}} \tag{4.20}$$

The procedure described above constitutes a large set of ODEs which can be solved using standard methods in Matlab.

For a more elaborate analysis, the full phase space-dependence of the neutron flux is required and the computational effort increased by orders of magnitude.

4.5 Numerical methods

Several numerical and computational methods have been used and developed (Bell & Glasstone, 1979; Case & Zweifel, 1967; Lewis & Miller, 1984) to solve neutron diffusion and transport problems and in thermal hydraulics, structural mechanics and other areas (Hutton, 2004; Lewis, Nithiarasu, & Seetharamu, 2004; Rao, 2017) with the complexity of realistic engineering systems. These include the Finite Difference Method (FDM), the Finite Element Method (FEM), the Finite Volume Method (FVM), nodal methods, spectral methods, quadrature-based methods, and hybrid methods.

This section considers two widely used methods *viz* the FDM and the FEM.

4.5.1 The Finite Difference Method

In a discretization of the differential equations, the entire domain is divided into elements, or meshes, and the solution is obtained at the node points. In a 1-D model, the domain $a \le x \le b$ is divided into a finite number of elements N (e.g., $N = 3$) for which there will be M (e.g., $M = 4$) mesh points while in a 2-D model, for example, in the Cartesian $x - y$ domain the node points will be in the x and y dimensions as shown in Fig. 4.7 (Fig. 4.11).

The FDM is applied to the 2-D steady-state heat conduction equation

$$\frac{\partial^2 T}{\partial x^2} + \frac{\partial^2 T}{\partial y^2} + \frac{\dot{q}}{k} = 0$$

Applying the first-order forms for the first and second derivatives

$$\frac{\partial T}{\partial x}\Big|_{i-1/2,j} = \frac{T_{i,j} - T_{i-1,j}}{\Delta x}$$

$$\frac{\partial T}{\partial x}\Big|_{i+1/2,j} = \frac{T_{i+1,j} - T_{i,j}}{\Delta x}$$

The second derivative is thus

$$\frac{\partial^2 T}{\partial x^2}\Big|_{i-1/2,j} = \frac{\frac{\partial T}{\partial x}\big|_{i+1/2,j} - \frac{\partial T}{\partial x}\big|_{i-1/2,j}}{(\Delta x)^2}$$

written as a "three node" term

$$\frac{\partial^2 T}{\partial x^2}\Big|_{i,j} = \frac{T_{i+1,j} - 2T_{i,j} + T_{i-1,j}}{(\Delta x)^2}$$

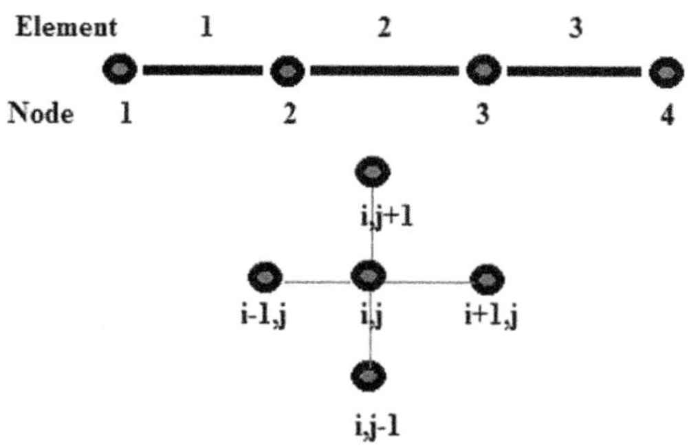

FIGURE 4.11 Elements and nodes.

Similarly, in the y-direction

$$\frac{\partial T}{\partial y}|_{i,j-1/2} = \frac{T_{i,j} - T_{i,j-1}}{\Delta y}$$

$$\frac{\partial T}{\partial y}|_{i,j+1/2} = \frac{T_{i,j+1} - T_{i,j}}{\Delta y}$$

The second derivative is thus

$$\frac{\partial^2 T}{\partial y^2}|_{i,j} = \frac{\frac{\partial T}{\partial x}|_{i+1/2,j} - \frac{\partial T}{\partial x}|_{i-1/2,j}}{(\Delta x)^2}$$

written as a "three node" term

$$\frac{\partial^2 T}{\partial y^2}|_{i,j} = \frac{T_{i,j+1} - 2T_{i,j} + T_{i,-1j}}{(\Delta y)^2}$$

the FDM is, for $\Delta x = \Delta y$,

$$\frac{T_{i+1,j} - 2T_{i,j} + T_{i-1,j}}{(\Delta x)^2} + \frac{T_{i,j+1} - 2T_{i,j} + T_{i,-1j}}{(\Delta y)^2} + \frac{\dot{q}}{k} = 0$$

which reduces to

$$T_{i+1,j} - 2T_{i,j} + T_{i-1,j} + T_{i,j+1} - 2T_{i,j} + T_{i,-1j} + \frac{\dot{q}}{k}(\Delta x)^2 = 0$$

or

$$T_{i,j} = \frac{1}{4}\left[\quad T_{i+1,j} + T_{i-1,j} + T_{i,j+1} + T_{i,-1j} + \frac{\dot{q}}{k}(\Delta x)^2\right]$$

In a square with temperatures given at the four sides and no heat generation ($\dot{q} = 0$), as shown in Fig. 4.12.

$$T_1 = \frac{1}{4}[\quad T_L + T_U + T_2 + T_3]$$

with $T_L = 300°\text{C}, T_U = 200°\text{C}, T_R = 400°\text{C}, T_D = 100°\text{C}$ the four simultaneous equations in matrix form are

$$\begin{bmatrix} 4 & -1 & -1 & 0 \\ -1 & 4 & 0 & -1 \\ -1 & 0 & 4 & -1 \\ 0 & -1 & -1 & 4 \end{bmatrix} \begin{bmatrix} T_1 \\ T_2 \\ T_3 \\ T_4 \end{bmatrix} = \begin{bmatrix} T_L + T_U \\ T_R + T_U \\ T_L + T_D \\ T_R + T_D \end{bmatrix} = \begin{bmatrix} 500 \\ 600 \\ 700 \\ 500 \end{bmatrix}$$

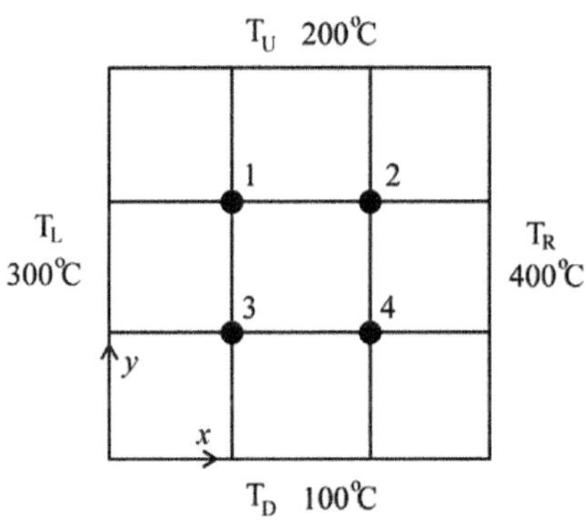

FIGURE 4.12 Square plate with Dirichlet boundary conditions.

Solving these equations in Matlab: $A = [4\ -1\ -1\ 0; -1\ 4\ 0\ -1; -1\ 0\ 4\ -1; 0\ -1\ -1\ 4]$; $B = [500;600;750;500]$; $T = (A\backslash B)$ gives the results

$$\begin{bmatrix} T_1 \\ T_2 \\ T_3 \\ T_4 \end{bmatrix} = \begin{bmatrix} 279.17 \\ 289.58 \\ 327.08 \\ 279.17 \end{bmatrix}$$

For a larger number of nodes, the surface temperature can be obtained from the Matlab program listed below, as shown in Fig. 4.9 for 400 (20×20) node points ($18 \times 18 = 324$ interior nodes) (Fig. 4.13) and compared with the exact solution given in Section (4.2.1).

The boundary conditions are visualized better from Fig. 4.10 (Fig. 4.14).

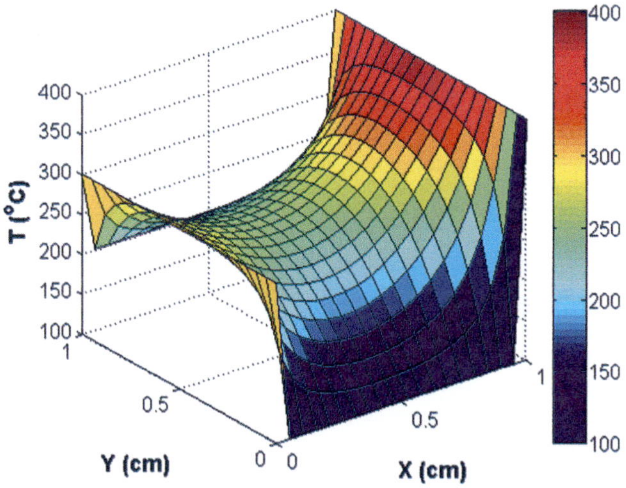

FIGURE 4.13 Temperature field $T(x,y)$ surface in a square plate.

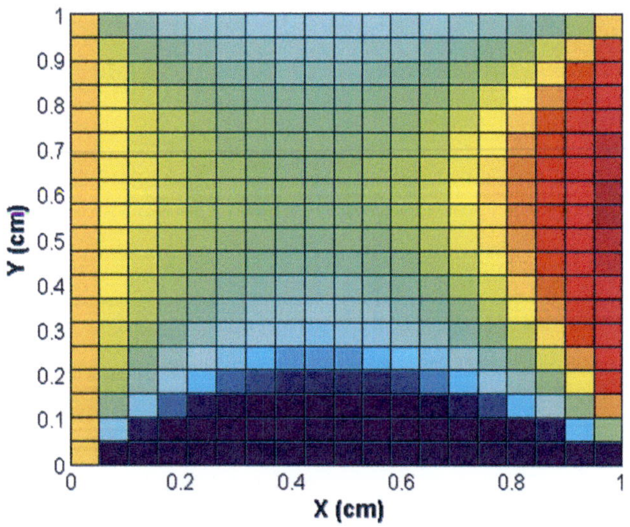

FIGURE 4.14 Temperature field $T(x,y)$ projection on a square plate.

4.5.1.1 Matlab program: Finite Difference Method

Steady-state heat conduction in a 2-D square plate with boundary temperatures specified.

```
% Ch4_FinDif2DSquarePlate.m
% Finite Difference
N = 20;% N-2 interior points
x = linspace(0,1,N);
dx = x(2)-x(1);
y = x; dy = dx;
eps = 1e-6; % convergence criterion
T  = zeros(N,N);
T(1,1:N) = 100;  %left
T(N,1:N) = 200;  %right
T(1:N,1) = 300;  %bottom
T(1:N,N) = 400;  %top
dt = dx^2/4;
Qdot=zeros(N,N); % heat generation rate
k=1; % thermal conductivity
err = 1; kk = 0;
while err > eps
kk = kk+1;
Told = T;
for i = 2:N-1
for j = 2:N-1
    a=(Told(i+1,j)-2*Told(i,j)+Told(i-1,j))/dx^2;
    b=(Told(i,j+1)-2*Told(i,j)+Told(i,j-1))/dy^2;
T(i,j) = dt* (a+b+Qdot(i,j)/k)+ Told(i,j);
end
end
error = max(max(abs(Told-T)));
end
x0=0;delx=1/(N-1);
y0=0;dely=1/(N-1);
for j=1:N
y(j)=y0+(j-1)*dely;
for i=1:N
x(i)=x0+(i-1)*delx;
end
end
surf(x,y,T)
set(gca,'FontSize',12)
xlabel('\bf X (cm)','fontsize',14)
ylabel('\bf Y (cm)','fontsize',14)
zlabel('\bf T ({}^oC)','fontsize',14)
```

Exercise 4.13: Steady-state head conduction (FDM).

With the given Matlab program, obtain the temperature profile in a 4×4 cm copper plate (thermal conductivity $k = 385$ W/(m-K)) with heat generation 2000 W at the center.

The greater the number of elements, the more will the computational effort generally resulting in more accuracy. For large 3-D models of reactor cores, for example, the FDM is used with refinements for both "structured" and "unstructured" grids. As problems become large, methods such as the FEM and FVM are used.

The resulting algebraic system can be solved by standard matrix methods such as the Gaussian elimination method, Choleski's method, Jacobi's method, the Gauss–Seidel method, and the Standard Over Relaxation method.

4.5.2 The Finite Element Method

In the FEM (Hutton, 2004; Lewis et al., 2004; Rao, 2017) the domain of interest is first discretized into a finite number of elements which fit into the specified (regular or irregular) geometry. For illustration, consider a 1-D domain consisting of three elements and four nodes (Fig. 4.15).

The field function, for example, $T(x)$, in an element with nodes i and j is expressed in terms of nodal values T_i and T_j and *shape functions* $N_i(x)$ and $N_j(x)$

$$T(x) = N_i(x)T_i + N_j(x)T_j = \begin{bmatrix} N_i(x) & N_j(x) \end{bmatrix} \begin{Bmatrix} T_i \\ T_j \end{Bmatrix}$$

where the shape functions are shown in Fig. 4.16.

The field equation is then reduced to a set of linear algebraic equations which are solved, by standard methods, for the nodal values. The equation for each *local* element is combined or *assembled* into a *global* system expressed mathematically as

$$[\mathbf{K}]\{\mathbf{T}\} = \{\mathbf{f}\}$$

where [**K**] is called the global stiffness matrix and {**f**} is called the force function. The temperature at the nodes is found as

$$\{\mathbf{T}\} = [\mathbf{K}]^{-1}\{\mathbf{f}\}$$

and from the nodal values, the temperature distribution is reconstructed using the shape functions. If the problem is in 2-D, the square and triangular elements shown in Fig. 4.13 can be used (Fig. 4.17).

Some commonly used elements in 1-D are the linear element (Fig. 4.11) with two nodes and the quadratic element with three nodes. In 2-D, the linear triangular element with three nodes (Fig. 4.13), the quadratic triangular element with six nodes, the quadrilateral element with four, eight or nine nodes, and the isometric element with eight nodes for curvilinear coordinates, that is, coordinates which change direction with position, are used. In 3-D, the linear tetrahedron element with four nodes, the linear hexahedron element with eight nodes, the linear prism element with six nodes and the quadratic tetrahedron element with ten nodes (Fig. 4.14) are used (Fig. 4.18).With $T(x) = a_1 + a_2x$ applied to both nodes of the element, coefficients a_1, a_2 can be found for the shape functions of Fig. 4.12.

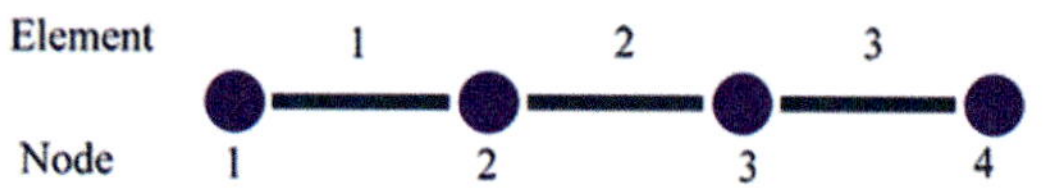

FIGURE 4.15 A 3-element, 4 node linear domain.

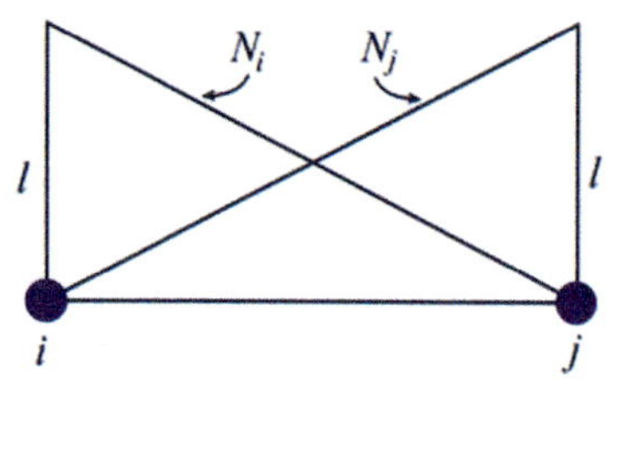

FIGURE 4.16 Shape functions $N(x)$.

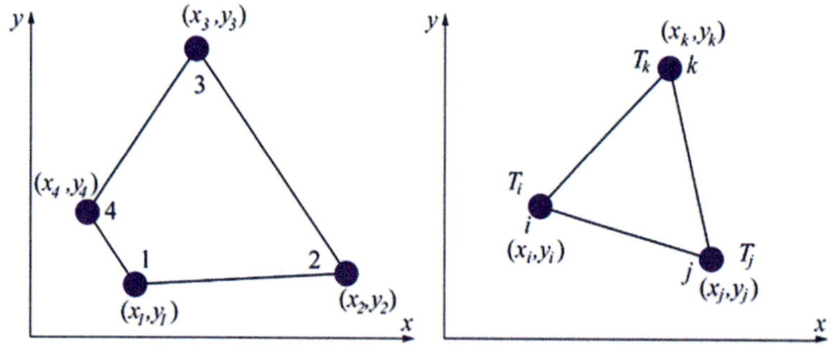

FIGURE 4.17 2-D quadrilateral and triangular elements.

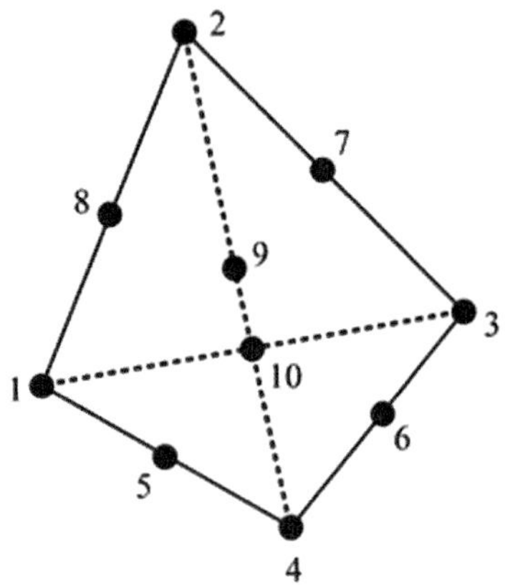

FIGURE 4.18 The 3-quadratic tetrahedron with ten nodes.

At $x = x_i$, $T_i = a_1 + a_2 x_i$, and at $x = x_j$, $T_j = a_1 + a_2 x_j$, giving the coefficients and the shape functions listed in Tables 4.3 and 4.4 respectively. The space-dependence of $T(x)$ is

$$T(x) = \frac{(T_j - T_i)}{(x_j - x_i)} x + \frac{(x_j T_i - x_i T_j)}{(x_j - x_i)}$$

For other elements, the coefficients and shape functions are listed in Tables 4.4 and 4.5 respectively.

The gradient variable **B** for the 1-D linear element is

$$\frac{dT}{dx} = g \equiv \begin{bmatrix} -\frac{1}{l} & \frac{1}{l} \end{bmatrix} \begin{Bmatrix} T_i \\ T_j \end{Bmatrix} = [\mathbf{B}][\mathbf{T}]$$

We first consider the procedure from a variational formulation where the stationary function of a functional (Lewis, Vrabie, & Syrmos, 2012) is the governing equation. In the case of the 2-D heat conduction equation, for example, the functional

$$I(T) = \frac{1}{2}\int_{\Omega}\left[k_x\left(\frac{\partial T}{\partial x}\right)^2 + k_y\left(\frac{\partial T}{\partial y}\right)^2 + k_z\left(\frac{\partial T}{\partial z}\right)^2 - 2GT\right]d\Omega + \int_{S_2} qT \quad ds + \int_{S_3} \frac{1}{2}h(T - T_a)^2 ds$$

is stationary for

$$\frac{\partial}{\partial x}\left(k_x \frac{\partial T}{\partial x}\right) + \frac{\partial}{\partial y}\left(k_y \frac{\partial T}{\partial y}\right) + \frac{\partial}{\partial z}\left(k_z \frac{\partial T}{\partial z}\right) + G = 0$$

with the following boundary conditions

$T = T_B$ on surface S_1

$k_x \frac{\partial T}{\partial x}\tilde{l} + k_y \frac{\partial T}{\partial y}\tilde{m} + k_z \frac{\partial T}{\partial z}\tilde{n} + q = 0$ on surface S_2

$k_x \frac{\partial T}{\partial x}\tilde{l} + k_y \frac{\partial T}{\partial y}\tilde{m} + k_z \frac{\partial T}{\partial z}\tilde{n} + h(T - T_a) = 0$ on surface S_3

where l, m, and n are surface normal, h is the heat transfer coefficient, k is the thermal conductivity and q is the heat flux.

The procedure is then as follows:

1. The temperature in each element with r nodes is written as

$$T^e = \sum_{i=1}^{r} N_i T_i = [\mathbf{N}]\{\mathbf{T}\}$$

2. Nodal values are to be selected to make the integral $I(T)$ stationary

$$\delta I(T) = \sum_{i=1}^{n} \frac{\partial I}{\partial T_i} = 0$$

thus for each element $\frac{\partial I}{\partial T_i} = 0$ for $i = 1, 2, \quad \ldots, n$

TABLE 4.4 Coefficients of linear and quadratic polynomials.

Element	Form of $T(x)$	a_1	a_2	a_3
1-D linear	$T(x) = a_1 + a_2 x$	$\frac{T_i x_j - T_j x_i}{x_j - x_i}$	$\frac{T_j - T_i}{x_j - x_i}$	–
1-D quadratic	$T(x) = a_1 + a_2 x + a_3 x^2$	T_i	$\frac{1}{l}(-3T_i + 4T_j - T_k)$	$\frac{2}{l^2}(T_i - 2T_j + T_k)$
2-D linear triangular	$T(x, y) = a_1 + a_2 x + a_3 y$	$\frac{1}{2A}\left[(x_j y_k - x_k y_j)T_i + (x_k y_i - x_i y_k)T_j + (x_i y_j - x_j y_i)T_k\right]$	$\frac{1}{2A}\left[(y_j - y_k)T_i + (y_k - y_i)T_j + (y_i - y_j)T_k\right]$	$\frac{1}{2A}\left[(x_k - x_j)T_i + (x_i - x_k)T_j + (x_j - x_i)T_k\right]$

TABLE 4.5 Shape functions for linear and quadratic polynomials.

Element	Form of $T(x)$	N_i	N_j	N_k
1-D linear	$T(x) = a_1 + a_2 x$	$\left[\frac{x_j - x}{x_j - x_i}\right]$	$\left[\frac{x - x_i}{x_j - x_i}\right]$	–
1-D quadratic	$T(x) = a_1 + a_2 x + a_3 x^2$	$\left[1 - \frac{3x}{l} + \frac{2x^2}{l^2}\right]$	$\left[\frac{4x}{l} + \frac{4x^2}{l^2}\right]$	$\left[\frac{2x^2}{l^2} - \frac{x}{l}\right]$
2-D linear triangular	$T(x, y) = a_1 + a_2 x + a_3 y$	$\frac{1}{2A}\left[(x_j y_k - x_k y_j)T_i + (x_k y_i - x_i y_k)T_j + (x_i y_j - x_j y_i)T_k\right]$	$\frac{1}{2A}\left[(y_j - y_k)T_i + (y_k - y_i)T_j + (y_i - y_j)T_k\right]$	$\frac{1}{2A}\left[(x_k - x_j)T_i + (x_i - x_k)T_j + (x_j - x_i)T_k\right]$

3. For an element,

$$I^e = \frac{1}{2}\int_\Omega \left[k_x\left(\frac{\partial T}{\partial x}\right)^2 + k_y\left(\frac{\partial T}{\partial y}\right)^2 + k_z\left(\frac{\partial T}{\partial z}\right)^2 - 2GT^e\right]d\Omega + \int_{S_{2e}} qT \ ds + \int_{S_{3e}} \frac{1}{2}h(T^e - T_a)^2 ds$$

4. The first derivative with respect to nodal temperatures is

$$\frac{\partial I^e}{\partial[\mathbf{T}]} = \int_\Omega \frac{1}{2}2[\mathbf{B}]^T[\mathbf{D}][\mathbf{B}][\mathbf{T}]d\Omega - \int d\Omega_\Omega \frac{1}{2}2G[\mathbf{N}]^T[\mathbf{T}] + \int_{S_{2e}} q[\mathbf{N}]^T[\mathbf{T}] + \int_{S_{3e}} h[\mathbf{N}]^T[\mathbf{T}]ds - \int_{S_{3e}} h[\mathbf{N}]^T T_a ds = 0$$

5. The terms are written in compact form as

$$k_x\left(\frac{\partial T^e}{\partial x}\right)^2 + k_y\left(\frac{\partial T^e}{\partial y}\right)^2 + k_z\left(\frac{\partial T^e}{\partial z}\right)^2 = \{g\}^T[\mathbf{D}]\{g\}$$

where

$$\{g\}^T[\mathbf{D}]\{g\} = \left\{\frac{\partial T^e}{\partial x} \quad \frac{\partial T^e}{\partial y} \quad \frac{\partial T^e}{\partial z}\right\}\begin{bmatrix} k_x & 0 & 0 \\ 0 & k_y & 0 \\ 0 & 0 & k_z \end{bmatrix}\begin{Bmatrix} \frac{\partial T^e}{\partial x} \\ \frac{\partial T^e}{\partial y} \\ \frac{\partial T^e}{\partial z} \end{Bmatrix}.$$

The algebraic system obtained is

$$[\mathbf{K}]\{\mathbf{T}\} = \{\mathbf{f}\}$$

where

$$[\mathbf{K}] = \int_\Omega [\mathbf{B}]^T[\mathbf{D}][\mathbf{B}]d\Omega + \int_{S_3} h[\mathbf{N}]^T[\mathbf{N}]s$$

and

$$[\mathbf{f}] = \int_\Omega G[\mathbf{N}]^T d\Omega - \int_{S_2} q[\mathbf{N}]^T ds + \int_{S_3} hT_a[\mathbf{N}]^T ds.$$

This procedure is demonstrated for a three-element, four-node wall of material, with a Neumann boundary condition on the left wall and a convective boundary condition on the right wall, the various terms are:

$$\int_0^l d\Omega[\mathbf{B}]^T k[\mathbf{B}] = kA\int_0^l dx \begin{bmatrix} -\frac{1}{l} \\ \frac{1}{l} \end{bmatrix}\begin{bmatrix} -\frac{1}{l} & \frac{1}{l} \end{bmatrix} = \frac{kA}{l}\begin{bmatrix} 1 & -1 \\ -1 & 1 \end{bmatrix}.$$

Since the heat generation is zero, $G = 0$ and

$$[\mathbf{f}] = -\int_{S_2} q[\mathbf{N}]^T ds + \int_{S_3} hT_a[\mathbf{N}]^T ds$$

the first term will appear only in the first element at which the Neumann boundary condition applies; similarly, the second term will appear only in the third element since the convective heat loss condition appears at the last node.

The element matrices are:for the first element:

$$[K]_1 = \frac{kA}{x_1}\begin{bmatrix} 1 & -1 \\ -1 & 1 \end{bmatrix}, \{f\}_1 = \begin{Bmatrix} qA \\ 0 \end{Bmatrix}$$

for the second element:

$$[K]_2 = \frac{kA}{x_2}\begin{bmatrix} 1 & -1 \\ -1 & 1 \end{bmatrix}, \quad \{f\}_2 = \begin{Bmatrix} 0 \\ 0 \end{Bmatrix}$$

and for the third element:

$$[K]_3 = \frac{kA}{x_3}\begin{bmatrix} 1 & -1 \\ -1 & 1 \end{bmatrix}, \{f\}_3 = \begin{Bmatrix} 0 \\ hAT_a \end{Bmatrix}$$

giving a 4×4 global matrix

$$\frac{kA}{x}\begin{bmatrix} 1 & -1 & 0 & 0 \\ -1 & 2 & -1 & 0 \\ 0 & -1 & 2 & -1 \\ 0 & 0 & -1 & 1+hA \end{bmatrix}\begin{bmatrix} T_1 \\ T_2 \\ T_3 \\ T_4 \end{bmatrix} = \begin{Bmatrix} qA \\ 0 \\ 0 \\ hAT_a \end{Bmatrix}$$

which can be solved to compute the temperatures T_i at the nodes.

Exercise 4.14: For a rod of length 10 cm, the temperature at the left and right nodes is $T_i = 100$ C and $T_j = 200$ C. Using the shape functions above, sketch the temperature variation in the element and calculate the temperature at a point 6 cm from the left node of the bar.

Exercise 4.15: For the above rod and end temperatures, use a 1-D quadratic element for which i, j, k are at 0, 5 10 cm from the left node to calculate the temperature at a point 6 cm from the left node of the bar. Compare the two temperature profiles.

Exercise 4.16: For the triangular element specified below, write the expression for $T(x, y)$ and plot the surface temperatures.

Coordinate	T (C)	X (cm)	Y (cm)
i	100	1	2
j	200	2	1
k	300	3	3

From the above elements and shape functions, it is clear that if the nodal values are known, then the function can be found for the entire element. From a single element, describe how you would solve the 8-element square plate problem illustrated in Fig. 4.19.

For the neutron diffusion equation, the functional

$$J(\phi) = \int_a^b \left\{ -D(x)\left[\frac{d\phi}{dx}\right]^2 + \Sigma_a(x)| \quad \phi(x)|^2 - 2S(x)\phi(x) \right\} dx$$

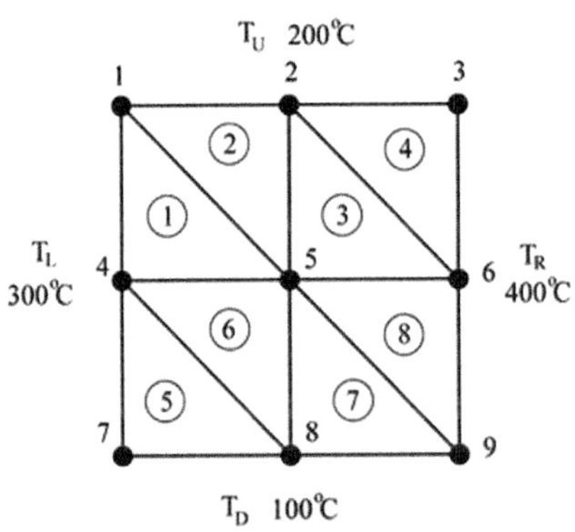

FIGURE 4.19 2-D triangular elements: square plate with Dirichlet boundary conditions.

gives an extremum that satisfies the governing equation. This extremum can be determined by varying ϕ to form the quantity $J(\phi+\delta\phi)-J(\phi)$. The procedure is, as described above, to express the flux in terms of "shape functions" $M(x)$, so that $\phi(x)=\sum_{i=1}^{N}A_iM_i(x)$. The unknown coefficients A_i are determined from the extremum condition

$$\frac{\partial J}{\partial A_i}=0$$

for all elements $i=1,2,3,\ldots N$.

An important equation in nuclear reactor theory is the eigenvalue equation

$$\hat{L}\phi=\frac{1}{k_i}\nu\Sigma_f\phi$$

which represent the loss terms on the left hand side and the gain terms on the right hand side; their are several values or modes for which the equation is balanced. The largest eigenvalue is the system multiplication k_{eff} of a reactor for which the corresponding eigenfunction is the stationary flux distribution.

In a simplified one-group model, the eigenvalue form of the neutron diffusion equation is

$$\left[\frac{d}{dx}\left(-D\frac{d}{dx}\right)+\Sigma_a\right]\phi(x)=\frac{1}{k}\nu\Sigma_f\phi(x)+S(x),\quad 0<x<L$$

with albedo boundary conditions

$$J(0)=-\alpha_L\phi(0),\quad J(L)=-\alpha_L\phi(L)$$

where

$$J=-D\frac{d\phi}{dx}.$$

By multiplying the above with Legendre polynomials for linear shape functions $P_l(N_i(x))$ and integrating over an element results in spatial balance equations for each element. The resulting system of simultaneous equations

$$\overline{\overline{A}}\overline{x}=\overline{B}$$

is solved for the vector x comprising nodal and cell-averaged fluxes and currents. The numerical results of this linear FEM have been determined to be more accurate for coarser grids than FDM for fixed-source problems (Brandao, Dominguez, & Iglesias, 2011).

In nuclear reactors with hexagonal cells, discretization with FEM or spectral methods is based on dividing hexagons into equilateral triangles and assuming that the neutron cross sections remain constant on each triangle. In the collocation method (González-Pintor, Ginestar, & Verdú, 2008) the flux on each triangle ϕ_e, in the eigenvalue equation, is expanded as

$$\phi_e(x,y)=\sum_{i,j}\phi_{e,i}g_{ij}(x,y)$$

where the spatial dependence is in the $g_{ij}(x,y)$ polynomial basis functions. The pseudo-spectral methods, based on modified Dubiner's polynomials for the basis functions, is applied to a 2-D IAEA reactor benchmark problem to compute eigenvalues and power distribution in the core.

Spectral methods can be used to obtain an approximate solution to differential equations in terms of a finite number of basis functions. Both spectral and FEMs are similar in the sense that some shape functions are used; the difference is that in spectral methods, the basis functions are global while in FEM, they are local, that is, within the nodes of an element.

One application of spectral methods is illustrated for the 2-D linear steady-state transport equation

$$\left[\mu\frac{\partial}{\partial x}+\sqrt{1-\mu^2}\frac{\partial}{\partial y}+\Sigma_t(x,y)\right]\phi(x,y,\mu)=\frac{1}{2}[\Sigma_s+\nu\Sigma_f]\int_{-1}^{1}d\mu'\phi(x,y,\mu')+S(x,\mu)$$

in a rectangular domain $-1\leq x,y\leq 1,\ -1\leq\mu\leq 1$ with given boundary conditions.

The angular flux is expanded in terms of Chebyshev polynomials of the first kind $T_i(y)$ which are orthogonal to a weight function $w(x)$

$$\int_{-1}^{1} T_i(x)T_j(x)w(x)dx = \begin{cases} 0. & i \neq j \\ \pi/2, & i = j \neq 0 \\ \pi, & i = j = 0 \end{cases}$$

The expansion is a function of space (y variable), $\phi(x, y, \mu) = \sum_{i=0}^{N} \phi_i(x, \mu)T_i(y)$ is then used, after some algebraic to reduce the 2-D equation into a set of N one-dimensional equations.

Another numerical approximation for the integral (Section 4.4.10) is the use of the quadrature rule, as in the discrete ordinates (S_N) method, to separate μ from the angular flux $\phi(x, y, \mu)$

$$\int_{-1}^{1} d\mu' \phi(x, y, \mu') \sim \sum_{i=1}^{N} w_n \phi_i(x, y)$$

The convergence of such a combined spectral-S_N method has been illustrated for the 2-D linear isotropic scattering transport equation (Asadzadeh & Kadem, 2006) which reduced to a system of 1-D equations.

For unstructured meshes, in the case of complex geometries, FEM and the FVM are used in 2-D and 3-D core calculations. In the FVM, the transported quantity is conserved within a volume; partial currents and surface fluxes are continuous at the boundaries of adjacent volumes. This method has been applied for 2-D and 3-D Light Water Reactors by Bernal, Miró, Ginestar, and Verdú (2014) demonstrating better accuracy and low computational times for 2-D results as compared to 3-D results.

4.6 Approximate methods

4.6.1 The Ritz method

The heat balance equation for a fin (Lewis et al., 2004) written as

$$\frac{d^2\theta}{d\zeta^2} - \mu^2\theta = 0$$

With boundary conditions: $x = 0, dT/dx = 0$ and at $x = L, T = T_b$, with $(T - T_s) = \theta$, $\zeta = x/L$, $\frac{hP}{kA = m^2}$ and $m^2L^2 = \mu^2$. At $\zeta = 0$, $d\theta/d\zeta = 0$ and at $\zeta = 1$, $\theta = \theta_0$ and with $(T - T_s) = \theta$ is solved with the approximation $\overline{T} = \sum_{i=1}^{n} a_i N_i(x)$ substitute into the equation and residual R.

Then a function $\theta(\zeta)$ which satisfies the b.c. is selected

$$\frac{\theta(\zeta)}{\theta_b} = 1 - (1 - \zeta^2)B$$

where B is an unknown parameter to be determined by requiring that

$$\int_0^1 \left(\frac{d^2\theta(\zeta)}{d\zeta^2} - \mu^2\theta \right) d\zeta = 0.$$

This determines the constant B

$$B = \frac{\frac{\mu^2}{2}}{1 + \frac{\mu^2}{3}}.$$

to yield the approximate solution

$$\frac{\overline{\theta}(\zeta)}{\theta_b} = 1 - \frac{3}{4}(1 - \zeta^2).$$

4.6.2 The Rayleigh–Ritz variational method

Consider the second-order ODE

$$\frac{d^2\theta(\zeta)}{d\zeta^2} - \mu^2\theta = 0$$

with boundary conditions

$$\frac{d\theta(0)}{d\zeta} = 0 \text{ and } \theta(1) = \theta_b.$$

The variational functional for the govening equation is

$$I = \int_0^1 \frac{1}{2}\left[\left(\frac{d\theta}{d\zeta}\right)^2 + \mu^2\theta^2\right] d\zeta$$

The θ profile that minimizes the integral I is the solution to the governing equation. The procedure is the same as above; a solution is assumed $\frac{\theta(\zeta)}{\theta_b} = 1 - (1 - \zeta^2)B$, and setting $\frac{\partial I}{\partial B} = 0$, gives

$$B = \frac{\frac{\mu^2}{2}}{1 + \frac{2}{5}\mu^2}$$

and the temperature profile

$$\frac{\theta(\zeta)}{\theta_b} = 1 - (1 - \zeta^2)\frac{\frac{\mu^2}{2}}{1 + \frac{2}{5}\mu^2}.$$

Note that the solution found is different from that with the Galerkin method.

4.6.3 The weighted residual method

With $T \approx \overline{T} = \sum_{i=1}^{n} a_i N_i(x)$, the coefficients a_i are determined by substituting the approximation and using the residual with weighting functions w_i to get

$\int_\Omega \omega_i(x) R dx = 0$ with $i = 1, 2, \quad \ldots, n$

Several weighting functions can be used, for example, in the collocation method $\omega_i = \delta(x - x_i)$, in the Galerkin method $\omega_i(x) = N_i(x)$ and in the least-squares method, $\omega_i = \partial R/\partial a_i$

Exercise 4.17: Use an approximate result for $\theta(x)$ from above in the Galerkin method to obtain a solution. Compare all three solutions for the given heat transfer equation.

4.7 The adjoint function

Mathematical operators have "adjoints." Thus an operator

$$\hat{L} \equiv \frac{\partial^2}{\partial x^2} + \frac{\partial}{\partial x} + 3$$

would have an adjoint operator written as

$$\hat{L}^+ \equiv \frac{\partial^2}{\partial x^2} - \frac{\partial}{\partial x} + 3$$

The adjoint of a derivative is its negative. A "forward" equation of the form

$$\hat{L}\phi(x) = S(x)$$

would have a "backward," or "adjoint" form expressed as

$$\hat{L}^{+}\phi^{+}(x) = S^{+}(x)$$

Both equations would be solved with the appropriate boundary conditions.

In nuclear engineering, ϕ^{+} is the adjoint neutron flux and is also called the "importance" function. For a critical reactor, the diffusion equation reads

$$\hat{L}\phi(x) = 0$$

where

$$\hat{L} \equiv \frac{d^2}{dx^2} + B^2$$

The adjoint operator $\hat{L}^{+}$ is defined as

$$\int_{-a/2}^{a/2} u\hat{L}v dx = \int_{-a/2}^{a/2} v\hat{L}^{+} u dx$$

where u and v are any two functions which vanish at the physical boundaries. The above is also written in the form of an inner product as

$$\left(u, \hat{L}v\right) = (v, \hat{L}^{+}u)$$

or equivalently as

$$\left(u, \hat{L}v\right) = (\hat{L}^{+}u, v)$$

For the second-order "diffusion" operator, it is readily shown, by carrying out an integration by parts, that the operator is self-adjoint, that is

$$\hat{L}^{+} = \hat{L}$$

The solutions, $(x)$and $\phi^{+}(x)$, to both the forward and backward homogeneous equations respectively, are thus proportional to each other.

4.8 Random processes, probability, and statistics

Modeling and simulation in nuclear engineering requires mathematical formulation of coupled systems with the central quantity being the neutron flux $\phi(\boldsymbol{P})$ in phase space $\boldsymbol{P}(r, E, \hat{\Omega}, t)$ where for a Cartesian coordinate system, $r = r(x, y, z)$ and $\hat{\Omega} = \hat{\Omega}(\theta, \varphi)$ and thus $\boldsymbol{P}$ is defined by seven variables. Thus, a neutron transport FDM, FVM or FEM computation with 10^3 points along each x, y, z axis, 10 angular bins, and 10 energy groups would require 10^{12} node point calculations in each time interval. Such intensive computational efforts led to alternate ways of modeling neutron transport. Another way of looking at nature is the probabilistic way where things happen randomly. The debate between scientists such as Einstein with his famous comment to Max Born that "God doesn't play dice" supported a deterministic model of nature an opposed to quantum physics where Planck, Heisenburg, Schrodinger, Dirac and others were creating a probabilistic interpretation. A similar view of transport phenomena took shape as a random process where particles interact in a probabilistic manner. The Monte Carlo (MC) method, based on laws of probability, was developed in the 1940s for the simulation of neutron transport and with advancements in computer hardware and computation, MC simulation has since been extended to several areas science, engineering, and the social sciences.

The following sections are intended to cover the concepts and mathematics as described for MC simulation of alpha particle transport (Section 4.1.61), electron transport (Section 4.1.62), gamma transport (Section 4.1.63), and neutron transport for criticality (Section 4.2.10). The later part of this book focuses on the applications of MC simulation in neutron and photon transport applied to nuclear reactors and systems.

The topics covered in these sections are random processes, random variables, probability distribution functions (PDFs), cumulative distribution functions (CDFs), random numbers, sampling methods, and accuracy and precision of MC methods.

4.8.1 Random processes

Random, or stochastic, processes are considered as an abstraction of events that happen and are observed as a collection of random variables such as Brownian motion (named after Robert Brown) of particles suspended in a liquid. Such processes, as shown below, are observed across science and engineering in signal processing, telecommunications, particle transport and in financial systems such as the fluctuations of the stock market.

Particle transport is considered to consist of random events as particles are "born" and travel in matter colliding with host nuclei, exchanging energy and undergoing nuclear interactions until they "die" by some capture or leakage process. The events are considered to be random as they are determined from probabilistic interaction data (Fig. 4.20).

4.8.2 Markovian processes

A Markovian process is stochastic in nature and has a future state that depends only on the present state without any dependence on its' past history. In the above Wiener process, for example, the "standard" Wiener continuous process $\mathscr{W}_t$ begins at zero, has independent increments in which the future is independent of the past, and has Gaussian increments with mean zero and variance 1 as will be explained in the following sections. In the above, the time step is taken $\Delta t = T/N$ where $T = 1\text{s}$ and $N = 10,000$ steps. With j as the time index and $\mathscr{W}_0 = 0$, $\mathscr{W}_t = \mathscr{W}_{t-1} + \xi\sqrt{\Delta t}$. The random variable ξ is independent and drawn from a normal distribution. Several other characterizations of $\mathscr{W}_t$ have been made for modeling Brownian motion. Processes with drift are also used to model fluctuations with a stochastic differential equation of the form $d\mathscr{X}(t) = a(t, \mathscr{X})dt + d\mathscr{W}(t)$.

For particle transport, this is a natural way of looking at events in history consisting of random events. In the stochastic sense, a history does not repeat itself and a random walk has a probability of occurrence. In this description of particle transport, nature is understood to behave in a stochastic rather than in a deterministic manner.

4.8.3 Population and sample

In a stochastic formulation, estimates are obtained from quantities from a finite sample of histories as it is not possible to consider the whole population which can be of the order of Avogadro's number of particles. When simulating particles in nuclear systems, it is typical to consider a sample of $\sim 10^6$ for which reasonably acceptable results are obtaineable. After reviewing some very powerful theorems of probability and statistics, it will be shown that the variance of a sample mean can be reduced in comparison with the population mean.

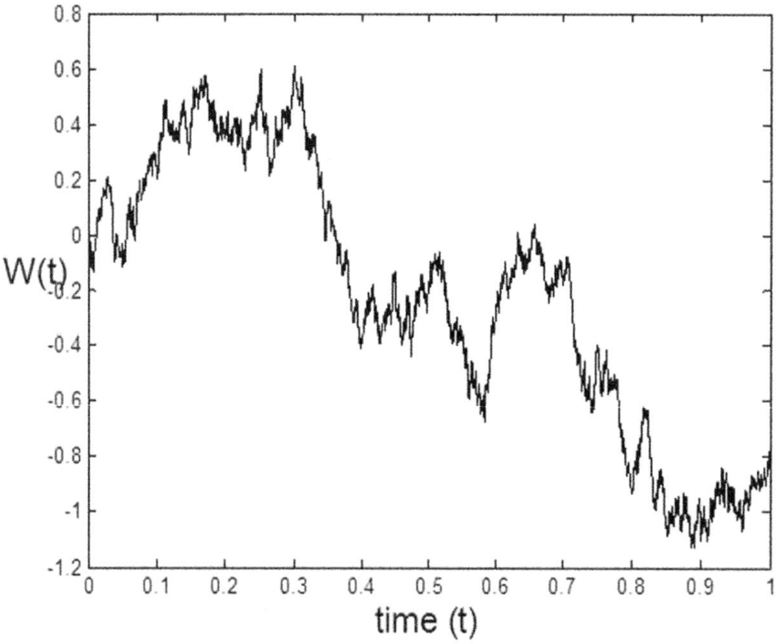

FIGURE 4.20 A random fluctuation depicted as a Wiener process.

4.8.4 Random variables, PDF, and CDF

In nuclear transport, modeled through stochastic simulations, the outcomes of an *experiment* are random; if the simulation is repeated the same result may not be obtained. It is thus considered a random event, just as radioactive transmutation and the compound nucleus formation described in Chapter 2 were based on one of possibly several outcomes.

Random variables are broadly classified as *discrete* and *continuous*; in the former, there are a finite number of realizations of a random variable while in the latter the relaizations are given by continuous functions.

For a discrete random variable X taking values x_i, $i = 1, 2, 3, \cdots, N$ with probability $p_i = P\{X = x_i\}$ has an expectation value $\langle X \rangle$

$$\langle X \rangle = E(X) = \sum_{i=1}^{N} p_i x_i = \mu. \tag{4.21}$$

Similar expressions can be written for powers of the random variable X as

$$\langle X^n \rangle = E(X^n) = \sum_{i=1}^{N} p_i x_i^n.$$

Consider for example, the second power, which is

$$\langle X^2 \rangle = E(X^2) = \sum_{i=1}^{N} p_i x_i^2.$$

If we want a metric for estimating how far each realization of the random variable is from its expectation value $\langle X \rangle$, it is useful to take the second moment

$$\langle (X-\mu)^2 \rangle = E\left((X-\mu)^2\right) = \sum_{i=1}^{N} p_i (x_i-\mu)^2$$

The right hand side can be further simplied as

$$\sum_{i=1}^{N} p_i (x_i-\mu)^2 = \langle X^2 \rangle - \langle X \rangle^2$$

which reads "the expectation value of X^2 minus the mean squared," a large value of this would mean that the values are far from the mean and thus vary to a large extent. This is the reason for this quantity to be called the variance, formally defined as

$$\mathrm{var}(X) = \sigma^2 = \langle X^2 \rangle - \langle X \rangle^2 \tag{4.22}$$

Eqs. (4.21) and (4.22) define the mean μ and variance σ^2 which are very important charactersitics of distributions of random variables.

As an example, consider the binomial distribution in which the outcome of an experiment can be zero or one. Thus in ten experiments, the outcome could vary from 0 to 10. If the probability of getting a 1 is p, then the probability of getting a 0 must be $1 - p$ since there are two possible outcomes and the probabilities therefore must add up to 1. Using the above analysis, the expectation values for an outcome is

$$\langle X \rangle = E(X) = \sum_{i=1}^{N} p_i x_i = p \times 1 + (1 - p) \times 0 = p$$

and

$$\langle X^2 \rangle = E(X^2) = \sum_{i=1}^{N} p_i x_i^2 = p \times 1^2 + (1 - p) \times 0^2 = p$$

giving the variance

$$\mathrm{var}(X) = \sigma^2 = \langle X^2 \rangle - \langle X \rangle^2 = p - p^2 = p(1 - p).$$

From the above, the value

$$S = \sum_{i-1}^{N} X_i$$

which is the sum of outcomes of N experiments, each of which has an outcome X_i is found if we knew how many of the exeriments had an outcome of 1; say n experiments had an outcome 1 and therefore $N-n$ would have outcomes 0. We can find the probability of having n outcomes as

$$P_n = \binom{N}{n} p^n (1-p)^{N-n}$$

If we carried out $N = 100$ experiments and $p = 0.2$, then

$$P_n = \binom{100}{n} 0.2^n 0.8^{100-n} = \frac{100!}{n!(100-n)!} 0.2^n 0.8^{100-n}$$

Thus $P_{20} = 0.0993$, $P_{40} = 2.316 \times 10^{-6}$,$P_{100} = 1.2677 \times 10^{-70}$. The probability P_n versus the number of outcomes n with score 1, plotted with the Matlab program CH4_BinomialPDF.m listed below, is shown in Fig. 4.21.

```
% Program name CH4_BinomialPDF.m
p=0.2;i=0;N=100;
for n=0:1:100
i=i+1;
a=factorial(N)/(factorial(N-n)*factorial(n));
P(i)=a*0.2^n*0.8^(100-n);
x(i)=n;
S(i)=(x(i)*P(i))*1;
end
OutcomeOfAllNexpts=sum(S)
```

We can thus find the total outcome of experiments

$$S = \sum_{i-1}^{N} X_i = nP_n = n\binom{N}{n} p^n (1-p)^{N-n} = Np$$

For $N = 100, p = 0.2$, $S = 20$ which is given from the Matlab program as OutcomeOfAllNexpts.

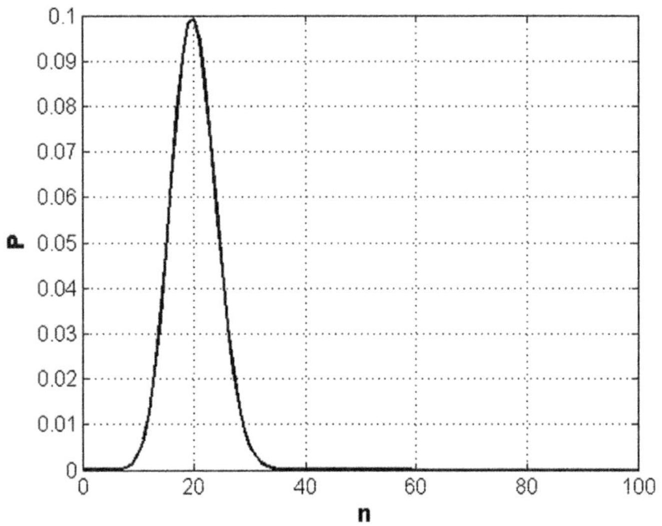

FIGURE 4.21 Probability of outcomes P_n versus the number of events n with score 1.

If we have N sets of two discrete random variables X and Y, and a joint distribution function $f_{X,Y}(x,y) = 1/N$, that is, each set is equiprobable, as an example, then the expectation (mean), variances, covariance and correlation can be obtained as:

$$E_{f_X}[X] = \sum_{i=1}^{N} x_i f_X(x_i)$$

$$E_{f_Y}[Y] = \sum_{i=1}^{N} y_i f_Y(y_i)$$

where

$$f_X(x_i) = \sum_{y} f_{X,Y}(x,y)$$

The variance, covariance, and correlation are defined as

$$\mathrm{Var}_{f_X}[X] = E_{f_X}\left[X^2\right] - \left[E_{f_X}\right]^2$$

$$\mathrm{Cov}_{f_{X,Y}}[X,Y] = E_{f_{X,Y}}[XY] - E_{f_X}E_{f_Y}$$

and

$$\mathrm{Corr}_{f_{X,Y}}[X,Y] = \frac{\mathrm{Cov}_{f_{X,Y}}[X,Y]}{\sqrt{\mathrm{Var}_{f_X[X]} \times Var_{f_X[X]}}}$$

The correlation coefficient lies between -1 and 1, the two limits at which the variables are uncorrelated and fully correlated.

Example 4.4: Given six pairs of values for two random variable, calculate their mean values, variances, covariance and correlation coefficient.

Let $(x,y) = \{(0,0),(1,2),(-1,2),(2,1),(-1,4),(2,3)\}$, $f_{X,Y}(x,y) = 1/6$ Given

$$f_X(x_i) = \sum_{y} f_{X,Y}(x,y) = \begin{cases} \frac{2}{6}, x = -1 \\ \frac{1}{6}, x = 0, 1 \\ \frac{2}{6}, x = 2 \end{cases}$$

the expectation values are

$$E_{f_X}[X] = \sum_{i=1}^{N} x_i f_X(x_i) = -1 \times \frac{2}{6} + 0 \times \frac{1}{6} + 1 \times \frac{1}{6} + 2 \times \frac{2}{6} = \frac{3}{6}$$

$$E_{f_X}\left[X^2\right] = \sum_{i=1}^{N} x_i^2 f_X(x_i) = 1 \times \frac{2}{6} + 0 \times \frac{1}{6} + 1 \times \frac{1}{6} + 4 \times \frac{2}{6} = \frac{11}{6}$$

and the variance is

$$Var_{f_X}[X] = E_{f_X}\left[X^2\right] - \left[E_{f_X}\right]^2 = \frac{11}{6} - \left(\frac{3}{6}\right)^2 = \frac{57}{36} = 1.5833$$

$$f_Y(y_i) = \sum_{x} f_{X,Y}(x,y) = \begin{cases} \frac{1}{6}, y = 0, 1, 3, 4 \\ \frac{2}{6}, y = 2 \end{cases}$$

$$E_{f_Y}[Y] = \sum_{i=1}^{N} y_i f_Y(y_i) = 0 \times \frac{1}{6} + 1 \times \frac{1}{6} + 2 \times \frac{2}{6} + 3 \times \frac{1}{6} + 4 \times \frac{1}{6} = \frac{12}{6} = 2$$

$$E_{f_Y}\left[Y^2\right] = \sum_{i=1}^{N} Y_i^2 f_X(y_i) = 0^2 \times \frac{1}{6} + 1^2 \times \frac{1}{6} + 2^2 \times \frac{2}{6} + 3^2 \times \frac{1}{6} + 4^2 \times \frac{1}{6} = \frac{34}{6}$$

The variance, covariance, and correlation coefficient are given by

$$\mathrm{Var}_{f_Y}[Y] = E_{f_Y}\left[Y^2\right] - \left[E_{f_Y}\right]^2 = \frac{34}{6} - 2^2 = 1.6667,$$

$$\mathrm{Cov}_{f_{X,Y}}[X, Y] = E_{f_{X,Y}}[XY] - E_{f_X}[X]E_{f_Y}[Y] = 0 \times \frac{1}{6} + 2 \times \frac{1}{6} - 2 \times \frac{1}{6} + 2 \times \frac{1}{6} + 6 \times \frac{1}{6} - 1 = = \frac{2}{6}$$

and

$$\mathrm{Corr}_{f_{X,Y}}[X, Y] = \frac{\mathrm{Cov}_{f_{X,Y}}[X, Y]}{\sqrt{Var_{f_X[X]} \times Var_{f_X[X]}}} = \frac{\frac{2}{6}}{\sqrt{1.5833 \times 1.6667}} = 0.2052.$$

For a random variable X that can take on a continuous set of realizations, the pdf $f(x)$ is a density function that represents the probability that X takes on a realization x. Thus, with the requirement that

$$\int_{-\infty}^{\infty} f(x)dx = 1,$$

as for discrete random variables, the probability P that X has value that lies between a and b is expressed in terms of an integral

$$P(a \le X \le b) = \int_a^b f(x)dx.$$

An associated quantity is the cdf

$$F(x) = P(X \le x) = \int_{-\infty}^{x} f(t)dt$$

Thus, $F(x)$ is a montonically increasing function, that is, it continues to increase.

An elementary discrete pdf is the uniformly distributed integer between 1 and n obtained from: $1 + \lfloor n(rnd) \rfloor$.

Some widely used continuous pdf's are the uniform, exponential, the binomial, and the Poisson distribution functions. Consider the uniform pdf (Fig. 4.22) $f(x) = 1, x \in (0, 1)$, and the exponential pdf (Fig. 4.23) $f(x) = 2e^{-2x}$ shown in Figs. 4.17 and 4.18 (Fig. 4.22).

The Statistics Toolbox of Matlab has GUI-based tools disttool for plotting distribution function (pdf's and cdf's) and randtool for randomly sampling from distribution functions. Of the over twenty distribution functions available, the ones used here is the uniform, exponential and normal distribution functions. Plots can also be obtained from the commands, for example, x = 0:0.1:3; y = exppdf(x,0.5); plot(x, y), mean = sum(x.*y)/sum(y)

The exponential pdf is $y = f(x|, \mu) = \frac{1}{\mu}e^{-x/\mu}$ with mean μ

The computed mean, obtained from the expression $\bar{x} = \int_a^b xy(x)dx / \int_a^b y(x)dx$, and $I_1 \equiv \int_a^b xy(x)dx \cong \sum_{i=0}^{M} x_i y_i$, $I_2 \equiv \int_a^b y(x)dx \cong \sum_{i=0}^{M} y_i$ is shown in Table 4.3. In the limit $b \to \infty$, $I_2 \to 1$. Table 4.2 shows that the computed mean tends to the exact mean $\mu = 1/2$ (Table 4.6).

The mean and variances for three continuous PDFs are given in Table 4.7.where

$$\mathrm{erf}(x) = \frac{2}{\sqrt{\pi}} \int_0^x e^{-t^2} dt.$$

The PDF and CDF of a normally distributed random variable, with $= 5, \sigma^2 = 2$, are shown in Fig. 4.24 and 4.25 respectively.

The Matlab program for Figs. 4.24 and 4.25 is listed below.

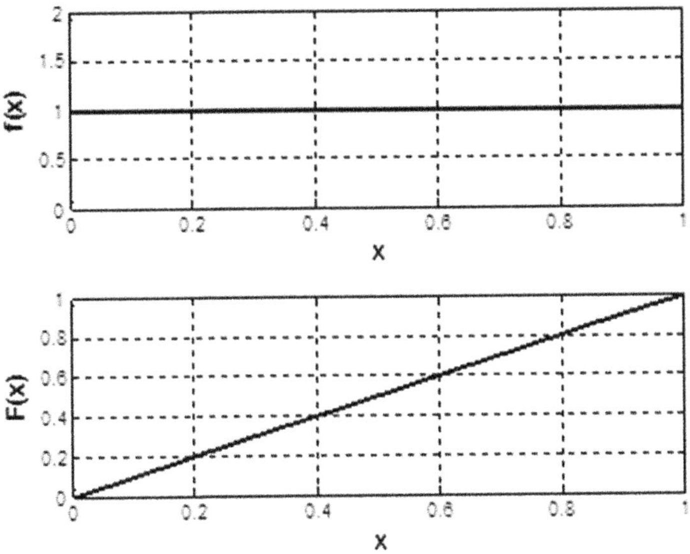

FIGURE 4.22 PDF and CDF of a uniformly distributed random variable.

TABLE 4.6 Estimated mean from an exponential distribution.

Range and intervals	Computed mean
0:0.10:3	0.4454
0:0.01:3	0.4877
0:0.001:3	0.4921
0:0.0001:3	0.4925
0:0.0001:3	0.4999

TABLE 4.7 PDF and CDF of the uniform, exponential and normal distribution functions.

Distribution	Uniform	Exponential	Normal
$f(x)$	$0, x < a, x > b \frac{1}{b-a}, a \leq x \leq b$	$0, x < 0 \lambda e^{-\lambda x}, x \geq 0$	$\frac{1}{\sigma\sqrt{2\pi}} e^{-\frac{(x-\mu)^2}{2\sigma^2}}$
μ	$\frac{a+b}{2}$	$\frac{1}{\lambda}$	μ
σ^2	$\frac{(b-a)^2}{12}$	$\frac{1}{\lambda^2}$	σ^2
$F(x)$	$0, \quad x < a \frac{x}{b-a}, \quad a \leq x \leq b \, 1, \quad x > b$	$0, \quad x < 0 \, 1 - e^{-\lambda x}, x \geq 0$	$\frac{1}{2}\left[1 + \text{erf}\left(\frac{x-\mu}{\sigma\sqrt{2}}\right)\right]$

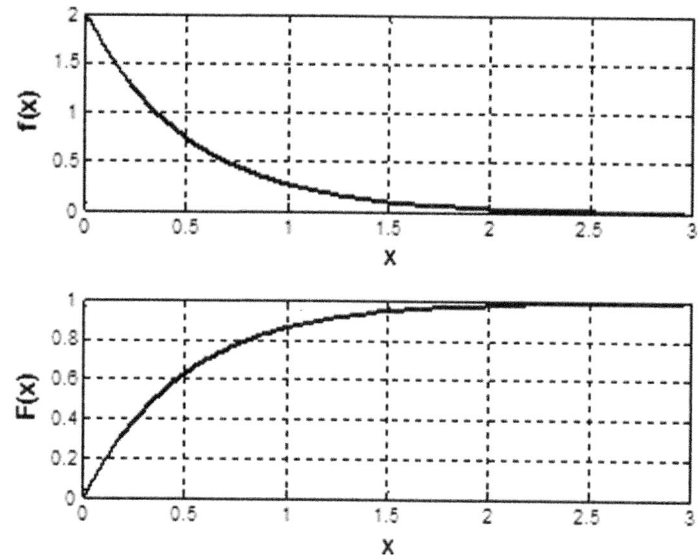

FIGURE 4.23 PDF and CDF of an exponentially distributed random variable.

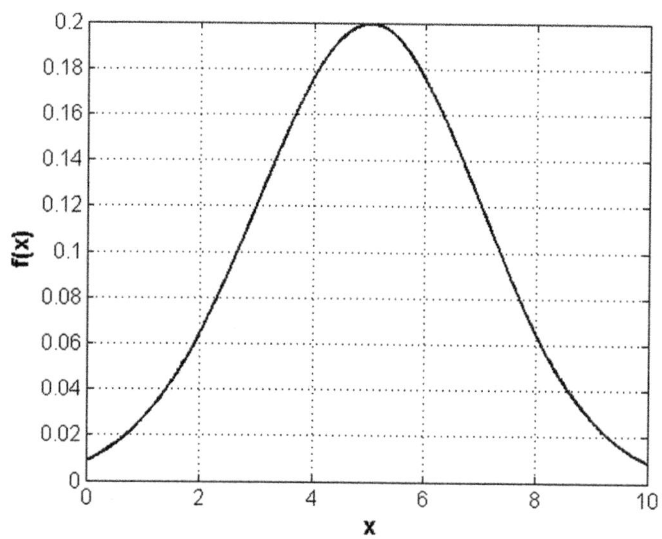

FIGURE 4.24 PDF of a normally distributed random variable.

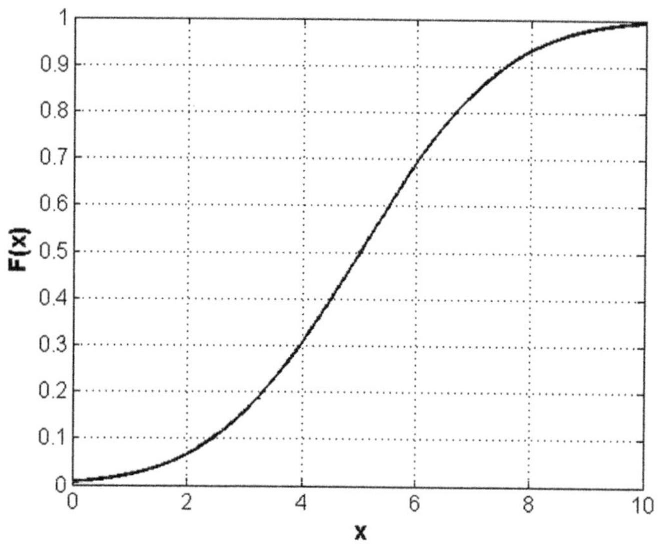

FIGURE 4.25 PDF of a normally distributed random variable.

```
% Program name CH4_NormalPDF.m
gid = fopen('out.txt','w');
mu=5;
sigma=2;
N=1000;
del=10/(N-1);
i=0; sum=0;sum2=0;
for x=0:del:10
    i=i+1;

PDF(i)=(1/(sigma*sqrt(2*pi)))*exp(-(x-mu)^2/(2*sigma^2));
CDF(i)= (1/2)*(1+erf((x-mu)/(sigma*sqrt(2))));
xx(i)=x;
sum=sum+x*PDF(i);
sum2=sum2+PDF(i);
end
mean=sum/sum2
```

The mean value of the random variable is computed in the program to verify the validity of sample size $N = 1000$ which was set arbitrarily.

4.8.5 Random numbers

Numbers drawn randomly from some underlying population distribution function, using an algorithm, are called pseudo-random numbers. Since the number of random numbers drawn from a population can be a finite, though large, sample, they need to satisfy statistical tests to ensure they are truly representative of the population from which they are drawn. For this, we will generate numbers and compute their means and variances. The correlation coefficient lies between -1 and 1, the two limits at which the variables are uncorrelated and fully correlated.

The Matlab functions generating uniform, exponential and normal random numbers are `rand`, `exprnd`, and `randn`.

One of the oldest and most reliable methods used for generating random numbers is the linear congruential random number generator (RNG)

$$S_{k+1} = (S_k g + c) \mathrm{mod} p$$

where $p = 2^m$, and the random number $\xi_k = S_k/p$. Here, S_0 is the seed, g and c are multipliers, and mod(p) implies the remainder after division by p; thus 4 mod 2 = 0. In Matlab the command is mod(4,2). The integers determine the period of the numbers and depend on the computer hardware; a large $N \sim 36$ ensures a large period. The Los Alamos Monte Carlo code uses $m = 48$ giving a period of $2^{46} \sim 7.03 \times 10^{13}$, $g = 5^{19}$, $c = 0$. These limits are dependent on the word length of a computer.

4.8.5.1 Matlab program

```
s(1)=1;c=47;p=100;N=100;Nbins=5;
for i=1:Nbins  count(i)=0; end
for i=1:N
t=c*s(i)+1;
s(i+1)=mod(t,p);
end
trand=s/p;
for i=1:N
    if (trand(i)<0.2)  count(1)=count(1)+1; end
    if ((trand(i)>=0.2) && (trand(i)<0.4))
        count(2)=count(2)+1; end
    if ((trand(i)>=0.4) && (trand(i) <0.6))
        count(3)=count(3)+1; end
    if ((trand(i)>=0.6) && (trand(i) <0.8))
        count(4)=count(4)+1; end
    if ((trand(i)>=0.8) && (trand(i) <=1))
        count(5)=count(5)+1; end
end
```

TABLE 4.8 Random number generation.

Seed	g	c	p	Period
1	47	1	100	20
1	2	0	9	6
3	2	0	9	2

The output is: count = 20 20 20 20 20 so that 20 random numbers fall into each bin.

When $c = 0$, the RNG is called a multiplicative congruential generator or the Lehmer RNG; when $c \neq 0$, the generator is called a mixed congruential generator. The period depends on the combination of values as shown below (Table 4.8).

The "holes" left by pseudo-random numbers in sampling space uniformly, is filled by quasi-random numbers which are used for numerical integration for sampling space more efficiently.

Exercise 4.18: Use the Matlab random number generator `rand` and generate 100 numbers and check their distribution. Using the tic and toc functions, get an idea of the time taken by your computer to generate 1,000,000 uniform random numbers.

Exercise 4.19: Generate a set of 1,000,000 uniform random numbers (ξ_1, ξ_2), test how many of these lie in the first quadrant of a unit circle and estimate π from the answer $\pi = 4 * \varpi$, where ϖ is the fraction of points falling inside the first quadrant. Now use the value from $\pi = 4 * \tan^{-1} 1$ to estimate the accuracy of a random number estimate of π.

4.8.6 Sampling from PDFs

4.8.6.1 Sampling from analytic PDFs

Sampling can be carried out from a discrete or continuous PDF by the direct method if the CDF can be analytically obtained, and by the rejection method otherwise.

For a discrete probability, the domain of interest is divided into a finite number of intervals, or bins, each bin having a probability f_i. As an example, consider the case of three bins. Since the sum of all probabilities is 1,

$$\sum_{i=1}^{3} f_i = 1.$$

If the bin probabilities are $f_1 = 0.2, f_2 = 0.7, f_3 = 0.1$, as shown in Fig. 4.26, the CDF is obtained by adding the bin probabilities. A uniform number ξ in the interval (0,1) is generated; as an example, if $\xi = 0.7$ as shown in Fig. 4.26, the dotted lines connect down into the second group.

The selection for a group is done by comparing ξ with the sums

$$\sum_{i=1}^{m-1} f_i < \xi \leq \sum_{i=1}^{m} f_i, m = 1, 2, 3$$

It is seen that $\xi = 0.7$ falls in the second interval as the sums are 0.2, 0.9, 1.0. Thus the second interval is selected. In transport simulation, discrete sampling is carried out to decide, for example, which event to select from a number of possible events, and to select a nuclide for a collision from a compound.

For an example of direct sampling, consider the uniform pdf for which the cdf is

$$F(x) = \begin{cases} x, 0 < x < 1 \\ 1, x \geq 1 \end{cases}$$

can be used to represent the uniform pdf. A simple proof is the "conservation of probability" as we move from one variable to another. Consider two random variables X and Y with pdf's $f(x)$ and $g(y)$. Since probability must be conserved regardless of the variable used, $f(x)dx = g(y)dy$, and if $X \equiv f, Y \equiv F$, then since $\frac{dF}{dx} = f$, the distribution function $g(F) = 1$, that is, the distribution of the cdf is uniform. It is for this reason that F can be used as a pseudo-random number.

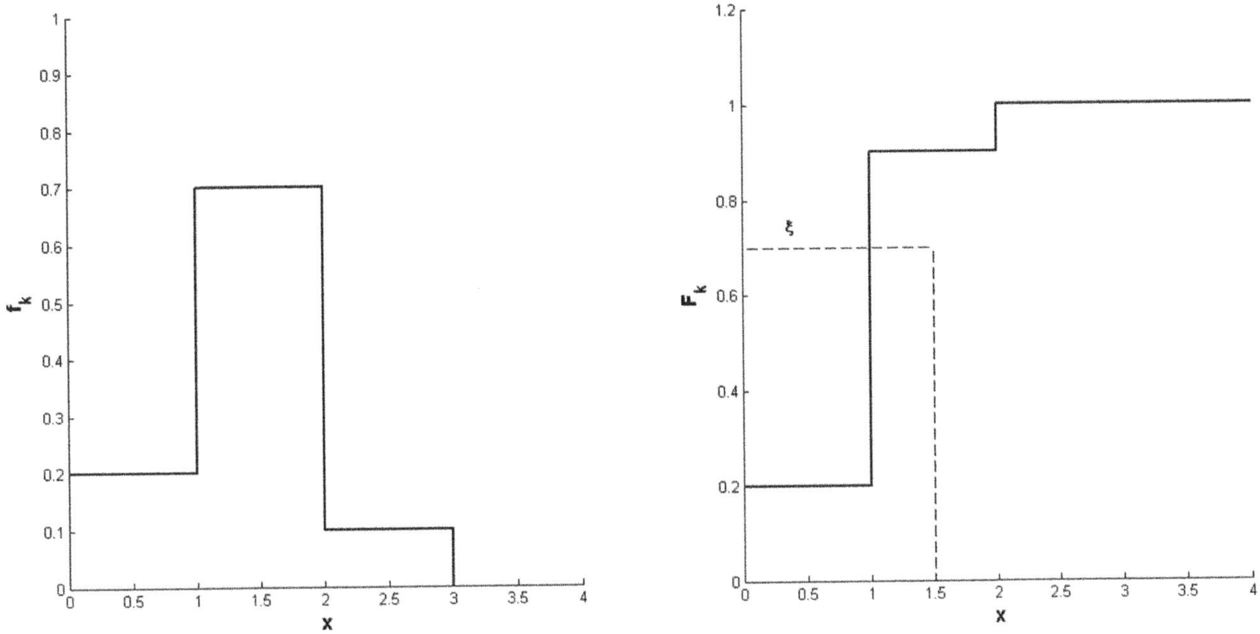

FIGURE 4.26 Sampling from a discrete distribution.

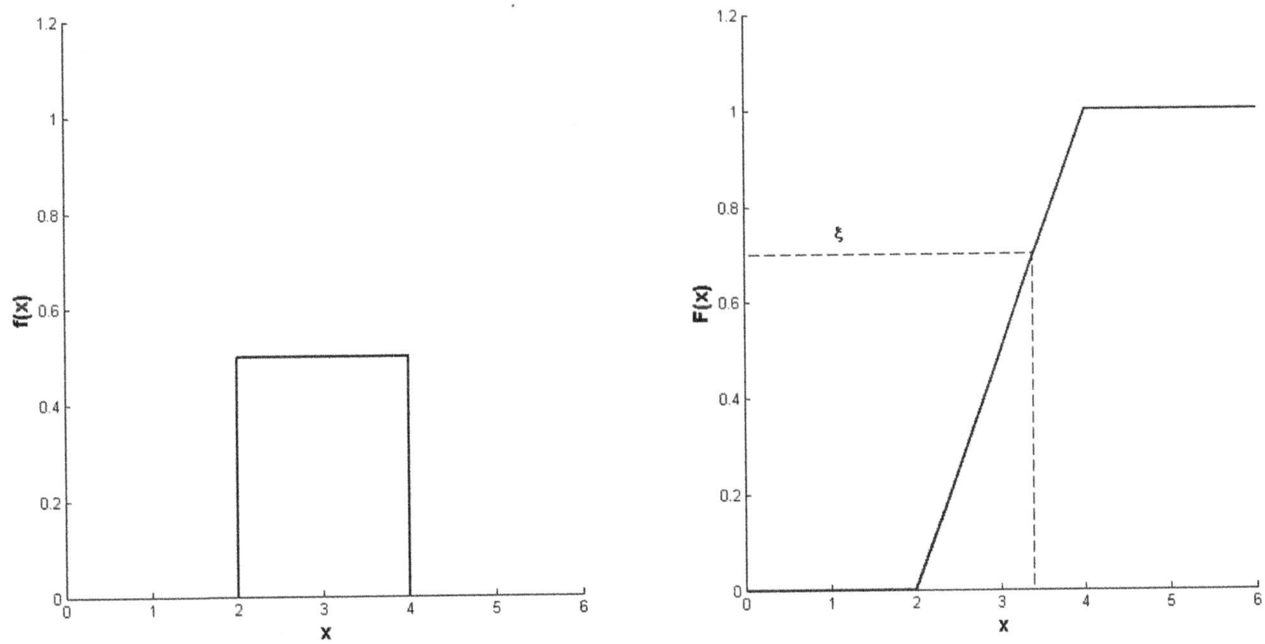

FIGURE 4.27 Sampling from a uniform distribution.

Fig. 4.27 shows the PDF and CDF for $f(x) = 1/2, 2 \le x \le 4$. The PDF is rectangular while the CDF is a monotonically increasing function. The dotted lines show that sampling for a random number ξ generated from a uniform distribution. If, for example, $\xi = 0.7$, then the intersection with the CDF is projected down to the $x-$ axis to get a sample $x = 3.40$.

For an exponential PDF, shown in Fig. 4.28, $f(x) = 2e^{-2x}$, $F(x) = \int_0^x f(x)dx = 1 - e^{-2x}$ the random variable X can be sampled using $x = F^{-1}(\xi)$, so that $x = (-1/2)\ln(1 - \xi)$ (Table 4.9).

From Fig. 4.27, for $\xi = 0.7$, the random variable sampled is $x = 0.6020$. This procedure is done on a computer where thousands or possibly millions of random numbers may be required depending on the problem.

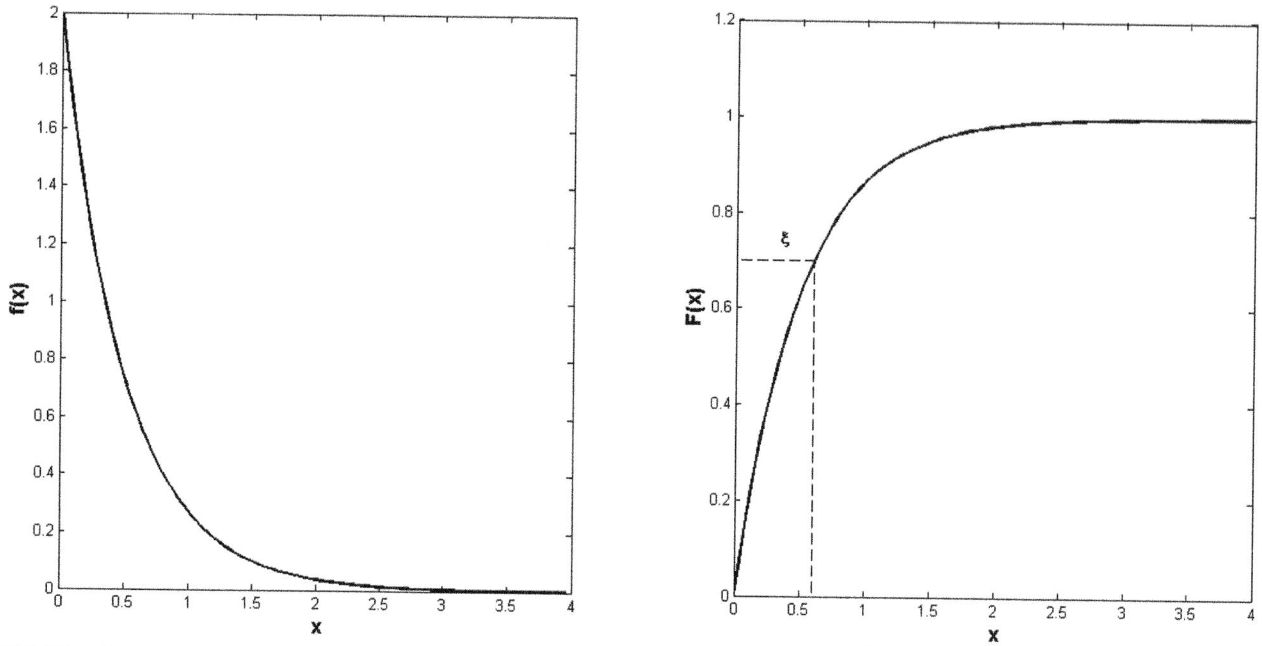

FIGURE 4.28 Sampling from an exponential distribution function.

TABLE 4.9 Sampling from uniform, exponential and normal PDFs.

Distribution	Uniform	Exponential	Normal
$f(x)$	$0, x<a, x>b \frac{1}{b-a}, a \le x \le b$	$0, x<0 \lambda e^{-\lambda x}, x \ge 0$	$\frac{1}{\sigma\sqrt{2\pi}} e^{-\frac{(x-\mu)^2}{2\sigma^2}}$
$F(x)$	$0, \quad x<a \frac{x}{b-a}, \quad a \le x \le b \; 1, \quad x>b$	$0, \quad x<0 \, 1-e^{-\lambda x}, x \ge 0$	$\frac{1}{2}\left[1+\text{erf}\left(\frac{x-\mu}{\sigma\sqrt{2}}\right)\right]$
x	$x=a+(b-a)\xi$	$-\frac{1}{\lambda}\log(1-\xi)$	$p=\sigma\sqrt{-2\log\xi_1}\theta=2\pi\xi_2\mu+p\sin\theta$

For the case of a linear PDF, $(x)=2x, 0 \le x \le 1$, the CDF is $F(x)=x^2$, for which x is sampled as $x=\sqrt{\xi}$.

Another useful case is the PDF $f(r)=re^{-r^2/2}$, which can be integrated easily to get the CDF $F(r)=1-e^{-r^2/2}$ from which the random variable can be sampled as $r=\sqrt{-2\log(1-\xi)}$.

An approximate method to sample from a Gaussian is to form the sum

$$Z=\xi_1+\xi_2+\cdots\xi_k, \text{large} k$$

from uniform random numbers ξ_i with mean and variance $E[\xi_i]=1/2, \sigma^2=1/12$. The mean and variance of Z are thus

$$E[Z]=k/2, \quad \sigma^2=k/12.$$

With the variable for a standardized normal (mean 0 variance 1), x and Z are related; here μ and σ are the mean and standard deviation of the Gaussian from which sampling is done. Then

$$\frac{x-\mu}{\sigma}=\frac{Z-k/2}{\sqrt{k/12}}$$

from which the sampled variable is

$$x \cong \sigma\left(\frac{Z-k/2}{\sqrt{k/12}}\right)+\mu,$$

This method works well for large ($\gtrsim$12).

4.8.6.2 *Sampling from nonanalytic PDFs*

Such a direct inversion is not possible for nonintegrable pdf's such as a rejection scheme in which a simple bounding function $g(x)$ is chosen $(f(x) \leq cg(x))$; the variable x is sampled from G and the value $\hat{x}$ is selected if $\xi_2 cg(\hat{x}) \leq f(\hat{x})$, otherwise it is rejected.

Exercise 4.20:

1. Use both direct and rejection schemes for sampling from the PDF: $f(x) = 2x,\ x \epsilon \left(0, \frac{1}{2}\right)$.
2. Use the rejection scheme for sampling from the normal PDF:

$$f(x) = \frac{1}{\sqrt{2\pi}\sigma} e^{-\frac{1}{2}\left(\frac{x-\mu}{\sigma}\right)^2}$$

4.8.7 Kullback–Leibler divergence for uniform random numbers

An important metric for estimating convergence in a sampling process is the Kullback–Leibler divergence,

$$D(p|(q) = \sum_x p(x) \log \frac{p(x)}{q(x)} = \sum_x p(x) \log p(x) - \sum_x p(x) \log q(x) \equiv -H(X) + H_c(X)$$

between a true pdf represented by p and an estimate q, where $H_c(X)$ is the cross-entropy which is a measure of the inefficiency. Ideally, the divergence $(p|(q)$, also called the relative entropy, should be zero so that no information is lost. Let us consider this metric for deciding how many uniform random numbers are enough for convergence. From Fig. 2.9, it is clear that less than $N = 100$ numbers is inadequate and $N \sim 500$ is a "good" sampling size.

The KL distance is seen in Fig. 4.26 to decrease as the sample size is increased. The convergence is also dependent on the number of bins where a larger sample size is required for 20 bins compared with that for 10 bins (Fig. 4.29).

4.8.8 The law of large numbers

For random numbers generated as independent and identically distributed, that is, each with the same mean μ then as their number increases the sample mean μ_s will tend to the theoretical mean μ.

Chebyshev's Inequality tells us that the deviation from the mean of a random variable more than k standard deviations is less than or equal to $1/k^2$. Thus the probability of being out of two standard deviations is ¼, and

$$P(|X - \mu| > k\sigma) \leq \frac{V(x)}{\varepsilon^2} = \frac{1}{k^2}$$

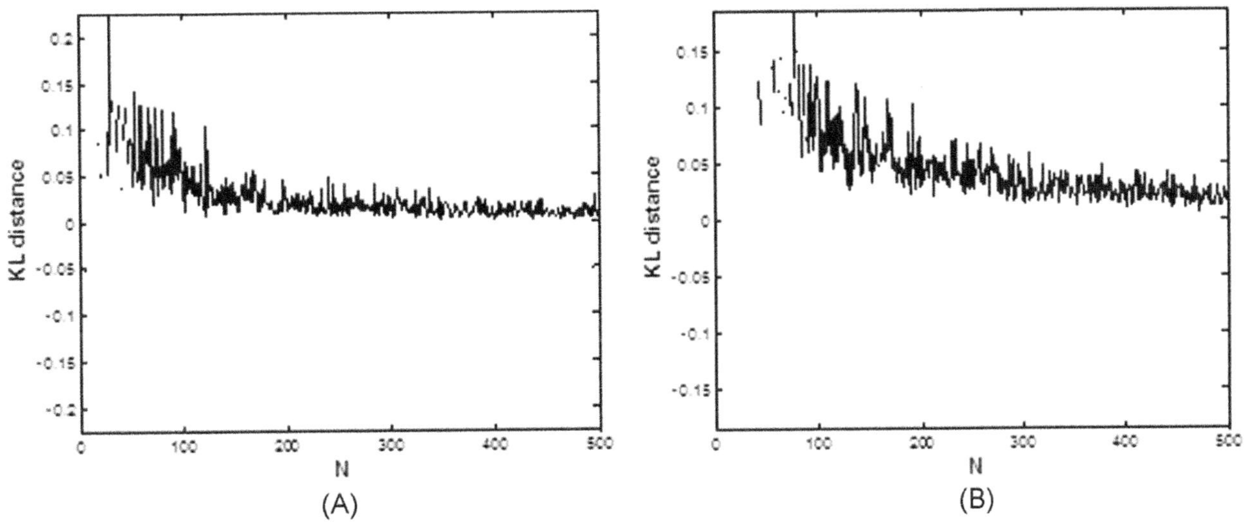

FIGURE 4.29 (A and B) KL distance for 10 and 20 bins.

According to the law of large numbers if $X_1, X_2, \ldots, X_n$ are independent trials, with an expectation $\mu = E(X_j)$ and finite variance $\sigma^2 = V(X_j)$, then for an average $\overline{S}_n \equiv \frac{S_n}{n} = \frac{X_1 + X_2 + \cdots X_n}{n}$, for any $\varepsilon > 0$

$$\lim_{n\to\infty} P\left(|\frac{\overline{S}_n}{n} - \mu| \geq \varepsilon\right) \to 0$$

$$\lim_{n\to\infty} P\left(\left|\frac{\overline{S}_n}{n} - \mu\right| < \varepsilon\right) \to 1$$

Example 4.5: Consider n Bernoulli trials processes and take a probability of success $p = 0.3$. If an outcome is a success, let the random variable be 1 otherwise it is zero. The expectation value $E(X_j) = 0.3$ and variance $V(X_j) = pq = 0.21$. Let $\overline{S}_n \equiv \frac{S_n}{n} = \frac{X_1 + X_2 + \cdots X_n}{n}$, then the expectation and variance of the average can be found as $< \overline{S}_n \geq 0.3$, and $V_{\overline{S}_n} = \frac{V_{S_n}}{n^2} = \frac{0.21}{n}$, and according to Chebyshev's Inequality, $P(|\overline{S}_n - 0.3| \geq 0.1) \leq 21/n(0.1)^2 = \frac{21}{n}$. If we have $n = 1000$ trials, then we can thus show that the probability that this estimate lies between 0.2 and 0.4 is greater than or equal to 0.979; or

$$P(0.2 < \overline{S}_{1000} < 0.4) \geq 0.979$$

Thus, for a very large number of trials the probability that the number will lie in the range 0.2 to 0.4 will be one. We know that the number should be 0.3 for a single trial; this states that it will tend to 0.3 for a large number of trials as well.

4.8.8.1 Application of the law of large numbers

For uniform random numbers between 0 and 1, the mean value and variance are ½ and 1/12 respectively. Thus if n numbers are drawn from this distribution, the $E(S_n/n) = 1/2$ and $V(Sn/n) = 1/12n$ so that

$$P\left(|S_n - \frac{1}{2}| \geq \varepsilon\right) \leq \frac{1}{12n\varepsilon^2}$$

For $\varepsilon = 0.1$, $P \leq 100/12n$, so that a sample size $n = 100$, would give $P \leq 8\%$, for $n = 1000$, $P < = 0.8\%$, so if want an error of less than 1%, we must have a sample size of $n \sim 1000$ (Grinstead & Snell, 1997). Thus Chebyshev's Inequality tells us the sample size required for the desired accuracy.

4.8.9 The central limit theorem

Chebyshev's Inequality gives an idea of the sample size while the stronger Central Limit Theorem tells us the effect of increasing sample size on the range of estimates.

Let $S_n = X_1 + X_2 + X_3 \cdots X_n$ be the sum of n discrete independent random variables with common distribution having expected value μ and variance σ^2. Then for $a < b$,

$$\lim_{n\to\infty} P\left(a < \frac{S_n - n\mu}{\sqrt{n\sigma^2}} < b\right) = \frac{1}{\sqrt{2\pi}} \int_a^b e^{-x^2/2} dx$$

$$\lim_{n\to\infty} P\left(a < \frac{S_n/n - \mu}{\sqrt{\sigma^2/n}} < b\right) = \frac{1}{\sqrt{2\pi}} \int_a^b e^{-x^2/2} dx \equiv \mathcal{N}(0,1)$$

In words, the above says that whatever the population distribution of the random variable X, with mean and variance, the sampled mean $\overline{X}$ will be normal as the sample size becomes very large. The sampled mean will have a mean equal to the population mean; however its variance will have reduced to σ^2/n. Thus, as the sample size becomes larger, the range becomes narrower.

Example 4.6: Population distribution of a random variable X is normal μ_p, σ_p;sample size N, sample of the mean has $\overline{\mu}_s, \sigma_s$; sample mean $\overline{X}$ is normally distributed with $\overline{\mu}_s = \mu_p$ and $\sigma_s = \sigma_p/\sqrt{N}$

4.8.9.1 Matlab program: mean and variance

```
mu=27;std=12;
fid=fopen('OutSample.txt','w');
N=36;
fprintf(fid,'\n  N=%8.0f',N)
for i=1:100
ct=0; % count people less than 30 years old
for j=1:N
    age(j)= mu+std*randn;
    if (age(j)<30)
        ct=ct+1;
    end
end
count(i)=ct;
SampleAvg(i)=mean(age);
SampleStd(i)=sqrt(var(age));
fprintf(fid,'\n    %6.0f    %6.0f  %8.4f
%8.4f',i,count(i),SampleAvg(i),SampleStd(i))
end

% now find the standard error of the SampledMean
AvgSampleMean=mean(SampleAvg);
StdSampledMean=sqrt(var(SampleAvg));
fprintf(fid,'\n    %8.4f    %8.4f',AvgSampleMean,StdSampledMean)
```

Table 4.10 shows results for N batches with sample sizes 36, 3600, and 360,000, that is, an increase of 100 in each computation. The decrease in the standard deviation can be seen to be by a factor of 10 while the sample mean tends to the population mean ($\mu_p = 27, \sigma_p = 12$).

4.9 Evaluation of integrals

There are several methods for evaluating integrals; the most basic of ideas is to break the region into a large number of rectangles and sum the areas of the rectangles. This would give an error at the curves of the function under which the area is being evaluated. Such methods (Abramowitz & Stegun, 1964) include the trapezoidal rule, Simpson's rule, Euler–Maclaurin formulas, Newton–Cotes formulas, and Gauss' formula (based on Legendre polynomials) for 1-D integrals. As the dimensionality increases, these methods become inefficient and random sampling begins to look attractive. In this section, the basics of the sampling methods are described.

A definite integral for a function f(x) is thus expressed as a finite sum

$$I = \int_{-1}^{1} f(x)dx \quad \approx \sum_{i=1}^{n} c_i f(x_i),$$

The reactangles are shown in Fig. 4.30.

TABLE 4.10 Sample mean $\overline{\mu}_s$ and standard deviation σ_s.

N (batches)	36	3600	360,000
100	26.8202(1.8477)	26.9864(0.2020)	27.0013(0.0189)
1000	27.1001(1.9830)	26.9950(0.2005)	27.0008(0.0192)

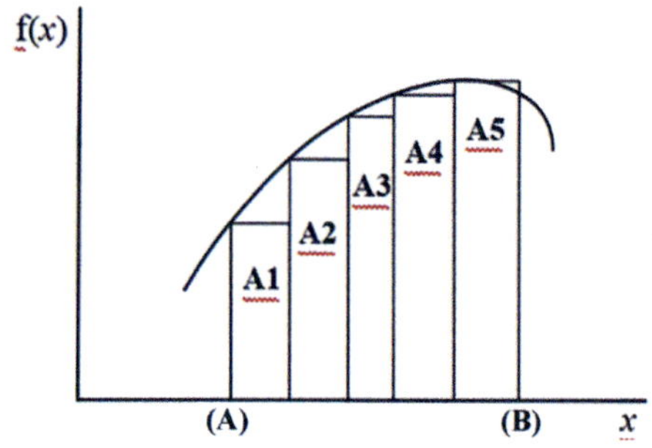

FIGURE 4.30 Numerical integration.

TABLE 4.11 Abscissa and weights of the Gaussian quadrature.

n	x_i	c_i
2	0.5773502691896257	1.0000000000000000
	− 0.5773502691896257	1.0000000000000000
3	0.7745966692414834	0.5555555555555556
	0	0.8888888888888888
	− 0.7745966692414834	-0.555555555555556
4	0.8611363115940525	0.3478548451374544
	0.3399810435848563	0.6521451548625460
	− 0.3399810435848563	0.6521451548625460
	− 0.8611363115940525	0.3478548451374544

In the Gaussian quadrature, the abscissa and weights arise out of requiring exact results for $(n + 1)$th order polynomials. For the case $n = 2$, an integral I is evaluated at to points; thus

$$I = \int_a^b f(x)dx = \sum_{i=1}^{2} w_i f(x_i) = w_1 f(x_1) + w_2 f(x_2).$$

For a third-order polynomial

$$f(x) = a_0 + a_1 x + a_2 x^2 + a_3 x^3$$

the integration is carried out resulting in four simultaneous linear equations for the four unknowns w_1, w_2, x_1, x_2; the solution gives

$$I = \frac{b-a}{2} f\left[\frac{b-a}{2}\left(-\frac{1}{\sqrt{3}}\right)\right] + \frac{b+a}{2} + \frac{b-a}{2} f\left[\frac{b-a}{2}\left(\frac{1}{\sqrt{3}}\right)\right] + \frac{b+a}{2}$$

Thus, if $a = -1, b = 1$

$$I = f\left[\left(-\frac{1}{\sqrt{3}}\right)\right] + f\left[\left(\frac{1}{\sqrt{3}}\right)\right]$$

which requires a change of limits on the integral

$$I = \int_{-1}^{1} f(x)dx = w_1 f(x_1) + w_2 f(x_2)$$

yielding the weights and abscissae of Table 4.11

$$w_1 = 1, \quad w_2 = 1, \quad x_1 = -\frac{1}{\sqrt{3}}, x_2 = \frac{1}{\sqrt{3}}.$$

Similarly, higher order polynomials are used to determine quadrature parameters for n = 2, 3 and 4 .

Example 4.7: Sketch the Integrals $I_1 = \int_0^5 x^2 e^{-x} dx$, $I_2 = \int_0^{2\pi} e^{-x} \sin 4x dx$, and $I_3 = \int_0^{10} e^{-x^2} dx$, and evaluate them using Simpson's Rule, and Gaussian Quadrature.

4.9.1 The Monte Carlo method for numerical integration

The computational effort for quadrature methods discussed above increases with the dimensionality of the integral thus a 1000 point evaluation in each dimension requires one billion evaluation points. For such large problems, Monte Carlo methods are more efficient. An integral evaluation of

$$I = \int_a^b g(x) f(x) dx$$

with the Monte Carlo Method (Dunn & Shultis, 2012; Kalos & Whitlock, 2008; Reiher, 1966) is carried out by sampling from the PDF $f(x)$ and estimating $g(x)$ as

$$I \cong \langle g(x) \rangle = \frac{1}{N} \sum_{i=1}^{N} g(x_i)$$

With the Central Limit Theorem, there is justification to anticipate that as the sample size $N \to \infty$, that is, a large number, then the domain of the result will become smaller and smaller with the sample mean tending to the population mean and the variance reducing inversely with the sample size N.

1. Sketch theiIntegrals, $I_1 = \int_0^2 dx \quad e^{-x}$, $\quad I_2 = \int_0^2 x e^{-x} dx$ and $I_3 = \int_0^2 e^{-x^2} dx$
2. Obtain an exact solution for each of the above ($I_3 = \int_0^2 \exp(-x^2) dx = \frac{\sqrt{\pi}}{2} \text{erf}(2) = 0.8821$)
3. With a calculator, taking $N = 10$ points, estimate the integrals and the standard deviations
4. Write a program which can be used to estimate the integrals using the MC method.

The functions are plotted in Fig. 4.31.

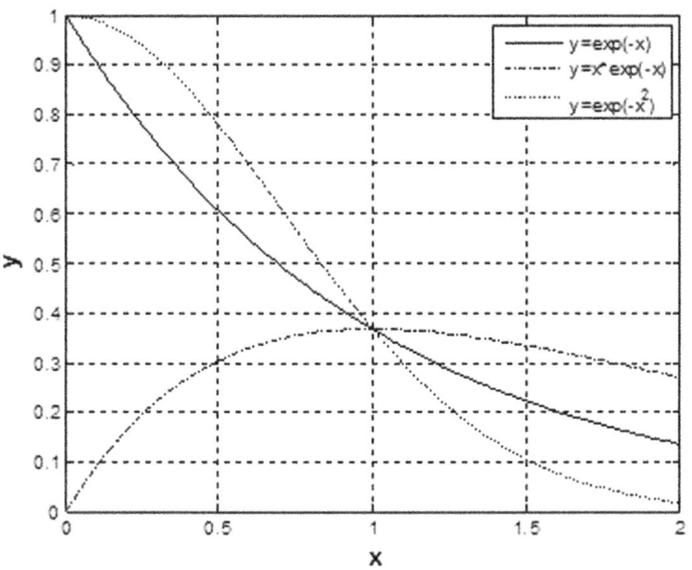

FIGURE 4.31 Functions for MC integration.

4.9.1.1 Matlab program

```
fid=fopen('outMain.txt','w');
x=0:0.001:2;
y1=exp(-x);
y2=x.*exp(-x);
y3=exp(-x.*x);
plot(x,y1,'-k','LineWidth',1.5)
hold on
plot(x,y2,'-.k','LineWidth',1.5)
hold on
plot(x,y3,':k','LineWidth',1.5)
h = legend('y=exp(-x)','y=x*exp(-x)','y=exp(-x^2)',1);
grid on
xlabel('\bf x','FontSize',12)
ylabel('\bf y','FontSize',12)
N=100;
% integral 1
F=@(x)exp(-x);
Q=quad(F,0,2);
sum=0.0; sum2=0.0;
for i=1:N
    t = rand;
    r = exp(-2.0*t);
    sum = sum + r ;
    sum2=sum2 + r^2;
end
Int1 = 2.0*sum/N;
var1 = sum2/N - (sum/N)^2;
std1 = sqrt(var1);
PercErr = 100.0*abs((Int1 - Q)/Q);
fprintf(fid,'\n Integral 1');
fprintf(fid,'\n N= %6.0f',N);
fprintf(fid,'\n    quad      Int1       var1');
fprintf(fid,'\n %8.4f %8.4f %12.4e ',Q,Int1,std1);
fprintf(fid,'\n Percentage Error = %8.4f',PercErr);
% integral 2

F=@(x)x.*exp(-x);
Q2=quad(F,0,2);
sum=0.0; sum2=0.0;
```

```
for i=1:N
    t = rand;
    r = t*exp(-2.0*t);
    sum = sum + r ;
    sum2=sum2 + r^2;
end
Int2 = 4.0*sum/N;
var2 = sum2/N - (sum/N)^2;
std2 = sqrt(var2);
PercErr = 100.0*abs((Int2 - Q2)/Q2);
fprintf(fid,'\n Integral 2');
fprintf(fid,'\n N= %6.0f',N);
fprintf(fid,'\n    quad      Int1       std2');
fprintf(fid,'\n %8.4f %8.4f %12.4e ',Q2,Int2,std2);
fprintf(fid,'\n Percentage Error = %8.4f',PercErr);

% integral 3
F=@(x)exp(-x.*x);
Q3=quad(F,0,2);
sum=0.0; sum2=0.0;
for i=1:N
    t = rand;
    r = exp(-4.0*t*t);
    sum = sum + r ;
    sum2=sum2 + r^2;
end
Int3 = 2.0*sum/N;
var3 = sum2/N - (sum/N)^2;
std3 = sqrt(var3);
PercErr = 100.0*abs((Int3 - Q3)/Q3);
fprintf(fid,'\n Integral 3');
fprintf(fid,'\n N= %6.0f',N);
fprintf(fid,'\n    quad      Int3       std3');
fprintf(fid,'\n %8.4f %8.4f %12.4e ',Q3,Int3,std3);
fprintf(fid,'\n Percentage Error = %8.4f',PercErr);
fclose(fid);
```

Results from MC evaluation of integrals are given in Table 4.12, giving the estimate and relative error with reference to the exact.

In Monte Carlo simulation, we say an estimate is accurate if its mean is close to the mean of the population, which is generally unknown. Precision is a measure of the spread or variance of the estimates from which the mean is computed. An estimate is not "good" if it is inaccurate but precise, or accurate and not precise; it is desirable to have an answer that is both accurate and precise. The "best" MC estimate would have the "correct" mean and "zero-variance."

TABLE 4.12 Values of integrals from Gaussian Quadrature and random sampling.

Integral	Matlab (quad)	Exact	$N = 100$	$N = 100{,}000$
$I_1 = \int_0^2 dx e^{-x}$,	0.8647	0.8647	0.9176[b] 2.4359×10^{-1} 6.13%	0.8649 2.4176×10^{-1} 0.027%
$I_2 = \int_0^2 x e^{-x} dx$	0.5940	0.5940	0.5924 4.4340×10^{-2} 0.2656%	0.5944 4.1886×10^{-2} 0.0656%
$I_3 = \int_0^2 e^{-x^2} dx$	0.8821	0.8821[a]	0.8827 3.2023×10^{-1} 0.0651%	0.8792 3.4455×10^{-1} 0.322%

[a] $I_3 = \int_0^2 \exp(-x^2) dx = \frac{\sqrt{\pi}}{2} \text{erf}(2) = 0.8821$.
[b] *Mean, standard deviation and percentage error (with quad taken as ref).*

There are several methods to obtain better and faster results such as importance sampling and other variance reduction techniques which will be discussed in the sequel in the context of simulations for nuclear systems.

Problems

1. Show that for a system in equilibrium with $n(t) = n_0$, subject to a step input ρ_0 at $t = 0$, the response $n(t)$ is given by

$$n(t) = \frac{n_o}{\omega_1 - \omega_2} \left[\left(\frac{\rho_o}{l} - \omega_2 \right) e^{\omega_1 t} + \left(\omega_1 - \frac{\rho_o}{l} \right) e^{\omega_2 t} \right]$$

For $\rho_0 > 0$ plot the response and comment on the effect of the roots on the response. Use the data for thermal fission U235.

$$\lambda = 0.0767 \quad \mathrm{s}^{-1}, \beta = 0.00650, \quad \Lambda = \frac{\beta}{\lambda} = 0.0847\mathrm{s}$$

2. Solve the steady-state heat conduction equation to plot the temperature in a fuel rod using data of Chapter 3. The volumetric heat generation is

$$q''' = q'''_{\max} \cos\left(\frac{\pi z}{\tilde{H}}\right).$$

From Chapter 3, use data for fuel rod diameter, clad thickness, fuel rod length, centerline fuel temperature.

3. Solve the steady-state heat conduction equation

$$\frac{d^2T}{dr^2} + \frac{1}{r}\frac{dT}{dr} + \frac{q'''}{k_f} = 0$$

with boundary conditions (1) T is nonsingular within the rod, (2) $T(0) = T_m$. Show that the temperature within the rod is

$$T = T_m - \frac{q''' r^2}{4k_f}$$

Now solve for the temperature in the clad. Show that the heat flow is

$$q = \frac{2\pi H k_c (T_s - T_c)}{\ln(1 + b/a)}$$

where $T(a) = T_s$, $T(a + b) = T_c$

4. Given the advection-diffusion equation

$$u\frac{dT}{dx} - k\frac{d^2T}{dx^2} = Q(x)$$

in the domain $a \le x \le b$, with the boundary conditions $T(a) = T_a$ and $T(b) = T_b$, write a finite difference equation and describe the computational strategy to solve for the temperature $T(x)$.

In the above equation, write the finite difference equation for the first two nodes for a given heat flux boundary condition at the boundary $x = a$.

5. Solve the Fredholm integral equation $(x) = 1 + \lambda \int_0^{\pi} \sin(x + t) y(t) dt$
6. Show that the resolvent kernel for $y(x) = 1 + \lambda \int_0^1 (1 - 3xt) y(t) dt$ I $R(x, t; \lambda) = \frac{4}{4 - \lambda^2}\left[1 + \lambda - \frac{3\lambda x}{2} - 3t\left\{x + \frac{\lambda}{2} - \lambda x\right\}\right]$ and the solution is $y(x) = \frac{4 + \lambda(2 - 3x)}{4 - \lambda^2}$, for $|\lambda| < 2$
7. Estimate the integral $I = \int_0^5 x^2 e^{-x} dx$ with x^2 as the estimator and e^{-x} as the pdf.
8. Use the linear congruential RNG $S_{k+1} = S_k g + c \bmod p$ with $g = 20$, $c = 1$, $S_o = 1$, $p = 100$ to generate random numbers.
9. In the previous question, increase the value of g to $g = 40$ and comment on the period of the random numbers. What general conclusion can be drawn?
10. Sample random numbers from a linear pdf $f(x) = \frac{2}{7}(2 + 3x)$ in the range (0, 1).

Nomenclature

English lower case

d	extrapolation distance
e	internal energy
f	probability distribution function
{f}	the force function
g	acceleration due to gravity
f_{xy}	joint distribution function
h	heat transfer coefficient
k	thermal conductivity
k	effective multiplication
k_c	thermal conductivity of cladding
k_{eff}	effective multiplication
k_f	thermal conductivity of fuel
n	number density
p	probability
p	pressure
q	heat flow (W)
q'''	volumetric heat generation (W/m^3)
$\overline{u}$	velocity vector

English upper case

A_i	area
B	buckling
C_i	concentration of precursor
D	diffusion coefficient
$\hat{D}$	derivative operator
D	Kullback–Leibler divergence
E	energy
F	force
F	cumulative distribution function
$H(\mu)$	Chandrasekhar's H function
J	variational functional
K	stiffness matrix
K	kernel
L	diffusion length
$\hat{L}^+$	adjoint operator
N_i	atomic density of the ith nuclide
N_i	shape function in the ith element
P_l^m	associated Legendre functions
R_c	cladding resistance
R_f	fuel resistance
S	source
S_k	k^{th} random number
T	absolute temperature
T_m	moderator temperature
T_C	clad temperature

Greek lower case

α	thermal diffusivity
α_F	temperature coefficient of reactivity of fuel
α_M	temperature coefficient of reactivity of moderator
β	separability constant

β	delayed neutron fraction
$\tilde{\beta}$	delayed neutron fraction for a mixture
γ	separability constant
γ_{ji}	production fraction of the ith isotope due to fission of isotope j
ε	emissivity
η	separability constant
λ	decay constant
μ	cosine of angle of scattering
μ	expectation, mean
ξ	random number
ρ	reactivity
ρ	density
σ	Stefan–Boltzmann constant
σ	standard deviation
τ	stress
ϕ	flux
ϕ_C	complementary solution
ϕ_P	particular solution
χ	fission spectrum

Greek upper case

Λ	neutron generation time (s)
Ω	solid angle
$\sum_a$	macroscopic absorption cross section

Abbreviations

CDF	cumulative distribution function
ODE	ordinary differential equation
PDE	partial differential equation
PDF	probability distribution function

References

Abramowitz, M., & Stegun, I. A. (1964). *Handbook of mathematical functions with formulas, graphs and mathematical tables*. Washington, DC: National Bureau of Standards, U.S. Department of Commerce.

Asadzadeh, M., & Kadem, A. (2006). Chebyshev spectral-SN method for the neutron transport equation. *Computers and Mathematics with Applications*.

Baker, G. (2016). Systems of differential equations. In *Differential equations as models in science and engineering*.

Bell, G., & Glasstone, S. (1952). *Nuclear reactor theory*. van Nostrand Reinhold Company.

Bell, G. I., & Glasstone, S. (1979). *Nuclear reactor theory*. New York: Robert E. Krieger Publishing Co., Inc..

Bernal, Á., Miró, R., Ginestar, D., & Verdú, G. (2014). Resolution of the generalized eigenvalue problem in the neutron diffusion equation discretized by the finite volume method. *Abstract and Applied Analysis*.

Brandao, M. A., Dominguez, D. S., & Iglesias, S. M. (2011). Linear finite element method for one-dimensional diffusion problems. In *International Conference on Mathematics and Computational Methods Applied to Nuclear Science and Engineering* (M&C 2011).

Case, K. M., & Zweifel, P. F. (1967). *Linear transport theory*. Reading, MA: Addison-Wesley Pub. Co..

Chandrasekhar, S. (1960). *Radiative transfer*. New York: Dover Publications.

Dunn, W. L., & Shultis, J. K. (2012). *Exploring Monte Carlo Methods*.

Filipov, S., & Faragó, I. (2018). Implicit euler time discretization and fdm with Newton method in nonlinear heat transfer modeling. arXiv.

González-Pintor, S., Ginestar, D., & Verdú, G. (2008). Continuous pseudo spectral method for the neutron diffusion equation on a triangular mesh. In *International Conference on the Physics of Reactors 2008, PHYSOR 08*.

Grinstead, C. M., & Snell, J. L. (1997). Introduction to probability: Second revised edition. *American Mathematical Society*.

Hetrik, D. (1993). *Dynamics of nuclear reactors*. La Grange Park, IL: American Nuclear Society.

Hutton, D. V. (2004). *Fundamentals of finite element analysis* (ISBN: 0072395362, ISSN: 1098-6596). Textbook (Important).

IAEA. (2008). *Thermophysical properties of materials for nuclear engineering: A tutorial and collection of data*. Vienna, Austria: IAEA.

Johnson, M., Lucas, S., & Tsvetkov, P. (2010). *Modeling of reactor kinetics and dynamics*.

Kalos, M. H., & Whitlock, P. A. (2008). *Monte Carlo methods*. Wiley-VCH Verlag GmbH & Co. KGaA.

Kanwal, R. P. (2013). *Linear integral equations: Theory and technique.*
Lamarsh, J. R., & Baratta, A. J. (2001). *Introduction to nuclear engineering* (3rd ed., p. 783)Prentice Hall.
Lewis, E. E., & Miller, W. F. Jr (1984). *Computational methods of neutron transport.* New York: John Wiley and Sons.
Lewis, F. L., Vrabie, D. L., & Syrmos, V. L. (2012). *Optimal control* (3rd ed.).
Lewis, R. L., Nithiarasu, P., & Seetharamu, K. N. (2004). *Fundamentals of the finite element method for heat and fluid flow.* John Wiley & Sons, Inc.
Ozisik, M. N. (1993). *Heat conduction* (2nd edn.). John Wiley & sons, Inc.
Polyanin, A. D., & Manzhirov, A. V. (2006). *Handbook of mathematics for engineers and scientists.*
Polyanin, A. D., & Manzhirov, A. V. (2008). *Handbook of integral equations* (2nd edn.).
Polyanin, A. D., & Zaitsev, V. F. (2004). *Handbook of nonlinear partial differential equations.*
Rao, S. S. (2017). *The finite element method in engineering.*
Reiher, W. (1966). *Hammersley, J. M., D. C. Handscomb: Monte Carlo Methods.* Methuen & Co., London, and John Wiley & Sons, New York, 1964. VII + 178 S. Preis: 25 s, Biom. Z.
Yeih, W. (2020). Solving the nonlinear heat equilibrium problems using the local multiquadric radial basis function collocation method. *Mathematics.*
Zwillinger, D. (1993). Handbook of differential equations. *Mathematics of Computation.*

Chapter 5

The neutron diffusion equation

A simple yet representative model of neutron transport is the neutron diffusion equation (NDE) for which analytical solutions can be readily found for nuclear systems modeled as regular geometries (rectangular, spherical, and cylindrical). For irregular geometries, numerical methods such as the Finite Difference Method (FDM), the Finite Element Method (FEM) and Finite Volume Method (FVM) are used. In reactor calculations, the Nodal Expansion Method (NEM) is used with the NDE to calculate the flux and power distribution over a full core. In such calculations, the NDE is easier to use with less computational effort than the neutron transport equation or the Monte Carlo (MC) method.

In this chapter, neutron diffusion theory is introduced for nuclear systems with nuclear fuel materials such as U^{235} and Pu^{239} as well as nonfissile materials such as aluminum, beryllium, boron, carbon (graphite), and iron used for structural applications.

The NDE is developed for one-dimensional regular geometries, and extended to three-dimensional geometries, for obtaining the neutron flux and associated reaction rates (absorption, scattering, etc.) in systems.

5.1 The conservation equation

Deterministic models that describe neutron behavior in a system are based on the diffusion equation (Jevremovic, 2009; Lamarsh, 2005), the integro-differential Boltzmann equation, and the integral equation. An elementary form of the diffusion equation can be derived from first principles by writing a conservation equation in a volume element assuming that neutrons do not lose energy during interactions with matter. Consider the volume element $\Delta V = \Delta x \Delta y \Delta z$ with net inward and outward currents in the x direction $J_{\text{in}} = J_x$ and $J_{\text{out}} = J_{x+\Delta x}$ shown in Fig. 5.1.

The change in neutron population $n(x,t)$, for this one-group model, can be written from the conservation equation expressing the net boundary crossings and losses and gains in the volume element. With the net boundary crossings taken only in the x-direction $(\overline{J}_{\text{in}} - \overline{J}_{\text{out}})A_{\perp}$ n/s, the absorptions $\Sigma_a \phi(x,t)\Delta V$ n/s and the source production $S(x,t)\Delta V$, the rate of change of neutron population is:

$$\frac{\partial n(x,t)}{\partial t}\Delta V = (\overline{J}_{\text{in}} - \overline{J}_{\text{out}})A_{\perp} - \Sigma_a \phi(x,t)\Delta V + S(x,t)\Delta V. \tag{5.1}$$

Using a Taylor series first-order expansion for the current, we can write

$$\overline{J}_{\text{out}}A_{\perp} \approx \overline{J}_{\text{in}}A_{\perp} + \nabla \cdot \overline{J}\Delta V$$

so that the diffusion equation, with $\phi = \text{nv}$, becomes

$$\frac{1}{v}\frac{\partial \phi(x,t)}{\partial t} = -\nabla \cdot \overline{J} - \Sigma_a \phi(x,t) + S(x,t).$$

Using Fick's law $\overline{J} = -D\overline{\nabla}\phi$, the diffusion equation reads

$$\frac{1}{v}\frac{\partial \phi(x,t)}{\partial t} = D\nabla^2\phi - \Sigma_a \phi(x,t) + S(x,t). \tag{5.2}$$

The diffusion coefficient D is given by

$$D = \frac{1}{3}\lambda_{tr} \equiv \frac{1}{3\Sigma_{tr}}$$

Nuclear Engineering. DOI: https://doi.org/10.1016/B978-0-323-90618-0.00005-3

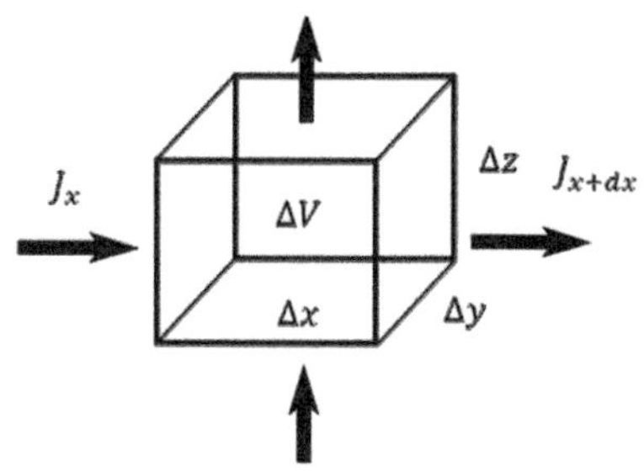

FIGURE 5.1 A volume element.

in terms of the macroscopic transport cross-section

$$\Sigma_{tr} = \Sigma_t - \mu_0 \Sigma_s.$$

where the average scattering angle $\hat{\Omega} \cdot \hat{\Omega}'$ for isotropic scattering in the center of mass system is $\mu_0 = 2/(3A)$.

The loss terms are the leakage $\nabla \cdot \bar{J}$ and the removal $\Sigma_r \phi$ while the gain terms are the direct source S and the fission production $\nu \Sigma_f \phi$; thus in general the steady-state diffusion equation can be expressed as

$$\nabla \cdot \bar{J} + \Sigma_r \phi = S + \nu \Sigma_f \phi.$$

In a two-group model, the neutrons produced by fission (per cm^3/s)

$$\nu_1 \Sigma_{f,1} \phi_1 + \nu_2 \Sigma_{f,2} \phi_2$$

would contribute to both groups in proportion to the fission spectrum $\chi_i, i = 1, 2$.

For steady-state, the one-group balance equation reduces to

$$D\nabla^2 \phi - \Sigma_a \phi + S = 0 \tag{5.3}$$

with absorption being the only removal process. If a two-group model is considered, then the processes have to be modeled in the groups separately. Thus the net leakage rate of the group-1 neutrons would be balanced by the removal rate due to capture in group-1 and to out-scattering to group-2 $\Sigma_{R,1}\phi_1 = \Sigma_{C,1}\phi_1 + \Sigma_{S,1\to2}\phi_1$ as well as to the source production in group 1. The two-group equations are

$$D_1 \nabla^2 \phi_1 - \Sigma_{r,1} \phi_1 + S_1 = 0 \tag{5.4}$$

$$D_2 \nabla^2 \phi_2 - \Sigma_{r,2} \phi_2 + S_2 = 0 \tag{5.5}$$

where S_1 could be a "direct" source and a fission source $\chi_1(\nu_1 \Sigma_{f,1}\phi_1 + \nu_2 \Sigma_{f,2}\phi_2)$ and $S_2 = p\Sigma_{S,1\to2}\phi_1 + \chi_2(\nu_1 \Sigma_{f,1}\phi_1 + \nu_2 \Sigma_{f,2}\phi_2)$, a "down-scattered" contribution plus a fission contribution. The eigenvalue form of the diffusion equation, in a 1-D slab of thickness L, as briefly discussed in Section 4.5, is written as

$$\left[\frac{d}{dx}\left(-D\frac{d}{dx}\right) + \Sigma_r\right]\phi(x) = \frac{1}{k}\nu \Sigma_f \phi(x) + S(x), 0 < x < L$$

where k is the eigenvalue that balances the loss and gain terms.

The multigroup form of the diffusion equation for energy group i is

$$-\nabla \cdot D_i(r)\nabla \phi_i(r) + \Sigma_i(r)\phi_i(r) = \sum_{i' \neq i} \Sigma^{(s)}_{i' \to i} \phi_{i'}(r) + \chi_i \sum_{i'} \nu \Sigma_{f,i'} \phi_{i'}(r) + S_i(r) \tag{5.6}$$

where $S_i(r)$ is the independent source.

The forms taken by the $\nabla \cdot D_i(r)\nabla$ were discussed in Chapter 4;for Cartesian

$$\left[\frac{\partial}{\partial x}D\frac{\partial}{\partial x} + \frac{\partial}{\partial y}D\frac{\partial}{\partial y} + \frac{\partial}{\partial z}D\frac{\partial}{\partial z} - \Sigma_r\right]\phi(x, y, z) + \nu \Sigma_f \phi(x, y, z) + S = 0,$$

cylindrical

$$\left[\frac{1}{r}\frac{\partial}{\partial r}rD\frac{\partial}{\partial r} + \frac{1}{r^2}\frac{\partial}{\partial \varphi}D\frac{\partial}{\partial \varphi} + \frac{\partial}{\partial z}D\frac{\partial}{\partial z} - \Sigma_r\right]\phi(r, \varphi, z) + \nu \Sigma_f \phi(r, \varphi, z) + S = 0,$$

and spherical coordinates

$$\left[\frac{1}{r^2}\frac{\partial}{\partial r}r^2D\frac{\partial}{\partial r}+\frac{1}{r^2\sin\theta}\frac{\partial}{\partial\theta}\sin\theta\, D\frac{\partial}{\partial\theta}+\frac{1}{r^2\sin\theta}\frac{\partial}{\partial\varphi}D\frac{\partial}{\partial\varphi}-\Sigma_r\right]\phi(r,\theta,\varphi)+\nu\Sigma_f\phi(r,\theta,\varphi)+S=0.$$

5.2 The one-group diffusion equation

The time-dependent diffusion Eq. (5.2), for a one-group energy model, can be solved analytically for simple problems in slab, cylindrical and spherical geometry. In this section, some elementary steady-state solutions are obtained to illustrate the flux distributions.

5.2.1 Nonmultiplying systems

Although the one-group model is too simple for practical applications, it gives considerable insight into important neutron characteristics of nuclear systems. For systems which are predominantly "one-group," that is, when the neutron spectrum is predominantly "fast" or "thermal," it is used with the appropriate "averaged" nuclear data to obtain the neutron flux and associated quantities such as criticality parameters and thermal power.

The NDE is a second-order ordinary differential equation (ODE) which can be solved given two boundary conditions: Neumann, Dirichlet or mixed, depending on the physical conditions. A bare assembly, for example, will have the condition of vanishing flux at the extrapolated boundary (Dirichlet boundary condition) while a reflecting surface will have a zero flux-gradient, or zero current, (Neumann boundary condition) condition. An interface will have both flux and current continuity conditions.

The NDE, based on Poisson's equation, is mathematically much simpler than the integro-differential transport equation and hence is used for simple problems. Even then, numerical solutions are required for large problems such as "whole-core" reactor design. In the multigroup form, the diffusion equation is used to obtain "first estimates" and can serve as a useful step for providing guesses to full transport calculations. A limitation of diffusion equation is near sources and boundaries where the angular flux requires detailed consideration as provided, for example, in the discrete ordinates and spherical harmonics methods.

5.2.1.1 *Finite slab*

In a graphite slab of thickness $2a$ in the x-direction and infinite in the y and z directions, with uniformly distributed sources emitting S neutrons/cm^3/s the neutron flux is given by

$$\phi(x)=\frac{S}{\Sigma_a}\left(1-\frac{\cosh\frac{x}{L}}{\cosh\left(\frac{a+d}{L}\right)}\right). \tag{5.7}$$

Another method for obtaining the flux for a distributed source is by expanding in eigenfunctions and using the orthogonality property to determine the coefficients. This formulations results in a Green's function which is useful for more general source distributions. From Eq. (5.7), the net current in the slab, from Fick's Law, is

$$J(x)=-D\nabla\phi(x)=SL\frac{\sinh x/L}{\cosh\left(\frac{a+d}{L}\right)}.$$

From $J(x)$, the escape probability $P_{es}(a)$, defined as the fraction of neutrons leaking out of the system across the surface $x=a$, is

$$P_{es}(a)=\frac{J(x=a)}{aS}=\frac{L}{a}\frac{\sinh a/L}{\cosh\left(\frac{a+d}{L}\right)}.$$

The total escape probability from the system is $P_{es}=P_{es}(-a)+P_{es}(a)$; due to the symmetry of the system

$$P_{es}=\frac{2J(x=a)}{2aS}=\frac{L}{a}\frac{\sinh a/L}{\cosh\left(\frac{a+d}{L}\right)}$$

From L'Hopital's rule it is readily seen that in the limit $a\to 0$, $P_{es}\to 1$ (for $d=0$) that is, all the neutrons escape from an intesimally thin slab.

The diffusion length defined as

$$L^2 = \frac{D}{\Sigma_a}$$

was discussed in Chapter 2 in the context of the slowing down of neutrons.

Example 5.1: .

Calculate the escape probability in a graphite slab of thickness 1 mean free path (mfp).
The first step is to find the atomic density N of graphite:

$$N = \frac{\rho N_{av}}{A} = \frac{1.60 \text{ X } 0.6023 \ \ 10^{24}}{12.01115} = 0.0802 \times 10^{24} \text{atoms cm}^{-3}$$

The next step is to determine the macroscopic cross-sections Σ_a, Σ_s, Σ_{tr}:

$$\Sigma_a = N\sigma_a = 0.0802 \text{ X } 0.0034 = 2.7268 \ \ 10^{-4} \text{ cm}^{-1}, \quad \Sigma_s = N\sigma_s = 0.0802 \times 4.75 = 0.3809 \text{ cm}^{-1}, \quad \Sigma_{tr} = N\sigma_{tr} = N\sigma_s(1-\overline{\mu}_o) = N\sigma_s\left(1 - \frac{2}{3A}\right) = 0.3599 \text{ cm}^{-1}$$

These are used to find the diffusion coefficient D and the diffusion length L:

$$D = \frac{1}{3\Sigma_{tr}} = \frac{1}{3(0.3599)} = 0.9262\text{cm}, \quad L = \sqrt{\frac{D}{\Sigma_a}} = \sqrt{\frac{0.9262}{2.7268 10^{-4}}} = 58.2804\text{cm}$$

and extrapolation distance $d = 0.71\lambda_{tr} = \frac{0.71}{0.3599} = 1.9728$cm. The mean free path λ, is found as

$$\lambda = \frac{1}{\Sigma_t} = \frac{1}{\Sigma_s + \Sigma_a} = \frac{1}{0.3812} = 2.6233 \text{ cm.}$$

For a graphite slab of thickness 1 mfp, the escape probability is

$$P_{es} = \frac{L \ \sinh a/L}{a\cosh\left(\frac{a+d}{L}\right)} = \frac{58.2804 \ \sinh 1.3116/58.2804}{2.6233/2 \ \ \cosh\frac{1.3116+1.9728}{58.2804}} = 0.9985.$$

Thus, in this case there is a 99.85% probability that a neutron will escape from the surfaces.

Fig. 5.2 shows the escape probability in the above graphite slab as a function of its thickness. The probability is "high" since graphite is a weak absorber and is thus used for moderating neutrons in a reactor.

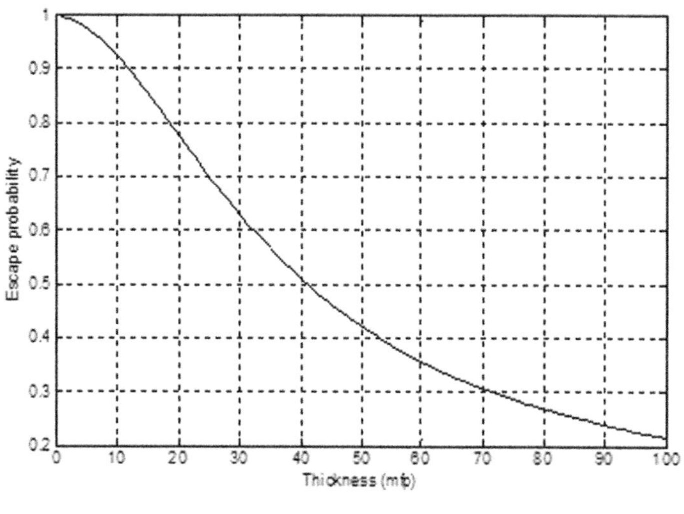

FIGURE 5.2 Escape probability in a graphite slab.

5.2.1.2 Finite cylinder

The governing equation

$$\left[\frac{1}{r}\frac{\partial}{\partial r}rD\frac{\partial}{\partial r}+\frac{\partial}{\partial z}D\frac{\partial}{\partial z}-\Sigma_r\right]\phi(r,z)+S=0$$

can be solved to obtain exact solutions. For an infinitely long cylinder with a uniformly distributed source as for the finite slab case, the solution is similar to Eq. (5.7) with the modified Bessel function of the first kind I_0 replacing the cosh x function. From the method of eigenfunctions, the flux in a finite cylinder can be found in the form

$$\phi(r,z)=\sum_{m,n}A_{mn}J_0\left(\frac{x_n r}{R}\right)\cos\frac{m\pi z}{H},n,m=1.23,\ldots \tag{5.8}$$

in terms of Bessel functions, eigenvalues and eigenfunctions as described in Chapter 4. The coefficients are then determined from the diffusion equation using the orthogonality condition.

5.2.1.3 Point source in an infinite medium

The flux in a finite sphere with a uniformly distributed source is of the same form as shown for the slab case in Eq. (5.7) with the sinh function replacing the cosh function as shown in Table 5.1.

For a point source in an infinite medium, the (thermal) neutron flux is

$$\varphi(r)=\frac{Se^{-r/L}}{4\pi Dr} \tag{5.9}$$

where $L\equiv L_T$, the thermal diffusion length.

Table 5.1 lists some solutions with source conditions, boundary conditions and one-group fluxes for nonmultiplying systems. Source conditions for infinite plane, point-like and infinite linear sources can also be derived from Gauss' divergence theorem (Carrillo, 2001) using Fick's law.

5.2.2 Multiplying systems

When the source is given by $S(r)=k_\infty\Sigma_a\varphi(r)$, with the infinite system multiplication factor $k_\infty=\varepsilon p\eta f_T$, the diffusion equation can be written as

$$\nabla^2\phi+B_m^2\phi=0 \tag{5.10}$$

where the material buckling B_m is

$$B_m^2=\frac{k_\infty-1}{L^2}.$$

Some elementary solutions are given here for slab, cylindrical, and spherical critical reactors.

5.2.2.1 Slab reactor

The 1-D slab reactor equation

$$\frac{d^2\phi}{dx^2}+B_m^2\phi(x)=0 \tag{5.11}$$

for $-a/2\le x\le a/2$ with the boundary conditions

$$\phi\left(-\frac{\tilde{a}}{2}\right)=\phi\left(\frac{\tilde{a}}{2}\right)=0,\frac{d\phi}{dx}|_0=0, \tag{5.12}$$

is critical for the flux

$$\phi(x)=A\cos Bx \tag{5.13}$$

TABLE 5.1 Exact solutions: 1-group nonmultiplying media.

Medium	Source	Boundary conditions	Flux
Slab			
Infinite	Planar source at $x = 0$ S n/cm²/s	Finite flux $\phi(x)$ Source condition $\lim_{x\to 0} J(x) = S/2$	$\phi(x) = \frac{SL}{2D} e^{-x/L}$
Infinite slab (in y and z) of thickness $2a$	Planar source at $x = 0$ S n/cm²/s	$\phi(a+d) = \phi(-a-d) = 0$ Source condition $\lim_{x\to 0} J(x) = S/2$	$\phi(x) = \frac{SL}{2D}\frac{\sinh[(a+d-\lvert x\rvert)/L]}{\cosh\left[\frac{a+d}{L}\right]}$
Infinite slab (in y and z) of thickness $2a$	Uniformly distributed sources S n/cm³/s	$\phi(a+d) = \phi(-a-d) = 0$ $\frac{d\phi}{dx} = 0$ at $x = 0$	$\phi(x) = \frac{S}{\Sigma_a}\left(1 - \frac{\cosh x/L}{\cosh\left(\frac{a+d}{L}\right)}\right)$
Cylinder			
Infinite cylinder (in z axis) finite radius R	as given	$\phi(R) = 0$ $\frac{d\phi}{dr} = 0$ at $r = 0$	$\sum_m A_n J_0\left(\frac{x_n r}{R}\right), n = 1.23, \ldots$
Finite cylinder radius R, height H		$\phi(R) = \phi(R) = 0$ $\phi'_{r=0} = \phi'_{z=H} = 0$	$\sum_{m,n} A_{mn} J_0\left(\frac{x_n r}{R}\right)\cos\frac{m\pi z}{H}, n, m = 1.23, \ldots$
Sphere			
Infinite	Point isotropic source at $r = 0$ S n/s	Finite flux $\phi(r)$ Source condition $\lim_{r\to 0} 4\pi r^2 J(r) = S$	$\phi(r) = \frac{Se^{-r/L}}{4\pi Dr}$
Finite sphere of radius R	Point isotropic source at $r = 0$ S n/s	Finite flux $\phi(r)$ and $\phi(\tilde{R}) = 0$	$\phi(r) = \frac{S}{4\pi D\sinh\left(\frac{R+d}{L}\right)}\frac{\sinh\kappa(R+d-r)}{r}$
Finite sphere of radius R	Uniformly distributed sources S n/cm³/s	Finite flux $\phi(r)$ and $\phi(\tilde{R}) = 0$	$\phi(r) = \frac{S}{\Sigma_a}\left(1 - \frac{R+d}{r}\frac{\sinh\kappa r}{\sinh\kappa(R+d)}\right)$

where the material and geometrical buckling are

$$B_m = \frac{k_\infty - 1}{L^2}, B_g = \frac{\pi}{a+d}.$$

5.2.2.2 *Cylindrical reactor*

For an infinite cylindrical reactor (i.e., infinite in height) which is critical and has radius R, the flux is space-dependent. The reactor equation

$$\frac{1}{r}\frac{\mathrm{d}}{\mathrm{dr}}r\frac{\mathrm{d}\phi}{\mathrm{dr}} + B^2\phi = 0 \tag{5.14}$$

has two independent solutions $J_0(\mathrm{Br})$ and $Y_0(\mathrm{Br})$ which are ordinary Bessel functions of the first- and second-kind. The general solution is

$$\phi = AJ_0(\mathrm{Br}) + CY_0(\mathrm{Br}) \tag{5.15}$$

where A and C are constants. Since $Y_0(x)$ is infinite at $x = 0$, while $J_0(0) = 1$, for ϕ to remain finite, C must be taken to zero and the flux is

$$\phi(r) = AJ_0(\mathrm{Br}). \tag{5.16}$$

The boundary condition therefore takes the form

$$\phi(\tilde{R}) = AJ_0(\mathrm{Br}) = 0. \tag{5.17}$$

The function $J_0(x)$ is zero at a number of values of x, so that ϕ is satisfied if B has eigenvalues

$$B_n = \frac{x_n}{\tilde{R}}. \tag{5.18}$$

In a critical system, the lowest eigenvalue has physical significance; thus

$$B_1^2 = \left(\frac{x_1}{\tilde{R}}\right)^2 = \left(2.405\tilde{R}\right)^2.$$

The one-group flux then becomes

$$\phi = AJ_0\left(2.405r\tilde{R}\right). \tag{5.19}$$

The constant A is determined after using from the reactor power given by

$$P = E_R\Sigma_f\int\phi(r)dV.$$

For a cylinder, $dV = 2\pi rdr$ and

$$P = 2\pi E_R\Sigma_f\int_0^R\phi(r)rdr = 2\pi E_R\Sigma_f A\int_0^R J_0\left(\frac{2.405r}{R}\right)rdr$$

which is evaluated using $\int J_0\left(x'\right)x'dx' = xJ_1(x)$ to get

$$P = \frac{2\pi E_R\Sigma_f R^2 AJ_1(2.405)}{2.405} = 1.35E_R\Sigma_f R^2 A.$$

After rearranging the equation for A, the flux is

$$\phi(r) = \frac{0.738P}{E_R\Sigma_f R^2}J_0\left(\frac{2.405r}{R}\right). \tag{5.20}$$

From Section 3.1, the flux in a critical finite cylindrical reactor of radius R and height H is

$$\phi(r,z) = AJ_0\left(2.405r\tilde{R}\right)\cos\left(\frac{\pi z}{\sim H}\right) \tag{5.21}$$

where normalization gives

$$A = \frac{3.63P}{E_R\Sigma_f V}.$$

For a reactor such as the pressurized water reactor (PWR) AP1000 described in Chapter 3, the average value of the fission cross-section in the core of a reactor can be obtained as

$$\overline{\Sigma_f|_c} = \frac{\overline{\Sigma_f|_r}n_r V_r}{V_c}$$

where $\overline{\Sigma_f|_c}$ is the average value of thermal fission cross-section in the core, $\overline{\Sigma_f|_r}$ is the average value of thermal fission cross-section in a fuel rod, n_r is the number of fuel rods, V_r is the volume of a fuel rod, and V_c is the core volume

The flux, plotted using the Matlab program listed below, is shown in Fig. 5.3 for a cylindrical reactor; the maximum power is at the origin of the reactor decreasing axially for a given radius. Physically, the flux signifies the power distribution in the reactor core. Thus, fuel rods at the center of a core are much "hotter" than those at the periphery.

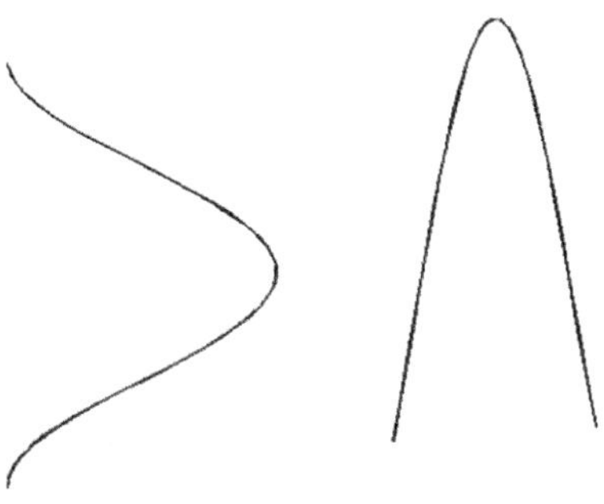

FIGURE 5.3 Flux in a critical finite cylindrical reactor of radius R and height.

Matlab program for flux in a cylindrical reactor

```
gid = fopen('out.txt','w');
C:\Users\Administrator\Desktop\Elsevier2020\PROGRAMS\Ch5_FluxPlotCyl.m
R=1.52;H=4.267/2;% half height
[X,Y] = meshgrid(0:0.1:R,0:0.1:H);
Z = besselj(0,2.405*X/R).*cos(pi*Y/H);
figure(1)
set(gca,'FontSize',12)
surf(X,Y,Z)
hold on
xlabel('\bf Radius (m)','fontsize',14)
ylabel('\bf Height (m)','fontsize',14)
ylim([0,H])
%legend('\bf n(E), n_0 kT','\bf n(v), n_0 {(m/(2kT))}^{1/2}')
grid on
fclose(gid)
```

Exercise 5.1: .

Plot the neutron flux in an assembly with $H = 2R = 5$ft. and S = 10^7 neutrons/s, $k_\infty = 0.980$, $p = 0.83$, $L_T^2 = 2.8$ cm^2, and $\overline{\Sigma}_a = 0.16$/cm.

5.2.2.3 Spherical reactor

In a critical spherical reactor, the flux can be obtained by solving

$$\left[\frac{1}{r^2}\frac{\partial}{\partial r}r^2\frac{\partial}{\partial r} + B^2\right]\phi(r) = 0 \tag{5.22}$$

and applying the boundary condition $\phi(\tilde{R}) = 0$ and the requirement of finite flux inside the reactor. The critical reactor flux is

$$\phi(r) = A\frac{\sin Br}{r} \tag{5.23}$$

where the geometric buckling $B = B_g = \pi/\tilde{R}$, since $\phi(\tilde{R}) = 0$. The constant A, as in the cylindrical case, is determined from the power of the reactor P:

$$A = \frac{P}{4E_R\Sigma_f R^2}.$$

Exercise 5.2: .

Using Eq. (5.23), obtain the (1) average flux in the reactor, and (2) the neutron current at the surface of the reactor. Calculate and plot the quantity $4\pi r^2\varphi(r)$ for $r/R \in (0, 1)$ and describe what it physically represents.

Example 5.2: .

Consider a spherical reactor of radius $R = 8.7407$ cm operating at a power of 100 W. Find the constant A in the expression for neutron flux.

Use the data: $\Sigma_f = 0.0672\text{cm}^{-1}, \tilde{R} = R + d, d = 1/3\lambda_{tr} = 1.0212\text{cm}$. The recoverable energy from fission E_R is assumed to be 200 MeV. Then

$$A = \frac{100}{20010^6 1.610^{19} 0.0672}\frac{1}{4(8.7407)^2} = 1.521710^{11} \quad \frac{n}{\text{cm}^2 \cdot \text{s}}$$

and since

$$B_g = \frac{\pi}{R+d}$$

For $B_m = B_g = B$, the flux is

$$\phi(r) = 1.521710^{11}\frac{\sin Br}{r}\ \frac{n}{\text{cm}^2\cdot\text{s}}.$$

Fig. 5.4 shows the flux for two power levels 100 and 200 W; the magnitude is determined by the operating power while the shape is determined by the "buckling," or the curvature.

Table 5.2 lists some interesting source conditions, boundary conditions and one-group fluxes for multiplying systems.

5.2.3 One-group criticality

From an elementary consideration that all neutrons have the same energy and are therefore classified as neutrons of one group, it is possible to estimate the criticality of slab, cylindrical and spherical reactors. These are inaccurate due to the assumptions made but are nonetheless useful to generate an idea of the size of critical systems.

For criticality, equating the material and geometrical bucklings $B_m = B_g = B$, gives

$$B^2 = \left(\frac{\pi}{R}\right)^2 = \frac{k_\infty - 1}{L^2}$$

in a spherical system from which

$$R_c = \pi\sqrt{\frac{L^2}{k_\infty - 1}}$$

gives an estimate for the critical radius R_c. Rearranging the above into a form used for the leakage probability, the effective multiplication for a critical system $k_{\text{eff}} = 1$ is

$$k_{\text{eff}} = \frac{k_\infty}{1 + B^2L^2} = 1. \tag{5.24}$$

Eq. (5.24) can be used, in an elementary analysis, to calculate the critical composition of a slab, cylinder or sphere.

Example 5.3: .

Given a U^{235}-Na fast spherical reactor with 1% weight fraction U^{235} and $k_\infty = 1.49$, determine the critical radius and the amount of U^{235} required.

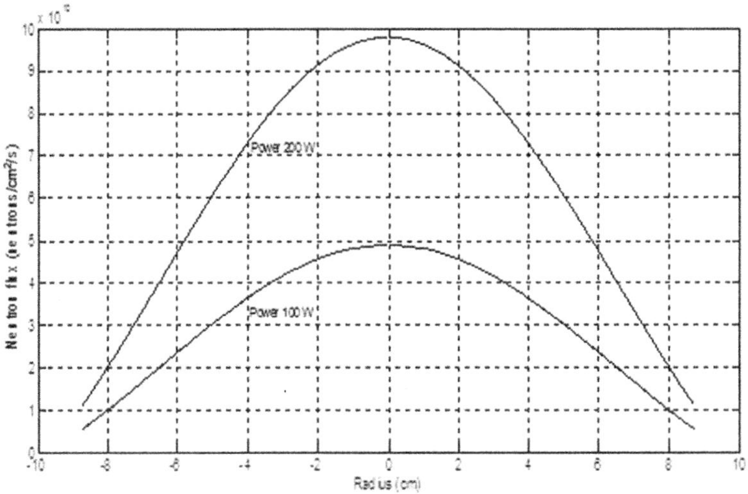

FIGURE 5.4 Neutron flux in a bare spherical reactor.

TABLE 5.2 Exact solutions: 1-group diffusion multiplying media.

Medium	Source	Boundary conditions	Flux
Slab			
Infinite critical slab reactor of thickness a, and $d \ll a$	Bare critical reactor	$\phi(a+d)=\phi(-a-d)=0$ $\frac{d\phi}{dx}=0$ at $x=0$	$\phi(x)=\frac{\pi P}{2aE_R\Sigma_f}\cos\frac{\pi x}{a}$
Bare cubical reactor of sides a	Bare critical reactor	$\phi(a+d)=\phi(-a-d)=0$ $\tilde{a}\equiv a+d$ $\frac{d\phi}{dx}=0$ at $x=y=z=0$	$\phi_T(x,y,z)=A\cos\frac{\pi x}{\tilde{a}}\cos\frac{\pi y}{\tilde{a}}\cos\frac{\pi z}{\tilde{a}}$
Cylinder			
Finite cylinder	Bare critical reactor	$\phi(\tilde{R},z)=0$ $\phi(r,\tilde{H})=0$	$\phi(r,z)=AJ_0\left(\frac{2.405r}{\tilde{R}}\right)\cos\left(\frac{\pi z}{\tilde{H}}\right)$
Infinite cylinder	Bare critical reactor	Finite flux $\phi(r)$ and $\phi(\tilde{R})=0$	$\phi(r)=\frac{0.738P}{E_R\Sigma_f R^2}J_0\left(\frac{2.405r}{R}\right)$
Sphere			
Bare spherical reactor of radius R	Bare critical reactor	Finite flux $\phi(r)$ and $\varphi(\tilde{R})=0$	$\phi(r)=\frac{\frac{P}{4E_R\Sigma_f R^2}\sin\pi r/\tilde{R}}{r}$

P is the reactor power, E_R is the energy recoverable from fission ($\sim$200 MeV).

The microscopic transport and absorption cross-sections are $\sigma_{tr,F}=6.8b, \sigma_{tr,M}=3.3b, \sigma_{a,F}=1.65b, \sigma_{a,M}=0.0008b$, With $B=\pi/\tilde{R}$, the critical radius $R_c=\tilde{R}-d$ is

$$R_c=\sqrt{\frac{\pi^2L^2}{k_\infty-1}}-d=\frac{\pi}{\sqrt{3\Sigma_{tr}\Sigma_a(k_\infty-1)}}-\frac{1}{3\Sigma_{tr}}.$$

Step 1: Calculate the atomic densities of fuel N_F and moderator N_M

From the mass ratio it is possible to find the atomic ratio:

$$\frac{N_M}{N_F}=\frac{\rho_M}{\rho_f}\cdot\frac{A_F}{A_M}=1011.7.$$

To get another equation for N_F and N_M calculate the atomic weight of the "mixture" $\overline{A}$ from the atomic fractions $\alpha_U=1/1012.7$ and $\alpha_{Na}=1011.7/1012.7$

$$\overline{A}=\alpha_F A_F+\alpha_M A_M=\frac{1}{1012.7}235.04+\frac{1011.7}{1012.7}23=23.2094.$$

Calculate the mixture density ρ_{mix} from the given weight fractions

$$\frac{1}{\rho_{mix}}=\frac{w_F}{\rho_F}+\frac{w_M}{\rho_M}=\frac{0.01}{18.7}+\frac{0.99}{0.97}=1.0212,$$

$\rho_{mix}=0.9792\text{g}\cdot\text{cm}^{-3}$, and calculate the atomic density of the mixture

$$N_{mix}=\frac{\rho_{mix}A_v}{\overline{A}}=\frac{0.9792\ 0.6023\ 10^{24}}{23.2094}=0.0254\ 10^{24}\text{atoms}\cdot\text{cm}^{-3}.$$

From this the individual atomic densities can be determined using the atomic fractions

$$N_F=\frac{1}{1012.7}N_{mix}=2.508110^{19}\text{atoms}\cdot\text{cm}^{-3}$$

$$N_M=\frac{1011.7}{1012.7}N_{mix}=2.540010^{22}\text{atoms}\cdot\text{cm}^{-3}$$

Step 2: Calculate the macroscopic cross-sections Σ_{tr}, Σ_a

$$\Sigma_{tr} = N_F\sigma_{tr,F} + N_M\sigma_{tr,M} = \left(2.508110^{-5}6.8\right) + \left(2.5410^{-2}3.3\right) = 0.0840\text{cm}^{-1}$$

and the extrapolation distance is $d = 0.71\lambda_{tr} = 0.71/(0.0840) = 8.3333\text{cm}$; the fuel and moderator macroscopic absorption cross-sections are:

$$\Sigma_{a,F} = N_F\sigma_{a,F} = \left(2.508110^{-5}1.65\right) = 4.138410^{-5} \quad \text{cm}^{-1}$$

$$\Sigma_{a,M} = N_M\sigma_{a,M} = \left(2.540010^{-2}0.0008\right) = 2.03210^{-5} \quad \text{cm}^{-1}$$

from which the mixture absorption cross-section is

$$\Sigma_a = N_F\sigma_{a,F} + N_M\sigma_{a,M} = 6.170410^{-5} \quad \text{cm}^{-1}.$$

Thus

$$L^2 = \frac{1}{3\Sigma_{tr}\Sigma_a} = \frac{1}{3(0.0840)(6.170410^{-5})} = 0.6431 \times \quad 10^5 \quad \text{cm}^2$$

and

$$R_c = \sqrt{\frac{\pi^2 L^2}{k_\infty - 1}} - d = \frac{253.5942\pi}{\sqrt{(1.49-1)}} - 8.3333 = 1130 \text{ cm}.$$

The volume of this spherical reactor is 6044 m^3. With the mixture density calculated above, this has a weight of about 6000 tons, of which about 60 tons is U^{235} and the rest, ~5940 tons, is Na.

5.3 The two-group diffusion equation

5.3.1 Nonmultiplying systems

The two-group NDE for the energy-averaged "fast" and "thermal" flux $\phi_1(\mathrm{r})$ and $\phi_2(\mathrm{r})$ are

$$D_1(r)\nabla^2\phi_1(\mathrm{r}) - \Sigma_1(\mathrm{r})\phi_1(\mathrm{r}) = 0 \tag{5.25}$$

$$D_2(\mathrm{r})\nabla^2\phi_2(\mathrm{r}) - \Sigma_2(\mathrm{r})\phi_2(\mathrm{r}) + \Sigma_1(\mathrm{r})\phi_1(\mathrm{r}) = 0. \tag{5.26}$$

For uniform material distribution, that is, constant macroscopic cross-sections, exact solutions can be obtained, from the above, in both finite and infinite media.

These are illustrated by writing a conservation equation and then going through a detailed derivation of the solution using complimentary and particular solutions for the above coupled second-order system.

To obtain the two-group fluxes in a nonmultiplying system, we begin with the elementary conservation of neutrons depicted in Fig. 5.5. In this simple illustration, there is a source of group-1 neutrons at the center of the sphere giving rise to a group-1 neutron flux, as well providing the slowing-down "source" of neutrons into group-2 in the form of

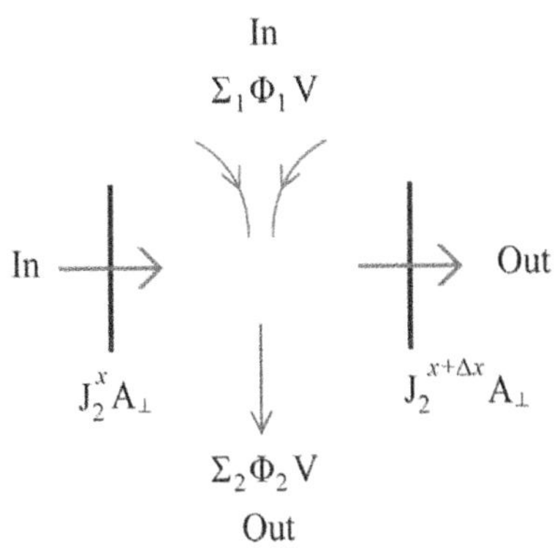

FIGURE 5.5 Conservation of neutrons in a two-group model.

slowed-down neutrons $\Sigma_1\Phi_1 V$ and neutrons crossing into the volume element $J_2{}^x A_1$. The "loss" terms of group-2 neutrons are the boundary crossings $J_2{}^{x+\Delta x}\ A_1$ and the absorptions $\Sigma_2\Phi_2 V$.

For group-1 neutrons,

$$(D_1\nabla^2 - \Sigma_1)\Phi_1(x) = 0 \tag{5.27}$$

with the boundary conditions

$$S = \left(4\pi r^2 \frac{d\Phi_1}{dr}\right)_{r\to 0},\ \Phi_1\ (R + d_1) = 0$$

The balance equations for group-2 neutrons is

$$J_2{}^x A_1 + \Sigma_1\Phi_1 V = \Sigma_2\Phi_2 V + J_2{}^{x+\Delta x}\ A_1$$

which, expanded into a first-order Taylor series

$$J_2{}^x A_1 + \Sigma_1\Phi_1 V = \Sigma_2\Phi_2 V + [J_2{}^x + \frac{d}{d_x} J_2{}^x \Delta x]A_1$$

simplifies to

$$\frac{d}{d_x} J_2{}^x V = \Sigma_1\Phi_1 V - \Sigma_2\Phi_2 V.$$

Using Fick's law, this becomes

$$\frac{d}{d_x}\left(-D_2\frac{d}{d_x}\Phi_2\right)V = \Sigma_1\Phi_1 V - \Sigma_2\Phi_2 V$$

Expressed, for constant diffusion coefficient, in the form

$$D_2\frac{d^2}{dx^2}\Phi_2 - \Sigma_2\Phi_2 + \Sigma_1\Phi_1 = 0 \tag{5.28}$$

or in operator form

$$D_2\nabla^2\Phi_2 - \Sigma_2\Phi_2 + S_2 = 0.$$

The boundary conditions are

$$-D_2 4\pi r^2 \Phi_2{}'|_0 = 0,\ \Phi_2(\bar{R}_2) = 0$$

In a spherical coordinate system, with angular symmetry

$$\nabla^2 = \frac{1}{r^2}\frac{\partial}{\partial r} r^2 D \frac{\partial}{\partial r} = \frac{2}{r}\Phi' + \Phi''$$

With the substitution $w = r\Phi$, $w' = r\Phi' + \Phi$, and $w'' = (r\Phi'' + \Phi') + \Phi'$

$$w'' = r\left[\Phi'' + \frac{2}{r}\Phi'\right]$$

so that the group-1 equation

$$\nabla^2\Phi_1(r) - \frac{1}{\tau}\Phi_1(r) = 0,$$

where $\tau = D_1/\Sigma_1$, becomes

$$w'' - \frac{1}{\tau}w = 0$$

with the solution

$$w(r) = A_1 e^{\frac{r}{\sqrt{\tau}}} + A_2 e^{-\frac{r}{\sqrt{\tau}}}$$

giving the group-1 flux

$$\Phi_1(r) = \frac{A_1}{r} e^{\frac{r}{\sqrt{T}}} + \frac{A_2}{r} e^{-\frac{r}{\sqrt{T}}}.$$

With the identities

$$\cosh x = \frac{1}{2}(e^x + e^{-x}), \cosh x + \sinh x = e^x$$

$$\sinh x = \frac{1}{2}(e^x - e^{-x}), \cosh x - \sinh x = -e^x$$

we can write

$$\Phi_1(r) = \frac{A_1}{r}\left(\cosh\frac{r}{\sqrt{T}} + \sinh\frac{r}{\sqrt{T}}\right) + \frac{A_2}{r}\left(\cosh\frac{r}{\sqrt{T}} - \sinh\frac{r}{\sqrt{T}}\right),$$

$$\Phi_1(r) = (A_1 + A_2)\frac{\cosh\frac{r}{\sqrt{T}}}{r} + (A_1 - A_2)\frac{\sinh\frac{r}{\sqrt{T}}}{r},$$

or

$$\Phi_1(r) = B_1\frac{\cosh\frac{r}{\sqrt{T}}}{r} + B_2\frac{\sinh\frac{r}{\sqrt{T}}}{r}$$

For the group-2 flux,

$$D_2(r)\nabla^2\Phi_2(r) - \Sigma_2\Phi_2(r) = -\Sigma_1\Phi_1(r),$$

We seek complimentary and particular solutions

$$\Phi_2(r) = \Phi_2{}^c(r) + \Phi_2{}^p(r).$$

For the complimentary solution,

$$D_2(r)\nabla^2\Phi_2(r) - \Sigma_2\Phi_2(r) = 0$$

gives

$$\Phi_2{}^e(r) = B_3\frac{\cosh\frac{r}{L}}{r} + B_4\frac{\sinh\frac{r}{L}}{r}.$$

For the particular solution,

$$\nabla^2\Phi_2(r) - \frac{1}{L^2}\Phi_2(r) = \frac{-\Sigma_1\Phi_1(r)}{D_2},$$

and in operator form

$$\Phi_2{}^p(r) = -\frac{1}{\hat{D}^2 - \frac{1}{L^2}} \quad \frac{\Sigma_1}{D_2}\Phi_1(r)$$

$$= \frac{L}{2}\left(\frac{1}{\hat{D} - \frac{1}{L}} - \frac{1}{\hat{D} + \frac{1}{L}}\right) \quad \frac{-\Sigma_1(r)}{D_2(r)}\Phi_1(r)$$

$$= -\frac{L^2}{2}\left[(1 - L\hat{D})^{-1} + (1 + L\hat{D})^{-1}\right] \quad \frac{-\Sigma_1(r)}{D_2(r)}\Phi_1(r)$$

$$= -\frac{L^2}{2}\left[1 + L\hat{D} - L^2D^2 + \ldots + 1 - LD + L^2D^2 + \ldots\right] \quad \frac{-\Sigma_1(r)}{D_2(r)}\Phi_1(r)$$

$$\Phi_2{}^p(r) = -L^2\left[1 + L^2\hat{D}^2 + L^4\hat{D}^4 + \ldots..\right]\frac{-\Sigma_1(r)}{D_2(r)}\Phi_1(r)$$

For constant cross-sections, the infinite series

$$\Phi_2{}^P(r) = -L^2 \left[1 + \frac{L^2}{\tau} + \left(\frac{L^2}{\tau}\right)^2 + \ldots\right] \frac{\mathrm{L}_1}{\mathrm{D}_2} \Phi_1(r),$$

is written in compact form, for $L^2 \ll \tau$,

$$\Phi_2{}^P(r) = L^2 \left[1 - \frac{L^2}{\tau}\right]^{-1} \frac{\mathrm{L}_1}{\mathrm{D}_2} \Phi_1(r).$$

Simplifying further,

$$\Phi_2{}^P(r) = \frac{\frac{\Sigma_1}{\Sigma_2}}{1 - \frac{L^2}{\tau}} \Phi_1(r).$$

Thus

$$\Phi_2(r) = \mathrm{B}_3 \frac{\cosh r/L}{r} + B_4 \frac{\sinh r/L}{r} + \frac{\frac{\sum_1}{\sum_2}}{1 - \frac{L^2}{\tau}} \Phi_1(r).$$

We now apply the boundary conditions to get the four coefficients B_1, B_2, B_3, B_4. The current boundary condition gives

$$\frac{d\Phi_1}{dr} = B_1 \frac{\frac{r}{\sqrt{\tau}} \sinh\frac{r}{\sqrt{\tau}} - \cosh\frac{r}{\sqrt{\tau}}}{r^2} + B_2 \frac{\frac{r}{\sqrt{\tau}} \cosh\frac{r}{\sqrt{\tau}} - \sinh\frac{r}{\sqrt{\tau}}}{r^2},$$

$$\left(4\pi r^2 \frac{d\Phi_1}{dr}\right)_{r\to 0} = [(B_1 4\pi)(-1) + 4\pi B_2(0)](-D_1),$$

$$S = -B_1(4\pi)(-D_1),$$

so that

$$B_1 = \frac{S}{(4\pi)(D_1)}.$$

Equating the group-1 flux at the extrapolated boundary condition to zero, $\Phi_1(R + d_1) = 0$, gives

$$B_1 \cosh\frac{R + d_1}{\sqrt{\tau}} + B_2 \sinh\frac{R + d_1}{\sqrt{\tau}} = 0$$

$$B_2 = -\frac{\cosh\frac{R+d_1}{\sqrt{\tau}}}{\sinh\frac{R+d_1}{\sqrt{\tau}}} B_1.$$

Simplifying,

$$\Phi_1(r) = B_1 \left[\frac{\cosh\frac{r}{\sqrt{\tau}}}{r} - \frac{\cosh\frac{R+d_1}{\sqrt{\tau}}}{\sinh\frac{R+d_1}{\sqrt{\tau}}} \frac{\sinh\frac{r}{\sqrt{\tau}}}{r}\right],$$

gives

$$\Phi_1(r) = \frac{B_1}{r} \left[\frac{\cosh\frac{r}{\sqrt{\tau}} \sinh\frac{R+d_1}{\sqrt{\tau}} - \sinh\frac{r}{\sqrt{\tau}} \cosh\frac{R+d_1}{\sqrt{\tau}}}{\sinh\frac{R+d_1}{\sqrt{\tau}}}\right],$$

which, upon using the identities,

$$\cosh(A \pm B) = \cosh A \cosh B \pm \sinh A \sinh B$$

$$\sinh(A \pm B) = \sinh A \cosh B \pm \cosh A \sinh B$$

$$\therefore \cosh\frac{r}{\sqrt{\tau}}\sinh\frac{R+d_1}{\sqrt{\tau}} - \cosh\frac{R+d_1}{\sqrt{\tau}}\sinh\frac{r}{\sqrt{\tau}} = \sinh\frac{\tilde{R}-r}{\sqrt{\tau}}$$

Similarly, applying the boundary condition to the group-2 flux,

$$\begin{aligned} \Phi_1(r) &= \alpha\frac{\sinh\frac{\overline{R}_1-r}{\sqrt{\tau}}}{r} \\ \Phi_2(r) &= \beta\frac{\sinh\frac{\overline{R}_2-r}{\sqrt{\tau}}}{r} + \gamma\Phi_1(r) \end{aligned} \tag{5.29}$$

where

$$\begin{cases} \alpha = \dfrac{S}{4\pi D_1}\dfrac{1}{\sinh\tilde{R}_1/\sqrt{\tau}} \\ \beta = \dfrac{S}{4\pi D_2}\dfrac{1}{\sinh\tilde{R}_2/L}\dfrac{L^2}{\tau - L^2} \\ \gamma = \dfrac{\Sigma_1}{\Sigma_2}\dfrac{\tau}{\tau - L^2} \end{cases}$$

n a finite medium, the boundary conditions, for the "fast" group: $\lim_{r\to 0} 4\pi r^2 J_1(r) = S$, and $\phi_1(\tilde{R}) = 0$ yield

$$\phi_1(r) = \frac{S}{4\pi D_1 r}\frac{\sinh(\tilde{R}-r)/\sqrt{\tau_T}}{\sinh\tilde{R}}/\sqrt{\tau_T}. \tag{5.30}$$

For the "thermal" group, there is no "direct" source so that the boundary conditions: $4\pi r^2 J_2(r)|_{r=0} = 0$, and $\phi_2(\tilde{R}) = 0$ yield the solution

$$\phi_2(r) = \frac{S}{4\pi D_2 r}\frac{L^2}{\tau - L^2}\frac{\sinh(\tilde{R}-r)/L}{\sinh\tilde{R}}/L + \frac{\Sigma_1}{\Sigma_2}\frac{\tau}{\tau - L^2}\phi_1(r). \tag{5.31}$$

For an infinite medium, it can readily be shown that the fluxes are

$$\phi_1^{\infty}(r) = \frac{S\exp\left(-\frac{r}{\sqrt{\tau_T}}\right)}{4\pi D_1 r} \tag{5.32}$$

and

$$\phi_2^{\infty}(r) = \frac{SL_2^2}{4\pi r D_2(L_2^2 - \tau_T)}\left(e^{-r/L_2} - e^{-r/\sqrt{\tau_T}}\right). \tag{5.33}$$

5.3.1.1 Computer programming example

Eqs. (3.26) and (3.27) for a slab of thickness, with an incident fast neutron source $S/2$ neutrons/cm^3/s on the left, are written as

$$D_1\frac{d^2\phi_1}{dx^2} - \Sigma_{r1}\phi_1 = 0$$

$$D_2\frac{d^2\phi_2}{dx^2} - \Sigma_{r2}\phi_2 + \Sigma_{r1}\phi_1 = 0$$

with boundary conditions

$$D_1 \frac{d\phi_1}{dx}|_{x=0} = S/2$$
$$\phi_1(x=T) = \phi_2(x=0) = \phi_1(x=T) = 0$$

and the following data:

$$D_1 = 1.1302 \text{ cm} \quad \Sigma_{r1} = 0.0418 \text{ cm}^{-1} \quad S = 1 \text{ cm}^{-3}\text{s}^{-1}$$
$$D_2 = 0.1107 \text{ cm} \quad \Sigma_{r2} = 0.0167 \text{ cm}^{-1}$$

These yield the exact solutions:

$$\phi_1(x) = A_1 e^{x/\sqrt{\tau}} + A_2 e^{-x/\sqrt{\tau}}$$

and

$$\phi_2(x) = A_3 e^{x/L} + A_4 e^{-x/L} + \beta\phi_1(x)$$

where

$$\tau = \frac{D_1}{\Sigma_{r1}}, L^2 = \frac{D_2}{\Sigma_{r2}}, \beta = \frac{\Sigma_{r1}}{\Sigma_{r2}} \frac{\tau}{\tau - L^2}$$

and $\tau = 27.0080 \quad \text{cm}^2, L^2 = 6.6128 \quad \text{cm}^2, \beta = 3.3100$.

The constants found from the boundary conditions are: $A = [-0.0010; 2.2981; 0; -7.6034]$.

Fluxes are evaluated with the Matlab program listed below:

```
den=1.0;MolWt=18;AvNo=0.6022e24;
No=den*AvNo/MolWt; %number density of water
sigr1=1.2508e-24; sigtr1=8.8158e-24; D1=1/(3*No*sigtr1);
sigr2=0.5004e-24; sigtr2=90.000e-24; D2=1/(3*No*sigtr2);
S=1;
Sigr1=No*sigr1;Sigr2=No*sigr2;tau=D1/Sigr1;Lsq=D2/Sigr2;L2=sqrt(Lsq);
beta=(Sigr1/Sigr2)*(tau/(tau-Lsq))
L=20; % slab thickness (cm)
AA=[  1                    -1                   0           0;
    exp(L/sqrt(tau)) exp(-L/sqrt(tau)) 0           0;
    beta                 beta                 1           1;
    beta*exp(L/sqrt(tau)) beta*exp(-L/sqrt(tau)) exp(L/L2) exp(-L/L2)]
B= [-S*sqrt(tau)/(2*D1);0;0;0]
A= inv(AA) * B
 flux1=@(x) A(1)*exp(x/sqrt(tau))+A(2)*exp(-x/sqrt(tau));
flux2=@(x) A(3)*exp(x/L2)+A(4)*exp(-x/L2)+
beta*(A(1)*exp(x/sqrt(tau))+A(2)*exp(-x/sqrt(tau)));
x=0:0.1:20;
FF1=feval(flux1,x); % group 1 flux
FF2=feval(flux2,x); % group 2 flux
figure(1)
plot(x,FF1,'k-','LineWidth',2);
hold on
plot(x,FF2,'k--','LineWidth',2)
grid on
xlabel('\bf x (cm)','fontSize',14)
ylabel('\bf Flux (n cm^{-2} s^{-1})','fontSize',14)
legend('\phi_1','\phi_2','Location','best','fontSize',12)
```

The resulting fluxes $\phi_{1,2}(x)$ are shown in Fig. 5.6. As expected, the fast flux $\phi_1(x)$ gradually decreases as neutrons undergo collisions in the host medium giving rising to the thermal flux $\phi_2(x)$ which subsequently rises and decays. Both

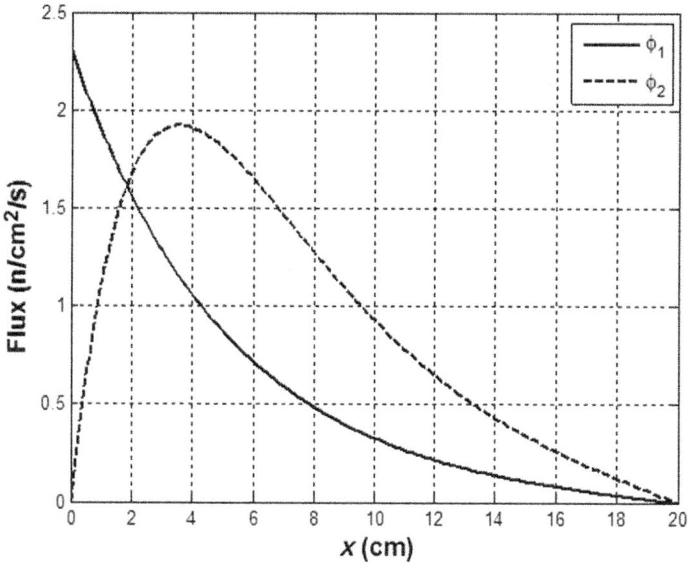

FIGURE 5.6 Two-group fluxes in a slab.

fluxes fall to zero at the physical (rather than the extrapolated) boundary as required by the specified boundary conditions.

Exercise 5.3: .

Estimate the time a person can stand a distance R from a Cf-252 point source of given strength S n/s shielded by water. The maximum permissible dose is 5 rem/y ($\sim$100 mrem/week). Use the infinite medium solutions Eqs. (5.32) and (5.33) and the flux-to-dose conversion factors (Fig. 2.13); justify any other assumptions you make.

5.3.1.2 Temperature effects on neutron flux

As discussed in Chapter 2, thermal properties of materials, such as diffusion coefficient and diffusion length, are temperature-dependent and corrections must be made to account for this dependence. For the thermal data given in Table 5.3, the corrections are made with the expressions

$$\overline{D}(\rho, T) = \overline{D}(\rho_0, T_0)\left(\frac{\rho_0}{\rho}\right)\left(\frac{T}{T_0}\right)^m$$

to evaluate the properties at any desired temperature.

Similarly, the diffusion length is evaluated as $L_T^2(\rho, T) = L_T^2(\rho_0, T_0)\left(\frac{\rho_0}{\rho}\right)^2\left(\frac{T}{T_0}\right)^{m+1/2}$ with $m = 0.470$ for H_2O and $m = 0.112$ for D_2O and zero otherwise.

5.3.2 Multiplying systems

In general, multiplying systems are designed in a way that neutron losses are minimized; this leads to efficient designs. Thus bare assemblies are surrounded by some scattering material, like water or beryllium, from which some of the leaking neutrons are reflected back into the core.

In such a core-reflector system, the two-group equations for the core are

$$D_{1c}\nabla^2\phi_{1c} - \Sigma_{1c}\phi_{1c} + \eta\Sigma_{2c}\phi_{2c} = 0 \tag{5.34}$$

and

$$D_{2c}\nabla^2\phi_{2c} - \Sigma_{2c}\phi_{2c} + \Sigma_{1c}\phi_{1c} = 0. \tag{5.35}$$

TABLE 5.3 Thermal data (at 20°C) for diffusion theory neutron flux: infinite medium.

Moderator	Density	$\overline{D}$	Σ_a	L_T
	g/cm[3]	cm	cm	cm
H_2O	1.00	0.16	0.0197	2.85
D_2O[a]	1.10	0.87	9.3×10^{-5}	97
Be	1.85	0.50	1.04×10^{-3}	21
Graphite	1.60	0.84	2.4×10^{-4}	59

[a]D_2O containing 0.25 weight/percent H_2O.

In the reflector, the equations are

$$D_{1r}\nabla^2\phi_{1r} - \Sigma_{1r}\phi_{1r} = 0 \tag{5.36}$$

and

$$D_{2r}\nabla^2\phi_{2r} - \Sigma_{2r}\phi_{2r} + \Sigma_{1r}\phi_{1r} = 0 \tag{5.37}$$

where the subscripts c, r refer to core and reflector, respectively. The boundary conditions are continuity of flux and current at the interface

$$\phi_{1c} = \phi_{1r} \tag{5.38}$$

$$\phi_{2c} = \phi_{2r} \tag{5.39}$$

$$D_{1c}\phi'_{1c} = D_{1r}\phi'_{1r} \tag{5.40}$$

$$D_{2c}\phi'_{2c} = D_{2r}\phi'_{2r} \tag{5.41}$$

and at the boundaries, for a spherical system,

$$-D_{1c}\phi'_{1c}|_{r=0} = -D_{2c}\phi'_{2c}|_{r=0} = 0 \tag{5.42}$$

and

$$\phi_{1c}(\tilde{R}) = \phi_{2c}(\tilde{R}) = 0. \tag{5.43}$$

We follow the solution procedure given in Lamarsh (1966), where two coupled second-order ODEs are converted into one fourth-order ODE.

From Eq. (5.34), ϕ_{2c} can be obtained in terms of ϕ_{1c} to get a fourth-order ODE for ϕ_{1c}

$$(\nabla^2 + \mu^2)(\nabla^2 - \lambda^2)\phi_{1c} = 0 \tag{5.44}$$

with

$$\mu^2 = \frac{1}{2\tau_1 L^2}\left[-(\tau_1 + L^2) + \sqrt{(\tau_1 + L^2)^2 + 4(k_\infty - 1)\tau_1 L^2}\right] \tag{5.45}$$

and

$$\lambda^2 = \frac{1}{2\tau_1 L^2}\left[(\tau_1 + L^2) + \sqrt{(\tau_1 + L^2)^2 + 4(k_\infty - 1)\tau_1 L^2}\right]. \tag{5.46}$$

For a uniformly distributed fuel, the coefficients in the above equations are constant and solutions are obtained by expressing the core equations as two 4th-order ODEs to yield the flux

$$\phi_{1c} = AX + CY \tag{5.47}$$

$$\phi_{2c} = AS_1X + CS_2Y \tag{5.48}$$

$$\phi_{1r} = FZ_1 \tag{5.49}$$

$$\phi_{2r} = S_3\phi_{1r} + GZ_2 \tag{5.50}$$

where the "coupling coefficients" S_1, S_2, S_3 are

$$S_1 = \frac{\Sigma_{1c}}{\Sigma_{2c}} \frac{1}{1+\mu^2 L_c^2}, \tag{5.51}$$

$$S_2 = \frac{\Sigma_{1c}}{\Sigma_{2c}} \frac{1}{1-\lambda^2 L_c^2}, \tag{5.52}$$

and

$$S_3 = \frac{D_{1r}}{\Sigma_{2r}} \frac{1}{\tau_T - L_r^2}. \tag{5.53}$$

The core and reflector functions X, Y for slab, cylinder and spherical geometries are given in Table 5.4 for the core and in Table 5.5 for the reflector.

After applying the interface conditions, the resulting equations are four linear homogenous equations, for which the only nontrivial solution, by Cramer's Rule, will be found when the determinant vanishes, that is,

$$\begin{vmatrix} X & Y & -Z_1 & 0 \\ D_{1c}X' & D_{1c}Y' & -D_{1r}Z_1' & 0 \\ S_1X & S_2Y & -S_3Z_1 & -Z_2 \\ D_{2c}S_1X' & D_{2c}S_2Y' & D_{2r}S_3Z_1' & -D_{2r}Z_2' \end{vmatrix} = 0. \tag{5.54}$$

Eq. (5.54) is called the two-group critical determinant. When the material properties are given, the critical dimension can thus be determined; conversely, when the dimension is given, the material properties (such as enrichment) can be determined. Consider a U^{235}-H_2O dilute homogenous solution with fuel-water ratio 1:500 and two-group data as given below (Table 5.6).

TABLE 5.4 Two-group core functions.

Geometry	X	Y
Infinite slab	$\cos\mu x$	$\cosh\lambda x$
Infinite cylinder	$J_0(\mu r)$	$I_0(\lambda r)$
Sphere	$\frac{\sin\mu r}{r}$	$\frac{\sinh\lambda r}{r}$

TABLE 5.5 Two-group reflector functions.

Geometry	Z	
	Reflector thickness b	Reflector thickness infinite
Infinite slab	$\sinh\kappa\left(\frac{a}{2}+b-\lvert x\rvert\right)$	$e^{-\kappa\lvert x\rvert}$
Infinite cylinder	$I_0(\kappa r)K_0[\kappa(R+b)] - I_0[\kappa(R+b)]K_0(\kappa r)$	$K_0(\kappa r)$
Sphere	$\frac{\sinh\kappa(R+b-r)}{r}$	$\frac{e^{-\kappa r}}{r}$

TABLE 5.6 Two-group data for homogeneous spherical reactor.

Data	Group 1	Group 2
D (cm)	1.13	0.16
Σ_c (cm)	0.0419	0.060
σ_{aW} (10^{-24} cm^2)	–	0.664
$\sigma_{a2,5}$ (10^{-24} cm^2)	–	678

For uniform fissile distribution, that is, constant values for D and Σ, the fluxes obtained in the core and reflector for the data in are

$$\phi_{1c}(r) = A\left(\frac{\sin 0.113r}{r} - 4.6710^{-8}\frac{\sinh 0.651r}{r}\right) \tag{5.55}$$

$$\phi_{2c}(r) = 0.676A\left(\frac{\sin 0.113r}{r} + 3.6410^{-7}\frac{\sinh 0.651r}{r}\right) \tag{5.56}$$

and

$$\phi_{1r} = 39.6A\frac{e^{-0.192r}}{r} \tag{5.57}$$

$$\phi_{2r} = 120A\left(\frac{e^{-0.192r}}{r} - 21.0\frac{e^{-0.351r}}{r}\right) \tag{5.58}$$

The core and reflector fluxes are shown in Fig. 5.7. The group-1 *fast* flux decreases steadily from the core into the reflector while the group-2 *thermal* flux also decreases within the core but has a hump at the core-reflector interface due to the reflection of slowed-down neutrons from the reflector.

For a uniform fissile distribution, with the flux shown in Fig. 5.8, the critical radius and mass obtained from diffusion theory are 21.9 cm and 1.15 kg U^{235}, respectively.

5.3.3 Two-group criticality

The two-group critical determinant results in a transcendental equation from which the critical radius R is determined. This is written in the form

$$\zeta_X = \frac{\alpha_1\zeta_Y\zeta_{Z2} + \alpha_2\zeta_Y\zeta_{Z1} + \alpha_3\zeta_{Z1}\zeta_{Z2}}{\alpha_4\zeta_Y + \alpha_5\zeta_{Z1} + \alpha_6\zeta_{Z2}} \tag{5.59}$$

where

$$\zeta_X \equiv \frac{X'}{X} = -\mu\left(\frac{1}{\mu R} - \cot\mu R\right)$$

$$\zeta_Y \equiv \frac{Y'}{Y} = \lambda\left(\coth\lambda R - \frac{1}{\lambda R}\right)$$

$$\zeta_Z \equiv \frac{Z'}{Z} = -\kappa\left(\frac{1}{\kappa R} + \coth\kappa b\right)$$

and

$$\alpha_1 = D_{1c}D_{2r}(S_3 - S_1), \alpha_2 = D_{1r}D_{2c}S_2 - D_{1c}D_{2r}S_3, \alpha_3 = D_{1r}D_{2r}(S_1 - S_2)$$
$$\alpha_4 = D_{1c}D_{2c}(S_2 - S_1), \alpha_5 = (D_{1r}D_{2c}S_1 - D_{1c}D_{2r}S_3), \alpha_6 = D_{1c}D_{2r}(S_3 - S_2).$$

The roots $\mu R = 2.472, R = 21.91$ cm and $\mu R = 5.61, R = 49.72$ cm are determined from the Matlab program listed below and shown in Fig. 5.8. The critical radius is thus established to be 21.91 cm, in agreement with Lamarsh.

The Matlab program uses the function

$$\zeta_Z \equiv \frac{Z'}{Z} = -\kappa\left(\frac{1}{\kappa R} + 1\right)$$

for an infinite reflector.

```
% Lamarsh Spherical Reflected Critical Reactor
% LamarshSphereCriticalityBookReactor.m
fid=fopen('Out.txt','w');
Nav=0.6023;denW=1.0;MolWt=18.02;AtDenW=denW*Nav/MolWt;
AtDen235=(1/500)*AtDenW;
fprintf(fid,'\n AtDen235=%12.4e  AtDenW=%12.4e',AtDen235,AtDenW)
MuFunc     = @(a,b) 0.5* ( -a + sqrt(a^2 + 4*b));
LambdaFunc = @(a,b) 0.5* (  a + sqrt(a^2 + 4*b));

% data
sigaW=0.664; sigaTh5=678; Sig1c=0.0419;Sig1r=0.0419;
Tauc=27; Taur=27;D1c=1.13; D2c=0.16;D1r=1.13; D2r=0.16;
p=1.0;
Sig2c=0.886*(AtDen235*sigaTh5+AtDenW*sigaW);
Sig2r=0.886*AtDenW*sigaW;
LsqC=D2c/Sig2c;eta=2.07;
LsqR=D2r/Sig2r; K1r=1/sqrt(Taur);K2r=1/sqrt(LsqR);
f=AtDen235*sigaTh5/(AtDen235*sigaTh5+AtDenW*sigaW);
kinf=eta*f;
a=LsqC+Tauc;b=(kinf-1)*Tauc*LsqC;
Mu2    =(1.0/(LsqC*Tauc))*MuFunc(a,b); Mu= sqrt(Mu2);
Lambda2=(1.0/(LsqC*Tauc))*LambdaFunc(a,b); Lambda  = sqrt(Lambda2);
S(1)=(p*Sig1c/Sig2c)/(1+Mu2*LsqC);S(2)=(p*Sig1c/Sig2c)/(1-Lambda2*LsqC);
S(3)=(D1r/Sig2r)/(Taur-LsqR);
Rhs1=D1c*D2r*(S(3)-S(1));
Rhs2=D1r*D2c*S(2)-D1c*D2r*S(3);
Rhs3=D1r*D2r*(S(1)-S(2));
Rhs4=D1c*D2c*(S(2)-S(1));
Rhs5=D1r*D2c*S(1)-D1c*D2r*S(3);
Rhs6=D1c*D2r*(S(3)-S(2));
fprintf(fid,'\n sigaW= %8.5f b  Tauc=%8.4f cm^2  sigaTh5=%8.4f b
Sig1c=%8.4f D1c=%8.4f cm  D2c=%8.4f cm ',sigaW,Tauc,sigaTh5,Sig1c,D1c,D2c)
fprintf(fid,'\n                       Taur=%8.4f cm^2
Sig1r=%8.4f D1r=%8.4f cm  D2r=%8.4f cm ',Taur,Sig1r,D1r,D2r)
fprintf(fid,'\n Rhs1=%8.4f  Rhs2=%8.4f Rhs3=%8.4f',Rhs1,Rhs2,Rhs3)
fprintf(fid,'\n Rhs4=%8.4f  Rhs5=%8.4f Rhs6=%8.4f\n',Rhs4,Rhs5,Rhs6)
fprintf(fid,'\n Sig2c= %8.4f cm^-1   Sig2r=%8.4f cm^-1   LsqC= %8.4f cm^2
LsqR=%8.4f   K1r=%8.4f cm^-1   K2r=%8.4f  cm^-1
',Sig2c,Sig2r,LsqC,LsqR,K1r,K2r)
fprintf(fid,'\n f= %8.4f   kinf= %8.4f Mu= %12.4e  Lambda= %12.4e
',f,kinf,Mu,Lambda)
fprintf(fid,'\n p= %8.4f ',p)
fprintf(fid,'\n S(1)= %12.4e  S(2)=%12.4e   S(3)=%12.4e ',S(1),S(2),S(3))
```

```
%b=22.5; % ref thickness
fprintf(fid,'\n Reflector thickness = %8.4f cm',b)
fprintf (fid,'\n\n        i     MuR          -R*LHS           -R*RHS
Diff ')
Rmin=0;Rmax=60;Nsteps=1000;del=(Rmax-Rmin)/Nsteps;
R=Rmin;RCrit=1e10;MinDif=1e10;
for i=1:Nsteps
R=R+del;
XprimeOverX= Mu*(cot(Mu*R)-1/(Mu*R));
YprimeOverY= Lambda*(coth(Lambda*R)-1/(Lambda*R));
% for a finite reflector
%Z1primeOverZ1= -K1r*coth(K1r*b);
%Z2primeOverZ2= -K2r*coth(K2r*b);
% for an infinite reflector
Z1primeOverZ1= -K1r*(1+1/(K1r*R));
Z2primeOverZ2= -K2r*(1+1/(K2r*R));
%
LHS=XprimeOverX;
r1=Rhs1*YprimeOverY*Z2primeOverZ2;
r2=Rhs2*YprimeOverY*Z1primeOverZ1;
r3=Rhs3*Z1primeOverZ1*Z2primeOverZ2;
RHSNum=r1 + r2 + r3;
r4=Rhs4*YprimeOverY;
r5=Rhs5*Z1primeOverZ1;
r6=Rhs6*Z2primeOverZ2;
RHSDen=r4+r5 + r6 ;
RHS=RHSNum/RHSDen;

X(i)=Mu*R;
Y(i)=-R*LHS;
Z(i)=-R*RHS;
Difference=Z(i)-Y(i);
Diff(i)=Difference;
if (abs(Difference)<MinDif)
    MinDif=abs(Difference);
    RCrit=R; %this is the critical radius found
end
%fprintf(fid,'\n  %6.0f  %8.4f  %12.4e  %12.4e
%12.4e',i,R,LHS,RHS,Diff(i))
fprintf(fid,'\n  %6.0f  %8.4f  %12.4e  %12.4e
%12.4e',i,X(i),Y(i),Z(i),Diff(i))
end
```

```
figure(1)
set(gca,'FontSize',14)
plot(X,Y,'k-','LineWidth',2)
hold on
plot(X,Z,'k-','LineWidth',2)
xlabel('\bf \mu R','fontSize',14)
ylabel('\bf ','fontSize',14);
text(4,8,'\bf LHS')
text(2.0,12.5,'\bf RHS')
text(5.0,12.5,'\bf RHS')
XL1=[2.472 2.472];YL1=[-20 4.12];
line(XL1,YL1,'LineWidth',0.5,'Color','k','LineStyle','--')
XL2=[5.61 5.61];YL2=[-20 8.08];
line(XL2,YL2,'LineWidth',0.5,'Color','k','LineStyle','--')
grid off
xlim([0 6.2])
ylim([-20 20])
```

Partial Output

```
AtDen235=  6.6848e-05  AtDenW=  3.3424e-02
sigaW= 0.66400 b Tauc= 27.0000 cm^2 sigaTh5=678.0000 b Sig1c= 0.0419
D1c=  1.1300 cm  D2c=  0.1600 cm
Taur= 27.0000 cm^2  Sig1r=  0.0419 D1r=  1.1300 cm  D2r= 0.1600 cm
Rhs1=  0.4283  Rhs2= -1.5021 Rhs3= 1.0738Rhs4= -1.0738 Rhs5= -0.4283
Rhs6=  1.5021

Sig2c=0.0598 cm^-1  Sig2r=0.0197 cm^-1  LsqC=2.6747 cm^2 LsqR=8.1369
K1r=  0.1925 cm^-1   K2r=  0.3506  cm^-1
f=0.6713   kinf=1.3896 Mu=1.1284e-01  Lambda=6.5088e-01 p=  1.0000
S(1)=   6.7737e-01  S(2)= -5.2617e+00   S(3)=  3.0465e+00
```

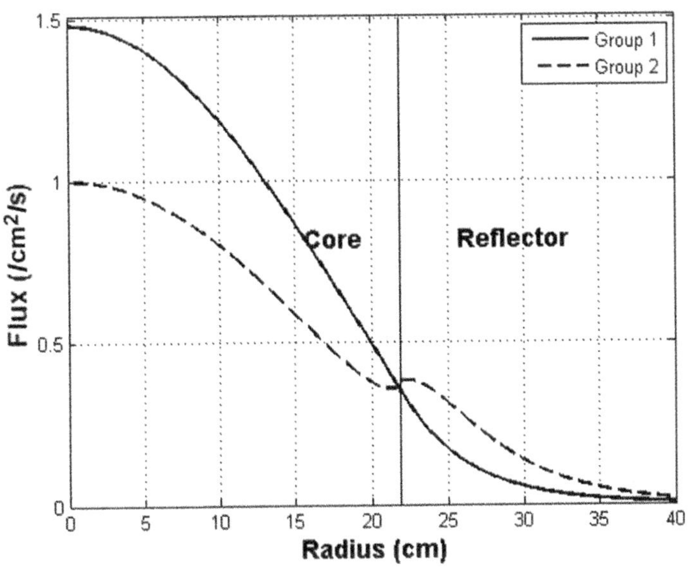

FIGURE 5.7 Two-group fluxes in a spherical reactor.

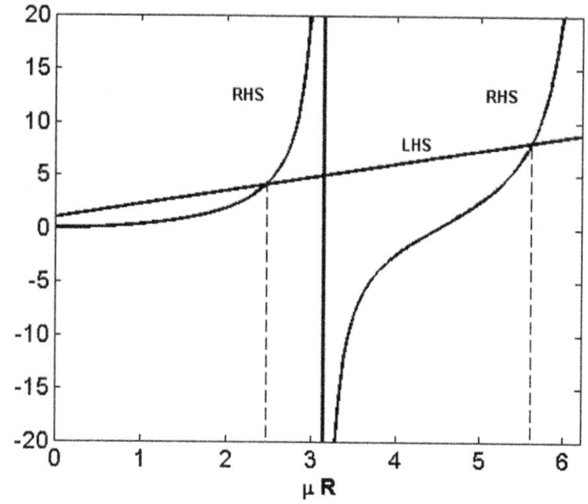

FIGURE 5.8 Transcendental equation for two-group criticality in a reflected spherical reactor.

The two-group expression for the system multiplication k_{eff} with nonleakage probabilities is an extension of the one-group expression; here the fast and thermal probabilities are both incorporated, based on Fermi's slowing-down equation where the neutron age τ_T, with units of cm^2 is introduced as the area "traveled" by a neutron as it slows down from source energy to thermal energy. Thus

$$k_{eff} = \frac{k_\infty e^{-B^2\tau_T}}{1+B^2L_T^2} = 1. \tag{5.60}$$

In order to find the critical radius, we can set $x \equiv B^2\tau_T$, and solve the transcendental equation

$$k_\infty e^{-x} = 1 + \left(\frac{L_T^2}{\tau_T}\right)x \tag{5.61}$$

to find x, and hence B from which the critical radius R is found.

Eq. (5.61), from Eq. (1.27), compares with

$$k_{eff} = k_\infty(1-P_F)(1-P_T)$$

where $1-P_F$ and $1-P_T$ were defined to be the fast and thermal nonleakage probabilities, from the finite system, respectively. Thus

$$P_F = e^{-B^2\tau_T} \cong \frac{1}{1+B^2\tau_T}$$

where the approximation holds for large reactors for which $B^2\tau_T < 1$, and

$$P_T = \frac{1}{1+B^2L_T^2}$$

An alternate expression for large reactors, called the modified one-group critical equation, is

$$1 = \frac{k_\infty}{1+B^2M_T^2} \tag{5.62}$$

where $M_T^2 = L_T^2 + \tau_T$ is called the thermal migration area.

We now have three expressions to compute critical compositions from one-group theory Eq. (5.24), two-group theory Eq. (5.60) and a modified one-group theory Eq. (5.62).

5.4 The multigroup diffusion equation

When a two-group formulation is inadequate to represent a system, such as for the case of modeling fast, epithermal and thermal regions, or a larger number of groups, a multigroup formulation is used.

Neutron energy during transport may vary from a few MeV down to a fraction of an electron-volt (thermal energy $E_T \sim kT \sim 0.025$ eV). This indeed is a broad range spanning over six orders of magnitude. It is thus convenient to divide the entire range into a finite number of groups within which neutronic properties are averaged and the neutron "group" equations are written from which group flux are subsequently obtained. While two-group calculations are extensively used for preliminary computations, elaborate MC simulations can be carried out for as many as 262 groups. Clearly, the more the number of groups, the better the results as fine details, such as cross-sections in the resonance region, may be represented. For the critical radius and corresponding mass, a detailed Monte Carlo simulation, with the Monte Carlo N-Particle (MCNP) code, for the 1:500 U^{235}-H_2O homogeneous solution described in the previous section gives a radius of 22.5 cm and mass of 1.2477 kg U^{235}respectively. Thus, the two-group results can be seen to be fairly accurate.

As an example, in a 6-group diffusion formulation to compute the neutron flux and critical radius of a bare sphere of Pu^{239} the group-constants data are given in Table 5.7 (Lamarsh, 1966).

For realistic neutronic analysis, a large number of groups is generally used for which data cross-section libraries are "generated" using group flux ϕ_i and associated reaction rates $<\sigma_x\phi>$ (where the inner product $<\ldots>$ implies integration over the energy group of interest), and setting the group cross-section, for group i, as $\sigma_x^{(i)} = <\sigma_x\phi> / \phi_i$. The existing methodology for obtaining multigroup cross-sections is based on reading ENDF pointwise cross-section data, by processing codes such as NJOY (LANL) and MC^2 -3 (ANL) to produce binned cross-sections for use in multigroup deterministic codes. This is achieved by the group flux-weighting mentioned above and accounting for resonances and self-shielding.

5.4.1 Numerical solution of the multigroup diffusion equations

In case of nonuniform material loading, the flux can be obtained by a numerical solution of the diffusion equations. A first step for a numerical procedure is the FDM for which a general iterative form for the multigroup diffusion equations, for ϕ_g,

$$-\nabla \cdot D_g(r)\nabla\phi_g^{(i)}(r) + \Sigma_{r,g}(r)\phi_g^{(i)}\left(r\right) - \sum_{i' \neq i} \Sigma_{s,g' \to g}\phi_{g'}^{(i)}(r) = S_g^{(i-1)}(r) \tag{5.63}$$

where, $D_g(r) =$ diffusion coefficient for energy group g at spatial position r, $\phi_g^{(i)} =$ "scalar" flux for energy group g at spatial position r at the ith iteration, $\Sigma_{r,g}(r) =$ removal cross-section for energy group g at spatial position r at the ith iteration, $\Sigma_{s,g' \to g} =$ 'in-scattering' cross-section for energy group g from groups at the ith at spatial position r, $S_g^{(i-1)}(r) =$ source for energy group g at spatial position r at the $(i-1)$ iteration.

The source term consists of a fission contribution and the contribution from an independent source $S_0(r)$

$$S_g^{(i-1)}(r) = \chi_g \sum_{g'} \nu\Sigma_{f,g'}(r)\phi_g^{(i-1)}(r) + S_0(r). \tag{5.64}$$

TABLE 5.7 Multigroup cross-sections for Pu^{239}.

g	E MeV	ν	χ	σ_{tr} (b)	σ_f (b)	σ_r (b)	$\sigma(g \to h)$				
							$h=2$	3	4	5	6
1	3.0–∞	3.48	0.204	4.25	1.90	0.03	0.20	0.27	0.45	0.31	0.04
2	1.4–3.0	3.09	0.344	4.50	1.95	0.05	–	0.18	0.50	0.35	0.05
3	0.9–1.4	2.99	0.168	4.80	1.83	0.07	–	–	0.45	0.30	0.06
4	0.4–0.9	2.93	0.180	5.70	1.70	0.11	–	–	–	0.29	0.05
5	0.1–0.4	2.88	0.090	8.40	1.67	0.17	–	–	–	–	0.05
6	0.0–0.1	2.86	0.014	12.0	2.05	0.50	–	–	–	–	–

Source: From Lamarsh J. R., Introduction to Nuclear Reactor Theory. Addison-Wesley Publishing Company, 1966, p. 368; ANL-5800.

In this iterative procedure, the first iteration gives $\phi_g^{(0)}$ from the source distribution $S_0(r)$, that is, by solving the equation

$$-\nabla\cdot D_g(r)\nabla\phi_g^{(0)}(r)+\Sigma_{r,g}(r)\phi_g^{(0)}(r)-\sum_{i'\neq i}\Sigma_{s,g'\to g}\phi_{g'}^{(0)}(r)=S_0(r). \tag{5.65}$$

With this initial distribution the iterations for $i=1,2,3,\cdots,N$ are carried out from the fission source

$$-\nabla\cdot D_g(r)\nabla\phi_g^{(i)}(r)+\Sigma_{r,g}(r)\phi_g^{(i)}(r)-\sum_{i'\neq i}\Sigma_{s,g'\to g}\phi_{g'}^{(i)}(r)=\chi_g\sum_{g'}\nu\Sigma_{f,g'}(r)\phi_{g'}^{(i-1)}(r)$$

Thus for $i=1$

$$-\nabla\cdot D_g(r)\nabla\phi_g^{(1)}(r)+\Sigma_{r,g}(r)\phi_g^{(1)}(r)-\sum_{i'\neq i}\Sigma_{s,g'\to g}\phi_{g'}^{(1)}(r)=\chi_g\sum_{g'}\nu\Sigma_{f,g'}(r)\phi_{g'}^{(0)}(r)$$

and for $i=2$

$$-\nabla\cdot D_g(r)\nabla\phi_g^{(2)}(r)+\Sigma_{r,g}(r)\phi_g^{(2)}(r)-\sum_{i'\neq i}\Sigma_{s,g'\to g}\phi_{g'}^{(2)}(r)=\chi_g\sum_{g'}\nu\Sigma_{f,g'}(r)\phi_{g'}^{(1)}(r)$$

and so on. The group flux $\phi_g(r)$ is then

$$\phi_g(r)=\sum_{i=0}^{N}\phi_g^{(i)}(r), \tag{5.66}$$

which converges, for a critical system for some large N. The system multiplication k_{eff} is determined by balancing the lloss and gain terms from the eigenvalue equation

$$-\nabla\cdot D_g(r)\nabla\phi_g^{(i)}(r)+\Sigma_{r,g}(r)\phi_g^{(i)}(r)-\sum_{i'\neq i}\Sigma_{s,g'\to g}\phi_{g'}^{(i)}(r)=\frac{\chi_g}{k_{\text{eff}}}\sum_{g'}\nu\Sigma_{f,g'}(r)\phi_{g'}^{(i-1)}(r)$$

where

$$k_{\text{eff}}{}^{(i)}=\frac{\int\sum_{g'}\nu\Sigma_{f,g'}(r)\phi_{g'}^{(i)}dr}{\int\sum_{g'}\nu\Sigma_{f,g'}(r)\phi_{g'}^{(i-1)}dr}. \tag{5.67}$$

The flux is normalized at each iteration, to satisfy the power output of a reactor;

$$\phi_{g'}^{(i)}(r)=\frac{1}{k_{\text{eff}}{}^{(i)}}\phi_{g'}^{(i-1)}(r). \tag{5.68}$$

can be written in the discretized form by equating the currents at the interfaces allowing for different materials in each mesh interval (Sekimoto, 2007) from Fig. 5.9

$$-\nabla\cdot D\nabla\phi|_k=c_{k-1}\phi_{k-1}+d_k'\phi_k+e_k\phi_{k+1} \tag{5.69}$$

for meshes numbered $k=1,2,3,\ldots,M$. With

$$\left(-e_1^{(1)}+\Sigma_{1\to2}^{(s)}\right)\phi_1^{(1)}+e_1^{(1)}\phi_2^{(1)}=1$$

the coefficients are given as

$$c_{k-1}=-\frac{2(p+1)r_{k-\frac{1}{2}}^{p}}{\left(r_{k+\frac{1}{2}}^{p+1}-r_{k-\frac{1}{2}}^{p+1}\right)\left[\frac{\Delta r_{k-1}}{D_i^{(k-1)}}+\frac{\Delta r_k}{D_i^{(k)}}\right]},$$

$$e_k=-\frac{2(p+1)r_{k+\frac{1}{2}}^{p}}{\left(r_{k+\frac{1}{2}}^{p+1}-r_{k-\frac{1}{2}}^{p+1}\right)\left[\frac{\Delta r_k}{D_i^{(k)}}+\frac{\Delta r_{k+1}}{D_i^{(k+1)}}\right]},$$

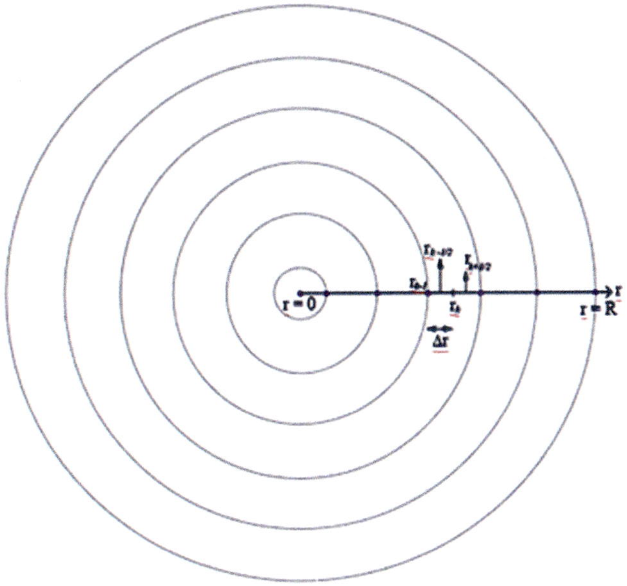

FIGURE 5.9 Mesh intervals for finite difference computation.

and

$$d'_k = -c_{k-1} - e_k$$

where $p = 0, 1, 2$ for slab, spherical and cylindrical geometries, respectively. In the above, the origin is at $r_{1/2}$ and the outer radius is $R = r_{M+1/2}$. The radii at the midpoints of each mesh interval are $r_1, r_2, r_3, \cdots r_K$ so that the mesh width of the kth mesh is: $\Delta r_k = r_{k+1/2} - r_{k-1/2}$.

Further, $c_0 = 0$, and for the first mesh, the point source boundary condition yields the equation

$$\left(-e_1 + \Sigma_{1\to 2}^{(s)}\right)\phi_1 + e_1\phi_2 = 1$$

where subscripts indicate energy and superscripts indicate the mesh interval. At the extrapolated boundary, it can be shown (Park et al., 2020) that $e_i^{(K)}$ is

$$e_i^{(K)} = -\frac{2(p+1)r^p_{K+\frac{1}{2}}}{\left(r^{p+1}_{K+\frac{1}{2}} - r^{p+1}_{K-\frac{1}{2}}\right)\left[\frac{\Delta r_K}{D_i^{(K)}} + 4.26\right]}$$

The procedure is as follows

1. Discretize Eq. (5.63) to obtain a form for $\phi_{k,g}$ for mesh interval k and energy group g.
2. Apply boundary conditions.
3. Write the equations in matrix form

$$\bar{\bar{A}}\bar{\phi} = \frac{1}{k_{\text{eff}}}\overline{B\phi}. \tag{5.70}$$

4. The total number of mesh intervals for a one-dimensional problem is K; the surrounding nodes for the kth element being $x_{k-1/2}$ and $x_{k+1/2}$. Thus element 1 is from $x_{1/2}$ to $x_{3/2}$. The column vector $\bar{\phi}$ is of size $K \times G$; thus for 100 mesh intervals and 2 groups, there are 200 fluxes to compute.
5. The matrix $\bar{\bar{A}} = \bar{\bar{L}} - \bar{\bar{S}}$ represents the three terms

$$-\nabla \cdot D_g(r)\nabla\phi_g^{(i)}(r) + \Sigma_{r,g}(r)\phi_g^{(i)}(r) - \sum_{i'\neq i}\Sigma_{s,g'\to g}\phi_{g'}^{(i)}(r)$$

the first two terms, the "loss terms," have element matrices $\bar{\bar{L}}_g$, $g = 1, 2, 3 \ldots G$ as diagonal elements while the third terminal $\bar{\bar{S}}$ is the scattering-in (lower triangular) matrix since down-scattering occurs from higher energy groups. Each element matrix $\bar{\bar{L}}_g$ is a tridiagonal matrix of size $K \times K$.

6. The matrix $\overline{B}$ represent the RHS fission term

$$\chi_g \sum_{g'} \nu\Sigma_{f,g'}(r)\phi_{g'}^{(i-1)}$$

7. In the power method, for eigenvalues $k_1, k_2, k_3, \ldots N$, the solution at the ith iteration is

$$\phi^{(i)} = \frac{1}{\prod_{i'=1}^{i-1} k^{(i)}} \left(\bar{\bar{A}} - 1\overline{B}\right)^{(i)} \phi^{(0)} \tag{5.71}$$

as $i \to \infty$, the flux $\phi^{(i)}$ converges to the flux ϕ and the largest eigenvalue is the value of k_{eff}.

Finally, the matrix can be solved by standard methods such as Gaussian elimination, Jacobi or Gauss-Siedel methods, or to reduce the memory advantage is taken of the sparse matrix and Choleski's method is more efficient. Computation is accelerated by the use of the Standard Overrelaxation Method (SOM) or a number of other methods.

As an example, for K mesh intervals and 2 groups, the matrices are

$$\bar{\bar{A}} = \begin{bmatrix} \overline{A}_1 & 0 \\ 0 & \overline{A}_2 \end{bmatrix}$$

in terms of the matrices $\overline{A}_g$, where the "group" coefficients $c, d, e\left(d_k = d'_k + \Sigma_k\right)$ are evaluated for each mesh interval. The group matrices $\overline{A}_g$ are

$$\begin{bmatrix} d_1 & e_1 & & & & 0 & \\ c_1 & d_2 & e_2 & & & & \\ & & \cdot & \cdot & & & \\ & & \cdot & \cdot & \cdot & & \\ & & & c_{k-1} & d_k & e_k & \\ & & & & \cdot & \cdot & \cdot \\ & 0 & & & & \cdot & \cdot \\ & & & & & c_{K-1} & d_K \end{bmatrix} \tag{5.72}$$

The $K \times 2$ flux, for each group and mesh interval, are

$$\overline{\phi} = \left[\phi_{1,1} \quad \phi_{2,1} \cdots \phi_{K,1} \quad \phi_{1,2} \quad \phi_{2,2} \cdots \phi_{K,2}\right]^T \tag{5.73}$$

and the "source" term is a vector of length 2K (for the case of 2 groups) and $\overline{B}$ are the fission terms.

5.5 Effect of fuel concentration on critical mass

A research area of interest in nuclear engineering has been the effect of nonuniform fuel distribution on the critical mass of nuclear systems.

The concept of minimum critical mass (MCM) by Goertzel (2003) is a classical and seminal problem in nuclear engineering; over the last few decades, it has retained a strong interest and led to further work on the requirement of flat thermal flux (FTF) as a necessary condition. Starting from a multigroup diffusion model by Goertzel and extending to a two-group transport model, Williams (2003) has shown that the condition for MCM is that the integral of the product of the thermal angular flux and a function related to the adjoint function is a constant and that an unsatisfactory aspect of the Goertzel theorem (that of adding delta functions at the core-reflector interface) disappears in transport theory. A number of models in reflected systems have been considered by Lewins (2004) in one- and two-group diffusion formulations to further establish maximum and minimum loading conditions.

There is also an anomaly (Van Dam, 2015) that "the critical size of a MCM system with continuous in-core fuel distribution can take the same value for two different reflector thicknesses." The thermal flux flattening has also been demonstrated for heterogeneous systems of thin fuel "foils" in a moderator (Van Dam, 2013).

The objective is to distribute the fissile material in such a way as to achieve criticality with minimum fissile material. Theoretical works have necessarily been based on assumptions and artifacts that need to be validated with elaborate simulations; hence the motivation to include this work in this chapter.

5.5.1 Goertzel's theorem

The two-group uniformly distributed fuel spherical reactor considered in Section 5.3.2 had a critical radius of 21.9 cm with a critical mass 1.15 kg U-235. Goertzel's theorem states that the critical mass can be reduced if a FTF is achieved in the core. This corresponds to a nonuniform fuel distribution which is covered in this section for slab and spherical reflected reactors. Calculations are compared with MC simulations.

5.5.2 Nonuniform fuel distribution: a slab model

Two-group diffusion theory is used for obtaining flux and criticality estimates for both uniform fuel loading and for loading resulting in a FTF which corresponds to MCM. An intriguing question regarding the "delta function" at the core-reflector interface and critical conditions arising out of analytical studies based on one- and two-group diffusion as well as two-group transport theory are illustrated for a "practical" core in which both minimum and maximum fuel loadings are required for achieving a FTF which corresponds to MCM.

Here, MC simulation, with MCNP5 (Briesmeister, 2000) is used for validation of two-group diffusion results.

Consider a 1D slab of a dilute U^{235}-H_2O solution reflected by water. For an exact solution with uniform fuel loading, followed by a FTF requirement, two-group diffusion equations are used. An "optimal" fuel loading is thus obtained and used in a MC simulation. The criticality conditions are obtained for both cases and used to reverse the larger core size by the addition of fuel at the core periphery.

These are discussed with Goertzel's result that more fissile material should be added to the central regions and progressively less material toward the end of the core region. The work by Williams (2003) concludes that "the major difference between diffusion and transport theories arises from the origin of the fuel mass required at the edges," and that "without the *ad hoc* addition of delta functions, diffusion theory would fail. In the transport equation on the other hand these features arise naturally." A detailed MC simulation is thus a natural extension of this study to compare the validity of both diffusion and transport models with results obtained from a detailed simulation where none of their assumptions, such as constant scattering, are used.

In Eqs. (5.34)–(5.37) if a FTF is assumed, that is, $\phi_{2c} = \phi_O$, then Eq. (5.35) is simplified to $-\Sigma_{2c}\phi_o + \Sigma_{1c}\phi_{1c} = 0$.

In the MCM problem, the objective is to find the fuel distribution $N_F(x)$ which will result in minimum fuel mass $\sim m$

$$\sim m = \alpha \int_0^{a/2} N_F(x) dV \tag{5.74}$$

where $\alpha = 2A_{235}/N_{\text{av}}$.

For a uniformly distributed fuel, described by Eqs. (5.34)–(5.37) the coefficients in Eq. (5.35) are constant and solutions are obtained by expressing the core equations as two 4th-order ODEs to yield the fluxes (Lamarsh & Baratta, 1955) with the resulting (transcendental) criticality equation for half-thickness a_o

$$\mu \tan \mu a_o/2 = \frac{\alpha_1 \lambda \tanh \lambda a_o/2 + \alpha_2}{\alpha_3 \lambda \tanh \lambda a_o/2 + \alpha_4} \tag{5.75}$$

where

$$\alpha_1 = -\kappa_{2r} D_{1c} D_{2r}(S_3 - S_1) - \kappa_{1r}(D_{1r} D_{2c} S_2 - D_{1c} D_{2r} S_3)$$

$$\alpha_2 = \kappa_{1r}\kappa_{2r} D_{1r} D_{2r}(S_1 - S_2)$$

$$\alpha_3 = -\kappa_{1r} D_{1c} D_{2c}(S_2 - S_1)$$

$$\alpha_4 = -\kappa_{2r} D_{1c} D_{2r}(S_3 - S_2) - \kappa_{1r}(D_{1r} D_{2c} S_1 - D_{1c} D_{2r} S_3)$$

where and μ^2 and λ^2 and the coupling coefficients are defined in are defined in Section 5.3.2.

For FTF ϕ_O in the core, Eqs. (5.34) and (5.35) yield

$$\phi_{1c}(x) = C_1 \cos \gamma x + \eta \frac{\Sigma_{2Mc}}{\gamma^2 D_{1c}} \phi_o \tag{5.76}$$

$$\phi_{1r}(x) = \phi_{1r}(a/2) e^{-\kappa_{1r}(x - |a/2|)} \tag{5.77}$$

$$\phi_{2r}(x) = fe^{-\kappa_{2r}(x-|a/2|)} + g\phi_{1r}(x) \tag{5.78}$$

where the constant C_1 in Eq. (5.76) can be found from the reflector interface condition

$$C_1\cos\gamma a = \phi_{1r}(a/2) - \eta\frac{\Sigma_{2Mc}}{\gamma^2 D_{1c}}\phi_o.$$

Here

$$\phi_{1r}(a/2) = \frac{D_{2r}}{\Sigma_{1r}}(\kappa_{1r} + \kappa_{2r})\kappa_{2r} \tag{5.79}$$

and

$$f = \frac{\kappa_{1r}}{\kappa_{1r} - \kappa_{2r}}\phi_o \quad \text{and} g = \frac{\Sigma_{1r}}{D_{2r}}\frac{1}{\kappa_{2r}^2 - \kappa_{1r}^2} \tag{5.80}$$

with $a/2$ being the half-width of the core.

The criticality condition is then

$$\gamma\kappa_{1r}S\tan\gamma a = 1 + \frac{\kappa_{1r}}{\kappa_{1r}} \tag{5.81}$$

where

$$S = \frac{\kappa_{1r}}{\kappa_{1r}} - \frac{1}{\eta - 1}.$$

Thus the critical half-thickness is

$$a = \frac{2}{\gamma}\cot^{-1}\left(\frac{\gamma D_{1c}}{D_{1r}\kappa_{1r}} - \frac{\gamma\tau_c\kappa_{1r}\gamma\Sigma_{2r}}{(\eta-1)D_{2r}\kappa_{2r}(\kappa_{1r}+\kappa_{2r})}\right) \tag{5.82}$$

leading to fuel loading

$$\Sigma_{2F}(x) = \Sigma_{2r}\left(\frac{\kappa_{1r}+\kappa_{2r}}{\kappa_{2r}}\right)\frac{\cos\gamma x}{\cos\gamma a/2} + \frac{\Sigma_{2Mc}}{\eta - 1}. \tag{5.83}$$

The fuel mass for both cases, uniform loading and FTF, are then

$$m_{F,o} = \alpha\Sigma_{F,o}\frac{a_o}{2} \tag{5.84}$$

and

$$m_F = \frac{\alpha\Sigma_{2r}}{\eta - 1}\left(\frac{\kappa_{1r}}{\kappa_{2r}}(\kappa_{1r}+\kappa_{2r})\tau_c + \frac{a}{2}\right). \tag{5.85}$$

The two-group data for uniform loading used is given in Table 5.6.

With a H_2O-U^{235} ratio 500–1 for uniform fuel loading ($N_5 = 6.6848 \times 10^{19}$ atoms/cm^3, $N_W = 3.342 \times 10^{22}$ molecules/cm^3) the criticality equation is

$$\tan\mu a_o/2 = \frac{0.8011\tanh\lambda a_o/2 + 0.6416}{0.6989\tanh\lambda a_o/2 + 0.4442} \tag{5.86}$$

($\mu = 0.1128, \lambda = 0.6509$) from which the critical thickness was found to be 16.1319 cm (half-thickness 8.06595 cm) with the two-group flux shown in Fig. 5.10 (mass 0.2106 g /cm^2).

Subsequently, for a slab of thickness 8.06 cm, height and width 20 cm with reflecting surfaces, a MCNP5 simulation with $N_5 = 6.7 \times 10^{19}$, $N_H = 6.7 \times 10^{22}$ and $N_O = 3.35 \times 10^{22}$ atoms/cm^3 (total 0.100567×10^{24}) in kcode with 1000 histories per cycle and 500 cycles with 10 skip cycles and tallies accumulated in two groups: group 1 (0.625 eV–1 MeV) and group 2 (0–0.625 eV) gives the fluxes shown in Fig. 5.11.

The result for this simulation gives $k_{\text{eff}} = 0.99544(0.00080)$ with a simulation time of 7.00 min on an Intel (R) Core (TM) i7–2620M CPU @2.70 GHz processor, 32-bit Operating System.

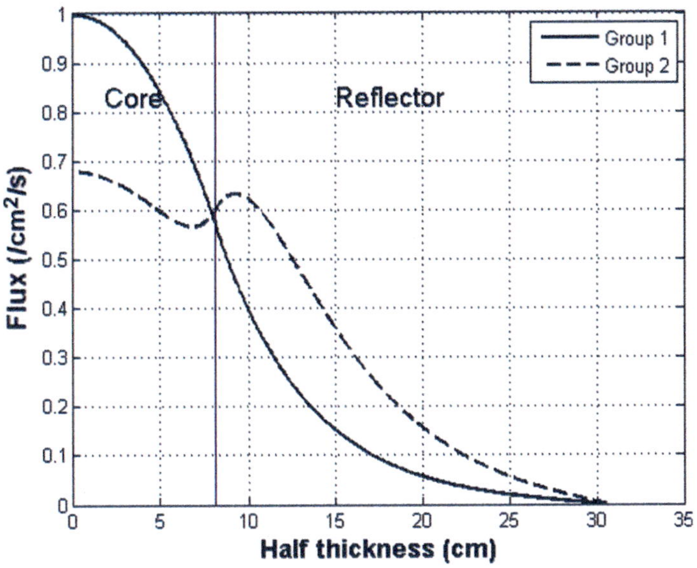

FIGURE 5.10 Two-group diffusion flux for a critical slab.

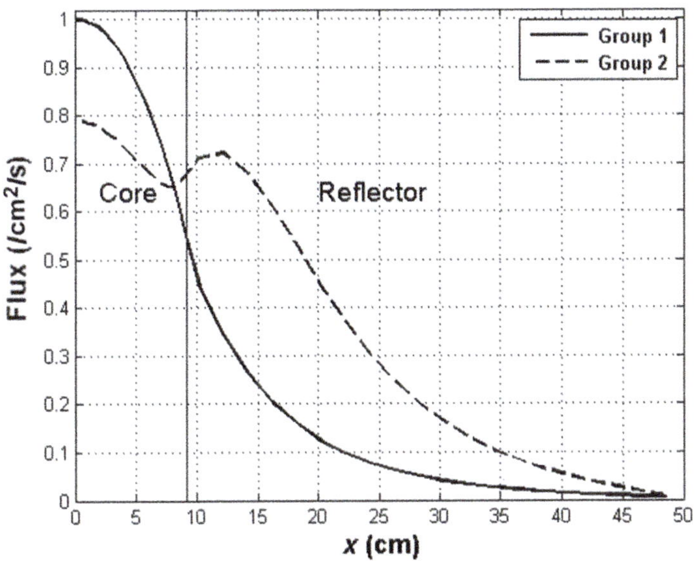

FIGURE 5.11 MCNP Two-group fluxes for uniform fuel distribution in a slab reactor. *MCNP*, Monte Carlo N-Particle.

The atom and mass fractions of U^{235} in the solution are 6.66223×10^{-4} and 2.54368×10^{-2} respectively. Each core mesh has a volume of 32.56 cm^3, and a mass of 33.4736 g, so that for 99 mesh intervals, the mass is 3.3139 kg. The total fuel mass is thus 84.2947 g, equivalent to 0.2107 g/cm^2, which matches very well with the diffusion estimate.

For FTF, the critical half-thickness, from the criticality equation

$$\tan\gamma a/2 = -3.8833 \tag{5.87}$$

was found to be 9.1567 cm with flux shown in Fig. 5.12. Thus the ratio of critical dimensions for uniform and FTF can be estimated as

$$\frac{a_o}{a} \approx \frac{\gamma}{\mu}\frac{0.9007}{\pi - 1.3188} = 0.8719 \tag{5.88}$$

The core region (half-thickness 9.1567 cm) is divided into 20 mesh intervals of equal width (0.4578 cm) and the reflector of thickness 22.5 cm is also divided into 20 mesh intervals of equal thickness (1.1250 cm). The decrease in Σ_{2F} is from 4.9002×10^{-2} to 1.3522×10^{-2}/cm over the core.

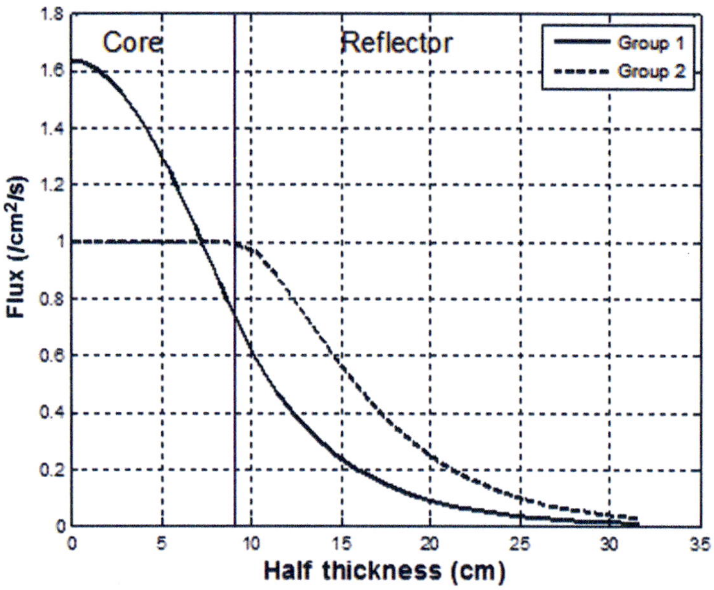

FIGURE 5.12 Two-group flux for flat thermal flux.

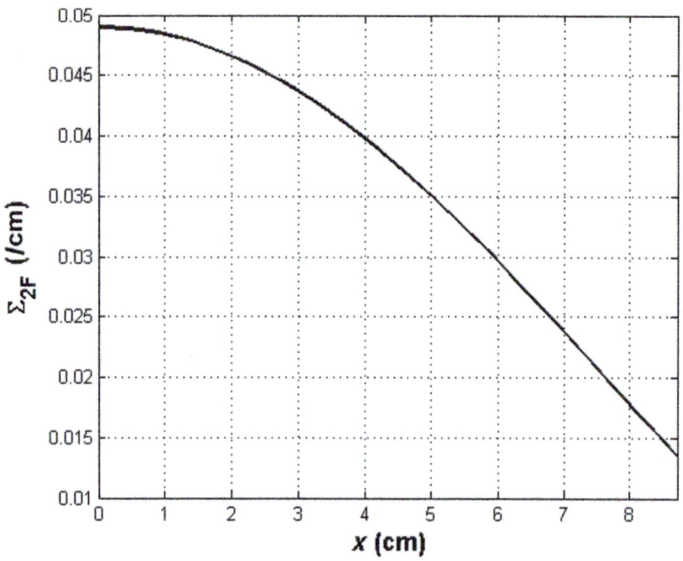

FIGURE 5.13 Fuel absorption cross-section.

The atomic density of U^{235}, N_5, is found from $N_5 = \Sigma_{2F}/\sigma_{a5}$ where the thermal fission cross-section for U^{235} is taken to be $\sigma_{f,5} = 0.886\sigma_{a,Th5}b$. The decrease in N_5 is from 8.1574×10^{19} to 2.2510×10^{19} atoms/cm^3 per mesh while the H_2O-to-U^{235} ratio increases from ~410 to ~1485 at the periphery. With this, the fuel mass decreases per mesh from 14.572 to 4.02 mg/cm^2 with a total fuel mass 0.2118 g/cm^2 compared with a fuel mass of 0.2106 mg/cm^2 for the case of uniform loading; hence no significant mass reduction.

With a cross-section area (depth times height) 1 cm^2 the volume of the core would be 4.0787 cm^3 so that the fuel concentration would be 3.3589 mg/cm^3 or 3.3589 g/L. These results are shown in Figs. 5.13–5.16.

A MCNP5 simulation was carried out with fuel loading of Fig. 5.16 shown in Fig. 5.17.

For a slab of thickness 9.1567 cm, with reflector thickness 22.5 cm, height and width 20 cm with reflecting surfaces, with varying N_5, $N_H = 6.7 \times 10^{22}$ and $N_O = 3.35 \times 10^{22}$ atoms/cm^3 with 1000 histories per cycle and 500 cycles with 10 skip cycles. Tallies were accumulated in two groups: group 1 (0.625 eV–1 MeV) and group 2 (0–0.625 eV) for which the flux are shown in Fig. 5.17. The result for this simulation was: $k_{\text{eff}} = 1.00282(0.0008)$, with a simulation time of 4.52 min on an intel processor.

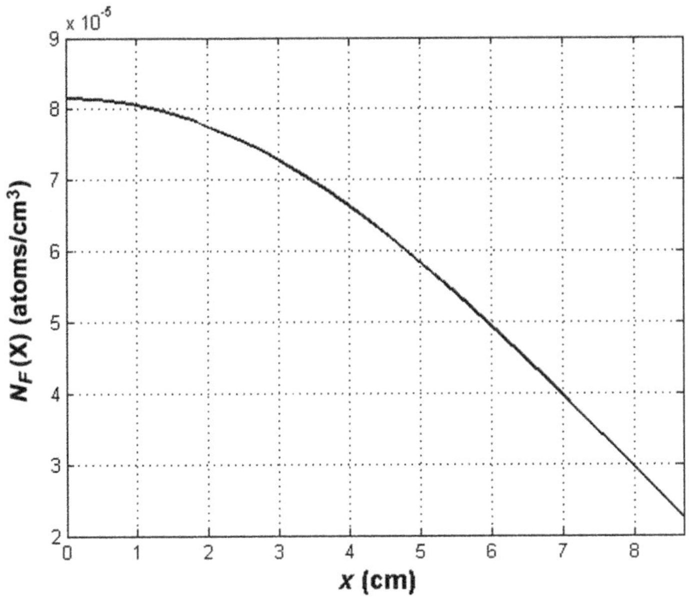

FIGURE 5.14 Fuel atomic density.

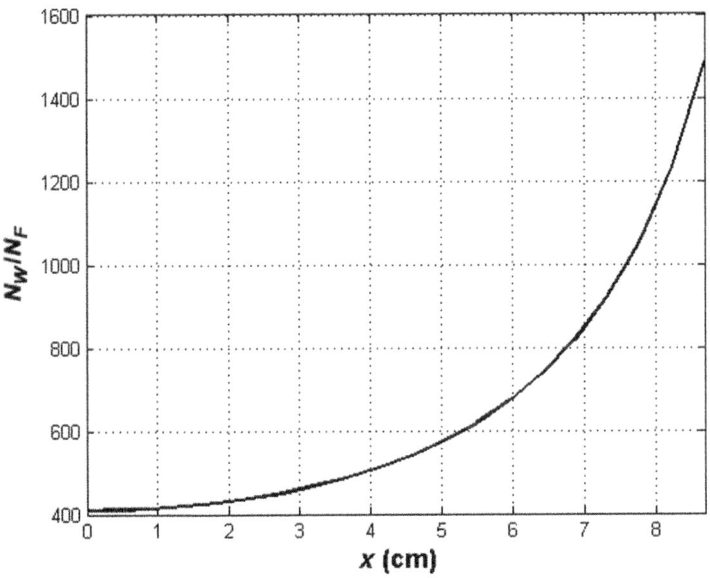

FIGURE 5.15 Water-to-fuel ratio.

The total fuel mass is 84.5594 g, equivalent to 0.2114 g/cm^2, which matches very well with the diffusion estimate.

The effect of reducing the core size was found with the truncated flat flux fuel distribution taken till 7.7832 cm, then continued for mesh intervals at 8.0, 8.02, and 8.06 cm. The mass reduction was 4.5066 g resulting in $k_{\text{eff}} = 0.98171(0.00087)$. Addition of material toward the end is thus required. With the atomic density shown in Fig. 5.18, that is, with a total fuel mass of 84.6945 g equivalent to 0.2117 g/cm^2, the system has a multiplication $k_{\text{eff}} = 1.00074(0.00083)$ with flux shown in Fig. 5.19.

The addition of fuel at the edges are estimated by MC reruns which is inefficient and recommended to be carried out by sampling for derivatives in a MC perturbation simulation (Koreshi & Lewins, 1990; Rief, 1984).

Thus, a uniform fuel distribution gave a critical half-thickness and fuel mass of 8.06595 cm, 0.2107 g/cm^2 with diffusion and 8.06 cm, 0.2106 g/cm^2 $k_{\text{eff}} = 0.99544(0.00080)$ with MCNP5 with the characteristic "hump" in the thermal flux at the core-reflector interface. With FTF in the core, it was found that the critical half-thickness and fuel mass were 9.1567 cm, 0.2118 g/cm^2 with diffusion and for the same thickness, fuel mass was 0.2106 g/cm^2 $k_{\text{eff}} = 1.00282(0.00080)$ with MCNP5.

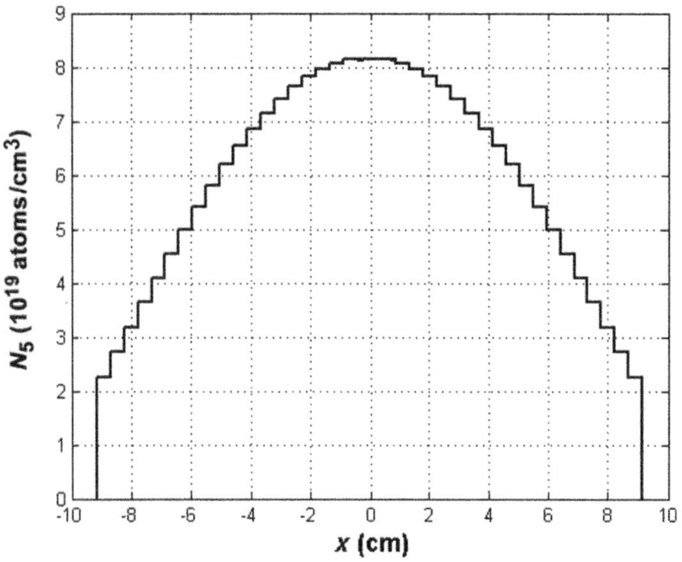

FIGURE 5.16 Fuel loading for flat thermal flux.

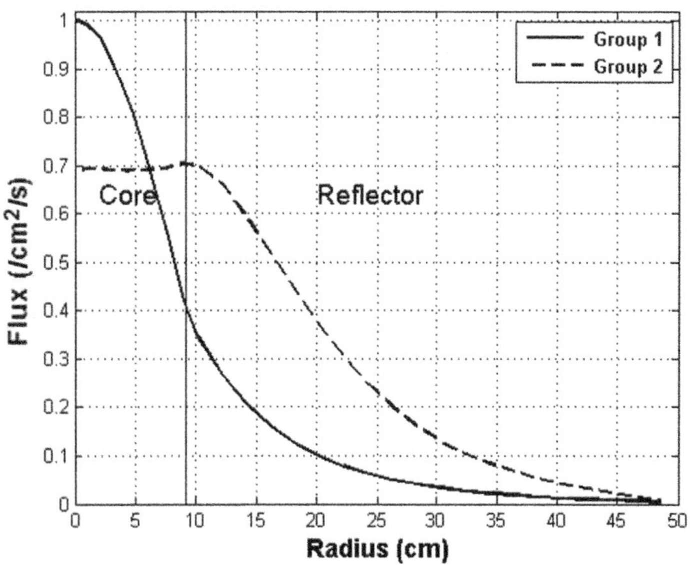

FIGURE 5.17 MCNP flux with flat thermal flux fuel loading. *MCNP*, Monte Carlo N-Particle.

5.5.3 Nonuniform fuel distribution: a spherical model

In a spherical homogenous U^{235}-H_2O core reflected with water, two-group diffusion theory is used for estimating the critical size given an initial composition. Subsequently, solutions are obtained for a FTF in the core to obtain a fuel distribution which gives the diffusion estimate of MCM. These estimates are validated with MC simulations using the MCNP5 code which shows the requirement of a "δ-source" at the core-reflector interface. In practical terms, the amount of fuel mass to be added in a finite region at the interface is estimated by sensitivity coefficients sampled from MC perturbation derivatives. These are used to add fuel at the interface to estimate the MCM. Diffusion theory with uniform composition gives a critical core of 21.0 cm radius with 1.15 kg fuel while a flat thermal flux gives 15.7668 cm compared with MCNP which gives a critical mass of 718.067 g with a "corrected" fuel distribution. While Goertzel's original work found the MCM to be 30% less than that for uniform fuel distribution, detailed simulations find a reduction by almost ~38%.

This section considers a spherical core surrounded with water reflector. The core is a dilute U^{235}-H_2O solution initially with a water-to-fuel atomic ratio 500 to 1. For an exact solution with uniform fuel loading, followed by a flat

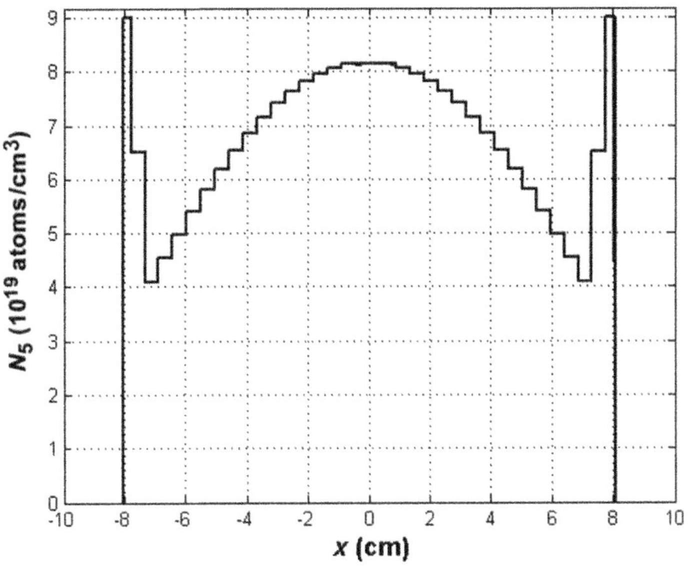

FIGURE 5.18 Atomic density with addition at periphery.

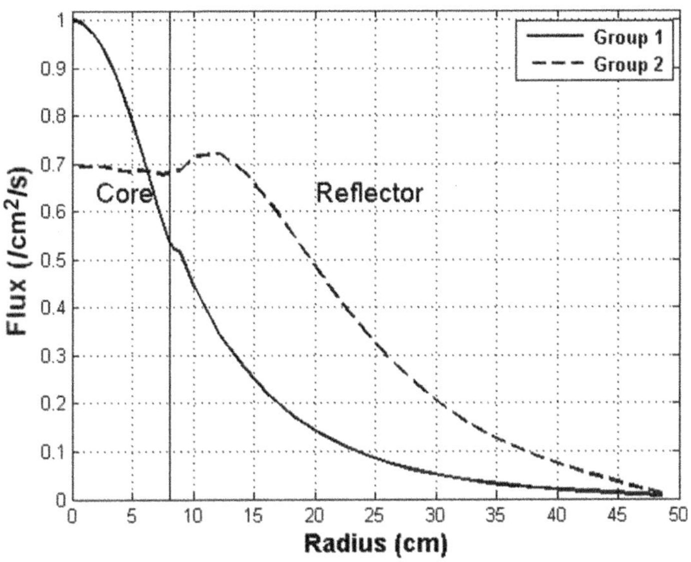

FIGURE 5.19 Thermal flat flux in core with reduced size.

thermal flux (FTF) requirement, two-group diffusion equations are used. An "optimal" fuel loading is thus obtained and used in a MC simulation.

For a uniformly distributed fuel, Section 5.3.2, the transcendental criticality equation for radius R *is*

$$\zeta_X = \frac{\alpha_1\,\zeta_Y\zeta_{Z2} + \alpha_2\zeta_Y\zeta_{Z1} + \alpha_3\zeta_{Z1}\zeta_{Z2}}{\alpha_4\zeta_Y + \alpha_5\zeta_{Z1} + \alpha_6\zeta_{Z2}} \tag{5.89}$$

with

$$\zeta_X \equiv \frac{X'}{X} = -\mu\left(\frac{1}{\mu R} - \cot\mu R\right)$$

$$\zeta_Y \equiv \frac{Y'}{Y} = \lambda\left(\coth\lambda R - \frac{1}{\lambda R}\right)$$

$$\zeta_Z \equiv \frac{Z'}{Z} = -\kappa\left(\frac{1}{\kappa R} + \coth\kappa b\right)$$

and

$$\alpha_1 = D_{1c}D_{2r}(S_3 - S_1), \alpha_2 = D_{1r}D_{2c}S_2 - D_{1c}D_{2r}S_3, \alpha_3 = D_{1r}D_{2r}(S_1 - S_2)$$
$$\alpha_4 = D_{1c}D_{2c}(S_2 - S_1), \alpha_5 = (D_{1r}D_{2c}S_1 - D_{1c}D_{2r}S_3), \alpha_6 = D_{1c}D_{2r}(S_3 - S_2).$$

For flat thermal flux $\phi_{2c} = \phi_O$ in the core

$$\phi_{1c}(r) = AX + \eta\frac{\Sigma_{2Mc}}{\gamma^2 D_{1c}}\phi_o \tag{5.90}$$

where

$$X = \frac{\sin\gamma r}{r}.$$

The constant A in Eq. (5.90) can be found from the core-reflector interface conditions

$$AX + \eta\frac{\Sigma_{2Mc}}{\gamma^2 D_{1c}}\phi_o = FZ_1 \tag{5.91}$$

and

$$D_1C_1X' = D_{1r}FZ_1'. \tag{5.92}$$

Also, $F = \alpha_1 G, \text{and} G = \alpha_2\phi_o$ where

$$\alpha_1 = -\frac{Z_2\,\zeta_{Z2}}{Z_1\,\zeta_{Z1}}$$

and

$$\alpha_2 = \frac{Z_1}{Z_2}\quad\frac{\zeta_{Z1}}{\zeta_{Z1} - \zeta_{Z2}}.$$

The criticality condition is then

$$R = \frac{1}{\gamma}\tan^{-1}\frac{\gamma D_{1c}}{D_{1c} + D_{1r}f(R)} \tag{5.93}$$

where

$$f(R) = \frac{\zeta_{Z1}\zeta_{Z2}}{\beta\zeta_{Z1} + (1-\beta)\zeta_{Z2}} \tag{5.94}$$

and

$$\beta = S_3\eta\frac{\Sigma_{2Mc}}{\gamma^2 D_{1c}}$$

leading to a fuel loading

$$\Sigma_{2F}(r) = A\frac{\sin\gamma r}{r} + \frac{\Sigma_{2Mc}}{\eta - 1}. \tag{5.95}$$

The fuel mass for both cases, uniform loading and FTF, are then

$$m_{F,o} = \alpha\Sigma_{F,o}V_c \tag{5.96}$$

and

$$m_F = \frac{\alpha}{\eta - 1}\left(\frac{-4\pi D_{1r}}{\phi_o}A\sin\gamma Rf(R) + \Sigma_{2Mc}V_c\right). \tag{5.97}$$

The two-group data for uniform loading used is: $D_{1c} = 1.13$ cm, $\Sigma_{1c} = 0.0419$/cm, $D_{2c} = 0.16$ cm, $\Sigma_{2c} = 0.060$/cm, $\sigma_{aW} = 0.664$ b, $\sigma_{a25} = 678$ b, $\tau_c = 27$ cm^2, $\eta_T = 2.07$. It is also assumed that $D_{1c} = D_{1r}$, $D_{2c} = D_{2r}, \tau_c = \tau_r$. Then, $\mu = 0.1128, \lambda = 0.6509$.

All MC simulations were performed on an Intel (R) Core (TM) i7–2620M CPU @2.70 GHz processor, 32-bit Operating System with 1000 histories per cycle and 500 cycles with 10 skip cycles. Tallies were accumulated in two energy groups: group 1 (0.625 eV–1 MeV) and group 2 (0–0.625 eV).

With a H_2O-U^{235} ratio 500:1 for uniform fuel loading ($N_5 = 6.6848 \times 10^{19}$ atoms/cm^3, $N_W = 3.342 \times 10^{22}$ molecules/cm^3) the criticality equation gives critical radius 21.9 cm with the two-group flux shown in Fig. 5.8 for which the fuel mass is 1.15 kg.

Subsequently, for a core of radius 22.5 cm with a 22.5 cm thick water reflector a MCNP5 simulation with $N_5 = 6.7 \times 10^{19}$, $N_H = 6.7 \times 10^{22}$ and $N_O = 3.35 \times 10^{22}$ atoms/cm^3 for which the flux are shown in Fig. 5.20. The result for this simulation was: $k_{\text{eff}} =$ 1.001364 (0.0011) with a simulation time of 6.52 min.

For a uniform fissile distribution, the critical radius and mass obtained from diffusion theory are close to the estimates obtained from MCNP: radius 22.5 cm and mass 1.2477 kg U^{235} (0.0262 kg/L).

5.5.4 Critical core with flat thermal flux loading

For FTF, the critical radius, from the criticality equation was found to be 15.7668 cm with flux shown in Fig. 5.21.

The core and reflector regions are divided into 20 mesh intervals of equal volume in-core and reflector, respectively. The decrease in Σ_{2F} is from 0.1725 to 0.0266/cm over the core.

The atomic density of U^{235}, N_5, is found from $N_5 = \Sigma_{2F}/\sigma_{a5}$ where the thermal fission cross-section for U^{235} is taken to be $\sigma_{f,5} = 0.886\sigma_{a,Th5}b$. The decrease in N_5 is from 2.8721×10^{20} to 4.4355×10^{19} atoms/cm^3 per interval while the H_2O-to-U^{235} ratio increases from ~116 to ~754 (~1081 at the interface) at the periphery. These results are shown in Figs. 5.22–5.25.

MC simulation with MCNP5 with the above fuel loading discretized as shown in Fig. 5.26 gives the fluxes shown in Fig. 5.27.

For a sphere of radius 15.7668 cm, with reflector thickness 22.5 cm, with varying N_5, the flux are shown in Fig. 5.9. For this simulation, $k_{\text{eff}} = 0.99447(0.0011)$, with a simulation time of 4.02 min. The value of k_{eff} indicates a requirement of more fuel in the core.

The total fuel mass is 702.8952 g, or 0.0428 kg/L.

Now consider the effect on the criticality of addition of fuel toward the interface to correct the k_{eff}. To estimate the amount of fuel to be added at the interface, sensitivity coefficients are obtained for a relative mass change of $\delta m/m_o = \pm 5\%$ with the uniform fuel distribution taken as the "reference" design point. The relative importance of the intervals, for a mass increase, is maximized at the center and steadily decreases toward the interface where it is seen to increase, suggesting the addition of mass in the last interval, the δ-source addition. For a 5% fractional increase, in the interface interval, the increase in system multiplication is $\delta k_{\text{eff}} \sim 9.66 \times 10^{-4}$.

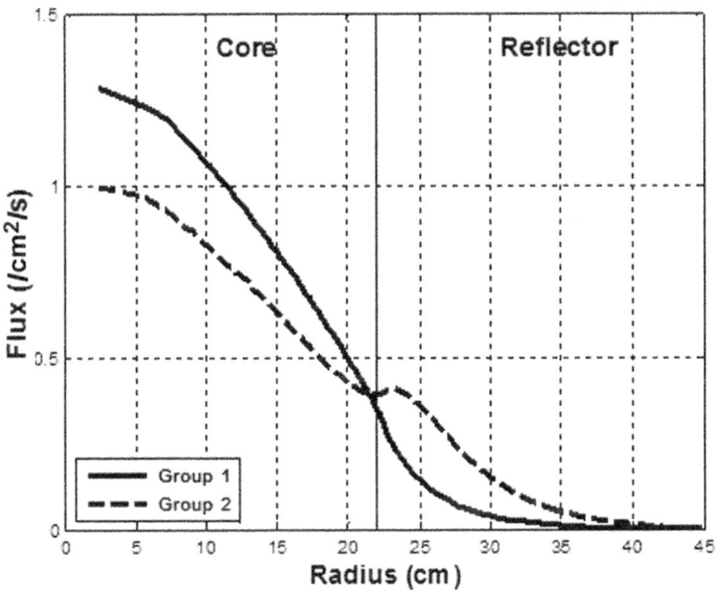

FIGURE 5.20 MCNP flux uniform distribution. *MCNP*, Monte Carlo N-Particle.

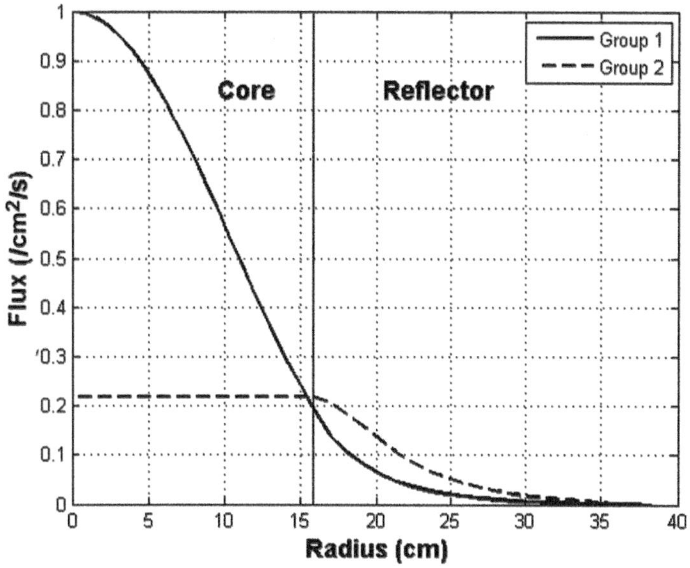

FIGURE 5.21 Two-group flux for flat thermal flux (spherical reactor).

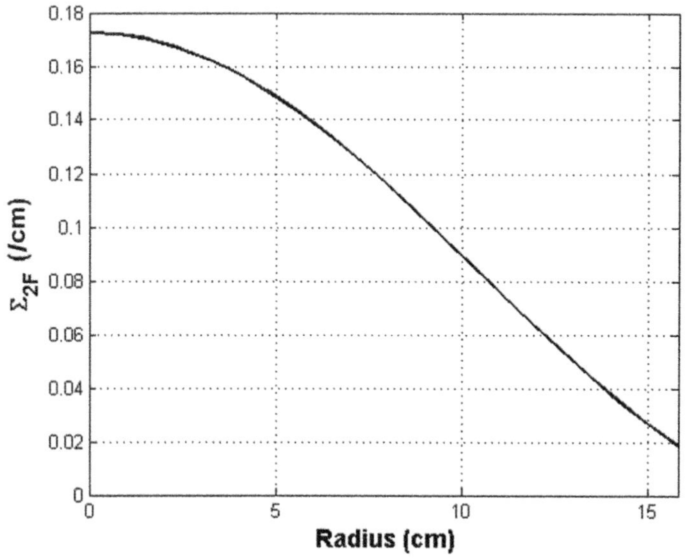

FIGURE 5.22 Fuel absorption cross-section.

The addition of 15.1695 g fuel in the last interval increases the critical mass to 718.067 g U^{235} and the system multiplication from $k_{\text{eff}} = 0.99447(0.0011)$ to 0.999495 (0.0013), that is, $\delta k_{\text{eff}} \sim 5 \times 10^{-3}$ which corresponds to a ~25% fractional increase in the mass of the interval at the interface.

It is also reported in the literature that for "optimum moderation" the smallest critical mass in a water reflected spherical U^{235} aqueous solution is 0.784 kg. Table 5.8 shows the results from diffusion and MC simulation; with MC giving 1.2477 kg U-235 compared with 1.150 kg from two-group diffusion theory. The flat thermal flux (FTF) reduces the critical mass to 0.7029 kg U-235.

5.6 The two-group adjoint diffusion equations

For the one-group diffusion equation which is a second-order ODE the operator $\hat{L}$ is self-adjoint, that is, $\hat{L}^+ = \hat{L}$. In contrast, the two-group equations, with $\chi_1 = 1, \chi_2 = 0, \Sigma_{f,1} = 0$ are

$$\left(-\nabla \cdot D_1 \nabla + \Sigma_{r,1}\right)\phi_1 - \nu\Sigma_{f,2}\phi_2 = 0 \tag{5.98}$$

$$\left(-\nabla \cdot D_2 \nabla + \Sigma_{r,2}\right)\phi_2 - \Sigma_{s,12}\phi_1 = 0 \tag{5.99}$$

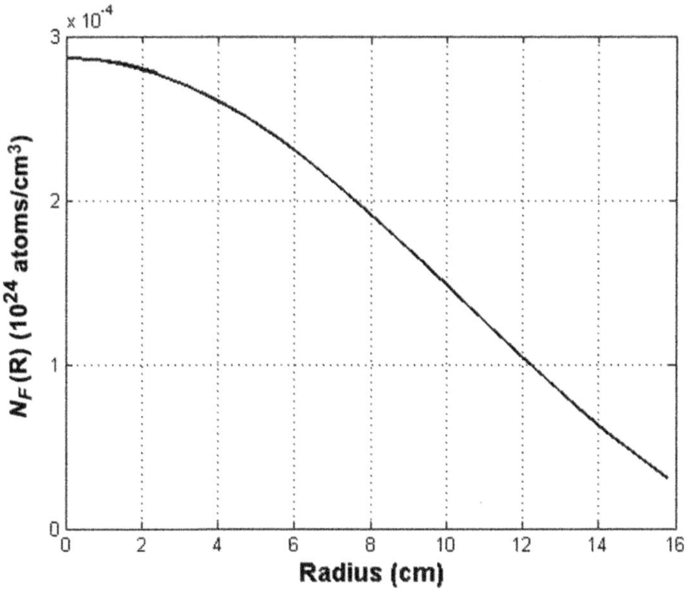

FIGURE 5.23 Fuel atomic density.

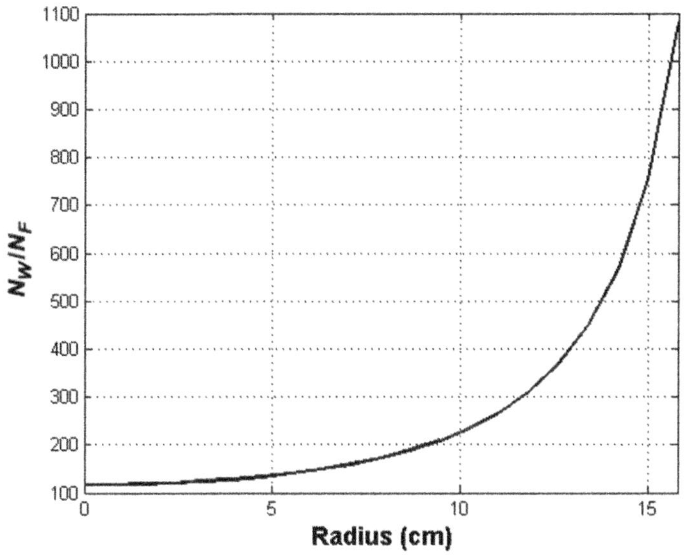

FIGURE 5.24 Water-to-fuel ratio (spherical reactor).

and with $\Sigma_{s,12} = p\Sigma_{s1}$, the above are written as

$$\begin{bmatrix} \text{div} \quad D_1 \quad \text{grad} - \Sigma_{r,1} & \nu\Sigma_{f,2} \\ p\Sigma_{s1} & \text{div} \quad D_2 \quad \text{grad} \end{bmatrix} \begin{pmatrix} \phi_1 \\ \phi_2 \end{pmatrix} = 0 \tag{5.100}$$

or, in matrix form,

$$\bar{\bar{\text{M}}}\bar{\phi} = 0 \tag{5.101}$$

In this case, it is observed that the adjoint operator $\bar{\bar{\text{M}}}^+$ is the transpose of $\bar{\bar{\text{M}}}$, that is, the rows and columns are interchanged. Thus

$$\bar{\bar{\text{M}}}^+ = \begin{bmatrix} \text{div} \quad D_1 \quad \text{grad} - \Sigma_{r,1} & \nu\Sigma_{f,2} \\ p\Sigma_{s1} & \text{div} \quad D_2 \quad \text{grad} \end{bmatrix}^T = \begin{bmatrix} \text{div} \quad D_1 \quad \text{grad} - \Sigma_{r,1} & p\Sigma_{s1} \\ \nu\Sigma_{f,2} & \text{div} \quad D_2 \quad \text{grad} \end{bmatrix} \tag{5.102}$$

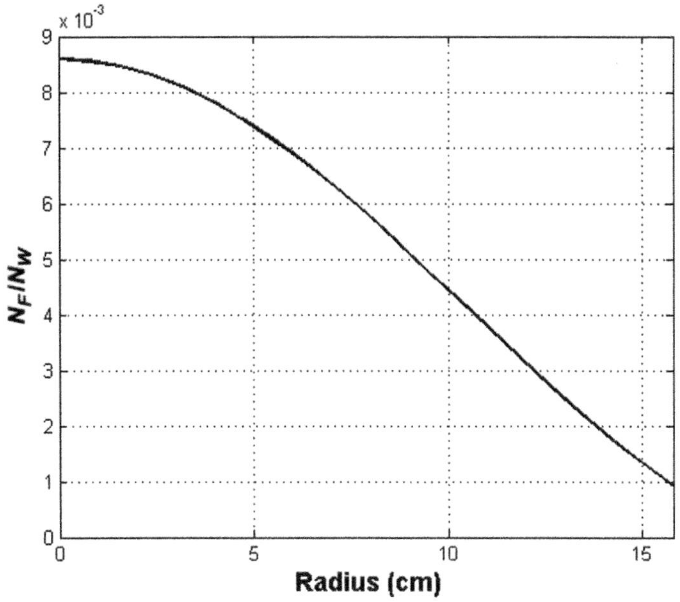

FIGURE 5.25 Fuel-to-water ratio.

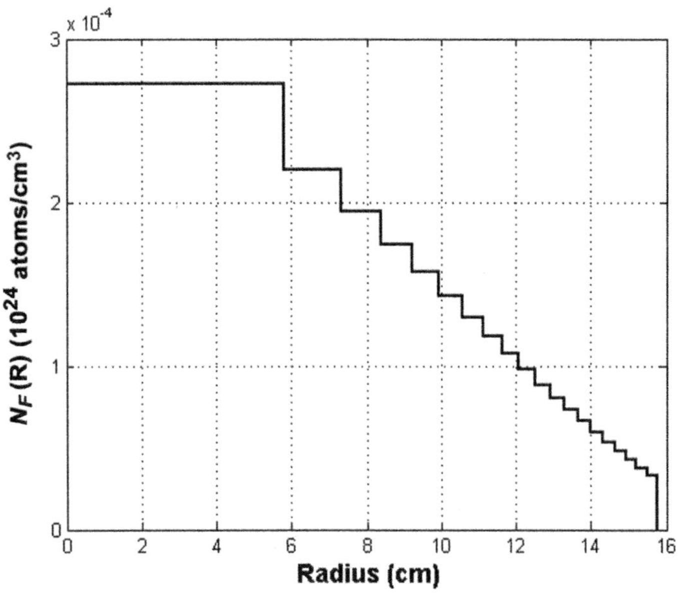

FIGURE 5.26 Fuel loading for flat thermal flux (spherical reactor).

and the set of equations is thus

$$\begin{bmatrix} \mathrm{div}D_1\mathrm{grad} - \Sigma_{r,1} & p\Sigma_{s1} \\ \nu\Sigma_{f,2} & \mathrm{div}D_2\mathrm{grad} \end{bmatrix} \begin{pmatrix} \phi_1^+ \\ \phi_2^+ \end{pmatrix} = 0 \tag{5.103}$$

The adjoint equations are thus, like the forward equations, coupled ordinary second-order differential equations which satisfy the same boundary conditions as the flux, that is, flux and current continuity at a physical interface and vanishing flux at the extrapolated boundary. The adjoint flux is an *importance function* which has physical significance in sensitivity analysis as will be demonstrated for variational formulations used in sensitivity analysis and optimization.

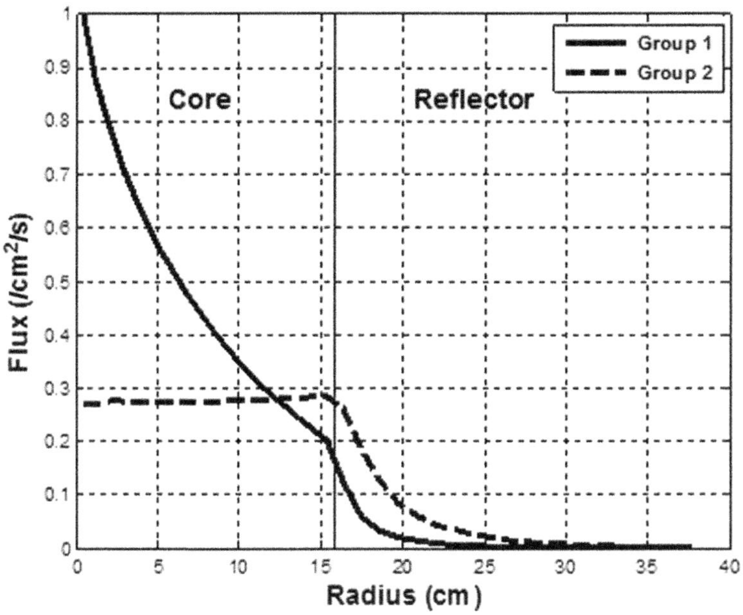

FIGURE 5.27 MCNP flux with flat thermal flux fuel loading (spherical reactor). *MCNP, Monte Carlo N-Particle.*

TABLE 5.8 Criticality estimates for a spherical reactor.

Criticality	MCNP			2-group diffusion
	FTF δ − source	Flat thermal flux (FTF) fuel dist.	Uniform	Uniform
R_c (cm)	15.7668	15.7668	22.5	21.9
ΔR_r (cm)	22.5	22.5	22.5	18.1
k_{eff}	0.999495 (0.0013)	0.99447 (0.0011)	1.001364 (0.0011)	1.0
M (kg)	0.718067	0.7028952	1.2477	1.150

5.7 Core neutronics with diffusion equations

Core neutronics calculations are performed over the whole core consisting of fuel assemblies which have fuel rods as well as control rods in the presence of a moderator or coolant. The most basic unit of an assembly is the unit lattice cell. This cell is representative of the fuel and moderator; as an entity it repeats itself within an assembly.

For the AP1000 reactor (Table 3.3) the core, shown in Fig. 5.28 has an equivalent diameter 3.04 m and an active core height 4.267 m.

Within the core are 157 assemblies with 17 × 17 rods in an arrangement represented in Fig. 5.29.

Each assembly, Fig. 5.30, is of dimensions 21.40204 × 21.40204 cm.

Within an assembly a *unit cell* of dimensions 1.26 × 1.26 cm, shown in Fig. 5.31, typically consists of fuel, IFBA (integral fuel burnable absorber), a helium gap, and cladding.

The fuel pellet is of radius 4.09575 mm, the IFBA is of thickness 0.00518 mm, the helium gap is of thickness 0.59202 mm, and the clad thickness is 0.05715 mm. Thus the outer fuel radius is 4.7475 mm (diameter 0.94950 cm). The unit cell represents basic nuclear characteristics of the fuel assembly; fluxes and reaction rates are determined within the unit cell. Lattice parameter programs are written to carry out calculations within this cell. The cell is typically divided into three regions, namely the fuel, the gap and the coolant. Within a cell, detailed calculations are performed to obtain its overall parameters which, in the second stage are used to carry out whole-core calculations. If fuel elements are placed at the corners of squares, the lattice is called a square lattice. In some configurations, hexagonal lattices are modeled when fuel elements are placed at vertices of a hexagon. Heterogeneity within a reactor core, for example when strong absorbers are placed within a

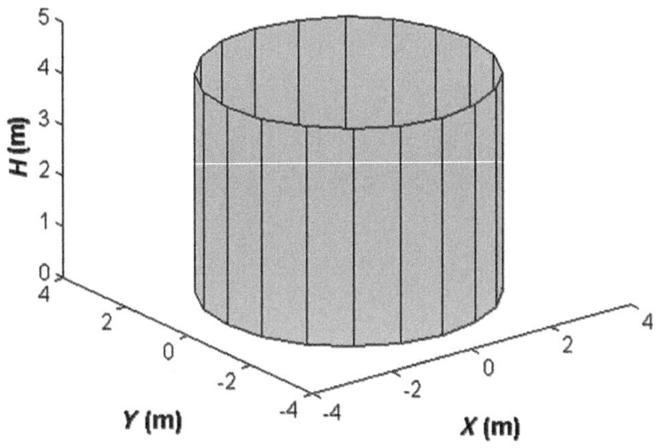

FIGURE 5.28 AP1000 core (equivalent radius 3.04 m, active core height 4.267 m).

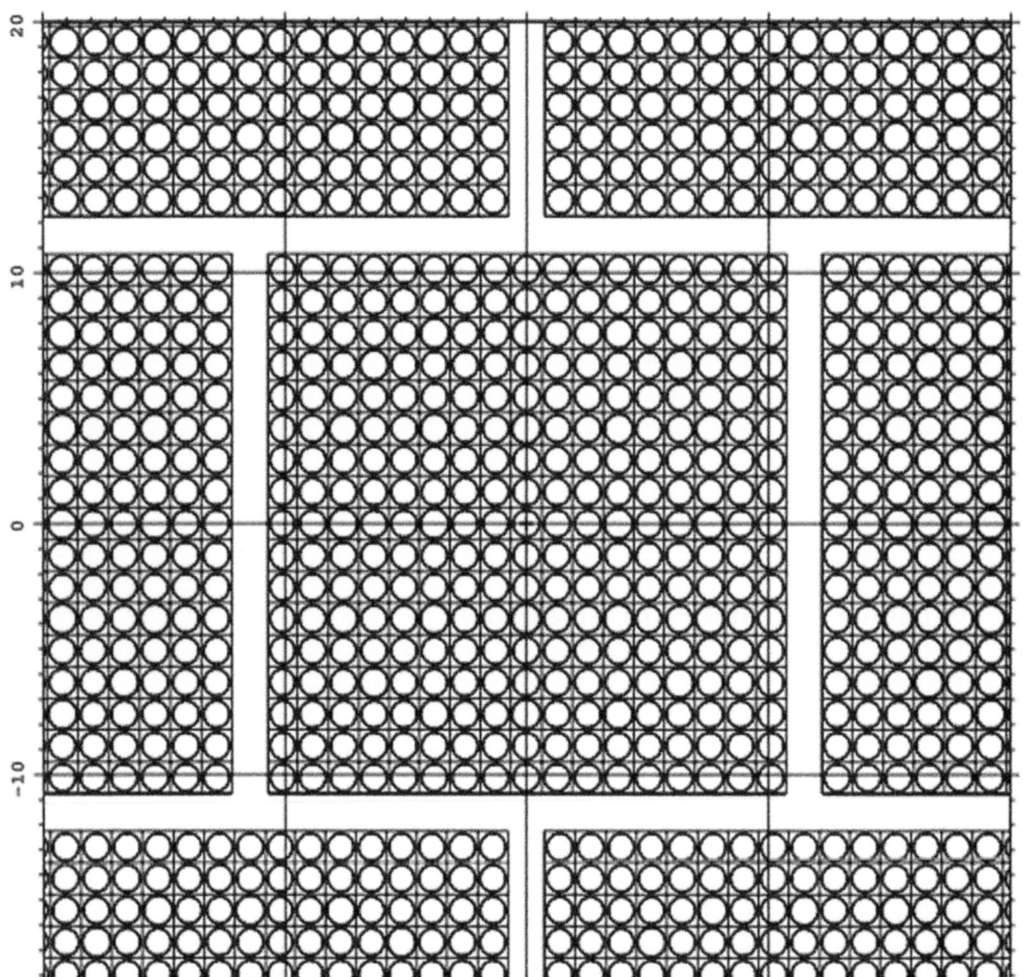

FIGURE 5.29 A typical PWR assembly with 17 × 17 rods. *PWR*, pressurized water reactor.

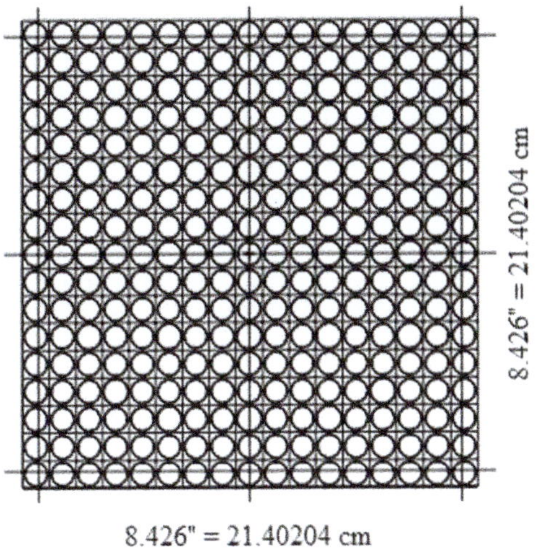

FIGURE 5.30 Fuel assembly of AP1000.

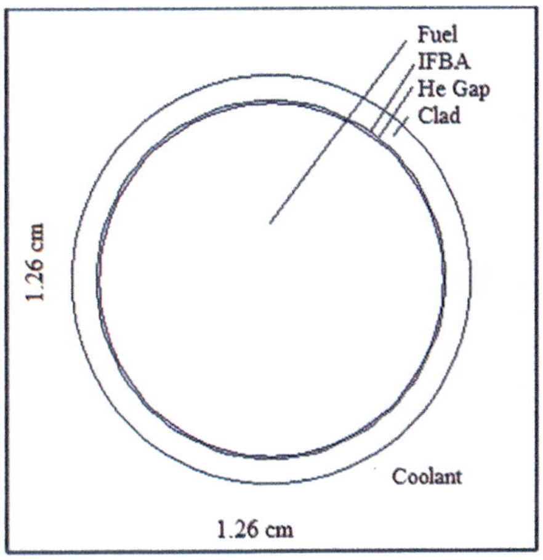

FIGURE 5.31 Unit cell of AP1000.

core, is modeled by the use of macro cells. A detailed transport calculation is carried out for a cell and in the second stage, a quarter model of the core, as shown in Fig. 5.32, is used with a less cumbersome diffusion model.

Codes produce lattice parameters to study the sensitivity of burnup, for example, to changes in operating conditions.

In a detailed calculation, starting from an initial configuration, a transport equation solution normalized to the given reactor power, is used for small time scales using the point reactor kinetics equations and for the long time scale using the burnup equation to calculate the isotopes produced (Section 4.4), and subsequently the new overall reactivity. A code such as WIMSD (Deen & Woodruff, 1995) is a typical reactor lattice code which at the first level, calculates the pin cell parameters in a 69-group computation. These are used for few-group calculations with detailed geometry models, followed by burnup calculations in an iterative procedure. In WIMSD, unit cell calculations are carried out with the integral transport equation using collision probability methods. One such code is Serpent (Leppänen, Pusa, Viitanen, Valtavirta, & Kaltiaisenaho, 2015), a multipurpose 3D continuous energy MC particle transport code distributed by OECD/NEA for neutron/photon transport with burnup calculation capability and coupled multiphysics simulations to generate homogenized few-group reaction cross-sections among several other parameters.

Here, an insight is given on the use of the NDE to obtain core fluxes and power distribution for which methods include the FDM, the coarse mesh FDM, and the NEM for structured meshes. In the case of unstructured meshes required for complex geometries, FEM and the FVM are used.

In the FDM methods for the NDE, the DIF3D code developed at Argonne National Laboratory and used for decades comes upgraded with diffusion and transport solvers (Nelson, Smith, & Heidet, 2021). In the nodal methods, for example, a coarse mesh analysis is carried out, with one node representing a fuel assembly; this is preferred over the use of FDM and FEM due to the computational efficiency of nodal methods though at the expense of detailed information especially if the heterogeneity is large, that is, when the number of control rods in an assembly is large. Some of the widely used nodal codes developed for thermal reactors are KIKO3D (Keresztúri et al., 2003), DYN3D (Rohde et al., 2016) and RAST-K (Park et al., 2020).

In one approach (Trkov, Najžer, & Škerget, 1990), the 2D (xy) diffusion equation for the quarter core of Fig. 5.32 is integrated over the y axis to obtain a 1D equation for each energy group. The nodal fluxes and currents are then obtained analytically. In another approach, the Green;s function is used for a solution of the NDE in which continuity conditions result in a system of equations giving nodal fluxes and currents.

The multigroup NDE for energy group g

$$-\nabla \cdot D_g(r)\nabla\phi_g(r) + \Sigma_{r,g}(r)\phi_g(r) = f_g(r) \tag{5.104}$$

with the "source" term

$$f_g(r) = \sum_{g' \neq g} \Sigma_{s,g' \to g}\phi_{g'}(r) + \frac{\chi_g}{k_{\text{eff}}}\sum_{g'} \nu\Sigma_{f,g'}(r)\phi_{g'}(r). \tag{5.105}$$

in 2D rectangular homogeneous zones, are integrated over the transverse direction y and using Fick's law, a set of one-dimensional equations are obtained for the quantities

$$\phi_{x(j)}(x) = \frac{1}{h_{(j)}}\int_{V_{j-1}}^{V_j} \phi(x,y)dy. \tag{5.106}$$

For a node $V_{(i,j)}$ for column i, $x_{i-1} \le x \le x_i$ width h_i and row j, $y_{j-1} \le y \le y_j$ height h_j, integration gives the transverse leakage

$$L_{y(j)}(x) = \frac{1}{h_{(j)}}\int_{V_{j-1}}^{V_j} -\frac{\partial^2}{\partial y^2}\phi(x,y)dy = \frac{n_y}{h_{(j)}}\left[\frac{J(x,y_j) - J(x,y_{j-1})}{D_{(j)}}\right] \tag{5.107}$$

resulting in the set of equations

$$-D_{(ij)}\frac{d^2\phi_{x(j)}}{dx^2} + \Sigma_{(ij)}\phi_{x_{(j)}} = f_{x(j)} - D_{(ij)}(r)L_{y(j)}. \tag{5.108}$$

Eq. (5.108) are then solved to obtain the nodal fluxes $\phi_{x_{(j)i}}$ and current $J_{x(j)i}$ on the node boundaries i. These give the average leakage in the x direction

$$L_{x(ij)} = \frac{1}{h_{(i)}}\left[\frac{J_{x(j)i} - J_{x(j)i-1}}{D_{(ij)}}\right]. \tag{5.109}$$

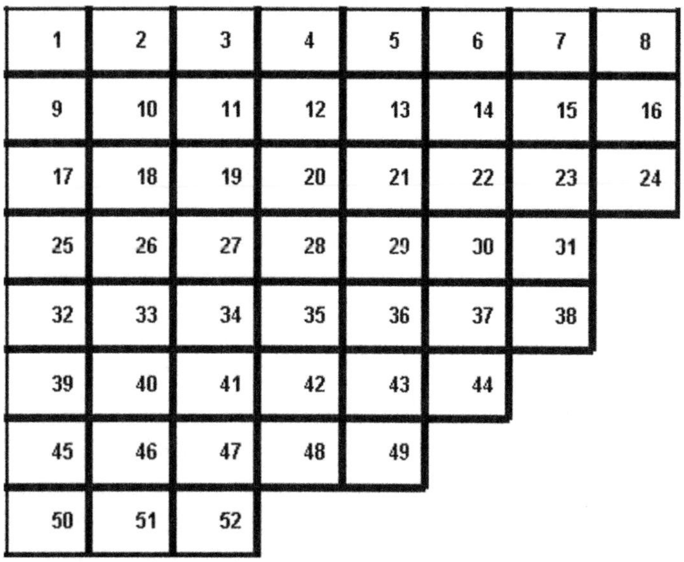

FIGURE 5.32 One-fourth model of a reactor core with 52 assemblies.

The integrations are performed in the x direction over a node, with a weighting function $w(x)$ which is the Green's function to the problem. This reduces the integrals to an algebraic expression involving the flux ϕ_{i-1}, ϕ_i and average current J_{i-1} and J_i in the node and at boundaries. All such equations for a a row along the x direction, applying continuity conditions for interfaces and external boundaries, are assembled into a global matrix and solved for fluxes and currents. The same procedure is repeated by integrating over the transverse direction.

In this method, there are no trial functions or a priori assumptions about the shape of the neutron flux.

These methods have been demonstrated for IAEA 2D and 3D PWR BSS-11 and BIBLIS benchmarks with a 20 cm coarse mesh giving overall system multiplication and power distribution within 4%–5%.

An application of an intermediate mesh method, such as FEM, with a coarse method, such as the Spectral Green's Function (SGF) Nodal Method (Rocha, Dominguez, Iglesias, & de Barros, 2016), has been demonstrated to solve the one-group multiplying media NDE for solving a benchmark problem with a high degree of heterogeneity. The equations

$$\frac{dJ(x)}{dx} + \Sigma_a(x)\phi(x) = \frac{1}{k_{\text{eff}}}\nu\Sigma_f(x)\phi(x) \tag{5.110}$$

and

$$J(x) = -D(x)\frac{d\phi(x)}{dx} \tag{5.111}$$

are discretized over a spatial domain $0 \le x \le L$, with albedo boundary conditions $J_{1/2} = -\alpha_L\phi_{1/2}$, and $J_{I+1/2} = \alpha_R\phi_{I+1/2}$ on both ends of a grid of I mesh intervals. The above balance equations are multiplied by the Legendre polynomial

$$P_l\left(\frac{2(x-x_i)}{h_i}\right)$$

integrated over an element $x_{i-1/2} \le x \le x_{i+1/2}$ to get averaged flux $\overline{\phi}_i, \hat{\phi}_i$ and current $\overline{J}_i, \hat{J}_i$ equations for the zeroth and first moments

$$\overline{\phi}_i = \frac{1}{h_i}\int_{x_{i-1/2}}^{x_{i+1/2}} \phi(x)dx, \tag{5.112}$$

$$\hat{\phi}_i = \frac{3}{h_i}\int_{x_{i-1/2}}^{x_{i+1/2}} \left(\frac{2(x-x_i)}{h_i}\right)\phi(x)dx, \tag{5.113}$$

$$\overline{J}_i = \frac{1}{h_i}\int_{x_{i-1/2}}^{x_{i+1/2}} J(x)dx, \tag{5.114}$$

and

$$\hat{J}_i = \frac{3}{h_i}\int_{x_{i-1/2}}^{x_{i+1/2}} \left(\frac{2(x-x_i)}{h_i}\right)J(x)dx. \tag{5.115}$$

The spectral parameters α_i, β_i are incorporated into the flux and current formulation through the FEM linear functions

$$\phi(x) = \alpha_i\overline{\phi}_i + \frac{2\beta_i}{h_i}(x-x_i)\hat{\phi}_i \tag{5.116}$$

and

$$J(x) = \alpha_i J_i + \frac{2\beta_i}{h_i}(x-x_i)\hat{J}_i. \tag{5.117}$$

The resulting algebraic equations contain the spectral parameters which are found by analytic solutions first proposed by Case & Zweifel (1967). The linear system of equations for $(6I+2)$ unknowns, that is, I values each of $\overline{\phi}_i, \hat{\phi}_i, \overline{J}_i, \hat{J}_i, \alpha_i, \beta_i$ and two end points, are reduced to $(2I+2)$ unknowns, and the same number of equations, by eliminating $\overline{\phi}_i, \hat{\phi}_i, \overline{J}_i, \hat{J}_i$, since they are expressed in terms of the eigenvalues and α_i, β_i are subsequently determined. Thus neutron flux distribution is calculated by direct or iterative techniques and the system multiplication k_{eff} is calculated by the power method described in the previous section.

Rocha et al. (2016) applied the hybrid FEM-SGF method described above to a six-region benchmark and compared results with a reference 960 element FEM calculation to find that, in the presence of strong heterogeneities, the accuracy in flux distribution was within 0.01% for 30 elements in a coarse mesh. This method appears to be promising for multigroup and higher dimensional eigenvalue calculations.

For unstructured meshes, the FVM has been demonstrated (Bernal, Miró, Ginestar, & Verdú, 2014) for 2D and 3D two-group NDE applicable to LWRs with the requirement of complex geometry modeling. For 2D reactor models, the computational time is reported to be in seconds while for 3D models, the time is in hours for fine meshes. A full 3D Nodal Diffusion Code for PWRs (Park et al., 2020) performs steady-state and transient analyses with generated microscopic cross-sections for 37 isotopes, and can handle a depletion chain with 22 actinides (uranium, plutonium. neptunium, americium and curium isotopes) and 15 fission products (xenon, iodine, samarium, promethium, and neodymium) and burnable poisons. The multigroup NDE are also solved with parallel computing using the two-node nodal method with the Coarse Mesh FDM (Song, Yu, & Kim, 2018) reducing the time for quarter-core and whole-core calculations. For multiphysics full 3D calculations, the computer code, RAST-K-v2, performs calculations with four nodes for a fuel assembly with 24–46 axial nodes. It can carry out core design, incorporating neutronics, thermal hydraulics, critical heat flux, departure from nucleate boiling (DNBR) and depletion calculations for prediction of isotope inventory in spent nuclear fuel. The RAST-K-v2 has been verified and validated for several Korean PWRs including the OPR-1000, APR-1400 and Westinghouse 2-loop and 3-loop reactors. The diffusion equation models are thus valuable for carrying out realistic whole-core calculations with parallel computing capabilities.

Problems

1. Consider slabs of aluminum, boron, beryllium, carbon, and iron of thickness 3 m.f.p each with a planar source at $x=0$ emitting 10^6 neutrons/s. Plot the flux and estimate the surface leakages from their sides (Table 5.9).
2. Show that the neutron flux in a sphere of radius R with a point isotropic source at $r=0$ emitting S neutrons/s is given by

$$\varphi = \frac{S}{4\pi D \sinh\left(\frac{R+d}{L}\right)} \frac{\sinh\kappa(R+d-r)}{r}$$

Find the leakage probability from the surface of a graphite sphere of radius 2 m.f.p. with a 1 Ci ^{226}Ra-Be point source at its center emitting 10^7 neutrons/s.

3. Compare the leakage probability of the above sphere if instead of graphite, the material was boron.
4. What is the magnitude of the flux at the center of a bare cubical reactor of sides 3 m operating at a power of 100 kW(th)?
5. Calculate the geometric buckling for a cylinder of radius 1 m and height 2 m. What information does this give on the infinite multiplication of the system?
6. How would you calculate the number of absorption and scattering reactions in a graphite sphere if the atomic density of the graphite is $N(r) = N_o(1 - r/R)$ where N_o is the nominal density?

TABLE 5.9 Thermal cross-sections.

Nuclide	Atomic weight	Density (g/cm^3)	$\sigma_a(b)$	$\sigma_s(b)$
Aluminum	26.9815	2.70	0.235	1.4
Boron	10.811	2.30	759	4.0
Beryllium	9.0122	1.85	0.0092	6.14
Carbon (graphite)	12.01115	1.60	0.0034	4.75
Iron	55.847	7.87	2.53	11.0

Nomenclature

English lower case

d	extrapolation distance
k_{eff}	effective multiplication
k_{∞}	infinite multiplication
n	number density
q'''	volumetric heat generation (W/m^3)

English upper case

A_i	area
B_g	geometrical buckling
B_m	material buckling
C_i	concentration of precursor
D	diffusion coefficient
$\hat{D}$	derivative operator
J	neutron current
L	diffusion length
$\hat{L}^+$	adjoint operator
N_i	atomic density of the ith nuclide
N_i	shape function in the ith element
P_{es}	escape probability
P_l	Legendre functions
R_c	critical radius
S	source

Greek lower case

λ	decay constant
μ	cosine of angle of scattering
μ_0	average cosine of scattering angle
ϕ	flux
ϕ^+	adjoint flux
ϕ_C	complementary solution
ϕ_P	particular solution
τ	neutron age
ν	number of neutrons produced per fission
χ	fission spectrum

Greek upper case

Ω	solid angle
$\sum_a$	macroscopic absorption cross-section
$\sum_f$	macroscopic fission cross-section
$\sum_r$	macroscopic removal cross-section
$\sum_s$	macroscopic scattering cross-section
$\sum_{tr}$	macroscopic transport cross-section
$\sum_t$	macroscopic total cross-section

Abbreviations

FD	finite difference method
FEM	finite element method
FVM	finite volume method
IFBA	integral fuel burnable absorber
NDE	neutron diffusion equation
NEM	nodal expansion method

References

Bernal, Á., Miró, R., Ginestar, D., & Verdú, G. (2014). Resolution of the generalized eigenvalue problem in the neutron diffusion equation discretized by the finite volume method. *Abstract and Applied Analysis*.

Briesmeister, J. F. (2000). MCNP – A general Monte Carlo N-particle transport code," Los Alamos National Laboratory.

Carrillo, H. R. V. (2001). The source condition in neutron diffusion theory. Revista Mexicana de física.

Case, K. M., & Zweifel, P. F. (1967). *Linear transport theory*. Reading, MA: Addison-Wesley Pub. Co.

Deen, J. R., & Woodruff, W. L. (1995). *WIMS-D4M user manual Rev.0*. Illinois: Champaign-Urbana.

Goertzel, G. (2003). Minimum critical mass and flat flux. *Journal of Nuclear Energy*.

Jevremovic, T. (2009). *Nuclear principles in engineering* (2nd ed.).

Keresztúri, A., Hegyi, G., Marázcy, C., Panka, I., Telbisz, M., Trosztel, I., & Hegedüs, C. (2003). Development and validation of the three-dimensional dynamic code—KIKO3D. *Annals of Nuclear Energy*, *30*(1), 93–120. Available from https://doi.org/10.1016/S0306-4549(02)00043-9, ISSN 0306-4549.

Koreshi, Z. U., & Lewins, J. D. (1990). Two-group Monte Carlo perturbation theory and applications in fixed-source problems. *Progress in Nuclear Energy*, *24*, 27–38.

Lamarsh, J. R. (1966). *Introduction to nuclear reactor theory*. Addison-Wesley Publishing Company.

Lamarsh, J. R. (2005). Introduction to nuclear reactor theory. *Interactions*.

Lamarsh, J. R., & Baratta, A. J. (1955). Introduction to nuclear engineering. Prentice Hall.

Leppänen, J., Pusa, M., Viitanen, T., Valtavirta, V., & Kaltiaisenaho, T. (2015). The Serpent Monte Carlo code: Status, development and applications in 2013. *Annals of Nuclear Energy*.

Lewins, J. (2004). Minimum critical mass and flat flux in a 2-group model, Annals of Nuclear Energy, 31, pp. 541-576.

Nelson, A. G., Smith, M. A., & Heidet, F. (2021). Verification of the diffusion and transport solvers within DiF3D for 3D hexagonal geometries. In *Proceedings of the* EPJ web of conferences.

Park, J., Jang, J., Kim, H., Choe, J., Yun, D., Zhang, P., Cherezov, A., & Lee, D. (2020). RAST-K v2—Three-dimensional nodal diffusion code for pressurized water reactor core analysis. *Energies*, *13*(23), 6324. Available from https://doi.org/10.3390/en13236324.

Rief, H. (1984). Generalized Monte Carlo perturbation algorithms for correlated sampling and a second-order Taylor series approach. *Annals of Nuclear Energy*.

Rocha, R. V. M., Dominguez, D. S., Iglesias, S. M., & de Barros, R. C. (2016). Finite Element Method with spectral Green's function in slab geometry for neutron diffusion in multiplying media and one energy group. *TEMA (São Carlos)*.

Rohde, U., Kliem, S., Grundmann, U., Baier, S., Bilodid, Y., Duerigen, S., ... Mittag, S. (2016). The reactor dynamics code DYN3D – models, validation and applications. *Progress in Nuclear Energy*, *89*, 170–190. Available from https://doi.org/10.1016/j.pnucene.2016.02.013, ISSN 0149-1970.

Sekimoto, H. (2007). *Textbook nuclear reactor theory*. Tokyo Institute of Technology.

Song, S., Yu, H., & Kim, Y. (2018). One-node and two-node hybrid coarse-mesh finite difference algorithm for efficient pin-by-pin core calculation. *Nuclear Engineering and Technology*.

Trkov, A., Najžer, M., & Škerget, L. (1990). Variant of Green's function nodal method for neutron diffusion. *Journal of Nuclear Science and Technology*, *27*, 766–777.

Van Dam, H. (2013). Minimum critical mass in a heterogeneous thermal system. *Annals of Nuclear Energy*.

Van Dam, H. (2015). On the anomaly in sizes of minimum critical mass cores. *Annals of Nuclear Energy*.

Williams, M. M. R. (2003). Some further considerations on Goertzel's minimum critical mass problem. *Annals of Nuclear Energy*.

Chapter 6

The neutron transport equation

The neutron transport equation (NTE) is a mathematical statement on the conservation of neutrons for the angular, rather than scalar, neutron flux in phase space. It is a more accurate description of the underlying phenomena of neutron transport than the neutron diffusion equation (NDE) described in Chapter 5.

There are several mathematical formulations of the NTE such as the integral form (Section 4.3) and the integro-differential form (Section 4.4) extensively covered in standard nuclear engineering text books (Bell & Glasstone, 1979; Case & Zweifel, 1967; Clark & Hansen, 1964; Davison, 1957; Duderstadt & Hamilton, 1976; Henry, 1975; Williams, 1971). The NTE can be interpreted as the zeroth moment of the Boltzmann transport equation which Boltzmann had derived for the kinetic theory of gases (Cercignani, 1988; Harris, 1971; Robson, White, & Hildebrandt, 2017) and is also applicable to radiation transport including charged particles, photons and neutrons (Section 1.6).

This chapter begins with a derivation of the integro-differential form of the NTE and shows some simple forms of the NTE, such as the one-speed equation with isotropic scattering, the one- and two-group transport equations similar to the two-group diffusion equations, the integral form, the collision and transport formulations, and the multigroup equations.

In Section 6.2, some classical exact, analytical, closed-form solutions obtained in the early days of neutron transport, the 1950s–70s, under idealized conditions with the Fourier and Laplace transforms are discussed. Classical solutions for the for infinite, semiinfinite, and some finite media were found with simplified scattering models. These are described here for the purpose of familiarizing the reader with the early approaches for obtaining the neutron flux and eigenvalues.

The compact exact solutions are then followed by the development and application of numerical methods described in Section 6.3. The most extensively used methods are the discrete ordinates (S_N) method, and the spherical harmonics (P_N) method. These are followed by the double P_N, or the DP_N method, the B_N method and Finite Element Method (FEM) described in Section 4.5, nodal and hybrid methods. Such methods are used extensively for neutron transport computations (Kang & Hansen, 1973; Lewis & Miller, 1984) for the design of nuclear reactor cores by calculations on lattice cells, in fuel assemblies and in whole-core neutronic analyses as discussed in the context of the NDE (Section 5.6).

In Section 6.4, two widely used methods for reactor calculations, namely the Collision Probability Method (CPM) and the Method of Characteristics (MOC), are described for lattice cell calculations and for whole-core eigenvalue calculations for the design of nuclear systems.

Current research and applications of transport theory are focused on the development of efficient computational methods and algorithms on multiprocessors to carry out multidimensional core neutronics calculations for the design of Gen III + and Gen-IV nuclear reactors described in Chapter 3.

It is important to realize that neutron transport contains some of the most elegant mathematics applied in nuclear engineering, and there are several methods and techniques, each impressive in its own right. It is not possible to include a more detailed introduction in a single chapter; indeed several books are devoted to neutron transport. The reader is advised to look into the research publications in journals such as Progress in Nuclear Energy, Annals of Nuclear Energy and Nuclear Science and Engineering for the latest trends and developments. The body of research contains new developments on making transport calculations more robust, applicable and fast. The gap between deterministic and stochastic computation is thus bound to become narrower and large reactor design calculations will become easier in the years to come.

Whatever numerical and computational advancements take place, the role of elegant mathematics will never end, as the foundations of neutron transport will always lie in the compact analytical expressions of the early days which will serve as standards for huge computational codes.

Nuclear Engineering. DOI: https://doi.org/10.1016/B978-0-323-90618-0.00006-5

An understanding of neutron diffusion, neutron transport, and in the later chapters, Monte Carlo simulation are the bedrock upon which nuclear engineering will always stand. It is with this objective that the following sections must be read to understand the equations.

6.1 Structure of the neutron transport equation

There are at least eight forms in which the NTE, in addition to its most basic integro-differential form, is used (Ganapol, 2008) viz (1) integral, (2) even-odd parity, (3) slowing down kernel, (4) multiple collision, (5) invariant embedding, (6) singular integral, (7) pseudoflux, and (8) Green's function, in addition to the Monte Carlo form which, though a stochastic simulation of the transport phenomena, can be interpreted to be a Neumann series solution of the integral equation. Each one of these forms has mathematical properties that enable a class of solutions.

In this chapter, two of these eight (deterministic) forms will be considered in some detail.

6.1.1 An integro-differential form of the neutron transport equation

For a derivation of the NTE from first principles, consider a neutron with energy E at time t at some position $\vec{r} = x\hat{i} + y\hat{j} + z\hat{k}$ where $\hat{i},\hat{j},\hat{k}$ are unit vectors along the x, y, z axes moving in a direction defined by two angles θ, φ such that the unit vector is $\hat{\Omega} = \Omega_x\hat{i} + \Omega_y\hat{j} + \Omega_z\hat{k}$, where $\Omega_x = \sin\theta\cos\varphi$, $\Omega_y = \sin\theta\sin\varphi$, $\Omega_z = \cos\theta$. These parameters define its "position" $P(\mathbf{r},\mathbf{\Omega}, E, t)$ in phase space. Let's say in the next Δt seconds, it moves a distance $v\Delta t$, during which the total number of interactions it is likely to have is $\Sigma_t v\Delta t$; conversely, the number of interactions it is likely not to have are $1 - \Sigma_t v\Delta t$, which means that if there were $N(\mathbf{r},\mathbf{\Omega}, E, t)dVdEd\Omega$ neutrons in a packet initially, there are $N(\vec{r}, \Omega, E, t)[1 - \Sigma_t(\vec{r}, E, t)v\Delta t]dVdEd\Omega$ neutrons left in the packet (Fig. 6.1).

The change in the number of neutrons is thus

$$Change = N(\mathbf{r} + v\mathbf{\Omega}\Delta t,\mathbf{\Omega}, E, t + \Delta t) - N(\mathbf{r},\mathbf{\Omega}, E, t).$$

This can be written as

$$\text{Change} = N(\mathbf{r} + v\mathbf{\Omega}\Delta t,\mathbf{\Omega}, E, t + \Delta t) - N(\mathbf{r},\mathbf{\Omega}, E, t + \Delta t) + N(\mathbf{r},\mathbf{\Omega}, E, t + \Delta t) - N(\mathbf{r},\mathbf{\Omega}, E, t)$$

where the term $N(\mathbf{r},\mathbf{\Omega},E,t + \Delta t)$ has been added and subtracted to be able to express the four terms as a space derivative and a time derivative, respectively. In the limit as $\Delta t \to 0$, the first term is $v\mathbf{\Omega}\cdot\nabla N(\mathbf{r}, \mathbf{\Omega}, E, t)$; the change in the packet, per unit volume interval, per unit energy interval and per unit solid angle interval, is thus

$$v\cdot\mathbf{\Omega}\nabla N(\mathbf{r},\mathbf{\Omega}, E, t) + \frac{\partial}{\partial t}N(\mathbf{r},\mathbf{\Omega}E, t).$$

We can write $v\cdot\mathbf{\Omega}\nabla N(\mathbf{r},\mathbf{\Omega}, E, t)$ as $\nabla\cdot\mathbf{\Omega}vN(\mathbf{r},\mathbf{\Omega}, E, t)$ and with angular flux $\phi(\mathbf{r},\mathbf{\Omega}, E, t)$ and the neutron current $J(\mathbf{r},E, t)$ defined as

$$\phi(\mathbf{r},\mathbf{\Omega}, E, t) = vN(\mathbf{r},\mathbf{\Omega}, E, t) \tag{6.1}$$

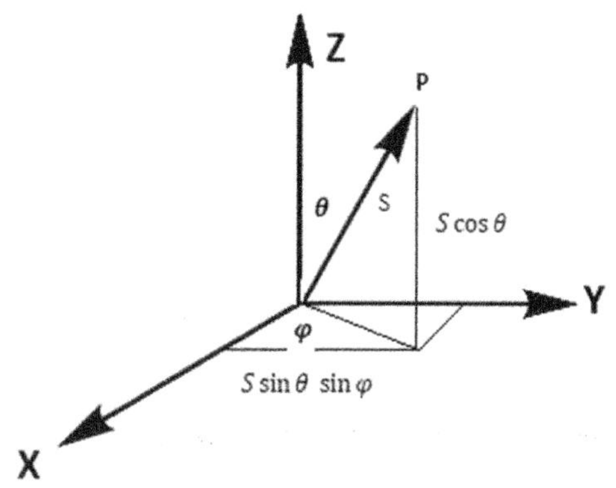

FIGURE 6.1 Coordinates in three-dimensional Cartesian axes.

and

$$J(\mathbf{r},E,t) = \int_0^{4\pi} \boldsymbol{\Omega}\phi(\mathbf{r},\boldsymbol{\Omega},E,t)d\boldsymbol{\Omega} \tag{6.2}$$

the leakage term is

$$\nabla\cdot\int_0^{4\pi} \boldsymbol{\Omega}\phi(\mathbf{r},\boldsymbol{\Omega},E,t)d\boldsymbol{\Omega} = \nabla\cdot J(\mathbf{r},E,t). \tag{6.3}$$

The change in the neutron population is due to the gains and losses expressed as

$$\text{Change} = \text{Gains} - \text{Losses} \tag{6.4}$$

The gains are due to in-scattering, fissions, an independent source, and any other gains due to inelastic scattering or other nuclear reactions:

$$\int_0^\infty dE' \int d\Omega' \Sigma_s(\mathbf{r},\boldsymbol{\Omega}'\cdot\boldsymbol{\Omega},E')\phi(\mathbf{r},\boldsymbol{\Omega}',E',t) + \frac{1}{4\pi}\chi(E)\int_0^\infty dE' \int d\Omega' \nu\Sigma_f(E')\phi(\mathbf{r},\boldsymbol{\Omega}',E',t) + S(\mathbf{r},\boldsymbol{\Omega},E,t) \tag{6.5}$$

Eq. (6.5) can be modified to separate the prompt and delayed fission neutrons by expressing the fission contribution as

$$\frac{1}{4\pi}\chi_p(E)\int_0^\infty dE' \int d\Omega'[1-\beta(E')]\nu\Sigma_f(E')\phi(\mathbf{r},\boldsymbol{\Omega}',E',t) + \frac{1}{4\pi}\sum_{i=1}^{6}\chi_j(E)\lambda_j C_j(t). \tag{6.6}$$

The losses due to the leakage of neutrons and their scattering out from the phase space of interest $P(\mathbf{r},\boldsymbol{\Omega},E,t)$ are

$$\nabla\cdot J(\mathbf{r},\boldsymbol{\Omega},E,t) + \Sigma_t(\mathbf{r},E)\phi(\mathbf{r},\boldsymbol{\Omega},E,t). \tag{6.7}$$

Now with the gain and loss terms substitued into Eq. (6.4) gives the time-dependent balance equation

$$\begin{aligned}\frac{1}{v}\frac{\partial}{\partial t}\phi(\mathbf{r},\boldsymbol{\Omega},E,t) &= \int_0^\infty dE' \int d\Omega' \Sigma_s(\mathbf{r},\boldsymbol{\Omega}'\cdot\boldsymbol{\Omega},E')\phi(\mathbf{r},\boldsymbol{\Omega}',E',t)\\ &+ \frac{1}{4\pi}\chi_p(E)\int_0^\infty dE' \int d\Omega'[1-\beta(E')]\nu\Sigma_f(E')\phi(\mathbf{r},\boldsymbol{\Omega}',E',t)\\ &+ \frac{1}{4\pi}\sum_{i=1}^{6}\chi_j(E)\lambda_j C_j(t) - [\nabla\cdot J(\mathbf{r},\boldsymbol{\Omega},E,t) + \Sigma_t(\mathbf{r},E)\phi(\mathbf{r},\boldsymbol{\Omega},E,t)] + S(\mathbf{r},\boldsymbol{\Omega},E,t).\end{aligned} \tag{6.8}$$

Eq. (6.8) is written with leakage and out-scattering on the LHS and gains on the RHS; rearranging gives the familiar form of the Linear Transport Equation (Bell & Glasstone, 1979)

$$\begin{aligned}\frac{1}{v}\frac{\partial}{\partial t}\phi(\mathbf{r},\boldsymbol{\Omega},E,t) + \nabla\cdot J(\mathbf{r},\boldsymbol{\Omega},E,t) + \Sigma_t(\mathbf{r},E)\phi(\mathbf{r},\boldsymbol{\Omega},E,t) &= \int_0^\infty dE' \int d\Omega' \Sigma_s(\mathbf{r},\boldsymbol{\Omega}'\cdot\boldsymbol{\Omega},E')\phi(\mathbf{r},\boldsymbol{\Omega}',E',t)\\ + \frac{1}{4\pi}\chi_p(E)\int_0^\infty dE' \int d\Omega'[1-\beta(E')]\nu\Sigma_f(E')\phi(\mathbf{r},\boldsymbol{\Omega}',E',t) &+ \frac{1}{4\pi}\sum_{i=1}^{6}\chi_j(E)\lambda_j C_j(t) + S(\mathbf{r},\boldsymbol{\Omega},E,t)\end{aligned} \tag{6.9}$$

where (Section 6.2) the scattering cross section is written in terms of the pre-collision energy and a function based on the collision dynamics.

$$\Sigma_s(\mathbf{r},\boldsymbol{\Omega}'\cdot\boldsymbol{\Omega},E') = \Sigma_s(\mathbf{r},E')f(\boldsymbol{\Omega}'\rightarrow\boldsymbol{\Omega},\mathrm{E}'\rightarrow E)).$$

Since scattering depends on the dot products of the two directions, rather than the angles themselves, a further simplification is

$$f(\boldsymbol{\Omega}'\rightarrow\boldsymbol{\Omega},\mathrm{E}'\rightarrow E) = f(\boldsymbol{\Omega}'\cdot\boldsymbol{\Omega},\mathrm{E}'\rightarrow E)$$

with the normalization

$$\int_0^\infty dE \int d\Omega f(\Omega' \cdot \Omega, E' \to E) = 1,$$

where integration is over the solid angle 4π steradians. The conservation of energy and momentum is expressed as

$$f(\Omega' \cdot \Omega, E' \to E) = p(\mu_0, E')\,\delta\left(E - \frac{1}{2}[(1+\alpha) + (1-\alpha)\cos\theta_c]E'\right)$$

with $\mu = \cos\theta_c$ in the center-of-mass system.

In Eq. (6.9), the time-dependent precursor concentration equation (Section 6.2) has a decay term as well as a flux-dependent production term due to the delayed fission neutrons; for the jth precursor, the concentration is

$$\frac{dC_j(t)}{dt} = -\lambda_j C_j(t) + \frac{1}{4\pi}\chi_j(E) \int_0^\infty dE' \int d\Omega' \beta(E') \nu \Sigma_f(E') \phi(\mathbf{r}, \Omega', E', t). \tag{6.10}$$

To look at a simpler form, we neglect delayed neutrons, for which the source term is

$$q(\mathbf{r}, \Omega, E, t) = \int_0^\infty dE' \int d\Omega' \Sigma_s(\mathbf{r}, \Omega' \cdot \Omega, E') \phi(\mathbf{r}, \Omega', E', t) + \frac{1}{4\pi}\chi(E) \int_0^\infty dE' \int d\Omega' \nu \Sigma_f(E') \phi(\mathbf{r}, \Omega', E', t) + S(\mathbf{r}, \Omega, E, t) \tag{6.11}$$

which can be justified since delayed neutrons account for less than one percent of the fission neutrons. For this model, the transport equation can be written in terms as

$$\frac{1}{v}\frac{\partial}{\partial t}\phi(\mathbf{r}, \Omega, E, t) + \Omega \cdot \nabla \phi(\mathbf{r}, \Omega, E, t) + \Sigma_t(\mathbf{r}, E)\phi(\mathbf{r}, \Omega, E, t) = q(\mathbf{r}, \Omega, E, t) \tag{6.12}$$

With the usual boundary conditions of flux and current continuity at an interface and appropriate surface conditions.

Thus, for distance s traveled in the direction $s\hat{\Omega}$, the quantity N must be constant across an interface from both sides, i.e.

$$N(\mathbf{r} + s\Omega, E, t + s/v)|_- = N(\mathbf{r} + s\Omega, E, t + s/v)|_+. \tag{6.13}$$

Similarly for a non reentrant or free surface denoted by a boundary b, and a free surface condition

$$N(\vec{r}, E, t)|_b = 0 \text{ for } \hat{n} \cdot \hat{\Omega} < 0 \tag{6.14}$$

that is, zero neutrons returning from a boundary.

Consider now the forms of $\Omega \cdot \nabla$ for slab, spherical and cylindrical geometry. In a one-dimensional slab, the distance traveled s is related to x by $x = s\cos\theta$, so that

$$\Omega \cdot \nabla = \mu \frac{\partial}{\partial x} \tag{6.15}$$

This is extended to rectangular geometry as

$$\Omega \cdot \nabla = \sqrt{1-\mu^2}\left(\cos\varphi \frac{\partial}{\partial x} + \sin\varphi \frac{\partial}{\partial y}\right) + \mu \frac{\partial}{\partial z}, \tag{6.16}$$

while in spherical geometry $\Omega \cdot \mathbf{r} = \mu$ and

$$\Omega \cdot \nabla = \frac{d}{ds} = \frac{\partial}{\partial r}\frac{dr}{ds} + \frac{\partial}{\partial \mu}\frac{d\mu}{ds}.$$

With $r = s\cos\theta = s\mu$,

$$\frac{d\mu}{ds} = \frac{d\mu}{d\theta}\frac{d\theta}{ds} = \frac{\sin^2\theta}{r} \equiv \frac{1-\mu^2}{r}$$

so that

$$\boldsymbol{\Omega}\cdot\nabla = \mu\frac{\partial}{\partial r} + \frac{1-\mu^2}{r}\frac{\partial}{\partial \mu}. \tag{6.17}$$

In cylindrical geometry,

$$\boldsymbol{\Omega}\cdot\nabla = \sqrt{1-\mu^2}\cos\chi\frac{\partial}{\partial r} + \frac{\sqrt{1-\mu^2}}{r}\sin\chi\left(\frac{\partial}{\partial \varphi} - \frac{\partial}{\partial \chi}\right) + \mu\frac{\partial}{\partial z} \tag{6.18}$$

where the vector $\boldsymbol{r}$ in the Cartesian coordinates has height z azimuthal angle φ and χ is the angle between the planes of $\boldsymbol{\Omega}$ and z vectors. The direction of the neutron $\boldsymbol{\Omega}$ is $\mu = \boldsymbol{\Omega}\cdot z$ and azimuthal angle χ. Thus, for an infinite (in height) cylinder with azimuthal symmetry

$$\boldsymbol{\Omega}\cdot\nabla = \sqrt{1-\mu^2}\cos\chi\frac{\partial}{\partial r} - \frac{\sqrt{1-\mu^2}}{r}\sin\chi\frac{\partial}{\partial \chi}.$$

Another useful relationship between the flight path and variables $x(s), t(s)$ is the change when expressed in terms of ds i.e.

$$\frac{d}{ds} = \frac{\partial}{\partial t}\frac{dt}{ds} + \frac{\partial}{\partial x}\frac{dx}{ds}$$

which is obtained with the relations (given initial conditions) $x = x_0 + s\mu$, and $t = t_0 + s/v$ as

$$\frac{d}{ds} = \frac{1}{v}\frac{\partial}{\partial t} + \mu\frac{\partial}{\partial x}. \tag{6.19}$$

With the above forms for the leakage term, the NTE can be explicitly written for regular geometry. Now consider another simplified form of the NTE known as the *one-speed* form, for which the steady state one-dimensional angular flux $\phi(x,\mu,E)$ in a slab would reduce the terms of Eq. (6.12) as follows:

$$\mu\frac{\partial}{\partial x}\phi(x,\mu,E) \rightarrow \mu\frac{\partial}{\partial x}\phi(x,\mu)$$

$$\Sigma_t(x,E)\phi(x,\mu,E) \rightarrow \Sigma_t(x)\phi(x,\mu)$$

$$\frac{1}{2}\int_0^\infty dE'\,\Sigma_s(x,E')\int_{-1}^{1} d\mu'\,\phi(x,\mu',E') \rightarrow \frac{1}{2}\Sigma_s(x)\int_{-1}^{1} d\mu'\,\phi(x,\mu')$$

and

$$\frac{1}{2}\chi(E)\int_0^\infty dE'\,\nu\Sigma_f(x,E')\int_{-1}^{1} d\mu'\,\phi(x,\mu',E') \rightarrow \frac{1}{2}\nu\Sigma_f\int_{-1}^{1} d\mu'\,\phi(x,\mu').$$

The steady-state one-dimensional transport equation is then

$$\left[\mu\frac{\partial}{\partial x} + \Sigma_t(x)\right]\phi(x,\mu) = q(x,\mu) \tag{6.20}$$

where

$$q(x,\mu) = \int d\mu'\,\Sigma_s(x,\mu')\phi(x,\mu') + \frac{1}{2}\int d\mu'\,\nu\Sigma_f\phi(x,\mu') + S(x,\mu). \tag{6.21}$$

Notice here that $\chi(E)$ has disappeared as it is understood that neutrons emerging from fission will fall in this energy group (for which the equation is being simplified). Mathematically what we are saying is that $\chi(E) = 1$; there is no other energy into which an emerging fission neutron can go.

With the above, a one-dimensional (azimuthal symmetry) one-energy NTE in plane geometry is

$$\mu\frac{\partial}{\partial x}\phi(x,\mu)+\Sigma_t(x)\phi(x,\mu)=\frac{1}{2}[\Sigma_s+\nu\Sigma_f]\int_{-1}^{1}d\mu'\,\phi(x,\mu')+S(x,\mu). \tag{6.22}$$

With the quantity c defined as the mean number of secondary neutrons emerging from a collision by any process, for example, scattering (c = 1), fission ($c=\overline{\nu}$), the (n,2n) reaction for which $c=2$,

$$c=\frac{\Sigma_s+\overline{\nu}\Sigma_f+\Sigma_{n,2n}+\cdots}{\Sigma_t}. \tag{6.23}$$

Eq. (6.22) is written as

$$\mu\frac{\partial}{\partial z}\phi(z,\mu)+\phi(z,\mu)=\frac{c}{2}\int_{-1}^{1}d\mu'\,\phi(z,\mu')+S(z,\mu) \tag{6.24}$$

where the unit of distance has changed to $z=\Sigma_t x$, the "optical distance," that is, unit of distance measured in terms of the mean free path $\lambda=1/\Sigma_t$.

In another simplified form of the one-speed transport equation the angular flux is expanded in spherical harmonics with Legendre polynomials (Section 4.2)

$$\phi(z,\mu)=\sum_{l=0}^{N}\frac{2l+1}{2}\phi_l(z)P_l(\mu) \tag{6.25}$$

where the moments $\phi_l(x)$ are defined as

$$\phi_l(z)=\int_{-1}^{1}d\mu\phi(z,\mu)P_l(\mu). \tag{6.26}$$

Expanding to $N=1$, Eq. (6.25) gives the angular flux in terms of the zeroth and first moments:

$$\phi(z,\mu)=\frac{1}{2}\left[\phi_0(z)+3\mu\phi_1(z)\right] \tag{6.27}$$

defined as

$$\phi_0=\int_{-1}^{1}\phi(z,\mu)d\mu,\ \phi_1=\int_{-1}^{1}\mu\phi(z,\mu)d\mu. \tag{6.28}$$

These equations, for a plane isotropic source $\delta(z)$, are

$$\frac{d\phi_1(x)}{dx}+(1-c)\phi_0(z)=\delta(z) \tag{6.29}$$

and

$$\frac{d\phi_0(z)}{dz}+3\phi_1(z)=0. \tag{6.30}$$

Eq. (6.30) is recognized as Fick's law and Eq. (6.29) is written as

$$\frac{dJ}{dx}+\Sigma_a\phi=S \tag{6.31}$$

or, in the form of the NDE (Section 4.1)

$$D\frac{d^2\phi}{dx^2}-\Sigma_a\phi+S=0. \tag{6.32}$$

We can also see an equivalence for the diffusion length (with $D=1/3$) in units of optical thickness,

$$L=\frac{1}{\sqrt{3(1-c)}}.$$

For a nonmultiplying medium, this gives

$$L = \frac{1}{\sqrt{3\Sigma_a/\Sigma_t}} = \sqrt{\frac{D}{\Sigma_a}}.$$

Thus, for fission

$$\int_0^\infty dE\chi(E)\int_0^{4\pi} d\Omega \frac{1}{4\pi} = 1$$

and to consider, for a moment, the steady-state equation, we have

$$\Omega\cdot\nabla\phi(\vec{r},\Omega,E) + \Sigma_t(\vec{r},E,t)\phi(\vec{r},\Omega,E) = \int_0^\infty dE' \int d\Omega' \Sigma_s(\vec{r},E',\Omega'\cdot\Omega)\phi(\vec{r},\Omega',E')$$

$$+ \frac{1}{4\pi}\chi(E)\int_0^\infty dE' \int d\Omega' \nu\Sigma_f(E')\phi(\vec{r},\Omega',E') + S(\vec{r},\Omega,E) \tag{6.33}$$

The NTE in the above form is an integro-differential equation; it is linear in the sense that superposition applies and Green's functions can be used. This means that if the flux is ϕ_1 for a source S_1 and ϕ_2 for a source S_2, then it must be $\phi_1 + \phi_2$ for $S_1 + S_2$; which in turn means that a Green's function $G(\boldsymbol{r}_0, \boldsymbol{\Omega}_0, E_0 \rightarrow \boldsymbol{r}, \boldsymbol{\Omega}, E)$ obtained for the steady-state transport equation with a point isotropic source $\delta(\boldsymbol{r}-\boldsymbol{r}_0)\delta(\boldsymbol{\Omega}-\boldsymbol{\Omega}_0)\delta(E-E_0)$ would give the flux for any source from

$$\phi\left(\boldsymbol{r},\boldsymbol{\Omega},E\right) = \int dV_0 \int d\boldsymbol{\Omega}_0 \int dE_0 S(\boldsymbol{r}_0, \ \boldsymbol{\Omega}_0, \ E_0)\, G\left(\boldsymbol{r}_0, \boldsymbol{\Omega}_0, E_0 \rightarrow \boldsymbol{r}, \ \boldsymbol{\Omega}, E\right). \tag{6.34}$$

6.1.2 The two-group transport equation

The two-group form of the NTE, similar to the two-group NDEs described in Section 5.3 are obtained from Eq. (6.9).

The neutrons are divided into two energy groups; for group-1 and group-2 energies E are in the range $E_1 \le E \le E_0$ and $E_2 \le E \le E_1$ where $E_2 = 0$ with the assumption that there is no up-scattering. The group-averaged neutron flux is ϕ_g and the group-averaged neutron current is J_g defined as

$$\phi_1(\boldsymbol{r}) = \int d\boldsymbol{\Omega}\int_{E_1}^{E_0} \phi(\boldsymbol{r},\boldsymbol{\Omega},E)dE, \phi_2(\boldsymbol{r}) = \int d\boldsymbol{\Omega}\int_0^{E_1} \phi(\boldsymbol{r},\boldsymbol{\Omega},E)dE, \tag{6.35}$$

and

$$J_1(\boldsymbol{r}) = \int d\boldsymbol{\Omega}\int_{E_1}^{E_0} \boldsymbol{\Omega}\phi(\boldsymbol{r},\boldsymbol{\Omega},E)dE, J_2(\boldsymbol{r}) = \int d\boldsymbol{\Omega}\int_0^{E_1} \boldsymbol{\Omega}\phi(\boldsymbol{r},\boldsymbol{\Omega},E)dE. \tag{6.36}$$

The fission contribution in group 1 is $\chi_1\left(\nu\Sigma_{f1}\phi_1(\boldsymbol{r}) + \nu\Sigma_{f2}\phi_2(\boldsymbol{r})\right)$. Assuming an isotropic source $S_{0,1}(\boldsymbol{r})$ in group 1 defined as

$$S_{0,1}(\boldsymbol{r}) = \int d\boldsymbol{\Omega}\int_{E_1}^{E_0} \mathrm{S}(\boldsymbol{r},\boldsymbol{\Omega},E)dE$$

the balance equation is

$$\nabla\cdot J_1(\boldsymbol{r}) + \Sigma_{t0,1}(\boldsymbol{r})\phi_1(\boldsymbol{r}) = \chi_1\left(\nu\Sigma_{f1}\phi_1(\boldsymbol{r}) + \nu\Sigma_{f2}\phi_2(\boldsymbol{r})\right) + S_{0,1}(\boldsymbol{r}) \tag{6.37}$$

with the flux-current relation Fick's law as follows:

$$\nabla\phi_1(\boldsymbol{r}) + 3\Sigma_{t,1}(\boldsymbol{r})J_1(\boldsymbol{r}) = 0. \tag{6.38}$$

In the above, the group-averaged total and fission macroscopic cross-sections are defined as

$$\Sigma_g(\boldsymbol{r}) = \frac{\int \Sigma(\boldsymbol{r},\mathrm{E})\,\phi(\boldsymbol{r},E)\,dE}{\int \phi(\boldsymbol{r},E)\,dE}. \tag{6.39}$$

Similarly for Group 2, the balance equation

$$\nabla\cdot J_2(\boldsymbol{r})+\Sigma_{t0,2}(\boldsymbol{r})\phi_2(\boldsymbol{r}) = \Sigma_{s,1\to2}\phi_1(\boldsymbol{r}) + \chi_2\left(\nu\Sigma_{f1}\phi_1(\boldsymbol{r}) + \nu\Sigma_{f2}\phi_2(\boldsymbol{r})\right) \tag{6.40}$$

has an in-scattering contribution from group 1 neutrons, The scattering cross-section is also defined as a flux-averaged quantity

$$\Sigma_{g'\to g} = \frac{\int dE'\,\phi(\boldsymbol{r},E')\int \Sigma(\boldsymbol{r},E'\to\mathrm{E})dE}{\int \phi(\boldsymbol{r},E')dE'} \tag{6.41}$$

where it is assumed that scattering is isotropic. The Fick's law statement for Group 2 is then

$$\nabla\phi_2(\boldsymbol{r}) + 3\Sigma_{t,2}(\boldsymbol{r})J_2(\boldsymbol{r}) = 0. \tag{6.42}$$

We will see this form of the equation obtained in the spherical harmonics form of the multigroup transport equation used for obtained numerical solutions for the angular flux.

6.1.3 The integral form of the transport equation

There are two ways possible to readily convert the integro-differential form; one is the MOC, using the derivative terms of Eq. (6.19) in which a PDE is converted into an ODE, and the other is an integration of the PDE.

with the derivative terms of the integro-differential transport equation in Cartesian geometry

$$\hat{\mathscr{L}}\phi(\boldsymbol{r},\boldsymbol{\Omega},E,t) \equiv \left(\frac{1}{v}\frac{\partial}{\partial t} + \Omega_x\frac{\partial}{\partial x} + \Omega_y\frac{\partial}{\partial y} + \Omega_z\frac{\partial}{\partial z}\right)\phi(\boldsymbol{r},\boldsymbol{\Omega},E,t) \tag{6.43}$$

which can be expressed as a total derivative

$$\frac{d\phi}{ds} = \frac{\partial\phi}{\partial t}\frac{dt}{ds} + \frac{\partial\phi}{\partial x}\frac{dx}{ds} + \frac{\partial\phi}{\partial y}\frac{dy}{ds} + \frac{\partial\phi}{\partial z}\frac{dz}{ds}$$

from which the "characteristics" are identified as $t = t_0 + s/v$, and $\boldsymbol{r} = \boldsymbol{r_0} + s\boldsymbol{\Omega}$. In three-dimensional Cartesian geometry,

$$x = s\sin\theta\cos\varphi = s\Omega_x$$
$$y = s\sin\theta\sin\varphi = s\Omega_y$$
$$z = s\cos\theta = s\Omega_z$$

so that

$$\frac{dt}{ds} = \frac{1}{v}, \frac{dx}{ds} = \Omega_x, \frac{dy}{ds} = \Omega_y, \frac{dz}{ds} = \Omega_z$$

then

$$\hat{\mathscr{L}}\phi(\boldsymbol{r},\boldsymbol{\Omega},E,t) = \frac{d\phi}{ds}.$$

The transport equation can thus be expressed as a first-order partial differential equation

$$\left[\frac{d}{ds} + \Sigma_t(\boldsymbol{r_0} + s\boldsymbol{\Omega}, E)\right]\phi(\boldsymbol{r_0} + s\boldsymbol{\Omega}, \boldsymbol{\Omega}, E, t_0 + s/v) = q(\boldsymbol{r_0} + s\boldsymbol{\Omega}, \boldsymbol{\Omega}, E, t_0 + s/v) \tag{6.44}$$

and integrated to yield the integral form of the transport equation (Bell & Glasstone, 1979)

$$\phi(\mathbf{r}, \boldsymbol{\Omega}, E, t) = \int_0^{\infty} e^{-\int_0^{s'} \Sigma_t(\mathbf{r}-s''\boldsymbol{\Omega},E)ds''} q\left(\mathbf{r} - s'\boldsymbol{\Omega}, \boldsymbol{\Omega}, E, t - s'/v'\right)ds'. \tag{6.45}$$

If we put the explicit form for $q(\vec{r}, \Omega, E, t)$ in the above, we get the integral equation expressed in the form of a scattered contribution and a direct contribution from the extraneous source. Thus, with

$$q(\mathbf{r}, \boldsymbol{\Omega}, E, t) = \int_0^{\infty} dE' \int d\boldsymbol{\Omega}' \Sigma_s(\mathbf{r}, E', \boldsymbol{\Omega}'\cdot\boldsymbol{\Omega})\phi(\mathbf{r}, \boldsymbol{\Omega}', E') + \frac{1}{4\pi}\chi(E)\int_0^{\infty} dE' \int d\boldsymbol{\Omega}' \nu\Sigma_f(E')\phi(\mathbf{r}, \boldsymbol{\Omega}', E') + S(\mathbf{r}, \boldsymbol{\Omega}', E') \tag{6.46}$$

we can write the terms iteratively as $\phi_0 = S'$, $\phi_1 = \mathrm{K}\phi_0, \ldots, \phi_{n+1} = \mathrm{K}\phi_n$. The interpretation of these terms is clear: ϕ_0 represents the "uncollided" flux, that is, the flux of source neutrons, multiplied by the attenuation factor, ϕ_1 is the flux of neutrons that have had one collision, and ϕ_n is the flux of neutrons that have had n collisions. The Neumann series $\sum_{n=0}^{\infty} \phi_n$ is a solution to the integral equation if it converges.

In this section, the integral formulation of the NTE for collision density $\psi(\mathbf{r}, \boldsymbol{\Omega}, E, t) = \Sigma_t(\mathbf{r}, E)\,\phi(\mathbf{r}, \boldsymbol{\Omega}, E, t)$ is expressed as

$$\psi(P) = S(P) + \int K(P' \to P)\psi(P')\,dP' \tag{6.47}$$

where $\psi(P)$ refers to the collision density in phase space P. This form is obtained by integrating the differential form of the transport equation described in the previous section. It is also "obvious" intuitively by considering that the collision density in some phase space P can come from either a direct source into P or by a collision density at some other phase space P' that makes a transition into P. This integral form is amenable to computational methods and is "naturally" suited for analog simulation as will be explained in Chapter 7 (The Monte Carlo Method).

We will later see that this approach is followed in the Monte Carlo simulation of neutrons as they are transported from the source in a successive series of events. Thus we can write (Rief, 1984)

$$\psi(P) = \sum_{n=1}^{\infty} \int \cdots \int K(P, u_n)\, K(u_n, u_{n-1}) \cdots K(u_2, u_1)\, S_0(u_1, u_0)\, du_n \ldots du_0 \tag{6.48}$$

where the collision density is the sum of collision densities from the direct source, those that have had one collision, two collisions and so on, up to n collisions

$$\psi(P) = \psi_0 + \psi_1 + \psi_2 + \cdots \psi_n.$$

The direct contribution from the source is the integral over all source neutrons that contribute to the phase space of interest P

$$\psi_0 = \int S(P, u_0)du_0,$$

The once-collided contribution is the source neutrons born in phase space u_0 that have a collision that takes them to phase space u_1 and then to phase space P

$$\psi_1 = \int K(P, u_1)du_1 \int S(u_1, u_0)du_0,$$

Thus, a source particle born at $u_0(\mathbf{r_0}, \boldsymbol{\Omega}_0, E_0, t_0)$ moves to $(\mathbf{r_1}, \boldsymbol{\Omega}_0, E_0, t_1)$ where it has its first collision that changes its energy and angle into phase space $(\mathbf{r_1}, \boldsymbol{\Omega}_1, E_1, t_1)$; we thus express the change as a kernel $K(\mathbf{r_0}, \boldsymbol{\Omega}_0, E_0, t_0 \to \mathbf{r_1}, \boldsymbol{\Omega}_1, E_1, t_1)$ in terms of a transport process $T(\mathbf{r_0}, t_0 \to \mathbf{r_1}, t_1; \boldsymbol{\Omega}_0, E_0)$ followed by a collision process $C(\boldsymbol{\Omega}_0, E_0 \to \boldsymbol{\Omega}_1, E_1; \mathbf{r_1}, t_1)$, or simply

$$K_{i-1} = C_{i-1} \cdot T_{i-1}.$$

Eq. (6.40) can be written as

$$\psi(P) = \sum_{n=1}^{\infty} \int \cdots \int \left(\prod_{i=1}^{n} C_i \cdot T_i du_i \right) S_0 du_0 \tag{6.49}$$

with the scattering kernel K, expressed as the product of a collision operator C, in which the precollision energy and angle will transform to the postcollision parameters, and a transport operator T.

Thus a collision will be represented by the term $K\phi \equiv C \cdot T\phi$ where the collision and transport operators $\hat{C} \equiv \hat{C}(r'; E' \to E, \hat{\Omega}' \to \hat{\Omega})$ $\hat{T} \equiv \hat{T}(r' \to r; E', \hat{\Omega}')$ represent the scattering process.

6.1.4 Multigroup form of the integral transport equation

The integral form of the transport equation can be "solved" in a continuous way which we can call *analog simulation*, or in a discrete way, by dividing the energy spectrum in a finite number of energy groups as was shown for diffusion equation in Chapter 3. In a multigroup form, where the continuous energy spectrum is discretized into a finite number of groups, as done in the MORSE code (Emmett, 2000), the integral form of the transport equation is obtained from the multigroup form of Eq. (6.12):

$$\left[\frac{1}{v_g} \frac{\partial}{\partial t} + \boldsymbol{\Omega} \cdot \nabla + \Sigma_t^{(g)}(\boldsymbol{r}) \right] \phi^{(g)}(\boldsymbol{r}, \boldsymbol{\Omega}, t) = q^{(g)}(\boldsymbol{r}, \boldsymbol{\Omega}, t) \tag{6.50}$$

to yield the collision density integral form

$$\psi^{(g)}\left(\vec{r}, \hat{\Omega}, t\right) = \sum_{g'=g}^{1} \int d\Omega' \Sigma_s^{g' \to g}(r, \hat{\Omega}' \to \hat{\Omega}) \psi^{(g')}\left(\vec{r}, \hat{\Omega}, t\right) + S^{(g)}\left(\vec{r}, \hat{\Omega}, t\right) \tag{6.51}$$

and with the collision and transport operators:

$$\psi^{(g)}\left(\vec{r}, \hat{\Omega}, t\right) = \hat{C}^{g' \to g}\left(\vec{r}; \hat{\Omega}' \to \hat{\Omega}\right) \hat{T}^{g'}\left(r' \to r; \hat{\Omega}'\right) \psi^{(g')}\left(r', \hat{\Omega}', t\right) + S^{(g)} \tag{6.52}$$

where the collision operator $\hat{C}$ is

$$\hat{C}^{g' \to g}\left(\vec{r}; \hat{\Omega}' \to \hat{\Omega}\right) = \sum_{g'=g}^{1} \int d\Omega' \frac{\Sigma_s^{g' \to g}(r, \hat{\Omega}' \to \hat{\Omega}}{\Sigma_t^{g'}(r)} \tag{6.53}$$

and the transport operator $\hat{T}$ is defined as

$$\hat{T}^{g'}\left(r' \to r; \hat{\Omega}'\right) = \int_0^{\infty} dR\, \Sigma_t^{(g)}(r)\, e^{-z^{(g)}(r, r - R\Omega)}. \tag{6.54}$$

In the integral form, the analog simulation this begins from a source neutron being transported to a collision site, where the collision process results in "postcollision" energy and angle of travel. The collision density is tallied and the simulation continues until a "history" ends due to an escape from the system or some other termination criteria as will be shown in the next chapter.

6.2 Exact solutions of the transport equation

The complexity of the transport equation makes it impossible to obtain exact solutions for all but idealized configurations such as infinite medium and regular geometry. A number of exact solutions are given in Bell and Glasstone (1979), and the following benchmarks by Ganapol (Ganapol, 2008) are available for comparing with numerical solutions: (1) Infinite medium slowing down, (2) Slowing down/Laplace transform solution in the B_L approximation, (3) Slowing down/the multigroup solution in the B_L approximation, (4) Slowing down and thermalization in an infinite medium/the embedded multigroup approximation, (5) Monoenergetic transport in an infinite medium/the Fourier transform solution, (6) Monoenergetic transport in a semiinfinite medium/the Laplace transform solution, (7) Monoenergetic transport in a 1-D slab/the Fn solution, (8) Monoenergetic transport in a 1-D cylinder/the Fn solution, (9) Multigroup transport in infinite media/the Fourier transform solution, (10) Multigroup slab transport/the Green's function method,

(11) Monoenergetic transport in a two-dimensional semiinfinite medium/the searchlight problem (SLP), and (12) Multgroup transport in a three-dimensional infinite medium/the point kernel method.

For the *source-free infinite medium* case, analytical solutions can be obtained, assuming separability $\phi(z,\mu) = \chi(z)\psi(\mu)$, in terms of eigenfunctions $\psi_\nu(\mu)$:

$$\phi_\nu(z,\mu) = e^{-z/\mu}\,\psi_\nu(\mu). \tag{6.55}$$

It can then be shown (Bell & Glasstone, 1979) that for all values of $\mu\in(-1,1)$ for ν not both real and in the interval $(-1,1)$, that

$$\psi_\nu(\mu) = \frac{c}{2}\frac{\nu}{\nu-\mu} \tag{6.56}$$

where the eigenvalues satisfy the transcendental equation

$$1 = c\nu_0 \tan h^{-1}\frac{1}{\nu_0} = \frac{c\nu_0}{2} ln\frac{\nu_0+1}{\nu_0-1}. \tag{6.57}$$

The above uses the relation

$$\tan h^{-1} x = \frac{1}{2} ln\frac{1+x}{1-x}.$$

When $c<1$ the roots are real (Section 2.10), as shown in Fig. 6.2, while for $c>1$, they are imaginary.

The value of ν_o determines the rate of decrease of the asymptotic flux in a source-free infinite medium and is therefore called the asymptotic relaxation length.

$$\phi_o^{\pm}(x,\mu) = e^{\mp x/\upsilon_0}\frac{c\nu_0}{2(\nu_0\mp\mu)}.$$

When c is near unity,

$$\frac{1}{\nu_0{}^2} = \frac{1}{\sqrt{3(1-c)}}\left[1+\frac{2}{5}(1-c)+\cdots\right]$$

For weakly absorbing media, this relaxation length is the same as the diffusion length L of diffusion theory

$$L = \frac{1}{\sqrt{3\Sigma_t\Sigma_a}} = \frac{1}{\sqrt{3\Sigma_t(1-c)}}.$$

Comparisons between diffusion and transport theory, and Monte Carlo simulation yet to be covered, will be demonstrated for some simple cases.

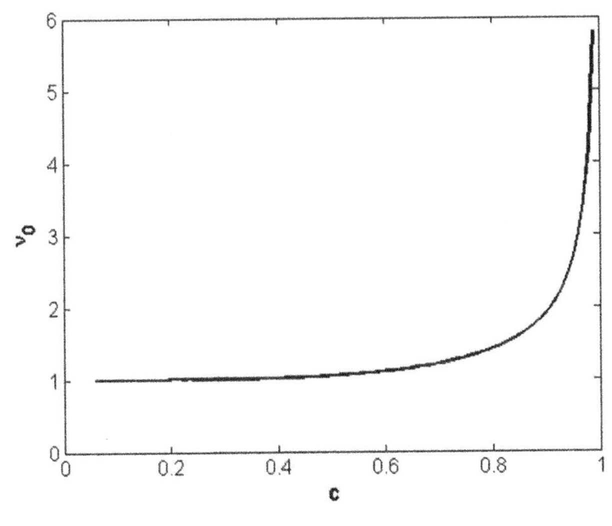

FIGURE 6.2 Relaxation length in m.f.p.'s for isotropic scattering.

It was shown above that for a source-free infinite medium for the asymptotic flux, for real values of ν_o is of the form

$$\phi_o^{\pm}(x,\mu) = e^{\mp x/\upsilon_0}\frac{c\nu_0}{2(\nu_0 \mp \mu)}$$

For the monoenergetic transport solution in a semiinfinite medium, for example, Ganapol (Ganapol, 2008) has obtained a "classical solution" for the scalar flux $\phi(x)$ in terms of the inverse Laplace transform

$$\phi(x) = \frac{2}{c}L_x^{-1}\left[\frac{\phi(0,-1/s)}{s}\right] \tag{6.58}$$

as

$$\phi(x) = \frac{2}{c}L_x^{-1}\begin{cases} \dfrac{c}{2}\dfrac{\mu_0}{1+s\mu_0}H(\mu_0)H(1/s), \text{Beam source} \\ \dfrac{\left[1-\sqrt{1-c}H(1/s)\right]}{s}, \text{Isotropic source} \end{cases} \tag{6.59}$$

Chandrasekhar's H function satisfies the integral equation

$$H(\mu) = 1 + \mu H(\mu)\int_0^1 \frac{\psi(\mu')}{\mu+\mu'}H(\mu')d\mu'$$

where $\int_0^1 \psi(\mu)d\mu = 1$. In case of isotropic scattering, $\psi(\mu) = c/2$. The equation is solved iteratively (Matlab Program 6.1) with

$$H_m^{k+1/2} = \left[1 - \frac{c}{2}\mu_m \sum_{m'=1}^{N} \omega_{m'} \frac{H_{m'}^k}{\mu_m + \mu_{m'}}\right]^{-1}$$

$$H_m^{k+1} = \frac{\alpha_0}{\alpha_0^{k+1/2}} H_m^{k+1/2}$$

with $\alpha_0^{k+1/2} \equiv \sum_{m'=1}^{N} \omega_{m'} H_{m'}^{k+1/2}$, and a normalization $\alpha_0 \equiv \int_0^1 d\mu H(\mu) = \frac{2}{c}\left[1-\sqrt{1-c}\right]$.

This will be referred to in the following sections, especially for exact solution for finite media configurations.

6.2.1 The classic albedo problem

The classic albedo problem, of determining reflectivity of a surface, was addressed originally in radiative transfer (Chandrasekhar, 1960) and exact solutions have been obtained (Ganapol, 2008) using the Laplace transform. For emerging radiation, $\mu < 0, x' \to \infty$, the solution is

$$\phi(x,-|\mu|) = \frac{c}{2|\mu|}\int_x^{\infty} dx' e^{-\frac{x-x'}{\mu}}\phi(x') \tag{6.60}$$

which, at the surface $x = 0$, reduces to

$$\phi(0,-|\mu|) = \frac{c}{2|\mu|}\int_0^{\infty} dx' e^{-\frac{x'}{\mu}}\phi(x'). \tag{6.61}$$

The exact solution, for an isotropic medium with a point isotropic source at the surface separating the half-space from the vacuum, in terms of Chandrasekhar's H functions is

$$\phi(0,-\mu) = 1 - \sqrt{1-c}H(\mu). \tag{6.62}$$

Fig. 6.3 shows the albedo for monoenergetic neutrons incident on a surface $x = 0$ as a function of c.

PROGRAM 6.1 Chandrasekhar's *H* function.

```
% Program C:\Users\User1\Documents\MATLAB\TrTheory2012\Hfunction
% give a range of values of c
% prog calculates H(mu) vs mu and plots
% open output file
resl=fopen('outTT.txt','w');

N=32;
if (N==32)
t(1)=0.0483076656877383162348126;    w(1)=0.0965400885147278005667648;
t(2)=0.1444719615827964934851864;    w(2)=0.0956387200792748594190820;
t(3)=0.2392873622521370745446032;    w(3)=0.0938443990808045656391802;
t(4)=0.3318686022821276497799168;    w(4)=0.0911738786957638847128686;
t(5)=0.4213512761306353453641194;    w(5)=0.0876520930044038111427715;
t(6)=0.5068999089322293900237475;    w(6)=0.0833119242269467552221991;
t(7)=0.5877157572407623290407455;    w(7)=0.0781938957870703064717409;
t(8)=0.6630442669302152009751152;    w(8)=0.0723457941088485062253994;
t(9)=0.7321821187402896803874267;    w(9)=0.0658222227763618468376501;
t(10)=0.7944837959679424069630973;   w(10)=0.0586840934785355471452836;
t(11)=0.8493676137325699701336930;   w(11)=0.0509980592623761761961632;
t(12)=0.8963211557660521239653072;   w(12)=0.0428358980222266806568786;
t(13)=0.9349060759377396891709191;   w(13)=0.0342738629130214331026877;
t(14)=0.9647622555875064307738119;   w(14)=0.0253920653092620594557526;
t(15)=0.9856115115452683354001750;   w(15)=0.0162743947309056706051706;
t(16)=0.9972638618494815635449811;   w(16)=0.0070186100094700966004071;
j=16;
for jj=1:16
    tt(jj)=-t(j); ww(jj)=w(j);
    j=j-1;
end
k=1;
for kk=17:32
tt(kk)=t(k); ww(kk)=w(k);
k=k+1;
end
end

%%% Begin
Ncases=6; Niters=50;
cc=[0.1;0.2;0.4;0.6;0.8;0.9];

for icase=1:Ncases
    c=cc(icase);
fprintf (resl,'\n c = %6.4f ',c);
Alpha0=(2.0/c)*(1.0-sqrt(1.0-c));
fprintf ('\n Alpha0  = %6.4f ',Alpha0);

for ih1 = 1:N
    mu(ih1)=0.5*(tt(ih1)+1.0);
    Hf(ih1) = 1.0;
end
```

(Continued)

PROGRAM 6.1 (Continued)

```
for ih6 = 1:Niters
for ih2 = 1:N
    sum =0.0;
    for ih3 = 1:N
    sum = sum + ww(ih3)*( Hf(ih3)/(mu(ih2)+ mu(ih3)));
    end
    sum = 0.5*sum;
    Hf(ih2) = 1.0/( 1.0 - (c/2.0)*mu(ih2)*sum );
end

Alpha0k=0.0;
for ih4=1:N
    Alpha0k = Alpha0k + ww(ih4)*Hf(ih4);
end
Alpha0k=0.5*Alpha0k;

Error = Alpha0-Alpha0k;

 for ih5=1:N
     Hf(ih5)=(Alpha0/Alpha0k)*Hf(ih5);
 end

end
%keep the H values for this case to be plotted later
for ih7=1:N
HH(icase,ih7)=Hf(ih7);
end

end
    for muVal=1:N
    H1(muVal)=HH(1,muVal);
    H2(muVal)=HH(2,muVal);
    H3(muVal)=HH(3,muVal);
    H4(muVal)=HH(4,muVal);
    H5(muVal)=HH(5,muVal);
    H6(muVal)=HH(6,muVal);
    end

  figure(1)
plot(mu,H1,':k','LineWidth',1.5)
hold on
plot(mu,H2,':k','LineWidth',1.5)
hold on
plot(mu,H3,':k','LineWidth',1.5)
hold on
plot(mu,H4,':k','LineWidth',1.5)
hold on
plot(mu,H5,':k','LineWidth',1.5)
hold on
plot(mu,H6,'-k','LineWidth',1.5)
xlabel('\mu','FontSize',12)
ylabel('H(\mu)','FontSize',12)
text(0.4,1.7,'c','FontSize',12)
```

(Continued)

PROGRAM 6.1 (Continued)
```
h = legend('c = 0.1','c = 0.2','c = 0.4','c = 0.6','c = 0.8','c = 0.9',2);
grid on

figure(2)
% plot albedo here
 for muVal=1:N
     H1(muVal)=(1-sqrt(1-cc(1)))*HH(1,muVal);
     H2(muVal)=(1-sqrt(1-cc(2)))*HH(2,muVal);
     H3(muVal)=(1-sqrt(1-cc(3)))*HH(3,muVal);
     H4(muVal)=(1-sqrt(1-cc(4)))*HH(4,muVal);
     H5(muVal)=(1-sqrt(1-cc(5)))*HH(5,muVal);
     H6(muVal)=(1-sqrt(1-cc(6)))*HH(6,muVal);

     end
plot(mu,H1,':k','LineWidth',1.5)
hold on
plot(mu,H2,':k','LineWidth',1.5)
hold on
plot(mu,H3,':k','LineWidth',1.5)
hold on
plot(mu,H4,':k','LineWidth',1.5)
hold on
plot(mu,H5,':k','LineWidth',1.5)
hold on
plot(mu,H6,'-k','LineWidth',1.5)
xlabel('\mu','FontSize',12)
ylabel('H(\mu)','FontSize',12)
text(0.4,1.7,'c','FontSize',12)
h = legend('c = 0.1','c = 0.2','c = 0.4','c = 0.6','c = 0.8','c = 0.9',2);
grid on

fclose(resl);
```

6.2.2 Infinite medium with a plane isotropic source

The steady-state transport equation exact solution for a plane isotropic source located at x_0 has been obtained by Case's method and by the Fourier Transform method (Bell & Glasstone, 1979). The angular flux, for $x_0 = 0$, is given by

$$\phi(x,\mu) = \frac{1}{4\pi}\left[\frac{\psi_0^+(\mu)e^{-\frac{x-x_0}{\nu_0}}}{N_0^+} + \int_0^1 \frac{\psi_\nu(\mu)e^{-\frac{x-x_0}{\nu}}}{N_\nu}d\nu\right] \tag{6.63}$$

for $x > 0$, where

$$\psi_0^+ = \frac{c}{2}\frac{\nu_0}{\nu_0 - \mu}$$

$$N_0^+ = \frac{c}{2}\nu_0^3\left[\frac{c}{\nu_0^2 - 1} - \frac{1}{\nu_0^2}\right]$$

$$\psi_\nu(\mu) = \frac{c}{2}\mathrm{P}\frac{\nu}{\nu - \mu} + \lambda(\nu)\delta(\mu - \nu)$$

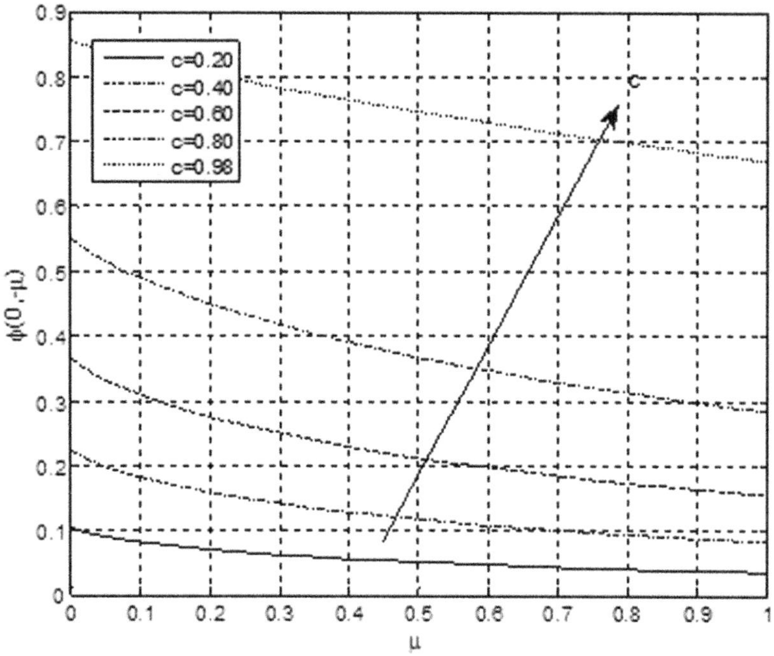

FIGURE 6.3 Albedo for monoenergetic neutrons for isotropic medium.

$$N_\nu = \nu\left[\lambda^2(\nu) + \frac{\pi^2 c^2}{4}\nu^2\right]$$

$$\lambda(\nu) = 1 - c\nu\tanh^{-1}\nu.$$

The angular fluxes, calculated from Matlab Program 6.2, are shown in Figs. 6.4–6.6 for the cases $c = 0.2, 0.4, 0.8$, that is, increasing scattering in the medium. In all cases, the "forward" flux ($\mu > 0$) is larger in magnitude than the "backward" flux ($\mu < 0$) while the magnitude of the angular flux increases as the material becomes less absorbing (higher c).

Integrating Eq. (6.63) gives the total flux

$$\phi(x) = \frac{1}{2}\left[\frac{e^{-\frac{x-x_0}{\nu_0}}}{N_0^+} + \int_0^1 \frac{e^{-\frac{x-x_0}{\nu_0}}}{N_\nu} d\nu\right]. \tag{6.64}$$

The flux given in Eq. (6.64) is the sum of an asymptotic ϕ_{as} and a transient part ϕ_{tr} with the asymptotic part representing the collision equilibrium component. The ratio of the asymptotic flux to the total flux $\phi = \phi_{as} + \phi_{tr}$ is shown in Fig. 6.7 where a high-scattering medium (high c) achieves equilibrium faster than low scattering media. A pure absorber ($c = 0$) would thus never achieve "collisional" equilibrium. The calculation for Fig. 6.7 is given in Matlab Program 6.3.

6.2.3 Finite sphere with a point isotropic source

For a sphere with a point isotropic source, an exact solution has been obtained by Erdmann and Siewert (1968), and by (Siewert & Grandjean, 1979) as

$$\phi(r) = \phi_\infty(r) - \frac{1}{4\pi r}\left[E(\nu_0)e^{-R/\nu_0}\sin h\left(\frac{r}{\nu_0}\right) + \int_0^1 E(\nu)e^{-R/\nu_0}\sin h\left(\frac{r}{\nu}\right)d\nu\right] \tag{6.65}$$

where

$$\phi_\infty = \frac{1}{4\pi r}\left[\frac{1}{\nu_0 N(\nu_0)}\exp\left(-\frac{r}{\nu_0}\right) + \int_0^1 \frac{1}{\nu N(\nu)}\exp\left(-\frac{r}{\nu}\right)d\nu\right] \tag{6.66}$$

PROGRAM 6.2 Angular flux in an infinite medium with a planar isotropic source.

```
% Program C:\Users\User1\Documents\MATLAB\TrTheory2012\AngFluxInfMed

% open output file
resl=fopen('outTT.txt','w');


N=32;

if (N==2)
t(1)=0.5773502691896257645091488;
t(2)=-t(1);
w(1)=1.0;
w(2)=1.0;
end

if (N==4)
 t(1)=0.3399810435848562648026658;
 t(2)=-t(1);
 w(1)=0.6521451548625461426269361;
 w(2)=w(1);
 t(3)=0.8611363115940525752239465;
 t(4)=t(3);
 w(3)=0.3478548451374538573730639;
 w(4)=w(3);
end

if (N==8)
t(1)=0.1834346424956498049394761; t(2)=-t(1);
t(3)=0.5255324099163289858177390; t(4)=-t(3);
t(5)=0.7966664774136267395915539; t(6)=-t(5);
t(7)=0.9602898564975362316835609; t(8)=-t(7);

w(1)=0.3626837833783619829651504; w(2)=w(1);
w(3)=0.3137066458778872873379622; w(4)=w(3);
w(5)=0.2223810344533744705443560; w(6)=w(5);
w(7)=0.1012285362903762591525314; w(8)=w(7);
end

if(N==16)
t(1)=0.0950125098376374401853193;   w(1)=0.1894506104550684962853967;
t(2)=-t(1); w(2)=w(1);

t(3)=0.2816035507792589132304605;  w(3)=     0.1826034150449235888667637;
t(4)=-t(3); w(4)=w(3);

t(5)=0.4580167776572273863424194; w(5)= 0.1691565193950025381893121;
t(6)=-t(5);    w(6)=w(5);

t(7)=0.6178762444026437484466718; w(7)= 0.1495959888165767320815017;
t(8)=-t(7);    w(8)=w(7);

t(9)=0.7554044083550030338951012; w(9)= 0.1246289712555338720524763;
t(10)=-t(9); w(10)=w(9);
```

(Continued)

PROGRAM 6.2 (Continued)

```
t(11)=0.8656312023878317438804679; w(11)=   0.0951585116824927848099251;
t(12)=-t(11);  w(12)=w(11);
t(13)=0.9445750230732325760779884; w(13)=   0.0622535239386478928628438;
t(14)=-t(13); w(14)=w(13);

t(15)=0.9894009349916499325961542; w(15)=   0.027152459411754094;
t(16)=-t(15); w(16)=w(15);
end

if (N==32)
t(1)=0.0483076656877383162348126;   w(1)=0.0965400885147278005667648;
t(3)=0.1444719615827964934851864;   w(3)=0.0956387200792748594190820;
t(5)=0.2392873622521370745446032;   w(5)=0.0938443990808045656391802;
t(7)=0.3318686022821276497799168;   w(7)=0.0911738786957638847128686;
t(9)=0.4213512761306353453641194;   w(9)=0.0876520930044038111427715;
t(11)=0.5068999089322293900237475;  w(11)=0.0833119242269467552221991;
t(13)=0.5877157572407623290407455;  w(13)=0.0781938957870703064717409;
t(15)=0.6630442669302152009751152;  w(15)=0.0723457941088485062253994;
t(17)=0.7321821187402896803874267;  w(17)=0.0658222227763618468376501;
t(19)=0.7944837959679424069630973;  w(19)=0.0586840934785355471452836;
t(21)=0.8493676137325699701336930;  w(21)=0.0509980592623761761961632;
t(23)=0.8963211557660521239653072;  w(23)=0.0428358980222266806568786;
t(25)=0.9349060759377396891709191;  w(25)=0.0342738629130214331026877;
t(27)=0.9647622555875064307738119;  w(27)=0.0253920653092620594557526;
t(29)=0.9856115115452683354001750;  w(29)=0.0162743947309056706051706;
t(31)=0.9972638618494815635449811;  w(31)=0.0070186100094700966004071;
for jj=2:2:32
    t(jj)=-t(jj-1); w(jj)=w(jj-1);
end
end

%%% Begin
c=0.8;
fprintf (resl,'\n c = %6.4f ',c);
% step 1   calc Nu0
TNu1=0.5; TNu2=10.0; NPTS=50000; del=(TNu2-TNu1)/NPTS;
TNu = TNu1;
for ip=1:NPTS
rt(ip)=TNu;
FN(ip) = 1.0-c*(TNu/2)*log((TNu+1)/(TNu-1));
a(ip)=sign(FN(ip));
TNu=TNu+del;
end
% find the root where the sign changes
root = 1.0;
for ir = 1:NPTS-1
    if (a(ir)~=a(ir+1))
        root = rt(ir);
    end
end
fprintf (resl,'\n nu0 = %6.4f ',root);
```

(Continued)

PROGRAM 6.2 (Continued)

```
% step2    calc N0+
N0Plus = 0.5*c*root^3*( (c/(root^2-1.0)) - (1/root^2));
fprintf (resl,'\n N0Plus = %12.4e ',N0Plus);

%xLo = 1.0; xHi = 10.0; DelX=1.0; Nx=10;
xLo = 0.1; xHi = 10.0;  Nx=100; DelX=(xHi-xLo)/Nx;
muLo=-1.0; muHi=1.0;  DelMu=0.4; Nangles=6;
keyX=0; iX=0; countX=0;
for iX=1:Nx
    x=xLo+(iX-1)*DelX;

    xx(iX)=x;

for iM=1:Nangles
    mu=muLo+(iM-1)*DelMu;

    mm(iM)=mu;

% step 3  calc psi0+
psi0Plus = 0.5*c * root/(root-mu);
phi1 = (psi0Plus*exp(-x/root))/N0Plus; % first term of angular flux

% step 4   calc lambda and N
sum=0.0;
for i=1:N
    nu=0.5*(1+t(i)); ww=w(i);
lambda=1-c*nu*atanh(nu);
Nnu = nu*(  lambda^2  + (1.0/4.0)*c^2*nu^2*3.14159^2 );
psiNu=0.5*c*(nu/(nu-mu));
value=(psiNu*exp(-x/nu))/Nnu;
sum = sum + ww*value;
end
beta=0.5*sum;

gamma=0.0;
if (mu>0)
% step 5   calc 2nd term part 2(gamma)
lambdaMu =1-c*mu*atanh(mu);
NnuMu = mu *(  lambdaMu^2  + (1.0/4.0)*c^2*mu^2*3.14159 );
gamma = (lambdaMu/NnuMu)*exp(-x/mu);
end

phi2=beta+gamma;% second term
phi(iX,iM)=(1.0/(4.0*pi))*(phi1+phi2);
end
end

% print the results
% there are countX values of x   and   countM values of mu
fprintf(resl,'\n x values are\n');
for ix = 1:Nx
    fprintf(resl,'\n x(%2.0f)=%6.4f',ix,xx(ix));
```

(*Continued*)

PROGRAM 6.2 (Continued)

```
end
fprintf(resl,'\n mu values are\n');
for im = 1:Nangles
    fprintf(resl,'\n mm(%2.0f)=%6.4f',im,mm(im));
end
fprintf (resl,'\n flux values are');
for ix = 1:Nx
    fprintf(resl,'\n flux at x = %6.4f ',xx(ix));
    for im = 1:Nangles
        fprintf(resl,'\n phi(%2.0f  %2.0f)=%12.4e',ix,im,phi(ix,im));
    end
end
% now collect, for a single angle, all the fluxes as a function of x
    for ix1=1:Nx
    phiMu1(ix1)=phi(ix1,1);
    phiMu2(ix1)=phi(ix1,2);
    phiMu3(ix1)=phi(ix1,3);
    phiMu4(ix1)=phi(ix1,4);
    phiMu5(ix1)=phi(ix1,5);
    phiMu6(ix1)=phi(ix1,6);
    end

    figure(1)
set(gca,'FontSize',12)
semilogy(xx,phiMu1,'-k','LineWidth',2)
hold on
%semilogy(xx,phiMu2,'-.k','LineWidth',1.5)
%hold on
semilogy(xx,phiMu3,':k','LineWidth',1.0)
hold on
semilogy(xx,phiMu4,':k','LineWidth',1.0)
hold on
%semilogy(xx,phiMu5,'-k','LineWidth',1.5)
%hold on
semilogy(xx,phiMu5,'-k','LineWidth',2)
hold on
xlabel('\bf x (mfp)','fontsize',14)
ylabel('\bf Angular Flux (cm^{-2} s^{-1} steradian^{-1})')
%h = legend('\mu=-1.0','\mu = -0.6','\mu = -0.2','\mu = 0.2','\mu =
0.6','\mu=1',2);
%h = legend('\mu=-1.0','\mu = -0.2','\mu = 0.2','\mu=1',2);
grid off
xlim([1 10])
ylim([1e-5 2e-1])
%      etime(clock,t0)
%      t=cputime;
%      fprintf(resl,'\n\n  CPU Time is %9.2f \n\n',t)
text(8.5,1e-1,'\bf c=0.8','fontsize',14)
text(1.3,1e-2,'\mu= -1')
text(1.1,3e-2,' -0.2')
text(1.8,3.5e-2,'0.2')
text(2,1e-1,'\mu=+1')
fclose(resl);
```

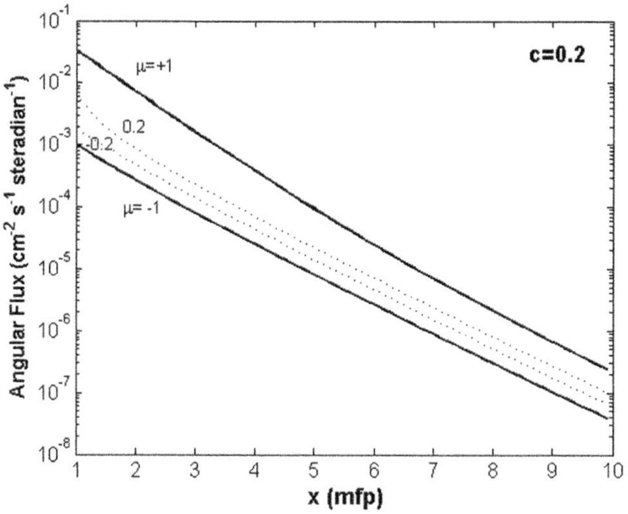

FIGURE 6.4 Angular flux for $c = 0.2$, infinite medium.

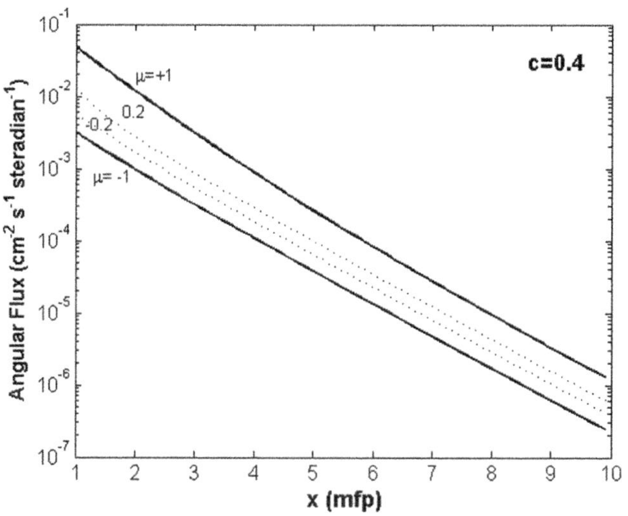

FIGURE 6.5 Angular flux $c = 0.4$, infinite medium.

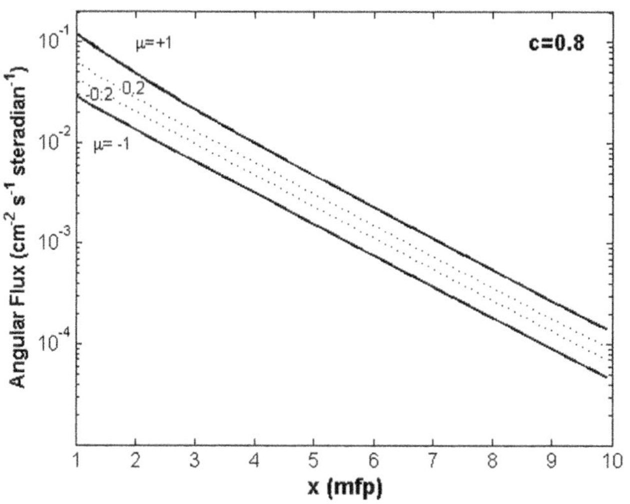

FIGURE 6.6 Angular flux $c = 0.8$, infinite medium.

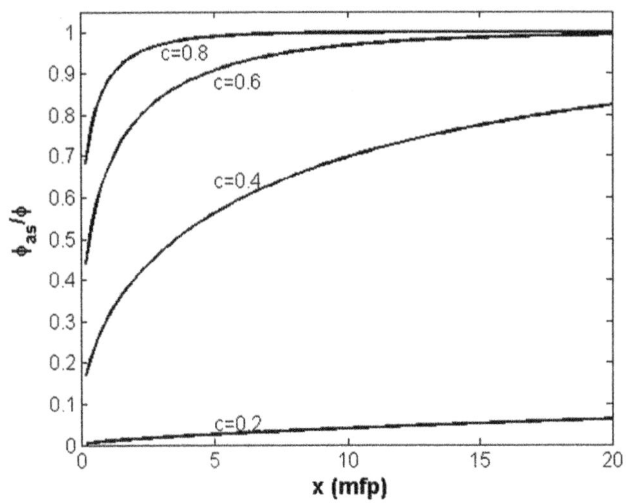

FIGURE 6.7 Total flux, infinite medium isotropic source.

is the infinite medium solution.

The expansion coefficient $E(\nu_0)$ is given by $E(\nu_0) = G(\nu_0) + \int_0^1 K(x)E(x)dx$ where $E(\nu)$ satisfies the Fredholm equation (Section 4.3) $E(\nu) = G(\nu) + \int_0^1 K(x \longrightarrow \nu)E(x)dx$. The four known functions G and K are:

$$\left\{1 - e^{-2\left(\frac{R+z_0}{\nu_0}\right)}\right\}G(\nu_0) = \frac{2}{\nu_0 N(\nu_0)}e^{-\left(\frac{R+2z_0}{\nu_0}\right)} + \frac{c\nu_0}{H(\nu_0)N(\nu_0)}\int_0^1 \frac{\exp\left(-\frac{R}{\nu}\right)}{H(\nu)N(\nu)}\frac{d\nu}{\nu+\nu_0} \tag{6.67}$$

$$\left\{1 - exp\left[-\frac{2(R+z_0)}{\nu_0}\right]\right\}K(x) = \frac{cx\nu_0}{2(\nu_0 + x)H(\nu_0)N(\nu_0)H(x)}\exp\left(-2R/x\right) \tag{6.68}$$

$$H(\nu)N(\nu)G(\nu) = \frac{c\nu\nu_0}{2(\nu+\nu_0)H(\nu_0)}\exp\left(-R/\nu_0\right)\left[G(\nu_0)\exp(-R/\nu_0) + \frac{2}{\nu_0 N(\nu_0)}\right] + c\nu\int_0^1 \frac{1}{(x+\nu)H(x)N(x)}\exp\left(-R/x\right)dx \tag{6.69}$$

and

$$H(\nu)N(\nu)K(x \longrightarrow \nu) = \frac{c\nu\nu_0}{2(\nu+\nu_0)H(\nu_0)}\exp(-2R/\nu_0)K(x) + \frac{c\nu x}{2(\nu+x)H(x)}\exp\left(-2R/x\right) \tag{6.70}$$

with

$$N(\nu_0) \equiv N_0^+ = \frac{c}{2}\nu_0^3\left[\frac{c}{\nu_0^2 - 1} - \frac{1}{\nu_0^2}\right]$$

$$N_\nu = \nu\left[\lambda^2(\nu) + \frac{\pi^2 c^2}{4}\nu^2\right]$$

$$\lambda(\nu) = 1 - c\nu \tan h^{-1}\nu$$

and z_0, the extrapolation distance, given by

$$z_0 = \frac{1}{2}\nu_0 ln\left[\frac{4N(\nu_0)H^2(\nu_0)}{c\nu_0}\right]. \tag{6.71}$$

PROGRAM 6.3 Angular flux in an infinite medium with a planar isotropic source.

```
% C:\Users\User1\Documents\MATLAB\TrTheory\TotFluxInfMed
% infinite medium  exact solution
% give 4 c values;
% get the total flux and plot

resl=fopen('outTT.txt','w');

N=32;
if (N==2)
t(1)=0.5773502691896257645091488;
t(2)=-t(1);
w(1)=1.0;
w(2)=1.0;
end

if (N==4)
 t(1)=0.3399810435848562648026658;
 t(2)=-t(1);
 w(1)=0.6521451548625461426269361;
 w(2)=w(1);
 t(3)=0.8611363115940525752239465;
 t(4)=t(3);
 w(3)=0.3478548451374538573730639;
 w(4)=w(3);
end

if (N==8)
t(1)=0.1834346424956498049394761; t(2)=-t(1);
t(3)=0.5255324099163289858177390; t(4)=-t(3);
t(5)=0.7966664774136267395915539; t(6)=-t(5);
t(7)=0.9602898564975362316835609; t(8)=-t(7);

w(1)=0.3626837833783619829651504; w(2)=w(1);
w(3)=0.3137066458778872873379622; w(4)=w(3);
w(5)=0.2223810344533744705443560; w(6)=w(5);
w(7)=0.1012285362903762591525314; w(8)=w(7);
end

if(N==16)
t(1)=0.0950125098376374401853193;    w(1)=0.1894506104550684962853967;
t(2)=-t(1); w(2)=w(1);

t(3)=0.2816035507792589132304605;  w(3)=     0.1826034150449235888667637;
t(4)=-t(3); w(4)=w(3);

t(5)=0.4580167776572273863424194; w(5)= 0.1691565193950025381893121;
t(6)=-t(5);    w(6)=w(5);

t(7)=0.6178762444026437484466718; w(7)= 0.1495959888165767320815017;
t(8)=-t(7);    w(8)=w(7);

t(9)=0.7554044083550030338951012; w(9)= 0.1246289712555338720524763;
t(10)=-t(9); w(10)=w(9);
```

(Continued)

PROGRAM 6.3 (Continued)

```
t(11)=0.8656312023878317438804679; w(11)=   0.0951585116824927848099251;
t(12)=-t(11);  w(12)=w(11);
t(13)=0.9445750230732325760779884; w(13)=   0.0622535239386478928628438;
t(14)=-t(13); w(14)=w(13);

t(15)=0.9894009349916499325961542; w(15)=   0.02715245941175409 4;
t(16)=-t(15); w(16)=w(15);
end

if (N==32)
t(1)=0.0483076656877383162348126;    w(1)=0.0965400885147278005667648;
t(3)=0.1444719615827964934851864;    w(3)=0.0956387200792748594190820;
t(5)=0.2392873622521370745446032;    w(5)=0.0938443990808045656391802;
t(7)=0.3318686022821276497799168;    w(7)=0.0911738786957638847128686;
t(9)=0.4213512761306353453641194;    w(9)=0.0876520930044038111427715;
t(11)=0.5068999089322293900237475;  w(11)=0.0833119242269467552221991;
t(13)=0.5877157572407623290407455;  w(13)=0.0781938957870703064717409;
t(15)=0.6630442669302152009751152;  w(15)=0.0723457941088485062253994;
t(17)=0.7321821187402896803874267;  w(17)=0.0658222227763618468376501;
t(19)=0.7944837959679424069630973;  w(19)=0.0586840934785355471452836;
t(21)=0.8493676137325699701336930;  w(21)=0.0509980592623761761961632;
t(23)=0.8963211557660521239653072;  w(23)=0.0428358980222266806568786;
t(25)=0.9349060759377396891709191;  w(25)=0.0342738629130214331026877;
t(27)=0.9647622555875064307738119;  w(27)=0.0253920653092620594557526;
t(29)=0.9856115115452683354001750;  w(29)=0.0162743947309056706051706;
t(31)=0.9972638618494815635449811;  w(31)=0.0070186100094700966004071;
for jj=2:2:32
    t(jj)=-t(jj-1); w(jj)=w(jj-1);
end
end

%%% Begin
Ncases=4;
cc=[0.2;0.4;0.6;0.8];

for ic =1:Ncases
c=cc(ic);
fprintf (resl,'\n c = %6.4f ',c);
% step 1   calc Nu0
TNu1=0.5; TNu2=10.0; NPTS=50000; del=(TNu2-TNu1)/NPTS;
TNu = TNu1;
for ip=1:NPTS
rt(ip)=TNu;
FN(ip) = 1.0-c*(TNu/2)*log((TNu+1)/(TNu-1));
a(ip)=sign(FN(ip));
TNu=TNu+del;
end
% find the root where the sign changes
root = 1.0;
for ir = 1:NPTS-1
    if (a(ir)~=a(ir+1))
        root = rt(ir);
    end
```

(*Continued*)

PROGRAM 6.3 (Continued)

```
end
Nu0(ic)=root;
fprintf (resl,'\n nu0 = %6.4f ',root);

% step2    calc N0+
N0Plus = 0.5*c*root^3*( (c/(root^2-1.0)) - (1/root^2));
fprintf (resl,'\n N0Plus = %12.4e ',N0Plus);
xLo = 0.2; xHi = 20.0; Nx = 100; DelX=(xHi-xLo)/Nx;
Nx=Nx+1;
keyX=0; iX=0; countX=0;
for iX=1:Nx
    x=xLo+(iX-1)*DelX;
    xx(iX)=x;
% step 3  calc psi0+
phi1 = (exp(-x/root))/N0Plus; % first term
% step 4   calc lambda and N
sum=0.0;
for i=1:N
    nu=0.5*(1+t(i)); ww=w(i);
lambda=1-c*nu*atanh(nu);
Nnu = nu*(  lambda^2  + (1.0/4.0)*c^2*nu^2*3.14159^2 );
value=(exp(-x/nu))/Nnu;
sum = sum + ww*value;
end
beta=0.5*sum;
gamma=0.0;

phi2=beta+gamma;% second term

phi(iX)=(1.0/(2.0))*(phi1+phi2);
phiAs(iX)=(1.0/2.0)*phi1;
ratio(iX)=phiAs(iX)/phi(iX);
end

% print the results

% now calc the total flux
fprintf(resl,'\n Total Flux  Asymp FLux  Ratio PhiAs/Phi \n\n');
fprintf(resl,'\n    i      x         Phi(x)      PhiAs(x)  Ratio ');
for k = 1:Nx
        fprintf(resl,'\n %4.0f  %8.4f  %12.4e %12.4e
%12.4e',k,xx(k),phi(k),phiAs(k),ratio(k));
        phiC(ic,k)=phi(k); % for each c value
        phiAsC(ic,k)=phiAs(k);
end


% now write all these values for this value of c
for k1 = 1:Nx
    ratioC(ic,k1)=ratio(k1);
end

end
```

(Continued)

PROGRAM 6.3 (Continued)

```
fprintf(resl,'\n\n     c      Nu0');
for in=1:Ncases
    fprintf(resl,'\n %6.4f %6.4f',cc(in),Nu0(in));
end

iplot=2;

if(iplot==1)
    set(gca,'FontSize',12)
semilogy(xx,phi,'-k','LineWidth',2)
hold on
semilogy(xx,phiAs,'-r','LineWidth',2)
xlabel('\bf x (mfp)''fontsize',14)
ylabel('\bf Total Flux cm^{-2} s^{-1}','fontsize',14)
h = legend('\phi','\phi_{as}',2);
grid on
end

if(iplot==2)
    set(gca,'FontSize',12)
    for ip=1:Nx
       ratio1(ip)=ratioC(1,ip); ratio2(ip)=ratioC(2,ip);
ratio3(ip)=ratioC(3,ip); ratio4(ip)=ratioC(4,ip);
    end
plot(xx,ratio1,'-k','LineWidth',1.5)
hold on
plot(xx,ratio2,'-k','LineWidth',1.5)
hold on
plot(xx,ratio3,'-k','LineWidth',1.5)
hold on
plot(xx,ratio4,'-k','LineWidth',1.5)
xlabel('\bf x (mfp)','FontSize',14)
ylabel('\bf \phi_{as}/\phi ','FontSize',14)
xp1=5; yp1= 0.05 ;text(xp1,yp1,'c=0.2','FontSize',12);
xp2=5; yp2= 0.64 ;text(xp2,yp2,'c=0.4','FontSize',12);
xp3=5; yp3= 0.88 ;text(xp3,yp3,'c=0.6','FontSize',12);
xp4=3; yp4= 0.95 ;text(xp4,yp4,'c=0.8','FontSize',12);
%h = legend('c=0.2','c=0.4','c=0.6','c=0.8',2);
grid off
ylim([0 1.05])
end
fclose(resl);
```

The Chandrasekhar H-function satisfies the integral equation

$$H(\mu) = 1 + \mu H(\mu) \int_0^1 \frac{\psi(\mu')}{\mu + \mu'} H(\mu') d\mu'. \tag{6.72}$$

The above equations are solved with the MATLAB program SpherePtSrcMain giving the total flux shown in Figs. 6.8–6.10 for various values of c

The values of flux given in Table 6.1 agree well with the results for $c = 0.3$ and $c = 0.9$ given by (Siewert & Grandjean, 1979).

Fig. 6.11 shows that transport theory flux for values of c approaching the critical case ($c = 1$) from which the cosine buckling is observable.

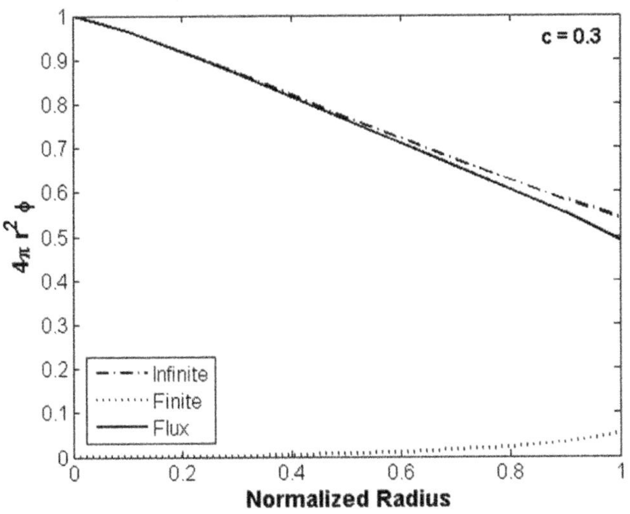

FIGURE 6.8 Transport theory flux in a finite sphere ($c = 0.3$).

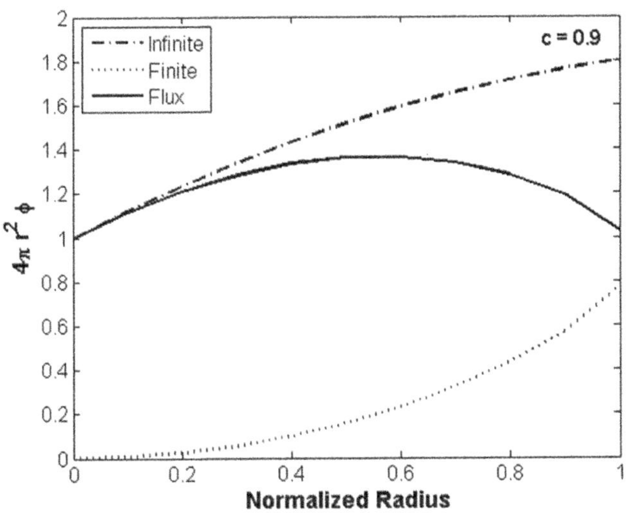

FIGURE 6.9 Transport theory flux in a finite sphere ($c = 0.9$).

6.3 Numerical methods for solving the transport equation

The first idealizations of infinite medium, semiinfinite medium and homogeneous half-spaces are useful for obtaining exact analytical solutions using Laplace and Fourier transforms, Green's functions, etc., but found very limited practical application. This was followed by semianalytical and numerical solutions methods such as S_N and P_N methods. Later, finite element (Section 4.5), boundary-element and finite volume methods were incorporated with tremendous numerical effort to extend the capabilities of deterministic codes to full 3D modeling.

6.3.1 The discrete ordinates method

The discrete ordinates method, first used by Wick and Chandrasekhar for radiative transfer analysis in stellar atmospheres (Bell & Glasstone, 1952), was developed for neutron transport by Carlson and Lathrop (1968). In the discrete ordinates, S_N, method, the angular distribution for solid angle Ω is discretized into a quadrature scheme as shown in Fig. 6.12 (Reuss, 2008) for the cases $N = 4.8$. Thus each octant has 3 and 10 elements, respectively.

The μ-interval $(-1,1)$ is divided into n intervals with $\mu_0 = -1, \mu_N = 1$. Thus, In the S_4 approximation, $\mu_0 = -1, \mu_1 = -1/2, \mu_2 = 0, \mu_3 = 1/2, \mu_4 = 1$.

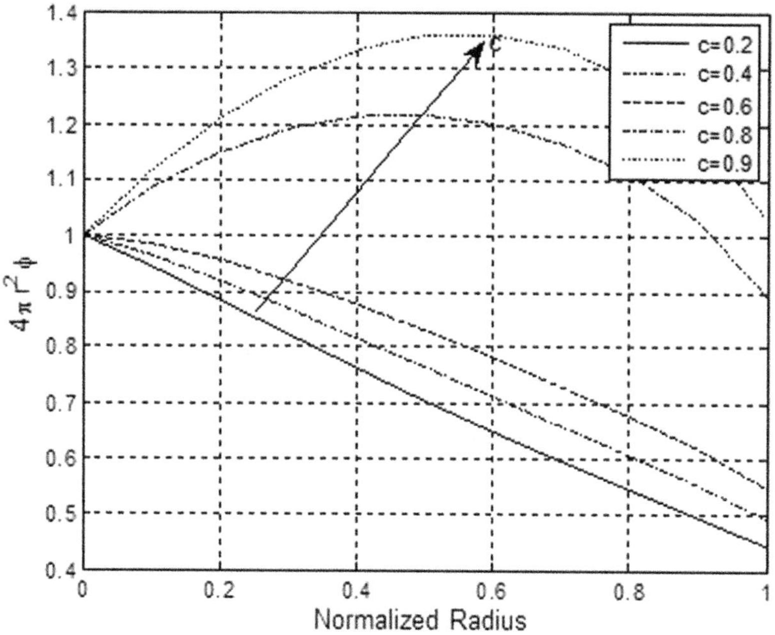

FIGURE 6.10 Transport theory flux in a finite sphere.

TABLE 6.1 Neutron flux in a finite sphere.

r	$c = 0.3$[a]		$c = 0.6$[b]		$c = 0.9$[c]	
	$4\pi r^2 \rho_\infty(r)$	$4\pi r^2 \rho(r)$	$4\pi r^2 \rho_\infty(r)$	$4\pi r^2 \rho(r)$	$4\pi r^2 \rho_\infty(r)$	$4\pi r^2 \rho(r)$
0	1.0	1.0	1.0	1.0	1.0	1.0
0.1	0.96427	0.96404	1.03384	1.03273	1.12076	1.11470
0.2	0.91944	0.91851	1.04871	1.04420	1.23423	1.20988
0.3	0.87072	0.86856	1.05026	1.03994	1.33831	1.28317
0.4	0.82055	0.81657	1.04180	1.02300	1.43252	1.33356
0.5	0.77035	0.76382	1.02562	0.99527	1.51681	1.36018
0.6	0.72103	0.71102	1.00350	0.95789	1.59138	1.36195
0.7	0.67318	0.65840	0.97682	0.91126	1.65659	1.33720
0.8	0.62717	0.60576	0.94672	0.85480	1.71285	1.28277
0.9	0.58325	0.55204	0.91410	0.78578	1.76064	1.19143
1.0	0.54155	0.49139	0.87972	0.68833	1.80046	1.02745

[a] $c = 0.3$, $v_0 = 1.0026$, $Z_0 = 2.50310$.
[b] $c = 0.6$, $v_0 = 1.1021$, $Z_0 = 1.19240$.
[c] $c = 0.9$, $v_0 = 1.9031$, $Z_0 = 0.78964$.

In the limits (μ_{j-1}, μ_j), the angular flux is approximated by a linear function

$$\phi(r,\mu) = \frac{\mu - \mu_{j-1}}{\mu_j - \mu_{j-1}}\phi(r,\mu_j) + \frac{\mu_j - \mu}{\mu_j - \mu_{j-1}}\phi(r,\mu_{j-1}). \tag{6.73}$$

This approximation is then substituted in the transport equation and after integration over μ and N equations are subsequently obtained. To obtain the final equations, an additional equation is required for $\mu = -1$.

For details on the mathematics and iterative algorithms on the S_N method, the reader is referred to Clark and Hansen (1964) and Bell and Glasstone (1979).

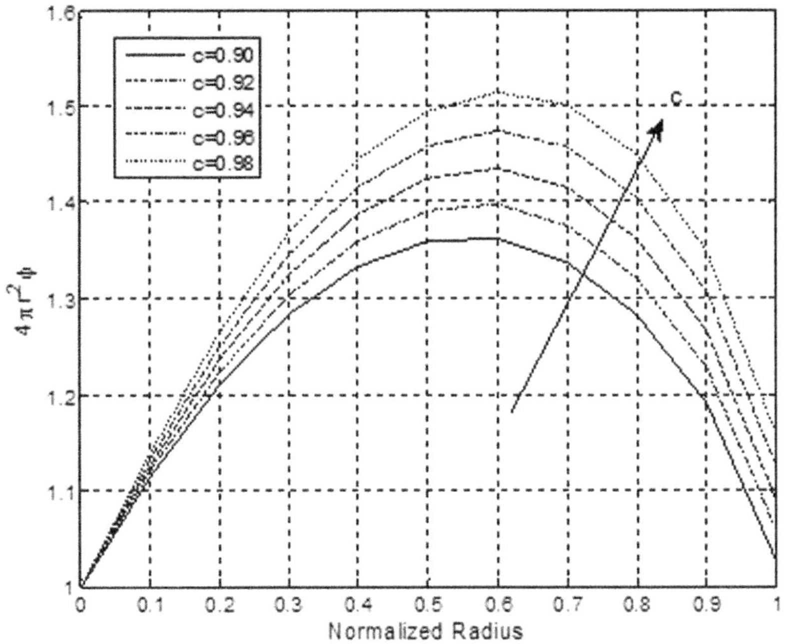

FIGURE 6.11 Transport theory flux for values approaching $c = 1$.

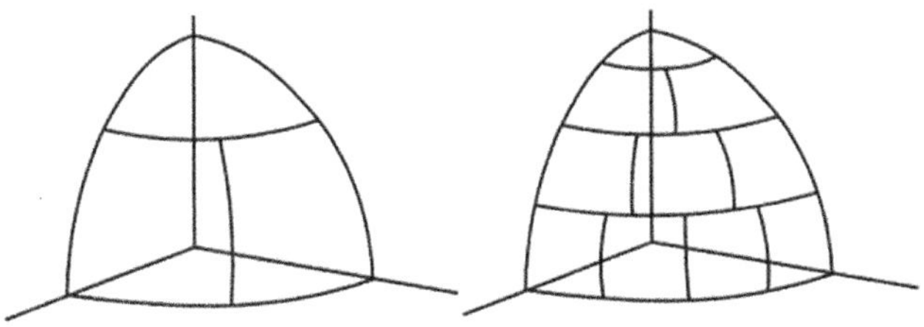

FIGURE 6.12 Discrete ordinates with $N = 4,8$.

In the transport equation, the integral term is represented by weighted angular flux at discrete directions. The representation is

$$\int_{-1}^{1} \phi(x, \mu')d\mu' \cong \sum_{i=1}^{N} w_i \phi(x, \mu_i) \tag{6.74}$$

where the quadrature weights, for a special choice of μ_i corresponding to $P_N(\mu_i) = 0$, given in Table 6.2 (Bell & Glasstone, 1979)

The discretized transport equation is

$$\mu \frac{\partial}{\partial z}\phi(z, \mu_j) + \phi(z, \mu_j) = \frac{c}{2}\sum_{i=1}^{N} w_i \phi(z, \mu_i) + S(z, \mu_j)$$

for angular bins specified for

$$j = 1, 2, 3, \cdots, N.$$

The discrete ordinates method is extensively applied in reactor calculations for lattice cells and whole-core eigenvalue calculations and is being made faster by using parallel algorithms in supercomputing environments.

6.3.2 The Spherical harmonics method

In this section, the spherical harmonics method, mentioned for the two-group transport equations, is described for a one-speed infinite slab with isotropic scattering followed by anisotropic scattering and then for a critical finite slab.

TABLE 6.2 Quadrature weights and angle for S_2, S_4, S_6 expansions.

N	w_i	μ_i
2	$w_1 = w_2 = 1.000$	$\mu_1 = -\mu_2 = 0.57735$
4	$w_1 = w_4 = 0.65215$	$\mu_1 = -\mu_4 = 0.33998$
	$w_2 = w_3 = 0.34785$	$\mu_2 = -\mu_3 = 0.86114$
6	$w_1 = w_6 = 0.46791$	$\mu_1 = -\mu_6 = 0.23862$
	$w_2 = w_5 = 0.36076$	$\mu_2 = -\mu_5 = 0.66121$
	$w_3 = w_4 = 0.17132$	$\mu_3 = -\mu_4 = 0.93247$

Consider now the case of an isotropic plane source $S(x) = \delta(x)/4\pi$. The one-speed equation from Eq. (6.22) is

$$\mu \frac{\partial}{\partial x}\phi(x,\mu) + \Sigma_t(x)\phi(x,\mu) = \frac{c}{2}\int_{-1}^{1} d\mu' \phi(x,\mu') + \frac{\delta(x)}{4\pi}. \tag{6.75}$$

Expanding the angular flux $\phi(x, \mu)$ in terms of Legendre polynomials $P_m(\mu)$

$$\phi(x,\mu) = \sum_{m=0}^{\infty} \frac{2m+1}{4\pi}\phi_m(x)P_m(\mu) \tag{6.76}$$

with moments defined by

$$\phi_m(x) = 2\pi\int_{-1}^{1} \phi(x,\mu)P_m(\mu)d\mu \tag{6.77}$$

so that the zeroth moment, as defined earlier, is the scalar flux

$$\phi_0(x) = 2\pi\int_{-1}^{1} \phi(x,\mu)d\mu \tag{6.78}$$

and the first moment is the current

$$\phi_1(x) = 2\pi\int_{-1}^{1} \mu\phi(x,\mu)d\mu = J(x). \tag{6.79}$$

In the spherical harmonics, P_N, method, the angles are written in terms of spherical harmonic functions which are orthogonal in the range $\mu\epsilon(-1, 1)$ where $\mu = \cos\vartheta$. The first few functions (Table 4.2) are

$$P_o(\mu) = 1, P_1(\mu) = \mu, P_2(\mu) = \frac{1}{2}(3\mu^2 - 1), \ldots$$

and orthogonality gives

$$\int_{-1}^{1} P_n(\mu)P_m(\mu)d\mu = \frac{2\delta_{n,m}}{(n+1)} \tag{6.80}$$

where

$$\delta_{n,m} = \begin{cases} 1, n = m \\ 0, n \neq m \end{cases}$$

Substituting Eq. (6.76) in Eq. (6.75)

$$\mu\sum_{m=0}^{\infty}(2m+1)\frac{d\phi_m}{dx}P_m(\mu) + \sum_{m=0}^{\infty}(2m+1)\phi_m(x)P_m(\mu) = c\phi_0(x) + \delta(x) \tag{6.81}$$

Using the recurrence relation for spherical harmonics:

$$(2m+1)\mu P_m(\mu) = (m+1)P_{m+1}(\mu) + mP_{m-1}(\mu) \tag{6.82}$$

and the orthogonality of Legendre polynomials, Eq. (6.81) reduces to a numerically useful form:

$$(n+1)\frac{d\phi_{n+1}(x)}{dx} + n\frac{d\phi_{n-1}(x)}{dx} + (2n+1)(1 - c\delta_{0n})\phi_n(x) = \delta_{0n}\delta(x) \tag{6.83}$$

from which moments can be computed. In Eq. (6.83), $\phi_{-1}(x) = 0$, and the Kronecker delta function $\delta_{0n} = \begin{cases} 1, n=0 \\ 0, n \neq 0 \end{cases}$.

Eq. (6.83) is truncated at n = N, and called the P_N approximation, with the condition

$$\frac{d\phi_{N+1}(x)}{dx} = 0.$$

An accurate answer is obtained for large N. The first step towards practical application is a slab finite in one dimension considered infinite in the other two dimensions. Such a configuration is an idealization of a critical slab for reactor calculations as described for the NDE in Chapter 5. This would require boundary conditions on both sides for incoming and outgoing flux or current. Spherical harmonics, based on Legendre polynomials have orthogonality for angles in the range $-1 \leq \mu \leq 1$ and thus are inadequate to represent a free boundary condition at the left face of a slab. The inward current at the left face of a slab ($x = 0$), $J^+(0)$ is zero if there are no incoming neutrons at this face. Such a condition is written as

$$J^+(0) = \int_0^1 \mu\phi(0,\mu)d\mu = 0. \tag{6.84}$$

The above is not to be confused with

$$J^-(0) = \int_{-1}^0 \mu\phi(0,\mu)d\mu \neq 0 \tag{6.85}$$

such that the net current

$$J(0) = J^+(0) - J^-(0) \neq 0. \tag{6.86}$$

Similarly on the right side of a critical slab ($x = a$), zero incoming current is expressed as

$$J^-(a) = \int_0^1 \mu\phi(a, -\mu)d\mu = 0, \tag{6.87}$$

while

$$J^+(a) = \int_0^1 \mu\phi(a,\mu)d\mu \neq 0, \tag{6.88}$$

and

$$J(a) = J^+(a) - J^-(a) \neq 0. \tag{6.89}$$

The generalized expression of such boundary conditions are the Marshak boundary conditions

$$\int_0^1 P_i(\mu)\phi(0,\mu)d\mu = \int_0^1 P_i(-\mu)\phi(a, -\mu)d\mu = 0, \tag{6.90}$$

for $i = 1, 3, 5, \ldots N$.

Exercise 6.1: Marshak conditions

Apply the Marshak boundary conditions for $i = 1$, and use Fick's Law (P_1 equations) to show that the linear extrapolation distance d for which $\phi(d) = 0$ is

$$d = \frac{2}{3}(\text{mfp}) \tag{6.91}$$

In the P_1 approximation, Marshak's boundary conditions for $c-1 \leq 1$, that is, $c \leq 2$, lead to the extrapolation distance

$$x_0 = \frac{2}{3}\left[1 - \frac{4}{9}(c-1) + \frac{16}{45}(c-1)^2 + \cdots\right]. \tag{6.92}$$

Recalling from Eq. (6.31) that c is the mean number of neutrons emerging from a collisions, assumed constant, $c=0$ for an absorber, $c=1$ for a pure scattering material, $c>1$ for neutron multiplication, $c=\overline{\nu}$ for fission. Thus, for the case $c=1$, $x_0=d$, as in the case for diffusion theory.

The Mark boundary conditions

$$\phi(0,\mu_i) = 0 \tag{6.93}$$

for $i = 1, 2, 3, \ldots (N+1)/2$, with N odd are another possibility for representing the boundary conditions at the two surfaces of a critical slab. Thus

$$\phi(0,\mu_1) = \phi(0,\mu_2) = \phi(0,\mu_3)\ldots = 0$$

where μ_1 are obtained from the roots of

$$P_{N+1}(\mu_i) = 0. \tag{6.94}$$

As an example, for $N=1$,

$$P_2(\mu_1) = \frac{1}{2}(3\mu_1^2 - 1) = 0$$

gives

$$\mu_1 = \frac{1}{\sqrt{3}}. \tag{6.95}$$

which is one of the two Gauss quadratures for $N=2$ used in numerical integration also (Section 6.4).

The extrapolation distance for the P_1 Mark boundary conditions are

$$x_0 = \frac{1}{\sqrt{3}}\left[1 - \frac{1}{3}(c-1) + \frac{1}{5}(c-1)^2 + \cdots\right]. \tag{6.96}$$

For $c=1$, the Marshak and Mark extrapolations length differ by about 12%.

The discrete ordinates method, using Gauss quadrature sets, and spherical harmonics method with Mark's boundary conditions are equivalent.

The preceding formulation for isotropic scattering has been extended to the case of anisotropic scattering

$$\mu\frac{\partial}{\partial x}\phi(x,\mu) + \phi(x,\mu) = c\int_0^{2\pi} d\varphi' \int_{-1}^{1} d\mu' f(\Omega' \to \Omega)\phi(x,\mu') + Q(x,\mu). \tag{6.97}$$

Since $f(\Omega' \to \Omega) = f(\Omega' \cdot \Omega) = f(\mu_0)$, a spherical harmonics expression

$$f(\Omega' \to \Omega) = f(\mu_0) = \sum_{l=0}^{\infty} \frac{2m+1}{4\pi} f_l P_l(\mu_0) \tag{6.98}$$

is used, with moments

$$f_l = 2\pi\int_{-1}^{1} f(\mu_0)P_l(\mu_0)d\mu. \tag{6.99}$$

$$f_0 = 2\pi\int_{-1}^{1} f(\mu_0)d\mu = 1 \tag{6.100}$$

$$\phi_m(x) = 2\pi\int_{-1}^{1} \phi(x,\mu)P_m(\mu)d\mu \tag{6.101}$$

With the addition theorem

$$P_l(\mu_0) = P_l(\mu)P_l(\mu') + 2\sum_{m=1}^{l}\frac{(l-m)!}{(l+m)!}P_l^m(\mu)P_l^m(\mu')\cos m(\varphi-\varphi') \tag{6.102}$$

where $\Omega = \Omega(\mu, \varphi)$. The angular flux and angular source $\phi(x, \mu)$ and $Q(x, \mu)$ are expanded as in Eq. (6.68)

$$Q(x, \mu) = \sum_{m=0}^{\infty}\frac{2m+1}{4\pi}Q_m(x)P_m(\mu) \tag{6.103}$$

with

$$Q_m(x) = 2\pi\int_{-1}^{1} Q(x, \mu)P_m(\mu)d\mu. \tag{6.104}$$

With the expansions (.) inserted into (..) and multiplying both sides by

$$\frac{1}{2}(2n+1)\int_{-1}^{1} P_m(\mu)d\mu* \tag{6.105}$$

the set of coupled ODEs is

$$(n+1)\frac{d\phi_{n+1}(x)}{dx} + n\frac{d\phi_{n-1}(x)}{dx} + (2n+1)\left(1-cf_n\right)\phi_n(x) = (2n+1)Q_n(x)n = 0, 1, 2, \ldots \tag{6.106}$$

For a P_3 solution, for illustration, the four first-order coupled ODEs for a one-dimensional slab are

$$\frac{d\phi_1(x)}{dx} + \left(1-cf_0\right)\phi_0(x) = Q_0(x) \tag{6.107}$$

$$2\frac{d\phi_2(x)}{dx} + \frac{d\phi_0(x)}{dx} + 3\left(1-cf_1\right)\phi_1(x) = 3Q_1(x) \tag{6.108}$$

$$3\frac{d\phi_3(x)}{dx} + 2\frac{d\phi_1(x)}{dx} + 5\left(1-cf_2\right)\phi_2(x) = 5Q_2(x) \tag{6.109}$$

and

$$4\frac{d\phi_4(x)}{dx} + 3\frac{d\phi_2(x)}{dx} + 7\left(1-cf_3\right)\phi_3(x) = 7Q_3(x) \tag{6.110}$$

These can be solved by standard methods, such as the finite difference method (Section 6.4) to obtain the moments in the expression

$$\phi(x, \mu) = \frac{1}{4\pi}\left[\phi_0(x)P_0(\mu) + 3\phi_1(x)P_1(\mu) + 5\phi_2(x)P_2(\mu) + 7\phi_3(x)P_3(\mu)\right] \tag{6.111}$$

The angular flux will then be

$$\phi(x, \mu) = \frac{1}{4\pi}\left[\phi_0(x) + 3\mu\phi_1(x) + \frac{5}{2}\left(3\mu^2-1\right)\phi_2(x) + \frac{7}{2}\left(5\mu^3-3\mu\right)\phi_3(x)\right] \tag{6.112}$$

Note that for a P_3 formulation, there are four coupled ODEs; in general there are $N+1$ equations with $N+1$ equations; thus, $N+1$ boundary and interface conditions are required. For the critical slab described above, free-surface boundary conditions, that is, zero incoming current at the left and right boundaries gives

$$J(0) = -\frac{1}{2}\phi(0), J(a) = \frac{1}{2}\phi(a) \tag{6.113}$$

which, using Fick's law can be expressed as

$$\phi + 2D\hat{\boldsymbol{n}}\cdot\nabla\phi = 0 \tag{6.114}$$

where $\hat{\boldsymbol{n}}$ is an outward unit normal to a boundary surface.

For a unit lattice cell, a reflective boundary condition, shown in Fig. 6.3, would imply zero (net) current at the point of reflection $\boldsymbol{x}_r$, i.e $J(\boldsymbol{x}_r) = 0$ (Fig. 6.13).

The general boundary condition is written in a mixed form, consisting of Dirichlet, Neumann or mixed boundary conditions.

In spherical geometry, the one-speed equation, using Eqs. (6.11), (6.12) and (6.17),

$$\left[\mu\frac{\partial}{\partial r} + \frac{1-\mu^2}{r}\frac{\partial}{\partial \mu} + \Sigma_t(\boldsymbol{r}, E)\right]\phi(\boldsymbol{r}, \boldsymbol{\Omega}, E, t) = \int_0^{2\pi} d\varphi' \int_{-1}^{1} d\mu' \Sigma_s(r, \mu_0)\phi(r, \mu') + Q(x, \mu) \tag{6.115}$$

and the spherical harmonics expansions (Bell & Glasstone, 1979) leads to the coupled ODEs

$$(n+1)\left(\frac{d}{dr} + \frac{n+2}{r}\right)\phi_{n+1}(r) + n\left(\frac{d}{dr} + \frac{n-1}{r}\right)\phi_{n-1}(r) + (2n+1)\Sigma_{s,n}\phi_n(r) = (2n+1)Q_n(r) \tag{6.116}$$

for $n = 0, 1, 2, \ldots$

$$\Sigma_s(r, \mu_0) = \sum_{m=0}^{\infty}\frac{2m+1}{4\pi}\Sigma_{s,m}(r)P_m(\mu_0) \tag{6.117}$$

As for the slab case, the $N+1$ equations require the same number of boundary conditions, for which half are specified at the outer boundary, as the free-surface conditions given above, and the other half requiring all fluxes $\phi_n(r=0)$ at the origin to be finite.

Another set of boundary conditions, as was applied in Chapter 5, is to use a free-surface boundary condition at the radius and a zero-current symmetry condition $\boldsymbol{J}(0) = 0$ at the origin.

Exercise 6.2: Spherical harmonics for an isotropic source

Show that the P_1 equations in spherical geometry for an isotropic source $Q_0(r)$ reduce to the diffusion equation

$$\frac{1}{r^2}\frac{d}{dr}\left[r^2 D\frac{d\phi(r)}{dr}\right] - \Sigma_{s,0}(r)\phi(r) + Q_0(r) = 0 \tag{6.118}$$

Exercise 6.3: The diffusion coefficient

From the P_1 equations, Eqs. (104) and (106), with an isotropic source, show that the diffusion coefficient is

$$D = \frac{1}{3(1-cf_1)}. \tag{6.119}$$

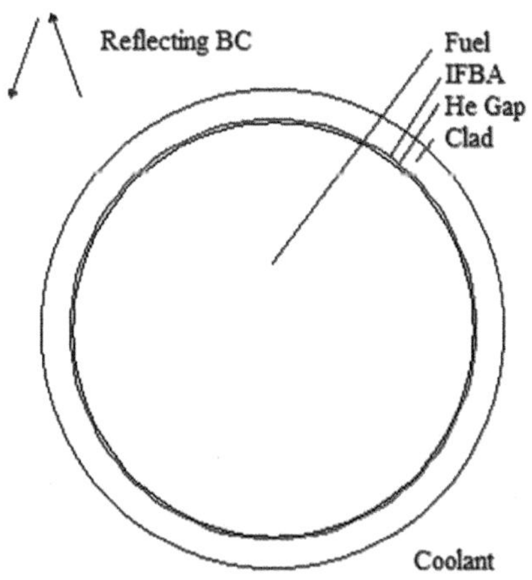

FIGURE 6.13 Reflective boundary condition on a unit lattice cell.

such that the transport cross-section is

$$\Sigma_{tr} = \Sigma_t \left(1 - cf_1\right). \tag{6.120}$$

The average cosine of the scattering angle $\overline{\mu}_0$ is

$$\overline{\mu}_0 = \frac{\int_{-1}^{1} \mu_0 f\left(\mu_0\right) d\mu_0}{\int_{-1}^{1} f\left(\mu_0\right) d\mu_0} = f_1 \tag{6.121}$$

For a scattering, nonfissile, material Eq. (6.120) is

$$\Sigma_{tr} = \Sigma_t - \Sigma_s \overline{\mu}_0. \tag{6.122}$$

The forms of $\nabla \cdot J$ for plane, infinetly long cylinder and sphere are (Section 5.1) of the form,

$$\nabla \cdot J = \frac{d}{dx} J_x,$$

$$\nabla \cdot J = \left(\frac{d}{dr} + \frac{2}{r}\right) J,$$

$$\nabla \cdot J = \left(\frac{d}{dr} + \frac{1}{r}\right) J_r,$$

which can be expressed as

$$\nabla \cdot J = \left(\frac{d}{dr} + \frac{n}{r}\right) J_r$$

with $n = 0, 1, 2$ for plane, cylinder and sphere. Similarly the diffusion equation, from the P_1 equations is

$$-\frac{1}{r^n} \frac{d}{dr} \left(r^n D \frac{d\phi(r)}{dr}\right) + \Sigma_t \phi(r) = Q_0(r). \tag{6.123}$$

6.3.3 The DP_N method

The double spherical harmonics DP_N method, proposed by Yvon (1957; Ziering & Schiff, 1958), gives an expansion of angular flux in the forward and backward hemispheres of angle, due to a singularity at $\mu = 0$, that is, at an angle of incidence $\theta = \pi/2$ on a free surface. The method gives good accuracy for a planar free surface or interface between different materials where the flux could be anisotropic.

The differences between the P_N and the DP_N methods are shown in Table 6.3. In the P_N method, the angular flux and source are expanded in terms of a complete set of Legendre polynomials with angular moments $\phi_l(x)$ and orthogonality in the range $\in (-1, 1)$. In the DP_N method, the expansion is for the half range at the left hemisphere $-1 \leq \mu \leq 0$

TABLE 6.3 Expansions, moments and orthogonality relations for P_N and DP_N methods.

P_N	DP_N
$\phi(x, \mu) = \sum_{l=0}^{N} \frac{2n+1}{2} P_n(\mu) \phi_n(x)$	$\phi(x, \mu) = \sum_{n=0}^{N} (2n+1) P_n(2\mu - 1) \phi_n^+(x), 0 \leq \mu \leq 1$
–	$\phi(x, \mu) = \sum_{n=0}^{N} (2n+1) P_n(2\mu + 1) \phi_n^-(x), -1 \leq \mu \leq 0$
$\phi_n(x) = \int_{-1}^{1} P_n(\mu) \phi(x, \mu) d\mu$	$\phi_n^+(x) = \int_0^1 P_n(2\mu - 1) \phi(x, \mu) d\mu$
–	$\phi_n^-(x) = \int_{-1}^{0} P_n(2\mu + 1) \phi(x, \mu) d\mu$
$\int_{-1}^{1} P_n(\mu) P_m(\mu') d\mu = \frac{2}{2n+1} \delta_{nm}$	$\int_0^1 P_n(2\mu - 1) P_m(2\mu - 1) d\mu = \frac{1}{2n+1} \delta_{nm}$
–	$\int_{-1}^{0} P_n(2\mu + 1) P_m(2\mu + 1) d\mu = \frac{1}{2n+1} \delta_{nm}$

and the half range at the right hemisphere $0 \leq \mu \leq 1$ of the angular domain with angular moments in the positive range, ϕ_l^+ and in the negative range ϕ_l^-. The orthogonality of the Legendre polynomials is separately for both angular domains.

When the flux is not heavily angle-dependent a few terms are sufficient in the spherical harmonics P_N method; conversely a high order P_N expansion is required for good accuracy when the angular dependence is prominent as could be the case for an interface between two very different materials.

Both the DP_0 and the P_1 are approximations of the diffusion equation. It is better to apply Yvon's method near the layer boundaries and for optically thin media while the P_N method would be expected to be better at large distance from boundaries or for optically thick media.

A finite difference form for the DP_N expansion can be obtained by mutilying Eq. (6.98) with $P_m(2\mu - 1)$ and integrating over the interval $0 \leq \mu \leq 1$ and then by $P_m(2\mu - 1)$ and integrating over $-1 \leq \mu \leq 0$. The first term is then

$$\int_0^1 d\mu \mu P_m(2\mu - 1)\frac{\partial \phi(x, \mu)}{\partial x}$$

which becomes

$$\frac{\partial}{\partial x}\sum_{n=0}^{N}(2n+1)P_n(2\mu - 1)\phi_n^+(x)\int_0^1 d\mu \mu P_m(2\mu - 1). \tag{6.124}$$

The integration over μ is carried out using the recurrence relation

$$\mu P_n(\mu) = \frac{1}{2n+1}[(n+1)P_{n+1}(\mu) + nP_{n-1}(\mu)]$$

So that Eq. (6.125) becomes

$$\frac{m}{2m+1}\frac{d\phi_{m-1}^+(x)}{dx} + \frac{m+1}{2m+1}\frac{d\phi_{m+1}^+(x)}{dx}$$

The full equation becomes

$$\frac{m}{2m+1}\frac{d\phi_{m-1}^{\pm}(x)}{dx} + \frac{m+1}{2m+1}\frac{d\phi_{m+1}^{\pm}(x)}{dx} \pm \frac{d\phi_m^{\pm}(x)}{dx} + 2\Sigma_t(x)\phi_m^{\pm}(x)$$
$$= \sum_{l=0}^{L}(2l+1)p_{lm}^{\pm}\Sigma_s(x)\sum_{n=0}^{N}\left(2n+1\right)\left[p_{lm}^+\phi_n^+(x) + p_{lm}^-\phi_n^-(x)\right] \tag{6.125}$$

where

$$p_{lm}^+ = \int_0^1 P_l(\mu)P_m(2\mu - 1)d\mu$$

and

$$p_{lm}^- = \int_{-1}^0 P_l(\mu)P_m(2\mu + 1)d\mu$$

so that $p_{0n}^{\pm} = \delta_{0n}$. Eq. (6.125) are a set of coupled ODEs which can be solved by standard methods.

The DP_N method has been applied by Ziering and Schiff and by Gelbard, Davis, and Pearson (1959) for plane geometry with four angular intervals to calculate the angular flux accurately. By the 1960s, the method had been extended to eigenvalue problems in two-dimensional geometries, and by the 1980s, it had been applied to PHWRs using the one-group and 27-group steady-state integral equation with isotropic scattering and constant source (Krishnani, 1982). A cylindrical lattice cell is used for modeling the 7-, 19- and 28-rod PHWR fuel clusters (Section 3.3) for hexagonal lattices with DP_1 and DP_2 expansions to calculate the flux distribution and overall k_{eff}. The DP_N method for slab geometry was incorporated into a multigroup transport code (Stepanek, 1981) and found to be more efficient than comparable S_N methods. Using a variational principle, the DP_N method has been extended to 2D plane geometry (Ghazaie, Zolfaghari, & Abbasi, 2017) and applied to several benchmarks including a 2D lattice cell and a miniature reactor problem.

6.3.4 The B_N method

In the B_N method, where the B denotes buckling (for spherical or cylindrical bare reactors for example, Section 5.2), the spatial part of the angular flux is assumed to be of a cosine form for a multiplying medium and an imaginary number for a nonmultiplying such as a reflector surrounding a core. The flux is assumed to be of the form

$$\phi(x,\mu,E) = e^{-iBx}\phi(\mu,E). \tag{6.126}$$

and similar expansions are assumed for the source term

$$Q(x,\mu,E) = e^{-iBx}Q(\mu,E)$$

For a bare reactor B is the buckling, and the real part of the above is

$$\phi(x,\mu,E) = \cos Bx\phi(\mu,E).$$

With the expansion given by Eq. (6.126), the "within group" flux equation becomes

$$[\Sigma_t(E) - iB\mu]\phi(\mu,E) = \sum_{n=0}^{\infty}\frac{2n+1}{2}P_n(\mu)\int dE'\,\Sigma_{s,n}(E'\rightarrow E)\int_{-1}^{1}d\mu'\,\phi(E',\mu')P_l(\mu') + \frac{1}{2}S(E) \tag{6.127}$$

Dividing by $[\Sigma_t - iB\mu]$ and multiplying the above by $P_l(\mu)$ and integrating, with

$$\int_{-1}^{1}d\mu P_n(\mu)\phi(\mu,E) = \phi_n(E)$$

and defining

$$\frac{1}{2}\int_{-1}^{1}\frac{P_l(\mu)P_n(\mu)}{[1-\mu g(E)]}d\mu = A_{l,n}$$

and

$$g(E) = iB/\Sigma_t(E)$$

gives

$$\Sigma_t(E)\phi_l(E) = \sum_{l=0}^{\infty}(2n+1)A_{l,n}\int dE'\,\Sigma_{s,n}(E'\rightarrow E)\phi_n(E') + S(E)A_{l,n}. \tag{6.128}$$

For an isotropic source $S_0(E)$,

$$\Sigma_t(E)\phi(\mu,E) = \sum_{l=0}^{\infty}(2n+1)A_{l,n}\int dE'\,\Sigma_{s,n}(E'\rightarrow E)\phi_n(E') + S_0(E)A_{l,0} \tag{6.129}$$

Then

$$A_{0,0}(g) = \frac{1}{2}\int_{-1}^{1}\frac{1}{[1-\mu g(E)]}d\mu = -\frac{1}{2g}\left(\ln|1-g| - \ln|1+g|\right) = \frac{1}{2g}\ln\frac{|1+g|}{|1-g|} = \frac{1}{g}\tanh^{-1}g$$

since

$$\frac{1}{2}\ln\left(\frac{x+1}{x-1}\right) = \tanh^{-1}x.$$

Also

$$A_{1,0} = \frac{1}{2}\int_{-1}^{1}\frac{\mu}{[1-\mu g(E)]}d\mu = \frac{1}{2g^2}\left(-2g - \ln|1-g| + \ln|1+g|\right) = \frac{1}{g}\left(A_{0,0}(g) - 1\right)$$

$$\frac{1}{g(E)}(2n+1)A_{l,n}(g) - (n+1)A_{l,n+1}(g) - A_{l,n-1}(g) = \frac{\delta_{l,n}}{g}.$$

Note that $A_{l,n} = A_{n,l}$.

Consider the B_0 approximation; with $l=0$, the scattering moments are truncated at $\Sigma_{s,0}(E' \to E)$ to give the integral equation

$$\Sigma_t(E)\phi_0(E) = A_{0,0}\int dE'\,\Sigma_{s,0}(E' \to E)\phi_0(E') + S_0(E)A_{0,0}. \tag{6.130}$$

For higher moments ϕ_l,

$$\Sigma_t(E)\phi_l(E) = \sum_{l=0}^{\infty}(2l+1)A_{l,n}\int dE'\,\Sigma_{s,l}(E' \to E)\phi_n(E') + S_0(E)A_{l,n}$$

and the angular flux $\phi(\mu, E)$ is easily found as

$$\phi(B,\mu,E) = \frac{A_{0,0}}{[\Sigma_t(E) - iB\mu]}\left[\int dE'\,\Sigma_{s,0}(E' \to E)\phi_0(E') + S_0(E)\right] \tag{6.131}$$

It is also possible to find all the $\phi_l(E)$ for $n=0$ from

$$\Sigma_t(E)\phi_l(E) = \sum_{n=0}^{\infty}(2n+1)A_{l,n}\int dE'\,\Sigma_{s,n}(E' \to E)\phi_n(E') + S_0(E)A_{l,0} \tag{6.132}$$

In the B_1 approximation, there will be two equations, one for $n=0$ and the other for $n=1$. Group constants, as defined in the two-group transport equation (Section 6.1) can be determined with this method. The general procedure for the B_N method is that a buckling is assumed for which a group-1 calculation is carried out. From this, a multigroup calculation gives the spatial distribution of the flux; if this differs considerably from the assumed distribution, then a series of iterations is carried out until the fluxes converge. When multigroup fluxes are found, the overall system criticality can be found. It is found that the converges of the B_N method is better than that of the P_N method.

6.3.5 The finite element method

The first and simplest numerical methods for computing fluxes were based on easy-to-use finite difference methods (Section 4.5.1) using spatial discretization but these were suited to square-type grids. These were refined in several ways, such as the FEM using "shape functions" (Section 4.5.2) to model irregular geometry as well as polynomial-type variations in the flux. The process of using these finite "elements" which are first solved at a unit level and then combined into a "global" system reduces the coupled differential equations into a system of linear algebraic equations of the form $\bar{\bar{A}}\bar{X} = \bar{B}$ which are subsequently solved by traditional methods of linear algebra. Techniques based on continuous and discontinuous Galerkin FEMs have also been used for improving the computational efficiency. For large problems, iterative schemes and convergence techniques have been used so that the FEM is widely used in reactor physics (Ackroyd, 1992; Kang & Hansen, 1973; Rothenstein & de Oliveira, 1991).

6.3.6 The nodal method with transport theory

In a typical 1000 MW PWR, there could be $\sim$3 M zones for which multiphysics computations would be required for carrying out stead state core analysis and other transient and safety studies. If in each zone a fine mesh structure would be used, the computational burden would become prohibitively large for supercomputers. Thus, in the first step a lattice cell calculation can be done using a fine mesh and reflective boundary conditions with a 2D transport code. Subsequently, in the second step, the resulting fluxes can be used to obtain spatially homogenized cross-sections for a simplified core model with nodes as each of the $\sim$3 M zones which could then be simulated with a less computationally expensive diffusion code (Smith, 1986). The nodal fluxes are then given a reconstructed shape from the nodal expansion function (Lawrence, 1986; Putney, 1986). High speedups with parallel algorithms have been demonstrated for Open multiprocessors for a pin-resolved 3D seven-group NuScale modular reactor core neutron transport calculations (Wang et al., 2019). Similarly, high accuracy and attractive computational speed has been demonstrated for pin power reconstruction by the high order triangle-based polynomial expansion nodal (TPEN) algorithm for a VVER-1000 water-water Russian energetic reactor (Safarzadeh, 2020).

6.3.7 Hybrid methods

There are several problems in nuclear engineering when a deterministic, or stochastic, approach in itself is not sufficient to give accurate answers within the limits of realistic computational effort. One such area is radiation shielding in ducts for which a hybrid computation, coupling a deterministic approach such as a 1D S_N transport computation with Monte Carlo simulation can be beneficial. An initial run of a deterministic code could give the biasing parameters favorable for a detailed Monte Carlo simulation to get "good" results which otherwise may not be possible due to large variances from analog simulations.

Another class of problems is the classical source-detector configuration where the "forward" or standard transport equation represents a solution at a detector given the source and conversely the "backward" or adjoint transport equation represents the importance function, that is, the importance of a particular phase space to the detector. Carrying out such coupled forward-backward computations can give the biasing parameters for a Monte Carlo simulation. These have also been used for design optimization in a variational formulation.

6.3.8 Criticality estimates

The single most important characteristic of a multiplying system is its criticality. In Chapter 2, the criticality of the Godiva and Jezebel assemblies was described, with critical radii given in Tables 2.19 and 2.22. The one-group criticality equation (Section 5.2.3) was the first expression given for estimating the k_{eff} of a bare slab, sphere or cylinder. The next step was to obtain a slightly better two-group criticality formula (Section 5.3.2) in which both fast and thermal leakages were incorporated. From a two-group, the formulation of a multigroup diffusion model was discussed in Section 5.4. The eigenvalue formulation of a steady-state neutron balance equation amounted to a problem in which the largest eigenvalue found had the physical significance of being related to the critical configuration. This was demonstrated for a two-group two-region spherical core surrounded by a reflector; a deterministic model resulted in the critical determinant from which the critical dimension was obtained. It was then understood that a criticality problem could be to determine the critical size of a system given its material composition, or alternately, to determine the material composition that would result in criticality for a certain shape. These were central to the description of nuclear power reactors, propulsion reactors and space reactors in Chapter 3. Whether it was PWRs, BWRs, PHWRs or any other reactor type, the most important design parameter was its critical configuration. We thus came to a description of neutron transport, in the present chapter. The improvement of the formulations in this chapter, over those in the previous chapter, was essentially in the description of the angular flux. This added a complexity to the mathematical description. We now realize that exact solutions of transport theory, done in the early 1960s and 1970s could also give criticality estimates, for example, from the asymptotic flux using the Mark boundary conditions (Eq. 6.96) and requiring the flux to be zero at the extrapolated distance gave the first estimates from this *end point method* of critical radius for a bare sphere as

$$a = \pi|\nu_0(c)| - x_0.$$

As an example, for $= 1.20$, $\nu_0(c) = 1.198$, These are listed in Table 6.4 for some values of c, the mean number of secondary neutrons emerging from a collision; the exact method rnsesults are from a transport theory solution by variational method.

From the exact solutions (Section 6.2) the description of numerical methods such as the S_N and P_N methods gave a better capability to solve eigenvalue multigroup transport problems to determine the systems criticality. Comparisons of transport theory estimates for the Godiva, Jezebel, Topsy and other assemblies (Bell & Glasstone, 1979; Bowen & Busch, 2005; Rawat & Mohankumar, 2011; Rowlands et al., 1999; Thompson et al., 2020) show good results between theory and experiments.

TABLE 6.4 Critical radius estimates of a sphere (in mean free paths); exact vs P_N method.

c	End point	Exact	P_1	P_3	P_5
1.05	7.277	7.2772	7.543	7.296	7.284
1.20	3.172	3.1720	3.513	3.204	3.181
1.60	1.476	1.4761	1.850	1.550	1.497

Source: Bell, G. I., & Glasstone, S. (1979). Nuclear *reactor theory*. New York: Robert E. Krieger Publishing Co., Inc., 1979, Table 2.7, p. 101.

It is seen that a multigroup (typically 6, 12, or 24 groups) S_4, P_3 calculation is sufficient to estimate the criticality of fast systems such as Godiva with certain data libraries. To close this chapter, we also describe methods, such as the MOC which have the capability of performing lattice cell calculations leading to whole-core neutronic eigenvalue calculations useful for the design of nuclear systems described in Chapter 3. Later in this book, we will cover the Monte Carlo method which is another powerful method as well as a simulation tool for the design of nuclear systems.

6.4 Transport theory for reactor calculations

The NDE, based on Poisson's equation, is mathematically much simpler than the integro-differential transport equation and hence is used for simple problems. Even then, numerical solutions are required for large problems such as "whole-core" reactor design. In the multigroup form, the diffusion equation is used to obtain "first estimates" and can serve as a useful step for providing guesses to full transport calculations. A limitation of diffusion equation is near sources and boundaries where the angular flux requires detailed consideration as provided, for example, in the discrete ordinates and spherical harmonics methods.

Reactor core calculations are carried out primarily to determine the overall core flux and distributions during steady-state operation as well as in the case of operational perturbations and in the determination of safety margins in case of accidents.

A typical 1000 MW PWR has an equivalent core diameter of ~3.37 m, height~ 3.66 m with 193 fuel assemblies, each 17 × 17 assembly containing 264 fuel rods. Thus there are 50,952 fuel rods in the core. If each fuel rod is partitioned, for computational purposes, into 50 zones axially then the total number of zones for which computation is required becomes 2,547,600. Within the assembly, we can identify a repeating pattern consisting of a cell with a fuel rod, clad and water coolant.

While it is desirable to carry out calculations for a full reactor core, it is generally quite difficult to do this due to the complicated geometry of the core, consisting of fuel pellets in fuel pins, cladding and structural material. Further, the heterogeneity of the core is due to various types of materials such as variable enrichment fuel, absorbers, and guide tubes. There is also the energy dependence of interaction cross-sections with resonances, as mentioned in Chapter 1. Thus, a two-step procedure is used. In the first step, lattice calculations over unit lattice cells, typically cylindrical, square, hexagonal (Fig. 6.4) or plate-type, are carried out with an ultra-fine and fine energy mesh structure for a fuel pin-cell with reflective boundary conditions. The purpose here is to generate energy- and space-averaged cross-sections, weighted over the neutron flux, from evaluated nuclear data files. In the second step, these "averages" are used as homogenized fuel assemblies with a coarse mesh in an integrated whole-core calculation (Fig. 6.14).

In obtaining "averages" for a whole-core calculation, a considerable amount of data processing and reactor physics is required to accurately obtain multigroup cross-sections particularly accounting for the resonances in the cross-section data. The number of groups used depends on the level of accuracy required; as an example, it is typical to take a 70-group structure for fast reactor cores and over 100 groups for thermal reactors which can be further group-collapsed into two-group cross-sections. The "computational burden" in whole-core calculations is reduced by using the diffusion equations for scalar flux instead of the transport equation for angular flux.

In deterministic methods, computations for the neutron flux are carried out by discretizing in space, energy and angle so that the entire domain is divided into a finite number of spatial zones, each of uniform nuclear properties, a

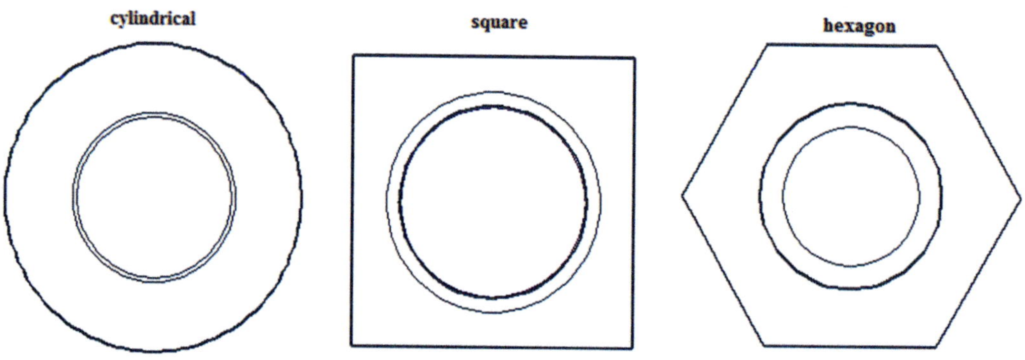

FIGURE 6.14 Cylindrical, square and hexagonal lattice cells.

finite number of energy groups and a finite number of angles. Thus, each discretization has its own limitations such as truncation errors.

In a nuclear reactor core there are typically several assemblies consisting of fuel, control rods, grid spacers, moderator, coolant and supporting structures. These assemblies are placed together in a lattice which could be rectangular as in the case of water-cooled reactors, and hexagonal as in the case of sodium cooled reactors. Thus, the generation of a reactor core mesh can become very complicated and could require large memory storage for a complete description. For this purpose, open-source mesh generation software is available for efficiently modeling realistic cores which could require millions of elements defined by mesh vertices (Jain & Tautges, 2014) requiring a mesh file size of a few GB. Modeling a fast reactor core, such as the Japanese MONJU reactor, comprising eight assembly types of 715 assemblies requires 101 million hexahedral elements generated by 712 processors in parallel. In realistic simulations involving the modeling of neutron transport with thermal hydraulics over possibly more than 1 billion hexahedral elements, the task of mesh generation is cumbersome as is the task of computation necessitating the most powerful supercomputers with hundreds of thousands of CPU cores.

6.4.1 Collision probability method

The collision probability $P_{ij,g}$ is defined as the probability that a neutron born, isotropically in the lab system and with a uniform spatial probability, in any region V_i of the lattice has its first collision in some region of interest V_j of a unit cell. The neutron flux and collision density are found from the spatially discretized multigroup integral form of the transport equation by tracking a neutron in a straight line to its next collision. After tracking a large number of trajectories the collision probabilities for all cells are computed and the scalar fluxes are then determined from a relationship of the form (Cacuci, 2010)

$$\phi_{j,g} = \frac{\sum_i Q_{i,g} V_i P_{ij,g}}{V_j \Sigma_{j,g}} \tag{6.133}$$

where $Q_{i,g}$ is the source term (including fission) for in-scattering into region i and energy group g, $\Sigma_{j,g}$ is the macroscopic total cross-section in region j and energy group g.

In a computer code such as DRAGON (Marleau, Hébert, & Roy, 2011), the integral form of the transport equation is written in terms of collision probabilities for which symmetry and reciprocity relations can be derived. Together with conservation relations (Stoke's theorem across boundaries), and with further assumptions of isotropic and uniform source, simple and compact expressions can be obtained for scalar flux and angular currents.

In the CPM, there are two main disadvantages: (1) full square matrices are produced, for example, a 20×20 matrix for 20 spatial regions, and (2) scattering is restricted to isotropic collisions in the L system. The CP method is thus preferred for problems where the number of regions is small and meshes are unstructured.

6.4.2 Method of characteristics

In the MOC, the solution is found for a fixed angle along a "straight path" across regions divided into uniform zones.

In the integral transport (Section 6.1.2)

$$\phi(\mathbf{r}, \boldsymbol{\Omega}, E, t) = \int_0^{\infty} e^{-\int_0^{s'} \Sigma_t(\mathbf{r}-s''\boldsymbol{\Omega},E)ds''} q\left(\mathbf{r} - s'\boldsymbol{\Omega}, \boldsymbol{\Omega}, E, t - s'/v'\right) ds' \tag{6.134}$$

with

$$q(\mathbf{r}, \boldsymbol{\Omega}, E, t) = \int_0^{\infty} dE' \int d\boldsymbol{\Omega}' \Sigma_s(\mathbf{r}, E', \boldsymbol{\Omega}' \cdot \boldsymbol{\Omega}) \phi(\mathbf{r}, \boldsymbol{\Omega}', E') + \frac{1}{4\pi} \chi(E) \int_0^{\infty} dE' \int d\boldsymbol{\Omega}' \nu \Sigma_f(E') \phi(\mathbf{r}, \boldsymbol{\Omega}', E') + S(\mathbf{r}, \boldsymbol{\Omega}', E') \tag{6.135}$$

the source term is written as

$$q_{i,j}^k = Q_{F,i}^k + Q_{S,i}^k + Q_i^k \tag{6.136}$$

where the indices i, j, k refer to the mesh, characteristic line (CL) and energy group respectively (Fig. 6.15).

A MOC computation flowchart, based on Eqs. (6.136–6.140), is shown in Fig. 6.16. At the first level, a unit lattice cell calculation begins with the selection of a representative cell such as the one shown in Fig. 6.15.

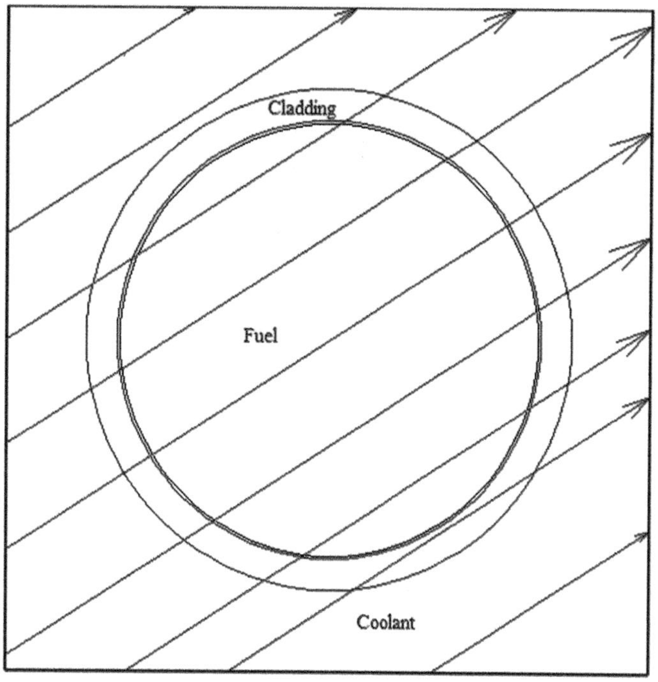

FIGURE 6.15 Characteristic lines in a lattice cell.

The cell is divided into zones within which the data is uniform, such as a fuel zone, a cladding zone and a coolant zone. The group cross-section data for each of these zones is given. The first step is an initialization of scalar fluxes, angular fluxes and the system multiplication k_{eff}.

A CL, represented by the index j, has a specified polar angle θ and a specified azimuthal angle φ; parallel lines are chosen by varying the initial point.

In the first step, with the initialized values, the fission source for zone i, energy group k. $Q_{F,i}^{k}$ is calculated from the initialized scalar fluxes $\varphi_i^{k'}$, as

$$Q_{F,i}^{k} = \frac{1}{4\pi}\frac{\chi(E)}{k_{\text{eff}}}\sum_{k'=1}^{G} \nu\Sigma_F^{k'}\phi_i^{k'}.$$

The fission source is calculated from Eq. (6.137) for all energy groups k in all meshes i.

In the second step, the scattering source $Q_{S,i}^{k}$ is calculated from all in-scattering cross-sections and scalar fluxes.

In the next step the outgoing angular flux $\phi_{i,j}^{\text{out},k}$ from a mesh is calculated from the incoming angular flux along a CL (j), $\phi_{i,j}^{\text{in},k}$

$$\phi_{i,j}^{\text{out},k} = \phi_{i,j}^{\text{in},k}e^{-\Sigma_i^k\Delta s} + \frac{Q_i^k}{\Sigma_i^k}\left(1 - e^{-\Sigma_i^k\Delta s}\right) \tag{6.137}$$

where $Q_i^k = Q_{F,i}^k + Q_{S,i}^k$ and Δs is the distance traveled along the CL across the mesh.

The average angular flux, along this CL is

$$\overline{\phi}_{i,j}^{k} = \frac{Q_i^k}{\Sigma_i^k} + \frac{\phi_{i,j}^{\text{in},k} - \phi_{i,j}^{\text{out},k}}{\Sigma_i^k\Delta s} \tag{6.138}$$

When all such CLs are processed, their average is the cell averaged group scalar flux $\overline{\overline{\phi}}_i^k$ from which the number of fissions, summed over all I meshes and G energy groups, is calculated as

$$\sum_{i=1}^{I}\sum_{k=1}^{G}\nu\Sigma_F^k\phi_i^k \tag{6.139}$$

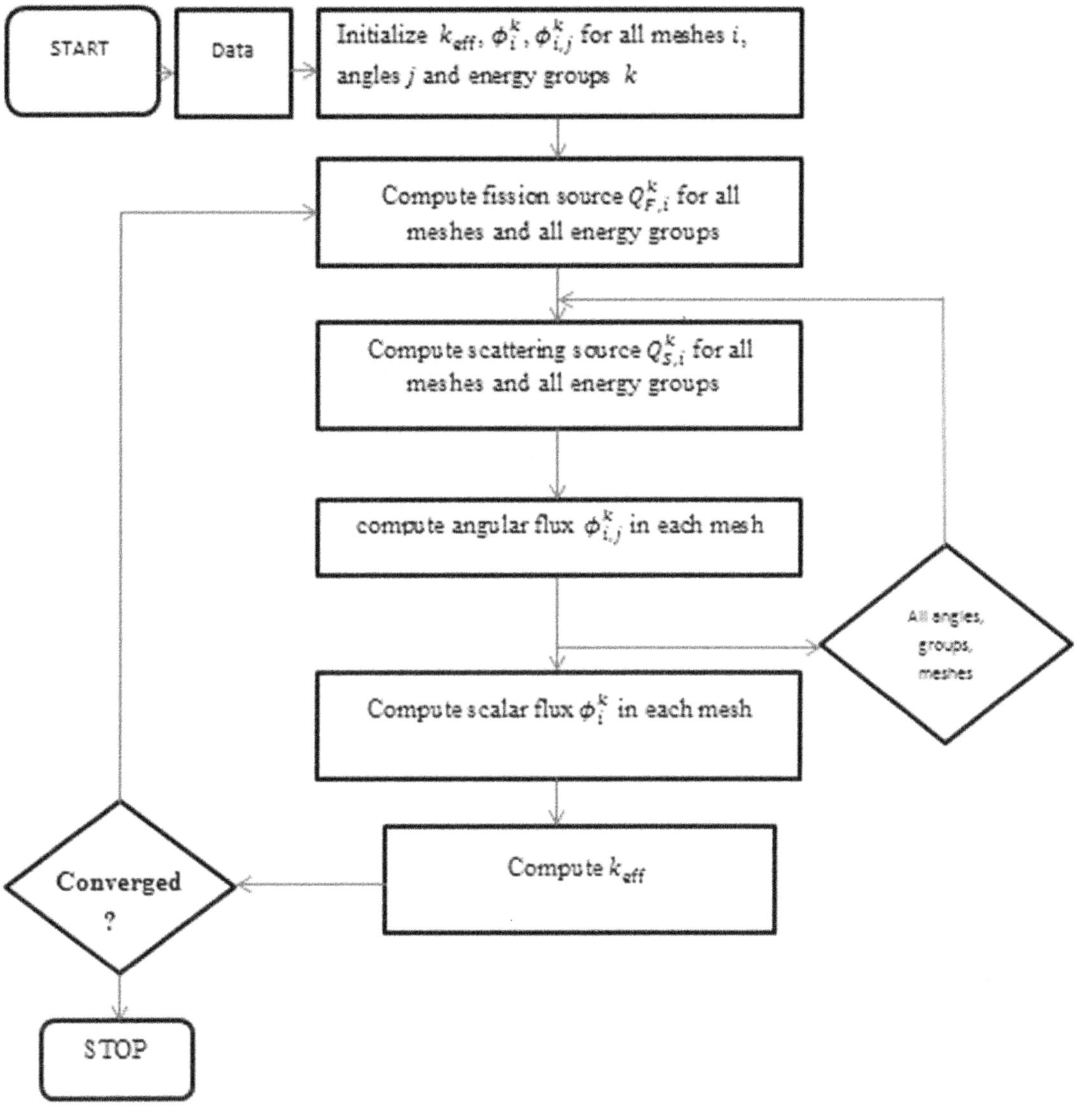

FIGURE 6.16 Flowchart for the method of characteristic in a lattice cell.

If the "new" scalar fluxes for the mth generation $\phi_i^k = \phi_i^{m,k}$ are used for the present *generation* and the "old" scalar fluxes $\phi_i^k = \phi_i^{m-1,k}$ are used for the previous, or $m-1$th, generation, then the ratio of the mth and $m-1$th generation of fissions neutrons would be the mth estimate of the system multiplication $k_{\text{eff}}^{(m)}$; thus

$$k_{\text{eff}}^{(m)} = \frac{\sum_{i=1}^{I}\sum_{k=1}^{G} \nu\Sigma_F^k \phi_i^{m,k}}{\sum_{i=1}^{I}\sum_{k=1}^{G} \nu\Sigma_F^k \phi_i^{m-1,k}} \tag{6.140}$$

Inner and outer iterations are carried out until the scalar fluxes and the $k_{\text{eff}}^{(m)}$ converge. This is the essence of an eigenvalue calculation at the level of the unit lattice cell.

There are several solvers used in MOC applications in reactor physics such as the cyclic tracking concept using infinite tracks with periodic characteristics and finite tracks with explicit boundary conditions. The latter are "faster solvers" and therefore more widely used in production codes. The numerical computation in the MOC is based on the ODE form, called the characteristic form, of the NTE and a track is computed as it traverses across homogenized spatial zones, crossing boundaries in a fixed solid angle, through the entire domain. Integrals are cast using quadrature schemes to compute scalar fluxes and boundary currents.

Several computation strategies have been developed and demonstrated for acceleration with advantages in solving unstructured 2D and structured 3D reactor core configurations. The MOC has been implemented in WIMS-E (Lindley et al., 2017) and DRAGON codes. The MOC has been used for fundamental lattice calculations in several applications such as in an ACR-type cell (Le Tellier & Hébert, 2005) consisting of a cluster geometry with light water coolant and heavy water moderator. Algorithms have been successfully implemented for full core simulations with modern single instruction multiple data (SIMD) computer architectures for high performance computing (Tramm et al., 2016).

Problems

1. Compare the neutron flux obtained by transport theory with that obtained by diffusion theory in a finite sphere with a point isotropic source at its center. Plot the flux for a source of 10^7 neutrons s^{-1}.
2. The asymptotic flux in a source-free sphere is given as $r\varphi_{as}(r) = A \sin \frac{r}{|\nu_o|}$. In end point theory an estimate for the critical radius R is obtained by putting the asymptotic flux equal to zero at the extrapolated boundary. This gives $R = \pi|\nu_o(c)| - d$, where d is the extrapolation radius. Using the data for Godiva estimate the critical radius and compare with that obtained with diffusion theory.
3. Compare the asymptotic flux of Problem 2 with the exact transport theory asymptotic flux and with the diffusion theory flux in Chapter 5.
4. Compare the reflectivity, or albedo, $\phi(0, -\mu) = 1 - \sqrt{1 - c}H(\mu)$ obtained from transport theory: with the corresponding estimate from diffusion theory and calculate the albedo of a slab of graphite 3 m.f.p. thick.

Nomenclature

English lower case

d	extrapolation distance
k_{eff}	effective multiplication
k_∞	infinite multiplication
n	number density

English upper case

A_i	area
B_g	geometrical buckling
B_m	material buckling
B_N	the B_N method
$\hat{C}$	collision operator
C_i	concentration of precursor
D	
$\hat{D}$	derivative operator
DP_N	the DP_N (double spherical harmonics) method
G	Green's function
J	neutron current
L	diffusion length
$\hat{L}^+$	adjoint operator
N_i	atomic density of the ith nuclide
N_i	shape function in the ith element
P_{es}	escape probability
P_l^m	associated Legendre functions
P_N	the P_N (spherical harmonics) method
R_c	critical radius
S	source
S_N	the S_N (discrete ordinates) method
$\hat{T}$	transport operator

Greek lower case

λ	decay constant
μ	cosine of angle of scattering
μ_0	average cosine of scattering angle
ϕ	flux
ϕ^+	adjoint flux
ϕ_C	complementary solution
ϕ_P	particular solution
τ	neutron age
ν	
χ	fission spectrum
ψ	collision density

Greek upper case

Ω	solid angle
$\sum_a$	macroscopic absorption cross-section
$\sum_f$	macroscopic fission cross-section
$\sum_r$	macroscopic removal cross-section
$\sum_s$	macroscopic scattering cross-section
$\sum_{tr}$	macroscopic transport cross-section
$\sum_t$	macroscopic total cross-section

Abbreviations

FD	finite difference method
FEM	finite element method
FVM	finite volume method
IFBA	integral fuel burnable absorber
MOC	method of characteristics
NDE	neutron diffusion equation
NEM	nodal expansion method

References

Ackroyd, R. T. (1992). Finite element methods for neutron transport based on maximum and minimum principles for discontinuous trial functions. *Annals of Nuclear Energy*.

Bell, G., & Glasstone, S. (1952). *Nuclear* reactor theo*ry*. van Nostrand Reinhold Company.

Bell, G. I., & Glasstone, S. (1979). *Nuclear* reactor the*ory*. New York: Robert E. Krieger Publishing Co., Inc..

Bowen, D. G., & Busch, R. D. (2005). Hand calculational methods for criticality safety – A primer. In *2005 NCSD topical meeting (American Nuclear Society Nuclear Criticality Safety Division)*.

Cacuci, D. G. (2010). *Handbook of Nuclear Engineering*, *2*.

Carlson, B. G., & Lathrop, K. D. (1968). *Transport theory the method of discrete ordinates. Computing methods in reactor physics*. .. .

Case, K. M., & Zweifel, P. F. (1967). *Linear* transport th*eory*. Reading, MA: Addison-Wesley Pub. Co..

Cercignani, C. (1988). *The Boltzmann* equation and its applica*tions*. Springer.

Chandrasekhar, S. (1960). *Radiative transfer*. Dover Publications.

Clark, M., & Hansen, K. F. (1964). *Transport methods. Numerical methods of reactor analysis*. .. .

Davison, B. (1957). *Neutron* transport theo*ry*. New York: Oxford University Press.

Duderstadt, J. J., & Hamilton, L. J. (1976). *Nuclear* reactor ana*lysis*. John Wiley and Sons.

Emmett, M. B. (2000). *MORSE: Present capabilities and future directions. Applied radiation and isotopes*. .. .

Erdmann, R. C., & Siewert, C. E. (1968). Green's functions for the one-speed transport equation in spherical geometry. *Journal of Mathematical Physics*.

Ganapol, B. D. (2008) *Analytical benchmarks for nuclear engineering applications case studies in neutron transport theory*.

Gelbard, E., Davis, J., & Pearson, J. (1959). Iterative solutions to the P 1 and double- P 1 Eq. 1. *Nuclear Science and Engineering*.

Ghazaie, S. H., Zolfaghari, A., & Abbasi, M. (2017). An extended half-range spherical harmonics method for first-order neutron transport equation based on variational treatment. *Progress in Nuclear Energy*.

Harris, S. (1971). *An introduction tot he theory of the Boltzmann equation*. New York: Dover Publications, Inc.

Henry, A. F. (1975). *Nuclear* reactor anal*ysis*. The M.I.T. Press.

Jain, R., & Tautges, T. J. (2014). Generating unstructured nuclear reactor core meshes in parallel. *Procedia Engineering*.

Kang, C. M., & Hansen, K. F. (1973). Finite element methods for reactor analysis. *Nuclear Science and Engineering*.

Krishnani, P. D. (1982). Interface current method for PHWR cluster geometry with anisotropy in the angular flux at interfaces. *Annals of Nuclear Energy*.

Lawrence, R. D. (1986). Progress in nodal methods for the solution of the neutron diffusion and transport equations. *Progress in Nuclear Energy*.

Le Tellier, R., & Hebert, A. (2005). Application of the characteristics method combined with advanced self-shielding models to an ACR-type cell. In *26th Annual Canadian Nuclear Society Conference and 29th CNS/CNA Student Conference*.

Lewis, E. E., & Miller, J. W. F. (1984). *Computational* methods of neutron transport. New York: John Wiley and Sons.

Lindley, B. A., et al. (2017). Current status of the reactor physics code WIMS and recent developments. *Annals of Nuclear Energy*.

Marleau, G., Hébert, A., & Roy, R. (2011). A user guide for DRAGON version 4. *Inst. génie nucléaire, Département génie mécanique, École Polytech. Montréal. (Report IGE–294)*.

Putney, J. M. (1986). A hexagonal geometry nodal expansion method for fast reactor calculations. *Progress in Nuclear Energy*.

Rawat, A., & Mohankumar, N. (2011). Benchmark results for the critical slab and sphere problem in one-speed neutron transport theory. *Annals of Nuclear Energy*.

Reuss, P. (2008). *Neutron* physics. Cedex, France: EDP Sciences.

Rief, H. (1984). Generalized Monte Carlo perturbation algorithms for correlated sampling and a second-order Taylor series approach. *Annals of Nuclear Energy*.

Robson, R. E., White, R. D., & Hildebrandt, M. (2017). *Fundamentals of charged particle transport in gases and condensed matter*.

Rothenstein, W., & de Oliveira, C. R. E. (1991). Discrete energy or multigroup finite element transport calculations in lattice physics. *Progress in Nuclear Energy*.

Rowlands, J., et al. (1999). *Intercomparison of calculations for Godiva and Jezebel*.

Safarzadeh, O. (2020). Development of the triangle-based nodal algorithm for reconstructing pin power distributions. *Progress in Nuclear Energy*.

Siewert, C. E., & Grandjean, P. (1979). Three basic neutron transport problems in spherical geometry. *Nuclear Science and Engineering*, *70*, 96–98.

Smith, K. S. (1986). Spatial homogenization methods for light water reactor analysis. *Progress in Nuclear Energy*.

Stepanek, J. (1981). The DP N surface flux integral neutron transport method for slab geometry. *Nuclear Science and Engineering*.

Thompson, N., et al. (2020). National Criticality Experiments Research Center (NCERC) - Capabilities and recent measurements. *EPJ Web of Conferences*.

Tramm, J. R., Gunow, G., He, T., Smith, K. S., Forget, B., & Siegel, A. R. (2016). A task-based parallelism and vectorized approach to 3D Method of Characteristics (MOC) reactor simulation for high performance computing architectures. *Computer Physics Communications*.

Wang, Y., Zhang, T., Lewis, E. E., Yang, W. S., Smith, M. A., & Wu, H. (2019). Three-dimensional variational nodal method parallelization for pin resolved neutron transport calculations. *Progress in Nuclear Energy*.

Williams, M. M. R. (1971). *Mathematical* methods in particle transport theory. New York: John Wiley & Sons.

Yvon, J. (1957). La diffusion macroscopique des neutrons une methode d'approximation. *Journal of Nuclear Energy*, *4*(3), 305–318.

Ziering, S., & Schiff, D. (1958). Yvon's method for slabs. *Nuclear Science and Engineering*.

Chapter 7

The Monte Carlo method

The Monte Carlo (MC) method (Dunn & Shultis, 2012; Kalos & Whitlock, 2008; McClarren, 2018) for radiation transport simulation is a powerful methodology for modeling realistic and often complex engineering systems. Though the method, with all its mathematical rigor and its strong foundations in the laws of probability, was fairly established, it was not until the Manhattan Project that it attracted the attention of nuclear scientists challenged with the estimation of the composition, dimensions and configurations of nuclear weapon systems.

The name "Monte Carlo" was given by Nicholas Constantine Metropolis (1915–99), of the Theoretical Physics Division of the Manhattan Project, in 1942, (Keys & Groves, 1963) to this statistical process of simulation (Metropolis & Ulam, 1949). His colleagues included Teller, John von Neumann, Stanislaw Ulam, and Robert Richtmyer. Those were the days of MANIAC, The Mathematical and Numerical Integrator and Computer, designed according to von Neumann's principle of the stored program (Anderson, 1986). It is said that Fermi used statistical sampling techniques as early as 1934, when he was working on neutron diffusion in Rome. When, in 1944, the development of the atomic-bomb project entered its final phase; the three problems at the center of the nuclear weapons program: neutron transport, behavior of materials under very temperatures and pressures, and fluid dynamics, required mathematical computation for design and prediction.

Earlier work was developed by Hammersley and Handscombe and Spanier and Gelbard (Hammersley & Handscomb, 1964; Hammersley, Handscomb, & Weiss, 1965; Spanier, Gelbard, & Bell, 1970). It does not per se involve the solution of any equation but can be interpreted to follow the integral form of the transport equation where the flux is expressed in terms of the uncollided and collided terms.

Several mathematical methods were developed, along with powerful computing hardware, that led to the successful testing of a nuclear weapon and the advent of the first electronic computer during World War II.

Since then, MC methods have been important in many areas of science and engineering, as well as to medicine and forecasting. In nuclear engineering, MC simulation is an essential component of the curriculum, starting preferably at the advanced undergraduate level. A picture of radiation transport in the mind of a student who has not been exposed to MC could be restrictive in several ways. Here, it is easier to relate to concepts such as flux, current, and the overall transport flow in complex geometries.

This chapter builds upon the fundamentals described for charged particles and gamma transport in Chapter 1, neutron interactions with matter in Chapter 2, mathematical foundations given in Chapter 4, and diffusion and transport modeling in Chapters 5 and 6. Here the simulation process is applied to the nuclear systems covered in Chapter 3 with emphasis on modeling fission and fusion systems.

7.1 Stochastic simulation

7.1.1 Markov processes

In Section 4.9.2, a Markovian process was described as a stochastic process in which the next state P_{i+1} depends on its present state $P_i$but there is no memory of the past.

The transport kernel (Section 6.1.4) $\hat{T}^{g}\left(\boldsymbol{r}'\to\boldsymbol{r};\hat{\Omega}'\right)$ represents a neutron at $\boldsymbol{r}',\hat{\Omega}',E'$ as shown in Fig. 7.1 being transported to position $\boldsymbol{r}$ where an interaction takes place represented by $\hat{C}^{g'\to g}\left(\boldsymbol{r};\hat{\Omega}'\to\hat{\Omega}\right)$ which changes the phase space of the neutron from $P(\boldsymbol{r}',\hat{\Omega}',E')$ to $P(\boldsymbol{r},\hat{\Omega},E)$; there is thus no dependence on any state prior to the state $P(\boldsymbol{r}',\hat{\Omega}',E')$. Since the interaction has no dependence on another state prior to $P\left(\boldsymbol{r}',\hat{\Omega}',E'\right)$, it is classified as a Markovian process.

7.1.2 Events in a random walk

As described in Chapter 1, a neutron source such as californium-252 can emit millions of neutrons per second. Each neutron is thus *born* as a source neutron at a position $\boldsymbol{r_0}$ with a certain energy E_0 at an angle Ω_0; it undergoes

Nuclear Engineering. DOI: https://doi.org/10.1016/B978-0-323-90618-0.00007-7

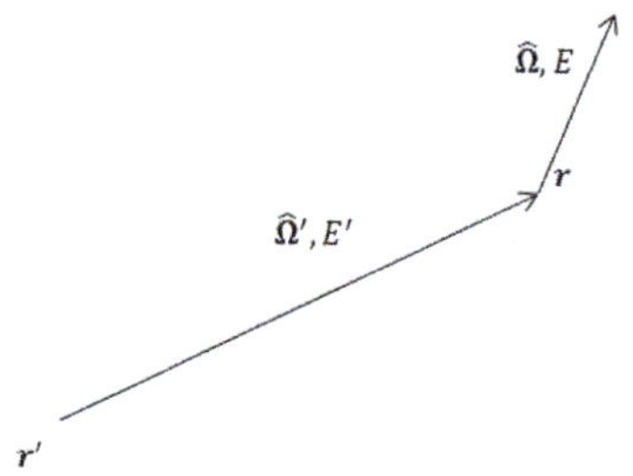

FIGURE 7.1 A neutron interaction.

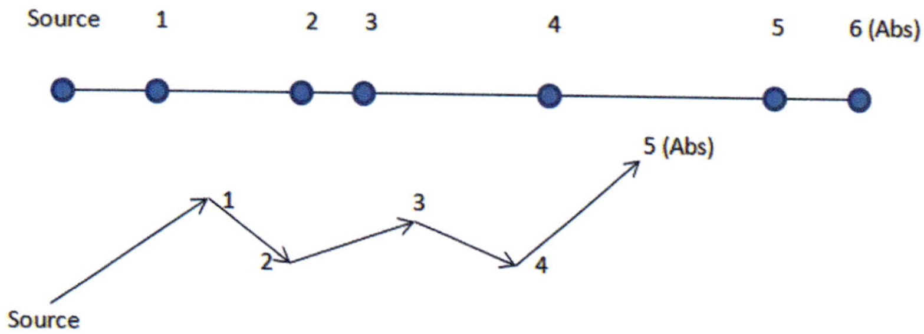

FIGURE 7.2 Random walks.

interactions in a *random walk* consisting of events. Two such random walks are shown in Fig. 7.2; in the first instance, a source neutron undergoes scattering collisions at locations 1–5 until it is absorbed at position 6. The second random walk shows a source neutron undergoing five interactions; the first four are scattering collisions with a change of angle (and of course energy) while the fifth interaction is an absorption which terminates its random walk. An absorption can be termed as the *death* of a neutron as it ends the random walk. In fact any event that takes a neutron out of the domain of interest, such as a leakage across the physical boundary of a nuclear system, is called a death. A random walk is a birth-to-death process with millions or several more such random walks considered in a simulation.

In Chapter 2, we discussed the several types of neutron interactions that are possible such as elastic scattering, inelastic scattering, radiative capture, fission and many more.

7.1.3 The physics of interactions

In MC simulation, the physics of interactions is modeled with very elaborate methods as it is crucial to have accurate knowledge of exactly what takes place in an interaction. As described in Section 2.2 (neutron interactions), radiative capture is understood to be a process in which a compound nucleus is formed for a very short while before the emission of a gamma ray. The details of such an interaction are part of a MC simulation for the calculation of postcollision parameters.

7.1.4 Nuclear interaction data

Data files such as the evaluated nuclear data file (ENDF/B) (Brown et al., 2018) briefly described in Chapter 6 contain elaborate neutron cross section data for $Z = 0 - 95$ (neutron to Amercium-241), neutron reaction sublibrary with 557 materials and photon data for 163 materials mostly with evaluations up to 140 MeV. There are fifteen sublibraries with photonuclear, photoatomic, radioactive decay, spontaneous fission yields and charged-particle cross sections. These libraries are used for the charged particle and gamma transport MC simulations described in Chapter 1 and neutron transport simulations described in Chapter 2.

ENDF/B files are part of the MCNP and SCALE packages but for other codes, group cross sections are prepared with codes such as NJOY and WIMSD as described in Chapters 5 and 6 for diffusion and transport codes. In MC simulation, the ENDF/B cross sections are used directly rather than going through the process of group-averaging.

7.1.5 How do we know an answer is good?

In MC simulation, quantities such as flux and current are estimated from an average of several random walks or histories. Generally, the larger the number of histories simulated, the greater will be the confidence level of an estimate, as predicted by the Central Limit Theorem. However, for a tally that is not well-behaved, such as in a rare event simulation, it is not necessary that the confidence level will increase with the number of histories. It is thus important to understand how reliable an estimate is. Two statistical quantities play a fundamental role here namely the mean of an estimate $\overline{x}$ and its variance σ^2. Actually we know the sample mean and the sample variance rather than the population mean and the population variance. The relationship between the sample mean and population mean is through the Law of Large Numbers, which states that as the sample size $N \rightarrow \infty$, the sample mean $\overline{x}$ tends to the true mean, or the population mean.

So that leaves us with the question: how is the sample variance σ_s^2 related to the population variance σ^2?

The sample variance of the mean $\sigma_s^2(\overline{x})$ is another quantity that is possible to calculate; this is expected to get smaller as the sample size increases. Thus

$$\sigma_s^2(\overline{x}) = \frac{\sigma_s^2}{N},$$

which means that we get continually better estimates of the sampled mean.

Our objective, to have as small as possible a value of $\sigma_s^2(\overline{x})$, can be achieved if we reduce the sample variance or increase the sample size. The first is possible by variance reduction techniques such as importance sampling in which we bias a simulation to give favorable estimates. As an example, if a quantity is known to be reducing at an exponential rate, then we would give more importance to lower values in a simulation. Thus instead of using uniformly distributed random numbers, we would use exponentially distributed random numbers. A simple demonstration of such an importance sampling will be demonstrated for the evaluation of integrals in Section 7.5.

It is now important to define what we mean by the quality of an estimate. If we get a bad estimate with a small variance then the answer can be classified as inaccurate but precise. For example, if the true value of a quantity is 32, and our estimate is 26 ± 0.0001, then this estimate is inaccurate but in terms of standard deviation, it is a precise estimate. Increasing the sample size N would be futile. On the other hand, an estimate of 31 ± 2 is a better estimate as the mean is closer to the true mean, while the standard deviation is much higher than that of the previous estimate. Here, it would be reasonable to expect an improvement in the estimate by increasing the sample size.

In MC simulations, the underlying model uses physics as well as cross section data which are in-built and we have no control on them. The dynamics of a neutron interaction that determine, for example, compound nucleus formation, are based on quantum physics (Section 2.8). We thus determine Doppler broadening from a given model. The two factors in our control are the coding and modeling of a problem. Coding, especially in the case of large computer codes, relies heavily on mathematical techniques used such as finite-difference, FEM, etc., to solve governing ODEs, PDEs and integro-differential equations. The modeling of a problem is in the hands of a user; free choices include the use of a particular method in preference to another, the use of a variance reduction scheme and the number of histories simulated. These choices determine the precision of an estimate.

In Chapter 4, the Central Limit Theorem was used to show that the distribution of the estimated means $\overline{x}$ would be approximately normal; the area under such a distribution would be ~ 0.68 within one standard deviation and ~ 0.95 within two standard deviations. These would serve as confidence levels for estimates obtained from a MC simulation. A useful metric would then be a relative error defined as the ratio of the standard deviation of the mean divided by the mean

$$R = \frac{\sqrt{\sigma_s^2(\overline{x})}}{\overline{x}}.$$

As described in MCNP Volume I, in the best estimate, R would be zero, while in a bad estimate R would be as large as the estimated mean $\overline{x}$. The relative error would thus lie in the range zero to one. In MCNP, the relative error is classified as generally reliable when $R < 0.10$ for a detector other than a point detector, and $R < 0.05$ for a point detector.

From the definition of the mean and standard deviation, the relative error is

$$R = \left[\frac{\sum_{i=1}^{N} x_i^2}{\left(\sum_{i=1}^{N} x_i \right)^2} - \frac{1}{N} \right]^{1/2}$$

which means that if a simulation is carried out for which each x_i is the same $x_i = \mu$, then

$$R = \left[\frac{\mu^2 \sum_{i=1}^{N} 1}{\left(\mu \sum_{i=1}^{N} 1 \right)^2} - \frac{1}{N} \right]^{1/2} = 0.$$

An easy example is the integral

$$I = \int_0^1 e^{-x} dx$$

for which the PDF can be the uniform PDF $f(x) = 1$, and the estimator is $g(x) = e^{-x}$ (in which case x_i's are all different) or alternately, in which the PDF is $f(x) = e^{-x}$ and the estimator is $g(x) = 1$ (in which case each history gives the same estimate). This choice of a PDF-estimator combination gives a zero-variance result. So, even one sample would be sufficient to give the correct answer.

When a zero-variance result is not foreseen, the value of the relative error squared is expected to decrease with an increase of the sample size N while the computer time would increase more-or-less linearly with N. It would thus be fair to expect that the quantity R^2T would remain constant; this quantity, in MCNP, is listed in the output as a Figure of Merit, FOM $= 1/(R^2T)$, from which the relative error is given as

$$R = \frac{1}{\sqrt{\text{FOM} \times T}}.$$

Another useful indicator for the quality of the results given by the MCNP code, is the variance of the variance (VoV) which is the estimated relative variance of the relative error R; a value less than 0.1 is taken to be indicative of a good result. Again, the VoV is expected to decrease as the sample size N increases. For a much more rigorous discussion the user is referred to the MCNP Volume I.

7.2 Simulation of a random walk

The integral equation "simulated" in MC for the collision density χ in a multigroup formulation with the collision and transport operators as discussed in Chapter 6 is

$$\chi^{(g)}\left(\vec{r}, \hat{\Omega}, t\right) = \hat{C}^{g' \to g}\left(\vec{r}; \hat{\Omega}' \to \hat{\Omega}\right) \hat{T}^{g'}\left(r' \to r; \hat{\Omega}'\right) \chi^{(g')}\left(r', \hat{\Omega}', t\right) + S^{(g)} \tag{7.1}$$

where $\hat{C}$ and $\hat{T}$ are the collision and transport operators. The MC procedure consists of starting a "source" neutron, transporting it to a collision site, processing the collision, making the appropriate tallies and continuing the simulation until the neutron is lost by some terminating process. Many such neutrons are simulated, and at the end of the simulation, tallies are averaged with both mean and variance computations. The MC process, which can be understood to be a simulation of the integral formulation, is also known as a "Neumann series" solution as will be outlined in more detail in the next section.

7.2.1 Monte Carlo simulation

In the MC method, a "large sample" of neutrons, representative of the total "population" is simulated to infer on quantities of interest using laws of probability. The MC approach allows modeling complexity of geometry and collision physics and hence is not limited to idealizations of the sort that limit deterministic approaches. One of its' drawbacks is heavy computational effort; this can be addressed by powerful hardware and by efficiency-improving techniques which result in considerable "speed-up."

Fig. 7.3 shows the transport of neutrons in matter. Three starting neutrons, A, B, and C, are shown here though in actual situations there can be a large number such as 10^{14} neutrons in a nuclear power reactor. Neutron A collides with a host nucleus which undergoes the nuclear fission reaction resulting in the prompt emission of three "first generation" neutrons *A-1*, *A-2*, and *A-3* while the nucleus recoils to a new position. Neutron *A-1* is captured by a host nucleus, while *A-2* collides with another nucleus which undergoes fission and emission of two neutrons *A-2−1* and *A-2−2*. The third neutron *A-3* collides with a host nucleus and scatters "off" it, that is, the neutron transfers some of its' energy to the nucleus. Thus the number of first generation neutrons is three. Each of these neutrons continue to be transported and collide with host nuclei undergoing one of a large possibilities of nuclear reactions, until they are either captured or they escape from the physical domain by crossing the boundary. In a straightforward simulation of neutron transport, each neutron is followed from its' birth as a source neutron to its' death when it is captured or has escaped. The birth-to-death process is called a "history" and the straight paths in between interactions during a history is called a "random walk" as it each interaction is determined randomly from one of several possible interactions. The random walk is "stochastic" in nature, that is, it may not repeat itself, and it is Markovian in the sense that the next event is not related to

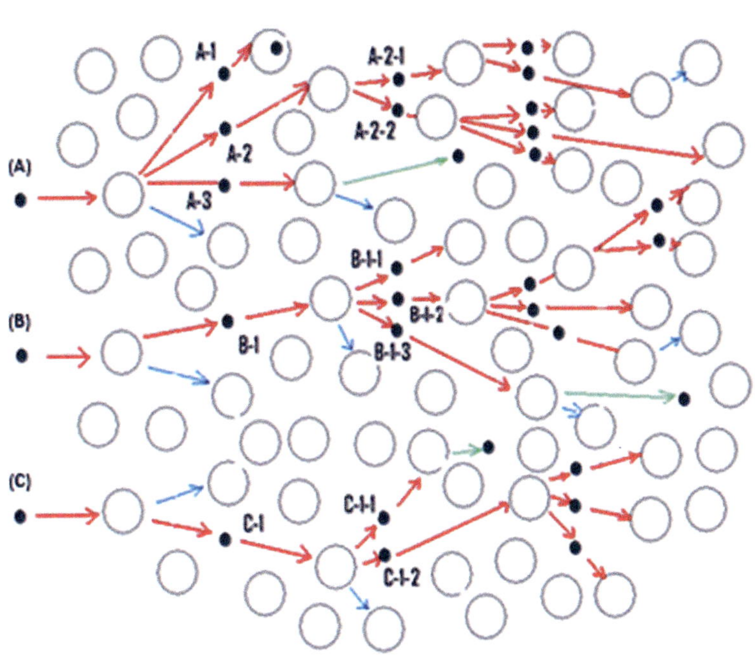

FIGURE 7.3 Neutron transport in matter.

the entire history of a neutron, just to its immediately preceding event. During a history, "tallies" are updated from which useful information on the neutron flux and subsequent reaction rates can be obtained.

In the MC simulation of neutron transport, several such histories are simulated, and their tallies are processed using laws of probability and statistics to infer on the averages of quantities obtained from the simulated sample. The usefulness of the MC method is largely due to the high-speed computing possible with modern processors capable of gigaflop computations.

As a result, nuclear reactors, nuclear weapons and other nuclear systems are designed with great precision using MC simulation. Since the 1940s MC applications in the Manhattan Project, the techniques have been adapted to a wide range of areas in computational mathematics and physics, econometrics and forecasting, to biology and medicine.

7.2.2 Estimators and tallies

The fundamental quantity of interest is the flux $\phi(\mathbf{r}, \boldsymbol{\Omega}, E, t)$ which can be obtained in exact analytical form from the transport equation for idealized conditions (Section 6.1) but in greater detail through numerical methods and simulation. In general, this quantity is obtained by dividing the phase space into a number of *bins* in which tallies are scored. In MC simulation, a result is obtained in a specified region of interest rather than over the whole domain as for diffusion and transport calculations. Thus the bin tallies amount to a multiangle multigroup approach which, as we will see, is a convenient approach for obtaining reliable estimates of the angular flux and from it, any reaction of interest.

Consider the three current and flux tallies (Werner, 2017)

$$J = \int dt \int dE \int d\Omega \int dA |\Omega \bullet \hat{n}| \quad \phi(\mathbf{r}, \boldsymbol{\Omega}, E, t), \tag{7.2}$$

the surface flux

$$\overline{\phi}_S = \frac{1}{A} \int dt \int dE \int d\Omega \int dA \qquad \phi(\mathbf{r}, \boldsymbol{\Omega}, E, t), \tag{7.3}$$

and the volume-averaged cell flux

$$\overline{\phi}_V = \frac{1}{V} \int dt \int dE \int d\Omega \int dA \quad \phi(\mathbf{r}, \boldsymbol{\Omega}, E, t). \tag{7.4}$$

The track-length and collision estimators (CE) are used to score the scalar neutron flux $\phi \equiv nv$, where n is the number density (neutrons/cm^3) and v is the neutron speed (cm/s). The flux, with units of neutrons/cm^2/s, represents the rate of neutrons crossing a unit area perpendicular to their direction of motion.

Typical tallies required in neutron/gamma transport calculations of nuclear fission reactors and systems are the system multiplication k_{eff}, neutron and gamma fluxes as well as their distribution in space and associated reaction rates such as radiative capture (gamma production), radiation dose, energy deposition. The flux loading determines the effect of radiation on materials in the reactor such as structural materials, coolants, moderators, control and instrumentation materials and wiring. The effect of radiation streaming through ducts and ports as well as the radiation environment external to the reactor/nuclear systems is an integral part of all neutron/gamma transport calculations.

For fusion reactors, such as ITER (International Thermonuclear Experimental Reactor), briefly described in Section 3.10, and the European DEMO reactor (Demonstration Fusion Power Reactor) beyond ITER, the tallies of interest are similar with little differences from nuclear fission reactors due to the differences in basic engineering design. In fusion reactors, very high temperatures and their loading on the vacuum vessel first wall (FW) as well as beyond the FW, the effect on superconducting coils, the tritium breeding in blanket are required in addition to the usual neutron/gamma transport calculations in fission systems. Radiation damage to steel affects its life and operation; it is calculated in terms of displacements per atom (dpa) per full power year (FPY).

Tallies for water in PWRs as well as in fusion reactor designs utilizing water as coolant, such as DEMO water-cooled lithium lead, are similar as they are based on the low water density at high temperatures and pressures.

Two main differences between fission and fusion power reactors are the neutron energy which is $\sim$ 1 MeV on the average from fission neutrons while a D-T fusion reactor would be a powerful source of 14 MeV neutrons. The other difference is that in a fusion reactor, the neutrons are born in the central vacuum chamber and stream around relatively easily due to gaps between surrounding blanket modules and more ducts and channels.

All tallies are based on the neutron/gamma flux, that is, on the current, surface fluxes, and volume-averaged fluxes [Eqs. (7.2)–(7.4)] and associated reaction rates R_x (Section 2.6) of the form

$$R_x = C\int dt \int dE\sigma_x(E) \quad \phi(\mathbf{r}, E, t). \tag{7.5}$$

where $\sigma_x(E)$ is a microscopic cross section for reaction type x (taken from a cross section library) and C is a normalization constant.

These reaction rates are calculated from a simulation using tally multiplier input commands as will be described for input files.

In Fig. 7.4 below, for a neutron history consisting of the four scattering events shown, the track lengths are d_1, d_2, d_3, d_4. For better efficiency, nonanalog MC uses a weight modification factor $p = \Sigma_s/\Sigma_t$, so that a history is continued after every collision with a reduced weight. The track-length estimator (TLE) for flux is

$$\phi_{\text{TLE}} = nv = (N.d)/(V.t) = wd/V \tag{7.6}$$

so that the flux estimate for this history is, for total starting weight W and volume V,

$$\phi_{\text{TLE}} = \left(\frac{1}{\text{WV}}\right)\sum_{i=1}^{4} w_i d_i \tag{7.7}$$

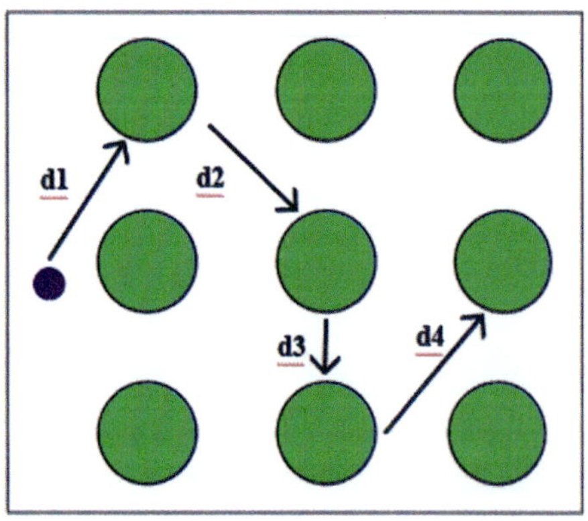

FIGURE 7.4 Random walk of a neutron.

TABLE 7.1 **Estimators for the system multiplication k_{eff}.**

Estimator	Tally
Collision estimator	$\frac{1}{N}\sum_{i=1}^{Ncol} W_i \left[\frac{\sum_{k=1}^{K} f_k \overline{\nu}_k \Sigma_{f,k}}{\sum_{k=1}^{K} f_k \Sigma_{t,k}}\right]$
Absorption estimator (analog simulation)	$\frac{1}{N}\sum_{i=1}^{Nabs} W_i \overline{\nu}_k \frac{\Sigma_{f,k}}{\Sigma_{c,k}+\Sigma_{f,k}}$
Absorption estimator (nonanalog simulation)	$\frac{1}{N}\sum_{i=1}^{Nabs} W_i' \overline{\nu}_k \frac{\Sigma_{f,k}}{\Sigma_{c,k}+\Sigma_{f,k}}$
Track-length estimator	$\frac{1}{N}\sum_{i=1}^{Npaths} W_i \rho d \sum_{k=1}^{K} f_k \overline{\nu}_k \Sigma_{f,k}$

The same quantity can be found using the collision estimator; in this case the estimate is made only at a collision site and not on an escape event, unlike the TLE.

$$\phi_{CE} = \left(\frac{1}{WV}\right)\sum_{i=1}^{4}\frac{w_i}{\Sigma_{t,i}} \tag{7.8}$$

Finally, the surface flux J_s, that is, the number of neutrons crossing the surface per unit area, can be estimated from the weight crossing a surface divided by the total starting weight and the volume to get

$$J_s = \left(\frac{1}{WV}\right)\sum_i w_i \tag{7.9}$$

The summation is over all the neutrons that cross the surface.

There are three commonly estimators namely the collision estimator (CE), the absorption estimator (AE) and the TLE with each having their own advantages such as when a large number of collisions are anticipated, the CE will give a good answer, that is, bot accurate and precise. When fewer collisions are anticipated, then nonanalog simulation is preferred since biasing a simulation, while preserving overall quantities, will give more tallies. Similarly, when the medium is optically thin then the TLE is bound to give the best estimate as the number of collisions in both analog and nonanalog simulations will be low or nonexistent while there will always be a track-length even in the absence of any collision.

Four estimators for the system multiplication k_{eff}
k_{eff}^{CE}, $k_{eff}^{A,as}$, $k_{eff}^{A,nas}$ and k_{eff}^{TLE} are listed in Table 7.1.

In the collision estimator, the estimate is made at each collision with the actual (analog) particle weight while in the AE the tally is updated at each absorption event. With implicit capture, a particle history is continued at an absorption but with reduced weight (equal to the survival probability). The TLE, in contrast to both, updates a tally regardless of a collision such as when it is crossing a region without undergoing a collision.

Exercise 7.1:

Assume a starting weight of one, $K=2$, (U-235 and U-238) with $f_5=0.2$, $f_8=0.8$, assume values for $\overline{\nu}_k$ and for cross sections, calculate values for a history with three collisions.

There is no fixed number for simulation parameters, but typical reactor simulations require a few million histories in over 100 cycles with thousands or hundreds of thousands of neutrons simulated per cycle for accurate results. Such a calculation takes several hours on multiprocessors with several cores.

Exercise 7.2:

From Fig. 7.5, estimate the TLE flux and the collision estimator flux. Are they the same? Also estimate the surface flux if line B represents the surface (assume unit surface area) (Table 7.2).

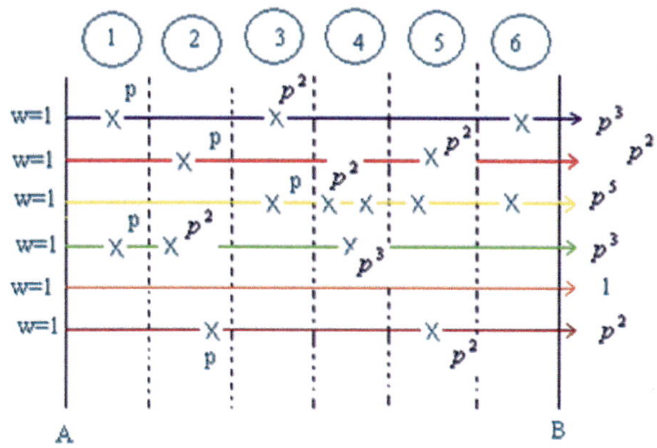

FIGURE 7.5 Forward scattering of neutrons in a slab.

TABLE 7.2 Tally estimators in MCNP.

Quantity	Estimator	Units
Current	W	Particles
Surface flux	$\frac{W}{\lvert\mu\rvert A}$	Particles/cm^2
Cell flux	$\frac{Wd}{V}$	Particles/cm^2
Point flux	$\frac{Wp(\Omega_p)e^{-\Sigma_t R}}{R^2}$	Particles/cm^2
Energy deposition	$\frac{\rho}{m}Wd\Sigma_t H(E)$	MeV/g
Heating	$\frac{\rho}{m}Wd\Sigma_f Q$	MeV/g
Pulse tally	W	Weight accumulated in energy bins

7.2.3 Sampling a source

The source in a simulation is specified according to the problem of interest, which are considered here in three categories:

1. fixed-source nonmultiplying medium,
2. criticality calculation in nuclear systems, and
3. mixed-mode neutron, photon, electron transport.

In the first category, the source would be specified in terms of its position energy, direction, and time. There could be one or several sources at fixed points or distributed over a surface or within a volume. A fusion source, for example, would have energy sampled from a fusion spectrum such as a Gaussian spectrum

$$p(E) = CE^{1/2}\exp\left[-\left((E-b)/a\right)^2\right]$$

where a is the spread in MeV and b is the average neutron energy in MeV; for DT fusion at 10 keV, $a = -0.01$ MeV and $b = -1$.

A fission energy spectrum, such the neutrons from a californium-252 source would be specified with a Watt spectrum

$$p(E) = C\exp\left(-E/a\right)\sinh\sqrt{bE}$$

where the parameters are given for the appropriate nuclide as shown for a few cases in Table 7.3.

In the second category, a criticality calculation starts by specifying a starting source at specified points or reading a source from a previous simulation. The system multiplication is calculated over a number of specified cycles for a specified number of source histories N, with some cycles skipped for using better estimates for calculating means and variances. The number of source histories is kept constant in each cycle by adjusting the weight of neutrons since one cycle may produce 1000 neutrons while another may produce 1200 neutrons from fission. The procedure for calculating k_{eff} is as dscribed in the previous section.

TABLE 7.3 Watt fission spectrum parameters.

Fission type	Nuclide	Energy (MeV)	a	b
Neutron induced	U-235	Thermal	0.988	2.249
		1	0.988	2.249
		14	1.028	2.084
	Pu-239	Thermal	0.966	2.842
		1	0.966	2.842
		14	1.055	2.383
Spontaneous	Cf-252	–	1.025	2.926
	Cm-244	–	0.906	3.848

In the third category, the simulation can be run as a neutron, photon, electron or as a mixed-mode simulation accounting for photon produced from nuclear interactions and secondary electrons produced from photons as described in Chapter 1.

MC simulations can handle a great variety of source specifications such as several point or distributed sources defined by discrete probabilities in energy bins. The source(s) can be biased in angle and energy as the problem may require. In shielding calculations, where very few tallies are scored in regions of interest, an MC simulation can be carried out to write a surface source where a good tally is scored. In a subsequent run, the surface source can be used to generate histories to score tallies in detector or regions of interest.

In this chapter, a fixed-source MC simulation is described. In a fission criticality problem, an initial fission source distribution is specified and the next generation fission points are obtained. This gives a first estimate of the system multiplication $k_{eff}^{(1)}$; these fission points are started in the second generation to give the next generation estimate and so on until convergence is reached. The energy of the source neutrons is sampled from a fission spectrum

For a point isotropic source of neutrons, the angles of emission are sampled using random numbers (Section 4.9)

$$\mu_i = 2\xi_i - 1, \varphi_i = 2\pi\xi_i \tag{7.10}$$

In Chapters 9 and 11, simulations will be carried out to demonstrate the versatility of MC simulation.

7.2.4 Sampling the "distance to collision"

The probability $\wp$ that a neutron at r_1, at time t_1, traveling with energy E_1 in the direction θ_1, φ_1 has its next interaction within dr_2 located at r_2 at a later time $t + dt$ is the product of two probabilities viz (1) the probability $\wp_1$ that it does not have an interaction between r_1 and r_2, and (2) the probability $\wp_2$ that, having reached r_2, it has its next collision in dr_2. Thus

$$\wp = \wp_1 \wp_2 \equiv \prod_{i=1}^{2} \wp_i = e^{-\Sigma_t |r_2 - r_1|} \Sigma_t dr_2 \tag{7.11}$$

For a one-dimensional case, the pdf $\wp$ is written as

$$f(x) = e^{-\Sigma_t x} \Sigma_t dx \tag{7.12}$$

and the cdf can be readily found as

$$F(s) = \int_0^s f(x) \mathrm{d}x = 1 - \mathrm{e}^{-\Sigma_t x} \tag{7.13}$$

The distance to collision is thus sampled as $s_i = \left(\frac{1}{\Sigma_t}\right) \ln\left(1 - \xi_i\right)$, which will have the same distribution as $s_i = \left(\frac{1}{\Sigma_t}\right) \ln\left(\xi_i\right)$ since ξ_i is uniformly distributed in (0,1) as will be $1 - \xi_i$.

7.2.5 Determining the type of event

At a collision site, determined by the mean free path, a MC simulation requires a determination of the type of event. This is selected by sampling from a possible chain of events as shown in Fig. 7.6. A random number ξ is used to determine whether the event is an absorption or a scattering event. In case of an absorption, another random number is used

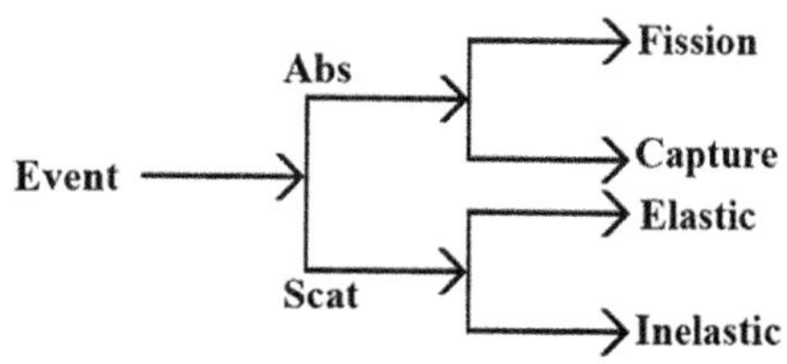

FIGURE 7.6 Chain of possible events.

to determine whether it is a fission or a capture. Similarly, in the case of a scattering event, a random number is used to determine whether it is elastic or inelastic scattering.

7.2.6 Determining the nuclide of interaction

In a MC simulation which handles nuclides independently, it is required to determine the nuclide with which a neutron collides at a collision site, for example, a neutron making a collision in water will collide with either a hydrogen atom or an oxygen atom. This decision is made from the probability of an interaction with nuclide i at incident energy E given as

$$p_i(E) = \frac{\Sigma_{t,i}(E)}{\Sigma_{t,5}(E) + \Sigma_{t,8}(E)}. \tag{7.14}$$

7.2.7 Processing a scattering event

In an elastic scattering collision, a neutron changes its direction with conservation of kinetic energy and momentum; there is no internal energy change or loss in friction.

The energy of a scattered neutron after collision E' can be found in terms of its energy before collision E, its angle of scattering in the laboratory system θ as

$$E' = \frac{E}{(1+A)^2}\left[\cos\theta + \sqrt{A^2 - \sin^2\theta}\right]^2 \tag{7.15}$$

The energy of a 1 MeV neutron colliding with a U^{235} nucleus is shown as a function of the scattering angle is shown in Fig. 7.7. It is seen that there is very little energy loss due to the high mass number ratio between the target and projectile. It is easy to understand that maximum energy loss will occur when the collision is between equal-mass particles; thus a neutron colliding with a hydrogen atom, in water for example, may easily lose most of its energy which essentially means that a "fast" (MeV) neutron can thermalize ($\sim$0.025 eV).

7.2.8 Processing a fission event

When a fission reaction (Section 2.9) takes place, two quantities needed are the number of prompt neutrons emitted and their associated energies. The average number of prompt neutrons emitted in fission is given by $\nu = \nu_o + \alpha E$ [Lamarsh (p. 95)] where the constants are given in Table 7.4.

From experiments carried out, it has been determined that neutron multiplication is a stochastic process and there is a probability distribution function assigned with $\overline{\nu}$ (Schmidt & Jurado, 2018). For U^{235}, the mean values measured were 2.41 for thermal fission, 2.49 for 0.5 MeV fission and 3.19 for 5.55 MeV fission.

The Gaussian distribution is used for the probability P_n, that n neutrons are emitted, given by

$$P_n = \frac{1}{\sqrt{2\pi\sigma^2}}\int_{n-1/2}^{n+1/2} \exp\left(\frac{(x-\overline{\nu})^2}{2\sigma^2}\right)dx \tag{7.16}$$

For the case $n = 0$, the lower limit on the integral is replaced by $-\infty$. Sampling is done from this normal distribution function using a rejection algorithm. The width of the distribution σ is taken as 1.088 for U^{235} and 1.116 for U^{238}. The mean $\overline{\nu}$ is energy dependent: for U^{235} it is $2.4 + 0.12\ E$ for $E < 5$ MeV and $2.2 + 0.16\ E$ for $E \geq 5$ MeV, while for U^{238} it is $2.4 + 0.0666\ E$ for $E < 3$ MeV and $2.1538 + 0.1654\ E$ otherwise. The returned value for ν is the rounded integer $\nu = \overline{\nu} + \sigma x\sqrt{-2\ln z/z}$.

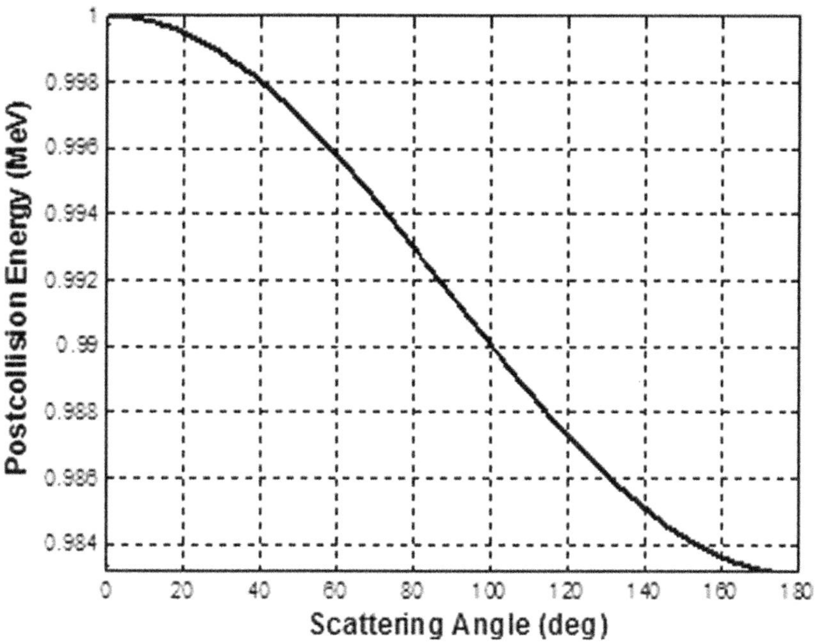

FIGURE 7.7 Postcollision energy for elastic scattering of a 1 MeV neutron-U235 nucleus.

TABLE 7.4 Neutrons emerging from fission in U235 and U238.

Isotope	ν_o	α (MeV^{-1})	E (MeV)
U^{235}	2.43	0.065	0<E<1
	2.35	0.150	E>1
U^{238}	2.30	0.160	All E

Another method is to estimate ν from the number of neutrons produced by fission $\nu\sigma_f/\sigma_t$ and to round off the number to the nearest integer.

The energy of fission neutrons can be sampled from the empirical, Maxwell or Cranberg fission spectra (Section 2) or others in the literature. The sampled values will be discussed in detail in a later chapter on MC simulation.

7.2.9 Processing a capture event

In analog simulation, a capture ends a history, as is done in the program written for this paper. However, a better way is to continue the history with a reduced "weight" until it is ended due to very low weight or escape from the system.

7.2.10 Processing an escape-from-system event

When a neutron escapes from the system, it is not added to the updated number of neutrons generated and thus does not contribute to the system multiplication.

7.2.11 Mean and variance

The mean and variance of an estimate (Section 4.8) is given as $\bar{\bar{x}} = \frac{1}{N}\sum_{i=1}^{N} x_i$, and the variance of the population is estimated from the sample as $\sigma^2 = \frac{1}{N}\sum_{i=1}^{N} x_j^2 - \bar{x}^2$, from which the variance of the mean is estimated as $\sigma_{\bar{x}}^2 = \frac{\sigma^2}{N}$.

7.2.12 Batch, history, random walk and events

The procedure for neutrons follows the same sequence as for alpha, beta and gamma transport (Section 1.6) depicted in the flow charts (Figs. 1.23, 1.24, and 1.27).

A simulation for a multiplying system begins with a batch of source neutrons N simulated for G generations for each source neutron.

The pseudocode is described below:

$i = 1$

1. Begin source neutron i
2. Initialize $n = 1$
3. Initialize neutron generation counter $n_g = 0, k = n = 1, j = 1$
4. Begin generation j
5. *Begin source neutron k for this generation*
6. Sample the distance to collision d
7. Sample the type of event
8. Process the event
9. Compute postcollision parameters
10. Determine if neutron still inside system
11. Update n_g
12. Update $k = k + 1$
13. If $k < n$ go to v (continue the same generation)
14. Comes here when all "mother neutrons" for this gen processed Update $n = n_g, k = k + 1$, for the next generation
15. Score for this source neutron i and generation j, M(i, j) $= n_g$
16. If $k < G + 1$ go to iv (begin a new generation)
17. Comes here when all generations processed
18. Update source neutron counter $i = i + 1$
19. If $i \leq N$ go to step I, begin a new source neutron
20. All neutrons have been processed

    Estimate k_{eff} from the following steps:

    a. Compute the number of neutrons produced in each generation: $P(j) = \sum_{i=1}^{N} M(i,j)$, for generations $j = 1, 2, 3, \ldots G$
    b. Compute the k_{eff} for each generation:

    $$k_{\text{eff}}(j) = P(j)/P(j-1),$$

    j = 2,3,. . .G, with $k_{\text{eff}}(1) = 1$

    c. Compute the error for each k_{eff}:

    $$\varepsilon(j) = k_{\text{eff}}(j)\sqrt{\frac{1}{P(j)} + \frac{1}{P(j-1)}} \tag{7.17}$$

    d. The estimate is

    $$k_{\text{eff}}(j) + \varepsilon(j) \tag{7.18}$$

7.3 Modeling the geometry

One of the strengths of the MC method for simulation of neutron transport in nuclear systems is its ability to model complex geometry typical of practical systems containing a variety of materials in regular and irregular shapes as well as ducting for instrumentation and biological shielding. An example of regular geometry is shown in Fig. 7.8 consisting of four boxes generated from the MCNP input file listed below. Another combination of primitive configurations is shown in Fig. 7.9.

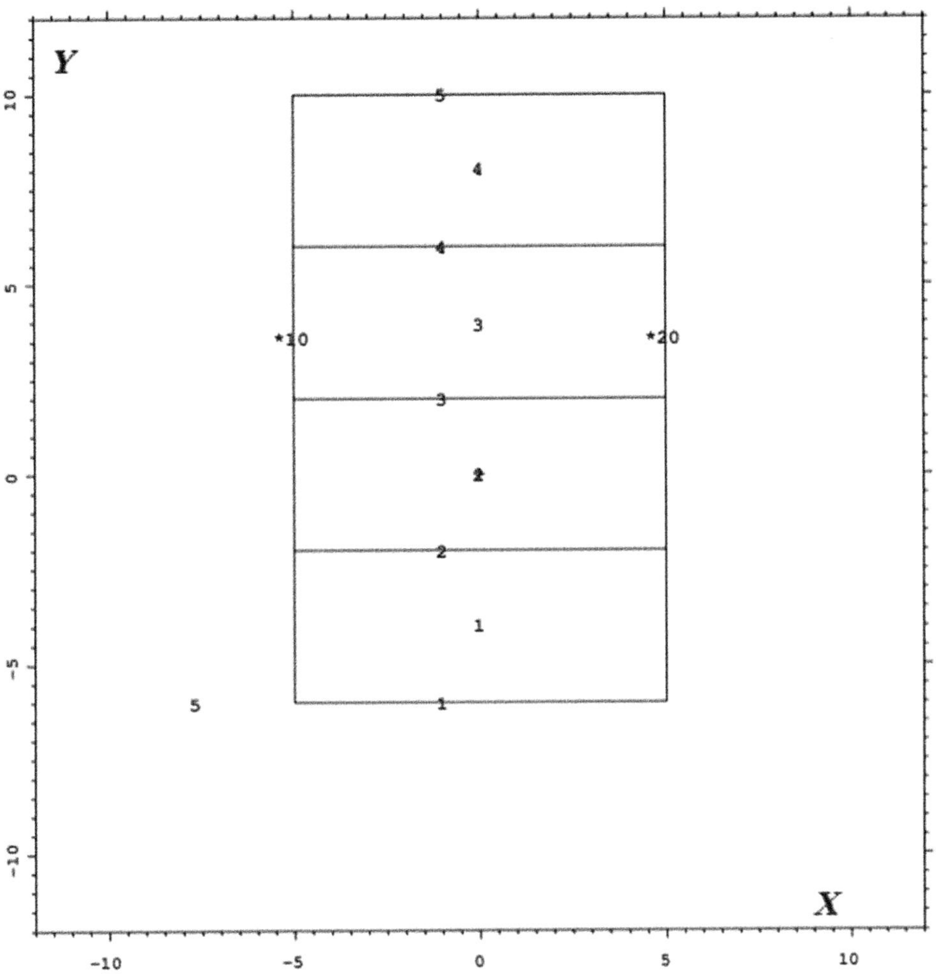

FIGURE 7.8 Four boxes with surface and cell numbers in MCNP.

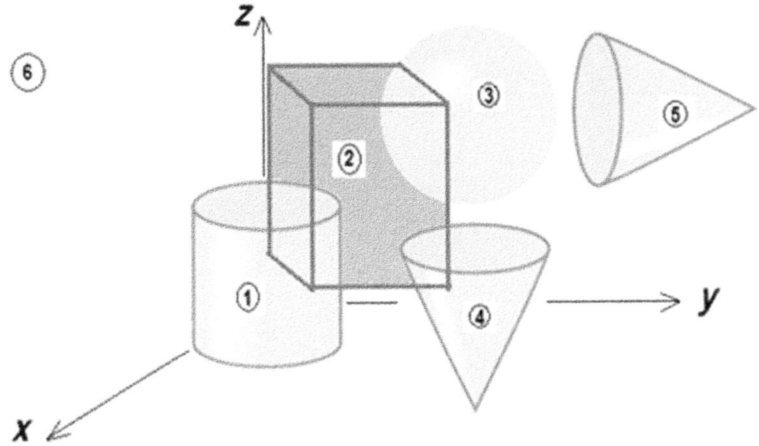

FIGURE 7.9 Unions and intersections of volumes.

Input file for Fig. 7.3

```
INPUT FILE BOOK06 FIG 7.3 Boxes
1    1 -1.0 1 -2 10 -20 30 -40
2    1 -1.0 2 -3 10 -20 30 -40
3    1 -1.0 3 -4 10 -20 30 -40
4   10 -2.7219e-03 4 -5 10 -20 30 -40
5    0  -1:5:-10:20:-30:40

1    py  -6
2    py  -2
3    py  2
4    py  6
5    py 10
*10  PX -5
*20  PX 5
*30  PZ -5
*40  PZ 5
```

MCNP commands

```
Mcnp5 ip inp=Book06
Pz 0 extent 12 12

Label 1 1 Scales 1
Color off file
```

A convenient way of describing a volume is in terms of its enclosing surfaces. Fig. 7.10 shows a cylinder and a truncated cone inside a sphere. The surface numbers from 10 to 16 represent the sphere of radius 8 cm centered at the origin, a cylinder of radius 3 cm on the z-axis within planes (surfaces 12 and 13) perpendicular to the z-axis, a cone with its vertex on the z-axis at the point (0,0,4) within its outside sheet, truncated by two surfaces perpendicular to the z-axis (surfaces 15 and 16). This way, large geometrical configurations can be modeled. Similarly Fig. 7.11 shows an *XY* view of a cylinder inside a sphere. Better visualizations are possible, and especially helpful for configurations such as the ITER design (Chapter 3) by the integration of CAD and MC input files.

Input file for Figs 7.5 and 7.6

```
Book7 Fig 7.5 (px 0) and 7.6 (pz=3)
1      0  -11 12 -13
2      0    -14 15 -16
3      0  -10 #1 #2
4      0    10

10    SO  8.00
11    CZ  3
12    PZ -2
13    PZ  2
14    KZ  4 1
15    PZ  2.5
16    PZ 3.5
```

MCNP commands

```
Mcnp5 ip inp=Book7
Pz 0 extent 10 10
Label 1 1 Scales 1

Color off file
```

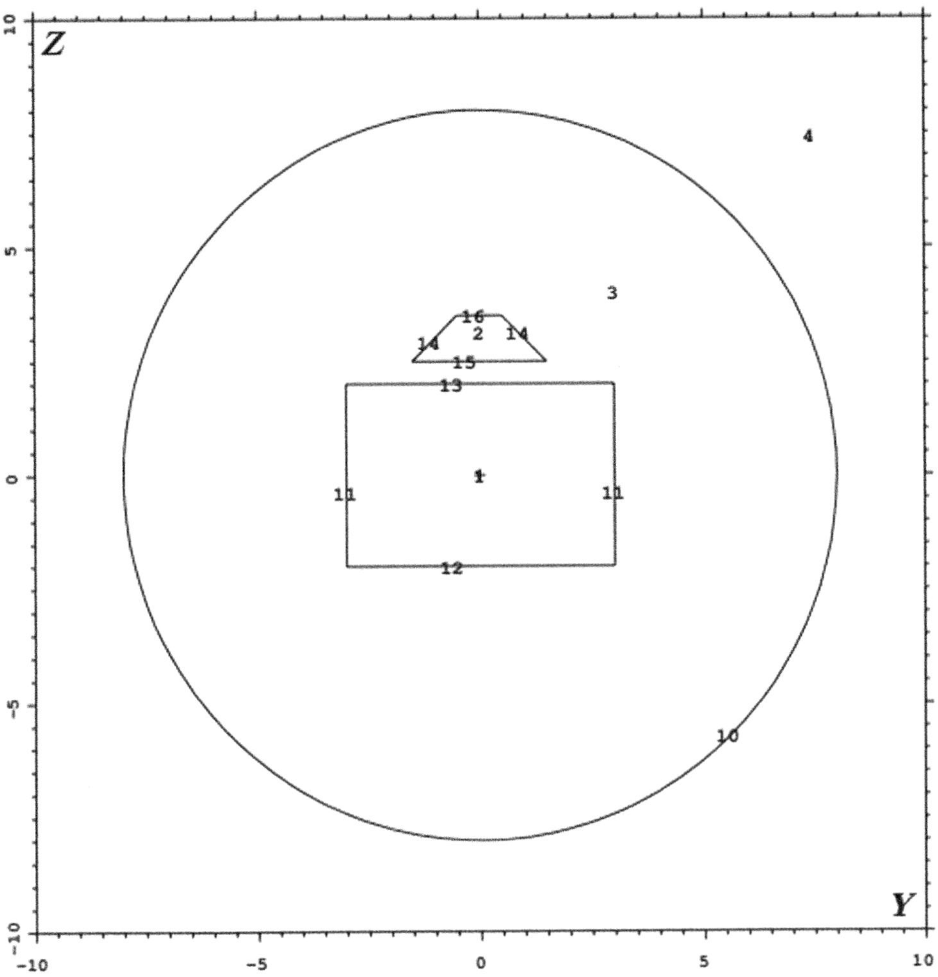

FIGURE 7.10 A cylinder and truncated cone within a sphere (YZ plane).

The above representations use constructive solid geometry to describe surfaces and regions through equations of *primitive* surfaces and combinatorial geometry based on the Boolean operators of intersection and unions.

Surfaces are defined by mnemonics and associated parameters or by specifying points explicitly. Some of the surfaces used in MCNP are listed in Table 7.5.

A *cell* is a physical region in space defined by combinatorial geometry using the Boolean operators for union and intersection. In Fig. 7.12, for two concentric spheres, the three cells are defined as:

Cell	Surface
1	-1
2	1 -2
3	2

In the above, the first cell is enclosed by surface 1, cell 2 is described as 1−2 which represents the intersection of a region with a *positive* with respect to surface 1 and a *negative* sense with respect to surface 2.

The concentric spheres centered at the origin are represented by the equation

$$f(x,y,z) \equiv x^2 + y^2 + z^2 - R^2 = 0$$

A *ray* is defined by its origin $R_o = (X_o, Y_o, Z_o)$ and its direction (unit) vector $\hat{\Omega} = \Omega_x \hat{i} + \Omega_y \hat{j} + \Omega_z \hat{k}$. The equation of a plane is $Ax + By + Cz + D = 0$. The normal vector $\overline{N}$ is (A, B, C). To find d, the distance from the origin of the ray to the plane where it intersects, we first need to find the point of intersection. Along the ray, the point is: $X_d = X_o + \Omega_x t$, $Y_d = Y_o + \Omega_y t$, $Z_d = Z_o + \Omega_z t$. The point $R_d = (X_d, Y_d, Z_d)$ lies on the plane; therefore $AX_d + BY_d + CZ_d + D = 0$. Thus the scalar t can be

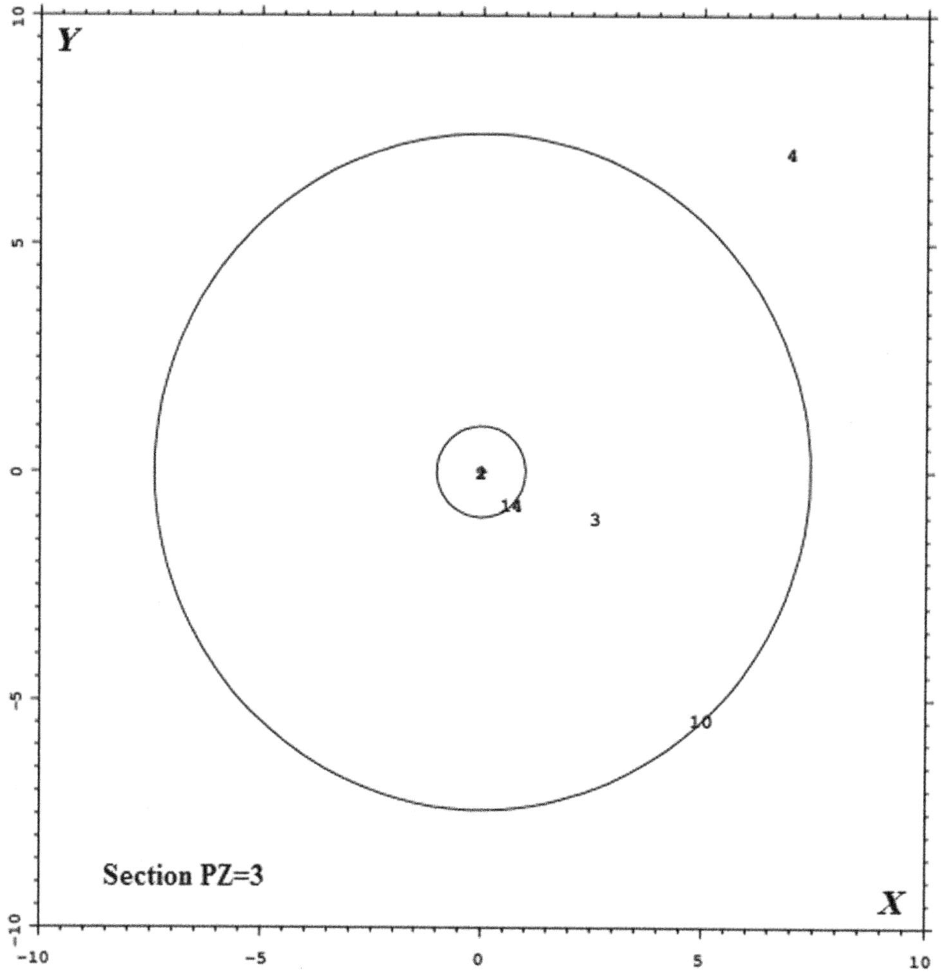

FIGURE 7.11 A cylinder within a sphere (*XY* plane).

found as

$$t = -\frac{(AX_o + BY_o + CZ_o + D)}{A\Omega_x + B\Omega_y + C\Omega_z} = -\frac{\overline{N} \bullet \overline{R}_o + D}{\overline{N} \bullet \hat{\Omega}}$$

as depicted in Fig. 7.13. If $\overline{N} \bullet \overline{R}_o = 0$, the ray and plane are parallel and there is no intersection; otherwise t is found from

$$t = -\frac{\overline{N} \bullet \overline{R}_o + D}{\overline{N} \bullet \hat{\Omega}}.$$

For $t < 0$, there is no intersection with the plane.

Exercise 7.3:

In Fig. 7.14, consider the ray at $R_o = (X_o, Y_o, Z_o) \equiv (2, 1, 3)$ in the direction $\hat{\Omega} = \Omega_x \hat{i} + \Omega_y \hat{j} + \Omega_z \hat{k}$ where the orthogonal angle is $\theta = 60^\circ$ and the azimuthal angle is $\varphi = 30^\circ$. Determine whether there is a point of intersection $R_d = (X_d, Y_d, Z_d)$ with the plane $y - 2 = 0$, and if there is, then find the distance to surface from R_o.

7.3.1 Geometries for illustration of Monte Carlo simulation

Some representative and practically important models considered in this chapter are fixed-source benchmarks, criticality benchmarks, a PWR unit lattice cell model and fusion reactor neutronics.

TABLE 7.5 Surface mnemonics for combinatorial geometry.

Mnemonic	Type	Description	Equation	Card entries
P	Plane	General	$Ax + By + Cz - D = 0$	ABCD
PX		Normal to X – axis	$x - D = 0$	D
PY		Normal to Y – axis	$y - D = 0$	D
PZ		Normal to Z – axis	$z - D = 0$	D
SO	Sphere	Centered at Origin	$x^2 + y^2 + z^2 - R^2 = 0$	R
S		General	$(x-\overline{x})^2 + (y-\overline{y})^2 + (z-\overline{z})^2 - R^2 = 0$	$\overline{x}\ \overline{y}\ \overline{z}R$
SX		Centered on X – axis	$(x-\overline{x})^2 + y^2 + z^2 - R^2 = 0$	$\overline{x}R$
SY		Centered on Y – axis	$x^2 + (y-\overline{y})^2 + z^2 - R^2 = 0$	$\overline{y}R$
SZ		Centered on Z – axis	$x^2 + y^2 + (z-\overline{z})^2 - R^2 = 0$	$\overline{z}R$
C/X	Cylinder	Parallel to X – axis	$(y-\overline{y})^2 + (z-\overline{z})^2 - R^2 = 0$	$\overline{y}\ \overline{z}R$
C/Y		Parallel to Y – axis	$(x-\overline{x})^2 + (z-\overline{z})^2 - R^2 = 0$	$\overline{x}\ \overline{z}R$
C/Z		Parallel to Z – axis	$(x-\overline{x})^2 + (y-\overline{y})^2 - R^2 = 0$	$\overline{x}\ \overline{y}R$
CX		on X – axis	$y^2 + z^2 - R^2 = 0$	R
CY		on Y – axis	$x^2 + z^2 - R^2 = 0$	R
CZ		on Z – axis	$x^2 + y^2 - R^2 = 0$	R
K/X	Cone	Parallel to X – axis	$\sqrt{(y-\overline{y})^2 + (z-\overline{z})^2} - t(x - \overline{x}) = 0$	$\overline{x}\ \overline{y}\ \overline{z}\ t^2 \pm 1$
		Parallel to Y – axis	$\sqrt{(x-\overline{x})^2 + (z-\overline{z})^2} - t(y - \overline{y}) = 0$	$\overline{x}\ \overline{y}\ \overline{z}\ t^2 \pm 1$
K/Z		Parallel to Z – axis	$\sqrt{(x-\overline{x})^2 + (y-\overline{y})^2} - t(z - \overline{z}) = 0$	$\overline{x}\ \overline{y}\ \overline{z}\ t^2 \pm 1$
KX		on X – axis	$\sqrt{y^2 + z^2} - t(x - \overline{x}) = 0$	$\overline{x}\ t^2 \pm 1$
KY		on Y – axis	$\sqrt{x^2 + z^2} - t(y - \overline{y}) = 0$	$\overline{y}\ t^2 \pm 1$
KZ		on Z – axis	$\sqrt{x^2 + y^2} - t(z - \overline{z}) = 0$	$\overline{z}\ t^2 \pm 1 \pm 1$ used only for one-sheet cone
SQ	Ellipsoid Hyperboloid Paraboloid	Axis parallel to X-, Y-, or Z-axis	$A(x-\overline{x})^2 + B(y-\overline{y})^2 + C(z-\overline{z})^2 + 2D(x - \overline{x}) + 2E(y - \overline{y}) + 2F(z - \overline{z}) + G = 0$	A B C D E F G $\overline{x}\ \overline{y}\ \overline{z}$
GQ	Cyl, Cone Ell, Hyp, Par	Axis not parallel to X-, Y-, or Z-axis	$Ax^2 + By^2 + Cz^2 + Dxy + Eyz + Fzx + Gx + Hy + Jz + K = 0$	A B C D E F G H J K
TX	Elliptical or circular torus	Axis parallel to X-, Y-, or Z-axis	$(x-\overline{x})^2/B^2 + \left\{\sqrt{(y-\overline{y})^2 + (z-\overline{z})^2} - A\right\}^2/C^2 = 1$	$\overline{x}\ \overline{y}\ \overline{z}$ A B C
TY			$(y-\overline{y})^2/B^2 + \left\{\sqrt{(x-\overline{x})^2 + (z-\overline{z})^2} - A\right\}^2/C^2 = 1$	$\overline{x}\ \overline{y}\ \overline{z}$ A B C
TZ			$(z-\overline{z})^2/B^2 + \left\{\sqrt{(x-\overline{x})^2 + (y-\overline{y})^2} - A\right\}^2/C^2 = 1$	$\overline{x}\ \overline{y}\ \overline{z}$ A B C

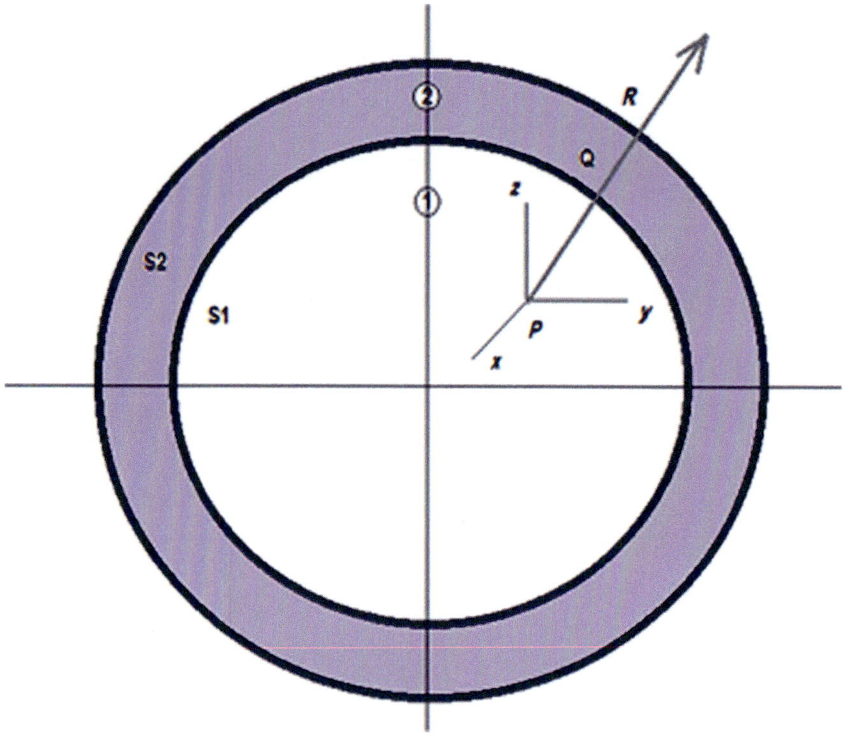

FIGURE 7.12 Concentric spheres.

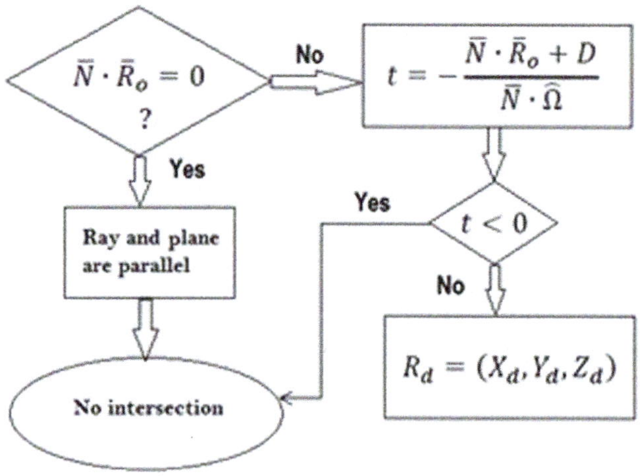

FIGURE 7.13 Flowchart for a ray-plane intersection.

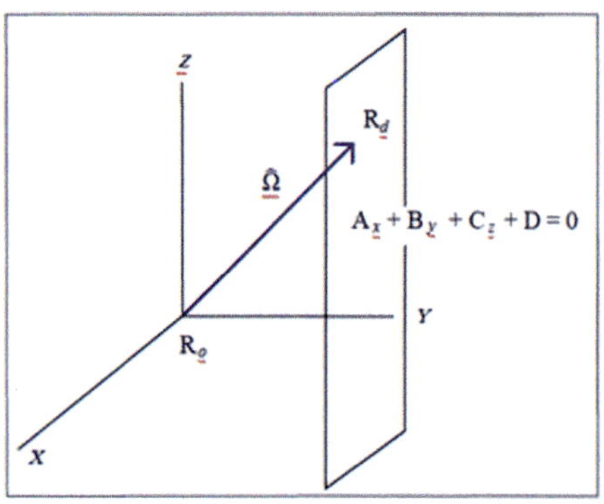

FIGURE 7.14 Ray-plane intersection.

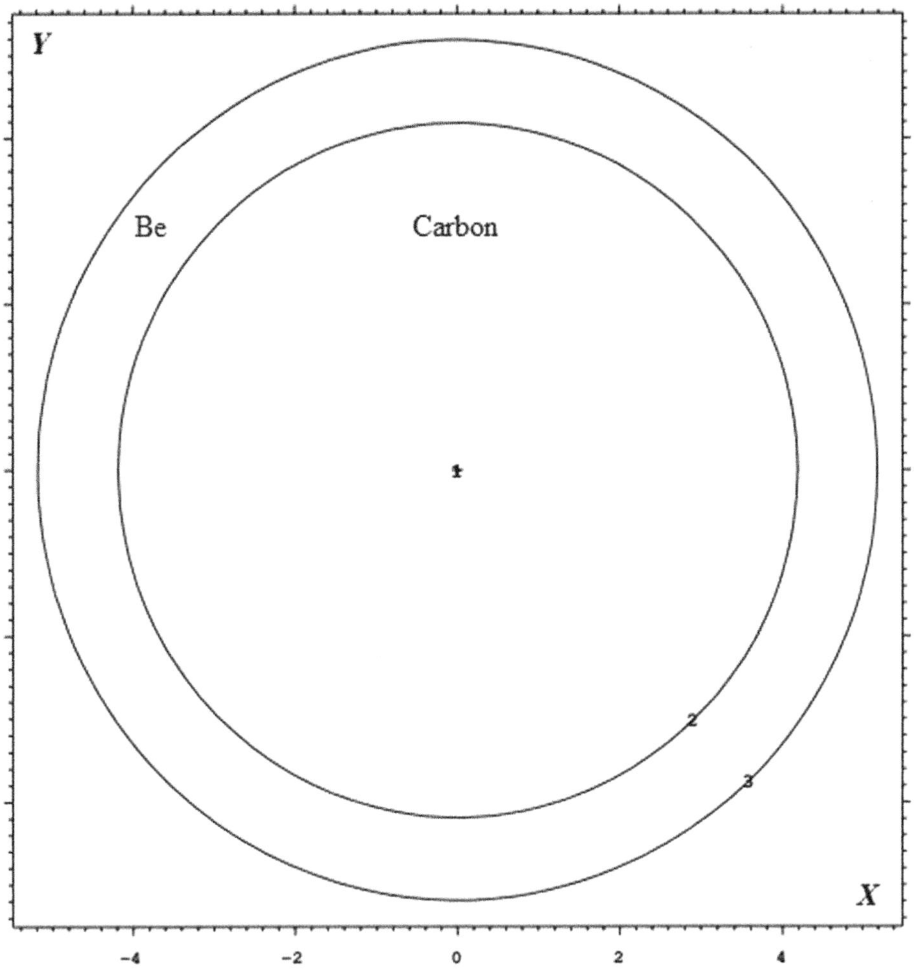

FIGURE 7.15 Carbon-Beryllium sphere (dimensions in cm).

Two simple fixed-source models, depicted in Fig. 7.15, considered are

1. a carbon bare sphere with a point isotropic source at its center, and
2. a carbon sphere reflected by a thin layer of beryllium.

The illustrative models for criticality benchmarks are bare spheres of the Godiva and Jezebel (Section 2.10, Table 2.21 for uranium and plutonium bare and reflected critical systems and Table 2.22 for Bare critical assemblies Godiva and Jezebel) assemblies. The MC simulation of critical assemblies is described in detail in Chapter 9.

For nuclear power reactors (Section 3.2), nuclear propulsion reactors (Section 3.3), SMRs and space reactors (Section 3.5), the overall design characteristics and core neutronics were described in Section 5.6 for neutron diffusion models and in Section 6.4 for neutron transport models for the unit lattice cell, fuel assembly, and the reactor core. The most basic unit for calculations in reactor core neutronics is the unit lattice cell for AP1000 (Section 3.2) shown in Fig. 7.16 represented by concentric cylinders in a bounding box with reflective surfaces.

The unit lattice cell is part of a fuel assembly and the reactor core as described for a PWR in Section 5.6. Its dimensions are very small as shown in the figure; the cell shown above has dimensions of 1.26 × 1.26 × 2 cm. This represents a part of a typical PWR fuel assembly containing 17 × 17 such cells of dimensions 21.40202 × 21.40204 cm with an active fuel height of 4.2 m.

For a Gen IV SMR (Section 3.5.3), Fig. 7.17 shows the core of the Toshiba 4S design (Koreshi & Hussain, 2014; Ueda et al., 2005) with eighteen hexagonal fuel assemblies and a single central control rod. Note the dimensions of the 4S core; the reflector outer radius is ~ 60 cm as compared to ~3 m for the AP1000. These differences arise out of different fuel enrichments and overall power requirements of both reactors.

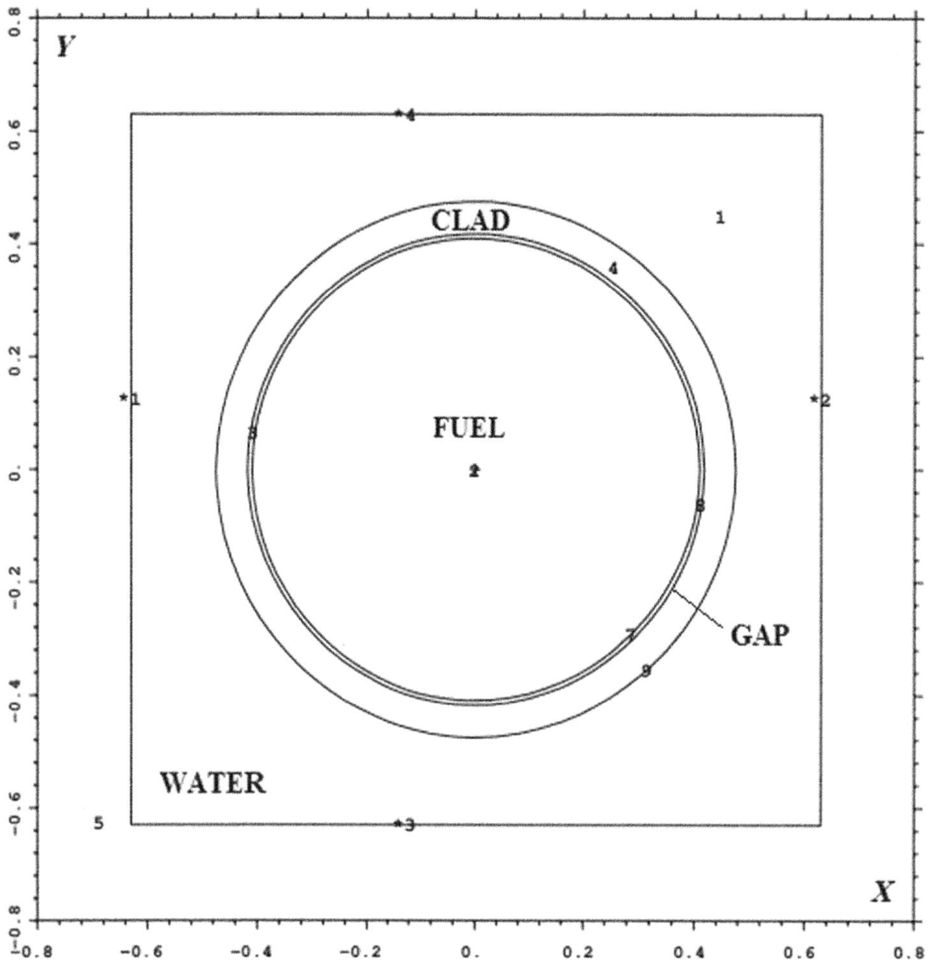

FIGURE 7.16 AP1000 unit lattice cell (dimensions in cm).

Fission neutronics give the following results

1. k_∞ in a unit lattice cell and fuel assembly
2. k_{eff} for full-core
3. Neutron flux distribution (n/cm^2/s)
4. Power density distribution (W/cm^3)
5. Peak power (W)
6. Average power (W)
7. Control rod worth (pcm)
8. Equilibrium xenon worth (pcm)
9. Boron coefficient of reactivity with/without xenon (pcm/ppm)
10. Doppler coefficients with/without xenon (pcm/K),
11. Cold shutdown (coolant heat removal after a reactor has shut down) calculation.

Typical calculations for an AP1000 (de Stefani, Losada Moreira, Maiorino, & Russo Rossi, 2019) are shown in Table 7.6.

Power density distribution is illustrated in a plot showing the power density (W/cm^3) in the fuel assemblies typically in a one-fourth core model.

Boron concentration is shown as a function of burnup.

Coupled analyses are carried out by integrating neutronics, activation and thermal hydraulics as described in Chapter 4 such as by combining MCNP (Werner, 2017), FISPACT (Sublet et al., 2017) and thermal hydraulics codes (Yang, Liu, Xiong, Chai, & Cheng, 2018).

An example of a reactor smaller than the Toshiba 4S reactor is the micronuclear reactor (MNR) shown in Fig. 7.18 (Aziz, Koreshi, Sheikh, & Khan, 2020; Sun et al., 2018) with overall physical design characteristics described in

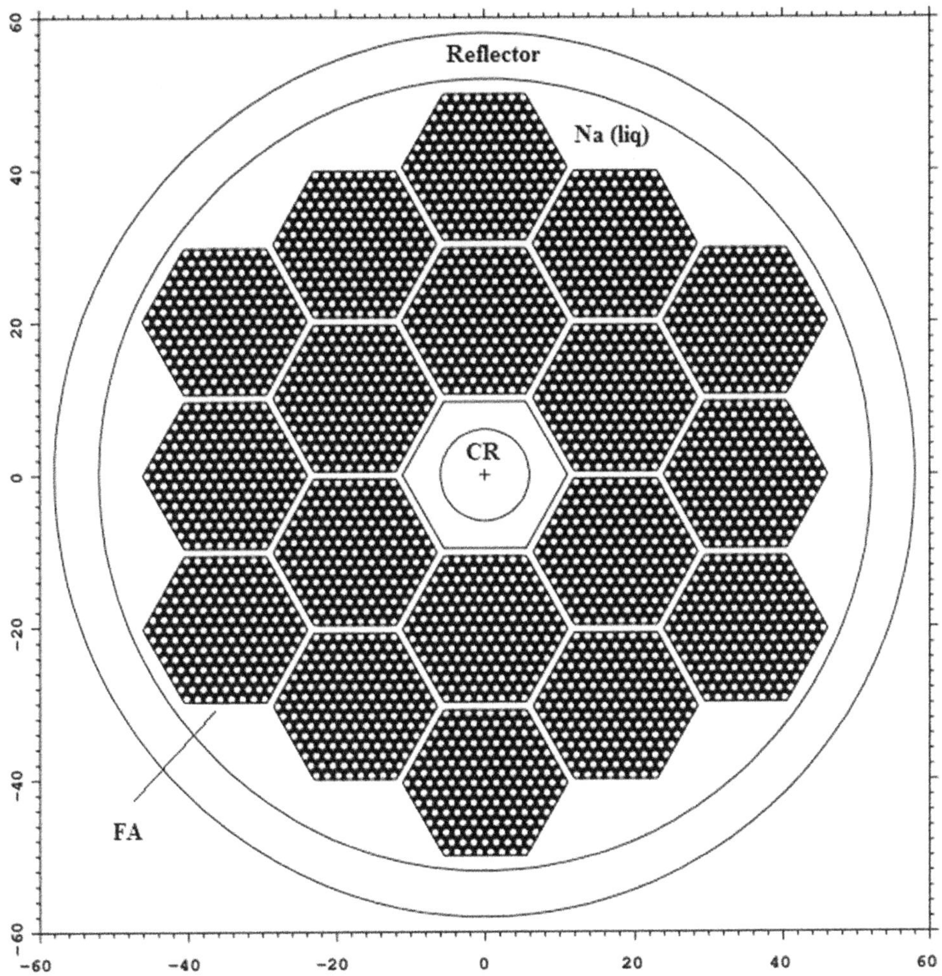

FIGURE 7.17 Core of the Gen IV Toshiba 4S reactor (FA = fuel assembly, CR = control rod, dimensions in cm).

TABLE 7.6 Some quantities of interest in fission neutronics.

Quantity	Result	Comments
Analyses at cold zero power (CZP), hot zero power (HZP), hot full power (HFP)	–	These are distinct *states* of a power reactor
k_∞ for unit cell and FA	R1.58 $1.21029 \pm 0,00003$, R2.35 $1.32863 \pm 0,00004$, R3.20 $1.40462 \pm 0,00004$, R3.40 $1.41697 \pm 0,00004$, R4.45 $1.46858 \pm 0,00003$,	Different fuel enrichments at CZP (defined T_f, T_m moderator specific mass and temperature in cross section library)
Neutron flux (n/cm^2/s)	$E \geq 1.0 MeV$	$1.37 \times 10^{14} \pm 0.09 \times 10^{14}$
	$5.53 keV < E < 1.0 MeV$	$1.93 \times 10^{14} \pm 0.09 \times 10^{14}$
	$0.625 eV < E < 5.53 keV$	$1.18 \times 10^{14} \pm 0.13 \times 10^{14}$
	$E \leq 0.625 eV$	$5.01 \times 10^{14} \pm 0.21 \times 10^{14}$
Power density distribution	Shown on a surface plot	
Power peak factor	2.71	
Peak power density	299.73 W/cm^3	
Average power density	110.6 W/cm^3	
Fuel depletion		
Radioactive waste production	Thermal output 2–20 kW/m^3	

Note: Uranium consumption, and plutonium build-up are shown as functions of burnup; at ~ 1 MWd/gU, 1 GWd would consume ~ 1 kgU.

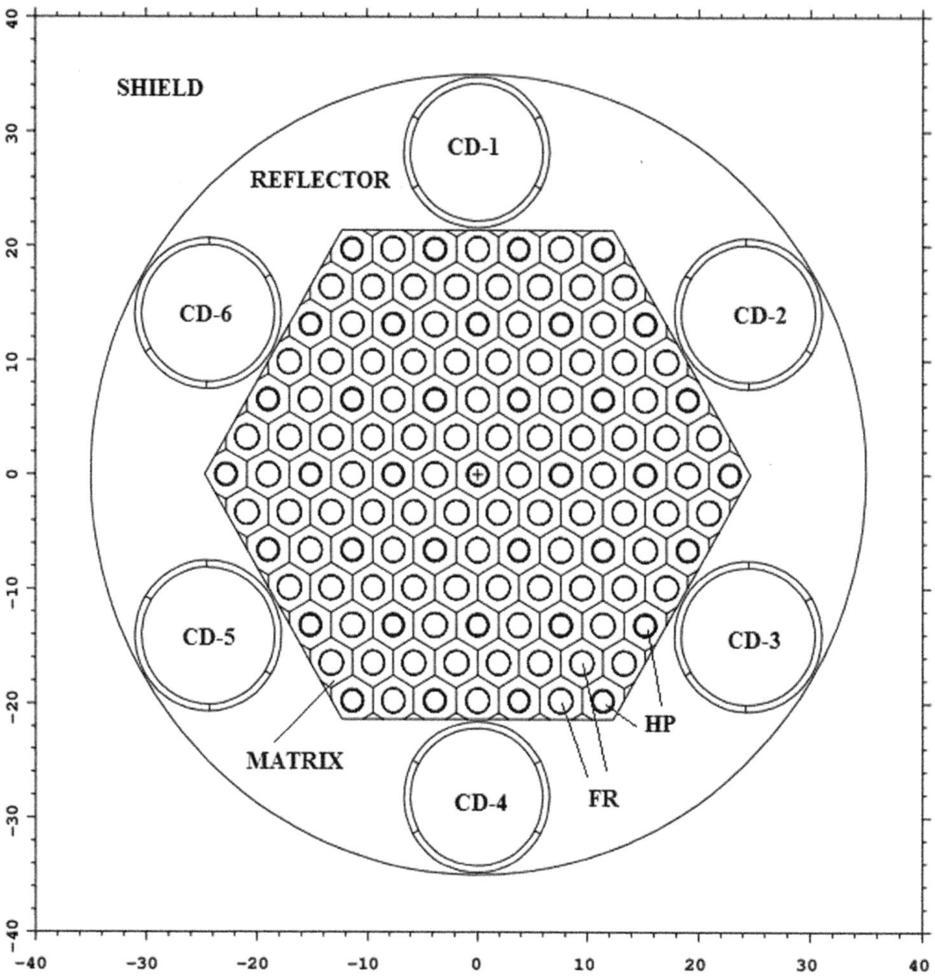

FIGURE 7.18 Core of a micronuclear reactor (FR = fuel rod, CD = control drum, dimensions in cm).

Section 3.8. Note again, the dimensions compared with the dimensions of the AP1000 and the 4S reactors. There are no fuel assemblies in the MNR due to the higher fuel enrichments and lower power requirements.

The geometry models for fusion Tokamak systems (Section 3.6) can be more complicated than for fission systems. ITER, for example is D-shaped torus with toroidal and poloidal field coils and considerable geometrical complexity of its systems.

In a Tokamak, with high neutron flux $\sim 10^{14}$ n/cm^2/s causing high loadings and subsequent materials challenges, some tallies of interest are

1. the FW loading (MW/m^2) to determine the rate of energy transfer through the wall (flux tally),
2. neutron (fast, $E>$ 0.1 MeV) fluence to toroidal field coils (TFC),
3. nuclear heating (W/m^3); over 80% of the DT energy is carried by a 14 MeV neutron and is deposited in the breeder blanket; the rest ($\sim$16%) in the FW, and tungsten armor, manifolds and surrounding structure,
4. the tritium breeding rate (flux multiplier tally)

   $01n + 36Li \rightarrow 24He + 13T + 4.8\text{MeV}$

   $01n + 37\text{Li} \rightarrow 24\text{He} + 13\text{T} + 01n - 2.5\text{MeV}$

   The TBR is calculated for candidate materials such as liquid metal, eutectics such as lithium leads, solids such as lithium oxide and ternary oxides (e.g., $LiAlO_2$, Li4SiO_4) and ceramics such as lithium titanate; design calculations are performed to optimize the lithium enrichment, mix of neutron multiplier such as beryllium and lead, coolant such as water, helium and molten salt, neutron reflector and structural material (steel),
5. helium production in steel structure and in reweldable zones; flux multiplier tally (n, α) reaction; calculated in ppm/FPY (atomic parts per million per FPY),
6. radiation damage (dpa) to copper (flux multiplier tally) and steel; since this degrades the material reducing its life; MC simulations thus give the dpa/FPY (displacements per atom per FPY),

7. radiation streaming from plasma to superconducting magnet TFC,
8. radiation dose (Gy) to epoxy in the winding pack facing the central cell,
9. radiation dose with reactor shutdown to permit maintenance,
10. radiation across ducts and shielding, and
11. activation of steel (effect of low activation materials in steel such as Eurofer), coolants and components to minimize activation and radioactive waste due to ~1021 neutrons per second during a power pulses.

The pressure and temperature operating conditions of tokamaks are of the same order as in PWRs, that is, 15.5 MPa, ~280°C; thus water density is significantly reduced.

Table 7.7 lists some estimates for an engineering study of the water-cooled lithium lead 1998 MW DEMO fusion reactor with net electrical power 500 MW (Moro et al., 2020) with 16 TFCs and a minor and major radius of 2.883 and 8.938 m, respectively, using mainly tungsten armor as the FW, LiPb, H_2O, and Eurofer steel as the breeder, and LiPb and H_2O manifolds.

Neutron studies lead to optimal designs, such as in the study quoted above (Moro et al., 2020), a coupled neutronics-thermal hydraulics analysis showed that FW cooling ws achievable by reducing the number of water channels which subsequently led to an increase in the Eurofer steel content and a resulting TBR increase to 1.138. The reason for such an increase is that water acts both as a moderator enhancing the Li_6 reaction and as a shield reducing the intensity of radiation.

Thus breeder design is an active optimization research area with the objective of enhancing tritium breeding from the above reactions of lithium (Del Nevo et al., 2019; Hernández & Pereslavtsev, 2018; Moro et al., 2020).

A simplified 1-D representative geometrical representation of a breeder blanket (Colling, 2016; Shimwell et al., 2016) of a spherical Tokamak reactor, is shown in Fig. 7.19.

A mockup experiment for tritium breeding in ITER is depicted in Fig. 7.20 (Jakhar et al., 2015; Mandal, Shenoi, & Ghosh, 2010) consists of a central D-T fusion source surrounded by lead (Pb) multiplier and reflected by high density polyethylene (HDPE) reflector.

When a geometrical model is made, the next step in MC simulation is the description of the physics of interactions, the material descriptions, the neutron/gamma source, tallies and simulation parameters.

TABLE 7.7 Some quantities of interest in fusion neutronics.

Quantity	Desired/limits	Objective
Average wall loading	1.04 MW/m^2	
TBR (desired)	>1.10	Highest TPR ~ 1.07×10^{12} tritium atoms/cm^3/s
TBR (assessed)	1.119	With reduced water density ($\rho = 0.725$g/cm3, TBR assessed is 1.118
Damage on vacuum vessel steel over 6 FPR (lifetime)	< 2.75 dpa	to ensure fracture toughness is reduced by no more than 30%)
Assessed radiation damage (I/B)	Eurofer FW ~ 8.1 dpa/FPY; SS316L VV inner shell ~ 8.67×10^{-3} dpa/FPY	–
Assessed Radiation damage (O/B)	Eurofer FW ~ 9.5 dpa/FPY; SS316L VV inner shell ~ 2.46×10^{-3} dpa/FPY	–
He concentration	< 1 ppm	To permit welding in reweldable zones
He concentration in FW I/B (assessed)	Eurofer in breeder blanket 94 ppm/FPY, VV inner shell 8.08×10^{-2} ppm/FPY, TFC 1.21×10^{-6} dpa/FPY	over a period of 6 FPY, 0.48 ppm
He concentration in FW O/B (assessed)	Eurofer in breeder blanket 114 ppm/FPY, VV inner shell 2.43×10^{-2} ppm/FPY, TFC 1.21×10^{-6} dpa/FPY	over a period of 6 FPY, 0.24 ppm
Fast neutron fluence on TFC	< 10^9 n/cm^2/s	shielding capability should ensure a fluence within the limits
Nuclear heating on winding pack	< 50 W/m^3	–
Total neutron (inboard) flux	5.13×10^{14} n/cm^2/s	–
Power deposition	W armor: ~ 3%, FW ~ 16%, Breeder zone ~ 75%, Manifolds ~ 3%	–
Heat load (O/B)	W armor ~ 28.4 W/cm^3, FW Eurofer ~ 9.59 W/cm^3	–

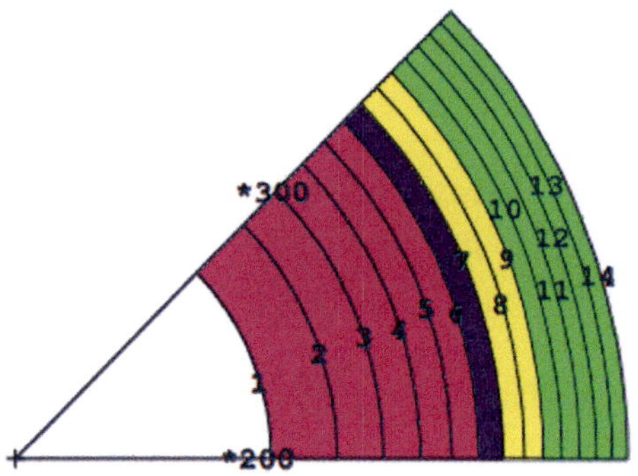

FIGURE 7.19 A simplified spherical model of a fusion reactor blanket for neutronics; vacuum, *purple* = breeder, *blue* = helium, *yellow* = beryllium multiplier, *green* = EUROFER steel structure.

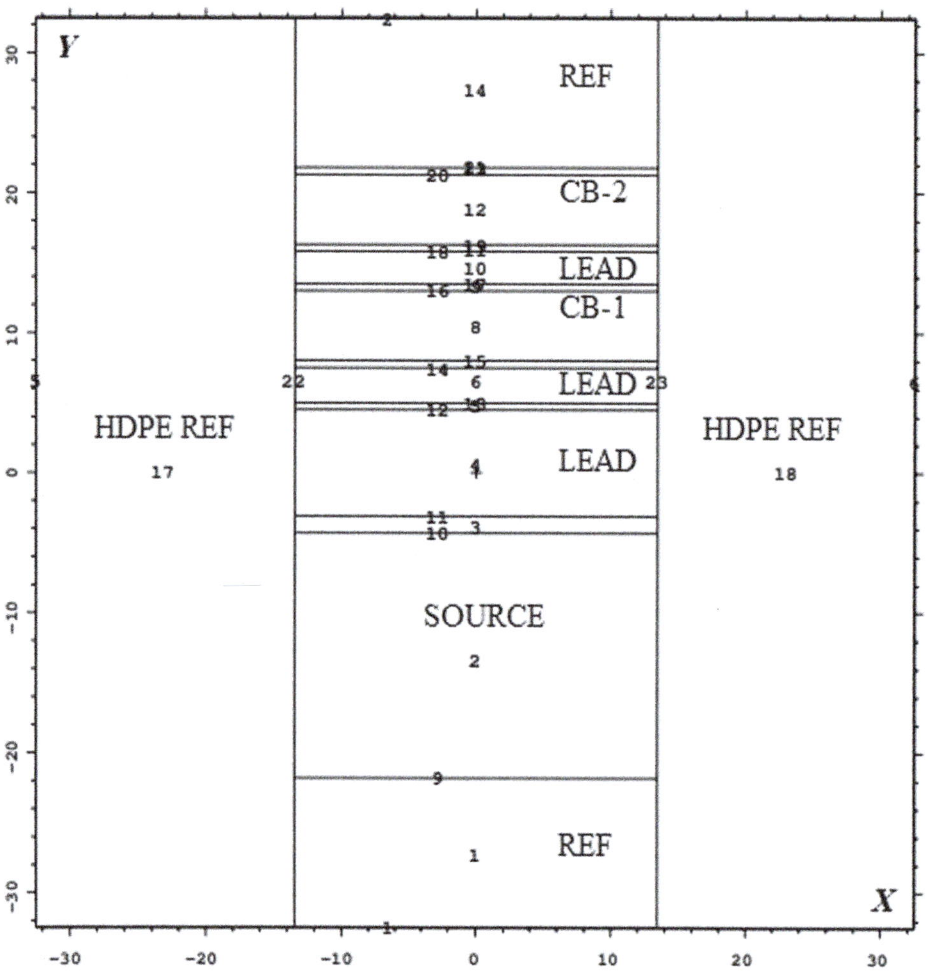

FIGURE 7.20 TBR ITER Mockup Experiment (dimensions in cm).

7.4 Demonstration

This section takes the reader through a fixed-source problem to introduce an input and output of a MC neutron transport simulation. Multiplying systems for simple critical assemblies and reactor lattices and nuclear reactor core calculations are described in chapters nine and eleven respectively.

Consider the carbon-beryllium sphere configuration shown in Fig. 7.15. There are three spheres in the model; the first sphere of radius 0.02 cm contains a point isotropic source of energy 1 MeV located at the origin. For such an idealized problem, exact analytical diffusion and transport solutions were described in Chapters 4 and 5. It is a good practice to learn MC simulation by first carrying out simulations for simple problems for which exact solutions are known. A time-of-flight benchmark for a carbon sphere is a useful starting point for comparing MC simulation results with experimental results.

Some benchmarks will be considered for critical systems in Chapter 9.

The MCNP input file (X-5 Monte Carlo Team, 2008) consists of three blocks separated by a blank line between blocks at the end of the third block. The first line of MCNP describes the simulation. In Block 1, the cells (volume regions) are defined in terms of their bounding surfaces and the material present in them.

In this example, Block 1 has four lines since there are four cells; source, carbon, beryllium, and the *outside world*.

The four lines, after the first description line, define the cell number followed by the material number, material density (negative sign meaning a gram density and positive meaning an atomic density) followed by the sense of the cell with respect to the surface number(s) given.

There are three surfaces defined in the second block, each is a sphere with its center at the origin; surfaces 1, 2 and 3 represent spheres of radii 0.02, 4.187, and 5.187 cm, respectively.

The first cell is defined to be within the region bounded by surface 1, that is, every point in cell 1 has a negative sense with respect to surface 1. Mathematically, this condition is stated as $f(x, y) = x^2 + y^2 + z^2 - (0.01)^2 < 0$ for a point x, y in cell 1 which contains material 1 with atomic density 0.001288 g/cm^3; the material entry is given in Block 3 of the input file. Similarly cells 2 and 3 are defined each between their two bounding surfaces. The last cell, cell 4, the outside world is everything beyond surface 3. A particle is considered to have left the system when it enters into the outside world and can not rerenter the system.

MCNP input file for a carbon-beryllium configuration

```
BK7_01 C sphere for Tallies F1,F2 and F4
C   Block 1
    1  1  -0.001288     -1
    2  2  -1.7           1 -2
    3  0          2 -3
    4  0                 3

C   Block 2
   1  SO  0.02
   2  SO  4.187
   3  SO  5.187

c  Block 3
c  this is a neutron only simulation
mode    n
imp:n    1 2R 0 $ each cell importance is 1 except for the outside world
c  which has zero importance
c
c   each cell has one material defined by its identifier
c   followed by its weight fraction (-ve entry) or atomic fraction
c   or atomic density (+ve entry)
c  --------------------- Air (Den 0.0012 gm/cm3)-------------
M1     7014.66c -0.75519 8016.66c -0.23179 6000.66c -0.00014
       18000 -0.01288
c  --------------------- Carbon PNNL 63 (Den 1.7 gm/cm3)----------------
M2     6000.66c 1
c  --------------------- Beryllium PNNL 24 (Den 1.848 g/cm3)--------------
M3     4009.66c 1
C
C    M4 and M5 are tallying materials
M4    6000.66c 1.0
M5    4009.66c 1.0
C
C  the bin boundaries for the simulation
C  1st bin E< 6.25e-7 MeV, 2nd bin between 6.25e-7 MeV and 1e-6 MeV
C  3rd bin between 1 eV and 1 MeV, 4th bin between 1 and 10 MeV
C  5th bin between 10 and 14 MeV (any neutrons above this energy
C  will not be tallied) you can define bins to get the desired output
```

```
e0 6.25e-7 1e-6 1 10 14
c  angular bins: bin 1 lies between cos values -1 and 0 (pi and pi/2)
c  bin 2 lies between cos values 0 1 i.e. between pi/2 and 0
c0  0 1
FC1  CURRENT ACROSS SURFACE 2 (CARBON SPHERE) eqn. 7.2
F1:N   2
fc2  SURFACE FLUX ACROSS SURFACE 2 (CARBON SPHERE) eqn. 7.3
F2:N   2
FC4  neutron flux in cell 2 (CARBON SPHERE) eqn. 7.4
F4:N   1
FC5   point detector of radius 0.5cm
f5:n   4.187 0 0 0.5
c  with the scalar flux, the reaction rates are calculated
FC24    reaction rates total x/s, (n.g) x/s in c sphere
c      R1=0.02 R2=4.187 V=(4/3)*pi*(R2^3-R1^3)=26.23e24
F24:N   2
FM24:n   (26.207 4 (-1) (-2) (-3) (102))
C  the source is defined as a point source located at 0,0,0 with
C  energy 1 MeV
sdef    pos = 0 0 0 erg=1
c
c  this simulation will run 10 million histories
NPS    10000000
C  all the output will be printed for results and diagnostics
print
```

The comment lines added to Block 3 describe the input lines for the mode of the simulation, the cell importance, the material descriptions, the bin structures for energy and angular distributions, tallies, tally multipliers, source description and simulation parameters.

The tally multiplier

```
FM24:n   (26.207 4 (-1) (-2) (-3) (102))
```

calculates quantities of the form described by Eq. (7.5). The FM24 follows a tally F24 which is a flux tally of type 4 (each succeeding tally is defined distinctly such as F4, F14, F24, F34,. . .). The form of this tally multiplier is

```
FM24:n   (C M (R1) (R2) (R3) (R4) . . .)
```

where the constant C is a multiplier, M is a material number and R1,. . .R4 are Reaction Numbers defined in the ENDF/B files.

In the present case, C is the atomic density of carbon multiplied by the volume of the carbon region and the Reaction Numbers are -1 (total cross section), -2 (absorption scattering cross section), -3 (elastic scattering cross section) and 102 (radiative capture cross section). In this demonstration only four reaction rates were specified; this depends on the user's requirements (limits are given in the MCNP manual).

Following an overall summary of the interactions and gains and losses, the required tallies are listed as shown in Table 7.8 for three simulations with 100, 10^4 and 10^6 histories.

The neutron current tally (F1) remains 0.99999 (0.0000) for all three simulations since the absorption is very small as shown in Table 7.8. The current is only in the forward cosine bin, that is, in the range $(0, \pi/2)$, in one energy bin (1 eV$-$1 MeV). This tally tells us that all postcollision energies lie in the range 1 eV$-$1 MeV; this range could have been made smaller to get finer detail.

Table 7.8 lists seven tallies per source neutron per second (surface flux, cell flux, point flux total and uncollided, total reactions, absorptions, elastic scatterings and radiative captures).

The results show that

1. For each tally, the quality of the results gets progressively better with the number of histories; this is due to the large number of collisions in the carbon sphere.

2. Relative error decreases by an order of magnitude as the simulations increase by two orders of magnitude.
3. The variance of the variance (VoV) also decreases in the same way as the relative error.
4. The (total) point flux improves with the number of histories while the uncollided point flux has zero relative error as it is a deterministic quantity.
5. Reactions are predominantly elastic scattering reactions; absorption is due to radiative capture alone. Since the absorptions include the (n, p), (n, t), (n, d), (n, He^3) and (n, γ) reactions, it is observed that absorption is only to the (n, γ) reactions.

Finally, the energy-dependent neutron flux, shown in Fig. 7.21 for one million histories, indicates that there is very little spectrum degradation due to the absence of any low-Z hydrogenous materials.

Such experiments, important as they are for an understanding of the simulation, will be left as exercises for the MC user.

TABLE 7.8 Tallies for a carbon sphere fixed-source Monte Carlo simulation.

Tally N	10^2	10^4	10^6
Surface flux (F2)	5.81407×10^{-3}	5.4334×10^{-3}	5.39525×10^{-3}
Relative error	0.0824	0.0055	0.0005
VoV	0.4931	0.0409	0.0004
Cell flux (F4)	1.74752×10^{-2}	1.80857×10^{-2}	1.80573×10^{-2}
Relative error	0.0329	0.0044	0.0004
VoV	0.0587	0.0018	0
Point flux (F5)	1.02127×10^{-2}	5.40671×10^{-3}	5.39618×10^{-3}
Relative error	0.3408	0.0428	0.0045
Point flux (F5) uncollided	1.81729×10^{-3}	1.81729×10^{-3}	1.81729×10^{-3}
Relative error	0	0	0
VoV	0.5535	0.1618	0.0028
Reactions/cm^3			
Total reactions	4.13757×10^{-3}	4.231955×10^{-3}	4.30813×10^{-3}
Relative error	0.0414	0.0056	0.0006
VoV	0.0748	0.0026	0
Absorptions	1.80589×10^{-9}	1.85912×10^{-9}	1.85759×10^{-9}
Relative error	0.0271	0.0036	0.0004
VoV	0.0748	0.0026	0
Elastic	4.13756×10^{-3}	4.31955×10^{-3}	4.30813×10^{-3}
Relative error	0.0414	0.0056	0.0006
VoV	0.0748	0.0026	0
Radiative capture	1.80589×10^{-9}	1.85912×10^{-9}	1.85759×10^{-9}
Relative error	0.0271	0.0036	0.0004
VoV	0.0748	0.0026	0

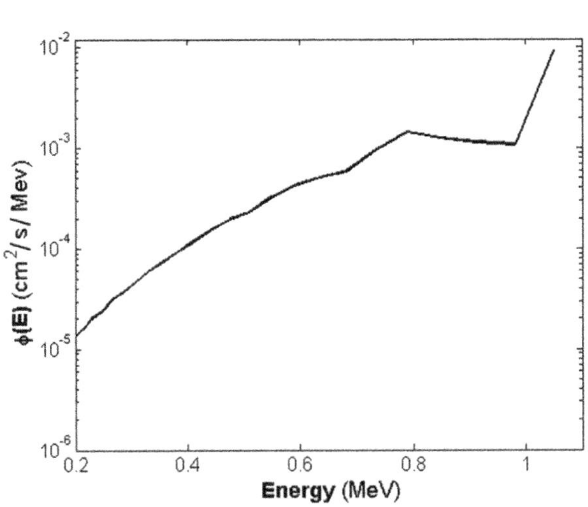

FIGURE 7.21 Neutron flux in a carbon sphere.

7.5 Variance reduction methods

In the previous section, a natural or straightforward MC simulation, called *analog simulation*, was carried out since it was anticipated, and demonstrated by simulation, that a large number of collisions took place in the carbon sphere. Subsequently, the results were both accurate and precise.

In nuclear fission reactor calculations, this is usually the case within the core where neutrons and gamma interactions take place frequently. However, as we go far from the source of activity the flux falls by orders of magnitude.

By design, radiation shielding has to be thick enough to attenuate the radiation so that very few neutrons or gammas are able to penetrate its thick layers. If, for example, one million neutrons are started as source neutrons and only two neutrons get through the shield to contribute to the tally of interest, then it will be an unreliable tally.

The validity of the Law of Large Numbers or the Central Limit Theorem would then become questionable and it would not make sense to compute a sample estimate or a variance.

As we saw in §7.1.5, there are two options for getting a better estimate; the first is to increase the histories to be simulated. An increase of 100 would reduce the standard deviation by 10. But simulating 100 million histories instead of 1 million would requires hours and hours of computer time, with no guarantees that a good enough tally would be possible. The second option is to force things to happen (bias the simulation) in such a way that more histories contribute to the tally.

This approach, of modifying the simulation in an unrealistic way would not be correct unless the overall weight is conserved. Thus in a small solid angle of interest, if there would have been 10 particles by geometrical considerations alone, then we can force 50 particles to travel toward the detector but with a weight reduced by a factor of five. Such a strategy, called a nonanalog simulation has several such *variance reduction methods* which help in obtaining reliable estimates.

Some methods are used in analog simulations to reduce the computational effort on particles which are likely to have a small contribution to a tally. For example, when the weight of a particle, after several interactions, has reduced below a small number such as 1.0e-6, then Russian Roulette is played to determine whether to let it continue, with increased weight. Say if the weight falls below the prescribed cutoff and it is given a small probability of survival, then a random number is generated to decide its fate. The probability that the random number would favor survival is also very low so in most cases, the particle would be terminated. However, there is small chance that it survives, in that case its weight is increased by the ratio of survival probability to nonsurvival probability and it continues. Conversely, if its weight becomes too large, then it is split into many particles of reduced weight. This process ensures that particles remain within a favorable weight window. Codes such as MCNP have a weight window generator which suggests best weights to specify. This is an art as well as a science as weight can be intuitively assigned as well.

Such splitting techniques can be carried out on the basis of space, energy and angle. When some regions are more *important* than others it is desirable to invest more time in simulating them as there is reason to believe that they will contribute toward the quality of a tally.

Similarly, when a particle moves into an important region, it is split into more particles, while conserving the overall weight. Conversely, when a particle moves into a region of less importance, a decision has to be made whether to let it continue at the expense of computational effort and time, or to terminate its history. This decision is based on a statistical process known as Russian Roulette. A random number is generated and with the prescribed probability of termination, the appropriate decision is made.

Geometry importances are assigned in a data file at the beginning of a MC simulation. If region A has importance one and region B has importance two, then the particle will be split into twice as many particles. In practice it is advisable to manually input cell importance and inspect the results to see whether they were set properly; this remains an art more than a science and is bound to rely heavily on intuition. A formal method is to run an adjoint problem but a code such as MCNP, which is regarded as industry-standard now, does not have this option in the "continuous energy" simulation mode.

For shielding problems where variance reduction can be necessary and most useful, a step-wise approach can be used with several runs, starting from an initial guess of cell importance and using generated weight windows progressively in regions closer to the shield.

Some more nonanalog techniques are exponential transform, forced collisions and source biasing where a probability distribution function is modified, and accounted for later, to give a large number of nonzero contributions from histories.

All these techniques can result in improved tallies, but judgment based mostly on intuition is the guide.

Importance biasing has mathematical foundations in the adjoint equation which is also called the backward equation as it represents a detector-to-source formalism rather than one based on the conventional source-to-detector formalism. The adjoint flux or "importance" function has then represents the importance of particles in phase space; these have also been called *contributions*. This amounts to knowing the solution and biasing it in such a way that only those particles are simulated that contribute to the tally.

A simple example of knowing the solution to carry out importance sampling is the integral evaluation deterministic problem described in Chapter 4. An altered estimator is used so that

$$I = \int_a^b \frac{f(x)}{g(x)} g(x)dx = \langle f/g \rangle \int_a^b dx = \langle f/g \rangle (b-a)$$

estimates $f(x)/g(x) \equiv h(x)$ while the PDF is $g(x)$. It can be shown that by a "suitable" choice of the "importance function" $g(x)$ this process leads to a "zero-variance" solution.

Consider the estimation of the integral $I = \int_0^1 e^{-x}dx$, first with the estimator $f(x) \equiv e^{-x}$ with sampling from the uniform PDF, and then with the estimator $h(x) = 1$, and of course the PDF normalization factor $(1 - e^{-1})$ outside the integral, with random number samped from the PDF

$$g(x) = \frac{e^{-x}}{1 - e^{-1}}$$

Clearly, this leads to

$$I = \int_0^1 g(x)dx$$

with a precise (zero-variance) answer of $(1 - e^{-1})$.

While it was easy to demonstrate a "zero-variance" result for the simple problem considered above, it is not always easy to "know" a priori the importance function $g(x)$ since the "best" $g(x)$ is the answer itself which we want to find.

Variance reduction based on intuition, or modified sampling or with the help of the adjoint equation has been found to be necessary in nuclear engineering for shielding and streaming problems to determine, with reasonable accuracy, the shutdown activation for example. In Chapter 9, variance reduction will be demonstrated for a representative problem.

7.6 Estimating perturbations with Monte Carlo simulation

In the design and operation of nuclear reactors and systems, it is often important to be able to estimate the effect of small material density perturbations arising out of the burnup of fuel or the insertion of control rods. Additionally, it may be important to estimate the effect of geometry perturbations due to moving components.

A drawback of MC methods is that the uncertainty of an estimate may be of the same order of magnitude as the perturbation itself and thus the effect of the design or operational change may become masked. This implies that the difference of two MC simulations may be inadequate to estimate perturbations. Thus other methods had been developed initially along the traditional route of deterministic formulations based on the adjoint function, or Lagrange multiplier, as a sensitivity coefficient. Such analyses have been used extensively to estimate perturbations from two MC runs viz a forward run and an adjoint run from codes such as MCNP and MORSE. New developments in this area came with "correlated tracking" and "derivative sampling" techniques (Rief, 1984) and implemented in KENO and TIMOC codes. The techniques were extended by Watkins for material perturbations in a research reactor and demonstrated for two-group transport (Koreshi & Lewins, 1990) and later extended to fusion reactor neutronics optimization studies. The perturbation feature is now part of several MC codes and is used extensively. Comparisons of the adjoint versus derivative approach have been discussed by Kiedrowski & Brown (2011). Applications of MC perturbation theory are given in Chapter 11 for estimating changes in k_{eff} due to material density pertrubations in a bare sphere such as Godiva. Such studies have also been carried out to quantify the sensitivity of estimates to uncertainties in cross section data. With reasonable confidence for small systems, perturbation can be carried out for design optimization in bigger simulations thus saving computational effort.

7.7 Conclusions

In this chapter, the fundamentals of the Monte Carlo (MC) method used for particle transport were presented. The use of random numbers, event probabilites, and probability distribution functions, given in Chapter 4, was described for simulating a random walk and tallying quantities using the collision, absorption, and track length estimators. The simulation for an elementary fixed-source problem was used to introduce the reader to prepare an input for carrying out a neutron transport simulation giving basic tallies and interpreting the estimated quantities.

Problems

1. Use the acceptance-rejection technique to sample from an exponential distribution and compare its efficiency by a straightforward analytical inversion.
2. Use the same acceptance-rejection technique to braw a bounding box around a normal distribution to get samples.
3. For Exercise 7.2, write a program to estimate neutron flux in a carbon slab of thickness 3 mfp with thermal cross sections.
4. Repeat the above for isotropic scattering.
5. Modify your program to use the collision estimator for flux rather than the TLE.
6. In the AP1000 unit lattice cell, using the models for alpha and beta transport (Chapter 1), what would be the significance of both radiations on the energy deposition and on the coolant activation?
7. Based on the topics covered in Chapter 4, how would the neutron fluxes calculated from a MC simulation be used to calculate the source term for spent nuclear fuel from a nuclear reactor?
8. With reference to quantities of interest in fission neutronics, list a few activation reactions in constituent elements of steel that could pose a long-term storage problem. How would these quantities be calculated using cross section fields?
9. With reference to Table 7.7 comment on the effect of water reduction leading to an increase in TBR.
10. In Section 7.5, for $I = \int_0^1 e^{-x}dx$, show that a zero-variance solution is obtained for an exponential PDF. Compare the estimate with a PDF representing the first quadrant of a circle.

Nomenclature

Greek lower case

ρ	density
θ	orthogonal angle
μ	cosine of orthogonal angle
μ	mean value
$\overline{\nu}_k$	average number of fission neutrons emerging from a fission
σ	standard deviation
σ_x	standard deviation of $\overline{x}$
τ	time constant
φ	azimuthal angle
$\overline{\phi}_S$	surface-averaged scalar flux
$\overline{\phi}_V$	volume-averaged scalar flux
χ	collision density

Greek upper case

Σ_t	macroscopic total cross section
Ω	solid angle

English lower case

d	distance
f_k	fraction of nuclide k
k_{eff}	system multiplication
m	mass
n	atomic density
p	resonance escape probability
p	probability
s	distance
ξ	random number
w	weight

English upper case

AE	absorption estimator
$\hat{C}$	collision kernel

Cf	californium-252
Cm	curium-244
H(E)	heating value as a function of energy
J	current
N	number of histories
Q	Q-value of a reaction
R	relative error
S	source
$\hat{T}$	transport kernel
T_f	fuel temperature
T_m	moderator temperature
ppm	atomic parts per million
dpa	displacements per atom
pcm	percent mille (10^{-5})

Abbreviations and Acronyms

CE	collision estimator
CZP	cold zero power
FOM	figure of merit
FPY	full power years
FW	first wall
HFP	hot full power
HZP	hot zero power
I/B	inboard
ENDF	evaluated nuclear data files
O/B	out-board
TBR	tritium breeding ratio
TFC	toroidal field coil
TLE	track-length estimator
TPR	tritium production rate
VV	vacuum vessel
MC	Monte Carlo

References

Anderson, H.L. (1986). Metropolis, Monte Carlo, and the MANIAC. Los Alamos Science.

Aziz, U., Koreshi, Z. U., Sheikh, S. R., & Khan, H. (2020). A study of criticality and thermal loading in a conceptual micronuclear heat pipe reactor for space applications. *Nuclear Technology and Radiation Protection*. Available from https://doi.org/10.2298/NTRP2003208A.

Brown, D. A., Chadwick, M. B., Capote, R., Kahler, A. C., Trkov, A., Herman, M. W., & Zhu, Y. (2018). ENDF/B-VIII.0: The 8th major release of the nuclear reaction data library with CIELO-project cross sections, New Standards and Thermal Scattering Data. *Nuclear Data Sheets*. Available from https://doi.org/10.1016/j.nds.2018.02.001.

Colling, B. R. (2016). *Blanket Performance and Radioactive Waste of Fusion Reactors: A Neutronics Approach*. England: Lancaster University.

Dunn, W. L., & Shultis, J. K. (2012). *Markov chain Monte Carlo. Exploring Monte Carlo Methods*. Available from https://doi.org/10.1016/b978-0-444-51575-9.00006-3.

Hammersley, J.M. & Handscomb, D.C. (1964). *Monte Carlo methods*. <https://doi.org/10.1007/978-94-009-5819-7>

Hammersley, J. M., Handscomb, D. C., & Weiss, G. (1965). Monte Carlo methods. *Physics Today*. Available from https://doi.org/10.1063/1.3047186.

Hernández, F. A., & Pereslavtsev, P. (2018). First principles review of options for tritium breeder and neutron multiplier materials for breeding blankets in fusion reactors. *Fusion Engineering and Design*. Available from https://doi.org/10.1016/j.fusengdes.2018.09.014.

Jakhar, S., Abhangi, M., Tiwari, S., Makwana, R., Chaudhari, V., Swami, H. L., & Chaudhuri, P. (2015). Tritium breeding mock-up experiments containing lithium titanate ceramic pebbles and lead irradiated with DT neutrons. *Fusion Engineering and Design*. Available from https://doi.org/10.1016/j.fusengdes.2015.04.003.

Kalos, M. H., & Whitlock, P. A. (2008). *Monte Carlo methods*. Wiley-Blackwell.

Keys, D. A., & Groves, L. R. (1963). Now it can be told. The story of the Manhattan project. *International Journal*. Available from https://doi.org/10.2307/40198796.

Kiedrowski, B. C. „ & Brown, F. B. (2011). Comparison of the Monte Carlo adjoint-weighted and differential operator perturbation methods. *Progress in Nuclear Science and Technology*. Available from https://doi.org/10.15669/pnst.2.836.

Koreshi, Z. U., & Hussain, S. Z. (2014). A neutronic pin cell and full-core design analysis of the 4 S reactor. *Annals of Nuclear Energy*, *72*, 145–150. Available from https://doi.org/10.1016/j.anucene.2014.05.015.

Koreshi, Z. U., & Lewins, J. D. (1990). Two-group Monte Carlo perturbation theory and applications in fixed-source problems. *Progress in Nuclear Energy*. Available from https://doi.org/10.1016/0149-1970(90)90020-6.

Mandal, D., Shenoi, M. R. K., & Ghosh, S. K. (2010). Synthesis & fabrication of lithium-titanate pebbles for ITER breeding blanket by solid state reaction & spherodization. *Fusion Engineering and Design*. Available from https://doi.org/10.1016/j.fusengdes.2010.06.018.

McClarren, R. G. (2018). Introduction to Monte Carlo methods. *Computational Nuclear Engineering and Radiological Science Using Python*. Available from https://doi.org/10.1016/b978-0-12-812253-2.00024-8.

Metropolis, N., & Ulam, S. (1949). The Monte Carlo method. *Journal of the American Statistical Association*. Available from https://doi.org/10.1080/01621459.1949.10483310.

Moro, F., Colangeli, A., Del Nevo, A., Flammini, D., Mariano, G., Martelli, E., & Villari, R. (2020). Nuclear analysis of the water cooled lithium lead DEMO reactor. *Fusion Engineering and Design*. Available from https://doi.org/10.1016/j.fusengdes.2020.111833.

Del Nevo, A., Arena, P., Caruso, G., Chiovaro, P., Di Maio, P. A., Eboli, M., & Martelli, E. (2019). Recent progress in developing a feasible and integrated conceptual design of the WCLL BB in EUROfusion project. *Fusion Engineering and Design*. Available from https://doi.org/10.1016/j.fusengdes.2019.03.040.

Rief, H. (1984). Generalized Monte Carlo perturbation algorithms for correlated sampling and a second-order Taylor series approach. *Annals of Nuclear Energy*. Available from https://doi.org/10.1016/0306-4549(84)90064-1.

Schmidt, K. H., & Jurado, B. (2018). Review on the progress in nuclear fission - Experimental methods and theoretical descriptions. *Reports on Progress in Physics*. Available from https://doi.org/10.1088/1361-6633/aacfa7.

Shimwell, J., Kovari, M., Lilley, S., Zheng, S., Morgan, L. W. G., Packer, L. W., & McMillan, J. (2016). A parameter study of time-varying tritium production in solid-type breeder blankets. *Fusion Engineering and Design*. Available from https://doi.org/10.1016/j.fusengdes.2016.01.011.

Spanier, J., Gelbard, E. M., & Bell, G. (1970). Monte Carlo principles and neutron transport problems. *Physics Today*. Available from https://doi.org/10.1063/1.3022338.

de Stefani, G. L., Losada Moreira, J. M., Maiorino, J. R., & Russo Rossi, P. C. (2019). Detailed neutronic calculations of the AP1000 reactor core with the Serpent code. *Progress in Nuclear Energy*. Available from https://doi.org/10.1016/j.pnucene.2019.03.030.

Sublet, J. C., Eastwood, J. W., Morgan, J. G., Gilbert, M. R., Fleming, M., & Arter, W. (2017). FISPACT-II: An advanced simulation system for activation, transmutation and material modelling. *Nuclear Data Sheets*. Available from https://doi.org/10.1016/j.nds.2017.01.002.

Sun, H., Wang, C., Ma, P., Liu, X., Tian, W., Qiu, S., & Su, G. (2018). Comnceptual design and analysis of a multipurpose micro nuclear reactor power source. *Annals of Nuclear Energy*, *121*, 118–127.

Ueda, N., Kinoshita, I., Minato, A., Kasai, S., Yokoyama, T., & Maruyama, S. (2005). *Sodium cooled small fast long-life reactor "4 S."*, . *Progress in Nuclear Energy* (47, pp. 222–230). . Available from https://doi.org/10.1016/j.pnucene.2005.05.022.

Werner, C. (ed.) (2017). MCNP (R) User's Manual Code Version 6.2.

X-5 Monte Carlo Team. (2008). MCNP - A general Monte Carlo N-particle transport code, Version 5 Volume I: Overview and theory. Los Alamos Nuclear Laboratory. <https://doi.org/LA-UR-03-1987>.

Yang, D., Liu, X., Xiong, J., Chai, X., & Cheng, X. (2018). Coupling of neutronics and thermal-hydraulics codes for the simulation of reactivity insertion accident for LFR. *Progress in Nuclear Energy*. Available from https://doi.org/10.1016/j.pnucene.2018.02.023.

Chapter 8

Computer codes

Computer codes have become the mainstay of nuclear engineering and are abundantly used for the design and analyses of nuclear systems. The use of codes, developed in universities, companies and government organizations, is indispensable as a first step in the design of a reactor and a requirement in regulatory processes. The USNRC has been using computer codes since the development of early nuclear reactors. Similarly other international efforts include the Institute de Radioprotection et de Surete Nucleaire (IRSN) in Europe and more recently, the Chinese nuclear regulatory institution through the National Nuclear Safety Administration (NNSA).

With the progression followed in this book, earlier chapters have covered the radiation transport processes covered charged-particle and gamma transport, neutron transport, a description of nuclear reactors, mathematical foundations, neutron diffusion followed by neutron transport and Monte Carlo simulation and optimization in the later chapters. All these are included in the computer codes used for the design and performance analyses of nuclear reactors and systems. In Chapter 4, the mathematical treatment of coupled analyses, such as neutronics coupled with thermal hydraulics, was introduced to demonstrate the interdisciplinary nature of nuclear engineering simulations. In these chapters, we saw that reactor core analyses requires such modeling of phenomena that take place within, for example, a reactor core typically consisting of several assemblies of fuel rods and control rods surrounded by moderator, coolant and protective structure. Thus mathematical formulations are necessarily coupled and the required computer codes are not restricted to any one field. The nuclear industry thus does not have a single code to handle all the requirements of nuclear systems analyses. Therefore, we will look separately at codes in the following areas which are fairly representative of the broader picture

1. Neutron and radiation transport,
2. Time-dependent reactor kinetics,
3. Thermal hydraulics,
4. Radiological protection, and
5. Performance and Safety analyses.

Over time, there has been a shift toward the usage of these codes utilizing the power of supercomputing architecture. The emphasis is thus on the development of algorithms and away from the basic mathematical rigor that was included earlier.

The strength of a computer code lies in its capability to model the interaction physics as well as the three-dimensional geometries of nuclear fission and fusion power reactors. The earlier codes, written for nuclear fission reactors and being restricted to "regular" geometry, that is, slab, circular or cylindrical systems were capable only of idealizing complex geometry features intrinsic to core internals and radiation as well as biological shielding systems surrounding the reactor. Over the years, with the development of better numerical schemes such as finite element methods in deterministic codes and combinatorial geometry in stochastic codes, it became possible to adequately describe many physical complexities. The same applies to complexities in the collision data, which was used in multi-group form rather than in continuous energy form, scattering models, fission spectra and so on.

With better capability in modeling real systems, the computational effort also increases and hence computer processing is a crucial requirement. Thus, computer codes have benefited from the processing speeds achievable by computer systems as they evolved from mechanical and electrical computers in the 1940s to electronic computers, mainframes such as IBM and CDC in the 1970s and 1980s, to workstations, RISC processors, Cray supercomputers with vector and parallel architecture, to modern-day portable and efficient computers (Schober, 2018). Looking to the future, it is possible that optical data communication may replace the present electronic circuitry making large computational codes even more attractive.

Nuclear Engineering. DOI: https://doi.org/10.1016/B978-0-323-90618-0.00008-9

The software and hardware features available in present-day computing permits simulation codes to cover complex features to a level sufficient for reliable design. A computational challenge in reactor neutronics has been to estimate "rare events" such as the transmission of radiation across the shielding.

Some of the widely used computer codes in nuclear systems are briefly described below.

8.1 Neutron and radiation transport codes

Early codes, which have been updated and are still used for reactor analysis including fuel depletion, include ANISN (Engle, 1967), CITATION, WIMS, and ORIGEN. The diffusion theory code CITATION (Fowler, Vondy, & Cunningham, 1971) released in 1972, solves the 3D neutron diffusion equation with the finite difference method and has been used together with WIMS-D and ORIGEN for calculating neutronics coupled with burn-up and depletion of fuel in nuclear systems. It is distributed by RSICC. WIMS (Winfrith Improved Multi-group Scheme) is a general code for reactor physics lattice cell calculations for nuclear reactor systems and has been used for the calculation of group constants (Section 6.1) for fuel assemblies such as for analysis of Loss of Flow transients (Noori-Kalkhoran, Minuchehr, Shirani, & Rahgoshay, 2014), for core optimization simulation of a PWR (Hussain, Xinrong, & Ahmed, 2008) and a wide range of neutronics.

The cross section data library has a number of resonance groups for thermal neutron analysis. WIMSD is distributed by the RSICC and it has been updated with several benchmarks (Lindley et al., 2017). ORIGEN (ORNL Isotope Degeneration and Depletion Code), a fuel cycle depletion code developed by ORNL, performs nuclide transmutation calculations based on neutron flux in a system (Bell, 1973; Yesilyurt, Clarno, Gauld, & Galloway, 2011). Other deterministic codes are described below.

8.1.1 ANISN

One of the first codes used for neutron/photon transport, ANISN (Anisotropic SN method), is based on an advanced discrete ordinates method proposed by Carlson in 1953 and developed by Wick and Carlson (Carlson & Lathrop, 1968) at Los Alamos Scientific Laboratory.

ANISN calculates angular flux in sufficient detail to solve deep-penetration problems in addition to regular calculations of flux and reaction rates in regions of nuclear systems. Older versions of ANISN-ORNL, which were operable on VAX and IBM mainframes, are now extended with improvements for Cray and IBM RISC versions as well as for Pentium IV computers under Windows XP operating systems.

8.1.2 DOT

The DOT code (Rhoades & Mynatt, 1973), as version DOT-4, calculates multi-group angular flux by solving the Boltzmann transport equation using the S_N method, diffusion theory or a special P_1 method, for two-dimensional geometric systems. It can solve for criticality also, but the principal application is to the more difficult deep-penetration transport of neutrons and photons for shielding problems. Cross sections are modeled using Legendre expansion to arbitrary order allowing for anisotropic scattering. A vectorized version, DOT-4/VE has increased computational efficiency. The CPU time required by DOT depends on the Flux Work Units (FWUs) which are dependent on the number of space mesh cells, energy groups, and the number of iterations per group. Typical of large deterministic codes, big problems could take over an hour on modern platforms.

8.1.3 TORT

TORT (2D, 3D) and DORT (1D, 2D) are Discrete-Ordinate Neutron and Photon transport codes (Kirk, 2009) with eigenvalue and deep-penetration calculation capabilities. TORT calculates the flux of particles due to particles incident upon the system from extraneous sources or generated internally as a result of interaction with the system. The Boltzmann transport equation is solved using the discrete ordinates method to treat the directional variable and the finite difference methods to treat spatial variables. Energy dependence is treated using a multi-group formulation. Time dependence is not treated.

8.1.4 PARTISN

PARallel **Ti**me-Dependent **SN** (Alcouffe, 2001), a successor of ONEDANT, TWODANT, and THREEDANT is a parallel time-dependent deterministic code that can solve three-dimensional problems in regular geometry using the discrete ordinates method.

Some of the extensively used Monte Carlo codes are described below.

8.1.5 MCNP

Monte Carlo methods for Neutrons and Photons (Pelowitz et al., 2013) were developed at Los Alamos National Labs (LANL) in the 1940s by the group that included von Neumann. Its' name later implied Monte Carlo *N*-Particle transport as it included charged particles. MCNP is used for a diverse range of problems involving neutron and radiation transport in nuclear reactor engineering with the capability of carrying out "whole-core" analyses, as well as medical physics, oil-well logging, and radiography. The first Monte Carlo code appeared as a Los Alamos report in 1963. By the early 1980s, MCNP begain to be widely distributed and used as MCNP3A and MCNP3B written in Fortran 77 with neutron and photon transport. The next version, MCNP4 had improvements such as electron transport capability as well as new tallies. It had parallel processing to speed up the computations. The perturbation capability was added in MCNP4B in the late 1990s. This was followed by MCNP4C which had some better physics such as the treatment of nuclear resonances. MCNP5 released in 2003 was re-written in Fortran 90 and had several improvements. The latest version, MCNP6.2 (2017) can transport 36 different particles from neutrons, photons, electrons to muons, anti neutrons various baryons, anti-particles, deuterons, tritons, alpha and heavy ions. The long list of improvements includes the modeling of physics, sources, data, tallies, unstructured meshes as well as code enhancements. This has led to a greatly increased versatility going far beyond the early neutron and photon transport to areas such as medical physics, accelerator-driven energy source research, high-energy dosimetry and futuristic areas such as charged-particle propulsion. Stochastic geometry has been incorporated to model the random arrangements in a geometry such as the high-temperature gas cooled reactor (HTGR) with its spherical fuel pebbles.

New Monte Carlo codes such as OpenMC (Romano et al., 2015) have improved combinatorial geometry models that have been demonstrated for photon simulation (Kargaran, Jafari, & Minuchehr, 2021) and are openly accessible.

MCNP benchmarks are available for neutron and problems (Whalen, Cardon, Uhle, & Hendricks, 1991), criticality safety problems (Wagner, Sisolak, & McKinney, 1992) and LWR criticality. MCNP has been used extensively for core neutronics; some representative latest work is the study on the performance analysis of thorium-based mixed oxide fuel in a PWR (Tucker & Usman, 2018), a conceptual design of a micronuclear reactor using heat pipes (Aziz, Koreshi, Sheikh, & Khan, 2020; Hao Sun et al., 2018) and the study of accident tolerant fuels in the upcoming NuScale SMR (Yu, Cai, He, & Li, 2021).

8.1.6 TART

TART 2005 (Kirk, 2009) is a coupled neutron photon Monte Carlo particle transport code written in FORTRAN and developed by Dermott E. Cullen (McKinley, Cullen, Latkowski, Procassini, & Skulina, 2006) and maintained by Lawrence Livermore Laboratories (LLL). It is time-dependent and uses combinatorial geometry allowing the modeling of large and complex problems. It is LLL's equivalent of LANL's MCNP and is said to be the fastest MC code.

8.1.7 MORSE

MORSE (Multi-group Oak Ridge Stochastic Experiment), succeeded by MONACO, is a Monte Carlo code developed by Oak Ridge National Laboratories (ORNL) (Emmett, 2000). Initially, it was also based on regular geometry and although it uses the same integral formulation of the transport equation as a continuous code such as MCNP, MORSE uses a multi-group approach. A neutron thus collides with an atom of the homogenized mixture prepared from the macroscopic cross sections rather than with individual atoms. Such a multi-group MC approach has served for quick comparisons with deterministic codes such as ANISN but for more elaborate computation, MCNP is preferred.

MORSE-SGC (Emmett, 1975) is run in stand-alone mode and is part of the SCALE licensing evaluation system. It uses the MARS (Multiple Array System) geometry package. The MORSE code is distributed by RSICC. The multi-group energy handling capability of MORSE has been useful for comparisons with ANISN and DOT computations. MORSE has also been coupled with discrete ordinates codes with the DOMINO-II code.

Dose rates from a cask array have been calculated with MAVRIC (MONACO with Automated Variance Reduction using Importance Calculations) for the safety of workers.

8.1.8 KENO

The KENO multi-group Monte Carlo code for criticality computations (Goluoglu, Bowman, & Dunn, 2007) was developed in the 1960s at ORNL. It is now part of the SCALE system and its use is as widespread as that of other MC codes. KENO is a distributed by the Nuclear Energy Agency (NEA) with generalized geometry capability allowing easy description of systems composed of cylinders, spheres, and cuboids (rectangular parallelepipeds) arranged in any order.

8.1.9 Other Monte Carlo codes

MONK and MCBEND (Long et al., 2015; Richards et al., 2015) are UKAEA's standard criticality and radiation shielding codes (Richards et al., 2015).

TRIPOLI (Nimal & Vergnaud, 1990, 2001; Vergnaud & Nimal, 1990) is a Monte Carlo code that was written at CEA, France, and used for "whole-core" neutronic reactor analyses, radiation shielding, and protection as well as for criticality calculations.

Other neutron/photon transport codes are VIM, a Monte Carlo code developed by Argonne National Labs, and SERPENT, a continuous energy MC code for neutron transport, criticality and reactor physics burn-up calculations developed by VTT Technical Research Center, Finland. Charged-particle codes that handle electron/photon transport include EGS5 and PENELOPE, while for high-energy transport codes include GEANT4, FLUKA and MCNPX.

8.2 Time-dependent reactor kinetics codes

PARCS (The Purdue Advanced Reactor Core Simulator), developed by the USNRC, solves for the transient neutron flux distribution and is used for the analysis of possible reactivity-initiated accidents in light water reactors. It can also be coupled to other USNRC thermal hydraulics codes such as RELAP.

8.3 Thermal hydraulics codes

Thermal hydraulics studies are carried out coupled with neutronics to study transients such as LOFA in WWER (Noori-Kalkhoran et al., 2014), and reactivity insertion accidents in Gen IV Liquid Metal Cooled Fast Reactor with the COBRA code (Yang, Liu, Xiong, Chai, & Cheng, 2018).

A thermal hydraulics code, such as RELAP, carries out a transient two-phase two-fluid hydrodynamic analysis for a nuclear reactor modeled in sufficient detail to analyze transient conditions for safety studies. Detailed studies are carried out in a suite such as the SCALE code system (Rearden & Jessee, 2016). A full-scale neutronic analysis for a sodium-cooled fast reactor core carried out using the lattice code HELIOS-2 with the Monte Carlo code Serpent to generate few-group constants used by the 3D deterministic code DYN3D was shown to give reliable results (Rachamin, Wemple, & Fridman, 2013).

For the analysis of transients and large/small LOCAs in PWRs and BWRs, TRACE/RELAP (**TRAC/RELAP** Advanced Computational Engine) is used. RELAP (The **R**eactor **E**xcursion and **L**eakage **A**nalysis **P**rogram) will be replaced by TRACE in a few years. Some representative calculations for new reactor systems are: a LOCA analysis for NuScale carried out to model an extremely severe accident caused by loss of decay heat removal (Skolik et al., 2021; Skolik, Trivedi, Perez-Ferragut, & Allison, 2019), and a study on the graphite oxidation behavior for reliability studies of a HTGR (Chai, Wu, & Okamoto, 2020).

Full 3D reactor core simulators, with the capability of neutronic-thermal hydraulics coupling are used to solve for the reactor power and flow field for steady-state as well as transient calculations. Free open source codes such as OpenMOC (Boyd, Shaner, Li, Forget, & Smith, 2014) and OpenMC (Romano et al., 2015) are available for carrying out deterministic computations with the method of characteristics and Monte Carlo simulations for neutronics and OpenFOAM (OpenFOAM, 2014) for computational fluid dynamics computations. Monte Carlo methods have been combined with the method of characteristics and demonstrated that the hybrid model is better than each one alone (Lee, Choi, & Lee, 2015). Coupling of neutronics with CFD has been carried out for Multiphysics simulations (Castagna et al., 2020; Seubert et al., 2012; Tuominen, Valtavirta, & Lepp, 2018; Tuominen, Valtavirta, Peltola, & Leppänen, 2016) of 3D deterministic as well as Monte Carlo codes for core analysis.

8.4 Radiological protection codes

For source term analysis (Section 3) and radionuclide releases under accident conditions, the USNRC codes RADTRAD (**RAD**ionuclide **T**ransport and **R**emoval **A**nd **D**ose Estimation) and RASCAL (**R**adiological **A**ssessment **S**ystems for **C**onsequence **A**na**L**ysis) are used.

8.5 Performance and safety analyses

For the performance of a single fuel rod under near-normal reactor operating conditions, codes developed by the USNRC are FRAPCON-3, for steady-state and mild transient analysis and FRAPTRAN for design basis accident analysis. These codes are very useful for the design of new SMRs, as described in Chapter 3. Two such examples are the work on a small modular 150 MWt PWR (Mirian & Ayoobian, 2020) and on studying aa accident tolerant fuel (He, Shirvan, Wu, & Su, 2019). In the first case, using FRAPCON, the effect of varying fuel diameter and pitch on the excess reactivity, neutron flux and burn-up were calculated by MCNP and coupled with FRAPCON to compute the fuel centerline temperature, stresses and pressures with the objective of enhancing the fuel cycle lifetime. In the second, using FRAPTRAN, a new silicon carbide (SiC) fuel was studied during a LOCA for its mechanical integrity.

For severe accident analysis the USNRC codes MELCOR for core meltdown analysis, such as for the modeling of Fukushima Daiichi Unit 1 (Herranz & López, 2020) and for a study on the AP1000 (Malicki, Pieńkowski, Skolik, & Trivedi, 2019) an integral severe accident analysis code, and SCDAP/RELAP5 are used.

To perform analyses on the possible atmospheric release of radionuclides with environmental consequences, the MACCS (The **MEL**CORE **A**ccident **C**onsequence **C**ode) Code has been widely used.

Probabilistic Risk Assessment: SAPHIRE (Systems Analysis Programs for Hands-on Reliability) developed by the USNRC

Typical simulation results from the codes described above include neutron and photon flux, criticality computations, and reaction rates in regions of interest. Deep-penetration calculations are required for the design of radiation and biological shielding around areas of high radiation.

For ITER, for example, elementary one-dimensional calculations are useful before proceeding to two-dimensional and full three-dimensional models. A simple 1D model of ITER in MCNP has been used to get energy-dependent neutron flux and dose rates.

Such 1D results are often useful to draw important engineering design parameters and limits. For example, in this 1D analysis, the authors conclude that "the magnitude of the dose rate on the outside hall of bioshield during normal ITER operation cannot be considered low in accordance with the result found in the simulation performed in this work, that is, 1 $\mu Sv/h$. It is important to remember that the contributions due to the presence of slits between blanket modules and also the various holes in the cryostat and bioshield were not considered in the calculation."

These are the very important findings of such simulations. Consider these with the acceptable radiation levels of the order of 0.05 Sv described in Chapter 1 and a preliminary estimate can be made to conclude that safety radiation levels will be exceeded by about five years. In the expression that "... the presence of slits between blanket modules ... were not considered in this calculation," it is clear that a 2D, or possibly a full 3D simulation is required.

To model geometrical complexity such as the ducts mentioned above, codes such as MCNP, KENO and DOT can be used to carry out two- or three-dimensional calculations for the streaming analysis.

A typical ducting problem that must be addressed in nuclear reactors *viz* the streaming of radiation in ports and ducts designed to place instrumentation at locations in the reactor. Results are found for levels of radiation that can subsequently be used to determine safety for workers and radiation damage to instrumentation.

8.6 Nuclear data

Computer codes come with nuclear data libraries which may need to be processed before they can be used. There are several nuclear data libraries such as ENDF/B (Evaluated Nuclear Data File, Brookhaven National laboratory) (Brown et al., 2018), ENDL (Evaluated Nuclear and Atomic Reaction Data Library, Lawrence Livermore Laboratory, USA), JEFF (European Joint Evaluated Fission and Fusion (JEFF) Library, OECD Nuclear Energy Agency), JENDL (Japanese Evaluated Nuclear Data Library), CENDL (Chinese Evaluated Nuclear Data Library) etc. which contain data obtained from experiments (Ge, Wu, Chen, & Xu, 2017; Herman & Trkov, 2010; Shibata et al., 2011).

These cross sections are available in the "old" formats with tabulations described below. The ENDF data is retrieved by specifying material numbers containing the atomic number Z and the atomic mass number A, the reaction of interest, the energy of interest, the version of the evaluation and several other details.

ENDF/B evaluations have a MAT number for all materials, an MF number for files, and a MT number for reaction type. For U^{238}, for example, the information file has the info: U-238 MAT = 9237 MF = 1 MT = 451 Library: ENDF/B-VII.0

Thus U^{238} evaluations have 10 information files *viz* MF = 1, 2, 3, 4, 5,6, 12, 13, 14 and 15, each of which carry information on the MT data available, for example, MT1 is the total neutron microscopic cross section σ_{tot} and MT16 is the $(n, 2n)$ data. A glance over the MF files is useful to appreciate the data "pointers", that is, the MT reaction data in each file.

MF = 1 General Information

MT = 452: total average neutron multiplicity per fission (total nubar)

MT = 455: average delayed neutron multiplicity per fission

MT = 456: average prompt neutron multiplicity per fission

MT = 458: energy release from fission

MF = 2 Resonance Parameters

MT = 151 resolved and unresolved resonance parameters

MF = 3 Neutron Cross sections

MT = 1: total cross section

MT = 2: elastic scattering cross section

MT = 3: nonelastic cross section

MT = 4: inelastic cross section

MT = 16: (n,2n) cross section

MT = 17: (n,3n) cross section

MT = 18: fission neutron cross section

MT = 19,20,21,38: multi-chance fission cross sections

MT = 37: (n,4n) cross section

MT = 51−71 discrete inelastic level cross sections

MT = 72−90: discrete inelastic level cross sections (direct reactions to groups of states) MT = 91: inelastic continuum neutron cross section

MT = 102: neutron radiative capture cross section

MF = 4 Angular Distributions of Secondary Particles

MT = 2: neutron elastic scattering angular distributions

MT = 51−53: discrete inelastic neutron angular distributions for the ground-state rotational band

MT = 54−71: discrete inelastic neutron angular distributions

MT = 72−90: discrete inelastic neutron angular distributions for direct reactions

MF = 5 Energy Distributions of Secondary Particles

MT = 18: prompt fission neutron spectrum matrix

MT = 455: delayed neutron emission spectra from fission

MF = 6 Product Energy-Angular Distributions

MT = 16,17,37: (n, xn) continuum distributions

MT = 91: (n, n′) continuum distributions

MF = 12 Photon Prod Multiplicities & Transition Probabilities

MT = 18,102: photon multiplicities from inelastic scattering, fission, and radiative capture

MF = 13 Photon Production Cross Sections

MT = 3: photon production cross sections from nonelastic reactions

MF = 14 Photon Angular Distributions

MT = 3,18,102: photon angular distributions from nonelastic, fission, and radiative capture reactions

MF = 15 Continuous Photon Energy Spectra

MT = 3,18,102: photon energy spectra from nonelastic reactions, fission and radiative capture

Reactions are also available with MT numbers in MF files.

In the table above, for MF = 1, the information in this file holds only five reaction types, that is, MT451, 452, 455, 456, and 458. MT451 is the INFO associated with the data, while MT452 holds $\overline{\nu}_t$, the average total number of neutrons (prompt plus delayed) neutrons released per fission event. More details are in the ENDF-6 Formats Manual on the BNL website. The first ten lines in this data file are

```
Retrieved by E4-Web: 2012/08/17,12:26:44                             1 0  0    0
 9.223800+4 2.360058+2          1          1          0          59237 1451    1
 0.000000+0 1.000000+0          0          0          0          69237 1451    2
 1.000000+0 3.000000+7          6          0         10          79237 1451    3
 0.000000+0 0.000000+0          0          0        675        1179237 1451    4
 92-U -238 ORNL,LANL+ EVAL-SEP06 Young,Chadwick,Derrien,Courcelle 9237 1451    5
                     DIST-DEC06 REV2-                             9237 1451    6
----ENDF/B-VII       MATERIAL 9237          REVISION 2            9237 1451    7
-----INCIDENT NEUTRON DATA                                        9237 1451    8
------ENDF-6 FORMAT                                               9237 1451    9
***************************************************************** 9237 1451   10
```

This section is reproduced from the "Formats Manual": To read the data, the position of the "records" has to be known. All records are one of six types: TEXT, CONT (has six special cases called DIR, HEAD, SEND, FEND, MEND, and TEND), LIST, TAB1, TAB2, and INTG. For example, SEND indicates a section end, FEND indicates a file end, MEND indicates a material end, and TEND indicates a "tape end." The counter NS is reset for every section in the file; the last is SEND which has NS = 99,999.

Every line toward the end has 92,237 which is the MAT number, followed by a one which is the file number, then 451 which is the MT (reaction) number, then the record number NS.

From the ENDF/B data, it can be verified that the first file has the first section (MT451) with 796 lines; the second section (MT452) has 07 lines; the third section (MT455) has 07 lines; the fourth section (MT456) has 07 lines; the fifth section (MT458) has 05 lines; then file 2 begins with MT151.

For specific applications where temperature-dependent data, for example, is required, nuclear data processing codes, such as NJOY, are used to process the cross sections in data libraries such as ENDF/B to prepare multi-group and mixture cross sections. Needless to state, data handling is itself a large field and professionals involved in the compilation, maintenance and processing of nuclear data most often stay within this domain.

As mentioned earlier, in the two step process, the first step is to use a lattice physics codes with a cross section generation code to provide the cross sections input to a neutronics code which computes the fluxes to get the power distribution. This power distribution is input to a thermal hydraulics code

One such system uses the GENPMAXS (Porhemmat, Hadad, & Faghihi, 2015) for the Generation of the Purdue Macroscopic XS set code module for the 3D Purdue Advanced Reactor Core Simulator PARCS. The generation code uses lattice codes such as HELIOS-2 (Wemple & Villarino, 2008), CASMO (Rhodes, Smith, & Lee, 2006), and TRITON (DeHart, 2006).

The use of codes, developed in universities, companies, and government organizations, is indispensable as a first step in the design of a reactor and a requirement in regulatory processes. The USNRC has been using computer codes since the development of early nuclear reactors. Similarly other international efforts include the Institute de Radioprotection et de Surete Nucleaire (IRSN) in Europe and more recently the Chinese nuclear regulatory institution through the National Nuclear Safety Administration (NNSA).

8.6.1. MCNP

The first Monte Carlo code appeared as a Los Alamos report in 1963. By the early 1980s, MCNP began to be widely distributed and used as MCNP3A and MCNP3B written in Fortran 77 with neutron and photon transport. The next version, MCNP4, had improvements such as electron transport capability as well as new tallies. It had parallel processing to speed up the computations. The perturbation capability was added in MCNP4B in the late 1990s. This was followed by MCNP4C which had some better physics such as the treatment of nuclear resonances. MCNP5 released in 2003 was rewritten in Fortran 90 and had several improvements. The latest version, MCNP6.2 (2017), can transport 36 different particles from neutrons, photons, electrons to muons, antineutrons various baryons, antiparticles, deuterons, tritons, alpha, and heavy ions. The long list of improvements includes the modeling of physics, sources, data, tallies, unstructured meshes, as well as code enhancements. This has led to a greatly increased versatility going far beyond the early neutron and photon transport to areas such as medical physics, accelerator-driven energy source research, high-energy dosimetry, and futuristic areas such as charged particle propulsion. Stochastic geometry has been incorporated to model the random arrangements in a geometry such as the high-temperature gas-cooled reactor with its spherical fuel pebbles.

8.7 Conclusion

Many of the computer codes used during the various stages of a nuclear reactor lifecycle are discussed in this chapter, from the concept and design stage to licensing, performance, regulatory, safety and accident analysis, to the decommissioning of a reactor. With advancements in computer hardware as well as the development of powerful algorithms, computer simulations and codes are most likely to play an increasingly important role in the nuclear industry leading to better, safer, and more competitive designs giving greater eminence to nuclear energy in the future.

Problems

1. Describe briefly how the ANISN code computes neutron flux in a cylinder. (a) What would be the difference in a computational strategy in ANISN for radiation shielding compared with a boron concentration criticality search? (b) How are neutron and photon dose rates obtained in ANISN?

2. For a discrete ordinates ANISN-S4 computation, what angular bins would be specified in a MCNP tally to compare results from both?

3. What is the fundamental difference between MORSE and MCNP (both Monte Carlo codes) in modeling neutron–nuclei interactions?

4. What is common between the deterministic code ANISN and the Monte Carlo code MORSE?

5. For a situation in which a pipe failure in the Spent Fuel Pool (SFP causes a loss of coolant accident (LOCA) describe what useful results the FRAPTRAN code could give. From the FRAPTRAN model, answer the following questions: (a) Why would the spent fuel rods stored in the SFP remain hot even though they have been removed from the reactor? (b) What model is used for thermal conductivity of uranium oxide fuel and how does it relate to the expression given in Chapter 4? (c) In the finite-difference scheme for the time-dependent conduction equation what quantities are obtained? (d) How is the stored energy (in a fuel rod) modeled in the code? (e) Which model for convection would be most appropriate to apply in SPF conditions? (f) What mechanical response does the code model? (g) What makes the modeling in the code multidisciplinary (as mentioned in Chapter 4)? (h) What would occur in the reactor building if the cladding were to burst?

6. How could the perturbation feature in MCNP be used for carrying out a boron criticality search in an LWR?

7. What makes the MC method particularly suitable for parallel computation? How would you carry out an MC simulation of particle histories using the parallel computing feature in Matlab and what computational speedup would you expect?

8. In an optimization study of a fusion reactor blanket design what factors could favor the use of a deterministic code such as ANISN over a Monte Code such as MORSE?

9. With reference to Chapter 4 and the description of the MCNP and ORIGEN codes given above, if the two codes are coupled for a burnup calculation in an LWR, (a) how will MCNP and ORIGEN be coupled, that is, what quantity will be given from MCNP to ORIGEN and which input data of MCNP will be modified for the next run in the loop, (b) which nuclide compositions will be of interest, and (c) when will a MCNP-ORIGEN coupling be favored over a ANISN-ORIGEN coupling?

10. In the MCNP−ORIGEN coupling discussed in 8.9, how would the reactor power be used in data transfer from MCNP to ORIGEN?

Nomenclature

Abbreviations and acronyms

ANISN	anisotropic SN method
CENDL	Chinese evaluated nuclear data library
DORT	discrete-ordinate neutron and photon transport
ENDF/B	evaluated nuclear data file, Brookhaven National Laboratory
ENDL	evaluated nuclear and atomic reaction data library
FORTRAN	formula translation
FRAPTRAN	fuel rod analysis program transient
GEANT	geometry and tracking
HTGR	high temperature gas cooled reactor
IBM	International Business Machines
ITER	International Thermonuclear Experimental Reactor
JEFF	European Joint Evaluated Fission and Fusion (JEFF) Library, OECD Nuclear Energy Agency
JENDL	Japanese evaluated nuclear data library
IRSN	Institute de Radioprotection et de Surete Nucleaire
LOCA	loss of coolant accident
LWR	light water reactor
MARS	multiple array system
MC	Monte Carlo
MCNP	Monte Carlo N-particle
MACCS	The **MEL**CORE **A**ccident **C**onsequence **C**ode
MF	material file
MORSE	Multigroup Oak Ridge Stochastic Experiment
MT	material type
NEA	Nucear Energy Agency
NNSA	National Nuclear Safety Administration
ORIGEN	**OR**NL **I**sotope De**gen**eration and Depletion Code
RELAP	The **R**eactor **E**xcursion and **L**eakage **A**nalysis **P**rogram
RADTRAD	**RAD**ionuclide **T**ransport and **R**emoval **A**nd **D**ose Estimation
RASCAL	**R**adiological **A**ssessment **S**ystems for **C**onsequence **A**na**L**ysis
RISC	reduced instruction set computer
RSICC	radiation safety information computational center
SAPHIRE	systems analysis programs for hands-on reliability
TORT	a three-dimensional discrete ordinates neutron/photon transport code
USNRC	United States Nuclear Regulatory Commission
WIMS	Winfrith improved multigroup scheme
WWER	water-water energetic reactor

References

Alcouffe, R. E. (2001). PARTISN calculations of 3D radiation transport benchmarks for simple geometries with void regions. *Progress in Nuclear Energy*.

Aziz, U., Koreshi, Z. U., Sheikh, S. R., & Khan, H. (2020). A study of criticality and thermal loading in a conceptual micronuclear heat pipe reactor for space applications. *Nuclear Technology & Radiation Protection*.

Bell, M. J. (1973). *ORIGEN—The ORNL isotope generation and depletion code*.

Boyd, W., Shaner, S., Li, L., Forget, B., & Smith, K. (2014). The OpenMOC method of characteristics neutral particle transport code. *Annals of Nuclear Energy*.

Brown, D. A., et al. (2018). ENDF/B-VIII.0: The 8th Major Release of the Nuclear Reaction Data Library with CIELO-project Cross Sections, New Standards and Thermal Scattering Data. *Nuclear Data Sheets*.

Carlson, B. G., & Lathrop, K. D. (1968). *Transport theory the method of discrete ordinates. Computing methods in reactor physics*. Gordon and Breach.

Castagna, C., et al. (2020). A Serpent/OpenFOAM coupling for 3D burnup analysis. *European Physical Journal - Plus*.

Chai, P., Wu, Y., & Okamoto, K. (2020). Improvement and validation of RELAP5/SCDAP code on evaluating the graphite behavior during oxidation process. *Progress in Nuclear Energy*.

DeHart, M. D. (2006). *Lattice physics capabilities of the SCALE code system using TRITON. PHYSOR-2006 - American Nuclear Society's topical meeting on reactor physics*. American Nuclear Society.

Emmett, M. (1975). *Morse Monte Carlo radiation transport code system*.

Emmett, M. B. (2000). MORSE: Present capabilities and future directions. In *Applied radiation and isotopes*.

Engle, W. W. J. (1967). *A users manual for ANISN: A one dimensional discrete ordinates transport code with anisotropic scattering*. Tennessee.

Fowler, T. B., Vondy, D. R., & Cunningham, G. W. (1971). *Nuclear reactor core analysis Code-CITATION*.

Ge, Z., Wu, H., Chen, G., & Xu, R. (2017). *CENDL project, the Chinese evaluated nuclear data library. EPJ web of conferences*. EDP Sciences.

Goluoglu, S., Bowman, S. M., & Dunn, M. E. (2007). *KENO monte carlo code capabilities. Transactions of the American Nuclear Society*. American Nuclear Society.

Hao Sun, G. S., Wang, C., Ma, P., Liu, X., Tian, W., & Qiu, S. (2018). Comnceptual design and analysis of a multipurpose micro nuclear reactor power source. *Annals of Nuclear Energy, 121*, 118–127.

He, Y., Shirvan, K., Wu, Y., & Su, G. (2019). Integrating a multi-layer deformation model in FRAPTRAN for accident Tolerant fuel analysis. *Annals of Nuclear Energy*.

Herman, M., & Trkov, A. (2010). *ENDF-6 formats manual*.

Herranz, L. E., & López, C. (2020). Challenges and sensitivities in the modelling of Fukushima Daiichi Unit 1 unfolding with MELCOR 2.2. *Annals of Nuclear Energy*.

Hussain, A., Xinrong, C., & Ahmed, R. (2008). *The conceptual design study of a TRISO fueled compact PWR core. International conference on nuclear engineering, proceedings, ICONE*. ASME.

Kargaran, H., Jafari, H., & Minuchehr, A. (2021). A novel approach based on improved combinatorial geometry (ICG) model for photon transport in Monte Carlo codes. *Progress in Nuclear Energy*.

Kirk, B. (2009). *Overview of neutron transport codes. 11th Neutron and ion dosimetry symposium-NEUDOS-11*. NEUDOS.

Lee, H., Choi, S., & Lee, D. (2015). A hybrid Monte Carlo/method-of-characteristics method for efficient neutron transport analysis. *Nuclear Science and Engineering: The Journal of the American Nuclear Society*.

Lindley, B. A., et al. (2017). Current status of the reactor physics code WIMS and recent developments. *Annals of Nuclear Energy*.

Long, D., et al. (2015). *MONK10: A Monte Carlo code for criticality analysis. ICNC 2015 - International Conference on Nuclear Criticality Safety*. ICNC.

Malicki, M., Pieńkowski, L., Skolik, K., & Trivedi, A. K. (2019). Development of MELCOR thermal hydraulic model of AP1000 and its verification for a DECL break. *Annals of Nuclear Energy*.

McKinley, S., Cullen, D. E., Latkowski, J. F., Procassini, R., & Skulina, K. M. (2006). *Application of MCNP, MERCURY and TART to calculation of the National Ignition Facility (NIF) shielding. Transactions of the American Nuclear Society*. American Nuclear Society.

Mirian, S. F., & Ayoobian, N. (2020). Investigations on a typical small modular PWR using coupled neutronic-thermal-mechanical evaluations to achieve long-life cycle-length. *Progress in Nuclear Energy*.

Nimal, J. C., & Vergnaud, T. (1990). TRIPOLI: A general Monte Carlo code, present state and future prospects. *Progress in Nuclear Energy*.

Nimal, J. C., & Vergnaud, T. (2001). *TRIPOLI-3: A neutron/photon Monte Carlo transport code. Advanced Monte Carlo for radiation physics, particle transport simulation and applications*. Springer.

Noori-Kalkhoran, O., Minuchehr, A., Shirani, A. S., & Rahgoshay, M. (2014). Full scope thermal-neutronic analysis of LOFA in a WWER-1000 reactor core by coupling PARCS v2.7 and COBRA-EN. *Progress in Nuclear Energy*.

OpenFOAM. (2014). *Open∇FOAM - The open source CFD toolbox - User guide*. OpenFOAM Found.

Pelowitz, D. B., et al. (2013). *MCNP6 user's manual*. La-Cp-13–00634.

Porhemmat, M. H., Hadad, K., & Faghihi, F. (2015). PARCS cross-section library generator; Part one: Development and verification. *Progress in Nuclear Energy*.

Rachamin, R., Wemple, C., & Fridman, E. (2013). Neutronic analysis of SFR core with HELIOS-2, Serpent, and DYN3D codes. *Annals of Nuclear Energy*.

Rearden, B. T., & Jessee, M. A. (2016). *SCALE code system*.

Rhoades, W. A., & Mynatt, F. R. (1973). *DOT III two-dimensional discrete ordinates transport code*.

Rhodes, J., Smith, K., & Lee, D. (2006). *CASMO-5 development and applications. PHYSOR-2006 - American Nuclear Society's topical meeting on reactor physics*. American Nuclear Society.

Richards, S. D., et al. (2015). MONK and MCBEND: Current status and recent developments. *Annals of Nuclear Energy*.

Romano, P. K., Horelik, N. E., Herman, B. R., Nelson, A. G., Forget, B., & Smith, K. (2015). OpenMC: A state-of-the-art Monte Carlo code for research and development. *Annals of Nuclear Energy*.

Schober, P. (2018). *Supercomputers. High-performance computing in finance: Problems, methods, and solutions*. Chapman and Hall/CRC.

Seubert, A., Sureda, A., Bader, J., Lapins, J., Buck, M., & Laurien, E. (2012). The 3-D time-dependent transport code TORT-TD and its coupling with the 3D thermal-hydraulic code ATTICA3D for HTGR applications. *Nuclear Engineering and Design*.

Shibata, K., et al. (2011). JENDL-4.0: A new library for nuclear science and engineering. *Journal of Nuclear Science and Technology*.

Skolik, K., Trivedi, A., Perez-Ferragut, M., & Allison, C. (2019). *LOCA-type scenario simulation for NuScale-SMR with RELAP/SCDAPSIM/MOD3.4. 18th International topical meeting on nuclear reactor thermal hydraulics, NURETH 2019*. American Nuclear Society.

Skolik, K., Allison, C., Hohorst, J., Malicki, M., Perez-Ferragut, M., Pieńkowski, L., & Trivedi, A. (2021). Analysis of loss of coolant accident without ECCS and DHRS in an integral pressurized water reactor using RELAP/SCDAPSIM. *Progress in Nuclear Energy*.

Tucker, L. P., & Usman, S. (2018). Thorium-based mixed oxide fuel in a pressurized water reactor: A burnup analysis with MCNP. *Annals of Nuclear Energy*.

Tuominen, R., Valtavirta, V., & Lepp, J. (2018). *Application of the serpent – openfoam coupled code system to the sealer reactor core. PHYSOR 2018: Reactor physics paving the way towards more efficient systems*. American Nuclear Society.

Tuominen, R., Valtavirta, V., Peltola, J., & Leppänen, J. (2016). *Coupling serpent and openfoam for neutronics - CFD multi-physics calculations. Physics of reactors 2016, PHYSOR 2016: Unifying theory and experiments in the 21st century*. American Nuclear Society.

Vergnaud, T., & Nimal, J. C. (1990). TRIPOLI capabilities proved by a set of solved problems. *Progress in Nuclear Energy*.

Wagner, J. C., Sisolak, J. E., & McKinney, G. (1992). *MCNP: Criticality safety benchmark problems, LA-12415*. New Mexico, United States.

Wemple, C. A., & Villarino, E. A. (2008). The HELIOS-2 lattice physics code. In *Physics*. College Park, MD.

Whalen, D. J., Cardon, D. A., Uhle, J. L., & Hendricks, J. S. (1991). *MCNP: Neutron benchmark problems*. LA−12212 DE92 004710.

Yang, D., Liu, X., Xiong, J., Chai, X., & Cheng, X. (2018). Coupling of neutronics and thermal-hydraulics codes for the simulation of reactivity insertion accident for LFR. *Progress in Nuclear Energy*.

Yesilyurt, G., Clarno, K. T., Gauld, I. C., & Galloway, J. (2011). Modular origen-S for multi-physics code systems. *American Nuclear Society*.

Yu, H., Cai, J., He, S., & Li, X. (2021). Analysis of neutron physics and thermal hydraulics for fuel assembly of small modular reactor loaded with ATFs. *Annals of Nuclear Energy*.

Chapter 9

Optimization and variational methods

9.1 Introduction

In engineering design three very important goals are safety, design optimization, and economic optimization.

This is true for nuclear technologies as nuclear power reactors are designed to operate safely, designed and built on the best knowledge of the day, and provide economically competitive energy. These features have been the hallmark of nuclear energy over the last 60 years of providing safe reliable and competitive energy (Section 3.1).

This last requirement will be most severely tested in the decades to come. Nuclear energy will have to compete with fossils and renewables to make a renaissance possible. The competitiveness will most likely come from the Gen IV reactor designs which have some very attractive features described in Chapter 3 (Section 3.5.3).

The present challenge for nuclear technology is optimization in proposed designs to win the licensing formalities and gain public acceptance to provide energy for the decades to come.

The main difference between design optimization in nuclear engineering and in other power systems is the high nuclear energy density that minimizes the tolerances. Both fission and fusion power systems contain some of the most sophisticated concepts about the way nature works as well as some of the most advanced technologies developed in the scientific community; and with it have come derivative benefits in the development of mathematical models and computing platforms.

Today, optimization in nuclear engineering is performed mostly with heuristic and *meta*-heuristic (meaning more efficient search techniques) methods with artificial intelligence expert knowledge-based systems to make a machine learn while it calculates, to mimic the workings of the human brain by neural networks rendering a machine towards a human way of thinking. It has thus become possible, for example, to optimize the fuel loading in 121 assemblies of a pressurized water reactor (PWR) core (Section 3.2.1) with the fuel of the right enrichment and the right burnup to be placed at the right place in the core at the right time of its power producing several-year schedule. A great achievement considering that the factorial of 121 is $\sim 10^{200}$ which means that out of so many possible solutions, the best must be found. How does one do that!

Reactor core fueling optimization is a huge search space even by modern computational capabilities, so a reduction in complexity is achieved by using a 1/8 core symmetric model to search from an optimal configuration from $15! \sim 10^{12}$ permutations and combinations, which becomes possible.

In nuclear engineering, optimization is applied in areas including overall reactor core design (Bae, Betzler, Chandler, & Ilas, 2021; Betzler, Chandler, Cook, Davidson, & Ilas, 2019), reactor operations (Kumar & Tsvetkov, 2015; Mousakazemi, 2021), safety-related management strategies, in-core fuel management (Chham et al., 2021), the design of criticality experiments, centrifuge cascade optimal configurations, radiation shielding design (Ahmad, Chang, Li, Yang, & Liu, 2021; Cai, Hu, Lu, & Jia, 2018), controller design (Mousakazemi, 2021; Mousakazemi, Ayoobian, & Ansarifar, 2018; Wan & Zhao, 2017), and many more.

Most of the present-day optimization is multi-objective and multi-modal and, in many cases, requires combinatorial optimizations made efficient by the use of machine learning algorithms and expert-based systems.

This chapter focuses on the applicable mathematics of optimization followed by a cursory review of some application areas.

Historically, this field has its origins in the works of Euler (1707–83) and Lagrange (1736–1813), following Newton's (1642–1726) optimization, and was further demonstrated by Bernoulli (1667–1748) in the classical brachistochrone problem. Two centuries later, Pontryagin (1908–88) developed what is now known and widely applied as the "bang-bang" control based on Pontryagin's Maximum Principle. Such variational methods were mainly focused on mechanics with considerable contributions by mathematicians including Hamilton (1805–65). The bedrock of optimization and optimal methods is thus considered to be Newton's optimization followed by Hamiltonian and Lagrangian

Nuclear Engineering. DOI: https://doi.org/10.1016/B978-0-323-90618-0.00009-0

formulations and the Euler-Lagrange equations as equivalent forms for the stationarity of a functional (Lewis, Vrabie, & Syrmos, 2012; Rao, 2009).

At the simplest level, one or many maxima or minima of a function are found by setting its first derivative to zero and then using the second derivative, or the Hessian matrix, to determine the nature of an extremum.

In the next step, with a quest to seek the best function a *new* calculus was developed by considering an independent function as a variable and variational calculus gave, in the simplest Euler-Lagrange equation, the capability to find the best function which minimized or maximized some functional. One common-sense illustration is that of a straight line corresponding to the shortest distance between two points on a flat surface (Section 9.2.2). Any other function would give a longer distance. Similarly, the function that gives the shortest distance between two points is a geodesic which lies on the larger circle on the surface of that sphere. The terminologies commonly used for the *best* value or function from calculus and variational calculus are optimization and optimal analysis.

Continuing the sequel in this book, with Chapter 4 covering the essential mathematics for a nuclear engineering basic degree curriculum, and Chapters 5–7 covering formulations in the diffusion and transport equations and the mathematics and tools for Monte Carlo simulation, the simple understanding is that nuclear, and in a wider sense, radiation transport is by no means easy to solve for realistic designs. Thus in Chapter 8, some illustrative and extensively used codes and databases were described.

With all the learning from Chapters 4 to 8, optimization would require several computations based on intuition and some learning. That would, of course, be a very inefficient way, maybe an impossible way, to obtain a nuclear reactor design. One can therefore imagine the large amount of experimentation that must have gone into the design of the first generation nuclear reactors described in Chapter 3.

Fortunately, strong and powerful computers were made in the 1960s and 1970s, notably the Control Data Corporation (CDC), International Business Machines (IBM), and Cray computers. With the hardware came vector and parallel processing capabilities and with all that the capability of performing huge computations became so simple. And that opened the door for stochastic and heuristic optimization techniques which are so widespread these days.

Traditional optimization methods are based on the adjoint function with a variational formulation and deterministic methods such as Dynamic Programming (DP) developed by Richard Bellman in the 1950s and widely used in optimal control as the Hamilton-Jacobi-Bellman equation (Bellman, 1952; Bertsekas, 2010). A major thrust however is on heuristic optimization methods using artificial intelligence, expert systems, neural networks, genetic algorithms (GAs), and several other methods (Lee & El-Sharkawi, 2007; Nissan, 2019; Pardalos & Resende, 2002). The optimization field is so vast and developed that it is not possible to consider more than just a few methods in this chapter, but the content here is sufficient to give a reasonable insight into the methods and applications in nuclear engineering.

9.2 Deterministic optimization

9.2.1 Deterministic optimization without constraints

The simplest method of finding the minima or maxima of a continuous function is from its first and second derivatives. As an example, for a two-variable function

$$f(x_1, x_2) = \frac{1}{2}x_1^2 + x_1x_2 + x_2^2 + x_2 \tag{9.1}$$

the partial derivatives are

$$\frac{\partial f}{\partial x_1} = x_1 + x_2, \quad \frac{\partial f}{\partial x_2} = x_1 + 2x_2 + 1, \text{and} \frac{\partial^2 f}{\partial x_1 \partial x_2} = \frac{\partial^2 f}{\partial x_2 \partial x_1} = 1. \tag{9.2}$$

Setting both first partial derivatives equal to zero and solving for the extremum $\boldsymbol{x}^*$, gives $(x_1^*, x_2^*) = (1, -1)$. The Hessian matrix is

$$\mathscr{H} = \begin{bmatrix} \dfrac{\partial^2 f}{\partial x_1^2} & \dfrac{\partial^2 f}{\partial x_1 \partial x_2} \\ \dfrac{\partial^2 f}{\partial x_2 \partial x_1} & \dfrac{\partial^2 f}{\partial x_2^2} \end{bmatrix} \tag{9.3}$$

with a determinant

$$|\mathscr{H}|^* = \left|\frac{\partial^2 f}{\partial x_1^2}\frac{\partial^2 f}{\partial x_2^2} - \frac{\partial^2 f}{\partial x_1 \partial x_2}\frac{\partial^2 f}{\partial x_2 \partial x_1}\right|^* = (1)(2) - (1)(1) = 1 \tag{9.4}$$

so that the extremum is a local minima. For a function of two independent variables, it is possible to plot the functions as shown in Fig. 9.1 for the surface $f(\boldsymbol{x})$ and in Fig. 9.2 for the contours illustrating that at the minima $f(\boldsymbol{x}^*) = 0$.

From Fig. 9.2, the minima is seen at $\boldsymbol{x}^* = (1, -1)$ for which $f(\boldsymbol{x}^*) = -0.5$.

9.2.2 Deterministic optimization with algebraic constraints

For optimizing a function $f(x)$ with N constraints $g_i(x) = 0$, we write a Lagrangian

$$\mathscr{L}(f, g, \boldsymbol{x}, \lambda) = f(\boldsymbol{x}) + \sum_{i=1}^{N} \lambda_i g_i(\boldsymbol{x})(i = 1, 2, 3 \cdots, N). \tag{9.5}$$

As an example of a two-variable function with a single algebraic constraint, consider the optimization problem: $\max V(R, H)$ subject to the constraint of available surface area A_s for a cylinder of radius R and height H. The volume is $V(R, H) = \pi R^2 H$ subject to a constraint of given surface area

$$A_s(R, H) = 2\pi R^2 + 2\pi RH = K. \tag{9.6}$$

An elementary consideration leads to a reduction of the two-variable problem to a single variable problem by substituting for the height of the cylinder

$$H(R) = \frac{K}{2\pi R} - R \tag{9.7}$$

when the volume can be written as

$$V(R) = \frac{1}{2}KR - \pi R^3. \tag{9.8}$$

The extremum can be found by differentiating the above to get the optimum for which the optimal radius and height are $R^* = \sqrt{K/6\pi}$, and $H^* = 2R_o$, respectively, and the volume is

$$V^* = 2\pi R^3 = 2\pi\left(\frac{K}{6\pi}\right)^{3/2}.$$

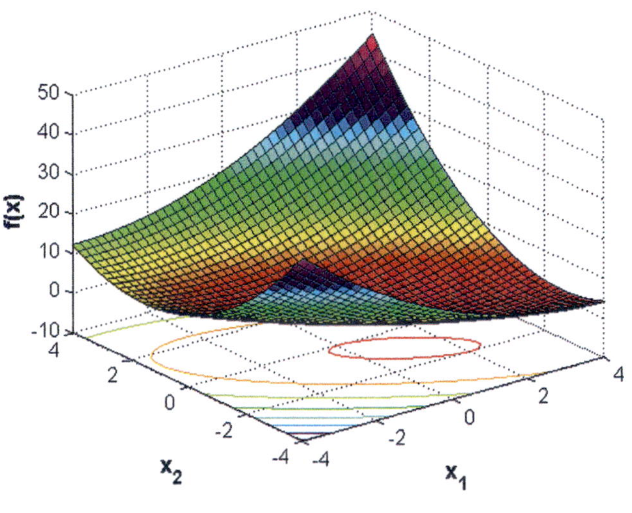

FIGURE 9.1 Surface plot of $f(x) = \frac{1}{2}x_1^2 + x_1x_2 + x_2^2 + x_2$ showing x_1, x_2 contours.

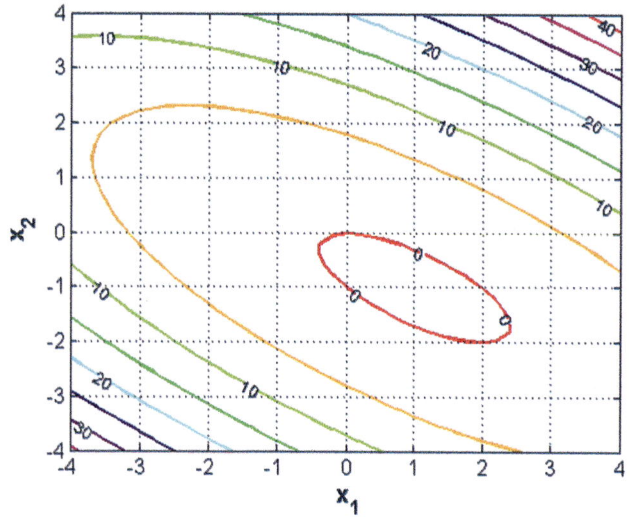

FIGURE 9.2 Contour plot of $\frac{1}{2}x_1^2 + x_1x_2 + x_2^2 + x_2 = C$.

With the Lagrangian (Eq. 9.5), $\mathscr{L}$, expressed as the function to maximize (or minimize) V, and the constraint condition $A_s - K = 0$, multiplied by a Lagrange multiplier λ (Lewis et al., 2012)

$$\overline{\nabla}\mathscr{L}|^* = \overline{\nabla}V|^* + \lambda\overline{\nabla}A_s|^* = 0 \tag{9.9}$$

since, at the extremum, $\overline{\nabla}\mathscr{L}|^* = 0$, so that the equations are

$$\frac{\partial V}{\partial R}|^* + \lambda\frac{\partial A_s}{\partial R}|^* = 0$$

and

$$\frac{\partial V}{\partial H}|^* + \lambda\frac{\partial A_s}{\partial H}|^* = 0.$$

This gives $H^* = 2R^*$, and the Lagrange multiplier

$$\lambda = -\frac{\overline{\nabla}V}{\overline{\nabla}A_s}|^* = -\frac{R^*}{2}. \tag{9.10}$$

Fig. 9.3 shows the permissible $R - H$ contours for surface constraints (almost straight lines) and the volume curves. In Fig. 9.2, there were no constraints but here the maximization or minimization must be performed within the specified constraints. The lines and curves move towards the right as values are increased.

Table 9.1 shows the optimum volumes V^* for values of $A_s = K = 10, 20, 30, 40,$ and $50\ \text{cm}^2$.

The optimal values R^*, H^* are indicated with crosses. Clearly, as the surface area increases, the optimal volume increases from 2.43 to 27.14 cm^3.

The area constraint, shown in Fig. 9.4, along with the volume contours athey move upwards for varying K show that the curves become tangential, that is, the normal vectors on both coincide with each other at the optimal point $R^* = 1.0301$ cm, $H^* = 2.0601$ cm with volume $V^* = 6.8671\ \text{cm}^3$ which is a maxima as shown in Fig. 9.5.

It is seen that the volume contours shift towards the right touching each other tangentially at the optimal value for the height for which the difference in the two values $H_V - H_A$ goes to zero at the optimal radius. This represents the geometrical interpretation of the Lagrange multiplier as the point where volume and surface area contours are tangential to each other.

9.2.3 Optimal solution with a system of first-order ordinary differential equation constraints

In the previous section, the best values of independent variables were found to maximize a function subject to an algebraic constraint.

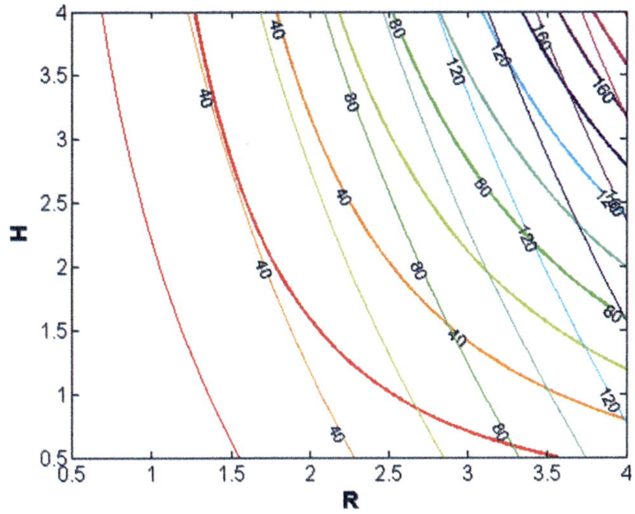

FIGURE 9.3 Objective function (volume) and constraint (area) curves.

TABLE 9.1 Constrained maximization of the volume of a cylinder.

K (cm^2)	10	20	30	40	50
R^* (cm)	0.7284	1.0301	1.2616	1.4567	1.6287
H^* (cm)	1.4567	2.0601	2.5231	2.9135	3.2574
V^* (cm^3)	2.4279	6.8671	12.6157	19.4231	27.1446

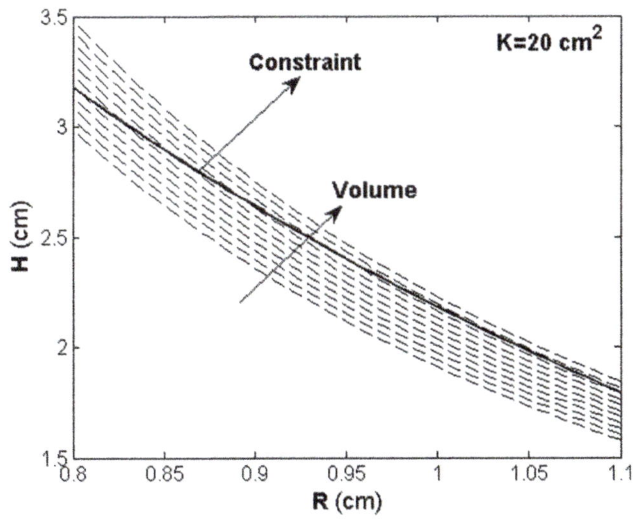

FIGURE 9.4 Surface area constraint for $A_s = 20$ cm^2 and a family of objective function (volume) curves.

Consider now the problem of finding a best function which maximizes an objective function

$$\varepsilon = \int_a^b \mathscr{F}(x, y(x), y'(x))dx \tag{9.11}$$

An example is to find a function that minimizes the distance between two points $A(x_a, y_a)$ and $B(x_b, y_b)$ on a flat surface. Clearly there are an infinite number of functions $g(x)$ that pass through these two points and satisfy the boundary

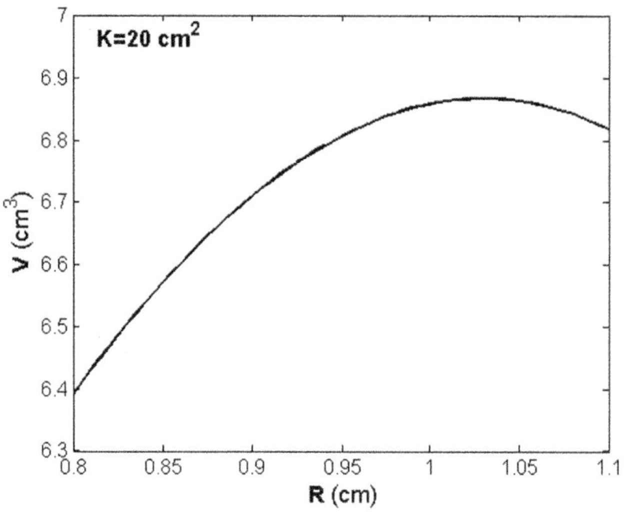

FIGURE 9.5 Objective function V versus radius R for a surface area constraint.

conditions at A and B. One such example is the function $g(x) = p + qx + rx^2$ where p, q, r are constants found by satisfying the boundary conditions. For an arc on $g(x)$ the objective function is

$$\mathscr{F}(x, y(x), y'(x)) = \sqrt{1 + (y')^2}. \tag{9.12}$$

Since the distance along the curve, using Pythagoras' Theorem for a right angle triangle relating the arc length ds to the other two sides of the triangle dx and dy, is

$$s = \int_a^b \mathscr{F}(x, y(x), y'(x))dx = \int_a^b \sqrt{1 + (y')^2}dx \tag{9.13}$$

Just like Newton's calculus for varying the independent variable to find the extremum of a function, here the function $y(x)$ is varied so that a new function considered is $g(x) = y(x) + \varepsilon h(x)$; then Eq. (9.11) is written as

$$\varepsilon = \int_a^b \mathscr{F}(x, g(x), g'(x))dx \tag{9.14}$$

In terms of the small parameter ε, differentiating ε with respect to ε gives

$$\frac{d\varepsilon}{d\varepsilon} = \int_a^b \frac{d\mathscr{F}}{d\varepsilon} \quad dx \tag{9.15}$$

which is

$$\frac{d\varepsilon}{d\varepsilon} = \int_a^b \left(\frac{\partial \mathscr{F}}{\partial x}\frac{dx}{d\varepsilon} + \frac{\partial \mathscr{F}}{\partial g}\frac{dg}{d\varepsilon} + \frac{\partial \mathscr{F}}{\partial g'}\frac{dg'}{d\varepsilon} \right) \quad dx \tag{9.16}$$

Note that for $\mathscr{F}(x, g(x), g'(x)) = \sqrt{1 + (g')2}$, the first two partial derivatives are zero while

$$\frac{\partial \mathscr{F}}{\partial g'} = \frac{g'}{\sqrt{1 + (g')2}}$$

and Eq. (9.16) becomes

$$\frac{d\varepsilon}{d\varepsilon} = \int_a^b \left(\frac{g'}{\sqrt{1 + (g')2}} h'(x) \right) \quad dx \tag{9.17}$$

Integrating by parts and setting the derivative $\varepsilon' = 0$,

$$\frac{d\varepsilon}{d\varepsilon} = \int_a^b \left(\frac{g'}{\sqrt{1+(g')^2}} \right) dh(x) = \left(\frac{g'}{\sqrt{1+(g')^2}} \right) h(x)|_a^b - \int_a^b h(x) d\left(\frac{g'}{\sqrt{1+(g')^2}} \right) = 0 \tag{9.18}$$

The first part of Eq. (9.18) can be made zero by choosing $h(x)$ such that $h(a) = h(b) = 0$ while the second term says that

$$\frac{d}{dx}\left(\frac{g'}{\sqrt{1+(g')^2}} \right) = 0 \tag{9.19}$$

which means that the term in the brackets is a constant in x. Thus $g'(x)$ is also a constant and a function that has a constant slope is a straight line. Therefore our assumption that $g(x) = p + qx + rx^2$ was a trial function could only be true if the constant $r = 0$.

The purpose of doing this exercise from Eq. (9.16) onwards for $g(x)$ considered as a second-order polynomial was to make the understanding easy. In general, a functional $\mathscr{F}$ can be written as in Eq. (9.11); for that case, Eq. (9.16) leads to the general form of Eq. (9.18) which is known as the Euler-Lagrange equation:

$$\frac{\partial \mathscr{F}}{\partial y} - \frac{d}{dx}\frac{\partial \mathscr{F}}{\partial y'} = 0 \tag{9.20}$$

To conclude: the Euler-Lagrange equation is the stationarity condition in variational calculus for an unconstrained optimization similar to the first derivative being zero in Newton's calculus.

9.2.4 Optimal solution with a system of first-order ordinary differential equation constraints

Now consider the next step in which the aim is to find a best function u of one or several variables to maximize a functional $J(u)$ subject to N first-order ordinary differential equation (ODE) constraints $\dot{x}_i(t) = f_i(\overline{x}, u, t)$ for $i = 1, 2, 3, \cdots, N$. This optimization of a functional rather than a function requires variational calculus and is classified as an optimal control problem.

A system is described by the equation of state

$$\dot{x} = f(x, u, t) \tag{9.21}$$

with a performance index (PI)

$$\varepsilon = \int_0^{\overline{t}} \mathscr{F}(x, u, t)dt. \tag{9.22}$$

A control $\overline{u}$ is implemented causing the system to move to a state $\overline{u}$; subsequently, a small change is made in the control δu, such that $|\delta u| \leq \varepsilon$ is made over the time domain $(0, t)$. This causes a change in x and $\dot{x}$ since the time rate of x changes from $d\overline{x}/dt$ to $d(\overline{x} + \delta x)/dt$ with a change $\delta \dot{x}$ which is related to the change in

$$\delta \dot{x} = \delta f = \frac{\partial f}{\partial x}\delta x + \frac{\partial f}{\partial u}\delta u + \mathcal{O}(\varepsilon) \tag{9.23}$$

causing a change in the PI

$$\delta\varepsilon = \int_0^{\overline{t}} \left[\frac{\partial \mathscr{F}}{\partial x}\delta x + \frac{\partial \mathscr{F}}{\partial u}\delta u \right] dt + \mathscr{F}(x, t)|_{t=\overline{t}}\delta t + \mathcal{O}(\varepsilon) \tag{9.24}$$

Multiplying Eq. (9.12) by $\lambda(t)$ integrating over t and adding to Eq. (9.13) gives

$$\delta\varepsilon = \int_0^{\overline{t}} \left[\left(\frac{\partial \mathscr{F}}{\partial x} + \lambda\frac{\partial f}{\partial x} \right)\delta x - \lambda\delta\dot{x} + \left(\frac{\partial \mathscr{F}}{\partial u} + \lambda\frac{\partial f}{\partial u} \right)\delta u \right] dt + \mathscr{F}|_{t=\overline{t}}\delta t + \mathcal{O}(\varepsilon) \tag{9.25}$$

Integrating $\lambda\delta\dot{x}$ by parts,

$$\delta\varepsilon = \int_0^{\bar{t}} \left[\left(\frac{\partial \mathcal{F}}{\partial x} + \lambda \frac{\partial f}{\partial x} + \dot{\lambda} \right) \delta x + \left(\frac{\partial \mathcal{F}}{\partial u} + \lambda \frac{\partial f}{\partial u} \right) \delta u \right] dt + \mathcal{F}|_{t=\bar{t}} \delta t - \lambda(\bar{t})\delta x(\bar{t}) + \lambda(0)\delta x(0) + \mathcal{O}(\varepsilon) \tag{9.26}$$

Therefore λ should satisfy the differential equation

$$\frac{\partial \mathcal{F}}{\partial x} + \lambda \frac{\partial f}{\partial x} + \dot{\lambda} = \frac{\partial H}{\partial x} + \dot{\lambda} = 0. \tag{9.27}$$

Then, the change in the PI is

$$\delta\varepsilon = \int_0^{\bar{t}} \left(\frac{\partial \mathcal{F}}{\partial u} + \lambda \frac{\partial f}{\partial u} \right) \delta u dt + \mathcal{F}|_{t=\bar{t}} \delta t - \lambda(\bar{t})\delta x(\bar{t}) + \lambda(0)\delta x(0) + \mathcal{O}(\varepsilon). \tag{9.28}$$

Consider now the boundary conditions.

When time t is specified then there is no uncertainty in the end points and $\delta t = 0$, so that the coefficient of δt does not need to be zero. In that case

$$\delta\varepsilon = -\lambda(\bar{t})\delta x(\bar{t}) + \lambda(0)\delta x(0) + \mathcal{O}(\varepsilon) \tag{9.29}$$

Further if the end-point is given, then $\delta x(\bar{t}) = 0$ and $\lambda(\bar{t})$ is not known. In case the end-point is free, then $\delta x(\bar{t}) \neq 0$ so that $\lambda(\bar{t}) = 0$.

When time t is not specified then $x(\bar{t} + \delta t) = x(\bar{t}) + \delta x(\bar{t}) + \mathcal{O}(\varepsilon)$. If x is free then $\mathcal{F}|_{t=\bar{t}} \delta t - \lambda(\bar{t})\delta x(\bar{t})$ can be simplified using the state equation $\dot{x} = f$ which can be understood to be $\delta x = f\delta t$, so that $\mathcal{F}|_{t=\bar{t}} \delta t - \lambda(\bar{t})\delta x(\bar{t}) = \left(\mathcal{F}|_{t=\bar{t}} - \lambda(\bar{t})f\right)\delta t$. Free and fixed boundary conditions are applied as specified in the problem (Table 9.2).

Eq. (9.18) becomes

$$\delta\varepsilon = \int_0^{\bar{t}} \left(\frac{\partial \mathcal{F}}{\partial u} + \lambda \frac{\partial f}{\partial u} \right) \delta u dt + \mathcal{O}(\varepsilon) \tag{9.30}$$

For $\delta\varepsilon \geq 0$, the value of δu can be chosen to be such that

$$\delta u = -\varepsilon \left(\frac{\partial \mathcal{F}}{\partial u} + \lambda \frac{\partial f}{\partial u} \right). \tag{9.31}$$

With a positive definite square term in the integral

$$\delta\varepsilon = -\varepsilon \int_0^{\bar{t}} \left(\frac{\partial \mathcal{F}}{\partial u} + \lambda \frac{\partial f}{\partial u} \right)^2 dt \geq 0. \tag{9.32}$$

The above condition is written as a minimum

$$\delta\varepsilon = \varepsilon \int_0^{\bar{t}} \left(\frac{\partial \mathcal{F}}{\partial u} + \lambda \frac{\partial f}{\partial u} \right)^2 dt \leq 0 \tag{9.33}$$

TABLE 9.2 Adjoint boundary conditions at initial and final time $(0, t)$.

x	Condition	Time	
		Specified	Unspecified
$x(t)$	Free	$\lambda(t) = 0$	$\lambda(t) = 0$
$x(t)$	Fixed	$\lambda(t)$ unspecified	$\mathcal{F}\vert_{t=\bar{t}} - \lambda(\bar{t})f$
$x(0)$	Free	$\lambda(0) = 0$	Initial time specified
$x(0)$	Fixed	$\lambda(0)$ unspecified	Initial time specified

and since $\delta\varepsilon$ cannot be negative, therefore the Hamiltonian function written as

$$H = \mathscr{F} + \lambda f \tag{9.34}$$

where the stationarity conditions are the state equation

$$\frac{\partial H}{\partial \lambda} = f \tag{9.35}$$

the adjoint, or co-state, equation

$$\frac{\partial H}{\partial x} = -\dot{\lambda} \tag{9.36}$$

and the optimal control equation

$$\frac{\partial H}{\partial u} = 0. \tag{9.37}$$

In nuclear engineering, an optimization problem could be the maximization of power for a given amount of fuel in some configuration. This would be as formulated as: maximize $P = \int \Sigma_f(x)\phi(x)dx$ subject to $N_o = \int N(x)dx$ and $\dot{x}_i(t) = f_i(\overline{x}, u, t)$ for $i = 1, 2$.

The Lagrange multiplier is a sensitivity coefficient and represents a powerful method to estimate the change in an optimal cost function per unit change in the constraint. It has thus been used for design sensitivity studies leading to optimization in nuclear reactor design.

The second-order neutron diffusion equation could be expressed as two first-order equations with state variables $x_1 = \phi, x_2 = \dot{\phi}$ giving two first-order ODEs.

$$\mathscr{L}(\phi, \lambda, x) = \langle \Sigma_f, \phi \rangle + \sum_i \lambda_i f_i. \tag{9.38}$$

Consider a system described by the coupled first-order ODEs

$$\dot{x}_1(t) = x_2(t) \equiv f_1(\overline{x}, u, t) \tag{9.39}$$

$$\dot{x}_2(t) = -x_2(t) + u(t) \equiv f_2(\overline{x}, u, t) \tag{9.40}$$

with boundary conditions $\overline{x}(0) = 0, \overline{x}(2) = [5 \quad 2]^T$, Find a function $u(t)$ such that the objective function, or *PI*,

$$J(u) = \frac{1}{2}\int_0^{t_f} u^2(t)dt \tag{9.41}$$

is maximized. The optimality analysis consists of the following steps:

1. Write the Lagrangian in terms of the objective function, the Lagrange multiplier(s), and the constraint(s),
2. Apply the stationarity conditions, and
3. Solve the equations to obtain the optimal control.

In this problem, the Lagrangian is

$$\mathscr{L}(\overline{x}, u, t) = \frac{1}{2}u^2(t) + \lambda_1 f_1 + \lambda_2 f_2 \tag{9.42}$$

with the stationarity conditions (state equations) giving the constraints

$$\dot{x}_1(t) = \frac{\partial \mathscr{L}}{\partial \lambda_1} = f_1 \tag{9.43}$$

$$\dot{x}_2(t) = \frac{\partial \mathscr{L}}{\partial \lambda_2} = f_2 \tag{9.44}$$

and the *adjoint* (Lagrange multipliers) or *co-state* equations

$$-\dot{\lambda}_1(t) = \frac{\partial \mathscr{L}}{\partial x_1} = 0 \tag{9.45}$$

$$-\dot{\lambda}_2(t) = \frac{\partial \mathscr{L}}{\partial x_2} = \lambda_1(t) - \lambda_2(t). \tag{9.46}$$

The above are solved to give

$$\lambda_1(t) = C_1 \tag{9.47}$$

$$\lambda_2(t) = C_1 + C_2 e^t \tag{9.48}$$

and the optimal control u^*

$$\frac{\partial \mathscr{L}}{\partial u} = u^* + \lambda_2(t) = 0. \tag{9.49}$$

Therefore, the optimal control is

$$u^* = -\lambda_2(t) = -\left(C_1 + C_2 e^t\right). \tag{9.50}$$

All that is required now is to find the constants C_1 and C_2. The optimal "trajectories" are then

$$x_1^*(t) = -C_1 t - \frac{C_2}{2} e^t - C_3 e^{-t} + C_4 \tag{9.51}$$

and

$$x_2^*(t) = -C_1 - \frac{C_2}{2} e^t + C_3 e^{-t}. \tag{9.52}$$

From the boundary conditions, the constants are:

$$C_1 = -7.2918, C_2 = 1.1870, C_3 = -6.6983, \text{ and } C_4 = -6.1048.$$

Then, the PI is found as

$$J(u^*) = \frac{1}{2}\int_0^{t_f} [u^*(t)]^2 dt = \frac{1}{2}\left[C_1^2 t_f + \frac{1}{2} C_2^2 \left(e^{2t_f} - 1\right) + 2C_1 C_2 \left(e^{t_f} - 1\right)\right] = 16.7507. \tag{9.53}$$

The optimal functions are plotted in Fig. 9.6; this represents the optimal solution.

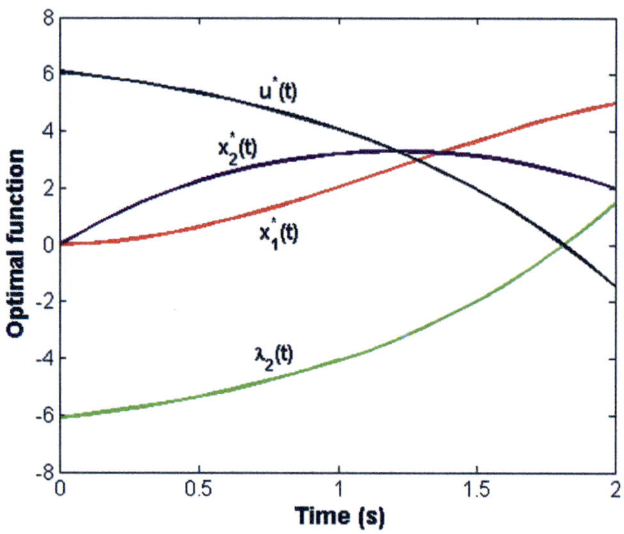

FIGURE 9.6 Optimal control $u^*(t)$, trajectories $x_1^*(t), x_2^*(t)$ and adjoint function $\lambda_2(t)$.

The same analysis is carried out using symbolic computing in MATLAB® using the dsolve function (Wang, 2007) with the steps below:

```
% step 1 declare the symbolic variables
syms x1 x2 p1 p2 u; % two equations so two Lagrange multipliers p1 p2% step 2 give the state equations
Dx1 = x2;
Dx2 = -x2 + u;
% step 3 write the cost function (inside the integral)
syms g;
g = 0.5*u^2;
% step 4 write the Hamiltonian
syms p1 p2 H;
H = g + p1*Dx1 + p2*Dx2;
% step 5    Adjoint (Co-state) equations
Dp1 = -diff(H,x1);
Dp2 = -diff(H,x2);
% step 6    solve for control u
du = diff(H,u);
sol_u = solve(du, 'u');
The solution has the following variables
sol_a =
   x2: [1x1 sym]
   x1: [1x1 sym]
   p1: [1x1 sym]
   p2: [1x1 sym]
and the optimal control u*is found as:           sol_u = -p2
% Substitute u in the state equations
Dx2 = subs(Dx2, u, sol_u);
% convert symbolic objects to strings for using 'dsolve'
eq1 = strcat('Dx1 =',char(Dx1));
eq2 = strcat('Dx2 =',char(Dx2));
eq3 = strcat('Dp1 =',char(Dp1));
eq4 = strcat('Dp2 =',char(Dp2));
sol_h = dsolve(eq1,eq2,eq3,eq4);
% BC x1(0) = x2(0) = 0; x1(2) = 5; x2(2) = 2;
conA1 = 'x1(0) = 0';
conA2 = 'x2(0) = 0';
conA3 = 'x1(2) = 5';
conA4 = 'x2(2) = 2';
sol_a = dsolve(eq1,eq2,eq3,eq4,conA1,conA2,conA3,conA4)
x1sol = sol_a.x1;
x2sol = sol_a.x2;
p2sol = sol_a.p2;
g1 = matlabFunction(x1sol);
g2 = matlabFunction(x2sol);
p2 = matlabFunction(p2sol);
%u = -p2;
x = 0:0.1:2;
xx1 = feval(g1,x); xx2 = feval(g2,x); pp2 = feval(p2,x); u = -pp2;
plot(x,xx1,'g','LineWidth',2);
hold on
plot(x,xx2,'b','LineWidth',2)
hold on
plot(x,u,'k','LineWidth',2)
legend('x1','x2','u')
grid on
```

The above program gives the functions x_1^*, x_2^*, u^* of Fig. 9.6.

Exercise 9.1: A first-order system is described as

$$\frac{d\theta(t)}{dt} = -a(\theta(t) - \theta_a) + bu(t)$$

where a and b are constants. Find the optimal control $u(t)$ which minimizes J

$$J = \frac{1}{2}\int_0^T u^2(t)dt$$

$\theta(t = 0) = \theta_a = 25°C$, and the final state is required to be $\theta(t = T) = 20°C$.

1. Sketch the optimal control $u^*(t)$ and the temperature $\theta^*(t)$, for $a = 0.25, b = 0.5$ and $T = 4$ s and calculate the optimal value J^*.
2. Sketch the Lagrange $\lambda(t)$ and explain how it can be used to find another optimal for different conditions.

Exercise 9.2: Maximize $J = \int_0^1 [x(t) + u(t)]dt$ subject to $\dot{x}(t) = 1 - u(t)^2, x(0) = 1$ by writing the Hamiltonian and applying the stationarity conditions. Sketch $u(t), \lambda(t), x(t)$ for the optimal solution and obtain the value of J.

9.2.5 Optimal discrete control (Pontryagin maximum principle)

In the previous section, the optimal control was a continuous function of the independent variable. When the control is discrete, the best control u^* is that which gives the maximum value of the Hamiltonian.

Given that the functional to be maximized is

$$J = \int_0^3 (x - u)dt \tag{9.54}$$

for a system described by

$$\dot{x} = x + u, \tag{9.55}$$

with initial and final conditions $x(0) = 2$, $x(3)$ free, with admissible controls (minimum and maximum) $u = [u_m \quad u_M] = [0 \quad 2]$

the Hamiltonian is written in the same way as before:

$$H = (x - u) + \lambda(x + u). \tag{9.56}$$

It is rearranged as $H = (1 + \lambda)x + (1 - \lambda)u$ to show that a sign change takes place for $\lambda > 1$. Thus in order to maximize H, u_m should be used for $\lambda > 1$. Note that the stationarity condition, Eq. (9.37), cannot be applied since the control is not continuous now.

Applying the stationarity condition for the adjoint equation

$$\frac{\partial H}{\partial x} = -\dot{\lambda} \tag{9.57}$$

gives the adjoint equation $1 + \lambda = -\dot{\lambda}$, which is easily solved to give $\lambda(t) = Ce^{-t} - 1$; applying the boundary condition $\lambda(3) = 0$, $C = \exp(3)$. This gives $\lambda(0) = 19.0855$, $\lambda(3) = -1$; at the switching point, the time T is found from $\lambda(T) = 1$, which gives $T = 2.3069$.

In the range $0 \leq t \leq T$, the control $u = u_M = 2$ and in $T \leq t \leq 3$, $u = u_m = 0$ would maximize J. This discrete optimal control solution is referred to as the bang-bang solution since one value is applied for some time and then the control is abruptly switched to another value.

Exercise 9.3: Show that the solutions for Eq. (9.44) $\dot{x} = x + u$ for $u = 0$ and $u = 2$ are $x(t) = 0.1991e^t$ and $x(t) = 2$ and calculate J.

An application of bang-bang control is in the placement of fuels of different enrichments in various spatial domains.

9.3 Controller design and optimization

Nuclear reactor control, achieved by the movement of control rods or by burnable poison distributed in the core, can be modeled by the point kinetic equations (PKE) introduced in Section2.1 which is a set of two non-linear first-order equations that can be combined into one second-order ODE. The reactor can be controlled by implementing reactivity changes which alter the neutron density $n(t)$ and bring the reactor back to the critical state. The manner in which a reactor is controlled has to ensure that no unsafe conditions are attained. Another situation is where a reactor is producing power according to some required load which may vary with time. In that case the reactor must be operated in a *load-following mode.* In both the situations, a controller must be designed and implemented accordingly.

In engineering systems, a control system would typically have some overshoot η and a finite (non-zero) response (settling) time τ_s, and will not therefore respond instantly. While an ideal controller cannot be designed, an *optimized controller* can be designed to give a response within acceptable safety limits.

In this section, the mass-spring-damper (MSD) equation, a second-order ODE, is considered to represent the PKE and the response is obtained for a given step (reactivity) input. The response is analyzed in terms of η and τ_s.

The solution of the MSD

$$m\frac{d^2x}{dt^2} + b\frac{dx}{dt} + kx = f(t) \tag{9.58}$$

for a step response $u(t) = 1, t \geq 0$, is found by taking the Laplace transform, so that Eq. (9.58) becomes

$$\left(ms^2 + bs + k\right)X(s) = \frac{1}{s}. \tag{9.59}$$

Now defining the natural frequency ω_n, the critical damping coefficient C_c and the damping coefficient ζ as

$$\omega_n = \sqrt{\frac{k}{m}}, \quad C_c = \sqrt{4km}, \zeta = \frac{b}{C_c}$$

Eq. (9.59) is written in terms of partial fractions and the coefficients in the numerator are then determined to get a useful form for inversion:

$$X(s) = \frac{1}{k}\left[\frac{1}{s} - \frac{s}{(s+\zeta\omega_n)^2 + \omega_n^2\left(1-\zeta^2\right)} + \frac{b}{m\omega_n^2\left(1-\zeta^2\right)}\frac{\omega_n^2\left(1-\zeta^2\right)}{(s+\zeta\omega_n)^2 + \omega_n^2\left(1-\zeta^2\right)}\right]. \tag{9.60}$$

$X(s)$ is readily inverted to give the compact time-dependent displacement (after some algebraic simplifications)

$$x(t) = \frac{1}{k}\left[1 - e^{-\zeta\omega_n t}\left\{\cos\omega_n\sqrt{1-\zeta^2 t} + \frac{\zeta}{\sqrt{1-\zeta^2}}\sin\omega_n\sqrt{1-\zeta^2 t}\right\}\right]. \tag{9.61}$$

To understand this solution, consider some special cases. First, for a mass-spring system without any damping $= 0$,

$$x(t) = \frac{1}{k}[1 - \{\cos\omega_n t\}] \tag{9.62}$$

which means that a unit step force would set the system into an oscillatory never ending motion as shown in the periodic function of constant amplitude in Fig. 9.7. For $\zeta < 1$, the displacement oscillates with an overshoot until it settles down to a steady gain as shown for the solutions for $\zeta = 0.2, 0.4, 0.8, 1.0$ in Fig. 9.7. Each of these cases has an overshoot η which decreases in magnitude as the damping ratio increases. Similarly, the settling time τ_s decreases with an increasing damping ratio. For higher damping when $\zeta > 1$ the square roots in Eq. (9.61) are complex and since $\sin i\theta$and $\cos i\theta$ are real numbers, there is no oscillatory motion. All these cases are shown in Fig. 9.7 for normalized spring constant k.

Exercise 9.4: A system is described by the state equations

$$\dot{n} = \frac{\rho - \beta}{l}n(t) + \lambda C(t)$$

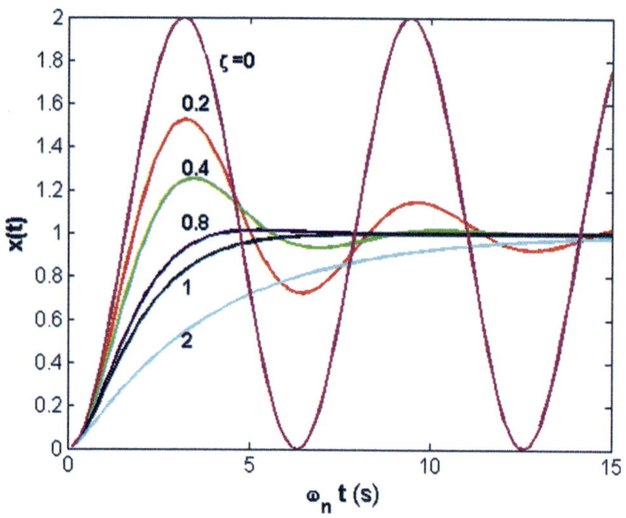

FIGURE 9.7 Response $x(t)$ versus $\omega_n t$ of a second-order system due to a unit step function.

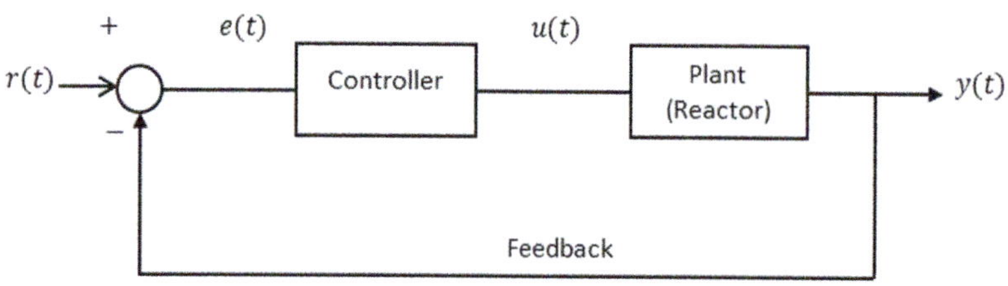

FIGURE 9.8 A proportional-integral-derivative controller.

$$\dot{C} = \frac{\beta}{l} n(t) - \lambda C(t)$$

where $n = n(t), c = c(t)$ β, l and λ are constants given below. The system is in equilibrium for $t \le 0$ when $c(t) = \beta n_0/\lambda l$. A step input $\rho(t) = \rho_0 > 0$, is applied at $t = 0$; $\rho(t) = 0, t \le 0$ and causing a jump in $n(t)$. Solve for the system response to compute $n(t)$ for $0 < t < 200$ s for $\rho_0(\rho_0/\beta = 0.1, 0.2, 0.5)$ and plot your results. The data for this exercise is: $\beta = 0.0079$, $l = 0.0001$ s, and $\lambda = 0.0767$ s^{-1}.

Now consider one of the simplest controllers, a proportional (P), integral (I), derivative (D), or proportional integral derivative (PID), controller which provides corrective action from the feedback of the plant (nuclear reactor) to take a feedback from the output signal $y(t)$, compare it with a desired signal $r(t)$ to produce an error signal $e(t)$ which is fed into the controller which subsequently produces a control signal $u(t)$ to drive the plant as desired, as depicted in Fig. 9.8. The plant output $y(t)$ is then optimized with respect to the control parameters.

The PID control signal

$$u(t) = K_P e(t) + K_I \int e(t)dt + K_D \frac{de}{dt} \tag{9.63}$$

has a transfer function (TF)

$$U(s) = \frac{K_D s^2 + K_P s + K_I}{s^2}$$

which modifies the TF (Eq. 9.59) to

$$\frac{X(s)}{R(s)} = \frac{K_D s^2 + K_P s + K_I}{s^3 + (b + K_D)s^2 + (k + K_P)s + K_I}. \tag{9.64}$$

The open-loop system response for $(m = 1, b = 3, k = 46.5, \zeta = 0.2)$ is shown in Fig. 9.9.

The objective is now to reduce the overshoot η, decrease the settling time τ_s, and increase the steady gain of the reactor (plant).

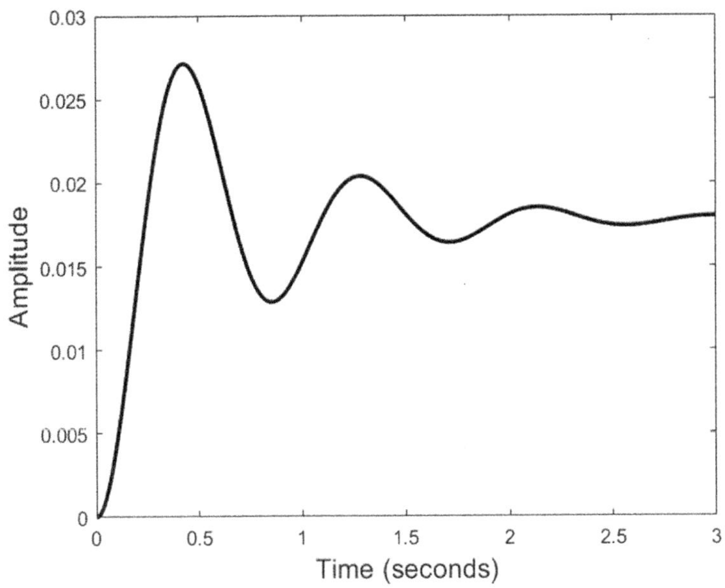

FIGURE 9.9 Open-loop system response for a unit step function.

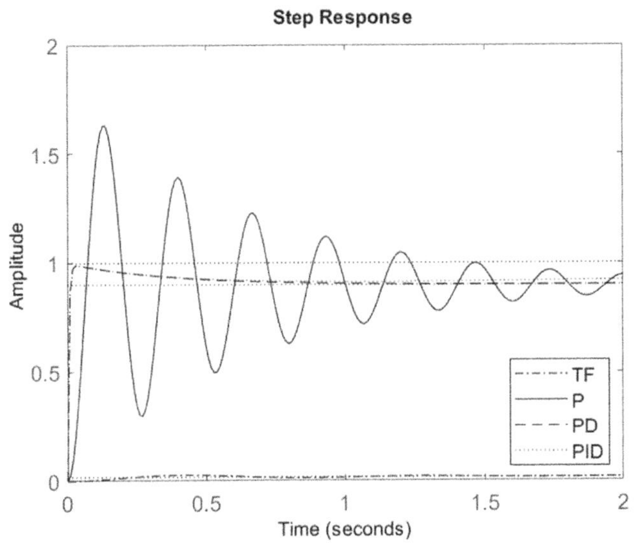

FIGURE 9.10 Unit step responses: open-loop response TF, P, PD, and PID responses.

For a preliminary analysis, the closed-loop parameters (Eq. 9.64) K_P, K_I and K_D are taken randomly. As K_P is increased, the overshoot increases and the steady value also increases; this is followed by increasing K_D which reduces τ_s. Finally, K_I is increased to improve the steady gain. For $K_P = 500, K_I = 100$ and $K_D = 200$, the rise time τ_r, the time at which the response first crosses the amplitude-one line, is 0.0111 s. All these cases namely, the open-loop response, TF and the responses for P, PD, and PID are shown in Fig. 9.10.

The PD response rises sharply to a response of ~1.0 then falls and stabilizes to a steady value of 1.0 by ~2 s. Even the PID response has not been able to rise to a steady value of 1.

There are several methods of tuning a PID controller, the oldest being the Nichols-Ziegler Method (Burns, 2001) which can readily give reasonably good response parameters. Here, the MATLAB tuning command is used to remove the overshoot and give the response shown in Fig. 9.11 with tuned parameters listed in Table 9.3.

The tuned response for the baseline parameters gave $K_P = 178.0441$, rise time (τ_r) 0.0735 s, settling time (τ_s) 2.51 s, overshoot (η) 73.4%, and peak amplitude 1.32. The final improvement gives the parameters listed in Table 9.3 with a steady peak amplitude 1.099 and rise and settling times $\tau_r = 0.944$s, settling time τ_s 2.27 s, and no overshoot.

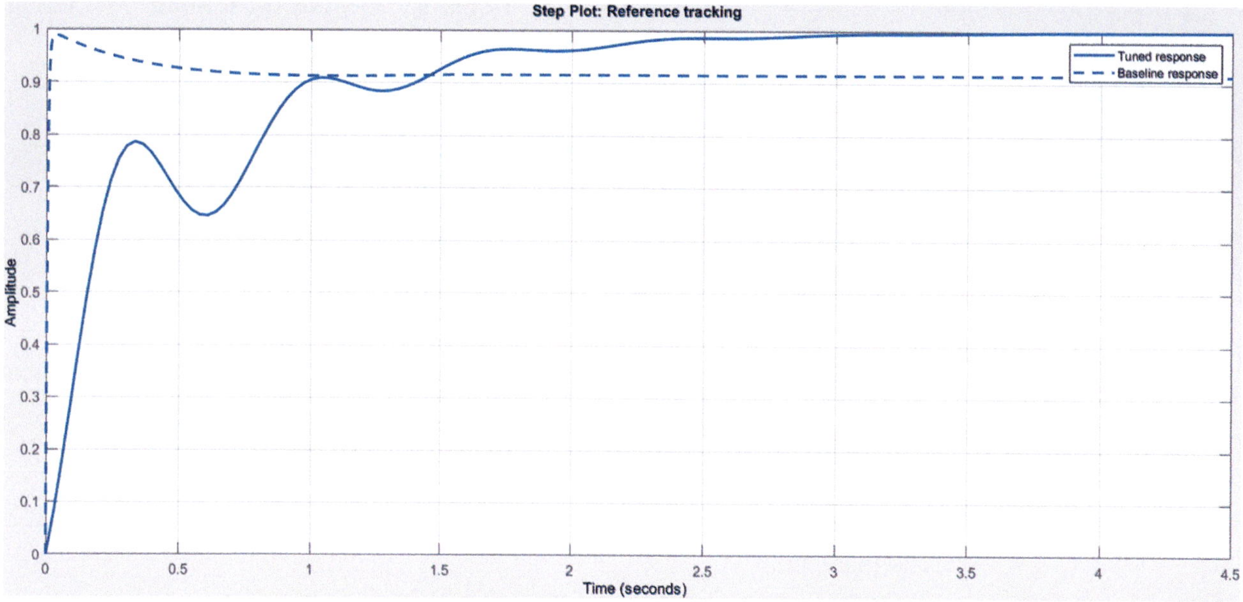

FIGURE 9.11 PID-tuned response compared with the baseline response.

TABLE 9.3 Proportional integral derivative controller tuning parameters.

Parameter	Tuned	Baseline
K_P	35.4172	500
K_I	129.7872	100
K_D	2.4162	200
τ_r (s)	0.944	0.0111
τ_s (s)	2.27	—
η (%)	0	0
Peak	1.099	NaN

Exercise 9.5: The closed-loop system for a PI controller based on an error signal $y(t)$ and input $g(t)$ is given by

$$\frac{d^2y}{dt^2} + \left(0.02 + k_p\right)\frac{dy}{dt} + k_i y = 10\frac{dg}{dt}.$$

When compared with a MSD

$$M\frac{d^2x}{dt^2} + D\frac{dx}{dt} + Kx = 0$$

for which the characteristic polynomial is $s^2 + 2\xi\omega_0 s + \omega_0^2$ where ξ and ω_0 denote relative damping and the undamped natural frequency

1. draw a block diagram of the system,
2. determine the controller parameters k_p, k_i for critical damping $\xi = 1$ with assumed values of $\omega_0 \in (0, 1)$, and
3. sketch the response and comment on the dynamic response of the system.

Another set of optimal tools is the linear quadratic controller (LQR) for optimizing the control when the plant model is linear and the PI is a quadratic function of the state variables and control. The plant can then can follow a reference trajectory (Burns, 2001; Lewis et al., 2012) as applied to a PWR with an optimal linear quadratic Gaussian (LQG) control with the robust integral sliding mode technique (Vajpayee et al., 2021a, 2021b) to track a reference trajectory with external disturbances and parametric uncertainties.

With optimized techniques on a PID, LQR, LQG a nuclear reactor can adjust itself to efficiently follow the load requirement.

9.4 Dynamic programming

DP is used in this chapter as a deterministic optimization method in which an optimal problem is structured into a number of multiple stages and considered as smaller sub-optimal problems. It is a powerful method used for deterministic problems, such as the Traveling Salesman Problem, to find an optimal policy and has been applied to decision-making under uncertainty as well.

In nuclear reactor refueling, the objective is to find the optimal refueling pattern at the end of each cycle when once- or twice- *burnt fuel* is replaced and thc rcmaining is shuffled. This has a bearing on the revenue requirement for producing energy in a cycle.

This section considers a simple example to illustrate DP; a more detailed description with relevance to fuel reloading in nuclear reactors is discussed in Section 13.2.

The reloading optimization problem can thus be expressed as

$$\min T_c = \min \sum_{i=1}^{N} C[x_i, \quad e_i, f_i] = \min \sum_{i=1}^{N} C_i \tag{9.65}$$

where C_i is the total discounted revenue requirement for producing energy E_i in the ith cycle, x_i e_i, and f_i are the state of the reactor, the reload enrichment and batch fraction at the beginning of the ith cycle.

In refueling optimization, a stage would represent a cycle at the end of which several possible states, defined by the fuel types (in the sense of enrichment and burnup) are available and a decision has to be made on the reloading pattern.

The essence of DP is that

1. A problem is configured into N stages consisting of various states $P_{i,k}$ where i can denote the fuel types available and k represents the stage number.
2. At each stage, a decision is made on which state $P_{i,k+1}$ to move to in the $(k+1)^{th}$ stage; this decision will influence the power distribution, criticality control (possibly by way of poison distribution), burnup behavior and cycle length in the next cycle. Of course, this decision will be based on several detailed coupled neutronics-depletion simulations, and hence will be computation-intensive.
3. An optimal path is found by starting at the end (the Nth stage), in an adjoint or backward manner with defined goals, moving to find the sub-optimal solution at the $(N-1)^{th}$ stage, and proceeding to the initial state.

The DP computation is thus expressed as a recursive algorithm; a simple example being to calculate the factorial of an integer or Fibonacci numbers.

```
function x = fact(n)
  if n< =1
    x = 1;
  else
    x = n.* fact(n-1);
  end
end
```

Consider the five-stage optimization problem, with the starting node as Stage 1 and the terminal node as Stage 5, with three intermediate stages as depicted in Fig. 9.12.

In this shortest path problem, the objective is to find the path from start to end which minimizes the total distance traveled. Such problems have been solved by Djikstra's algorithm by generating a *minimum path tree* by scanning all neighboring nodes from a starting node.

In this section, consider the recursive algorithm

$$f_s(i) = \min\{c_{i,j} + f_{s+1}(j)\} \tag{9.66}$$

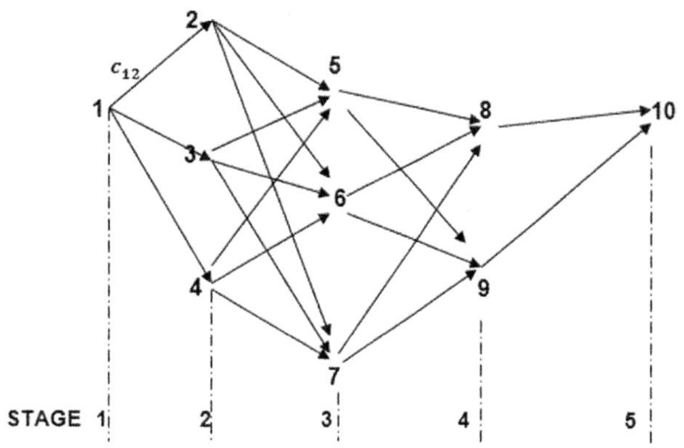

FIGURE 9.12 A five-stage forward network for finding the shortest route.

TABLE 9.4 Distances c_{ij} between nodes i and j in a network.

Node	1	2	3	4	5	6	7	8	9	10
1	0	2	5	6	—	—	—	—	—	—
2	—	0	—	—	3	4	6	—	—	—
3	—	—	0	—	3	2	4	—	—	—
4	—	—	—	0	4	3	5	—	—	—
5	—	—	—	—	0	—	—	1	4	—
6	—	—	—	—	—	0	—	6	3	—
7	—	—	—	—	—	—	0	3	3	—
8	—	—	—	—	—	—	—	0	—	3
9	—	—	—	—	—	—	—	—	0	4
10	—	—	—	—	—	—	—	—	—	0

Starting from the last stage N, the algorithm Eq. (9.64) gives the distances as.

where $f_s(i)$ is the cost associated with a transition from state i to the end state node N in the stage s and $c_{i,j}$ is the cost associated in taking the system from state i to state j, listed in Table 9.4. In this problem, the assumption is that only forward transitions are allowed.

Stage $s = 4$

$$f_4(8) = 3$$

$$f_4(9) = 4$$

Moving back another stage:

Stage $s = 3$

$$f_3(5) = \min\begin{cases} c_{5,8} + f_4(8) = 1 + 3 = \mathbf{4} \\ c_{5,9} + f_4(9) = 4 + 4 = 8 \end{cases}$$

$$f_3(6) = \min\begin{cases} c_{6,8} + f_4(8) = 6 + 3 = 9 \\ c_{6,9} + f_4(9) = 3 + 4 = \mathbf{7} \end{cases}$$

and

$$f_3(7) = \min\begin{cases} c_{7,8} + f_4(8) = 3 + 3 = \mathbf{6} \\ c_{7,9} + f_4(9) = 3 + 4 = 7 \end{cases}$$

At Stage 3, the minimum cost paths from states 5, 6 and 7 are: 5–8–10, 6–9–10, and 7–8–10, respectively. Moving now to Stage 2, the calculations are as follows:

For stage $s = 2$

$$f_2(2) = \min\begin{cases} c_{2,5} + f_3(5) = 3 + 4 = \mathbf{7} \\ c_{2,6} + f_3(6) = 4 + 7 = 11 \\ c_{2,7} + f_3(7) = 6 + 6 = 12 \end{cases}$$

At this stage, the minimum node is thus found to be 2–5 and 2–6 and 2–7 are rejected.

$$f_2(3) = \min\begin{cases} c_{3,5} + f_3(5) = 3 + 4 = \mathbf{7} \\ c_{3,6} + f_3(6) = 2 + 7 = 11 \\ c_{3,7} + f_3(7) = 4 + 6 = 10 \end{cases}$$

Similarly the route 3–5 is the best out of 3–5, 3–6, and 3–7.

$$f_2(4) = \min\begin{cases} c_{4,5} + f_3(5) = 4 + 4 = \mathbf{8} \\ c_{4,6} + f_3(6) = 3 + 7 = 10 \\ c_{4,7} + f_3(7) = 5 + 6 = 11 \end{cases}$$

This calculation again shows that the intermediate node (node 5) is the best from node 2.
Stage $s = 1$

$$f_1(1) = \min\begin{cases} c_{1,2} + f_2(2) = 2 + 7 = \mathbf{9} \\ c_{1,3} + f_2(3) = 5 + 7 = 12 \\ c_{1,4} + f_2(4) = 6 + 11 = 17 \end{cases}$$

This stage calculation says that the rout 1–2 is the best of all options. The *optimal* path is thus found to be 1–2–5–8–10 with a minimum cost of 9 units.

9.5 Stochastic optimization

Stochastic methods, equipped with the elaborate mathematical and physical foundations of transport phenomena, and capability of modeling realistic 3D configurations, have powerful techniques which efficiently search for an optimal solution from a very large search space. This section illustrates GAs for constrained optimization of a simple two-variable function (Section9.1) and describes some other methods and applications.

9.5.1 Genetic algorithms

The GA method is used as an optimization method for large problems where better computational efficiency, in comparison with deterministic methods, can be achieved.

The GA method (Haupt & Haupt, 2004; Holland, 1975) is based on the "survival of the fittest" biological evolutionary processes and has been adapted in computational algorithms that carry out a random-search for the best values of one or several variables that maximize (or minimize) an "objective" function. A typically large search space is scanned with the help of algorithms incorporating concepts such as genes (bits), chromosomes (bytes), fitness function (objective function), and "selection" tests of genetic crossover and mutation. Thus, "fitter" successive generations are obained to ultimately "converge" to an optimal as shown.

The cylinder volume optimization is repeated with the GA method using the MATLAB program given in Haupt and Haupt (2004).

It is useful to go through the calculation details for the first two iterations given below. The simulation parameters are listed in Table 9.5. Note that simulations will have a much larger number of variables, population size, number of iterations, and length of a chromosome.

After the simulation parameters are fixed, as shown in Fig. 9.13, the initial population of chromosomes is generated and evaluated.

The maximization problem (Section 9.2.2, Eq. 9.8) is converted to a minimization problem with a sign change and each chromosome is evaluated for its fitness. The chromosomes in binary form with 4 bits, are converted into real numbers in the search space for radius $R \in (0, 2)$; thus a chromosome 1111 with a value 15 would correspond to the real number 2.0. Thus for a chromosome in the range 0–15, the linear mapping in the range 0–2 would give the required

TABLE 9.5 Simulation parameters for genetic algorithm optimization.

Parameter	Value/description
Population size (no. of chromosomes)	10
No. of iterations	10
Length of a chromosome (bits)	4
Selection fraction	0.50
Crossover strategy	Single bit
Mutation rate μ	0.15
Objective function (Eq. 9.8)	$-\left(\frac{1}{2}KR-\pi R^3\right)$
Search range for radius R	$R\in(0,2)$

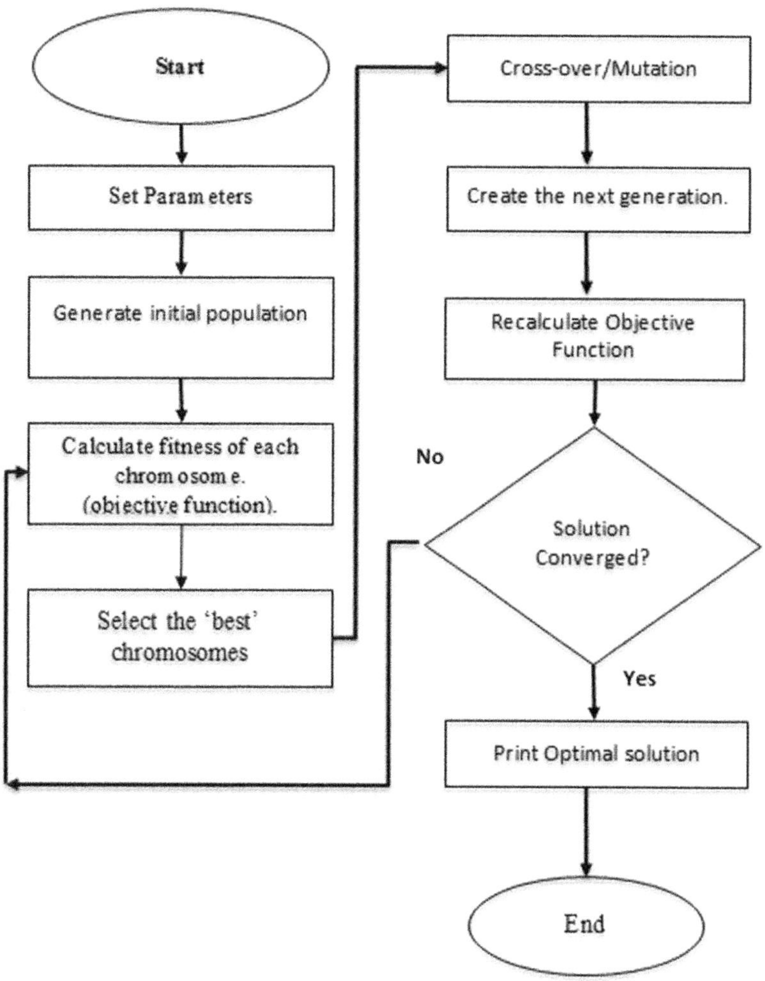

FIGURE 9.13 Flowchart of the genetic algorithms optimization method.

transformation giving a four digit accuracy. A 16-bit chromosome would give $2^{16}-1=65{,}535$ mapping with $R=2$ to give an accuracy 10^{-5}.

```
i   chromosome   number    cost fn
1     1010       1.3333    -5.8866e+00
2     1101       1.7333    -9.7285e-01
3     1010       1.3333    -5.8866e+00
4     1111       2.0000    -5.1327e+00
5     0101       0.6667    -5.7358e+00
```

```
6      0111    0.9333    -6.7791e+00
7      1000    1.0667    -6.8539e+00
8      1011    1.4667    -4.7551e+00
9      1111    2.0000     5.1327e+00
10     0101    0.6667    -5.7358e+00
```

The fitness of the chromosomes is sorted with the best being ranked first out of 10 chromosomes.

```
Sorted Cost   (lowest obj. fn. first)
i   number          obj. fn.
1     1.0667      -6.8539e+00
2     0.9333      -6.7791e+00
3     1.3333      -5.8866e+00
4     1.3333      -5.8866e+00
5     0.6667      -5.7358e+00
6     0.6667      -5.7358e+00
7     1.4667      -4.7551e+00
8     1.7333      -9.7285e-01
9     2.0000        5.1327e+00
10    2.0000       5.1327e+00
```

The chromosomes generated above give a minimum, or best value, -6.8539 and a mean value -3.2340.

The next step in the optimization is to improve upon the initial population by selected the best and creating the next generation by crossover and mutation. In this computation the number of matings is 3 obtained from ceil (popsize-keep)/2 where the ceiling of the difference between the population size and the chromosomes to keep (fraction 0.5 as set in Table 9.5). The probability density function (PDF), $f(i)$ for chromosome i is calculated by its value of the objective function divided by the sum of all values. From the PDF, the cumulative distributive function ($F(i)$)is calculated (Section 4.9) as shown below. For 10 chromosomes, 5 were kept and used to calculate PDF and cumulative distribution function (CDF) values.

```
i       f(i)        F(i)
1     0.3333      0.3333
2     0.2667      0.6000
3     0.2000      0.8000
4     0.1333      0.9333
5     0.0667      1.0000
```

Note, as stated before, that the CDF is a monotonically increasing function rising to the value one.

The mates are selected by three uniform random numbers ξ_1, ξ_2 generated in the range (0,1):

i	ξ_1	ξ_2
1	0.9941	0.6520
2	0.3977	0.9061
3	0.6533	0.1331

A mate is selected from the random numbers compared with the CDF values, for example, for the first value $\xi_1 = 0.9941$, the last bin of the CDF (chromosome 5) is selected.

```
i   Mate1   Mate2
1     5       3
2     2       4
3     3       1
```

The next generation of chromosomes are obtained from crossover and mutation:

```
Mating using single-point crossover

i  chromosome
   1000
   0111
   1010
   1010
```

```
   0101
Mutation using mutrate (μ)
i  chromosome
   1000
   0010
   1010
   1010
   0101
```

The new chromosomes are re-evaluated for their fitness:

```
i    chromosome   number    obj. fn.
1      1000       1.0667   -6.8539e+00
2      1000       1.0667   -6.8539e+00
3      1001       1.2000   -6.5713e+00
4      0110       0.8000   -6.3915e+00
5      1010       1.3333   -5.8866e+00
6      1010       1.3333   -5.8866e+00
7      1010       1.3333   -5.8866e+00
8      0101       0.6667   -5.7358e+00
9      1011       1.4667   -4.7551e+00
10     0010       0.2667   -2.6071e+00
Sorted Cost (lowest cost first)

i    number    obj. fn.
1    1.0667   -6.8539e+00
2    1.0667   -6.8539e+00
3    1.2000   -6.5713e+00
4    0.8000   -6.3915e+00
5    1.3333   -5.8866e+00
6    1.3333   -5.8866e+00
7    1.3333   -5.8866e+00
8    0.6667   -5.7358e+00
9    1.4667   -4.7551e+00
10   0.2667   -2.6071e+00
minimum cost of population minc(1) = -6.8539
mean cost of population meanc(1) = -5.7428
```

This process continues until the maximum number of iterations specified in the simulation parameters. Each iteration ends with a minimum cost and a mean cost (plotted in Figs. 9.14–16):

```
minimum cost of population    minc(1) = -6.8539
mean        cost of population meanc(1) = -5.3445
```

Table 9.5 shows GA results for $A_s = 20$ cm^2 for which the deterministic results (Table 9.1) are $R^* = 1.0301$ cm, $H^* = 2.0601$ cm, $V^* = 6.8671$ cm^3. For all the GA runs, the selection fraction was 0.5, mutation rate was 0.15 and each "chromosome" was of length 8 bits. The search space specified for the radius was (0, 2). The GA results for the optimal R and H are in good agreement with the 'exact' result and convergence is seen in the fourth generation while the mean fluctuates randomly. For the fourth row of results listed in Table 9.6, the GA result for optimum volume (6.8670) matches the deterministic exact result (6.8671) to three decimal places.

The "quality" of results from a GA simulation is dependent on the specified simulation parameters for which there is no formal method and the selection remains an "art" rather than a science.

Such random methods are feasible when the number of variables is large. In principle, GA is efficient in a "parallel" computing environment with benefits of computational speed-up.

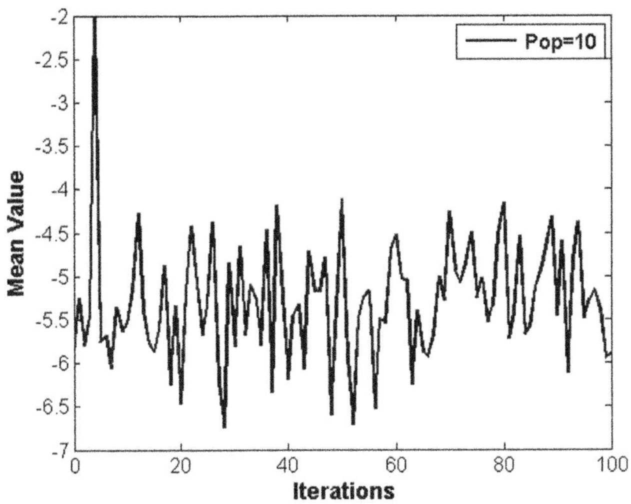

FIGURE 9.14 Genetic algorithm results for minimization ($N = 10$, $M = 100$, $N_{bits} = 4$).

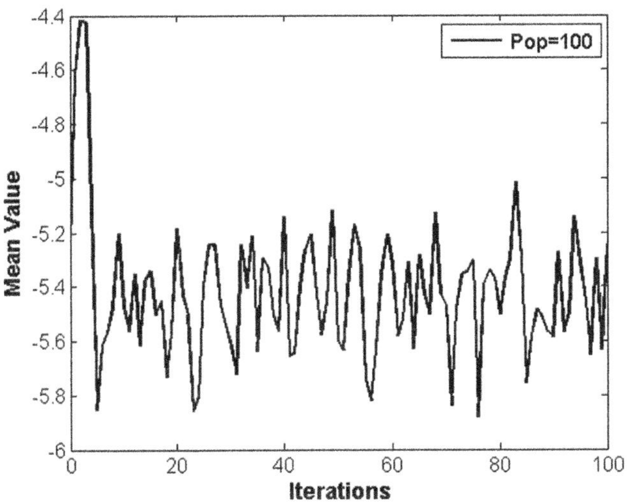

FIGURE 9.15 Genetic algorithm results for minimization ($N = 100$, $M = 100$, $N_{bits} = 4$).

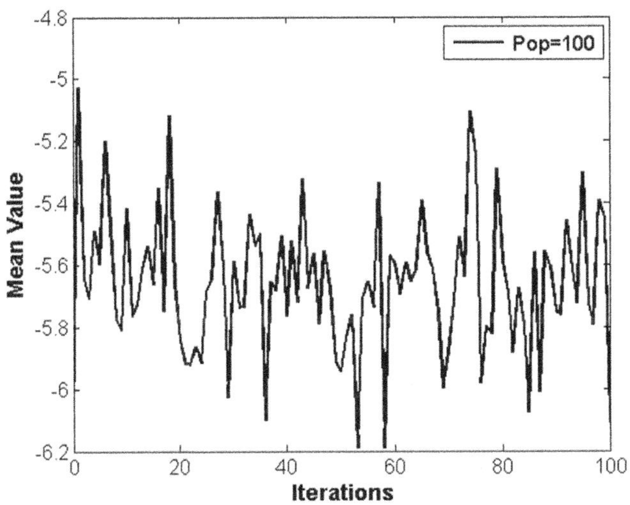

FIGURE 9.16 Genetic algorithm results for minimization ($N = 100$, $M = 100$, $N_{bits} = 8$).

TABLE 9.6 Genetic algorithm results for optimized cylinder volume ($V^*K = 20$) cm^2.

Population size	No. of bits	Iterations	(R^*, H^*, V^*)
10	4	100	1.0667, 2.1334, 6.8539
100	4	100	Same as above
100	8	100	1.0275, 2.0350, 6.8670

9.5.2 Particle swarm optimization

Particle swarm optimization (PSO), first introduced by Kennedy and Eberhart (Kennedy & Eberhart, 1995) in 1995 is a swarm intelligence (*meta*-heuristic) algorithm inspired by the movement of a flock of birds with an objective to survive by looking for a safe place to land with the availability of food. Since the time PSO was introduced, several variants of the method have been developed. In nuclear engineering the method has been extensively applied in-core design optimization, controller design optimization (Domingos, Schirru, & Pereira, 2006; Mousakazemi, 2021; Mousakazemi et al., 2018) and several other areas

The PSO algorithm, in the standard form, is a simple two step procedure; in the first, the position $\boldsymbol{x}$ and velocity $\boldsymbol{v}$ of particles in a swarm are initialized and evaluated using the objective function. The best values p_t^b for each particle in the tth generation and the best values for all particles in the tth and previous generations, p^g, are obtained. With this information and a set of weighting factors w_i, c_1, c_2 with randomness modeled by random numbers ξ, the velocity is updated for the next iteration as:

$$v_{t+1} = w_i v_t + c_1\xi_1\left(p_t^b - x_t\right) + c_2\xi_2(p^g - x_t). \tag{9.67}$$

It is worth mentioning that this is not truly a velocity in the sense of distance divided by time; it is given this name most probably since it is a parameter that travels to the next iteration where the position of each particle is updated as follows:

$$x_{t+1} = x_t + v_{t+1}. \tag{9.68}$$

The iteration continues until the objective function converges to some prescribed tolerance ε. The variables in Eq. (9.67) and Eq. (9.68) are formally defined as

$c_{1,2}$ correction factors
p^g global best (combination of values)
p_t^b particle best position at the tth iteration
v_t velocity at iteration step t
v_{t+1} velocity at iteration step $t+1$
w_i inertia weight
$\xi_{1,2}$ uniform random number $\epsilon(0, 1)$.

The PSO method is simple in its standard form and is best illustrated by an exercise.

Exercise 9.6: Consider a minimization problem for the objective function given as Eq. (9.1) with

$f(x_1, x_2) = \frac{1}{2}x_1^2 + x_1x_2 + x_2^2 + x_2$ for a search space $-4 \leq x_1, x_2 \leq 4$. For a hand calculation of two iterations, use the following values: Population size (Pop) = 5, $c_1 = c_2 = 1.5$, No. of iterations $= 10$, $w = 0.8$ and fill Table 9.7.

Steps for the first iteration

Step 1. Generate 5 random numbers ξ in the range (0, 1) first for v_1 then for v_2
Step 2. Generate 5 random numbers ξ in the range (0, 2) first for x_1 andx_2.
Step 3. For each pair x_1x_2 calculate $f(x_1, x_2)$
Step 4. For the first iteration of each pair $x_1.x_2$ is the best value since there are no values from the previous iteration.
Step 5. Find the global best pair, that is, the x_1x_2 pair with the lowest f value

TABLE 9.7 Optimization by the particle swarm optimization method (first iteration).

Particle	Velocity		Position		Obj. Fn.	Particle best	
	v_1	v_2	R	H	$f(R,H)$	p_1^b	p_2^b
1							
2							
3							
4							
5							

This completes the first iteration.

$$P_g^b = [\quad]$$

Steps for the second iteration

Step 1: update v_1, then v_2 from Eq. (9.2).
Step 2: update R from Eq. (9.3), then update H from Eq. (9.3).
Step 3: find the particle best value p_1^b from this and previous iteration, then find p_2^b
Step 4: find the best global value $P_g^b = [\quad]$ from this and the previous iteration.

The calculated values can be filled in a Table similar to Table 9.7.

The procedure in the PSO method is simpler than that in the GA method since there are no chromosomes, cross-overs, or mutations.

Other optimization methods used are Simulated Annealing and the algorithms based on the observation of ants, birds, fish and animals as they behave for survival: Ant Colony Optimization, Grey Wolf Optimization, and Cuckoo optimization.

9.6 Applications of optimization in reactors

In nuclear engineering, the refinement of mathematical models and the development of powerful algorithms enabling efficient computer simulation toward the design and operations of present and future nuclear power reactors makes it possible to optimize designs and strategies to a high level of accuracy.

This section, building upon the introduction given in Section 9.1 provides further insight into the areas where optimization methods are being applied and indeed, have considerable scope for refinement.

9.6.1 Multi-objective core optimization

Optimization for reactor design optimization is an extensive and challenging multi-discipline process involving coupled phenomena (Section 4.4, Section 4.8) and several interdependent parameters mostly requiring multi-objective optimization for a wide spectrum (Section 3.2) covering the primary system (core and associated systems), Balance of Plant, back-end fuel cycle, and safety analyses.

The optimization methods require a search for the optimal design parameters for which a very basic and elegant theory is provided by variational calculus especially Pontryagin's bang-bang control (Section 9.2.5) while large problems can be efficiently simulated by stochastic heuristic methods (Section 9.5) such as GA, PSO, and simulating annealing.

Optimization broadly falls into two categories namely design and operations. The design optimization process begins from the conceptual and feasibility studies as for vSMR's (Section 3.2.5) and Gen IV reactors (Section 3.5.3) going through the licensing procedures and freezing of a final design. This is followed by the operational phase from startup to normal operations, maintenance and unplanned shutdowns and periodic refueling.

For large operating baseload reactors mainly in the >700 MWe category (Section 3.1), optimization is performed for better understanding operating parameters such as Doppler broadening due to temperature (Section 2.8) improving operating parameters and to carry out safety-related simulations including source-term calculations (Section 3.5.4) such as for the Fukushima Boiling Water Reactors (BWRs). The most suitable techniques are stochastic heuristic coupled with numerical solutions such as nodal methods and the method of characteristics of the diffusion or transport equations (Section 5.4.1, Section 5.6).

Typical illustrations of operational areas for optimization are prediction of the critical heat flux (CHF) and moisture carryover (MCO) in a BWR. For both CHF and MCO prediction, the attractive optimization tools include arificial intelligence (AI) data-driven models, Support Vector Regression, and Gaussian Process Regression for producing lookup tables and AI-based predictive models in the face of non-linearity and uncertainty.

Several present-day optimization challenges need to be addressed for design changes in nuclear submarine reactors (Section 3.3), basic design configurations of micronuclear reactors and space propulsion reactors (Section 3.7) and nuclear fusion reactors (Section 3.6, Section 10.6). Some of these challenges for submarine reactors (Section 3.3.6) could be itemized as follows:

1. minimize the overall reactor size and weight,
2. design long "lifetime cores" without refueling for ~50 years,
3. design better, lighter and more effective neutron and gamma shielding,
4. increase the operating temperatures ~600°F–700°F for better thermodynamic efficiency,
5. explore the use of advanced high burnup fuels such as U-Zr, U-Al and ceramics,
6. optimize the power conversion system,
7. plan for overcoming dead time from xenon $54135Xe$ (half life 9.2 hours) poisoning, and
8. simulate off-normal operating conditions with a potential for accident scenarios.

9.6.2 Pressurized water reactor core pattern optimization

A fuel reloading strategy is necessary to remove spent fuel, add new fuel, and shuffle the remaining fuel within the core at the end of a cycle. This requires placement of fuels of differing enrichment as well as distributing the poison to ensure criticality with changing nuclide concentrations. Load Pattern Optimization (LPO) is best achieved by stochastic heuristic, and AI-driven *meta*-heuristic, optimization (Section 9.5) with GA (Section 9.4), PSO, Ant Colony Optimization, Evolutionary Algorithms, Differential Evolutionary Algorithms (DEA) and DP (Section 9.5.1).

AI-driven *meta*-heuristics for LPO have shown significant speedups for the first core of the 1000 MWe PWR with optimization to manage the radial power peaking factor (RPPF) under Xe equilibrium conditions. The maximum RPPF of the obtained loading pattern (LP) is decreased more than that of the LP by the designer.

By similar methods, automated and optimized control rod pattern design in a BWR has been demonstrated to calculate the axial power distribution to calculate k_{eff} and the shutdown margin.

9.6.3 Controller proportional integral derivative

An optimized controller design for nuclear power reactors is used to optimize the power output and follow a load pattern in the presence of external disturbances and statistical uncertainties. Some designs are based on the PID (Section 9.3), LQR, and LQG controllers.

9.6.4 Radiation shielding

Radiation shielding design (Section 10.4.2) optimization requires that the best combination and configuration of materials is found providing shielding from both neutron and gamma radiation. Thus, stochastic heuristic methods are again an attractive choice and have been used to give optimal designs.

Extensive work in shield design optimization has been carried out using GAs (Cai, Hu, Pan, Hu, & Zhang, 2018; Cai, Hu, Pan, Sun, & Yan, 2018) while multi-grid algorithms with better computational efficiency have been used by Asbury (Asbury, Holloway, Fleming, Gallimore, & Martin, 2012).

GA combined with diffusion theory has been used for shielding design optimization (Abdul Rahman, Lee, & Franceschini, 2018; Israeli & Gilad, 2018).

Simulation with the Monte Carlo N-Particle code has also been carried out in the area of detector shielding design, Boron Neutron Capture Therapy and neutron source imaging (Cai, Hu, Pan, Hu, et al., 2018; Hegazy, Skoy, & Hossny, 2018; Li et al., 2016; Williamson, 2010).

In other studies (Yadollahi, Nazemi, Zolfaghari, & Ajorloo, 2016), the effect of boron in concrete has been studied to find that the optimal mixture of concrete, used for neutron shielding, has a water–cement composition of 38%, cement content of 400 kg/m^3, 50% aggregate with Colemanite and 15% silica fume–cement. Similarly, the effect of another strong absorber, gadolinium, has been studied (Park, Kim, & Yi, 2017) to find that in concrete Gd content up to 10 at.% concrete composite (10 cm × 5 cm thick) shielding efficiency of around 86%. For multi-layered iron-water shields (Fuse, Yamaji, & Miura, 1970), it was found that the optimum arrangement was a composite iron-water-iron configuration, rather than a homogeneous mixture.

9.6.5 Some other applications of optimization

Other areas in nuclear engineering where optimization is applied is in gas centrifuge configurations, in the design of criticality, in condenser control, and in the optimal placement of instrumentation and sensors in a reactor core.

In a centrifuge hall, for example, the design of an optimal cascade which gives the minimum specific cost of enriched uranium, is a problem in the domain of combinatorial optimization.

As a final remark, optimization in nuclear engineering is an area of significant research output with emphasis towards AI-based *meta*-heuristic stochastic methods and machine learning based on classification and regression.

Problems

1. A recursive algorithm for the minimum value, in this case the distance traveled to node j after n intermediate points is $V_n(j) = \min\left[c_{ij} + V_{n-1}(i)\right], i = 1, 2, 3, \cdots, N$. Evaluate $V_1(4)$ = shortest distance from node 1 to node 4 with at most 1 intermediate point, the algorithm will use $V_1(j=4) = \min\left[c_{ij} + V_0(i)\right], j = 1, 2, 3, \cdots, N$.
2. Consider the well-known Bounded Knapsack problem where the weight w_i and value v_i of N items is given and the objective is to fill a bag with the maximum value of items while not going beyond a permissible weight W. It is also allowed to fill x_i copies of each item but not exceeding c copies This optimization problem can be written as max $\sum_{i=1}^{M} v_i x_i$ subject to $\sum_M w_i x_i \leq W$ and $0 \leq \quad x_i \leq c$. Comment on the classification of this problem in terms of computational complexity as a $NP-$complete or as a $NP-$hard problem. How does it compare with the use of DP for solving a $NP-$hard problem.
3. How would the shortest route problem be solved by the GA method described in Section 9.5.1? Would the results be the same if DP follows a backwards calculation while GA follows a forward calculation?
4. In Section 9.3 (Controller design and optimization) comment on the objective function and its relevance to tracking a given load requirement.
5. With the PSO method described in Section 9.5.2, maximize the two-variable function $f(R, H) = 2\pi R^2 + 2\pi RH$ subject to the constraint condition Eq. (9.6). The search space is $0 \leq R, H \leq 2$. Compare your results with the results of Section 9.5.1 obtained with the GA method.

Nomenclature

English lower case

b	damping coefficient
k	spring constant
l	neutron generation time
n	neutron density

English upper case

C	precursor concentration
C_c	critical damping coefficient
K_D	derivative gain
K_I	integral gain
K_P	proportional gain

Greek lower case

β	delay fraction
ε	error
ζ	damping ratio
η	overshoot
λ	decay constant
ϕ	neutron flux

ρ reactivity
τ_r rise time
τ_s settling time
ω frequency

Greek upper case

$\mathscr{F}$ objective function
Σ_f (macroscopic) fission cross section
H Hamiltonian
J integral performance index
$\mathscr{L}$ Lagrangian

Letter-like symbols

$\mathscr{E}$ integral performance index
$\mathscr{H}$ Hessian matrix

Abbreviations and acronyms

CDC Control Data Corporation
IBM International Business Machines
LQR linear quadratic control
PID proportional integral derivative

References

Abdul Rahman, F. B., Lee, J. C., & Franceschini, F. (2018). A deterministic method for PWR loading pattern optimization. *Annals of Nuclear Energy, 117*, 131–144.

Ahmad, S., Chang, B., Li, B., Yang, Q., & Liu, C. (2021). Mass optimization of the radiation shadow shield for space nuclear power system. *Progress in Nuclear Energy.*

Asbury, S. T., Holloway, J. P., Fleming, R. F., Gallimore, A. D., & Martin, W. R. (2012). Multi-grid genetic algorithms for optimal radiation shield design. PhD Thesis, Nuclear Engineering and Radiation Sciences, University of Michigan.

Bae, J. W., Betzler, B. R., Chandler, D., & Ilas, G. (2021). Automated fuel design optimization for high flux isotope reactor low enriched uranium core design. *EPJ Web Conference.*

Bellman, R. (1952). On the theory of dynamic programming. *Proceedings of the National Academy of Sciences of the United States of America.*

Bertsekas, D. P. (2010). (3rd ed.). *Dynamic programming and optimal control*, (Vol. II). Athena Scientific.

Betzler, B. R., Chandler, D., Cook, D. H., Davidson, E. E., & Ilas, G. (2019). Design optimization methods for high-performance research reactor core design. *Nuclear Engineering and Design.*

Burns, R. S. (2001). *Advanced control engineering*. Butterworth-Heinemann.

Cai, Y., Hu, H., Pan, Z., Sun, W., & Yan, M. (2018). Metaheuristic optimization in shielding design for neutrons and gamma rays reducing dose equivalent as much as possible. *Annals of Nuclear Energy.*

Cai, Y., Hu, H., Lu, S., & Jia, Q. (2018). Optimization of radiation shielding material aiming at compactness, lightweight, and low activation for a vehicle-mounted accelerator-driven D-T neutron source. *Applied Radiation and Isotopes: Including Data, Instrumentation and Methods for Use in Agriculture, Industry and Medicine.*

Cai, Y., Hu, H., Pan, Z., Hu, G., & Zhang, T. (2018). A method to optimize the shield compact and lightweight combining the structure with components together by genetic algorithm and MCNP code. *Applied Radiation and Isotopes: Including Data, Instrumentation and Methods for Use in Agriculture, Industry and Medicine.*

Chham, E., et al. (2021). Fuel reloads optimization for TRIGA research reactor using Genetic Algorithm coupled with neutronic and thermal-hydraulic codes. *Progress in Nuclear Energy.*

Domingos, R. P., Schirru, R., & Pereira, C. M. N. A. (2006). Particle swarm optimization in reactor core design. *Nuclear Science and Engineering: The Journal of the American Nuclear Society.*

Fuse, T., Yamaji, A., & Miura, T. (1970). The optimum arrangement of laminated iron-water shields. *Nuclear Engineering and Design.*

Haupt, R. L., & Haupt, S. E. (2004). *Practical genetic algorithms* (2nd ed). John Wiley & Sons.

Hegazy, A. H., Skoy, V. R., & Hossny, K. (2018). Optimization of shielding- collimator parameters for ING-27 neutron generator using MCNP5. *EPJ Web Conference, 177*, 02003.

Holland, J. H. (1975). *Adaptation in natural and artificial systems: An introductory analysis with applications to biology, control, and artificial intelligence*, University of Michigan Press.

Israeli, E., & Gilad, E. (2018). Novel genetic algorithm for loading pattern optimization based on core physics heuristics. *Annals of Nuclear Energy.*

Kennedy, J., & Eberhart, R. (1995). *Particle swarm optimization. IEEE International Conference on Neural Networks - Conference Proceedings.* IEEE.

Kumar, A., & Tsvetkov, P. V. (2015). A new approach to nuclear reactor design optimization using genetic algorithms and regression analysis. *Annals of Nuclear Energy*.

Lee, K. Y., & El-Sharkawi, M. A. (2007). *Modern heuristic optimization techniques: Theory and applications to power systems*. John Wiley & Sons.

Lewis, F. L., Vrabie, D. L., & Syrmos, V. L. (2012). *Optimal control* (3rd ed.). John Wiley & Sons.

Li, H., Zhang, G., You, Z., Li, H., Zhang, G., & You, Z. (2016). Optimization design and simulation of a multi-source energy harvester based on solar and radioisotope energy sources. *Micromachines*, *7*(12), 228.

Mousakazemi, S. M. H. (2021). Control of a pressurized light-water nuclear reactor two-point kinetics model with the performance index-oriented PSO. *Nuclear Engineering and Technology*.

Mousakazemi, S. M. H., Ayoobian, N., & Ansarifar, G. R. (2018). Control of the pressurized water nuclear reactors power using optimized proportional–integral–derivative controller with particle swarm optimization algorithm. *Nuclear Engineering and Technology*.

Nissan, E. (2019). An overview of AI methods for in-core fuel management: Tools for the automatic design of nuclear reactor core configurations for fuel reload, (re)arranging new and partly spent fuel. *Designs*.

Pardalos, P., & Resende, M. (2002). *Handbook of applied optimization*. Oxford University Press.

Park, J.-S., Kim, J., & Yi, S.-H. (2017). Effects of Gadolinium in Fe based amorphous ribbons with high boron contents on the neutron shielding efficiency. *Annals of Nuclear Energy*, *109*, 365–369.

Rao, S. S. (2009). *Engineering optimization: Theory and practice* (4th ed.). Wiley.

Vajpayee, V., Becerra, V., Bausch, N., Deng, J., Shimjith, S. R., & Arul, A. J. (2021a). LQGI/LTR based robust control technique for a pressurized water nuclear power plant. *Annals of Nuclear Energy*.

Vajpayee, V., Becerra, V., Bausch, N., Deng, J., Shimjith, S. R., & Arul, A. J. (2021b). Robust-optimal integrated control design technique for a pressurized water-type nuclear power plant. *Progress in Nuclear Energy*.

Wan, J., & Zhao, F. (2017). Optimization of AP1000 power control system setpoints using genetic algorithm. *Progress in Nuclear Energy*.

Wang, X. (2007). *Solving optimal control problems with Matlab - Indirect methods*. Raleigh, NC: North Carolina State University.

Williamson, M. R. (2010). Multivariate optimization of neutron detectors through modeling. *Philosophy (London, England)*.

Yadollahi, A., Nazemi, E., Zolfaghari, A., & Ajorloo, A. M. (2016). Optimization of thermal neutron shield concrete mixture using artificial neural network. *Nuclear Engineering and Design*.

Chapter 10

Monte Carlo simulation in nuclear systems

10.1 Introduction

Criticality is central to all nuclear designs from the first critical assemblies, Godiva and Jezebel, and the first reactor core calculations for nuclear power reactors (Section 3.1). Criticality calculations are relevant to the nuclear industry where fissile material is transported, processed, and stored in the form of gases, liquids and solids, and for the design of nuclear reactors and systems to determine the materials required, their amounts, and configurations.

This chapter takes the reader through representative and illustrative situations in the following areas:

1. criticality of bare fissile assemblies (Section 10.2),
2. criticality safety in the front-end of the fuel cycle (Section 10.3),
3. radiation moderation and shielding (Section 10.4),
4. fission neutronics (Section 10.5), and
5. fusion neutronics (Section 10.6).

A detailed simulation of the legacy Godiva bare assembly is demonstrated to give physical insight into its neutronics; the energy- and space dependence of the flux and associated reaction rates.

In the front-end of the nuclear fuel cycle, the ore is processed using uranyl nitrate solution $UO_2(NO_3)_2$ to produce hexafluoride (UF_6). In the back-end of the fuel cycle, decladded spent fuel is stored and processed in which it is dissolved in nitric acid to form uranyl nitrate used for reprocessing. MC simulations are carried out to determine the safety of storage configurations,

Nuclear thermal power reactors (Section 3.1) require uranium enriched to typically 2–5 weight % while fast reactors (Section 3.2) require higher enrichments and naval propulsion reactors could require highly enriched uranium (HEU) ($\sim$90% enriched) just as in weapons grade uranium (WPG). Uranium enrichment is therefore a central part of the nuclear fuel cycle for all applications of nuclear technology. From several demonstrated enrichment technologies, commercial processes use gaseous diffusion (mechanical flow through a porous membrane) and gas centrifuge (gas separation by centripetal force in a rotating cylinder). In an enrichment plant, uranium hexafluoride (UF_6) is transported and stored at various stages in its processing from gas with composition varying from that of natural uranium composition to possible HEU. In the *front-end* of the nuclear fuel cycle, uranium ore mining and milling takes place, followed by conversion from uranium oxide in the form of U_3O_8, commonly known as yellow cake to UF_6 in a form (solid, liquid, or gas) depending on the process conditions, followed by enrichment and fabrication.

UF_6 gas is used in an enrichment plant to separate $U^{235}F_6$ from $U^{238}F_6$ by a process such as gas diffusion or a gas centrifuge. In gaseous diffusion, the gas is heated and passed through a porous membrane in which $U^{235}F_6$ passes faster than $U^{238}F_6$ due to their weight difference. Similarly in a fast-spinning centrifuge, typically 1150 Hz, the heavier gas flows to the sides while the lighter gas stays closer. Each stage thus gives a small separation which, accumulated over a cascade with thousands of stages, results in significant enrichment. At room temperature, UF_6 is in the form of white crystalline material (density 5.09 g/cm^3); its triple point is at a pressure of about 1.5 times atmospheric pressure, that is, 22 psia and at 64.05°C. Thus at room temperature with normal atmospheric pressure, UF_6 is in solid phase and sublimes to the gas phase (directly). Its liquid phase occurs above the triple point so that in an enrichment plant, just the solid and gas phases are typically observed.

UF_6 is transported to a cascade facility in large steel containers with about 12 tons of material stored as liquid but cooled to solid form during transport. At the cascade facility, it is heated to the liquid phase and transferred through pipes to the enrichment site. Since UF_6 is corrosive, it is generally transported in nickel or stainless steel pipes. In cylinders during storage, it forms a layer which inhibits corrosion. In enriched form, the UF_6 is shipped out of the cascade

Nuclear Engineering. DOI: https://doi.org/10.1016/B978-0-323-90618-0.00010-7

facility in cylinders of varying capacity depending on the amount of enrichment. For a reactor fuel grade gas, the cylinders contain typically about two and half tons material.

Depleted UF_6 (waste stream from an enrichment process) is stored in large cylindrical tanks containing typically up to 14 tons of material, filled as a liquid which cools and solidifies, occupying about 65% of the cylinder volume. At the bottom of the cylinder, there is solid crystals and above it is the gas. When small cylinders are moved, the noise made by the crystals can be heard just like that of table salt.

During its transportation and storage, the issues of criticality safety and impact to environment are of vital importance.

Very large amounts of UF_6 are transported and stored in enrichment plants. For example, the amount of UF_6 required to produce 1 kg product gas to 4.4% U^{235} with feed assay 0.711% (natural uranium) and tails assay 0.23% is about 8.7 kg U as UF_6 requiring about 6.7 SWU. Thus a 100-ton production would require about 870 tons of reactor-grade enriched material to be transported and at 6700 SWU/t, a total of 670,000 SWU. Compare this with the total world enrichment capacity of 66,700,000 SWU/y which is sufficient for the fuel of 100 such reactors every year. To produce 1 kg 90% U (WGU) in UF_6 would require about 187 kg UF_6 at 0.711% feed and 0.23% tails, with about 215 SWU. Thus 20 kg WGU/y would require 4300 SWU/y; thus is each centrifuge gives 5 SWU, then about 1000 centrifuges are required, and the material flow for 20 kg HEU/y is about 3.7 t/y. The largest enrichment facilities in the world are Rosatom (27,654 SWU/y), Urenco (18,320 SWU/y), Orano (7500 SWU/y) and CNNC (6,750 SWU/y). All these facilities require criticality analyses.

The quantities of interest are the k_{eff} of a single cylinder or hundreds or thousands of cylinders in their storage facility, and the radiation dose. The system multiplication is a criticality calculation carried out in much the same way as for fissile materials; standard analytical and computational tools are based on the diffusion, transport and Monte Carlo methods.

MC simulations are carried out for the design of detection systems, for example, in Prompt Neutron Activation Analysis (PNAA) or in Thermal Neutron Activation Analysis (TNAA). A PNAA detection system is described in Section 10.4 for determining the best moderator resulting in an *optimal* energy spectrum and a *best* physical configuration to get a good signal and to protect the detector from radiation.

In radiation shielding, MC simulation is used for the design of shields from α, β, γ and neutron radiation. Alpha emitters such as U-238 and Pu-239 may contain trace impurities which could possibly emit other radiations. However, the range of α radiation is small and shielding is possible with a thin sheet of paper. Similarly the penetrating power of β radiation is not strong, typically just a few millimeters in tissue and hence they can be shielded with a few mm thickness of a low-Z material such as plastic. Gamma radiation pose a serious radiation risk due to their penetration power. Typical radiation shield materials in the nuclear industry are water, concrete, and iron. The half value layer (HVL), the thickness of a material that reduces the intensity of a radiation by half, is typically in the range 0.5–2 cm for a lead shield for 0.5–2 MeV gamma radiation. The next best shield is iron, followed by concrete and water. As described in Section 1.3, gamma radiation follows an exponential attenuation in a shield and a $1/R^2$ intensity reduction at a distance R. However, simple theoretical models use a buildup factor to account for multiple scattering which is best estimated by Monte Carlo simulation. Neutron radiation also poses a serious problem in the design of shields due to the penetration power of neutrons and the several possible reactions. A typical strategy used for the design of neutron shields is to first slow the fast neutrons with a hydrogenous material and then absorb the neutrons. This can produce gamma rays which then need to be shielded by lead, iron or concrete as discussed. In nuclear reactors, openings are kept for piping, ducts and instrumentation; this provides channels for radiation steaming which is a challenge for MC simulation as discussed in Section 7.5 (variance reduction).

In fission neutronics, MC simulations are very effective for core neutronics starting from the basic unit of a lattice cell to a fuel assembly followed by a full-core analysis.

In this chapter, the three core designs considered are

1. Westinghouse AP-1000
2. Toshiba Gen-IV 4S, and the
3. Micronuclear heat pipe reactor (MNR).

Section 10.4 describes unit lattice cell MC simulations, extended to infinite assembly simulations and whole-core simulations for these designs.

Finally, in Section 10.6, fusion neutronics are described with particular reference to the ITER design for which MC simulations are useful in determining several engineering design parameters, for example, the First Wall (FW) loading, the tritium breeding rate (TBR), and the displacements per atom (dpa) in the superconducting coils.

10.2 Bare critical assemblies

10.2.1 Godiva

A detailed Monte Carlo simulation for Godiva, as a follow-up to Section 2.10 and Section 6.3, is of great educational value as it gives physical insight into the flux and associated reaction rates. A problem generally associated with large simulations is that one can easily lose track of the physical significance of voluminous results.

The physical model of Godiva is a bare sphere of radius 8.741 cm with a composition of 1.02 wt.% U-234, 93.7 wt.% U-235, and 5.27 wt.%U-238, density 18.74 g/cm^3, mass 52.4254 kg, atomic density = 4.79838×10^{22}/cm^3. The total volume (2.7951×10^3 cm^3) is divided into 50 zones of equal volume.

The input file BK10Gdva is listed in Annex A. The weight fractions and atomic fractions of uranium nuclides are given in Table 10.1.

The nuclear data used is from endf66c, 92234.66c for uranium nuclides at temperature 293.6K. Simulation parameters are given in Table 10.2.

With 3000 particles per history, a total of 150 cycles out of which 60 cycles are skipped, and an initial guess $k_{\text{eff}} = 1$ for the purpose of estimating k_{eff} and the prompt removal lifetime τ_p. As described in Section 7.2, M is the number of source points generated in each simulation cycle, shown in Fig. 10.1.

Each cycle gives an estimate of the prompt removal lifetime $\tau_p^{(a)}$ (abs) and the source points generated: $\tau_p^{(a)} = 0.967664$ shakes (1 shake = 10^{-8} s), $\tau_p^{(c)} = 0.969156$ shakes, and $\boldsymbol{k}_{\text{eff}}$ shown in Table 10.3.

TABLE 10.1 Material composition of Godiva.

Composition/nuclide	U-234	U-235	U-238
Weight fraction	1.02010×10^{-2}	9.37094×10^{-1}	5.27053×10^{-2}
Atomic fraction	1.02511×10^{-2}	9.37677×10^{-1}	5.20719×10^{-2}

TABLE 10.2 MC simulation parameters.

M	N	K	K_s	k_{eff}
Varies (Fig. 10.1)	3000	150	60	1.0

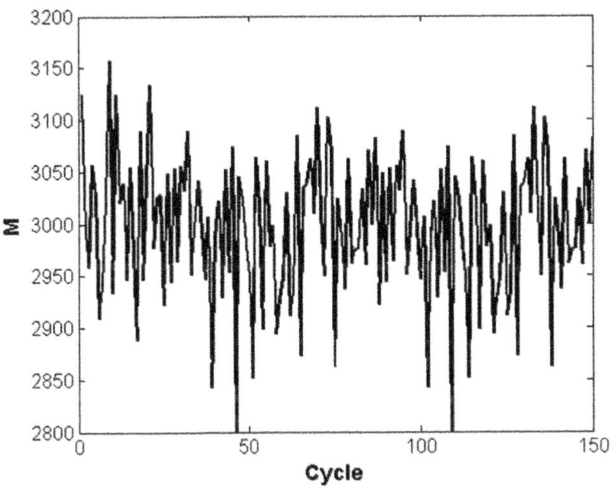

FIGURE 10.1 Fission points simulated in each cycle.

TABLE 10.3 Estimates of k_{eff} with collision-, absorption and track length-estimators.

Estimate	Average of 90 cycles
$k_{eff}^{(c)}$	0.994040 (0.0014)
$k_{eff}^{(a)}$	0.993941 (0.0014)
$k_{eff}^{(T)}$	0.994835 (0.0012)
$< k_{eff}^{(CAT)} >$	0.994272 (0.0012)

$k_{eff}^{(c)}$ = system multiplication (collision estimate), $k_{eff}^{(a)}$ = system multiplication (absorption estimate), $k_{eff}^{(T)}$ = system multiplication (track length estimate), and $< k_{eff}^{(CAT)} >$ = average system multiplication (collision, absorption, track length estimate).

The above results compare with $k_{eff} = 0.9976$ (0.0011) by Whalen et al. (Whalen, Cardon, Uhle, & Hendricks, 1991). Some important observations from the simulation results are:

1. The average neutron energy is 2.0621 MeV
2. Particles created per particle due to weight cutoff was 0.032885 at an energy 1.1819×10^{-2} MeV
3. The number of neutron collisions per source particle is 4.063
4. The random numbers generated are 43,579,980 (~44 million)
5. The source tracks were 269,862 each with a weight 1
6. Total source tracks 272,187 with weight 1.0383 energy 2.0776 MeV
7. Source multiplication (*n*, xn) was 2325 tracks weight per source particle 5.4134×10^{-3} and energy 3.7340×10^{-3} MeV.

The number of collisions in each zone is fairly large, varying from 8369 in the outermost cell to 43,082 in the innermost cell. About 93.6% of the collisions are with the U-235 nuclide; this is also the atomic abundance of U-235. Since the number of collisions is high in every cell, there is no need for variance reduction except for implementing a low weight cutoff to avoid "wasting" time on low weight particles.

The average track mean free path varies between 2.6251 cm in the innermost cell increasing to 2.6676 cm in the outermost cell. These numbers tell us that this is a simulation where we can expect good results due to the large number of nonzero histories.

To get an idea of the gains and losses in the system, Table 10.4 gives two contributions, namely the source itself and the small contribution from weight cutoffs. The total weight of 1.0383 must be balanced in the *losses* table.

Table 10.5 illustrates where the losses are taking place: the escapes, the weight cutoffs, captures, multiplication and fission treated as a loss mechanism.

The convergence of k_{eff} is seen in Fig. 10.2; for 150 cycles, we can see that it is a good idea to skip a few cycles at the beginning of a simulation.

After the first ~ ten cycles, Fig. 10.2 gives a satisfactory result. Of course, the number of skip cycles is more or less a guess given at the beginning of a simulation.

Deciding which estimator to rely on for the k_{eff} estimate, Fig. 10.3 shows that the absorption and collision estimators underestimate the tally; the collision estimator is above the absorption estimate and hence closer to the track length estimate. It would be fair to conclude that since this is not a strongly absorbing medium, the TLE and collision estimators would be expected to give more reliable results.

In the tallies, the standard and nonstandard ENDF/B reaction numbers for the reactions tallied in the simulation are listed below

standard ENDF/B reaction numbers

1 total cross-section
2 elastic scattering cross-section
16 n,2n reaction
17 n,3n
18 n,f
102 (n,γ)
103 (n,p)
104 (n,d)

TABLE 10.4 Gains per source particle.

Mechanism	Tracks	Weight	Energy (MeV)
source	269862	1.00000	2.06210
weight cutoff	–	0.032885	0.011819
Total	272187	1.03830	2.07760

TABLE 10.5 Losses per source particle.

Mechanism	Tracks	Weight	Energy (MeV)
Escape	231,765	0.574920	0.925520
Weight cutoff	39,260	0.033177	0.011845
Capture	0	0.044765	0.027477
Down scattering	0	0	0.525060
(*n*,xn)	1162	0.0027056	0.021057
(*n*,*f*)	0	0.382730	0.566660
Total	272,187	1.038300	2.077600

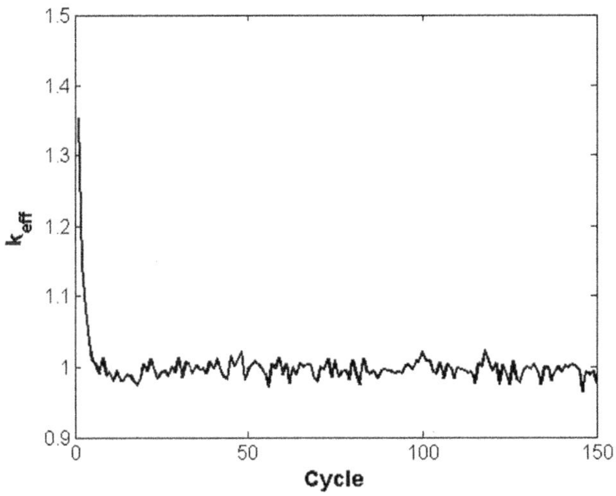

FIGURE 10.2 System multiplication k_{eff} (TLE).

105 (n,t)
106 (n,He3)
107 (n,α)

nonstandard ENDF/B reaction numbers

−2 absorption cross-section
−6 total fission cross-section
−7 fission nu

The tallies specified in the input file are the neutron current (F1), the neutron flux (F4), the energy deposition (F6), the fission energy deposition (F7) and the reaction rates (FMn); the tally multipliers are the product of the atomic number density and the volume. The units of the tally multipliers are thus *reactions/s*.

The simulation gives the number of neutrons escaping as (F1:N =) 5.75122 $\times$ 10^{-1} 0.0011, energy deposition (F6:N =) 1.23121 $\times$ 10^{-3} 0.0011 MeV/g and fission energy deposition (F7:N =) 1.32154 $\times$ 10^{-3} 0.0011 MeV/g (includes gamma ray heating; it is assumed that fission photons deposit energy locally).

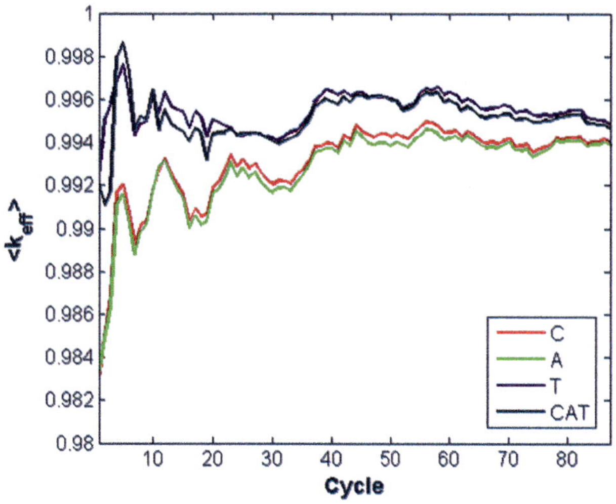

FIGURE 10.3 System multiplication (average) k_{eff} after 60 skip cycles.

TABLE 10.6 Tallies for Godiva.

Group/Tally	Thermal	Intermediate	Fast	Total
	$E < 0.625$ eV	0.625eV – 100keV	$E > 100$keV	
Total reactions	0	1.54931×10^{-1} 0.0094	2.49234 0.0012	2.64727 0.0012
Elastic scattering	0	0.126275 0.0094	1.53455 0.0014	1.66082 0.0014
Absorption(−2)	0	6.50478×10^{-3} 0.0096	3.82374×10^{-2} 0.0019	4.47422×10^{-2} 0.0021
Radiative capture (102)	0	6.50478×10^{-3} 0.0096	3.82374×10^{-2} 0.0019	4.47422×10^{-2} 0.0021
Fission reactions (−6)	0	2.04106×10^{-2} 0.0095	3.62598×10^{-1} 0.0012	3.83008×10^{-1} 0.0011
Fission neutrons (−6) (−7)	0	4.9529×10^{-2} 0.0095	9.45185×10^{-1} 0.0012	9.94714×10^{-1} 0.0011
(n,2n) + (n,3n)	0	0	2.59771×10^{-3} 0.0139	2.59771×10^{-3} 0.0139

The tallies in Table 10.6, for the thermal, intermediate, and fast neutron energy groups clearly show that most of the activity is in the fast group.

The flux falls off with radius as shown in Fig. 10.4, as a classic one-group flux.

The center of the sphere has the highest flux as shown in Fig. 10.5.

The same picture of flux is shown in Fig. 10.6 with the central values about five times higher than the peripheral values.

The value of the flux, normalized to the power P, is

$$\phi = \frac{1}{3.1724 \times 10^{-11}} \frac{\overline{\nu} P \phi_{F4}}{k_{eff}}$$

where the energy from fission is 3.1724×10^{-11}J and ϕ_{F4} is the F4 scalar flux tally of MCNP.

The energy-dependent flux, shown in Fig. 10.7, exhibits high values at near source energies and is therefore a *hard* spectrum. There is no Maxwellian at lower energies due to the absence of any hydrogenous medium.

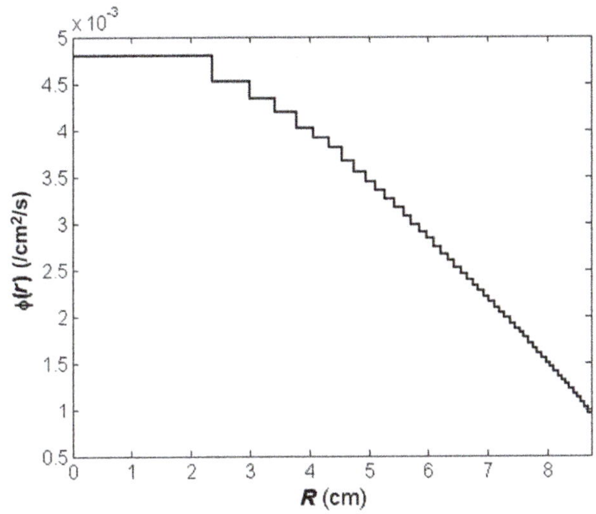

FIGURE 10.4 Neutron flux $\phi(r)$ versus radius r.

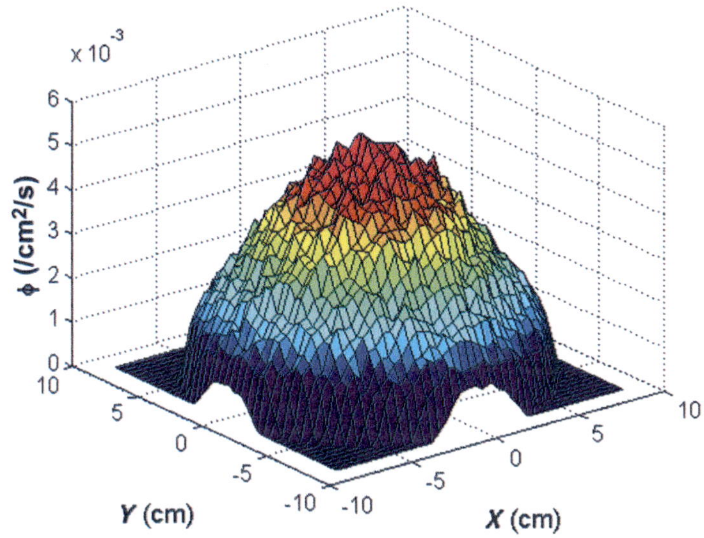

FIGURE 10.5 Neutron flux $\phi(r)$ versus $x$and y.

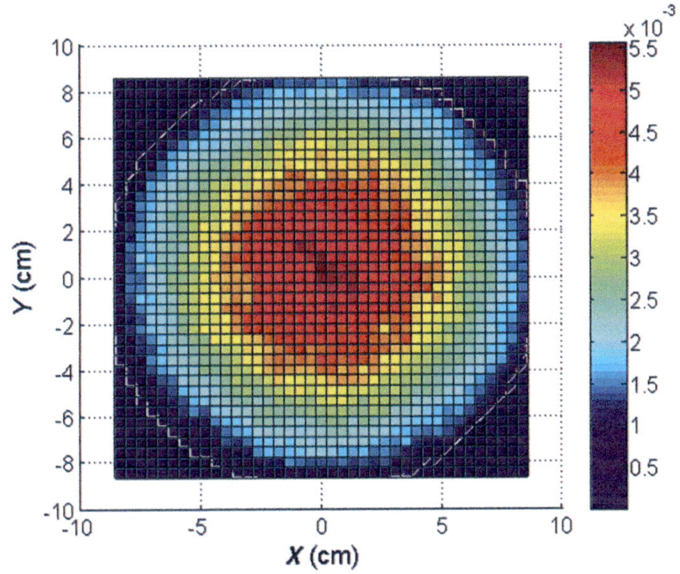

FIGURE 10.6 Neutron flux (r).

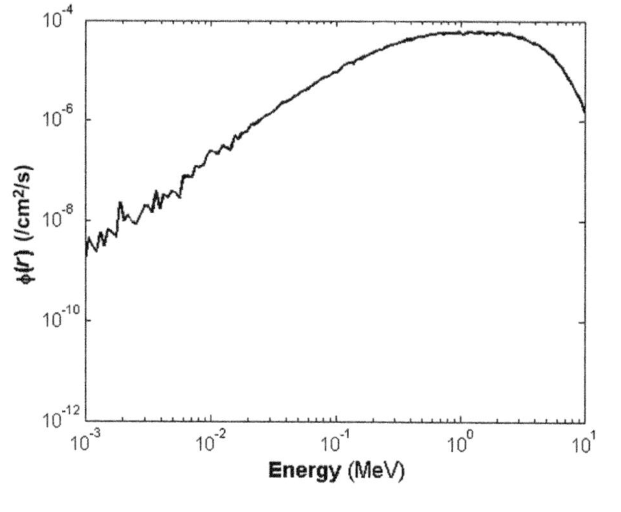

FIGURE 10.7 Neutron flux $\phi(E)$ versus energy E.

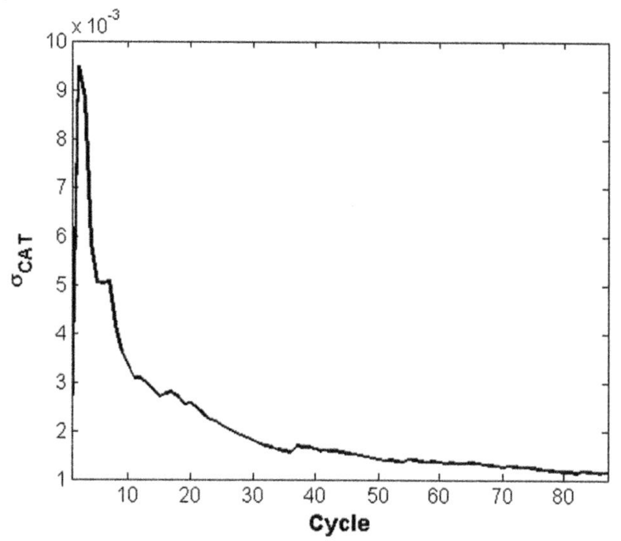

FIGURE 10.8 Standard deviation versus cycle number.

The quality of the simulation results is illustrated in Fig. 10.8; the standard deviation of the collision-absorption-TLE averaged k_{eff} decreases by a factor ~ 10 by the 90th active cycle.

In this simulation, the relative error was found to decrease as $1/\sqrt{N}$, the variance of the variance was found to decrease as $1/N$, and the FOM is constant as seen in Fig. 10.9.

The Godiva simulation of this section is meant to understand the qualitative aspect of the results as much as the quantitative aspects. In any nuclear system, it is most important to have an idea of the overall space- and energy-dependence of the flux. In a fast reactor, for example, the spectrum is similar to the Godiva spectrum while in a water reactor, two Maxwellian peaks will be found; one for the source neutrons and the other for thermal neutrons. There relative magnitudes will indicate the reaction rates when the cross-section behavior of the predominant nuclides is also understood.

10.2.2 Jezebel

The input file for the Jezebel (Chapter 2, Section 6.3) assembly (Rowlands et al., 1999) is listed in Annex B. The bare assembly is modeled as a plutonium sphere (95.5 wt.% Pu-239 and 4.5 wt.% Pu-240) with a radius = 6.385 cm, density = 15.61 g/cm^3 and atomic fractions shown in Table 10.7.

Table 10.8 shows the tallies for Jezebel for the same MC simulation parameters and energy bins as for Godiva indicating the same overall behavior, that is, hard spectrum and predominance of the fast neutrons.

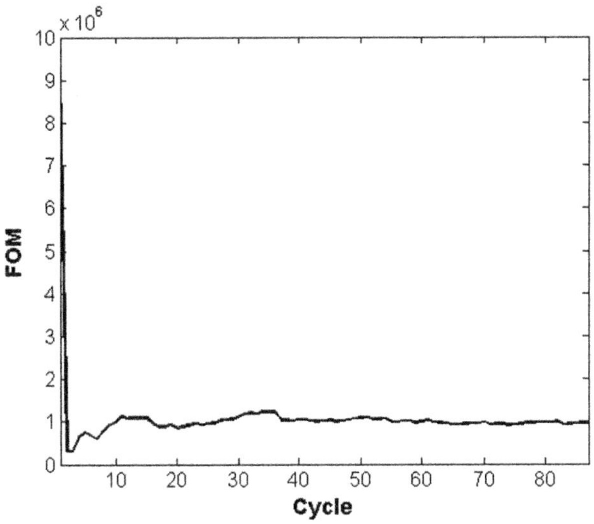

FIGURE 10.9 Figure of merit versus cycle number.

TABLE 10.7 Material composition of Jezebel.

Composition/nuclide	Pu-239	Pu-240
Weight fraction	0.955	0.045
Atomic fraction	0.955179	0.0448206

TABLE 10.8 Tallies for Jezebel.

Group/Tally	Thermal	Intermediate	Fast	Total
	$E < 0.625$ eV	0.625eV – 100keV	$E >$ 100keV	
Total reactions	0	5.90774×10^{-2} 0.0143	1.41514 0.0012	1.47422 0.0013
Elastic scattering	0	4.83311×10^{-2} 0.0143	8.22303×10^{-1} 0.0014	8.70634×10^{-1} 0.0014
Absorption (−2)	0	1.76497×10^{-3} 0.0165	8.76700×10^{-3} 0.0032	1.05320×10^{-2} 0.0037
Radiative capture (102)	0	1.76497×10^{-3} 0.0165	8.76700×10^{-3} 0.0012	1.05320×10^{-2} 0.0037
Fission reactions (−6)	0	6.78738×10^{-3} 0.0143	3.11815×10^{-1} 0.0011	3.18602×10^{-1} 0.0011
Fission neutrons (−6) (−7)	0	1.96806×10^{-2} 0.0143	9.86400×10^{-1} 0.0012	1.00608 0.0011
(n,2n) + (n,3n)	0	0	1.12502×10^{-3} 0.0140	1.12502×10^{-3} 0.0140

The tallies are: neutrons escaping from sphere (F1:N =) 6.73781×10^{-1} 0.0008, energy deposition (F6:N =) 3.26504×10^{-3} 0.0011 MeV/g, fission energy deposition (F7:N =) 3.54458×10^{-3} 0.0011 MeV/g (includes gamma ray heating; it is assumed that fission photons deposit energy locally).

The results of the simulation are: k_{eff} = 1.006261 0.0012 (combined average col/abs/tle), removal lifetime = 0.37120 0.0023 (combined average col/abs/tle).

10.3 Criticality safety

10.3.1 Storage of interacting units

Typical benchmarks for critical assemblies (Whalen et al., 1991) and their storage as interacting units are as follows:

1. Enriched uranium metal (93.2% enriched) cylinders $2 \times 2 \times 2$ unreflected array critical experiment (Wagner, Sisolak, & McKinney, 1992) with cylinders 10.765 cm in height and 11.496 cm in diameter. These cylinders are contained in cuboids referred to as a 2C8 unit. The spacing between the units is 2.244 cm in the x and y directions and 2.245 cm in the z direction. The TLE k_{eff} of this system was found to converge to 1.0000 with MCNP version 4.2 with 3000 particles per cycle in 200 cycles and 20 skip cycles.
2. In the second experiment, the 2C8 unit with increased separation of approximately 11.98 cm, is reflected on six sides of the cube by 15.24 cm of paraffin and gives a MC simulation result converged to 1.0000.
3. SP12: A composite array of four HEU (93.2%) cylinders and four cylindrical plexiglass containers filled with HEU (92.6%) uranyl nitrate solution $UO_2(NO_3)_2$ are critical.
4. SP16: an infinite number of slabs of uranyl fluoride solution UO_2F_2 contained in Pyrex glass and separated by borated UO_2F_2.

10.3.2 Storage of uranium hexafluoride cylinders

All uranium hexafluoride operations in nuclear facilities are carried out within the regulations of the International Atomic Energy Agency (IAEA), International Standards Organization (ISO), and United States Nuclear Regulatory Commission (USNRC). In the United States, operations are conducted under the Nuclear Regulatory Commission (Code of Federal Regulations, Section 10), American National Standards Institute (ANSI), Department of Transportation and the Department of Energy.

During the handling, transportation and storage of fissile material, subcriticality within margins of safety is maintained. These safety limits are derived on the basis of one of two types of criteria: criteria based on the value of k_{eff} for the system under analysis and criteria based on the critical value of one or more control parameters, such as mass, volume, concentration, geometry, moderation, reflection, interaction, isotopic composition and density, and with account taken of neutron production, leakage, scattering and absorption (IAEA, 2014).

Cylinder sizes vary (USEC (United States Enrichment Corporation), 1995) from the 1S Model with a diameter of 1.5 inches (minimum volume 0.15 L) for 100 w/o U-235 enrichment and a maximum shipping limit of 0.45 kg, to the 48 H Model with a diameter of 48 inches and minimum volume 3964 L for maximum 1 w/o enriched U-235 in UF_6 and maximum shipping limit 12,261 kg.

For radiation dose calculations, the *source term* is calculated from the modeling and simulation of transmutation of nuclides for both neutron and gamma sources. The direct neutron source depends on the enrichment and is assumed independent of the decay chain while the photon source depends on the decay (Su, 2015); the UF_6 neutron source term for a 48Y cylinder (12501 kg UF_6) with 0.711 weight percent U-235 is 4.978×10^5 n/s with $\sim$62% in the range 0.4 MeV- 1.42 MeV. The neutron source term decreases with decreasing tails U-235 enrichment. The photon source term is due to Th-231, Th-234, Pa-234, and Pa-234m in addition to the uranium nuclides. It is a factor $\sim 10^5$ times greater than the neutron source but the spectrum is towards lower energies $\sim$10 keV − 0.2 MeV increasing steadily over a two-year time period. Dose calculations were carried out using MCNP5 and ENDF/B-VII.0 with ANSI/ANS-6.1.1−1977 flux-to-dose conversion factors.

For a filled cylinder, with 0.711 wt. % U-235 the dose rate from both neutrons and photons is 5.317 mrem/h inside ($\sim$94% photon dose) while outside the cylinder it is 1.739 mrem/h ($\sim$88% photons). With distance, the dose rate falls off by about four orders of magnitude at 100 m and another two orders by the next 100 m. At the boundaries of such storage sites, the dose rates, are below the limits prescribed in 10 CFR Section 20.1301 (Dose Limits for individual members of the public). When such filled feed (natural uranium in UF^6) cylinders are stacked one or two layers up and thousands of cylinders are placed on the ground in a storage yard, their combined effect, calculated from MC simulations, is summarized as follows:

In a 10×10 array, the single, double and triple stack MC neutron (and photon) dose rates are 0.165, 0.183 and 0.191 (and 1.13, 1.19 and 1.21) mrem/h falling off in the radial direction to 2.58×10^{-3}, 4.01×10^{-3} and

TABLE 10.9 Atom density of UF_6 (atoms/b-cm).

Density (g/cc)	U-235	U-238	F	H
5.1	4.4168E-4	8.2860E-3	5.31340E-2	7.6800E-4

4.96×10^{-3} mrem/h (and 8.19×10^{-3}, 1.39×10^{-2} and 1.85×10^{-2}) mrem/h by 50 m and to 1.76×10^{-5}, 2.17×10^{-5} and 2.70×10^{-5} (and 3.70×10^{-5}, 5.14×10^{-5} and 7.54×10^{-5}) mrem/h by 395 m. These MC estimates, with no cell source have relative error within 1%, 1%–4% and 5%–8% for neutrons and within 1%–2%, 1%–3%, 3%–4% for photons at 0, 50, and 395 m respectively.

In a 100×100 array, the single, double and triple stack MC neutron (and photon) dose rates are 0.209, 0.248, and 0.258 (1.16, 1.25 and 1.27) mrem/h falling off in the radial direction to 4.02×10^{-2}, 5.07×10^{-2} and 5.56×10^{-2} (6.19×10^{-2}, 7.58×10^{-2} and 8.89×10^{-2}) mrem/h by 50 m and to 8.13×10^{-4}, 9.96×10^{-4} and 1.04×10^{-3} (1.22×10^{-3}, 1.45×10^{-3} and 1.55×10^{-3}) mrem/h by 395 m.

These MC estimates, with no cell source have relative error within 1%–2%, 1%–3% and 2%–4% for neutrons and within 1%, 3%–5%, 3%–5% for photons at 0, 50, and 395 m, respectively.

The effect of ground scattering is also considered for calculating radial and axial dose rates. This reduces the dose rate.

The above results show that the photon dose forms the major component of the total dose; its relative importance decreases with distance due to increased neutron skyshine (emitted radiation reaching a facility by scattering from the atmosphere).

These calculations can be used to design storage facilities in terms of arrays and stacking for estimating the distance to the boundary for safe and compliant radiation dose. A 10×10 array size with triple stacking, for example, would require a site boundary of 80 m while a 100×100 array size with the same triple stacking would require 270 m to the boundary.

Criticality calculations carried out for a 2½ ton UF_6 steel cylinder with 5% enrichment UF_6 (material composition listed in Table 10.9) at Oak Ridge National Laboratory using the SCALE system with ENDF/B-IV cross sections (Broadhead, 1991) give a single unit $k_{eff} = 0.453 \pm 0.003$ with an effectively infinite water reflector (SG = 1).

The k_{eff} varies with the specific gravity of water; from 0.72 at zero SG increasing to a maximum of ~0.82 for SG 0.02 then falling off to ~0.445 at 0.5 SG and slightly increasing to ~0.453 at SG 1.0.

For single 10- and 14-ton UF_6 cylinders in an infinite water-reflected array (SG = 1), $k_{eff} = 0.526 \pm 0.002$ and $k_{eff} = 0.533 \pm 0.003$.

Sensitivity studies are carried out to investigate the effect of temperature due to density and resonance-capture variations and fuel location patterns. It is reported that k_{eff} is insensitive to temperature.

These studies seek optimization to get the best separation distance and to estimate the density and temperature effects.

10.4 Radiation moderation and shielding

10.4.1 Radiation moderation for a neutron generator

In Prompt Gamma Neutron Activation Analysis (PGNAA), a sample is activated with a neutron source, characteristic gamma rays are subsequently emitted, and measurement is performed simultaneously (IAEA, 2012). In some applications, such as the detection of explosives, thermal neutron activation analysis (TNAA) is used to enhance radiative capture reactions.

The placement of the sample, between the source and the detector determines the quality of the signal. MC simulations are carried out to select the best moderator and to determine the minimum size and optimal configuration.

Powerful neutron sources are used such as californium-252 with a mixed energy spectrum and D-D fusion monoenergetic source of energy 2.5 MeV emitting 10^{11} n/s. In such cases, an important consideration is the protection of a detector since efficiency is degraded due to the effect of fast neutron bombardment on scintillators and germanium detectors.

When there is no medium in between a source and detector, the dose can be calculated by the geometrical attenuation of the neutron flux, for example, a 14 MeV flux of 1.7×10^8 n/cm^2/h gives a dose of 1 Sv/h (100 rem/h); compared with the highest permissible dose of 10 μSv/h (1 mrem/h).

TABLE 10.10 Comparison of moderators for obtaining maximum flux at a detector.

Moderator	Min. R (cm)	Max. ϕ_{th} 10^8 n/cm²/s
Light water	78	8.3
Polyethylene	63	12
Heavy water	85	4.08
Graphite	130	1.8

For attenuation of fast neutrons in the range 1–2.5 MeV, MC simulations show that the best shielding is by a hydrogenous material such as polyethylene and the worst would be by a high scattering material such as graphite. Table 10.10 shows results for three moderators to produce the lowest thermal flux on a detector with minimum thickness.

MC simulations have been used for the design optimization of a neutron source generator for PNAA (Ref) by repeated simulations to show that an optimal design with lowest fast neutron and photon fluence at a detector for a 10^{11} n/s D-D generator heavy water is more effective than light water, polyethylene, beryllium, and graphite and that the optimal radius of moderator of about 86 cm with a minimum lead thickness of 25 cm. This optimal design reduces the thermal neutron flux to 2.9×10^7 n/cm²/s.

10.4.2 Radiation shielding

For neutrons, as mentioned above, the best shield will be a hydrogenous material such as water, followed by lead, then low Z and high Z (water-lead) materials. Fast neutrons slowed down by water, for example, will produce 2.2 MeV gamma rays from the radiative capture reactions in hydrogen. These energetic gammas will require shielding by a high-Z material such as iron or lead.

The HVL for 0.5, 1.0, 1.5 and 2.0 MeV photons is 0.51, 0.76, 1.27, and 1.52 cm for lead, and 3.30, 4.57, 5.84, and 6.60 cm for concrete, respectively.

A simple and practical demonstration of shielding illustrated in Fig. 10.10 shows a concrete shell, labeled cell "2," located 10 ft. (304.8 cm) from a 14 MeV point isotropic D-D source at the center of the sphere labeled "S" emitting 10^9 neutrons per second. In a neutron transport MC simulation, MCNP input file BK10Shld (Annex C), the neutron current, surface flux, and radiation dose are estimated.

Results of the simulation for 400,000 histories and a lower energy cutoff 1 MeV are listed in Table 10.11.

For a void, the surface flux 5.37428×10^{-7} n/cm²/s from simulation is equal to the exact value at the surface labeled "1" in Fig. 10.10.

$$\phi(R) = \frac{1}{4\pi R^2}$$

at $R = 304.8$ cm which gives a dose of 2.79462×10^{-16} Sv for one source neutron. The neutron flux-to-dose coefficients are used from the ICRP conversion factors (ICRP, 1996). For 10^{11} n/s, the dose is 100.6063 mrem/h. With a concrete shell, this dose is reduced to 1.2008 and 0.5117 mrem/h for shield thickness 70 and 80 cm, respectively. The relative errors of the estimates are within 2.4%.

The calculation is done as follows:

1 rem = 0.01 Sv, D-D fusion source 14 MeV 10^9 n/s

Dose = $10^9\ \text{n/s} \times 2.79462 \times 10^{-16} \frac{Sv}{n/s} \times 3600 \frac{\text{s}}{\text{h}} \times 10^5 \frac{\text{mrem}}{\text{Sv}} = 100.6063$ mrem/h.

10.5 Nuclear fission applications

In this section, MC simulations are demonstrated for unit lattice cells and reactor core models of the AP1000, Toshiba 4S and the micronuclear heat pipe cooled reactor (MNR).

10.5.1 Unit lattice cell and fuel assembly of the AP1000 reactor

The AP1000 unit lattice cell (Fig. 7.16) has fuel of three enrichments as shown in Table 10.12.

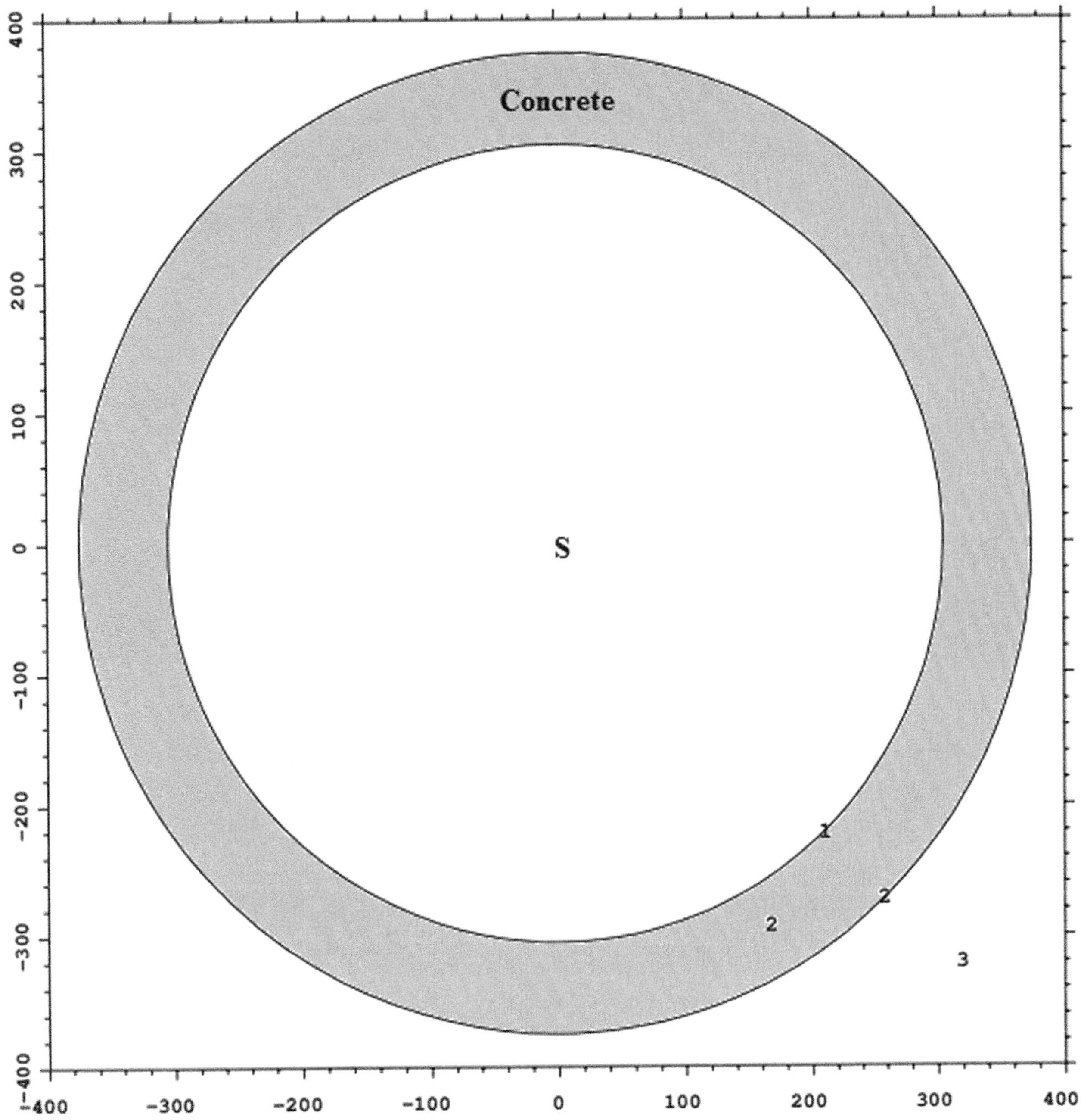

FIGURE 10.10 Concrete shell with surface and cell numbers.

The atomic densities and volume occupied by water (at 300K), fuel, helium gap and the cladding are listed in Table 10.13. The quantity *NV* is used in the MCNP tally multiplier to convert the product of microscopic cross-section and flux into a reaction rate.

In a PWR such as the AP1000, the water density at 600K decreases appreciably to ~70% of its value at 300K so that atomic densities used in a MC simulation are as given in Table 10.14.

The integral fuel burnable absorber (IFBA) consists of a mixture of zirconium and boron with atomic densities listed in Table 10.15 (Laranjo de Stefani, Losada Moreira, Maiorino, & Russo Rossi, 2019).

The MCNP input file for the unit lattice cell, BK10AP10 is listed in Annex D. The simulation is carried out for 5000 histories per cycle for 150 cycles with 5 skip cycles and an initial guess $k_{\text{eff}} = 1$. The neutron flux is fairly steady as seen in Fig. 10.11. This is due to the small dimensions of the lattice cell; otherwise the flux would have the same shape as for Godiva.

TABLE 10.11 Radiation dose across a concrete shield.

Tally	Units	Void	Concrete	
			70 cm	80 cm
Neutron current	Neutrons	1.0	9.7329×10^{-3} 0.0122	4.22892×10^{-3} 0.0180
Surface flux	n/cm²/s	5.37428×10^{-7}	8.08496×10^{-9} 0.0148	3.50463×10^{-9} 0.0244
Radiation dose	Sv	2.79462×10^{-16}	3.33562×10^{-18} 0.0146	1.42129×10^{-18} 0.0239
	mrem/h	100.6063	1.2008	0.5117

TABLE 10.12 AP1000 fuel atomic densities.

Enrichment w/o	2.35	3.4	4.45
Density (g/cm³)	10.47635	10.47635	10.47635
Mass fraction			
U-235	0.0207146	0.0299696	0.0392243
U-238	0.8607574	0.8514884	0.8422196
O-16	0.1185280	0.1185421	0.1185561
Total	1.0000000	1.0000000	1.0000000
MCNP input N			
U-235	5.56014e-04	8.04432e-04	1.05284e-03
U-238	2.28120e-02	2.25663e-02	2.23207e-02
O-16	4.67360e-02	4.67416e-02	4.67471e-02
Total	7.01040e-02	7.01123e-02	7.01206e-02

AP1000 atomic densities, are obtained from the MATLAB program % \Elsevier\Programs\Ch7_AP1000_MCNPinput.m.

TABLE 10.13 Atomic densities and volumes of cells in AP1000 unit lattice cell.

Region	Atomic density (*N*) atom/b	Volume (*V*)	*NV* (atoms cm²/b)
Water	1.00367×10^{-1}	3.51510	0.35280
Fuel	as in Table 10.12	2.10829	0.1478
Gap	1.00000×10^{-4}	8.52588×10^{-2}	8.5259e-6
Clad	4.34418×10^{-2}	6.41741×10^{-4}	0.0279

Fig. 10.12 shows the energy spectrum in the lattice cell; this is very different from the Godiva hard spectrum. In this case, thermalization occurs due to neutrons scattering with water, and two peaks of comparable magnitudes are observed. These correspond to thermal neutrons due to moderation and the fission spectrum due to source neutrons.

TABLE 10.14 AP1000 water atomic densities.

Water temp (K)	300	600
Density (g/cm^3)	1	0.7
Mass fraction		
H	0.1111111	0.1111111
O-16	0.8988889	0.8988889
Total	1.0000000	1.0000000
MCNP input N		
H	6.69111e-02	4.68378e-02
O-16	3.34556e-02	2.34189e-02
Total	1.00367e-01	7.02567e-02

TABLE 10.15 AP1000 IFBA (zirconium diboride) atomic densities. Material density 5.42 g.cm^3.

Isotope	Weight fraction	Atomic density (atoms/cm^3)
B-10	0.0187	5010
B-11	0.1713	5011
Zr-90	0.416745	40090
Zr-91	0.090882	40091
Zr-92	0.138915	40092
Zr-94	0.140778	40094
Zr-96	0.02268	40096

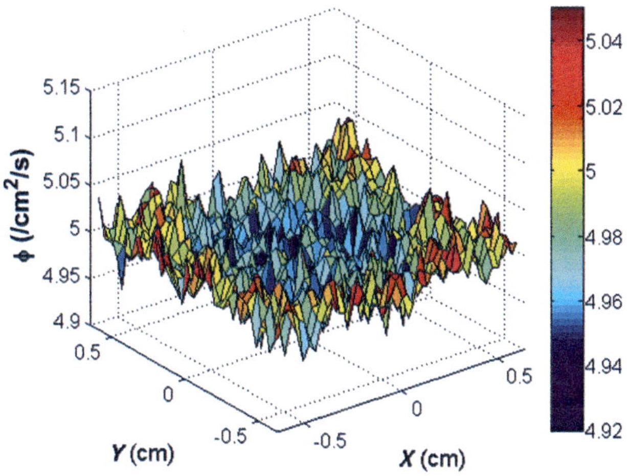

FIGURE 10.11 Neutron flux $\phi(r)$ versus x and y.

In the subsequent analysis, the following tallies listed in Table 10.16, are obtained in the fuel and water regions

1. Total reaction rate
2. Elastic scattering rate

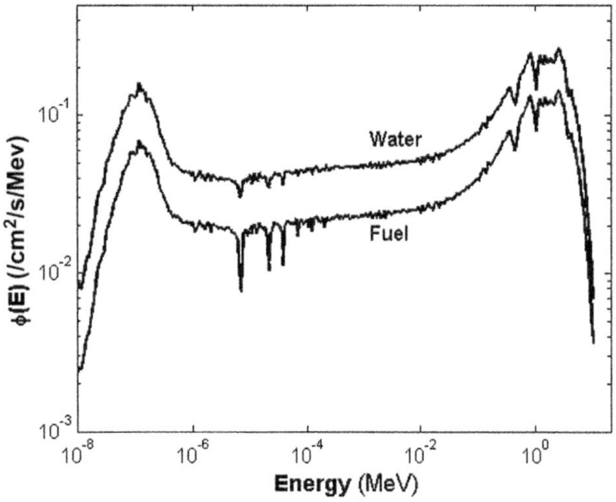

FIGURE 10.12 Neutron flux $\phi(E)$ versus E.

3. Absorption rate (in water and fuel regions)
4. Radiative capture

and the fission reaction rate and number of fission neutrons emitted in the fuel. Results are binned in three energy groups corresponding to thermal $E < 0.625$ eV, intermediate $0.625\text{eV} - 100\text{keV}$ and fast $E > 100\text{keV}$ energies.

Neutrons undergo about nine times more collisions in water than in fuel regions, with the energy distribution shown in Table 10.17. Elastic and absorption reactions are predominantly in the thermal and intermediate energy groups. In fact, radiative capture is *the* absorption mechanism in these two lower groups.

Table 10.17 shows that fissions are predominantly in the thermal group, as expected (since the fission cross-section is orders of magnitude higher at thermal energies). Subsequently, the number of fission neutrons emitted is also mainly in the thermal group. We see that the number of neutrons emitted per fission, obtained as

$$\overline{\nu} = \frac{\langle \nu \Sigma_f \phi \rangle}{\langle \Sigma_f \phi \rangle}$$

is 2.4367, 2.4342, and 2.7579 for the thermal, intermediate, and fast groups, respectively. The cell energy deposition tallies give 1.57493 (0.0012) MeV/g in water and 9.86552 (0.0011) MeV/g, in fuel. The fission energy deposition is 6.11183 (0.0011) MeV/g in the fuel region.

The unit lattice cell is part of a fuel assembly as shown in Fig. 10.13.

A detailed MCNP simulation for the AP1000 by Stefani et al. (Laranjo de Stefani et al., 2019) estimates k_∞ for the Beginning of Life (BOL) and End of Life of the core, and reactivity coefficients.

For a lattice with fuel rod including gap, Zirlo cladding and pitch, for various enrichments k_∞ is 1.21029 ± 0.00003 (1.58%), 1.32863 ± 0.00004 (2.35%), 1.40462 ± 0.00004 (3.20%), 1.41697 ± 0.00004 (3.40%) and 1.21029 ± 0.00003 (4.45%).

For the AP1000 assembly k_∞ is reported for a number of configurations; for 4.45% enriched fuel with 24 Pyrex and 72 IFBA burnable absorbers, $k_\infty = 1.26524 \pm 0.00008$.

10.5.2 The Toshiba 4S reactor

The 4S Gen-IV design (Section 3.5) is based on the idea of a small low-power high temperature metal-cooled fast reactor with no moving parts.

To realize such a design, enriched U-10Zr fuel is placed in hexagonal fuel assemblies in a liquid sodium pool moderator with a single central absorber rod and a beryllium reflector (Koreshi & Hussain, 2014; Tsuboi, Arie, Ueda, & Grenci, 2012).

In this thirty-year two-phase operation, with minimal hands-on maintenance, the first phase of fifteen years of operation is achieved by the gradual withdrawal of the central rod. Thus the beginning of cycle (BOC) reactivity should be high enough such that the reactor is subcritical with the rod fully inserted and sufficiently critical to produce power.

TABLE 10.16 Reaction rates in an AP1000 unit cell.

	Water			Fuel		
	Thermal	Intermediate	Fast	Thermal	Intermediate	Fast
	$E < 0.625$ eV	**0.625eV − 100keV**	**$E > 100$ keV**	**$E < 0.625$ eV**	**0.625eV − 100keV**	**$E > 100$ keV**
Total	3.75361E + 01 0.0012	3.91317E + 01 0.0005	1.40565E + 01 0.0008	2.89720E + 00 0.0012	4.29777E + 00 0.0007	3.45535E + 00 0.0009
Elastic	3.73608E + 01 0.0012	3.91139E + 01 0.0005	1.40468E + 01 0.0008	1.51791E + 00 0.0012	3.77071E + 00 0.0007	2.81159E + 00 0.0009
Abs	1.75349E-01 0.0013	1.77668E-02 0.0010	7.21758E-03 0.0053	3.06725E-01 0.0013	3.73804E-01 0.0019	2.48070E-02 0.0012
(n,γ)	1.75349E-01 0.0013	1.77668E-02 0.0010	9.11486E-05 0.0009	3.06725E-01 0.0013	3.73804E-01 0.0019	2.05502E-02 0.0009

TABLE 10.17 Fissions and multiplication in an AP1000 unit cell.

	Thermal	Intermediate	Fast
	$E < 0.625$ eV	0.625eV – 100keV	$E > 100$ keV
Fission $\langle \Sigma_f \phi \rangle$	1.07256E + 00 0.0013	1.47861E-01 0.0013	6.54692E-02 0.0016
Fission neutrons $\langle \nu \Sigma_f \phi \rangle$	2.61351E + 00 0.0013	3.59921E-01 0.0013	1.80569E-01 0.0017
$(n, 2n) + (n, 3n)$	0	0	2.58042E-03 0.0137

FIGURE 10.13 An AP1000 fuel assembly with 25 Pyrex burnable absorbers.

In the second phase, the gradual upward movement of the external beryllium reflector keeps the reactor critical. The movement is upwards, against gravity to ensure passive safety. In the event of power failure, the reflector would fall downwards and the reactor would return to a subcritical safe state.

The neutronic design challenge is thu s the BOC excess reactivity, the worth of the central absorber and the capacity of the beryllium reflector.

MC simulation is the perfect tool for carrying out such an analysis. Compared with the AP1000 in the previous section, the 4S is smaller and has fewer assemblies. Its electrical output is 10 MW in the uranium-fueled design and 50 MW in the plutonium-fueled design.

The movement of the liquid sodium metal is also upwards, by stationary electromagnetic pumps. This movement of the coolant is another passive safety design feature. Again, a failure of the electrical power in the plant will lead to subcriticality.

The issue in that case would be the heat removal from the core to prevent it from achieving melt-down conditions.

The 4S core model for a Monte Carlo code MCNP simulation, shown in Fig. 10.14 from the top and in Fig. 10.15 from the side, fits into a reflector cylinder with outer radius 60 cm and height 260 cm.

A standard MC simulation would thus carry out a preliminary pin-cell simulation to obtain the neutron spectrum, the infinite multiplication k_∞ and the effect of temperature and moderator density followed by a full-core simulation for the effective multiplication factor k_{eff} with a central absorber control rod.

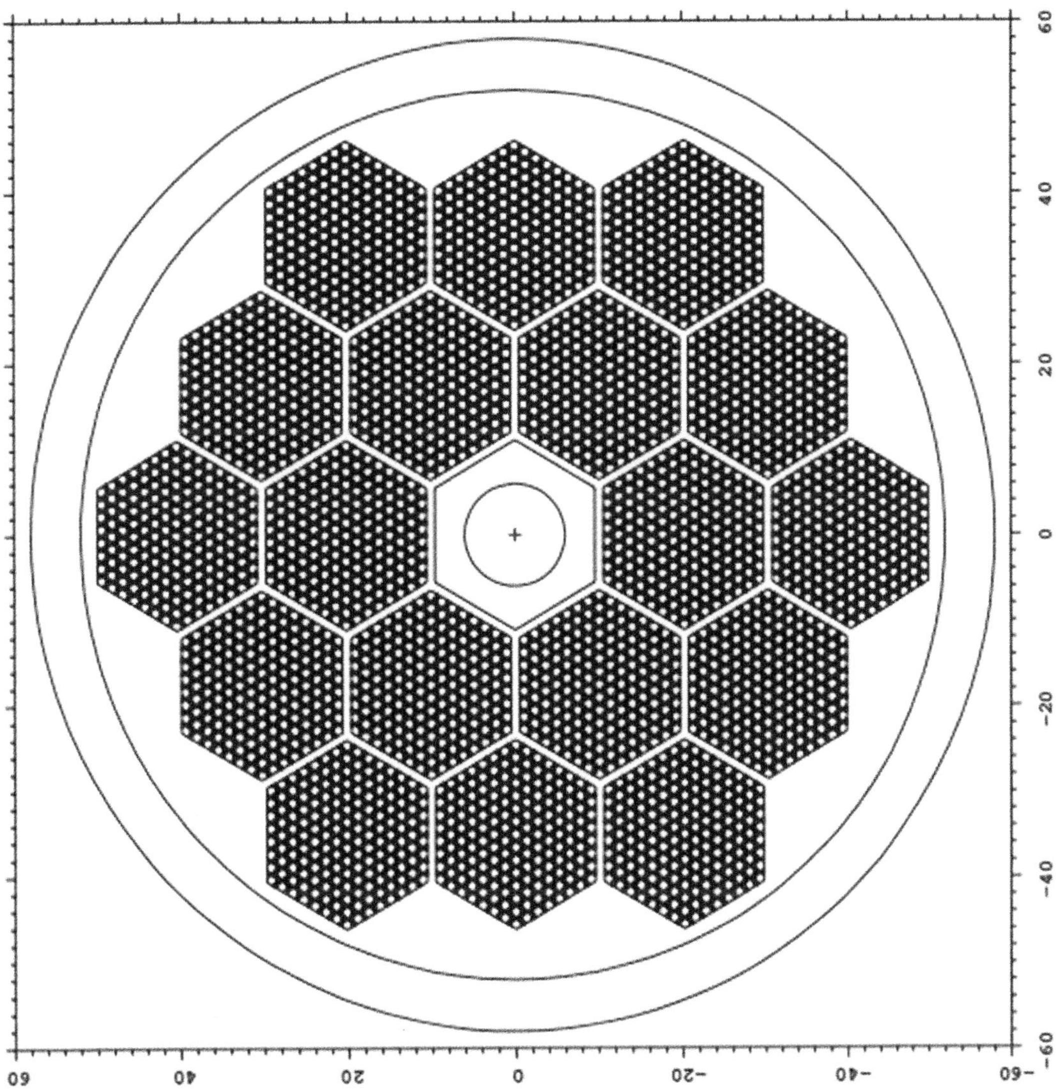

FIGURE 10.14 Top view of the 4S reactor core model.

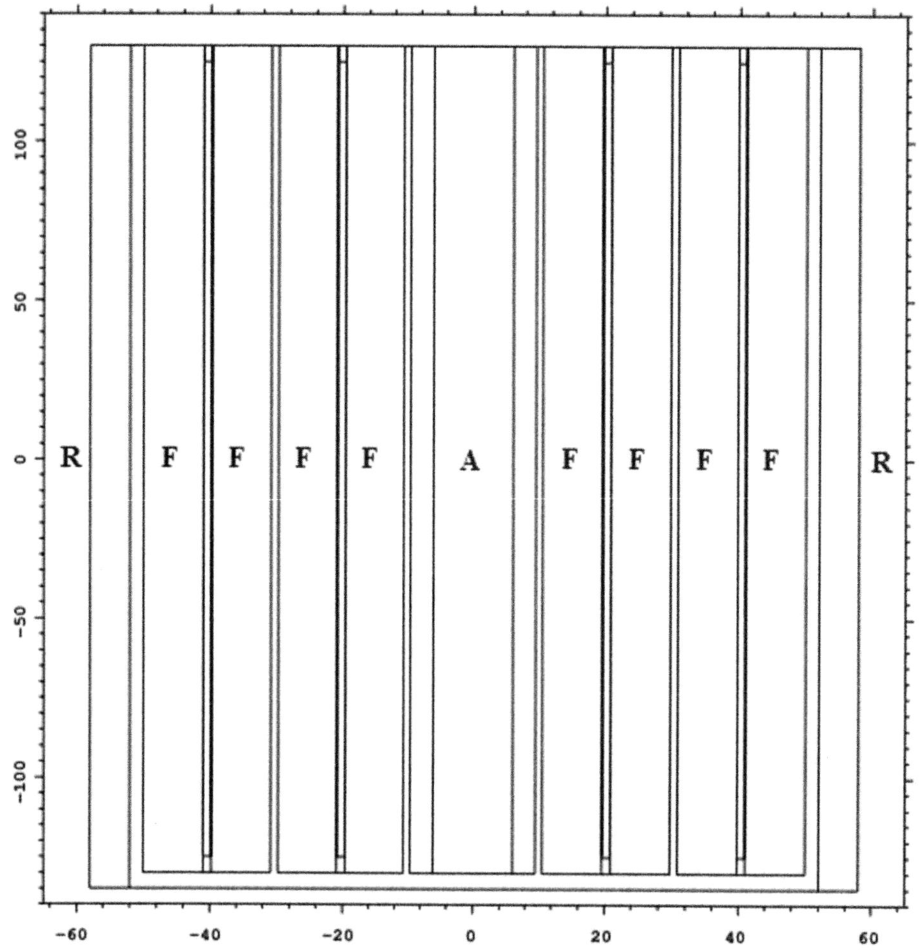

FIGURE 10.15 Side view of the 4S reactor core model (R = Reflector, F = Fuel, A = Absorber).

TABLE 10.18 Physical parameters of the 4S reactor core.

Core	U-10Zr	
	17%	19%
No. of assemblies	6	12
Mass of uranium (t)	3.0903	6.1805
Mass of uranium-235 (t)	0.5253	1.1743
Atomic density (10^{24} atoms/cm^3)		
U-235	6.1329×10^{-3}	6.8545×10^{-3}
U-238	2.9565×10^{-2}	2.8852×10^{-2}
Zr	1.0328×10^{-2}	1.0328×10^{-2}

The 4S 10 MW(e) design, Table 10.18, consists of a cylindrical core with active core height 2.5 m, and diameter 1.16 m. The fuel (diameter 10.50 mm, length 2.50 m) is metallic U/10% Zr 17%−19% enrichment, density 15.6425 g/cm^3 with 169 (13 × 13) fuel rods in a triangular fuel pin arrangement (hexagonal array) fuel assembly with pitch 1.5 cm, and 18 assemblies, with pitch 20.2073 cm, comprising a heavy metal (U) inventory of 9.2708 t and a fissile U^{235} inventory of 1.6996 t. The moderator/coolant is sodium metal (8*t*) operating in the temperature range 355−510 C.

Three unit-cells in the core configuration are:

1. the central cell with the control rod,
2. the 169-fuel pin 17% enriched fuel cell, and
3. the 169-fuel pin 19% enriched fuel cell.

Each lattice fuel cell has a hexagonal structure with fuel surrounded by liquid sodium moderator.

The source is sampled using the KSRC card at the center of each assembly. The KCODE card is used with 2000 histories and 510 cycles with 10 skip cycles. The MCNP cross-section files used were: for sodium, endf66a (ENDF-6.1); for zirconium, endf66b (ENDF-6.1); for natural hafnium endf60 (ENDF-6); for U^{235} and U^{238}, endf66c (ENDF-6.5). All evaluations used are at a temperature of 293.6K.

A preliminary *pin-cell simulation* for the 1693-fuel pin 17% enriched fuel cell gave $k_\infty = 1.49739(0.00067)$ for a fairly "hard" spectrum in the range $10^{-2} - 10$ MeV. For a change in temperature from 300K to 3000K, k_∞ reduces to 1.48583.

Similarly as moderator density varies from 0.93–0.83 g/cm^3 in the range 97.80°C–500°C, the change in k_∞ was from 1.50010 to 1.49415 (Table 10.19).

In a *full-core simulation* for the effective multiplication factor k_{eff}, two absorber materials are considered viz boron carbide (B_4C) and hafnium (Hf) to determine the corresponding fuel inventory to balance the BOC excess reactivity.

For the given fuel rod, k_{eff} decreases from 0.98208 (0.00186) without rod, to 0.89988(0.00150) with a fully inserted rod of 8 cm radius. Thus the core can not attain criticality even with the rod fully removed. The worth of a control rod is found from k_{eff} with and without the rod; when k_{eff} is 0.98274 (0.00061) without a control rod, a 1 cm radius fully inserted hafnium reduces k_{eff} to 0.98179 (0.00058) while for a similar B_4C control rod the change is to 0.98057 (0.00056).

It is found that for a single central absorber rod of radius 9 cm, the maximum fuel rod radius permissible is 6 mm for hafnium and 6.2 mm for B_4C control rods.

Without a reflector and 10000 histories, 210 cycles, 10 skip cycles, $k_{eff} = 1.00783(0.00043)$ for the same B_4C (21.89% B10) rod of radius 6 cm.

The flux is generally flat as shown in Fig. 10.16.

TABLE 10.19 Atomic densities of boron carbide.

AvNo (at./gm-atom)	0.6022 E24		
Density (g/cc)	2.54		
At. Wt. B10	10.01		
At. Wt. B11	11.01		
AtFrB10	0.199	0.2189	0.2587
AtFrB11	0.801	0.7811	0.7413
Abar B (g/g-atom)	10.811	10.7911	10.7513
At. Wt. C-12 (g/g-atom)	12		
Abar B4C [g/(g-mol)]	55.244	55.1644	55.0052
N B4C (atom/b-cm)	0.0276879	0.0277278	0.0278081
B10 enrich (%)	19.9	21.89	25.87
N B	0.1107514	0.1109112	0.1112322
N B10	2.2040E-02	2.4278E-02	2.8776E-02
N B11	8.8712E-02	8.6633E-02	8.2456E-02
N C	2.7688E-02	2.7728E-02	2.7808E-02
N total (atom/b-cm)	1.3844E-01	1.3864E-01	1.3904E-01

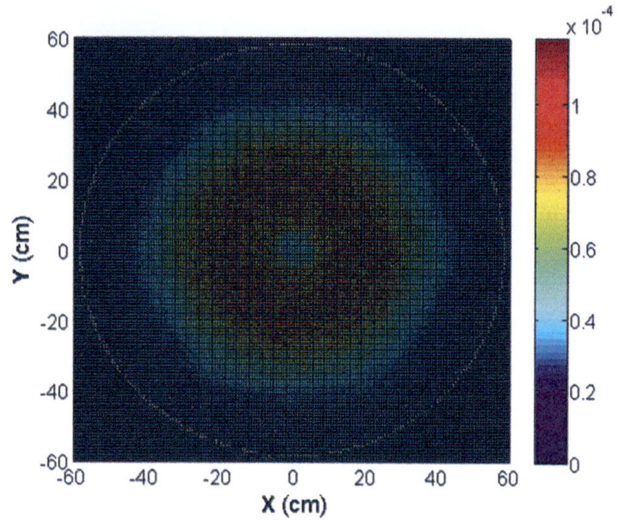

FIGURE 10.16 Neutron flux in the 4S reactor core.

10.5.3 Micronuclear reactor

With renewed interest in space applications, micronuclear reactors (Section 3.8, Section 7.3) with <100 kWe generation based on HEU (60%–70% enriched U-235) and heat pipes with liquid metals for high temperature operation are being considered (Aziz, Koreshi, Sheikh, & Khan, 2020; Chenglong, Sun, Simiao, Wenxi, & Suixheng, 2020; Hao Sun, Pan Ma, Wenxi Tian, & Suizheng Qiu, 2018).

The MNR described in shown in Figs. 3.18 and 7.18 with a compact core within a cylinder of radius of radius 35 cm and height 40 cm, listed in Table 3.20, is much smaller than the 4S reactor discussed in the previous section. This portable core with mass ~503 kg has 90 fuel rods and 37 heat pipes in a monolithic Nb-1Zr matrix. As described in Section 3.8, the total uranium fuel is ~149 kg containing ~104 kg U-235.

Monte Carlo whole-core simulations for such advanced compact fast reactors have been carried out, with materials and atomic densities listed in Table 10.20, to determine neutronic parameters such as fuel design, power distribution and control rod worth.

MC simulations with 1000 histories per cycle and 5000 cycles give $k_{\text{eff}} = 0.955739$ 0.0003 and 1.026546 0.0003 for absorbers in the front and back positions respectively. The source was sampled uniformly in each of the 90 fuel rods with a Watt fission spectrum. The delay fraction β from two simulations (with and without delayed neutrons)

$$\beta = 1 - \frac{k_p}{k_{\text{eff}}}$$

was found to be 0.006951. The maximum excess reactivity is thus $ 3.66 (dollars), and the shutdown margin is $ 5.445 (dollars); one dollar being the reactivity normalized with the delayed neutron fraction.

The fluxes in the core and water radial shield are shown in Fig. 10.17. The fluxes can be seen to be dominated by the higher energy range in the fuel and matrix in contrast to the fluxes in the water shield. Thus power in the core is expected to have more contribution from high energies.

With the 1/6th symmetry in the core, as shown in Fig. 10.18, the power distribution was estimated using the fission reaction tally.

Fig. 10.19 show the power distribution with absorber rods facing the core. The radial power peaking (RPP) is predominantly in the inner fuel cells showing a maximum of 11% with respect to the average power distribution. These decrease gradually towards the periphery and show the trend, as anticipated, of lowest RPP (0.86) in the fuel cells directly facing the absorber and 1.09 each in cells influenced by the reflector.

For rods facing the core and away from the core the power variations can be up to ~20% which requires appropriate heat removal from the heat pipes with sodium, potassium or lithium as heat transfer fluids. In this design, a heat pipe has the capacity to remove power in 14–17 kW range. Thus if the reactor operates at a thermal power of 2–3 MW, each fuel rod will be producing ~20–30 kWt which would exceed the capabilities of sodium and potassium. Even lithium with a maximum envisaged of ~18 kWt would be working at its maximum capacity. For a power

TABLE 10.20 Material data for micronuclear reactor simulation.

Material	Temp (K)	Density (g/cm^3)	Atoms/cm^3/atom fraction*
Fuel: UN (70%)	2000 K	13.6	U235 0.023025, U238 0.009743, N14 0.032768
Helium	–	3.7 × 10^{-5}	He 1
Lithium	1200	0.4	Li6 0.01 Li7 0.99
Lithium Vapor	–	0.001	Li6 0.01 Li7 0.99
Mo-14Re	–	12	Mo 0.064777 Re 0.0054332
Reflector BeO	900	3.01	Be9 −0.360320, O16 −0.639680
Absorber B_4C	–	2.52	B10 −0.782610, C12 −0.217390
Matrix Nb-1Zr	1200K	6.55	Nb93 4.2031e-2 Zr 4.3239e-4
Water	–	1.00	H1 2, O16 1
Tungsten	–	19.3	W74 1

when not specified, $T = 2.5300E\text{-}8$ MeV; negative fractions indicate weight fractions.

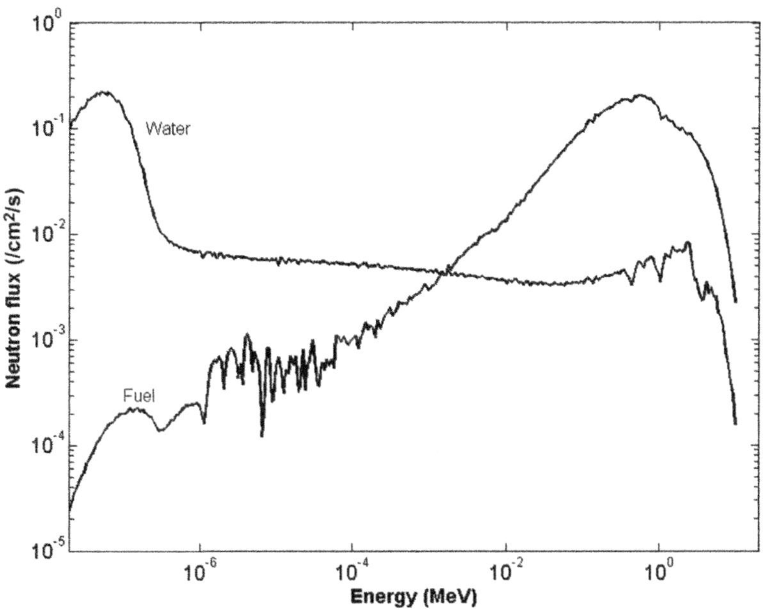

FIGURE 10.17 Core and shield fluxes.

conversion system, a thermo electric generator (TEG) operating at 5%−10% conversion efficiency would be able to generate typically ~150 kWe.

10.6 Nuclear fusion applications

Following a basic introduction to nuclear fusion and its confinement schemes given in Section 3.6 introducing the ITER design, and the introduction to the Monte Carlo method in Chapter 7, an overview of the parameters of interest in fission and fusion neutronics was given in Section 7.3 and Table 7.7. In this section, the geometrical complexity of the Tokamak design and requirements and challenges for MC applications are reviewed.

We first take a look at the overall ITER design in Fig. 10.20. The size and geometrical complexity are far more difficult to model than in fission power reactors.

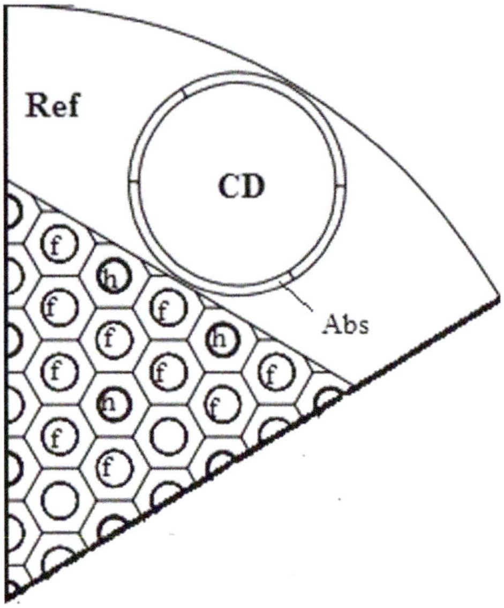

FIGURE 10.18 Micronuclear reactor core 1/6th model.

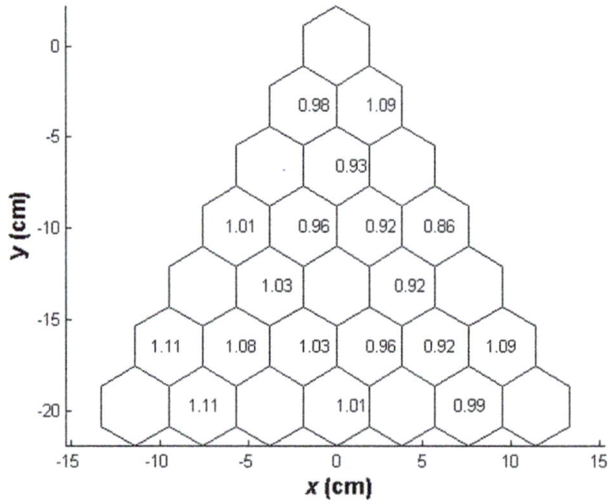

FIGURE 10.19 Power distribution for absorber rods facing core.

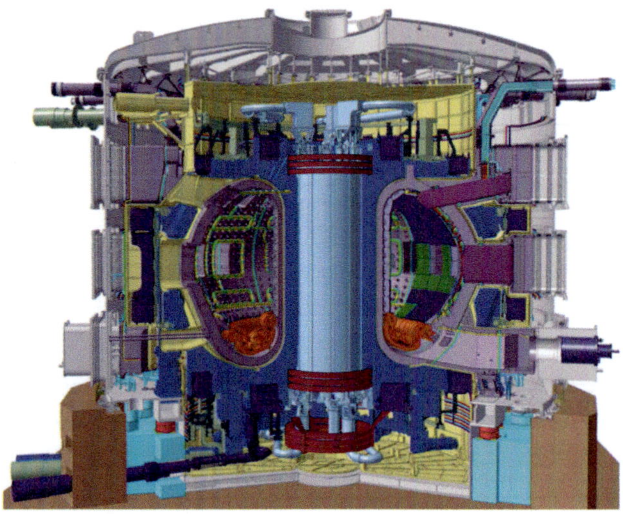

FIGURE 10.20 A view of international thermonuclear experimental reactor. *Courtesy http://www.iter.org.*

The ITER experiment has a fusion power of 500 MW and its main purpose is to demonstrate significant energy gain with pulses of the order of $\sim$400 s which are called *long* pulses due to its promise for the production of steady power in the next design. Presently, the low coolant temperature is not meant to produce electrical power. It is evident that special techniques would be required for accurate modeling of the complex geometry of ITER; for this purpose, the fusion community has combined the MCNP code with CAD software (Juarez et al., 2021; López-Revelles et al., 2018).

In Fig. 10.21, the Toroidal Field Coils represent a superstructure within ITER; they are huge, complex and provide the high magnetic field (5.3 T) necessary for plasma confinement.

There are 18 D-shaped superconducting toroidal field coils, and six poloidal field coils, operating at cryogenic temperatures cooled by supercritical helium; a TFC has coil height 16.5 m, width 9 m and weighs 310 t. Each TFC has 134 turns with $\sim$4 miles of conductor constructed from 40 tons of niobium-tin superconducting strands. The material niobium-tin (Nb_3Sn) in the magnets becomes superconducting at low temperatures down to 4K while the boiling point of liquid helium is 4.2K at atmospheric pressure. The TFCs weigh $\sim$6000 tons which is over one-fourth of the total weight of the Tokamak and have a total magnetic energy of 41 GJ.

The performance of TFCs is thus dependent on parameters such as heat deposition and radiation damage which a MC simulation should be capable of providing with confidence.

In addition to the TFCs, some other critical components are the FW, the divertor the blanket, the shielding and the various ducts and openings for cables and instrumentation. Another feature of the ITER Tokamak is its modularity which present added technical challenges for modeling and ensuring protection of equipment.

With these design features, some parameters needed from an elaborate MC simulation are:

1. FW loading ($\sim$MW/m^2) is calculated mainly by the fast fluence ($E > 0.1$ MeV). The FW receives very high neutron fluxes as well as the bombardment of energetic ions. MC simulations need to assess the performance of high melting point high-strength metals such as W, Re, Ta, Mo, Nb, V.
2. Total nuclear heating to the inboard leg of TFC; about 70% of the total heating is due to neutrons and photons. Heating causes thermal and mechanical stresses.
3. The prompt dose outside the biological shield needs to be estimated to ensure safety limits of occupational dose compliant with regulatory standards such as the CFR 10 Part 20; this is an enormous modeling challenge for MC simulations due to the large size and the strong shielding material. Since analog simulation would yield unreliable results, variance reduction (source biasing and WWG, automated procedures for variance reduction) and hybrid MC/deterministic techniques such as the Consistent Adjoint Driven Importance Sampling (CADIS) are used to optimize particle population for reliable tallies. This has shown attractive enhancement in the quality of the results with more precision and higher FOM (Ibrahim et al., 2011, 2015).
4. Radiation dose to materials due to neutrons and photons. The integrity of components is threatened by fast neutron radiation damage due to radiation displacing atoms from there sites. The dislocation of a Primary Knock-on Atom

FIGURE 10.21 Toroidal field coils of the international thermonuclear experimental reactor machine. *Courtesy http://www.iter.org.*

(PKA) creates a vacancy; the PKA moves about until it becomes an interstitial stationary atom at some location creating a *Frenkel pair*. The pair leads to embrittlement of structural materials such as iron and steel. MC simulations give the displacement reactions $R_{DPA} = N \int_{E_m}^{E_M} \sigma_D \phi(E) dE$ where the displacement per atom (DPA) is an important quantity to estimate the degradation of mechanical and electrical properties of materials such as copper in the electrical wiring (Mascitti & Madariaga, 2011; Rajput, Subhash, & Srinivasan, 2020; Valentine, Colling, Worrall, & Leppänen, 2021).

5. Tritium production in the blanket models. The TBR is required as an essential performance parameter.
6. Hydrogen and helium production. As described earlier, the presence of these gases makes welding difficult if the gas concentrations exceed ppm levels
7. Shutdown dose due to neutron and photons. A regulatory compliance and an ITER safety limit of 100μSv/h = 10 mrem/h twelve days after shutdown is a requirement for the shielding around ITER. The biological dose is calculated at the position of the person standing in Fig. 10.21.
8. The effectiveness of beryllium, lead or any other multiplier for enhancing the TBR.
9. The effectiveness of beryllium as a plasma-facing material being a low-*Z* material with low radiation losses and enhanced oxygen removal from the plasma.
10. Neutrons produce activation in coolants such as water by the reaction $O^{16}(n,p)N^{16}$.
11. High-strength materials such as molybdenum and niobium are high activation elements and are thus replaced by vanadium and tungsten.
12. Nb-93 (n,2n), Au-197 (n,γ), Al (n,α), Ni(n,p) foil.
13. Neutron/photon streaming through ducts.
14. Radiation damage to the tungsten tiles in the divertor assemblies at the bottom of the vacuum vessel.

The displacement cross-section for stainless steel decreases from 100 b to $\sim$0.015 b from 1×10^{-10} to $\sim 10^{-3}$ MeV then increases to $\sim$ 10^4 b till 14 MeV; thus fast neutrons produce the highest displacements in the spectrum.

Elementary 1-D models have been extensively used in early fusion design studies. A typical 1-D slab, cylindrical or spherical geometry simulation has two main parts of the configuration, the inboard and the outboard regions to model the TF Coils, FW, vacuum vessel, breeder blanket units shown in Fig. 10.22 (Section 7.3), and shielding.

Inboard side: TFC with layers of structural material (e.g.,316SS), coolant (e.g., H_2O), multiplier (Pb, Be, etc.) insulator, gaps and experimental ports.

In the simulation, the 14.1 MeV source neutrons are generated in the vacuum vessel.

FIGURE 10.22 International thermonuclear experimental reactor blanket modules. *Courtesy ITER: https://www.iter.org/mach/blanket.*

Better 3-D MC simulations model millions of cells and may require several hours of run-time on fast supercomputing platforms. MC simulation is carried out with codes including MCNP, OpenMC, KENO, TRIPOLI, Serpent and FLUKA for elaborate physical and geometrical modeling. Fusion-specific cross-section library such as the Fusion Evaluated Nuclear Data Library (FENDL) with continuous energy pointwise data with MCNP or with multigroup 46-neutron/21-gamma data which includes data from ENDF/B-VIII for neutrons; for photons the MC library MCPLIB is used.

In terms of neutronics the main difference in fission and fusion systems lies in the neutron spectrum; in nuclear fission the average neutron energy is $\sim$ 1 MeV while in fusion the DT reaction gives a 14.1 MeV neutron.

In terms of operating temperature, PWRs and ITER have lower temperatures than metal salt reactors (MSR) and Gas-Cooled Fast Reactors (GFR) and of course much lower than the Very High Temperature Reactors or the MNR using liquid metals win heat pipes.

Due to the very high plasma temperature in the central chamber of the Tokamak, special materials with compatible thermo-mechanical properties, such as niobium and tungsten (melting point 3414°c), are used in fusion reactors.

The radiation damage in fusion reactors will generally be higher than that in fission power rectors due to the harder spectrum in the former.

Due to the difference in neutron energies in fission and fusion, the primary knock-on atom (PKA) has an energy up to a few keV while in fusion the energy is up to $\sim$ 1 MeV. The radiation damage in fusion reactors due to the higher fluence is also expected to be significantly larger than in fission reactors.

Special problems in fusion reactors are the high flux environment in the FW and divertor giving rise to much higher mechanical and thermal stresses.

The radiation waste, handling and storage from fusion reactors will be much less of a challenge than that from fission reactors. MC simulations are capable of giving good estimates of activation to determine the levels of radiation in the waste and spent fuel in fission reactors.

Overall, the technical challenges in fusion are arguably much steeper than those for fission reactors.

Once, MC simulations are carried out and reliable estimates are obtained with all the computational effort described above, the greater challenge of optimization will remain.

Such studies are in progress as developments continue with proposed designs for better materials utilization and configurations. Parametric studies will require efficient tools for estimating the design sensitivity of replacing structural materials as in the case of low activation EUROFER structural steel, and tungsten and its alloys such as W-Re for divertor surfaces, compounds of lithium such as ceramic titanates for enhancing the TBR and optimized shielding designs for minimizing the biological dose.

Problems

10.1 In Sec. (10.2) for the Godiva simulation results briefly describe how the mean free path can be estimated from the total number of collisions.10.2 For the Godiva spectrum shown in Fig. 10.7 can you justify the use of a one-group diffusion model for estimating its critical radius? What advantages would one-speed transport have over a one-speed diffusion model?10.3 How would you model the simulation when the Godiva assembly is surrounded by people? What effects would you anticipate regarding the safety of such a configuration?10.4 For a uranium hexafluoride storage facility, how would you estimate the relative effects of the neutron and photon skyshine some distance away?10.5 In Sec. (10.4) how would you design a multi-layered low-Z high-Z shield for best attenuation? How would you expect this to compare with a composite shield of varying composition?10.6 In Sec. (10.5) what effect does the reduced density at high temperature have on the criticality?10.7 Compare the neutron flux spectra of Godiva (Fig. 10.7) and AP1000 (Fig. 10.12) to comment on the radiation dose from both systems.10.8 For the 4S reactor core in Sec. (10.5) what would be effect on criticality and power distribution if the positions of the low enriched (17%) and higher enriched (19%) fuel rods are interchanged?

Nomenclature

d	grain size/sphere diameter
g	acceleration due to gravity
h	wick thickness
k_{eff}	effective multiplication factor
k_p	multiplication factor with prompt neutrons

l_e	evaporator length
l_c	condenser length
l_a	adiabatic region length
l_t	total length of heat pipe
r_i	inner heat pipe radius

Greek letters

Ψ	tilt angle from vertical axis
ε	porosity
ϕ	flux
σ_{DPA}	microscopic cross-section for displacements per atom
τ_p	prompt removal lifetime

Abbreviations and acronyms

BOC	beginning of cycle
DPA	displacement per atom
FOM	figure of merit
HEU	highly enriched uranium
HVL	half value layer
IAEA	International Atomic Energy Agency
ISO	International Standards Organization
ITER	International Thermonuclear Experimental Reactor
Gen-IV	Generation IV
MNR	Micronuclear reactor
PFC	Poloidal field coil
PGNAA	Prompt gamma neutron activation analysis
RPP	Radial power peaking
SWU	Separative work units
TEG	Thermo electric generator
TFC	Toroidal field coil
TVL	Tenth value layer
USEC	United States Enrichment Corporation
WGU	Weapons grade uranium

References

Aziz, U., Koreshi, Z. U., Sheikh, S. R., & Khan, H. (2020). A study of criticality and thermal loading in a conceptual micronuclear heat pipe reactor for space applications. *Nuclear Technology and Radiation Protection*, *35*, 208–215. Available from https://doi.org/10.2298/NTRP2003208A.

Broadhead, B. L. (1991). *Criticality safety review of 2½-, 10-, and 14-ton UF6 cylinders*. Tennessee.

Chenglong, W., Sun, H., Simiao, T., Wenxi, T., & Suixheng, Q. (2020). Thermal-hydraulic analysis of a new conceptual heat-pipe cooled small nuclear reactor system. *Nuclear Engineering and Technology*, *52*, 19–26.

IAEA. (2012). *Neutron generators for analytical purposes*. IAEA Radiation Technology Reports Series.

IAEA. (2014). Criticality safety in the handling of fissile material, *Specific Safety Guide SSG-27*. Vienna, Austria.

Ibrahim, A. M., Mosher, S. W., Evans, T. M., Peplow, D. E., Sawan, M. E., Wilson, P. P. H., … Heltemes, T. (2011). Iter neutronics modeling using hybrid Monte Carlo/ deterministic and CAD-based Monte Carlo methods. In *Nuclear technology*. <https://doi.org/10.13182/NT175-251>.

Ibrahim, A. M., Wilson, P. P. H., Sawan, M. E., Mosher, S. W., Peplow, D. E., Wagner, J. C., & Grove, R. E. (2015). Automatic mesh adaptivity for hybrid Monte Carlo/deterministic neutronics modeling of difficult shielding problems. *Nuclear Science and Engineering*, *181*. Available from https://doi.org/10.13182/NSE14-94.

ICRP. (1996). ICRP Publication 74. Conversion coefficients for use in radiological protection against external radiation. In *Annals of the ICRP*.

Juarez, R., Pedroche, G., Loughlin, M. J., Pampin, R., Martinez, P., De Pietri, M., & Sanz, J. (2021). A full and heterogeneous model of the ITER tokamak for comprehensive nuclear analyses. *Nature Energy*. Available from https://doi.org/10.1038/s41560-020-00753-x.

Koreshi, Z. U., & Hussain, S. Z. (2014). A neutronic pin cell and full-core design analysis of the 4S reactor. *Annals of Nuclear Energy*, *72*, 145–150. Available from https://doi.org/10.1016/j.anucene.2014.05.015.

Laranjo de Stefani, G., Losada Moreira, J. M., Maiorino, J. R., & Russo Rossi, P. C. (2019). Detailed neutronic calculations of the AP1000 reactor core with the Serpent code. *Progress in Nuclear Energy*. Available from https://doi.org/10.1016/j.pnucene.2019.03.030.

López-Revelles, A. J., Catalán, J. P., Kolšek, A., Juárez, R., García, R., García, M., & Sanz, J. (2018). MCNP model of the ITER Tokamak complex. *Fusion Engineering and Design*. Available from https://doi.org/10.1016/j.fusengdes.2018.04.023.

Mascitti, J. A., & Madariaga, M. (2011). Method for the calculation of DPA in the reactor pressure vessel of Atucha II. *Science and Technology of Nuclear Installations*. Available from https://doi.org/10.1155/2011/534689.

Rajput, M., Subhash, P. V., & Srinivasan, R. (2020). Displacement damage study in tungsten and iron for fusion neutron irradiation. *Fusion Engineering and Design*. Available from https://doi.org/10.1016/j.fusengdes.2019.111370.

Rowlands, J., Bosq, J. C., Brinckman, J. J., Fort, E., Giffard, F., Jacqmin, R., ... Pelloni, S. (1999). *Intercomparison of calculations for Godiva and Jezebel*.

Su, S.-D. (2015). *Environmental radiation dose assessment and validation for uranium hexafluoride storage management*. University of New Mexico. <https://digitalrepository.unm.edu/ne_etds?utm_source = digitalrepository.unm.edu%2Fne_etds%2F11&utm_medium = PDF&utm_campaign = PDFCoverPages>.

Sun, H., Wang, C., Ma, P., Liu, X., Tian, W., Qiu, S., & Su, G. (2018). Conceptual design and analysis of a multipurpose micro nuclear reactor power source. *Annals of Nuclear Energy*, *121*, 118–127. Available from https://doi.org/10.1016/j.anucene.2018.07.025.

Tsuboi, Y., Arie, K., Ueda, N., & Grenci, T. (2012). Design of the 4S reactor. *Nuclear Technology*, *178*, 201–217.

USEC (United States Enrichment Corporation). (1995). *Uranium hexafluoride: A manual of good handling practices*.

Valentine, A., Colling, B., Worrall, R., & Leppänen, J. (2021). Benchmarking of the Serpent 2 Monte Carlo code for fusion neutronics applications. *EPJ Web of Conferences*. Available from https://doi.org/10.1051/epjconf/202124704015.

Wagner, J. C., Sisolak, J. E., & McKinney, G. (1992). MCNP: Criticality safety benchmark problems, LA-12415. New Mexico, USA.

Whalen, D. J., Cardon, D. A., Uhle, J. L., & Hendricks, J. S. (1991). *MCNP: Neutron benchmark problems*. LA- – 12212 DE92 004710.

Annex A MCNP listing for Godiva (Section 10.2.1)

```
BK10Gdva Godiva with 50 equi-volume regions
c      this gives f(r)
 1  1 -18.74    -1        imp:n=1
 2  1 -18.74     1     -2 imp:n=1
...
... lines removed for space
...
48  1 -18.74    47    -48 imp:n=1
49  1 -18.74    48    -49 imp:n=1
50  1 -18.74    49    -50 imp:n=1
51  0 50 imp:n=0

  1    SO     2.37267244
  2    SO     2.98937995
  3    SO     3.42198581
 ...
... lines removed for space
 46    SO     8.50139893
 47    SO     8.56256218
 48    SO     8.62286390
 49    SO     8.68233380
 50    SO     8.74100000

kcode 3000 1.0 60 150 $ criticality calculation
c       in kcode fission is treated as an absorption
ksrc  0 0 0
m1    92234 -1.02
      92235 -93.7
      92238 -5.27
m2    92234 -1.02 92235 -93.7 92238 -5.27
m3    92234 1
m4    92235 1
m5    92238 1
f1:n 1
fc4      tallies in bare sphere
f4:n  1  2  3  4  5  6  7  8  9 10
     11 12 13 14 15 16 17 18 19 20
     21 22 23 24 25 26 27 28 29 30
     31 32 33 34 35 36 37 38 39 40
     41 42 43 44 45 46 47 48 49 50
fmesh24:n GEOM=xyz ORIGIN = -8.741 -8.841 -0.5
          IMESH=8.741 IINTS=40 JMESH=8.741 JINTS=40 KMESH=0.5 KINTS=1
c
E34      0.625e-6 0.1 14
c        fc34      tallies in bare sphere (N=0.0479838,R=8.741,
c        V=2.7975e3, NV=134.2353)
```

```
f34:n  1  2  3  4  5  6  7  8  9 10
      11 12 13 14 15 16 17 18 19 20
      21 22 23 24 25 26 27 28 29 30
      31 32 33 34 35 36 37 38 39 40
      41 42 43 44 45 46 47 48 49 50
fc34     NV          nu f     n,xn    a    f     n,g
fm34   (134.2353 2  (-6 -7)  (16:17) (-2) (-6)  (102) )
c
c       fc44     tallies in bare sphere
f44:n 1
fc44     NV     m  n,g
fm44   (134.2353 4 (102))
c
c
FC54:N   Energy dependent scalar flux to plot f(E)
F54:N    1
e54   3.77e-9 4.05e-9 4.36e-9 4.69e-9 5.04e-9 5.42e-9 5.83e-9 6.27e-9
      6.74e-9 7.25e-9 7.79e-9 8.38e-9 9.01e-9 9.69e-9 1.04e-8 1.12e-8
....... lines removed for space
      3.01e-2 3.24e-2 3.48e-2 3.74e-2 4.02e-2 4.33e-2 4.65e-2
      5.e-2   5.38e-2 5.78e-2 6.22e-2 6.69e-2 7.19e-2 7.73e-2
      8.32e-2 8.94e-2 9.61e-2 1.03e-1 1.11e-1 1.20e-1 1.29e-1
      1.38e-1 1.49e-1 1.60e-1 1.72e-1 1.85e-1 1.99e-1 2.14e-1
      2.30e-1 2.47e-1 2.66e-1 2.86e-1 3.07e-1 3.30e-1 3.55e-1
      3.82e-1 4.10e-1 4.41e-1 4.75e-1 5.10e-1 5.49e-1 5.90e-1
      6.34e-1 6.82e-1 7.33e-1 7.89e-1 8.48e-1 9.12e-1 9.81e-1
      1.05    1.13    1.22    1.31    1.41    1.52    1.63
      1.75    1.88    2.03    2.18    2.34    2.52    2.71
      2.91    3.13    3.37    3.62    3.89    4.19    4.50
      4.84    5.20    5.60    6.02    6.47    6.96    7.48
      8.04    8.65    9.30    10
Print
```

Annex B MCNP input listing (Jezebel, Section 10.2.2)

```
BK10Jzbl Jezebel concentration
 1  1 -15.61    -1        imp:n=1
 2  0 1 imp:n=0

1   so    6.385

kcode 3000 1.0 60 150
ksrc  0 0 0
m1   94240 -4.5 94239 -95.5
m3   94239 1
m4   94240 1
m5   94240 1
f1:n 1
fc4     tallies in bare sphere
f4:n  1
fmesh24:n GEOM=xyz ORIGIN = -6.385 -6.385 -0.5
         IMESH=6.385 IINTS=40 JMESH=6.385 JINTS=40 KMESH=0.5 KINTS=1
c
E34     0.625e-6 0.1 14
c       fc34     tallies in bare sphere (N=0.0393163,R=6.385,
c       V=1.0904e3, NV=42.8691)
f34:n  1
fc34      NV          tot el  nu f      n,xn    a    f     n,g
fm34   (42.8691 2  (1) (2)(-6 -7)  (16:17) (-2) (-6)  (102) )
c        fc44     tallies in bare sphere
f44:n 1
fc44      NV     m  n,g
fm44   (42.8691 3 (102))
c
FC54:N   Energy dependent scalar flux to plot f(E)
F54:N    1
e54    3.77e-9 4.05e-9 4.36e-9 4.69e-9 5.04e-9 5.42e-9 5.83e-9 6.27e-9
       6.74e-9 7.25e-9 7.79e-9 8.38e-9 9.01e-9 9.69e-9 1.04e-8 1.12e-8
       1.20e-8 1.29e-8 1.39e-8 1.50e-8 1.61e-8 1.73e-8 1.86e-8
       2.e-8   2.15e-8 2.31e-8 2.49e-8 2.68e-8 2.88e-8 3.09e-8
       3.33e-8 3.58e-8 3.85e-8 4.13e-8 4.45e-8 4.78e-8 5.14e-8
       5.53e-8 5.94e-8 6.39e-8 6.87e-8 7.39e-8 7.95e-8 8.54e-8
       9.19e-8 9.88e-8 1.06e-7 1.14e-7 1.23e-7 1.32e-7 1.42e-7
       1.53e-7 1.64e-7 1.77e-7 1.90e-7 2.04e-7 2.19e-7 2.36e-7
       2.54e-7 2.73e-7 2.93e-7 3.15e-7 3.39e-7 3.65e-7 3.92e-7
       4.22e-7 4.53e-7 4.88e-7 5.24e-7 5.64e-7 6.06e-7 6.52e-7
       7.01e-7 7.54e-7 8.10e-7 8.71e-7 9.37e-7 1.01e-6 1.08e-6
       1.16e-6 1.25e-6 1.35e-6 1.45e-6 1.56e-6 1.67e-6 1.80e-6
       1.94e-6 2.08e-6 2.24e-6 2.41e-6 2.59e-6 2.78e-6 2.99e-6
       3.22e-6 3.46e-6 3.72e-6 4.e-6   4.30e-6 4.62e-6 4.97e-6
       5.35e-6 5.75e-6 6.18e-6 6.65e-6 7.15e-6 7.69e-6 8.26e-6
       8.89e-6 9.56e-6 1.03e-5 1.10e-5 1.19e-5 1.28e-5 1.37e-5
       1.48e-5 1.59e-5 1.71e-5 1.84e-5 1.97e-5 2.12e-5 2.28e-5
       2.45e-5 2.64e-5 2.84e-5 3.05e-5 3.28e-5 3.53e-5 3.79e-5
       4.08e-5 4.39e-5 4.72e-5 5.07e-5 5.45e-5 5.86e-5 6.31e-5
       6.78e-5 7.29e-5 7.84e-5 8.43e-5 9.06e-5 9.75e-5 1.05e-4
```

```
      1.13e-4 1.21e-4 1.30e-4 1.40e-4 1.51e-4 1.62e-4 1.74e-4
      1.87e-4 2.01e-4 2.17e-4 2.33e-4 2.50e-4 2.69e-4 2.89e-4
      3.11e-4 3.35e-4 3.60e-4 3.87e-4 4.16e-4 4.47e-4 4.81e-4
      5.17e-4 5.56e-4 5.98e-4 6.43e-4 6.91e-4 7.43e-4 7.99e-4
      8.60e-4 9.24e-4 9.94e-4 1.07e-3 1.15e-3 1.24e-3 1.33e-3
      1.43e-3 1.54e-3 1.65e-3 1.78e-3 1.91e-3 2.05e-3 2.21e-3
      2.37e-3 2.55e-3 2.75e-3 2.95e-3 3.17e-3 3.41e-3 3.67e-3
      3.95e-3 4.24e-3 4.56e-3 4.91e-3 5.28e-3 5.67e-3 6.10e-3
      6.56e-3 7.05e-3 7.58e-3 8.15e-3 8.77e-3 9.43e-3 1.01e-2
      1.09e-2 1.17e-2 1.26e-2 1.36e-2 1.46e-2 1.57e-2 1.68e-2
      1.81e-2 1.95e-2 2.09e-2 2.25e-2 2.42e-2 2.60e-2 2.80e-2
      3.01e-2 3.24e-2 3.48e-2 3.74e-2 4.02e-2 4.33e-2 4.65e-2
      5.e-2   5.38e-2 5.78e-2 6.22e-2 6.69e-2 7.19e-2 7.73e-2
      8.32e-2 8.94e-2 9.61e-2 1.03e-1 1.11e-1 1.20e-1 1.29e-1
      1.38e-1 1.49e-1 1.60e-1 1.72e-1 1.85e-1 1.99e-1 2.14e-1
      2.30e-1 2.47e-1 2.66e-1 2.86e-1 3.07e-1 3.30e-1 3.55e-1
      3.82e-1 4.10e-1 4.41e-1 4.75e-1 5.10e-1 5.49e-1 5.90e-1
      6.34e-1 6.82e-1 7.33e-1 7.89e-1 8.48e-1 9.12e-1 9.81e-1
      1.05    1.13    1.22    1.31    1.41    1.52    1.63
      1.75    1.88    2.03    2.18    2.34    2.52    2.71
      2.91    3.13    3.37    3.62    3.89    4.19    4.50
      4.84    5.20    5.60    6.02    6.47    6.96    7.48
      8.04    8.65    9.30    10
FC6    Energy deposition averaged over cell (MeV/g)
F6:N   1
FC7    Fission energy deposition averaged over cell (MeV/g)
F7:N   1
Print
```

Annex C MCNP input listing (BK10Shld, Section 10.5.1)

```
BK10Shld Concrete shell 30 cm thick
c Shield neutron 2 june 2021
c cell cards
   1      0 -1          imp:n=1
   2      1 -2.3 1 -2 imp:n=1   $ concrete shield
   3      0  2          imp:n=0

c surface cards
  1       so 304.8
  2       so 374.8

c geometry cards
mode    n
c      concrete (rho = 2.3 g/cm^3)
m1     1001  1.68765e-1
       8016  5.62493e-1
      11023  1.18366e-2
      12000  1.39951e-3
      13027  2.14316e-2
      14000  2.04076e-1
      19000  5.65495e-3
      20000  1.86720e-2
      26054  2.47295e-4
      26056  3.91067e-3
      26057  9.38014e-5
      26058  1.19384e-5
       6012  1.41730e-3
sdef    erg=14
f1:n    1 2
f2:n    1 2
E0      1 12i 14.0
f45:n    0    374.8 0 0
cut:n    j 1.0 $ cut neutrons at 1 MeV
f12:n    2
c ambient neutron dose equiv. H*(10mm) ICRP Sv cm^2
de12       2.500E-08 1.000E-07 1.000E-06 1.000E-05 1.000E-04 1.000E-03
           1.000E-02 2.000E-02 5.000E-02 1.000E-01 2.000E-01 5.000E-01
           1.000E+00 1.500E+00 2.000E+00 3.000E+00 4.000E+00 5.000E+00
           6.000E+00 7.000E+00 8.000E+00 1.000E+01 1.400E+01 1.700E+01
           2.000E+01
df12       8.000E-12 1.040E-11 1.120E-11 9.200E-12 7.100E-12 6.200E-12
           8.600E-12 1.460E-11 3.500E-11 6.900E-11 1.260E-10 2.580E-10
           3.400E-10 3.620E-10 3.520E-10 3.800E-10 4.090E-10 3.780E-10
           3.830E-10 4.030E-10 4.170E-10 4.460E-10 5.200E-10 6.100E-10
           6.500E-10
nps        400000
dd     0.1  1e100
print
```

Annex D MCNP input listing (BK10AP10, Section 10.5.1)

```
BK10AP10 Unit Lattice Cell AP1000  23 MAY 2021
c
1    1 -1.0 (1 -2 3 -4 5 -6) (7) imp:n=1 tmp=2.5300e-8 $ water
2    2 -10.47635 (-7 5 -6)       imp:n=1 tmp=2.5300e-8 $ fuel pin
3    3  1.0e-4 (7 -8 5 -6)       imp:n=1 tmp=2.5300e-8 $ gap
4    4  4.295e-2 (8 -9 5 -6)   imp:n=1 tmp=2.5300e-8 $ clad
5    0  (-1:2:-3:4:-5:6)         imp:n=1 tmp=2.5300e-8 $ outside world

C   surface cards
  *1  PX  -0.63
  *2  PX   0.63
  *3  PY  -0.63
  *4  PY   0.63
  *5  PZ   -2
  *6  PZ    2
   7  CZ  0.4096
   8  CZ  0.4178
   9  CZ  0.4750

mode   n
c  ---------WATER (den 1 g/cm3)
M1   8016.50C 1     $ rmccs 2.53e-8
     1001.50C 2     $ rmccs 2.53e-8
MT1  lwtr.01t       $ tmccs 2.53e-8
c --------- Fuel (den 10.4 g/cm3 4.5%)  endf66a 2.53e-8
c m2   8016.66c 4.64149e-2 92234.66c 8.49269e-6 92235.66c 1.05705e-3
c      92238.66c 2.21413e-2
c --------- Fuel (den 10.4 g/cm3 4.45%)  endf66a 2.53e-8
m2   8016.66c 4.67471e-2 92235.66c 1.05284e-3 92238.66c 2.23207e-2
c ---------- Helium4 Gap (den 0.1785e-3 g/cm3)
m3   2004.50c 1 $ rmccs  2.53e-8
c --------clad(6.506 g/cm^3) at den=4.295e-2  PNNL ----------
m4      40000.66c 1 $ endf66b
c
c tallies
m5   92235.66c 1
m6   92238.66c 1
c
FC4  neutron flux in cells 1 2 3 4
F4:N   1 2 3 4
c       sd4
E4     0.625e-6 0.1 14
VOL    1 1 1 1
c       ----------------
FC14    TALLIES IN WATER
C       ----------------
E14     0.625e-6 0.1 14
F14:N   1
C       NV          tot el     a     n,g
fm14   (0.3526 1  (1) (2)   (-2)   (102) )
C  fM14  (1 5 (-6))(1 6 (102))
C       ---------------
FC24    TALLIES IN FUEL
C       ---------------
E24     0.625e-6 0.1 14
F24:N   2
```

```
C          NV          tot el  nu f      n,xn     a    f    n,g
fm24    (0.1478 2  (1) (2)(-6 -7)  (16:17) (-2) (-6)  (102) )
C
C       ---------------------
FC34    TALLIES IN HELIUM GAP
C       ---------------------
E34     0.625e-6 0.1 14
F34:N   3
C          NV          tot el     a     n,g
fm34    (8.5259e-6 3 (1) (2)  (-2)   (102) )
C
C       ---------------
FC44    TALLIES IN CLAD
C       ---------------
E44     0.625e-6 0.1 14
F44:N   4
C          NV          tot el  nu f      n,xn     a    f    n,g
FM44    (0.0279 4   (1) (2)(-6 -7)  (16:17) (-2) (-6)  (102) )
fmesh54:n GEOM=xyz ORIGIN = -0.63 -0.63 -0.5
           IMESH=0.63 IINTS=40 JMESH=0.63 JINTS=40 KMESH=0.5 KINTS=1
FC64:N   Energy dependent scalar flux to plot f(E) in water (1) and fuel (2)
F64:N    1 2
e64     3.77e-9 4.05e-9 4.36e-9 4.69e-9 5.04e-9 5.42e-9 5.83e-9 6.27e-9
        6.74e-9 7.25e-9 7.79e-9 8.38e-9 9.01e-9 9.69e-9 1.04e-8 1.12e-8
        1.20e-8 1.29e-8 1.39e-8 1.50e-8 1.61e-8 1.73e-8 1.86e-8
        2.e-8   2.15e-8 2.31e-8 2.49e-8 2.68e-8 2.88e-8 3.09e-8
        3.33e-8 3.58e-8 3.85e-8 4.13e-8 4.45e-8 4.78e-8 5.14e-8
        5.53e-8 5.94e-8 6.39e-8 6.87e-8 7.39e-8 7.95e-8 8.54e-8
        9.19e-8 9.88e-8 1.06e-7 1.14e-7 1.23e-7 1.32e-7 1.42e-7
        1.53e-7 1.64e-7 1.77e-7 1.90e-7 2.04e-7 2.19e-7 2.36e-7
        2.54e-7 2.73e-7 2.93e-7 3.15e-7 3.39e-7 3.65e-7 3.92e-7
        4.22e-7 4.53e-7 4.88e-7 5.24e-7 5.64e-7 6.06e-7 6.52e-7
        7.01e-7 7.54e-7 8.10e-7 8.71e-7 9.37e-7 1.01e-6 1.08e-6
        1.16e-6 1.25e-6 1.35e-6 1.45e-6 1.56e-6 1.67e-6 1.80e-6
        1.94e-6 2.08e-6 2.24e-6 2.41e-6 2.59e-6 2.78e-6 2.99e-6
        3.22e-6 3.46e-6 3.72e-6 4.e-6   4.30e-6 4.62e-6 4.97e-6
        5.35e-6 5.75e-6 6.18e-6 6.65e-6 7.15e-6 7.69e-6 8.26e-6
        8.89e-6 9.56e-6 1.03e-5 1.10e-5 1.19e-5 1.28e-5 1.37e-5
        1.48e-5 1.59e-5 1.71e-5 1.84e-5 1.97e-5 2.12e-5 2.28e-5
        2.45e-5 2.64e-5 2.84e-5 3.05e-5 3.28e-5 3.53e-5 3.79e-5
        4.08e-5 4.39e-5 4.72e-5 5.07e-5 5.45e-5 5.86e-5 6.31e-5
        6.78e-5 7.29e-5 7.84e-5 8.43e-5 9.06e-5 9.75e-5 1.05e-4
        1.13e-4 1.21e-4 1.30e-4 1.40e-4 1.51e-4 1.62e-4 1.74e-4
        1.87e-4 2.01e-4 2.17e-4 2.33e-4 2.50e-4 2.69e-4 2.89e-4
        3.11e-4 3.35e-4 3.60e-4 3.87e-4 4.16e-4 4.47e-4 4.81e-4
        5.17e-4 5.56e-4 5.98e-4 6.43e-4 6.91e-4 7.43e-4 7.99e-4
        8.60e-4 9.24e-4 9.94e-4 1.07e-3 1.15e-3 1.24e-3 1.33e-3
        1.43e-3 1.54e-3 1.65e-3 1.78e-3 1.91e-3 2.05e-3 2.21e-3
        2.37e-3 2.55e-3 2.75e-3 2.95e-3 3.17e-3 3.41e-3 3.67e-3
        3.95e-3 4.24e-3 4.56e-3 4.91e-3 5.28e-3 5.67e-3 6.10e-3
        6.56e-3 7.05e-3 7.58e-3 8.15e-3 8.77e-3 9.43e-3 1.01e-2
        1.09e-2 1.17e-2 1.26e-2 1.36e-2 1.46e-2 1.57e-2 1.68e-2
        1.81e-2 1.95e-2 2.09e-2 2.25e-2 2.42e-2 2.60e-2 2.80e-2
        3.01e-2 3.24e-2 3.48e-2 3.74e-2 4.02e-2 4.33e-2 4.65e-2
        5.e-2   5.38e-2 5.78e-2 6.22e-2 6.69e-2 7.19e-2 7.73e-2
        8.32e-2 8.94e-2 9.61e-2 1.03e-1 1.11e-1 1.20e-1 1.29e-1
```

```
      1.38e-1 1.49e-1 1.60e-1 1.72e-1 1.85e-1 1.99e-1 2.14e-1
      2.30e-1 2.47e-1 2.66e-1 2.86e-1 3.07e-1 3.30e-1 3.55e-1
      3.82e-1 4.10e-1 4.41e-1 4.75e-1 5.10e-1 5.49e-1 5.90e-1
      6.34e-1 6.82e-1 7.33e-1 7.89e-1 8.48e-1 9.12e-1 9.81e-1
      1.05    1.13    1.22    1.31    1.41    1.52    1.63
      1.75    1.88    2.03    2.18    2.34    2.52    2.71
      2.91    3.13    3.37    3.62    3.89    4.19    4.50
      4.84    5.20    5.60    6.02    6.47    6.96    7.48
      8.04    8.65    9.30    10
FC6    Energy deposition averaged over cell (MeV/g)
F6:N   1 2 3 4 T
FC7    Fission energy deposition averaged over cell (MeV/g)
F7:N   1 2 3 4 T
kcode 5000 1 5 150
prdmp 150 150 150
ksrc  0 0 -1
      0 0  0
      0 0  1
print
```

Chapter 11

Comparisons: Monte Carlo, diffusion, and transport

11.1 Introduction

A university nuclear engineering curriculum typically places neutron diffusion in the early to middle part of undergraduate studies while transport theory is delayed to the final year. It is not uncommon for students to feel uncomfortable with the neutron transport equation and usually one does not muster the courage to go deeper and explore whether these two formulations have common ground.

By the time the final year is about to end, one gets to hear of Monte Carlo (MC) simulation but since it is not covered in sufficient detail, one typically graduates without seeing the connection.

It is in a Master's or PhD program that one takes a serious look at the transport equation and appreciates the power of MC simulation for a thesis and beyond. That power is often comforting, especially to see voluminous results presented in colorful pictures.

In the 1970s and early 1980s, professors and students would spend a great amount of effort and time in getting the mathematics right. The power of theory, its challenges, and elegance, were considered the power of academia. With the demands of industry, that focus on theory probably weakened.

This chapter is meant to present a *unified* picture of neutron diffusion, transport, and MC plus the realization that it comes under the more general scope of radiation transport. This picture comes towards the end of this book when the reader has covered all three with a fair amount of depth and breadth.

Once a unified picture has been made in the mind, it should be natural to view things as one; whether deterministic or stochastic, nuclear engineering as a profession will cherish both and see their strengths.

To present a unified picture, some elementary models are reviewed and solutions with diffusion, transport and MC simulation are presented for each.

This chapter comes after diffusion theory (DT) (Chapter 5), transport theory (Chapter 6), and the basics of MC method (Chapter 7) have been covered in a fair amount of detail.

A comparison must begin with benchmarks which are fortunately available in nuclear engineering to a good extent. In the preceding chapters, these benchmarks have been cited and used on occasions; in criticality, diffusion, transport and MC.

The early work, nicely covered by Bell and Glasstone (1979), Clark and Hansen (1964), Ganapol (2008) begin from the early work in radiative transfer and use powerful mathematics such as Fourier transforms, Laplace transforms, and the Green's functions. Pioneering work by Case and Zweifel, the Weiner-Hopf method, and Chandrasekhar's H-function are the solid bedrock upon which nuclear engineering stands firm.

The following four sections cover system multiplication in a bare sphere, 1D slab, and spherical systems and a comparison of fluxes for a one-group flux calculation.

11.2 Criticality in a bare sphere

11.2.1 One-group diffusion theory criticality

This section is based on Chapter 5 (The neutron diffusion equation). From Section 5.2.2, the one-group diffusion equation (Eq. 5.11) expressed as an eigenvalue equation is

$$D\nabla^2\phi - \Sigma_r\phi + \frac{1}{k}\nu\Sigma_f\phi = 0 \qquad (11.1)$$

gives the infinite medium equation, setting the leakage ($\nabla^2\phi$) term to zero, for which the infinite multiplication factor k_∞ is

Nuclear Engineering. DOI: https://doi.org/10.1016/B978-0-323-90618-0.00011-9

TABLE 11.1 One-group data for Godiva.

Nuclide	ν	σ_a (b)	σ_f (b)	σ_{tr} (b)
U^{235}	2.60	1.65	1.40	6.80
U^{238}	2.60	0.255	0.095	6.90

Source: Criticality Hand calculations LA14244-M p. 45/181.

TABLE 11.2 One-group macroscopic cross sections for Godiva.

Nuclide	N atoms/b-cm	Σ_a cm^{-1}	Σ_f cm^{-1}	Σ_{tr} cm^{-1}
U-235	4.489×10^{-2}	7.407×10^{-2}	6.285×10^{-2}	3.0530×10^{-1}
U-238	3.081×10^{-2}	7.857×10^{-4}	2.927×10^{-4}	2.130×10^{-2}

$$k_\infty = \frac{\nu\Sigma_f}{\Sigma_r} \tag{11.2}$$

and for the finite system

$$k_{\text{eff}} = \frac{k_\infty}{1 + L^2B^2}. \tag{11.3}$$

The corresponding neutron flux is

$$\phi(r) = A\frac{\sin Br}{r} \tag{11.4}$$

where A is a constant depending on the power P. The Godiva data in Tables 11.1 gives the macroscopic quantities listed in Table 11.2.

From the above, the diffusion coefficient is $D = 1/3\Sigma_{tr} = 1.0209$ cm, the diffusion length (squared) is $L^2 = D/\Sigma_a = 13.6382$ cm^2, from which $k_\infty = \nu\Sigma_f/\Sigma_a = 2.1932$.

The criticality equation gives $B^2 = 0.0875$ cm^2, from which the extrapolated radius is $\overline{R} = 10.6213$ cm. With $d = 2.13 \times D = 2.1745$ cm, the critical radius is $R = \overline{R} - d = 8.4468$ cm.

11.2.2 Two-group diffusion theory criticality

From Section 5.3.2, the two-group diffusion equations are

$$D_1\nabla^2\phi_1 - \Sigma_{1r}\phi_1 + \frac{1}{k}\left(\nu\Sigma_{f1}\phi_1 + \nu\Sigma_{f2}\phi_2\right) = 0 \tag{11.5}$$

and

$$D_2\nabla^2\phi_2 - \Sigma_{2r}\phi_2 + \Sigma_{s1\to2}\phi_1 = 0 \tag{11.6}$$

with the assumption that there is no fission contribution into group 2.

The two-group (core) fluxes are given by Eqs. (5.47) and (5.48) for functions X and Y and coupling coefficients.

For an infinite system, setting the leakage (∇^2..) terms to zero, gives k_∞ as

$$k_\infty = \frac{\Sigma_{2r}\nu\Sigma_{f1} + \Sigma_{s1\to2}\nu\Sigma_{f2}}{\Sigma_{1r}\Sigma_{2r}} \tag{11.7}$$

and the finite mutiplication as

$$k_{\text{eff}} = \frac{\nu\Sigma_{f1}}{D_1B^2 + \Sigma_{1r}} + \frac{\Sigma_{s1\to2}\nu\Sigma_{f2}}{(D_1B^2 + \Sigma_{1r})(D_2B^2 + \Sigma_{2r})}. \tag{11.8}$$

The critical size, given the composition, is found from the 4X4 criticality determinant (Eq. 5.54) with the appropriate two-group data as for the reflected spherical reactor in Section 5.3.2. A simpler expression, the six-factor formula, was obtained in Section 2.10 for a single generation, as

$$k_{\text{eff}} = 1 = \frac{k_\infty}{(1 + B^2\tau)(1 + L^2B^2)}. \tag{11.9}$$

In Eq. (11.9), the effects of the fast and thermal groups appear through the leakage terms with neutron age and diffusion length respectively.

11.2.3 One-speed transport theory criticality

To compare diffusion estimates with the transport result, consider the asymptotic flux (2.10, Section 6.3.8) in a sphere of radius r

$$\phi_{as}(r) = \frac{A}{r}\sin\frac{r}{|\nu_0|} \tag{11.10}$$

where ν_0 is a solution of the transcendental equation

$$1 = c\nu_0 \tanh^{-1}\frac{1}{\nu_0} = \frac{c\nu_0}{2}\ln\frac{\nu_0 + 1}{\nu_0 - 1}. \tag{11.11}$$

Recall that c is the number of neutrons emerging from an interaction such that $0 < c < 1$ for non-multiplying media and $c > 1$ for a multiplying medium such as Godiva. The sphere is approximately critical when the asymptotic flux is zero at the extrapolated radius $\overline{R}$; thus $\phi_{as}(\overline{R}) = 0$ at $\overline{R} = R_c + x_0 = \pi|\nu_0(c)|$.

The solutions of (Eq. 11.11) are real for $c < 1$ and complex for $c > 1$; an approximation for ν_0 is

$$\frac{1}{\nu_0{}^2} = \frac{3(1-c)}{c}\left[1 - \frac{9}{5}\frac{1-c}{c} - \cdots\right]. \tag{11.12}$$

For c near unity, another approximation is

$$\nu_0 = \frac{1}{\sqrt{3(1-c)}}\left[1 + \frac{2}{5}(1-c) + \cdots\right]. \tag{11.13}$$

To find the critical radius ν_0 is determined from the transcendental equation and the critical radius is found from $R_c = \pi|\nu_0| - x_0$, where the extrapolation distance x_0 can be calculated from the Mark P_1 boundary condition (Eq. 2.65).

In the spherical harmonics P_1 "diffusion approximation," the neutron flux is given by Eq. (2.54), re-written here for quick reference:

$$\phi(x) = \frac{1}{2}\sqrt{\frac{3}{1-c}}\; e^{-|x|/L}$$

where the diffusion length is

$$L = \nu_0 = \frac{1}{\sqrt{3(1-c)}}. \tag{11.14}$$

Values of ν_0 for $c < 1$ are plotted in Fig. 6.2; others are reproduced from Bell and Glasstone (Bell & Glasstone, 1979) (Table 11.3).

For a one-group calculation, the data of Table 2.17(a) (Bell and Glasstone) is used for U-235 (density 18.8 g/cm^3) assuming isotropic scattering:

$\nu = 2.50$, $\sigma_f = 1.3$ b, $\sigma_s = 4.0$ b. This gives $c = 1.3679$, for which Eq. (11.11) gives

$$|i\nu_0| = 0.8359.$$

Note that the transcendental equation for $c < 1$ is straightforward as the roots are real numbers while for $c > 1$, with complex roots, the MATLAB® program listed below was written.

The extrapolation radius, from the Mark P_1 boundary conditions is $x_0 = 0.5181$ cm. The critical radius, $R_c = \pi|\nu_0(c)| - x_0 = 8.2558$ cm which is to be compared with the computed value of the critical radius 8.710 cm with $k_{\text{eff}} = 0.9912$.

TABLE 11.3 Relaxation lengths for isotropic scattering (in mean free paths).

c	Exact (Eq. 11.11)	Approximation (Eq. 11.13)	Diffusion theory (Eq. 11.14)
$c<1$			
$\lvert\nu_0\rvert$			
0.99	5.797	5.797	5.774
0.90	4.116	4.115	4.083
0.80	1.408	1.394	1.291
0.50	1.044	0.979	0.816
0	1.000	0.808	0.577
$c>1$			
$\lvert i\nu_0\rvert$			
1.01	5.750	5.751	5.774
1.05	2.532	2.531	2.582
1.10	1.757	1.756	1.826
1.20	1.198	1.195	1.291
1.50	0.689	0.680	0.816

The relaxation length in one-speed transport theory P_1 diffusion approximation is ν_0, and it determines the shape of the flux just like the buckling in one group DT.

MATLAB program for computing the eigenvalue v_0 from Eq. (11.11)

```
% calculate eigenvalue from transcendental equation
% Bell and Glasstone p.126 q.11
nu=2.5; sigF=1.3;
sigC=0;sigS=4;sigT=sigF+sigC+sigS;A=235;mu0=2/(3*A);sigTR=sigS*(1-mu0);
c=(nu*sigF+sigS)/sigT; k_inf=nu*sigF/(sigT-sigS);
c=1.3679; % godiva
z= 0.8:0.0001:0.96; x=z*i; % convert to a complex number
f = @(x) (2./(c*x)) - log((x+1)./(x-1)); y=f(x); yabs=abs(y);
plot(z,yabs);

% find root
n=size(z);
ymin=10;
for j=1:n(2)
     if(yabs(j)<ymin)
          Nu0=z(j); ymin=yabs(j);
     end
end

% extrapolated radius Rc
x0=(1/sqrt(3))*(1-(1/3)*(c-1)+(1/5)*(c-1)^2-(1/7)*(c-1)^3); % extrap dist
rho=18.8;AvNo=0.6022e24;M=235;N=rho*AvNo/M;

SIGT=N*sigT*1e-24; lambda=1/SIGT;

Rc=pi*Nu0-x0; % critical radius in mfp
Rc_cm=Rc*lambda; % critical radius in cm
```

Exercise 11.1: One-speed transport theory critical radius.

Use Eq. (11.14) (DT approximation) to calculate the eigenvalue and the critical radius of Godiva with the data given in this section.

11.3 The classic albedo calculation

In the classic albedo calculation, the reflection is estimated at a surface due to an incident source as shown in Fig. 11.1.

In DT from the infinite slab one-group flux (Table 5.1), the albedo

$$\beta = \frac{J_{\text{out}}}{J_{\text{in}}} \tag{11.15}$$

is given by

$$\beta = \frac{1 - (2D/L)\coth a/L}{1 + (2D/L)\coth a/L}. \tag{11.16}$$

where a is the slab thickness.

Fig. 11.2 shows the albedos for light water (LW), beryllium (Be), aluminum (Al), graphite (C), and heavy water (HW) using thermal data listed in the MATLAB program CH11_AlbedoSlabDiffTh.m (Annex A). The highest albedo (0.7369) is for a water slab, saturated at a thickness ~10 cm, with progressively lower values for Be, Al, C and the lowest for heavy water.

In Section 6.2.1, the albedo from transport theory was given as

$$\phi(0, -\mu) = 1 - \sqrt{1 - c}H(\mu).$$

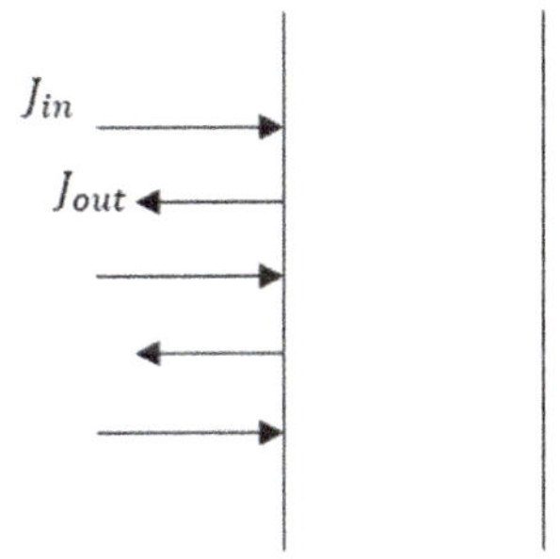

FIGURE 11.1 Reflection of an incident beam of neutrons.

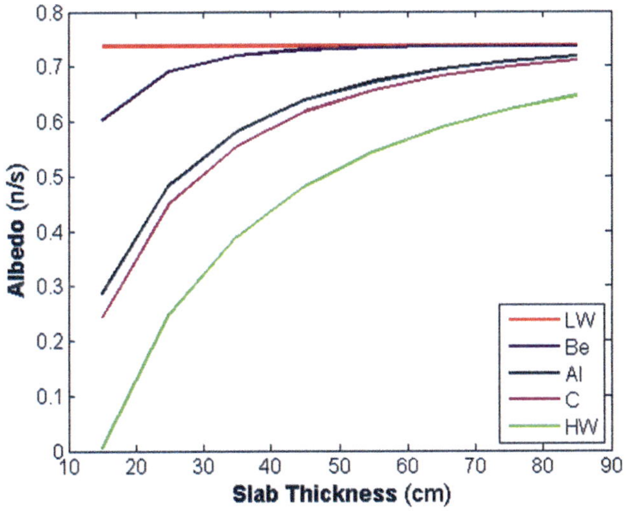

FIGURE 11.2 Albedos for light water, beryllium (Be), aluminum (Al), graphite (C), and heavy water (HW) due to an anisotropic source incident on the left face. [Annex A: CH11_AlbedoSlabDiffTh.m]

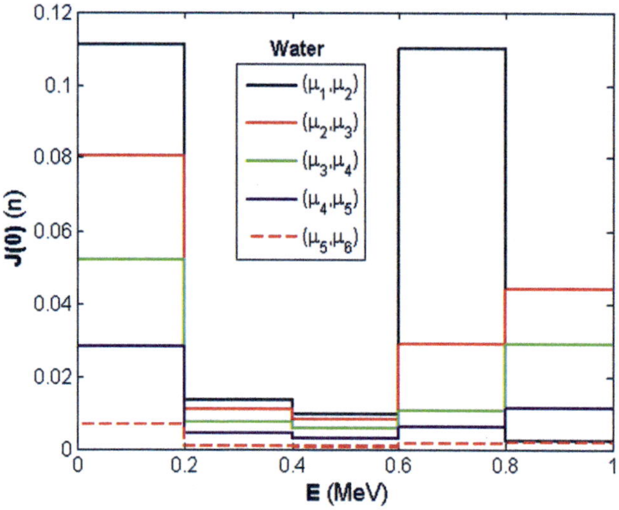

FIGURE 11.3 Albedo for 1 MeV neutrons incident on the left face of a water slab of thickness 10 cm. ($\mu_1 = -1, \mu_2 = -0.8, \mu_3 = -0.6, \mu_4 = -0.4, \mu_5 = -0.2, \mu_6 = -1$).

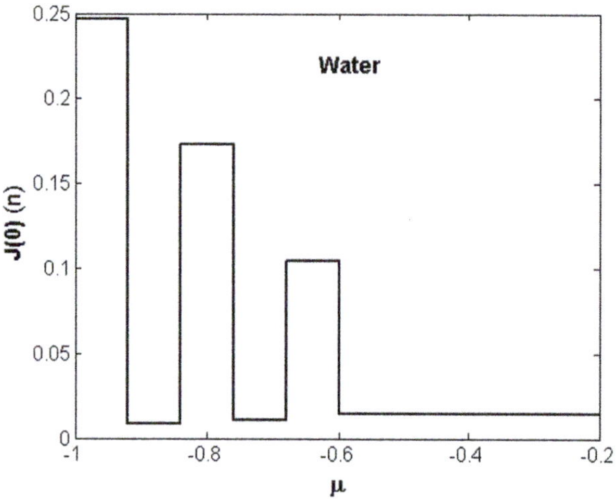

FIGURE 11.4 Albedo for a water slab of thickness 10 cm for 1 MeV incident neutrons.

The MC results for albedo were obtained from a MCNP simulation (input file ANNEX B: BK11albd) for 1 MeV mono-energetic neutrons incident on the left face of a slab of water of thickness 10 cm. The number of neutrons simulated was 400,000 giving a relative error of $<1\%$. The albedos obtained from a MCNP simulation are shown in Fig. 11.3 as a function of the reflection angle ($\mu_1 = -1, \mu_2 = -0.8, \mu_3 = -0.6, \mu_4 = -0.4,\ \mu_5 = -0.2,\ \mu_6 = -1$). The highest reflection in the range $-1 < \mu < -0.8$ is for low energy neutrons ($E < 0.2$ MeV) since higher energy neutrons penetrate deep into the medium.

The energy-integrated albedos in water are shown in Fig. 11.4; the large-angle reflection is most prominent falling off with decreasing angle.

Exercise 11.2:

From the steady-state one-group diffusion equation (Eq. 5.3), show that for an incident source S n/s on the left side of a slab of thickness a, and extrapolated distance d, the flux $\phi(x)$ and albedo β are:

$$\phi(x) = A \quad \cosh\frac{x}{L} + B\sinh\frac{x}{L} \tag{11.17}$$

where the constants A and B are

$$A = -\,\mathrm{Btanh}\frac{a+d}{L}$$

TABLE 11.4 Albedo for water and graphite slabs (MC simulations).

μ	Water	SS316
−1.0, −0.8	0.246371 0.0086	0.207984 0.0031
−0.8, −0.6	0.172610 0.0107	0.169504 0.0035
−0.6, −0.4	0.104693 0.0144	0.130964 0.0040
−0.4, −0.2	0.0529103 0.0208	0.086881 0.0051
−0.2, 0	0.0120528 0.0446	0.026080 0.0096

and the albedo is

$$\beta = \frac{J_{\text{out}}}{J_{\text{in}}} = \frac{p - \frac{2D}{L}\coth\frac{a+d}{L}}{p + \frac{2D}{L}\coth\frac{a+d}{L}},$$

where $p = \Sigma_s/\Sigma_t$.

Note: The incident and reflected currents are

$$J_{\text{in}}(0) = \frac{\Sigma_s}{4\Sigma_t}\phi(0) - \frac{D}{2}\frac{d\phi}{dx}|_{x=0}$$

and

$$J_{\text{out}}(0) = \frac{\Sigma_s}{4\Sigma_t}\phi(0) + \frac{D}{2}\frac{d\phi}{dx}|_{x=0}$$

and the net current is $J = J_{\text{in}} - J_{\text{out}}$.

The energy-integrated albedos are listed in Table 11.4; in water, the back-scattering is predominant for a mono-energetic anisotropic source.

Comparing the diffusion values of albedo in Fig. 11.2 with MC simulation values in Fig. 11.4 for 1 MeV neutrons, a large difference is observed. This is due to the energy of incident neutrons. For water, the albedo increases from 0.588636 (0.0040) to 0.74175 (0.0023) as the energy of the incident neutrons decreases from 1 MeV to 0.1 eV.

11.4 Flux in a slab

Three idealized and illustrative cases in slab geometry are infinite medium with an incident source, finite medium with incident source on the left boundary, and finite medium with an isotropic source.

11.4.1 Diffusion theory

Analytical solutions for infinite media fixed source problems are typically in the form of decaying exponentials while finite media have algebraic, trigonometric, or hyperbolic functions.

Some elementary cases for non-multiplying media in DT for slab geometry (Section 5.2.1; Table 5.1) are:

- infinite medium with an incident source S_0 neutrons/s, for which the boundary conditions of $\lim A_{x\to 0} J = S_0/2$ and $x \to \infty$ give the flux

$$\phi = \frac{S_o}{2\kappa D}\exp(-\kappa|x|), x \neq 0 \tag{11.18}$$

where $\kappa = \sqrt{\Sigma_a/D}$ the inverse of the diffusion length expressed in terms of the total cross-section

$$L = (3\Sigma_t\Sigma_a)^{-1/2} = \{3(1-c)\}^{-1/2}$$

which, is the same as the eigenvalue ν_0 given by the transcendental equation (Eq. 11.11) in Table 11.3.

- finite slab $x \in (0, \Delta)$, with an impinging flux at the left surface of a slab $J(0) = F_L$ and $\phi(\Delta) = 0$

- finite slab with an isotropic source at the center, the boundary conditions are $\phi(0) = \phi(\Delta) = 0$ and $\lim_{x \to 0} J = \lim_{x \to 0} \left(-D d\phi/dx\right) = S_0/2$. For the point isotropic source, the flux

$$\phi(x) = \frac{S_o L}{2D}\left[1 + e^{-\Delta/L}\right]^{-1}\left(e^{-|x|/L} - e^{(|x|-\Delta)/L}\right) \tag{11.19}$$

which can also be expressed in the following forms:

$$\phi(x) = \frac{S_o L}{2D}\frac{\sinh[(\Delta - 2|x|)/2L]}{\cosh(\Delta/2L)} \tag{11.20}$$

and

$$\phi(x) = \frac{2S_o}{\Delta\Sigma_a}\sum l_{odd}\frac{1}{1 + B_l^2 L^2}\cos\frac{l\pi x}{\Delta}. \tag{11.21}$$

The transport correction in the diffusion model is usually made by considering the slab to be of physical dimension $\Delta - 2d$ where the extrapolation distance is $d = 0.71\lambda_{tr}$. The extension of the eigenfunction form of the solution is easily made to the 3D case where the flux $\phi(x, y, z)$ can be found as

$$\phi = \frac{8S_o}{\Delta^3\Sigma_a}\sum l, m, n,_{odd}\frac{1}{1 + B_{l,m,n}^2 L^2}\cos\frac{l\pi x}{\Delta}\cos\frac{m\pi y}{\Delta}\cos\frac{n\pi z}{\Delta}. \tag{11.22}$$

where

$$B_{l,m,n}^2 = \left(l^2 + m^2 + n^2\right)\left(\frac{\pi}{a}\right)^2.$$

11.4.2 Transport theory

In transport theory, some elementary cases (Section 6.2) for slab geometry are

- source-free infinite medium, when separation of variables is assumed with eigenfunctions $\psi_v(\mu)$ and eigenvalues ν (Table 11.3) to give an asymptotic solution

$$\phi(x, \mu) = e^{-x/\nu}\psi_\nu(\mu) \tag{11.23}$$

based on the 1D transport equation with isotropic scattering.

$$\left[\mu\frac{\partial}{\partial x} + 1\right]\phi(x, \mu) = \frac{c}{2}\int_{-1}^{1} d\mu' \phi(x, \mu') \tag{11.24}$$

- infinite medium spherical harmonics method in the P_1 approximation

$$\phi(x) = \frac{1}{2}\sqrt{\frac{3}{1-c}}e^{-\sqrt{3(1-c)}x} \tag{11.25}$$

- infinite medium with plane isotropic source has a *transient* behavior (in the sense of collisional equilibrium) in the P_3 approximation, for which a Fourier transform of Eq. (11.23) for a plane source $\delta(x)$ is

$\phi(x) = \frac{1}{2\pi}\int_{-\infty}^{\infty}\frac{e^{ikx}}{(1-c) + g(k)}$where

$$g(k) = \frac{\frac{9}{7}k^4 + 15k^2}{\frac{55}{7}k^2 + 15}.$$

Evaluating the residues at the simple poles, yields

$$\phi(x) = \frac{7}{9}\left[\frac{-\frac{55}{7}m^2 + 15}{2m(n^2 - m^2)}e^{-m|x|} - \frac{-\frac{55}{7}n^2 + 15}{2n(n^2 - m^2)}e^{-n|x|}\right] \tag{11.26}$$

where m and n, and the eigenvalues, which are the relaxation lengths, $v_0 \equiv 1/m$ and $v_i \equiv 1/n$, are given in Table 11.5.

Spatial attenuation is, as for the cases above, defined by eigenvalues that include ν_0. Far from the source, the asymptotic part will dominate (Bell & Glasstone, 1979). Conversely, in strongly absorbing media, the transient part will dominate as there will hardly be any collisional equilibrium. The source appears as a normalization factor for collisional equilibrium while it dominates for weak equilibrium.

- finite 1D slab, Ganapol (2008) with the impinging (on left surface) and vacuum (on right surface) boundary conditions for: and $\mu > 0$; $\phi(0, \mu) = F_L(\mu)$ and $\phi(\Delta, \mu) = 0$. The exact F_N solution, based on the integral equations, for this case has been given as

$$\phi(x) = \int_0^1 d\mu F_L(\mu) e^{-x/\mu} + \sum_{\alpha=0}^{N-1} c_\alpha\left(x\right) \int_0^1 d\mu \phi_\alpha(\mu) \tag{1}$$

where $\phi_\alpha(\mu)$ are basis functions and the functions $c_\alpha(x)$ are determined from the procedure defined.

11.4.3 Monte Carlo simulation

The MC simulation uses the collision estimator (CE) to estimate the total number of collisions in a region, that is, the weight participating in collisions in a region s $\psi = \Sigma_t \varphi = \langle p_i \rangle$ so that $\varphi_{CE} = \langle p_i \rangle / \Sigma_t$, and the track-length estimator (TLE) For the TLE, in which the flux is the mean weighted track-length expressed as

$$\phi_{TLE} = nv = \frac{N}{V} * \frac{s}{t} \equiv \frac{1}{Vt} \langle p_i s_i \rangle .$$

11.4.4 Comparison

The simulation is carried out for strong absorbing materials, such as boron and gadolinium, to strong scattering materials such as iron and aluminum using one-group data (Lamarsh & Baratta, 1955) listed in Table 11.6.

For a unit isotropic source located in the center of a cube of 5 m.f.p. the thermal flux, obtained from DT (ANNEX C, MATLAB Program: ExactSolSlabJan03) is shown in Fig. 11.5.

It can be seen that the flux in a high scattering material such as aluminum will attenuate slower than that in a highly absorbing material. Both gold (Au) and boron (B) are highly absorbing and hence the flux drops rapidly.

The leakage, per unit source neutron, computed for the slabs is as follows: Al: 7.802×10^{-2}, Fe: 4.832×10^{-2}, U: 5.063×10^{-3}, Au: 5.146×10^{-4}, B: 4.648×10^{-4}, Gd: 3.532×10^{-4}.

TABLE 11.5 Infinite medium eigenvalues from P_3 transport theory.

c	0.5	0.6	0.8	0.9	0.95	0.99
v_0	1.0114	1.0838	1.4047	1.9029	2.6350	5.7971
v_i	0.4094	0.4271	0.4660	0.4866	0.4969	0.5051

TABLE 11.6 Thermal cross-section data.

Element	σ_a (b)	σ_s (b)	$p = \sigma_s/\sigma_t$
Aluminum	0.235	1.4	0.8563
Boron	759	4.0	0.0052
Iron	2.53	11.0	0.8130
Gold	98.8	9.3	0.0860
Gadolinium	46,000	4.0	8.695×10^{-5}
Uranium	7.5	8.3	0.5253

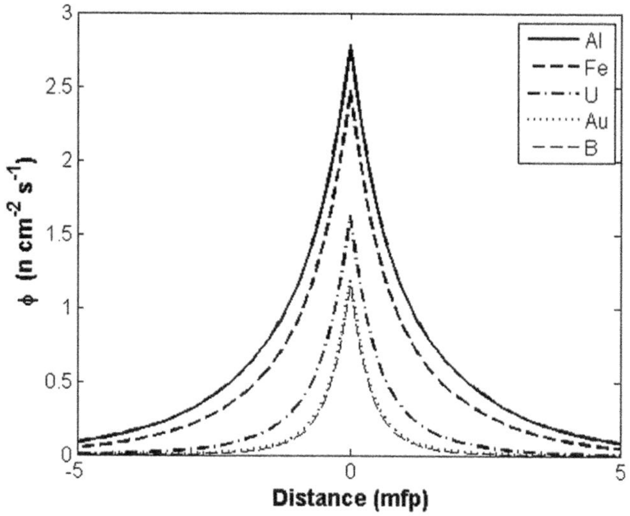

FIGURE 11.5 One-speed flux in a 1D slab.

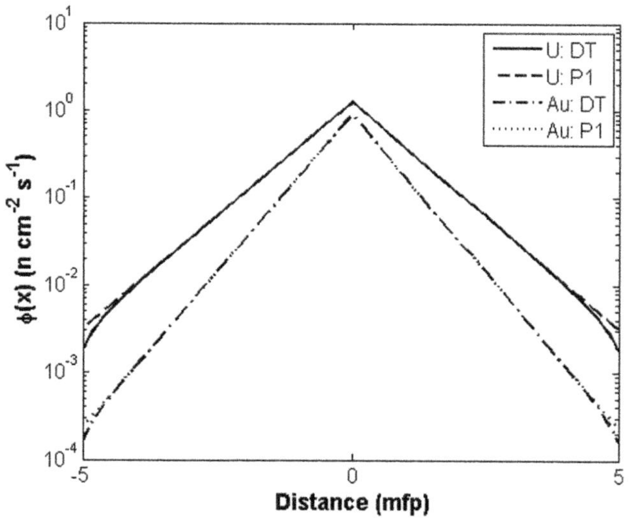

FIGURE 11.6 One-speed flux in 1D slab (DT vs P_1).

A comparison of DT and spherical harmonics P_1 approximation for U and Au is shown in Fig. 11.6. The significant difference is near the boundaries where the validity of Fick's law is doubtful.

Fig. 11.7. shows the scalar flux in a 1D slab of gadolinium with an isotropic source at the center. It is seen that for a strong absorber, there is significant difference between the P_3 result and the DT result near the source and near the boundaries. Thus, for a strong absorber, the sharper rise due to lack of collisional equilibrium is better modeled by the P_3 approximation.

For a slab with an incident source J_{in} on the left side of the surface (Fig. 11.8), the DT fluxes are plotted for light water (LW), beryllium (Be), aluminum (Al), graphite (C), and heavy water (HW). The sharpest attenuation is for water which is the most absorbing of the five materials considered; while heavy water has the slowest attenuation. Beyond a slab thickness of ~11 cm, the flux has become negligible due which the albedo in Fig. 11.2 saturates to a constant value.

To consider a more detailed flux behavior, a simulation carried out with MCNP (ANNEX B input file) for an anisotropic 1 MeV mono-energetic source incident from the left face of water and steel (SS-316) slabs, shows a rapid drop in Fig. 11.9. At the interface, the flux in both cases is high (~100 times higher than in the next 1 cm layer). The flux in water falls off faster than that in SS-316 (higher c than that for water).

It is also important to visualize the energy spectrum of the flux in a hydrogenous medium such as water and a high c material such as SS-316.

As shown in Fig. 11.10, the energy dependent flux in SS-316 has a hard spectrum while a Maxwellian is observed in water at thermal energies (~0.025 eV).

Now consider the MC collision density in a 5 m.f.p thick aluminum slab with a point isotropic source at the center.

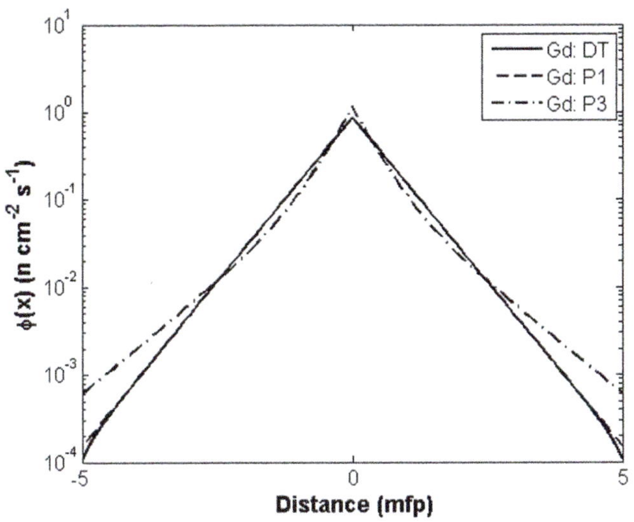

FIGURE 11.7 One-speed flux in Gd: DT, P_1, P_3 comparisons.

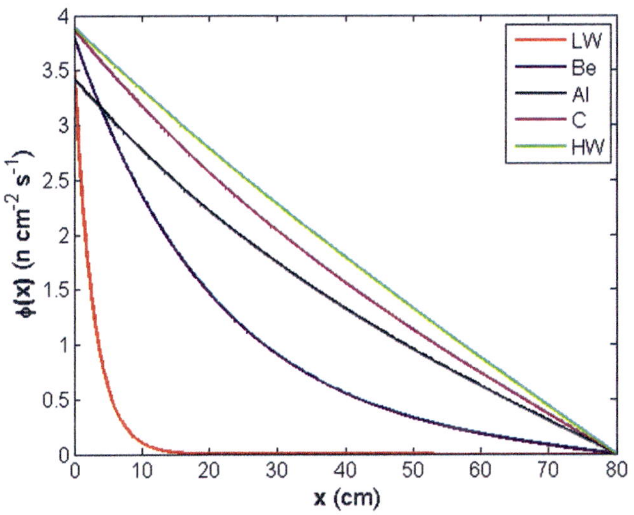

FIGURE 11.8 Neutron flux for light water, beryllium (Be), aluminum (Al), graphite (C), and heavy water (HW) due to an anisotropic source incident on the left face. [Program CH11_AlbedoSlabDiffTh.m].

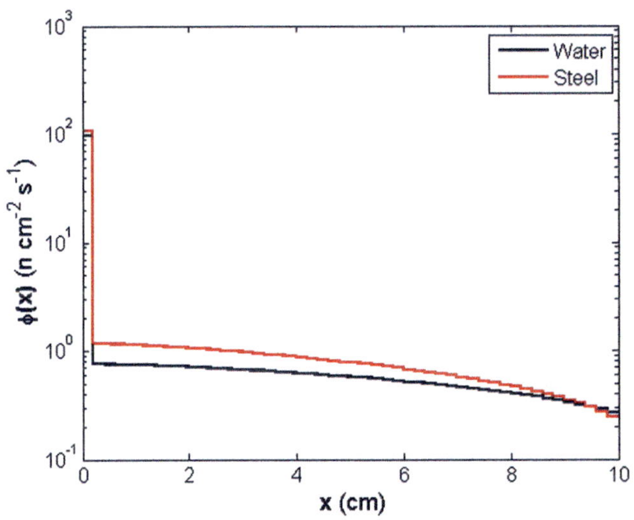

FIGURE 11.9 Neutron flux versus distance (x) in water and SS-316 slabs of thickness 10 cm due to a 1-MeV mono-energetic anisotropic source incident from the left face. (ANNEX B input file).

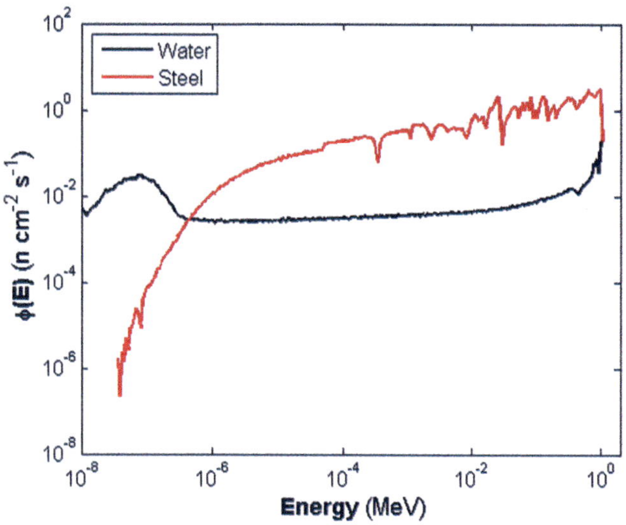

FIGURE 11.10 Neutron flux versus energy E in water and SS-316 slabs of thickness 10 cm due to a 1-MeV mono-energetic anisotropic source incident from the left face.

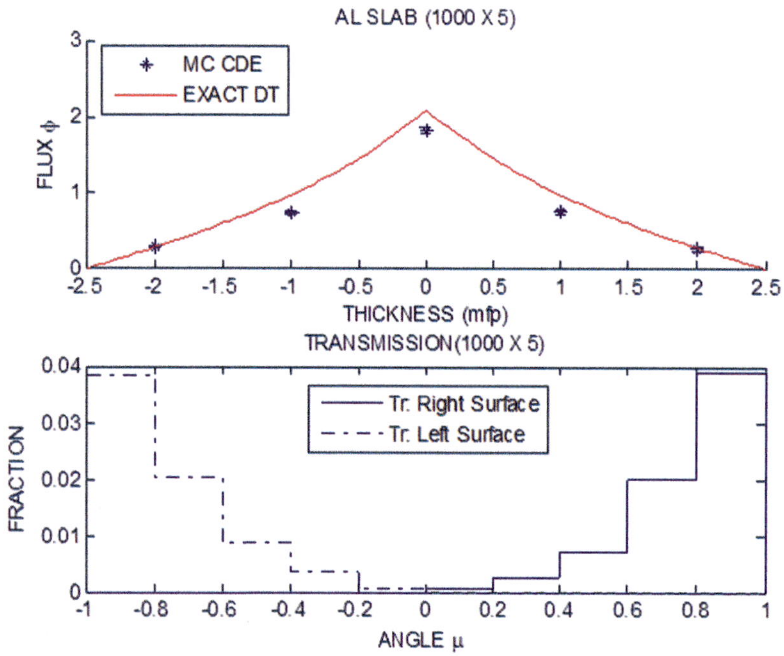

FIGURE 11.11 Comparison of collision density: Monte Carlo versus diffusion theory.

Fig. 11.11 shows the thermal flux computed with the collision density estimator (CDE) and the one-group diffusion expression. It can be seen that DT over-estimates the flux, especially near the source. However, the difference is smaller further from the source.

The difference between the CDE and TLE flux estimates for a sample size 1000×5 (1000 histories and 5 batches) is <2%. The transmission from the left (and right) surfaces is <4% for aluminum. For iron and boron the transmission was ~3.5% and 0.6.% respectively.

11.5 Flux in a finite sphere with a point isotropic source

11.5.1 Diffusion theory

In Section 5.2.2, the neutron DT flux is given by

$$\phi(r) = \frac{S}{4\pi Dr} \frac{\sinh\kappa(R+d-r)}{\sinh\kappa(R+d)}$$

and the leakage from the sphere is given by

$$L = 4\pi R^2 J(R) = -4\pi R^2 D\overline{\nabla}\phi = \frac{S\kappa(R+d)}{\sinh\kappa(R+d)}.$$

Consider an aluminum sphere of radius $R = 25$ cm, $\rho = 2.70$ g/cm^3 (mass = 176.7 kg) with the data: $\sigma_s = 1.4$ b, $\sigma_a = 0.235$ b, $N = 0.06027$ nuclei (b.cm)$^{-1}$, $\Sigma_t = 0.0985$ cm^{-1}.

With this data, the mean free path is $\lambda = 1/\Sigma_t = 10.1478$ cm, so that the sphere is of radius 2.4636 mean free paths.

From the data: $\mu_0 = 2/3A = 0.0247$, $\Sigma_{tr} = 0.0823\text{cm}^{-1}$, $d = 0.71/\Sigma_{tr} = 8.6275$ cm, $\overline{d} = 0.8502$ fp, $D = 4.0505$ cm, $L = \sqrt{D/\Sigma_a} = 16.9108$ cm, $\overline{L} \equiv L/\lambda = 1.6664$, $c \equiv \Sigma_s/\Sigma_t = 0.8563$.

From the one-speed transport theory spherical harmonics P_1 approximation, the quantity L_T is equivalent to the diffusion length (since $1-c$ is the macroscopic absorption cross-section) with the diffusion coefficient $D = 1/3$ in units of mean free paths. For this data, $L_T = 1/\left[\sqrt{3(1-c)}\right] = 1.5229$. However, transport theory gives the (normalized) relaxation length as $\nu_0 = 1.6115$. The DT flux shown in Fig. 11.12 becomes very large at $r = 0$ where the source is located.

Fig. 11.13 is the thermal flux (DT) for a point isotropic source at the center of a sphere; compared with Fig. 11.8 it has the same shape except that light water is at the top. This shows the effect of geometry on the attenuation of flux.

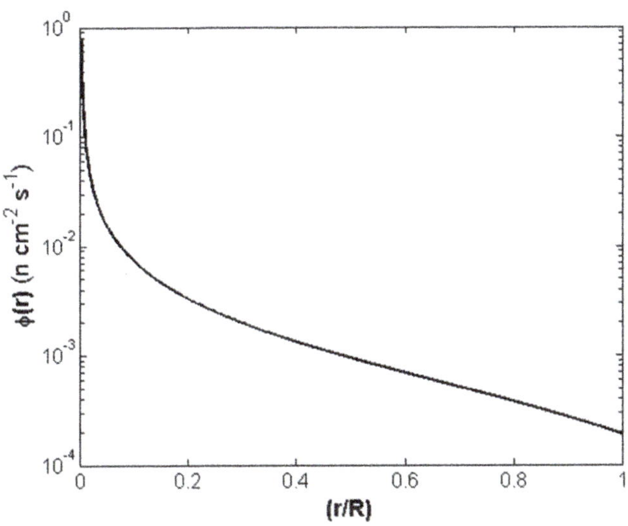

FIGURE 11.12 Diffusion Theory flux in an aluminum sphere.

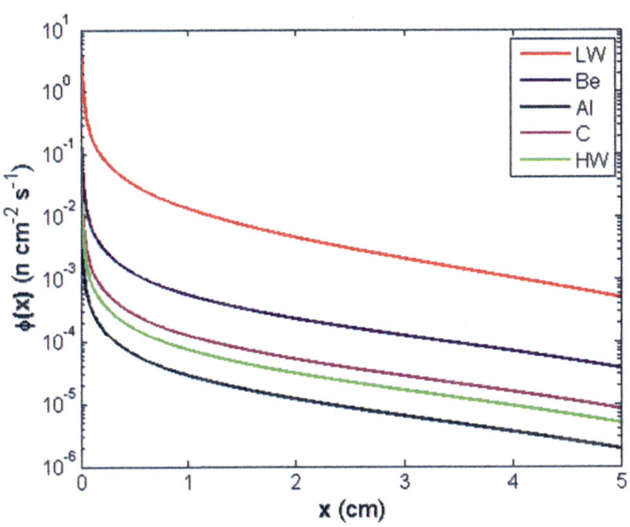

FIGURE 11.13 Thermal flux (diffusion theory) for a point isotropic source at the center of a sphere.

11.5.2 Transport theory exact solution

In Section 6.2, the flux in a sphere with point isotropic source equations is obtained with a program written in MATLAB using a Gaussian quadrature with $N = 32$. The expansion coefficients are obtained by solving the Fredholm equations iteratively. For $c = 0.8563$ (the one-group data for aluminum) (Lamarsh & Baratta, 2001), gives $\nu_0 = 1.6183$, $N(\nu_0) = 0.2669$, $H(\nu_0) = 1.903$, $G(\nu_0) = 1.1601$, $E(\nu_0) = 1.1627$, and the extrapolation distance, from Eq. (11.7), $z_0 = 0.83024$.

Table 11.7 lists the results, from DT and transport theory. Clearly, DT does not give good results near the source and boundary. The ratio of the transport-to-diffusion flux is close to unity in the range 0.74–2.2 m.f.p. This will have an important bearing when choosing a reference for the Kullback-Leibler estimates.

To better understand the fluxes in Fig. 11.13, a MCNP simulation was carried out to compare fluxes in water, uranium and aluminum. The simulation used 50,000 histories with a point isotropic mono-energetic source of 1 MeV located at the center of the sphere of radius 8.4710 cm (the same as for Godiva).

The neutron fluxes shown in Fig. 11.14 validate Fig. 11.13 with the scalar flux in water being higher than that for aluminum and uranium.

Although, in Fig. 11.14, the neutron flux is higher in water, the emergent spectrum, given in Table 11.8, shows that water moderates the high incident energy while aluminum and uranium do not result in comparable shielding of neutron radiation.

TABLE 11.7 Diffusion theory flux compared with transport theory.

r/R	r	$\bar{r}$	$\phi(r)$	$4\pi r^2\phi(r)$	
	cm	m.f.p	n/cm^2/s	DT	TT
0.0	0.0	0.0	—	—	1.0
0.1	2.5	0.2464	6.7341e-003	0.5289	1.1017
0.2	5.0	0.4927	2.8785e-003	0.9043	1.1823
0.3	7.5	0.7391	1.6353e-003	1.1559	1.2413
0.4	10.0	0.9854	1.0406e-003	1.3076	1.2788
0.5	12.5	1.2318	7.0198e-004	1.3783	1.2946
0.6	15.0	1.4782	4.8905e-004	1.3827	1.2882
0.7	17.5	1.7245	3.4613e-004	1.3321	1.2586
0.8	20.0	1.9709	2.4557e-004	1.2344	1.2032
0.9	22.5	2.2172	1.7214e-004	1.0951	1.1157
1.0	25.0	2.4636	1.1678e-004	0.9172	0.9643

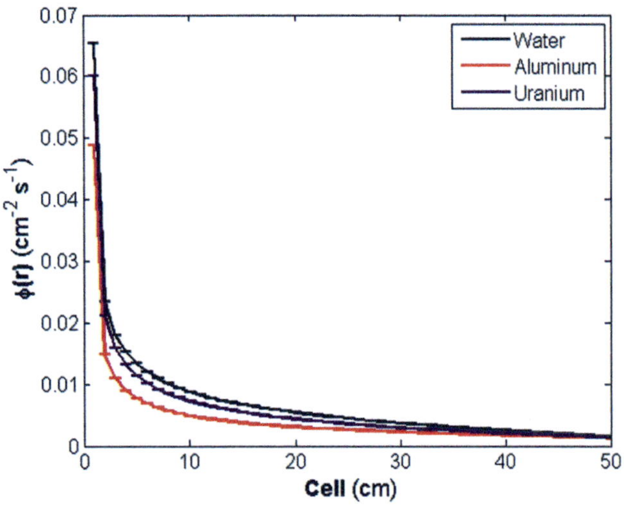

FIGURE 11.14 Neutron flux versus cell number in a sphere of radius 8.7410 cm divided into 50 equal-volume cells.

TABLE 11.8 Emerging current spectrum from a sphere of radius 8.471 cm.

Energy bin	Water	Aluminum	Uranium
0–1 eV	1.65416×10^{-1} 0.0092	0	0
1 eV–1 keV	1.70841×10^{-1} 0.0098	0	0
1 keV–1 MeV	6.26264×10^{-1} 0.0035	9.99463×10^{-1} 0.0000	8.99460×10^{-1} 0.0008
Total	9.62522×10^{-1} 0.0005	9.99463×10^{-1} 0.0000	8.99460×10^{-1} 0.0008

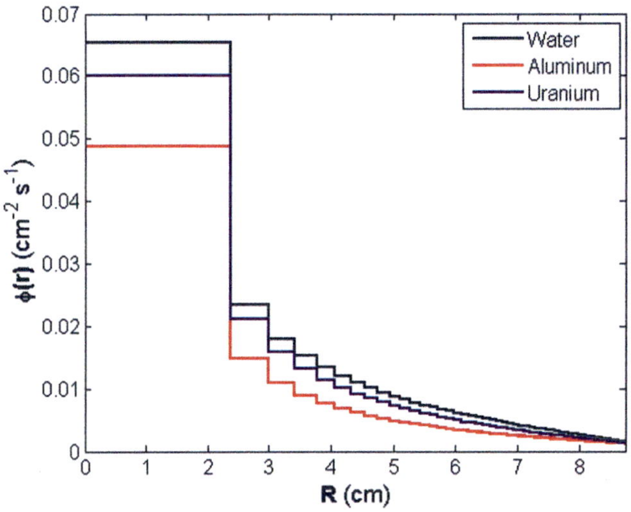

FIGURE 11.15 Neutron flux versus radius in a sphere of radius 8.7410 cm divided into 50 equal-volume cells.

A brief discussion on Fig. 11.9 in which a very high flux appears at the interface where source neutrons are incident is helpful to clarify this artefact. In both slab and spherical geometry, when equal volume regions are specified, the *first volume* has a larger radius (difference) and the fluxes appear to take a shape illustrated in Fig. 11.15 which is the same result as shown in Fig. 11.14. So, the magnitude of the flux is very high in a region adjacent to the source in both slab and spherical geometries.

11.5.3 Monte Carlo simulation

Fig. 11.16 shows a comparison of the exact DT flux with the TLE flux in the 30 equal-volume regions of an aluminum sphere of radius 25 cm (2.4636 m.f.p) with a mass of 176.7 kg. To comment on the accuracy of the results, we need to consider two parameters *viz* the mean free path λ and the scattering probability p_s. For this one-speed model, $\lambda \sim 10.15$cm and $p_s \sim 0.86$. We can thus expect that there will be a sufficient number of collisions for the CE flux estimate to agree well with the TL estimate. For this case, the average number of collisions per source neutron is ~ 1.88 in the first region gradually decreasing to ~ 0.17 in the last region which is about 2.4 m.f.p's away from the source. For a sample size $N \sim 10,000$ this means that we will have of the order of 19,000 collisions in the first region and ~ 1700 collisions in the last region. Similarly, the relatively high scattering probability makes this material far from an absorber. It is seen that these two factors contribute to the good agreement between DT and the MC estimate from the mid-point of the first region which is half a mean free path away. Similarly, good agreement is observed for the leakage, which is 0.54319 from the MC simulation, compared with 0.5548 from the exact solution.

Comparisons of diffusion and transport flux show good agreement far from the boundaries. The convergence can be verified using the Kullback-Leibler divergence indicating that N$\sim$10,000 is a sufficient sample size for simulation.

Consider now the individual cell balances shown in Table 11.9. The steady-state balance requires

$$J_{in}^{(c)} + S^{(c)} = J_{out}^{(c)} + A^{(c)}$$

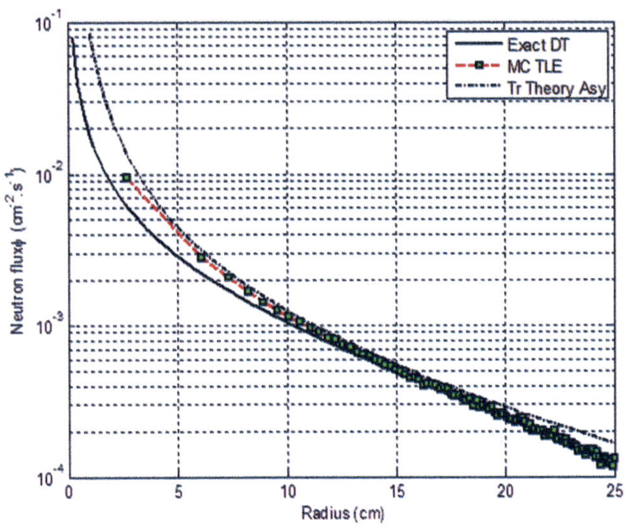

FIGURE 11.16 Monte Carlo compared with transport and diffusion theory.

TABLE 11.9 Cell balances using diffusion theory flux.

Cell	Sources			Sinks			Δ
	$S^{(c)}$	$J_{in}^{(c)}$	Total In	$A^{(c)}$	$J_{out}^{(c)}$	Total Out	
1	1	0	1.0000	0.1467	0.8533	1.0000	0
2	0	0.8533	0.8533	0.0590	0.7943	0.8533	0
3	0	0.7943	0.7943	0.0411	0.7532	0.7943	0
4	0	0.7532	0.7532	0.0317	0.7215	0.7532	0
5	0	0.7215	0.7215	0.0256	0.6958	0.7215	0
6	0	0.6958	0.6958	0.0213	0.6745	0.6958	0
7	0	0.6745	0.6745	0.0181	0.6564	0.6745	0
8	0	0.6564	0.6564	0.0155	0.6409	0.6564	0
9	0	0.6409	0.6409	0.0134	0.6276	0.6409	0
10	0	0.6276	0.6276	0.0116	0.6160	0.6276	0

The reaction rate $\mathscr{R}_k^{(c)}$ for a reaction k in a cell c between radii $r^{(c)}$ and $r^{(c+1)}$ $\mathscr{R}_k^{(c)} = \int \Sigma_k^c \phi(r) dV$ is

$$\mathscr{R}_k^{(c)} = \frac{S\Sigma_k}{D\sinh\kappa(R+d)}\left\{-\frac{r^{(c+1)}}{\kappa}\cosh\kappa\left(R+d-r^{(c+1)}\right)+\frac{r^{(c)}}{\kappa}\cosh\kappa\left(R+d-r^{(c)}\right)-\frac{1}{\kappa^2}\sinh\kappa\left(R+d-r^{(c+1)}\right)+\frac{1}{\kappa^2}\sinh\kappa\left(R+d-r^{(c)}\right)\right\}$$

$$\mathscr{R}_k^{(T)} = \frac{S}{\Sigma_a}\Sigma_k\left[\frac{\sinh\kappa(R+d)-\sinh\kappa d-\kappa R\cosh\kappa d}{\sinh\kappa(R+d)}\right]$$

where $\mathscr{R}_k^{(T)}$, obtained by setting $r^{(c)}=0, r^{(c+1)}=R$, is the total reaction rate of reaction type k in the sphere. With the leakage given by

$$\mathscr{L} = \int \nabla \bullet J \quad dV = -\int D\nabla^2\phi \quad dV = -4\pi D\left(r^2\frac{d\phi}{dr}\Big|_r^{r+dr}\right)$$

$$= 4\pi r_2^2 J(r_2) - 4\pi r_1^2 J(r_1) = J_{out}^{(c)} - J_{in}^{(c)}$$

where

$$J(r) = -D\frac{d\phi}{dr} = \frac{S}{4\pi r^2}\left[\frac{\kappa r\cosh[\kappa(R+d-r)]+\sinh[\kappa(R+d-r)]}{\sinh\kappa(R+d)}\right].$$

Elementary solutions and simulations show that

1. DT, based on Fick's law, is valid in the "interior" of a medium and not near a source or boundary;
2. the asymptotic relaxation length v_0 for a source-free infinite medium transport solution is identical to that from DT, when it is called the diffusion length L, for a pure scattering medium ($c = 1$), and
3. the relaxation length, expressed by the eigenvalues, has a "transient" and an "asymptotic" part which is a function of the medium properties (c).

Problems

1. In Section 11.1, the eigenvalues ν_o are defined by the transcendental equation (Eq. 11.11). Calculate ν_o for c = 1.0, 1.2, 1.4 and compare with values in Table 11.3.
2. Repeat Q.1 using values of for c = 0.2, 0.4, 0.6, 0.8, 1.0 and comment on your results, especially in the limits of low c and the case c = 1.
3. Explain how the Kullback-Leibler Divergence estimates can be used to comment on the quality of the results in Fig. 11.16 (MC compared with transport and diffusion estimates).
4. For the slab and spherical geometry results, use a different material from aluminum with a lower c value, such as boron, and again compare the results. Does MC sample size depend on the scattering power of the material?
5. Plot the relative errors for the above and again, comment on the trend you observe.

Nomenclature

English lower case

c	number of neutrons emerging from an interaction
d	extrapolation distance
k_{eff}	effective multiplication
k_∞	infinite multiplication
N	atomic number density

English upper case

B	buckling
D	diffusion coefficient (xx)
$\hat{D}$	derivative operator
H	Chandrasekhar H-function
J	neutron current
L	diffusion length
N_i	atomic density of the ith nuclide
R_c	critical radius
$\overline{R}$	extrapolated radius
S	source

Greek lower case

β	albedo
μ	cosine of angle of scattering
μ_0	average cosine of scattering angle
ν_0	eigenvalue
ν	number of neutrons produced per fission
ϕ	flux
ϕ_{as}	asymptotic flux
σ_a	microscopic absorption cross section
σ_f	microscopic fission cross section
σ_r	microscopic removal cross section
σ_s	microscopic scattering cross section
σ_{tr}	microscopic transport cross section

σ_t microscopic total cross section
τ neutron age
χ fission spectrum

Greek upper case

$\sum_a$ macroscopic absorption cross section
$\sum_f$ macroscopic fission cross section
$\sum_r$ macroscopic removal cross section
$\sum_s$ macroscopic scattering cross section
$\sum_{tr}$ macroscopic transport cross section
$\sum_t$ macroscopic total cross section

Abbreviations

m.f.p mean free path
CDE collision density estimate
DT diffusion theory
MC Monte Carlo
TLE track-length estimator
SS-316 Stainless Steel Type 316

References

Bell, G. I., & Glasstone, S. (1979). *Nuclear reactor theory*. New York: Robert E. Krieger Publishing Co., Inc..

Clark, M., & Hansen, K. F. (1964). *Transport methods. Numerical methods of reactor analysis*. Elsevier.

Ganapol, B. D. (2008). *Analytical benchmarks for nuclear engineering applications case studies in neutron transport theory*. Nuclear Energy Agency.

Lamarsh, J. R., & Baratta, A. J. (2001). *Introduction to nuclear engineering* (3rd ed., p. 783)*Prentice Hall.*

Lamarsh, J. R., & Baratta, A. J. (1955). *Introduction to nuclear engineering*. McGraw-Hill.

Annex A MATLAB Program AlbedoSlabDiffTh.m (Section 11.3)

```
% CH11_AlbedoSlabDiffTh.m
% Aluminum data Lamarsh (INE)
N=0.06027e23; siga=0.235e-24;Siga=N*siga; D=4.0505;
L=sqrt(D/Siga);kappa=1/L;a=80; % slab thickness
d=8.6275;den=2.70;S=1.0;
% Diffusion coefficient and thermal diffusion length (source L&B)
%    water                heavy water         beryllium        graphite
Dw=0.16;Lw=2.85;     Dhw=0.87;Lhw=97;  Dbe=0.50;Lbe=21;  Dc=0.84;Lc=59;

% slab thickness a
% S=1 incident from left side (J+)
% phi(x) =B1 cosh x/L + B2 sinh (x/L)

B2Al =(-4*S*cosh(a/L))/( (2*D/L)*cosh(a/L) + sinh(a/L)); B1Al =4*(S+(D/
(2*L))*  B2Al);
B2w  =(-4*S*cosh(a/Lw))/((2*Dw/Lw)*cosh(a/Lw) + sinh(a/Lw));B1w  =4*(S+(Dw/
(2*Lw))* B2w);
B2hw =(-4*S*cosh(a/Lhw))/((2*Dhw/Lhw)*cosh(a/Lhw) +
sinh(a/Lhw));B1hw=4*(S+(Dhw/(2*Lhw))*B2hw);
B2be =(-4*S*cosh(a/Lbe))/((2*Dbe/Lbe)*cosh(a/Lbe) +
sinh(a/Lbe));B1be=4*(S+(Dbe/(2*Lbe))*B2be);
B2c  =(-4*S*cosh(a/Lc))/( (2*Dc/Lc)*cosh(a/Lc) + sinh(a/Lc));B1c
=4*(S+(Dc/ (2*Lc))* B2c);

Steps=1000;Del=a/Steps;Len=0;
for i=1:Steps
    Len=Len+Del;    x(i)=Len;
    phi(i)   = B1Al*cosh(x(i)/L) +   B2Al *sinh(x(i)/L);
    phiw(i)  = B1w *cosh(x(i)/Lw) +  B2w  *sinh(x(i)/Lw);
    phihw(i) = B1hw*cosh(x(i)/Lhw) + B2hw *sinh(x(i)/Lhw);
    phibe(i) = B1be*cosh(x(i)/Lbe) + B2be *sinh(x(i)/Lbe);
    phic(i)  = B1c *cosh(x(i)/Lc ) + B2c  *sinh(x(i)/Lc);
end

for j=1:8
    SlabTh(j)=10*j+5;
ggAl=(2*D/L)*coth(SlabTh(j)/L);  albedoAl(j) = (1-ggAl)/(1+ggAl);
ggw =(2*D/L)*coth(SlabTh(j)/Lw); albedow(j)  = (1-ggw)/(1+ggw);
gghw=(2*D/L)*coth(SlabTh(j)/Lhw); albedohw(j) = (1-gghw)/(1+gghw);
ggbe=(2*D/L)*coth(SlabTh(j)/Lbe); albedobe(j) = (1-ggbe)/(1+ggbe);
ggc=(2*D/L)*coth(SlabTh(j)/Lc); albedoc(j) = (1-ggc)/(1+ggc);
end

figure(1)
set(gca,'FontSize',12)
plot(x,phiw,'r-','LineWidth',1.5)
hold on
plot(x,phibe,'b-','LineWidth',1.5)
hold on
plot(x,phi,'k-','LineWidth',1.5)
hold on
plot(x,phic,'m-','LineWidth',1.5)
hold on
plot(x,phihw,'g-','LineWidth',1.5)
hold on
```

```
legend('LW','Be','Al','C','HW',12,'Location','NorthEast')
%text(0.1,0.7,'\bf Water','fontsize',12)
%xlim([1e-8 2])
ylim([0 4])
%hold on
grid off
xlabel('{\bf x} (cm)','fontsize',14)
ylabel('{\bf \phi(x)} (n cm^{-2} s^{-1})','fontsize',14)

figure(2)
set(gca,'FontSize',12)
plot(x,phiw,'k--','LineWidth',1.5)
hold on
plot(x,phibe,'k:','LineWidth',1.5)
hold on
plot(x,phi,'k-','LineWidth',1.5)
hold on
plot(x,phic,'k-.','LineWidth',1.5)
hold on
plot(x,phihw,'k--','LineWidth',1)

legend('LW','Be','Al','C','HW',12,'Location','NorthEast')
%text(0.1,0.7,'\bf Water','fontsize',12)
%xlim([1e-8 2])
ylim([0 4])
%hold on
grid off
xlabel('{\bf x} (cm)','fontsize',14)
ylabel('{\bf \phi(x)} (n cm^{-2} s^{-1})','fontsize',14)

figure(3)
set(gca,'FontSize',12)
plot(SlabTh,albedow,'r-','LineWidth',1.5)
hold on
plot(SlabTh,albedobe,'b-','LineWidth',1.5)
hold on
plot(SlabTh,albedoAl,'k-','LineWidth',1.5)
hold on
plot(SlabTh,albedoc,'m-','LineWidth',1.5)
hold on
plot(SlabTh,albedohw,'g-','LineWidth',1.5)
hold on

legend('LW','Be','Al','C','HW',12,'Location','SouthEast')
%text(0.1,0.7,'\bf Water','fontsize',12)
%xlim([1e-8 2])
%ylim([1e-3 2e-1])
%hold on
grid off
xlabel('{\bf Slab Thickness} (cm)','fontsize',14)
ylabel('{\bf Albedo} (n/s)','fontsize',14)
```

```
figure(4)
set(gca,'FontSize',12)
plot(SlabTh,albedow,'k--','LineWidth',1.5)
hold on
plot(SlabTh,albedobe,'k:','LineWidth',1.5)
hold on
plot(SlabTh,albedoAl,'k-','LineWidth',1.5)
hold on
plot(SlabTh,albedoc,'k-.','LineWidth',1.5)
hold on
plot(SlabTh,albedohw,'k--','LineWidth',1)

legend('LW','Be','Al','C','HW',12,'Location','SouthEast')
%text(0.1,0.7,'\bf Water','fontsize',12)
%xlim([1e-8 2])
%ylim([1e-3 2e-1])
%hold on
grid off
xlabel('{\bf Slab Thickness} (cm)','fontsize',14)
ylabel('{\bf Albedo} (n/s)','fontsize',14)
```

Annex B MCNP Input File BK11Albd (Section 11.2)

```
BK Ch11 Albedo comparison with H function
1    12 -1.0    1 -2   3 -4 5 -6 imp:n=1
2    0 (-1:2:-3:4:-5:6) imp:n=0

1    py  0
2    py  10
*3   px  -0.5
*4   px   0.5
*5   pz  -0.5
*6   pz   0.5

c       BLOCK 3    ........................................
SDEF POS=0 0 0 AXS=0 1 0 EXT=0 RAD=d1 PAR=1 ERG=1 VEC=0 1 0 DIR=1
SI1 0 0.05 $ radial sampling range: 0 to Rmax (=0.05cm)
SP1 -21 1 $ radial sampling weighting: r^1 for disk
c
mode    n
c  STEEL(316)(7.92 g/cm3)
m3      14000 -0.010 24000 -0.170 25055 -0.020 26000 -0.655
        28000 -0.120 42000 -0.025
c  Water (1.0 g/cm3)
m12     1001 2 8016 2
mt12    lwtr.02t
FC1    neutrons crossing surface 1, 2
F1:N    1 2
E1     0.2 0.4 0.6 0.8 1.0
C1     -0.8 -0.6 -0.4 -0.2 0 1.0 $ six bins
c ENERGY SPECTRUM OF FLUX IN SLAB
fc64 energy spectrum of flux in slab
f64:n 1
e64    3.77e-9 4.05e-9 4.36e-9 4.69e-9 5.04e-9 5.42e-9 5.83e-9 6.27e-9
       6.74e-9 7.25e-9 7.79e-9 8.38e-9 9.01e-9 9.69e-9 1.04e-8 1.12e-8
       1.20e-8 1.29e-8 1.39e-8 1.50e-8 1.61e-8 1.73e-8 1.86e-8
       2.e-8   2.15e-8 2.31e-8 2.49e-8 2.68e-8 2.88e-8 3.09e-8
       3.33e-8 3.58e-8 3.85e-8 4.13e-8 4.45e-8 4.78e-8 5.14e-8
       5.53e-8 5.94e-8 6.39e-8 6.87e-8 7.39e-8 7.95e-8 8.54e-8
       9.19e-8 9.88e-8 1.06e-7 1.14e-7 1.23e-7 1.32e-7 1.42e-7
       1.53e-7 1.64e-7 1.77e-7 1.90e-7 2.04e-7 2.19e-7 2.36e-7
       2.54e-7 2.73e-7 2.93e-7 3.15e-7 3.39e-7 3.65e-7 3.92e-7
       4.22e-7 4.53e-7 4.88e-7 5.24e-7 5.64e-7 6.06e-7 6.52e-7
       7.01e-7 7.54e-7 8.10e-7 8.71e-7 9.37e-7 1.01e-6 1.08e-6
       1.16e-6 1.25e-6 1.35e-6 1.45e-6 1.56e-6 1.67e-6 1.80e-6
       1.94e-6 2.08e-6 2.24e-6 2.41e-6 2.59e-6 2.78e-6 2.99e-6
       3.22e-6 3.46e-6 3.72e-6 4.e-6   4.30e-6 4.62e-6 4.97e-6
       5.35e-6 5.75e-6 6.18e-6 6.65e-6 7.15e-6 7.69e-6 8.26e-6
       8.89e-6 9.56e-6 1.03e-5 1.10e-5 1.19e-5 1.28e-5 1.37e-5
       1.48e-5 1.59e-5 1.71e-5 1.84e-5 1.97e-5 2.12e-5 2.28e-5
       2.45e-5 2.64e-5 2.84e-5 3.05e-5 3.28e-5 3.53e-5 3.79e-5
       4.08e-5 4.39e-5 4.72e-5 5.07e-5 5.45e-5 5.86e-5 6.31e-5
       6.78e-5 7.29e-5 7.84e-5 8.43e-5 9.06e-5 9.75e-5 1.05e-4
       1.13e-4 1.21e-4 1.30e-4 1.40e-4 1.51e-4 1.62e-4 1.74e-4
       1.87e-4 2.01e-4 2.17e-4 2.33e-4 2.50e-4 2.69e-4 2.89e-4
       3.11e-4 3.35e-4 3.60e-4 3.87e-4 4.16e-4 4.47e-4 4.81e-4
       5.17e-4 5.56e-4 5.98e-4 6.43e-4 6.91e-4 7.43e-4 7.99e-4
       8.60e-4 9.24e-4 9.94e-4 1.07e-3 1.15e-3 1.24e-3 1.33e-3
```

```
      1.43e-3 1.54e-3 1.65e-3 1.78e-3 1.91e-3 2.05e-3 2.21e-3
      2.37e-3 2.55e-3 2.75e-3 2.95e-3 3.17e-3 3.41e-3 3.67e-3
      3.95e-3 4.24e-3 4.56e-3 4.91e-3 5.28e-3 5.67e-3 6.10e-3
      6.56e-3 7.05e-3 7.58e-3 8.15e-3 8.77e-3 9.43e-3 1.01e-2
      1.09e-2 1.17e-2 1.26e-2 1.36e-2 1.46e-2 1.57e-2 1.68e-2
      1.81e-2 1.95e-2 2.09e-2 2.25e-2 2.42e-2 2.60e-2 2.80e-2
      3.01e-2 3.24e-2 3.48e-2 3.74e-2 4.02e-2 4.33e-2 4.65e-2
      5.e-2   5.38e-2 5.78e-2 6.22e-2 6.69e-2 7.19e-2 7.73e-2
      8.32e-2 8.94e-2 9.61e-2 1.03e-1 1.11e-1 1.20e-1 1.29e-1
      1.38e-1 1.49e-1 1.60e-1 1.72e-1 1.85e-1 1.99e-1 2.14e-1
      2.30e-1 2.47e-1 2.66e-1 2.86e-1 3.07e-1 3.30e-1 3.55e-1
      3.82e-1 4.10e-1 4.41e-1 4.75e-1 5.10e-1 5.49e-1 5.90e-1
      6.34e-1 6.82e-1 7.33e-1 7.89e-1 8.48e-1 9.12e-1 9.81e-1
      1.05    1.13    1.22    1.31    1.41    1.52    1.63
      1.75    1.88    2.03    2.18    2.34    2.52    2.71
      2.91    3.13    3.37    3.62    3.89    4.19    4.50
      4.84    5.20    5.60    6.02    6.47    6.96    7.48
      8.04    8.65    9.30    10
print
c cutoff the neutrons at 0 MeV
cut:n  j 0.0
nps    400000
```

Annex C MATLAB Program CH11ExactSolSlabJan03.m (Section 11.4.4)

```
%Exact solution slab ZK January 2010
%  CH11_ExactSolSlabJan03.m
% C:\MATLAB7\work\ModSim\ExactSolSlabJan3.m
%
%  zafar koreshi
%  2010
%
% open output file
resl=fopen('out1.txt','w');
tic
% begin data and inputs
% material data
factor=1.0;

Msig_a(1)=0.235e-24;Msig_s(1)=1.40e-24;MMolWt(1)=26.980;Mden(1)=2.70;% Al Lamarsh
pp558-559
Msig_a(2)=2.53e-24;Msig_s(2)=11.0e-24;MMolWt(2)=55.847;Mden(2)=7.87;% Fe Lamarsh
pp558-559
Msig_a(3)=7.60e-24;Msig_s(3)=8.3e-24;MMolWt(3)=238.03;Mden(3)=19.1;% U Lamarsh
pp558-559
Msig_a(4)=98.8e-24;Msig_s(4)=9.3e-24;MMolWt(4)=196.967;Mden(4)=19.32;% Au Lamarsh
pp558-559
Msig_a(5)=759.0e-24;Msig_s(5)=4.0e-24;MMolWt(5)=10.811;Mden(5)=2.30;% B Lamarsh
pp558-559
Msig_a(6)=46000.0e-24;Msig_s(6)=4.0e-24;MMolWt(6)=157.25;Mden(6)=7.95;% Gd Lamarsh
pp558-559

AvNo=6.023e24;

for imat = 1:6
    fprintf (resl,'\n \n ********** BEGINNING MATERIAL NO: %3.0f \n',imat);
    sig_a=Msig_a(imat);
    sig_s=Msig_s(imat);
    MolWt=MMolWt(imat);
    den=Mden(imat);
NoDen = den*AvNo/MolWt;
Sig_a=NoDen*sig_a;
Sig_s=NoDen*sig_s;

Sig_t=Sig_a+Sig_s;%cm^-1
mubar = 2./(3.*MolWt);
Sig_tr = Sig_t*(1.-mubar);
ExtrapDist = 0.71/Sig_tr;
DiffCoeff = 1./(3.*Sig_tr);
DiffLength =sqrt(DiffCoeff/Sig_a);
Pscat = Sig_s/Sig_t; % scattering probability
%
NR=10;
for i=1:NR
    D(i)=i;
end
```

```
XOPT(1)=0.5*(0.0+D(1));
for ir1 = 2:NR
    ir2=ir1-1;
    XOPT(ir1)=0.5*(D(ir2)+D(ir1));
end
% exact answer for this problem
% slab with two free surfaces; isotropic source at center
So = 1.0; %unit source strength
CubeLength = 10.0; % mfp's
EX1 = So * DiffLength /(2.0 * DiffCoeff);
% use the extrpolated distance in EX2
EX2 = CubeLength/Sig_t + ExtrapDist ; % actual length of the cube + the
extrapolation distance
EX3 = ( 1.0 + exp(-EX2/DiffLength) );
EX4 = EX1/EX3 ;
fprintf (res1,'\n Exact Solution   1D Slab Pt Isotropic Source \n');
fprintf (res1,'  Xopu      X (cm)      FLUX      EigenFunction Flux     %error \n');
LeftBdry = -CubeLength/2; % optical path units
XPTS = 101;
DEL  = CubeLength/(XPTS-1);

% Transport P-1 approx which is the diffusion flux with the
% transport correction
TrFactor1 = 0.5*sqrt(3./(1.0-Pscat));
TrFactor2 = sqrt(3.0*(1.0-Pscat));

% % Transport solution P-3 approx; Tm, Tn are the eigenvalues
Tcc = Pscat;
fprintf (res1,'\n\n Transport P-3 Fourier Transform Infinite Med B&G\n');
fprintf (res1,' c = %8.4f \n',Pscat);
Ta=9/7;
Tb = (55/7)*(1-Tcc)+5;
Tc=15*(1-Tcc);
Tp=[Ta 0 Tb 0 Tc];
Tt=roots(Tp);
Tm=imag(Tt(3));
Tn=imag(Tt(1));
fprintf (res1,'\n   m  =  %12.4e     n =  %12.4e \n',Tm,Tn);
Tcoeff1=(-(55/7)*Tm*Tm+15)/( (2*Tm)  *  (Tm^2-Tn^2)   );
Tcoeff2=(-(55/7)*Tn*Tn+15)/( (2*Tn)  *  (Tn^2-Tm^2)   );
fprintf (res1,'\n Tcoeff1 = %12.4e    Tcoeff2 = %12.4e  \n',Tcoeff1,Tcoeff2);

fprintf (res1,'\n\n X (mfp)      X (cm)     CD Flux      EigenFn Flux     percentage
error');
for irE = 1:XPTS
    ColDenEstFlux_ex(irE)=0.0;
    XXO(irE) = (LeftBdry + (irE-1)*DEL );% optical path units in range -a/2 to a/2
    XX(irE)  = XXO(irE)/Sig_t; % actual units of distance (cm)
    EX5 = abs(XX(irE))/DiffLength;
    EX6 = (abs(XX(irE))-EX2)/DiffLength;
    %%%%%%%% DIFFUSION THEORY FINITE MEDIUM SOLUTION %%%%%%%%%%
    ColDenEstFlux_ex(irE) = EX4*(exp(-EX5) - exp(EX6));

    %%%%%%%%%% DIFFUSION THEORY INFINITE MEDIUM SOLUTION %%%%%%%%%
    DTFlux(irE) = (DiffLength/(2.*DiffCoeff))*exp(-EX5);

    %%%%%%%%% TRANSPORT CORRECTED DIFFUSION INFINITE MEDIUM AND P-1 SOLUION
    TrFlux(irE) = TrFactor1*exp( -TrFactor2*Sig_t*abs(XX(irE)));
    %
    %%%%%%%%%%%% TRANSPORT P-3 SOLUTION INFINITE MEDIUM %%%%%%%%%
```

```
    TrFluxP3(irE) = (7/9)*( -Tcoeff1*exp(-Tm*Sig_t*abs(XX(irE))) -Tcoeff2*exp(-
Tn*Sig_t*abs(XX(irE))));
    %Eigenfunction method numerical series computation of flux and leakage
FX1= (2./(EX2*Sig_a));
L=1;
sum = 0.0;
sum2 = 0.0;
for iseries = 1:10
    BL =  L*pi/EX2;
    TT = (1./(1.+BL*BL*DiffLength*DiffLength))*cos(BL*XX(irE));
    sum = sum + TT;

    % leakage
    CL = L*pi/2.0;
    EE = (1./(1.+BL*BL*DiffLength*DiffLength))*(sin(CL)/CL);
    sum2 = sum2+2.0*EE;
    %
    L=L+2; % L is always odd
end
EigenFnFlux(irE) = (2./(EX2*Sig_a))*sum;

PercErr = abs (100.*((EigenFnFlux(irE)-
ColDenEstFlux_ex(irE))/ColDenEstFlux_ex(irE)));

    fprintf (resl,'\n  %8.4f   %8.4f %12.4e %12.4e
%12.4e',XXO(irE),XX(irE),ColDenEstFlux_ex(irE),EigenFnFlux(irE),PercErr);
end
for irE = 1:XPTS
    CD(imat,irE)=ColDenEstFlux_ex(irE); % finite medium diffusion theory flux
    DT(imat,irE)=DTFlux(irE);           % infinite medium diffusion theory flux
    TF(imat,irE)=TrFlux(irE);           % infinite medium diffusion theory flux P1
correction
    TFP3(imat,irE)=TrFluxP3(irE);       % infinite medium transport theory flux P3
approx
end
% and now for the transport approx

% Leakage
LeakageExact =  1./cosh(EX2/(2.0*DiffLength));
fprintf (resl,'\n \n Exact Leakage is  %12.4e \n',LeakageExact);
%
LeakageEigenFn = 1.0 - sum2;
fprintf (resl,'\n Leakage from eigenfunction analysis is %12.4e
\n',LeakageEigenFn);
end
      toc
% FIG 1
% PLOT OF COLLISION DENSITY vs OPTICAL DISTANCE
%  ColDenMean(iregion)      vs   iregion   (colDenStDe is the sigma)
%  ColDen_ex(iregion)
% XOPT is the midpoint of each region in optical path units
```

```
        for irE = 1:XPTS
        % Al
        CDP_1(irE)   =CD(1,irE);
        DTh_1(irE)   =DT(1,irE);
        TF_1(irE)    =TF(1,irE);
        TFP3_1(irE) =TFP3(1,irE);
        Del_DT_TFP1_1(irE) = DTh_1(irE)-TF_1(irE);  % inf medium difference
        Del_DT_TFP3_1(irE) = DTh_1(irE)-TFP3_1(irE);% inf medium difference
        % FE
        CDP_2(irE)   =CD(2,irE);
        DTh_2(irE)   =DT(2,irE);
        TF_2(irE)    =TF(2,irE);
        TFP3_2(irE) =TFP3(2,irE);
        Del_DT_TFP1_2(irE) = DTh_2(irE)-TF_2(irE);  % inf medium difference
        Del_DT_TFP3_2(irE) = DTh_2(irE)-TFP3_2(irE);% inf medium difference
        % U
        CDP_3(irE)   =CD(3,irE);
        DTh_3(irE)   =DT(3,irE);
        TF_3(irE)    =TF(3,irE);
        TFP3_3(irE) =TFP3(3,irE);
        Del_DT_TFP1_3(irE) = DTh_3(irE)-TF_3(irE);  % inf medium difference
        Del_DT_TFP3_3(irE) = DTh_3(irE)-TFP3_3(irE);% inf medium difference
        % Au
        CDP_4(irE)   =CD(4,irE);
        DTh_4(irE)   =DT(4,irE);
        TF_4(irE)    =TF(4,irE);
        TFP3_4(irE) =TFP3(4,irE);
        Del_DT_TFP1_4(irE) = DTh_4(irE)-TF_4(irE);  % inf medium difference
        Del_DT_TFP3_4(irE) = DTh_4(irE)-TFP3_4(irE);% inf medium difference
        % B
        CDP_5(irE)   =CD(5,irE);
        DTh_5(irE)   =DT(5,irE);
        TF_5(irE)    =TF(5,irE);
        TFP3_5(irE) =TFP3(5,irE);
        Del_DT_TFP1_5(irE) = DTh_5(irE)-TF_5(irE);  % inf medium difference
        Del_DT_TFP3_5(irE) = DTh_5(irE)-TFP3_5(irE);% inf medium difference
        %Gd
        CDP_6(irE)   =CD(6,irE);
        DTh_6(irE)   =DT(6,irE);
        TF_6(irE)    =TF(6,irE);
        TFP3_6(irE) =TFP3(6,irE);
        Del_DT_TFP1_6(irE) = DTh_6(irE)-TF_6(irE);  % inf medium difference
        Del_DT_TFP3_6(irE) = DTh_6(irE)-TFP3_6(irE);% inf medium difference
        end

        fprintf (resl,'\n  Material No 1    \n');
        fprintf (resl,'\n  Pt      X        Diff Th        P-1          P-3');
        % write these results, DT, P-1, P-3 in the output file
        fprintf (resl,'\n I  X (cm)  FinMed DT  InfMedDT   InfMedTC_P1
InfMedTT_P3\n');
        % material 1  Al
        for irP = 1:XPTS
            fprintf(resl,'\n %3.0f %8.5f %12.4e %12.4e
%12.4e',irP,XX(irP),CDP_1(irP),DTh_1(irP),TF_1(irP),TFP3_1(irP));
        end
```

```
        %%%%%%%%%%%%%%%%%%%%%%%%%%%%%%%%%%%%%%%
        ThisPlot=4; % select the plot required
        %%%%%%%%%%%%%%%%%%%%%%%%%%%%%%%%%%%%%%%

       % if (ThisPlot==1)
       figure(1)

       set(gca,'FontSize',12)

%legend('LW','Be','Al','C','HW',12,'Location','SouthEast')
%text(0.1,0.7,'\bf Water','fontsize',12)
%xlim([1e-8 2])
%ylim([1e-3 2e-1])
%hold on
grid off
%xlabel('{\bf Slab Thickness} (cm)','fontsize',14)
%ylabel('{\bf Albedo} (n/s)','fontsize',14)
set(gca,'XTickLabel',{'-5','-4','-3','-2','-1','0','1','2','3','4','5'})

%subplot(2,2,1)
semilogy(XXO,CDP_1,'k-','LineWidth',1.5)
%plot(XXO,CDP1,'-')
hold on
semilogy(XXO,TF_1,'k-.','LineWidth',1.5)
%plot(XXO,TF1,'-.')
 hold on
semilogy(XXO,TFP3_1,'k:','LineWidth',1.5)
%plot(XXO,TFP3_1,'.')
 h = legend('Al: DT','Al: P1','Al: P3',2,'fontsize',12);
       % end

       figure(2)
        %if (ThisPlot==2)
         set(gca,'FontSize',12)
%subplot(2,2,1)
 hold on
 semilogy (XXO,CDP_2,'k:','LineWidth',1.5)
 hold on
 semilogy(XXO,TF_2,'k-.','LineWidth',1.5)
 hold on
 semilogy(XXO,TFP3_2,'k.','LineWidth',1.5)
h = legend('Fe: DT','Fe: P1','Fe: P3',2,'fontsize',12);
       % end
        figure(3)
        % FIGURE 5
        %if (ThisPlot==3)
         set(gca,'FontSize',12)
%subplot(2,2,2)
 %semilogy (XXO,CDP_3,'-r','LineWidth',1.5)
 semilogy (XXO,CDP_3,'-k','LineWidth',1.5)
 hold on
 %semilogy(XXO,TF_3,'-g','LineWidth',1.5)
 semilogy(XXO,TF_3,'--k','LineWidth',1.5)
  hold on
 %semilogy (XXO,CDP_4,'-b','LineWidth',1.5)
 semilogy (XXO,CDP_4,'-.k','LineWidth',1.5)
 hold on
```

```
  %semilogy(XXO,TF_4,'-c','LineWidth',1.5)
  semilogy(XXO,TF_4,':k','LineWidth',1.5)
 h = legend('U: DT','U: P1','Au: DT','Au: P1',2,'fontsize',12);
 % title 'NEUTRON FLUX'
      xlabel ('\bf Distance (mfp)','fontsize',14)
      ylabel ('\bf \phi(x) (n cm^{-2} s^{-1})','fontsize',14)
     % set(gca,'XTickLabel',{'-5','-4','-3','-2','-1','0','1','2','3','4','5'})

        %end

 figure(4)
% if (ThisPlot==4)
 set(gca,'FontSize',12)
% subplot(2,2,3)
 semilogy (XXO,CDP_5,'k-','LineWidth',1.5)
 hold on
 semilogy(XXO,TF_5,'k.','LineWidth',1.5)
 hold on
 semilogy(XXO,TFP3_5,'k.','LineWidth',1.5)
     h = legend('B: DT','B: P1','B: P3',2,'fontsize',12);
% end

 % FIGURE 6
 figure (5)
 %if (ThisPlot==5)
  set(gca,'FontSize',12)
 %   subplot(2,2,4)
    %semilogy(XXO,CDP_6,'-r','LineWidth',1.5)
    semilogy(XXO,CDP_6,'-k','LineWidth',1.5)
 hold on
 %semilogy (XXO,TF_6,'-g','LineWidth',1.5)
 semilogy (XXO,TF_6,'--k','LineWidth',1.5)
 hold on
 %semilogy(XXO,TFP3_6,'-b','LineWidth',1.5)
 semilogy(XXO,TFP3_6,'-.k','LineWidth',1.5)
    h = legend('Gd: DT','Gd: P1','Gd: P3',2,'fontsize',12);
     % title 'NEUTRON FLUX'
      xlabel ('\bf Distance (mfp)','fontsize',14)
      ylabel ('\bf \phi(x) (n cm^{-2} s^{-1}) ','fontsize',14)

% end

 % NOW PLOT ALL P-3 FLUXES INFINITE MEDIUM
 figure(6)
% if (ThisPlot==981)
 set(gca,'FontSize',12)
%plot(XXO,TFP3_1,'-r','LineWidth',1.5)
plot(XXO,TFP3_1,'k-','LineWidth',1.5)
hold on
%plot(XXO,TFP3_2,'-g','LineWidth',1.5)
plot(XXO,TFP3_2,'k--','LineWidth',1.5)
hold on
%plot(XXO,TFP3_3,'-b','LineWidth',1.5)
plot(XXO,TFP3_3,'k-.','LineWidth',1.5)
```

```
hold on
%plot(XXO,TFP3_4,'-c','LineWidth',1.5)
%plot(XXO,TFP3_4,'X','LineWidth',1.5)
hold on
%plot(XXO,TFP3_5,'-m','LineWidth',1.5)
plot(XXO,TFP3_5,'kx','LineWidth',1.5)

h = legend('Al','Fe','U','B',2,'fontsize',12);
%set(gca,'XTickLabel',{'-5','-4','-3','-2','-1','0','1','2','3','4','5'})

%h = legend('Al','Fe','U','Au','B',2);
% title 'NEUTRON FLUX'
      xlabel ('\bf Distance (mfp)','fontsize',14)
      ylabel ('\bf \phi(x) (n cm^{-2} s^{-1}','fontsize',14)
%hold on
%plot(XXO,TFP3_6,'-y')
%h = legend('Al','Fe','U','Au','B','Gd',2);
% end

  % NOW PLOT ALL FLUXES FINITE MEDIUM    FIGURE 3 in Paper
  figure(7)
 %if (ThisPlot==983)
  set(gca,'FontSize',12)
%semilogy(XXO,CDP_1,'-r','LineWidth',1.5)
semilogy(XXO,CDP_1,'-k','LineWidth',1.5)
hold on
%semilogy(XXO,CDP_2,'-g','LineWidth',1.5)
semilogy(XXO,CDP_2,'--k','LineWidth',1.5)
hold on
%semilogy(XXO,CDP_3,'-b','LineWidth',1.5)
%semilogy(XXO,CDP_3,'ok','LineWidth',1.5)
hold on
%semilogy(XXO,CDP_4,'-c','LineWidth',1.5)
%semilogy(XXO,CDP_4,'xk','LineWidth',1.5)
hold on
%semilogy(XXO,CDP_5,'-m','LineWidth',1.5)
semilogy(XXO,CDP_5,'-.k','LineWidth',1.5)
%h = legend('Al','Fe','U','Au','B',2); % for 5 graphs  in ICONE 18
h = legend('Al','Fe','B',2,'fontsize',12); % for 3 graphs  in NED

% title 'NEUTRON FLUX'
      xlabel ('\bf Distance (mfp)','fontsize',14)
      ylabel ('\bf\phi','fontsize',14)
%hold on
%plot(XXO,TFP3_6,'-y')
%h = legend('Al','Fe','U','Au','B','Gd',2);
% end

 % NOW PLOT ALL P-3 FLUXES INFINITE MEDIUM
 figure(8)
 %if (ThisPlot==984)
  set(gca,'FontSize',12)
plot(XXO,TFP3_1,'k-','LineWidth',1.5)
```

```
hold on
plot(XXO,TFP3_2,'k--','LineWidth',1.5)
hold on
plot(XXO,TFP3_3,'k-.','LineWidth',1.5)
hold on
plot(XXO,TFP3_4,'k:','LineWidth',1.5)
hold on
plot(XXO,TFP3_5,'k--','LineWidth',1.0)
h = legend('Al','Fe','U','Au','B',2,'fontsixe',12);
% title 'NEUTRON FLUX'
      xlabel ('\bf Distance (mfp)','fontsize',14)
      ylabel ('\bf \phi  (n cm^{-2} s^{-1})','fontsize',14)
%hold on
%plot(XXO,TFP3_6,'-y')
%h = legend('Al','Fe','U','Au','B','Gd',2);
 %end

 % Fig 4 in ICONE18 Paper  (Fig 3  in   NED paper)
  if (ThisPlot==985)

%semilogy(XXO,CDP_1,'-r')
semilogy(XXO,CDP_1,'-k')
hold on
%semilogy(XXO,TF_1,'-g')
semilogy(XXO,TF_1,'--k')
 hold on
%semilogy (XXO,CDP_2,'-b')
 semilogy (XXO,CDP_2,'-.k')
 hold on
%semilogy(XXO,TF_2,'-c')
 semilogy(XXO,TF_2,':k')

h = legend('Al: DT','Al: P1','Fe: DT','Fe: P1',2);
%title 'NEUTRON FLUX'
      xlabel ('DISTANCE (mfp)')
      ylabel ('FLUX \phi')
         end

 % plot differences for Infinite Medium Only
 % title 'DIFFERENCE IN INFINITE MEDIUM NEUTRON FLUX'
 if (ThisPlot==99)
     subplot(2,2,1)
     plot (XXO,Del_DT_TFP1_1,'-k')
     % XXO is in mfp
     hold on
     plot (XXO,Del_DT_TFP3_1,'-.k')
   %  title 'Aluminium'
      xlabel ('DISTANCE (mfp)')
      ylabel ('\delta\phi')
      % h = legend('DT-DC','DT-TTP3',2);
      hold on
      subplot(2,2,2)
```

```
    plot (XXO,Del_DT_TFP1_3,'-k')
    % XXO is in mfp
    hold on
    plot (XXO,Del_DT_TFP3_3,'-.k')
%    title 'Uranium'
     xlabel ('DISTANCE (mfp)')
     ylabel ('\delta\phi')
     % h = legend('DT-DC','DT-TTP3',2);
     hold on
     subplot(2,2,3)
    plot (XXO,Del_DT_TFP1_5,'-k')
    % XXO is in mfp
     hold on
     plot (XXO,Del_DT_TFP3_5,'-.k')
 %   title 'Boron'
     xlabel ('DISTANCE (mfp)')
     ylabel ('\delta\phi')
     % h = legend('DT-DC','DT-TTP3',2);
     hold on
      subplot(2,2,4)
    plot (XXO,Del_DT_TFP1_6,'-k')
    % XXO is in mfp
    hold on
    plot (XXO,Del_DT_TFP3_6,'-.k')
 %   title 'Gadolinium'
     xlabel ('DISTANCE (mfp)')
     ylabel ('\delta\phi')
     % h = legend('DT-DC','DT-TTP3',2);
end
     toc
fclose(resl);
```

Chapter 12

Exercises in Monte Carlo simulation

This chapter comprises hands-on exercises in:

1. random sampling from the Watt fission spectrum,
2. Monte Carlo (MC) simulation for the one-group neutron flux in a non-multiplying sphere,
3. search for a reflected spherical critical configuration,
4. reactor core modeling,
5. radiation shielding,
6. perturbation analysis, and
7. geometry modeling.

12.1 Sampling from a distribution function

In the MC simulation of a multiplying medium, the neutron source energy is sampled from a spectrum such as the Watt fission spectrum (Section 2.9). As described in Chapter 4, random sampling is straightforward when the probability distribution function (PDF) is simple and analytic; in that case the cumulative distributive function (CDF) is readily inverted. When the CDF is not easy to obtain, or when the PDF is tabulated, other methods, such as the acceptance-rejection method, must be used.

For an exponential PDF

$$f(E) = \lambda e^{-\lambda E} \tag{12.1}$$

the CDF is

$$F(E) = 1 - e^{-\lambda E} \tag{12.2}$$

and the energy E_i can be sampled as

$$E_i = -\frac{1}{\lambda}\ln\left(1 - \xi_i\right). \tag{12.3}$$

where ξ_i is a uniform random number.

In MATLAB®, the commands for Eq. (12.3) are

```
N=1000;
lambda=1; % for mean = 1
E=(-1/lambda)*log(1-rand(N,1));
```

and the mean energy can be found as

```
Ebar= sum(E)/N;
```

As an example, if $\lambda = 1$ MeV, with $N = 1000$ Ebar $= \overline{E} = 1.0398$ MeV.

Nuclear Engineering. DOI: https://doi.org/10.1016/B978-0-323-90618-0.00012-0

12.1.1 Sampling from a normal distribution

For neutron energy E distributed normally with mean μ and standard deviation σ

$$f(E) = \frac{1}{\sqrt{2\pi}\sigma} e^{-\frac{1}{2}\left(\frac{E-\mu}{\sigma}\right)^2}. \tag{12.4}$$

Fig. 12.1 shows a plot of $f(E)$ for $\mu =$ MeV and $\sigma =$ MeV. The total area of the curve is 1.0; in MATLAB this is easily found with the example below

```
fun = @(x,mu,sigma)(1/(sigma*sqrt(2*pi)))* exp(-0.5*((x-mu)./sigma).^2);
q = integral(@(x) fun(x,5,2),-15,15)
q= 1.0000
```

The MATLAB(R) function `normpdf` can also be used to plot the function

```
x=-10:0.5:10;
y=normpdf(x,5,2);
plot(x,y)
```

and the function `normcdf(x,5,2)` gives the CDF at $x = 2$.

The energy is sampled from

$$E_i = \sigma\xi_i^{(N)} + \mu \tag{12.5}$$

giving a sampled PDF $f_1(E)$ and

$$E_i = \mu + \sigma\sqrt{-2\ln\xi_{i,1}}\cos 2\pi\xi_{i,2} \tag{12.6}$$

giving $f_2(E)$, also shown in Fig. 12.1. The random numbers $\xi_{i,1}$ and $\xi_{i,2}$ are generated from a uniform distribution while $\xi_i^{(N)}$ are generated from a normal distribution.

The MATLAB program CH12_NormalSampling.m used for sampling from the normal PDF is listed in Annex A.

In Fig. 12.1, both sampled distributions $f_1(E)$ and $f_2(E)$ are for a sample size $N = 10,000$ which show acceptable agreement, except for at the ends, with the population distribution $f(E)$.

Exercise 12.1: From the matlab program given above, compute the sample standard deviation from each sampling scheme.

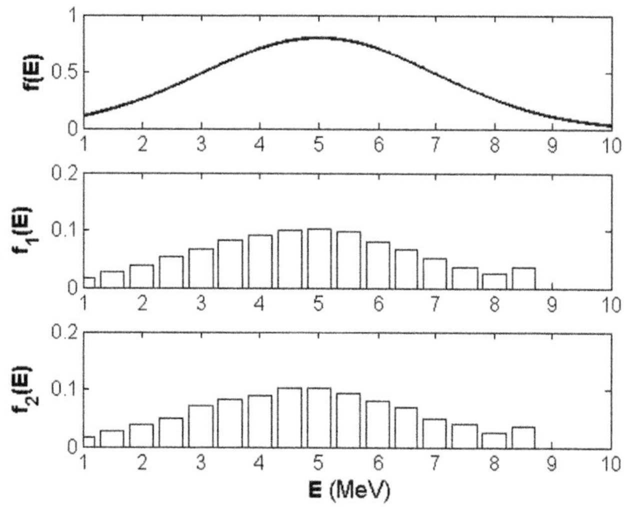

FIGURE 12.1 Sampling from a normal distribution with $\mu = 5, \sigma = 12$.

Exercise 12.2: Show that the CDF of the normal distribution is

$$F(E) = \frac{1}{2}\left[1 + \text{erf}\left(\frac{x-\mu}{\sigma\sqrt{2}}\right)\right] \tag{12.7}$$

12.1.2 Sampling from a Watt fission spectrum

The energy of the prompt neutron(s) emitted from fission, E, is found from the Watt Spectrum (Cullen, 2004; Froehner & Spencer, 1980; Monte Carlo Team, 2005)

$$W(a,b,E) = Ce^{-E/a}\sinh\sqrt{bE} \tag{12.8}$$

where

$$C = \left(\frac{\pi a^3 b}{4}\right)^{-1/2} e^{-ab/4}.$$

and the constants a, b are weak functions of the energy. For neutron-induced fission in U-235 at thermal and 1 MeV neutron energy $a = 0.988$ MeV and $b = 2.249\ \text{MeV}^{-1}$.

At low energies $E \lesssim 1\text{MeV}$

$$W(a,b,E) \sim AE^{1/2} \tag{12.9}$$

and at high energies $E \gtrsim 1$ MeV

$$W(a,b,E) = Ce^{\sqrt{bE}-E/a} \tag{12.10}$$

From Eq. (12.8), the Watt fission spectrum is shown, on a log-log scale, in Fig. 12.2. Figs. 12.2–12.4 are obtained from the MATLAB Program CH12_WattSampling.m listed in Annex B.

Energy is sampled from (Everett, Turner, & Cashwell, 1973)

$$K = 1 + \frac{ab}{8}, \quad L = a\left(K + \sqrt{K^2 - 1}\right), \quad M = \frac{L}{a} - 1$$
$$x = -\ln\xi_1, y = -\ln\xi_2 \tag{12.11}$$

If $(y - M(x+1))^2 \leq bLx$, then the numbers are accepted and the energy is $E = Lx$; otherwise a new set of random numbers is generated (program listed in Annex B).and from a Maxwellian in the center of mass system transformed to the Lab system (Brown; Froehner & Spencer, 1980)

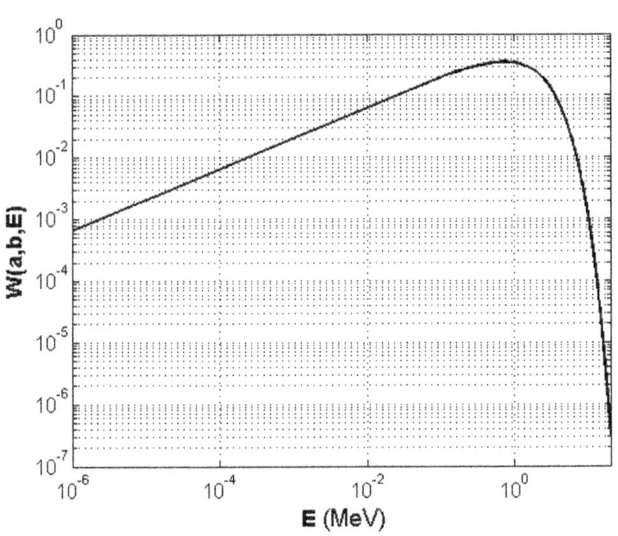

FIGURE 12.2 Watt fission spectrum for U-235 induced fission ($a = 0.988$ MeV, $b = 2.249\ \text{MeV}^{-1}$).

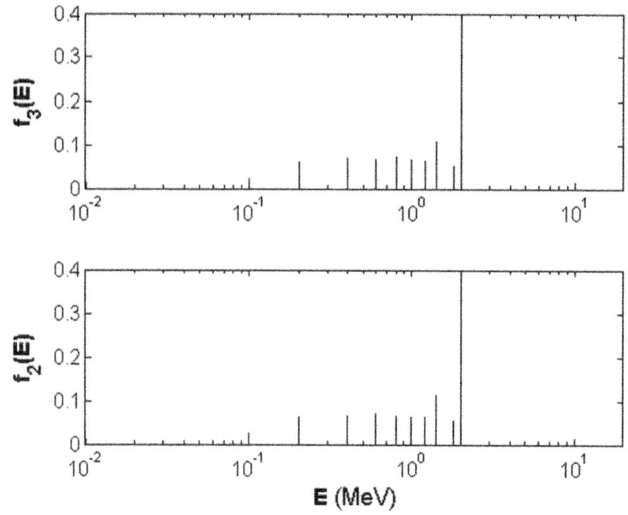

FIGURE 12.3 Sampled Watt fission spectra probabilities $f_3(E)$ and $f_4(E)$.

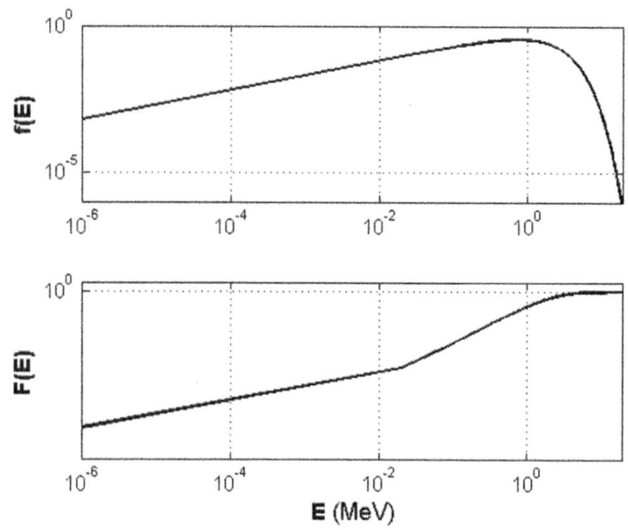

FIGURE 12.4 PDF and CDF of the U-235 Watt spectrum.

$$w \leftarrow a\left(-\ln\xi_1 - \ln\xi_2 \cos^2\frac{\pi}{2}\xi_3\right)$$

$$E \leftarrow w + \frac{a^2 b}{4} + (2\xi_4 - 1)\sqrt{a^2 b w} \qquad (12.12)$$

The sampled spectrum using Eq. (12.11), $f_3(E)$, and Eq. (12.12), $f_4(E)$, are shown in Fig. 12.3. For 100,000 points simulated, the sampled mean energies for $f_3(E)$ and $f_4(E)$ were 2.0343 and 2.0310 MeV respectively. The acceptance-rejection sampling scheme had a 75.9% acceptance.

The PDF $f(E)$ and CDF $F(E)$ are shown in Fig. 12.4. The CDF can be curve-fitted and used for indirect sampling. Alternatively, the acceptance-rejection scheme can be used for sampling.

Exercise 12.3: Run the above program for sample size N varying from 100 to 1,000,000 and note the accuracy and sampling time.

Exercise 12.4: For the low- and high-energy approximations given by Eqs. (12.2) and (12.3) find expressions for the cumulative distribution function to cary out analytical sampling. Again, comment on the accuracy of using these approximations.

Exercise 12.5: Apply the acceptance-rejection scheme for sampling the Watt energy spectrum. Recall from Chapter 4, the scheme for sampling from a difficult PDF f is based on considered a simple PDF g and choosing a constant c such that $f(y) \leq cg(y)$. For a random number ξ (uniform in 0, 1) the point is selected if

$$\xi \leq \frac{f(\xi)}{cg(\xi)}.$$

Exercise 12.6: Use Eqs. (12.9) and (12.10) to carry out direct sampling and compare the efficiencies.

12.2 Estimating the neutron flux in a non-multiplying sphere

This section considers a non-multiplying sphere of radius R with a point isotropic source located at the origin emitting S neutrons/s. Recall that in Section 11.5 the neutron flux was calculated from diffusion theory, transport theory, and MC simulations. Comparisons were also illustrated for diffusion theory and MC results obtained from a modest program MCFSoneProg.m which is listed here as a hands-on exercise.

The main program has 162 lines and calls three functions:

Function and arguments returned	Line called from
`[RG, RS, Vol] = MCFSShells(Rad_sphere, NV);`	6
`IREG = MCFSregion(POS);`	28, 34, 77
`[Ncols, FLUX] = MCFSsameR(FREG, WGT, DTC, Ncols, FLUX);`	38

This program could have been written in fewer lines; the purpose here is to illustrate the philosophy of having many subroutines and functions to do their bits separately rather than as one main program; this is the way large codes are written.

12.2.1 The simulation process

Step 1: Enter the basic data

The data entered is:

- Avogadro's number
- molecular weight
- density
- radius
- number of regions NREG
- nuclear data $(\sigma_s.\sigma_a, \sigma_t)$
- number of histories N to simulate.

Step 2: Initialize counters

The following counters are initialized

- scattering, absorption and collision counters
- weight participating in the collision WtPartCol
- collision estimator (CE) flux (CEflux)
- track-length estimator (TLE) flux (FLUX).

```
for irl=1:NREG
NScats(irl)=0; Nabs(irl)=0; Ncols(irl)=0;
FLUX(irl)=0.0;WtPartCol(irl)=0.0;CEflux(irl)=0.0;end
```

Step 3: Begin simulation

Source histories are simulated by generating source neutrons at the origin with a starting weight of one isotropically

$$\mu = 2\xi_1 - 1, \quad \theta = \cos^{-1}\mu, \quad \varphi = 2\pi\xi_2$$

The distance to collision (DTC) is used to calculate the final position.

$$DTC = \frac{1}{\Sigma_t}\ln(1 - \xi_3) \tag{12.13}$$

Exercise 12.7: What is the purpose of defining bounding surfaces in this program?

```
function [RG,RS,Vol]=MCFSShells(Rad_sphere,NV)
Volume = (4/3)*3.14159*Rad_sphere^3;
EachVolume = Volume/NV;
RS(1)=(3*EachVolume/(4*3.14159))^(1/3);
Vol(1)=(4/3)*3.14159*RS(1)^3;
for iv = 2: NV
    RS(iv)= ( RS(iv-1)^3 + (3/(4*3.14159))*EachVolume )^(1/3);
    Vol(iv)=(4/3)*3.14159*(RS(iv)^3-RS(iv-1)^3);
end

RS

% REGION      SURFACES
%RG(region no,surfaces)
RG(1,1)= 1;   RG(1,2)= 0; % BOUNDING SURFACES
RG(2,1)= 1;   RG(2,2)= 2;
RG(3,1)= 2;   RG(3,2)= 3;
RG(4,1)= 3;   RG(4,2)= 4;
RG(5,1)= 4;   RG(5,2)= 5;
RG(6,1)= 5;   RG(6,2)= 6;
RG(7,1)= 6;   RG(7,2)= 7;
RG(8,1)= 7;   RG(8,2)= 8;
RG(9,1)= 8;   RG(9,2)= 9;
RG(10,1)= 9;  RG(10,2)= 10;


for ir=2:NV
    RG(ir,1)=ir-1;
    RG(ir,2)=ir;
end
```

The function MCFSregion determines the regions IREG and FREG for the initial and final positions.

```
function IREG=MCFSregion(POS)
global RS
global NREG
% given the position, find the region

IREG=0; keyR=0;
while (keyR==0)
    for ir = 1:NREG-1
        if ((POS>=0)&&(POS<RS(1)))
            IREG=1;
            keyR=1;
        end
        if ((POS>=RS(ir))&&(POS<RS(ir+1)))
            IREG=ir+1;
            keyR=1;
        end
        if (POS>RS(NREG))
            IREG=1000; % escape from system
            keyR=1;
        end
    end
end
```

There are two possibilities viz (1) where the final and initial regions are the same, (IREG = FREG, then MCFSsameR is called) and (2) where the regions are different.

```
function [Ncols,FLUX,FLUXsq]=MCFSsameR(FREG,WGT,DTC,Ncols,FLUX,FLUXsq)
%fprintf('flux and Ncols updated in same region\n')
Ncols(FREG)=Ncols(FREG)+1;
FLUX(FREG) = FLUX(FREG) + WGT*DTC;
FLUXsq(FREG) = FLUXsq(FREG) + (WGT*DTC)^2;
```

Exercise 12.8: What is the purpose of updating Ncols, FLUX and FLUXsq in MCFSsameR?

In the former, a collision is tallied by updated the number of collisions in this region Ncols, the track-length flux FLUX, and the weight participating in the collision WtPartCol (which will give the collision estimate of flux). The second case, where the initial and final regions are different, has to determine the path of the neutron as it crosses intermediate regions to ultimately collide in the final region or when it escapes from the system.

A region number of 1000 is used to represent the region external to the sphere. To account for double precision, a number Small = 1.0e−6 is used to avoid small numbers which may misrepresent a neutron as not lying on a surface when it has been transported to a surface in its path. Note the distances DTC, DTS and Dleft which are used to update the distance left Dleft from the distance to collision DTC when a distance to surface (DTS) is used to determine whether the history continues or is ended in a final region. It is crucial to determine which of the bounding surfaces in a region a neutron will be transported to. For this, the MATLAB function `isreal` is used to reject a complex root. Similarly, a negative real root is not selected and thus only one real root for the parameter DTS is used to determine the surface of interest. During a surface crossing, only the track-length estimator (TLE) is updated while in a final region both the CE and TLE counters are updated. When a neutron escapes from the physical domain, that is, the sphere, the counters Nescapes, WGT_escaping and TotalWGTlost are updated to later contribute to the leakage estimates. In a collision occurring in the final region FREG, the weight is reduced in this non-analog simulation and the history is continued. The parameter key takes the value zero until a history is inside the domain, and one when it terminates due to an escape from the domain.

Step 4: Calculate mean values

The mean values are obtained by dividing the estimates by the number of neutrons started to get the value for one source neutron.

TABLE 12.1 Some of the variables used in the Monte Carlo program.

Quantity	Variable
Avogadro's number	AvNo
Radius	Rad_sphere
Nuclear data	sigma_s, sigma, sigma_t
Distance left (remaining)	Dleft
Distance to collision	DTC
Distance to surface	DTS
Number of histories	N
Present region	IREG
Final region	FREG
Number of regions	NREG
A small number (1×10^{-6})	Small
Weight participating in a collision	WtPartCol
Collision estimator flux	CEflux
Flux variable updated at each interaction	Flux
Number of scatterings	Nscats
Number of absorptions	Nabs
Number of collisions	Ncols
Number of escapes	Nescapes
Update total (statistical) weight of neutron	TotalWGT
Weight lost by escape	TotalWGTlost
Position x, y, z	XPOS, YPOS, ZPOS
Direction cosines	Uu, vv, ww
Particle weight	WGT

Step 5: Calculate exact flux

The exact flux is calculated (line 118) with $\mu_0 = 2/(3A)$, extrapolation distance $d = 0.71\lambda_{tr}$, $\Sigma_{tr} = \Sigma_s(1 - \mu_0)$, diffusion coefficient $D = 1/(3\Sigma_{tr})$, diffusion length $L = \sqrt{D/\Sigma_a}$ (Table 12.1).

12.2.2 MATLAB program for point source in a finite non-multiplying sphere

A MATLAB program is given below for the steps described in the previous section. It is recommended that these lines are "copied and pasted" into a MATLAB.m file and executed.

```
Step 1: Enter the basic data
Av_No = 0.6023e+24; % Avogadro's number
Rad_sphere = 25.0; %cm
NREG=10; % divide the sphere into 10 regions of equal volume
[RG,RS,Vol]=MCFSShells(Rad_sphere,NV);
% Aluminum thermal cross-sections
sigma_s = 1.4e-24; % cm^2
sigma_a = 0.235e-24;%
sigma_t = sigma_s+sigma_a;
N=10000; % simulate N neutrons
Step 2: Initialize counters
NEscapes=0; WGT_escaping=0.0; TotalWGTlost=0.0;
for ir1=1:NREG
NScats(ir1)=0; Nabs(ir1)=0; Ncols(ir1)=0;
FLUX(ir1)=0.0;WtPartCol(ir1)=0.0;CEflux(ir1)=9.0;end
Step 3: Begin simulation
for  J =1:N
% begin a source neutron
XPOS=0.0;YPOS=0.0;ZPOS=0.0;POS =sqrt(XPOS^2+YPOS^2+ZPOS^2);WGT=1.0;
StartingWeight(J)=WGT;
key=0;% key will be non zero when this particle escapes or history ends
while(key==0)
mu = 2.0*rand -1.0; theta=acos(mu); phi = 2.0*pi*rand;
DTC=(-(1.0/Big_sigma_t)*log(1.0-rand));
% store current position P
XPOS_P=XPOS; YPOS_P=YPOS; ZPOS_P=ZPOS; POS_P=POS;
IREG=MCFSregion(POS); % ireg can be from 1,2,3...NREG or 1000 for escape
% get direction cosines and get final position
uu = sin(theta)*cos(phi);vv=sin(theta)*sin(phi);ww=cos(theta);
XPOS_F=XPOS+DTC*uu ; YPOS_F=YPOS+DTC*vv; ZPOS_F=ZPOS+DTC*ww;
% now check if the updated position is inside or outside the sphere
POS_F =sqrt(XPOS_F^2+YPOS_F^2+ZPOS_F^2);
FREG=MCFSregion(POS_F);
% the neutron can be in another region
% first, check if it is in the same region
if (FREG==IREG)
[Ncols,FLUX]=MCFSsameR(FREG,WGT,DTC,Ncols,FLUX);
WtPartCol(IREG)=WtPartCol(IREG)+WGT;
end
% going to a different region
if (FREG~=IREG)
small=1.0e-6;
hitSurf=0;
Dleft=DTC;
uu = sin(theta)*cos(phi);vv=sin(theta)*sin(phi);ww=cos(theta);
while ((Dleft>0)&&(key==0))
% find next surface NS it could encounter;
% surfaces are RG(IREG,1) and RG(IREG,2)
nsurfs=0;
for is=1:2
SURF=RG(IREG,is);
if (SURF>0)
```

```
aa=1;bb=2.0*(XPOS_P*uu+YPOS_P*vv+ZPOS_P*ww);
cc=(XPOS_P^2+YPOS_P^2+ZPOS_P^2-RS(SURF)^2);
det = bb^2-4*aa*cc;
root1 = (-bb+sqrt(det))/(2*aa); root2= (-bb-sqrt(det))/(2*aa);
% check if roots are real
tf1=isreal(root1); tf2=isreal(root2);
if ((tf1==1)&&(root1>small))
NS = SURF; DTS = root1;
nsurfs=nsurfs+1;end
if ((tf2==1)&&(root2>small))
NS = SURF; DTS = root2;
nsurfs=nsurfs+1;
end  end end  end
  % will it hit NS or will it collide before it hits the surface
if (DTS<Dleft)
hitSurf=hitSurf+1; % will hit surface
FLUX(IREG) = FLUX(IREG) + WGT*DTS;
% update this temporary position in its flight
XPOS_P=XPOS_P+DTS*uu ; YPOS_P=YPOS_P+DTS*vv; ZPOS_P=ZPOS_P+DTS*ww;
% find the new region it is going to enter
XREGchk=XPOS_P+0.1*DTS*uu;YREGchk=YPOS_P+0.1*DTS*vv;
ZREGchk=ZPOS_P+0.1*DTS*ww;
POSchk = sqrt(XREGchk^2 + YREGchk^2 + ZREGchk^2);
IREG=MCFSregion(POSchk);
% WHAT IF THE NEXT REGION IS 1000 (a void) ?
if (IREG==1000)
key=1;end end
if (DTS>Dleft)
Ncols(IREG)=Ncols(IREG)+1; % final collision
WtPartCol(IREG)=WtPartCol(IREG)+WGT;
FLUX(IREG) = FLUX(IREG) + WGT*Dleft;
% update this final position in its flight
XPOS_F=XPOS_P+Dleft*uu ; YPOS_F=YPOS_P+Dleft*vv; ZPOS_F=ZPOS_P+Dleft*ww;end
% reduce DTC by the amount DTS, Dleft = DTC-DTS
Dleft = Dleft-DTS;end
% if the new region is IREG=1000, score an escape
if (IREG==1000)
NEscapes=NEscapes+1;
WGT_escaping=WGT;
TotalWGTlost=TotalWGTlost+WGT_escaping;
Dleft=-1E9;
% update this final position in its flight
XPOS_F=1.0E+5; YPOS_F=1.0E+5; ZPOS_F=1.0E+5;
key=1;end end
% final position now becomes the initial position
XPOS=XPOS_F;YPOS=YPOS_F;ZPOS=ZPOS_F;POS =sqrt(XPOS^2+YPOS^2+ZPOS^2);
% post-collision weight
% now alter the weight of the particle and continue the history
WGT = WGT*P_scat;
% if it goes on and on then use weight cutoff
end %this ends the WHILE loop
 % comes here when the particle history has ended
% comes here only if key is non zero
end %    end of J loop
```

<u>Step 4: Calculate mean values</u>

```
% normalize results to one source neutron
for i=1:NREG
    Ncols(i)=Ncols(i)/N;
  % collision estimator flux
    WtPartCol(i)= WtPartCol(i)/N;
    WtPartCol(i)= WtPartCol(i)/Big_sigma_t;
    CEflux(i)    = WtPartCol(i);
  % track-length flux
    FLUX(i)      = FLUX(i)/N;end
```

<u>Step 5: Calculate exact flux</u>

```
% exact solution
dext        = 0.0;
mu_0        = 2.0/(3.0*Mol_Wt);
Sig_tr      = Big_sigma_s*(1.0- mu_0);
dext        = 0.71*(1.0/Sig_tr);
DiffCoeff   = 1.0/(3.0*Sig_tr);
DiffLength = sqrt(DiffCoeff/Big_sigma_a);
Src         = 1.0;
tt1        = Src/(4.0*3.14159*DiffCoeff*sinh((Rad_sphere+dext)/DiffLength));
rad(1)=0.5*RS(1);
Flux(1)=tt1*sinh((Rad_sphere+dext-rad(1))/DiffLength);
Flux(1)  = Flux(1)/rad(1);
for irex = 2:NREG
    delR = RS(irex)-RS(irex-1);
    rad(irex)= RS(irex-1)+ 0.5*delR;
    Flux(irex)  = tt1*sinh((Rad_sphere+dext-rad(irex))/DiffLength);
    Flux(irex)  = Flux(irex)/rad(irex);end
WLPP=TotalWGTlost/NParts;
%exact leakage
LeakageEx=
Src*(Rad_sphere+dext)/(DiffLength*sinh((Rad_sphere+dext)/DiffLength));
```

<u>Step 6: Plot Track-length flux vs exact flux</u>

```
% show in a plot of TLE vs Exact
rad1 = 1; rad2 = Rad_sphere;
deltaR=(rad2-rad1)/100;
radius=0.0;
for i = 1: 100
    radius     = radius + deltaR;
    Fluxact = tt1*sinh((Rad_sphere+dext-radius)/DiffLength);
    Fluxact = Fluxact/radius;
    x(i)       = radius;
    y(i)       = Fluxact; end
for irSctPlot = 1:NREG
    xT(irSctPlot) = rad(irSctPlot);
    yT(irSctPlot) = fluxPv(irSctPlot); % TLE flux per unit vol
end
semilogy (x,y,'-k')
hold on
semilogy(xT,yT,'--
rs','LineWidth',2,'MarkerEdgeColor','k','MarkerFaceColor','g','MarkerSize',
10)
      etime(clock,t0)
      t=cputime;
      fprintf(resl,'\n\n  CPU Time is %9.2f \n\n',t)
```

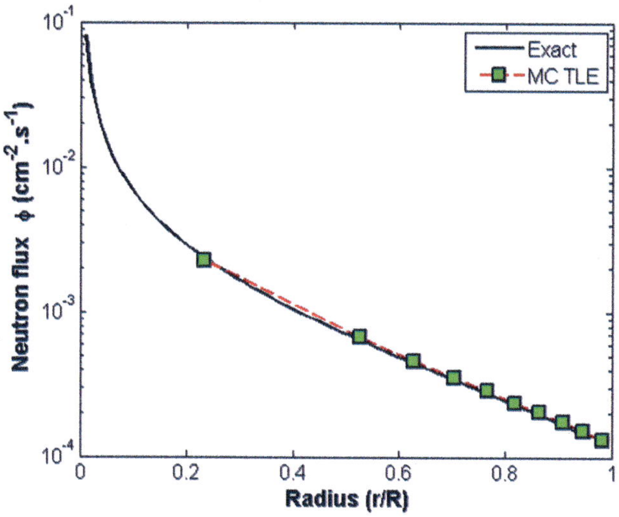

FIGURE 12.5 Neutron flux (Monte Carlo and diffusion theory) in a sphere.

12.2.3 Results

The MC TLE flux for $N = 1000$ histories is shown in Fig. 12.5. For each region, the TLE flux is estimated to be $<x>$ where

$$<x> \equiv \frac{1}{N}\sum_{i=1}^{N} w_i s_i$$

It is also necessary to estimate the standard deviation of $<x>$, written as $\sigma_{<x>}$. For a random variable x drawn from a population with PDF $f(x)$, the variance σ^2 is

$$\sigma^2 = \int (x - <x>)^2 f(x)dx$$

and subsequently, from the Central Limit Theorem, the variance of the estimated mean $\sigma^2_{<x>}$ is given by

$$\sigma^2_{<x>} = \frac{\sigma^2}{N}$$

An estimate of the variance of the estimated mean is thus

$$\sigma^2_{<x>} = \frac{\sigma^2}{N} = \frac{1}{N}\left(\overline{x^2} - \overline{x}^2\right)$$

Thus

$$\sigma_{<x>} = \sqrt{\frac{1}{N}\left(\overline{x^2} - \overline{x}^2\right)}$$

and another quantity, the relative standard deviation (RSD) $R_{<x>}$ of the estimated mean is

$$R_{<x>} = \frac{\sigma_{<x>}}{<x>}$$

Fig. 12.2 shows the neutron flux from MC with N = 1000 histories sampled (green box) compared with the exact diffusion theory solution (dashed line). In this case, the comparison is seen to be acceptable. It is good practice to compare a simulation result with an "exact" analytical result but this is not always possible, as has been stated in earlier chapters.

The output of the program given above is summarized below:

```
Radius = 25.0000 cm, Volume = 65449.7917 cm^3,
Regions = 10
 Reg    Radius    Volume
  1     11.6040 6544.9792
  2     14.6201 6544.9792
  3     16.7358 6544.9792
  4     18.4202 6544.9792
  5     19.8425 6544.9792
  6     21.0858 6544.9792
  7     22.1976 6544.9792
  8     23.2079 6544.9792
  9     24.1372 6544.9792
 10     25.0000 6544.9792

Mass= 176714.4375 g
sigma_s=  1.4000e-024 cm2 sigma_a= 2.3500e-025 cm2 No_Den= 6.0271e+022 cm-3
 1 mfp = 10.1478 cm
Radius of sphere is   2.4636 mfps
Scattering Probability P_scat 0.856269
Reg  avNcols  TLE (cm)        avFlux cm-2          CEflux cm-2          ExactFlux cm-2
1    1.8810  1.4917e+001     2.2792e-003          2.2982e-003          2.3575e-003
2    0.7037  4.3437e+000     6.6368e-004          6.6187e-004          6.4109e-004
3    0.5187  2.9640e+000     4.5287e-004          4.5032e-004          4.4485e-004
4    0.4256  2.2866e+000     3.4936e-004          3.5429e-004          3.4245e-004
5    0.3401  1.8250e+000     2.7884e-004          2.7560e-004          2.7683e-004
6    0.3153  1.6006e+000     2.4455e-004          2.4586e-004          2.3022e-004
7    0.2497  1.3210e+000     2.0184e-004          1.9081e-004          1.9494e-004
8    0.2214  1.1354e+000     1.7348e-004          1.6826e-004          1.6707e-004
9    0.1990  9.8959e-001     1.5120e-004          1.5464e-004          1.4433e-004
10   0.1656  8.6412e-001     1.3203e-004          1.2790e-004          1.2532e-004

 Total Weight lost= 5431.9003  Weight Loss per source neutron= 5.4319e-001
 Total escapes      =    10000
 LeakageEx from surface =    0.5548


  Reg  radMdPt      Fluxact
   1     5.8020    2.3575e-003
   2    13.1120    6.4109e-004
   3    15.6780    4.4485e-004
   4    17.5780    3.4245e-004
   5    19.1313    2.7683e-004
   6    20.4642    2.3022e-004
   7    21.6417    1.9494e-004
   8    22.7028    1.6707e-004
   9    23.6726    1.4433e-004
  10    24.5686    1.2532e-004
```

Consider now the estimate of the standard deviation of the estimated mean $\sigma_{<x>}$ and the relative estimated mean $R_{<x>} \equiv \sigma_{<x>}/<x>$ shown in Table 8.5.

The additions to the code are then an estimate of the square of the estimator FLUXsq at every event in the region, that is,

```
FLUX(reg)    = FLUX(reg)    +  WGT*DTC;
FLUXsq(reg)  = FLUXsq(reg)  + (WGT*DTC)^2;
```

and the standard deviation (STD) $\sigma_{<x>}$ is obtained for each region:

```
for ireg=1:NREG
  % track-length flux
    FLUX(ireg)=FLUX(ireg)/N;
    FLUXsq(ireg)=FLUXsq(ireg)/N;
    STD(ireg)= sqrt( (FLUXsq(ireg)-FLUX(ireg)*FLUX(ireg))/N);
end
```

from which the RSD $R_{<x>}$ is obtained for each region:

```
for ireg=1:NREG
    RSTD(ireg)=STD(ireg)/FLUX(ireg);
end
```

It is important to note that $R_{<x>}$ reduces an order of magnitude when N increases by two orders of magnitude in accordance with the variation of the variance following the central limit theorem: $\sigma^2 \sim 1/N$. It can be concluded that with a sample size $N = 100,000$ the flux have "converged" since the RSD is less than 0.5% for the "least precise" TLE flux estimate in region 10, farthest from the source (Table 12.2).

12.3 Reflected assemblies

In Section 10.1, MC simulation of the Highly Enriched Uranium (HEU) Godiva sphere with radius 8.741 cm and mass 52.4254 kg respectively gave $<k_{\text{eff}}^{(CAT)}> = 0.994272$ (0.0012) with the Monte Carlo N-Particle code (MCNP5) which close to the Los Alamos value of $k_{\text{eff}} = 0.9976$ (0.0011) (Whalen, Cardon, Uhle, & Hendricks, 1991).

An important exercise in nuclear engineering is to carry out simulations for reduction of fissile material to achieve criticality by the use of a reflector such as beryllium, graphite or water, as was demonstrated in Section 5.3.2 for a spherical solution of U-235 in water surrounded by a water reflector using two-group diffusion theory.

With MC simulations, much better designs in the sense of smaller size and efficient performance can be achieved.

In this section, simulations are demonstrated for reflected neutron systems with 93.5% U-235 surrounded by graphite and 97.6% U-235 surrounded by water (Whalen et al., 1991). There are several references for critical systems (Bowen & Busch, 2005; DeVolpi, 1982; Snood, Forster, & Parsons, 2003; Wagner, Sisolak, & McKinney, 1992) which can be used to test simulation results.

TABLE 12.2 Mean $<x>$ and relative standard deviation $R_{<x>}$ of the Monte Carlo TLE flux.

Region	$N = 100$		$N = 10,000$		$N = 100,000$	
	$<x>$	$R_{<x>}$	$<x>$	$R_{<x>}$	$<x>$	$R_{<x>}$
1	15.128	0.0000	14.853	0.0000	14.8950	0.0000
2	4.0574	0.0000	4.3861	0.0000	4.4056	0.0000
3	3.2868	0.0660	2.9959	0.0050	3.0262	0.0015
4	2.2036	0.0542	2.2855	0.0069	2.3241	0.0023
5	1.5420	0.0951	1.8662	0.0091	1.8830	0.0028
6	1.3609	0.0747	1.5436	0.0103	1.5610	0.0033
7	1.1447	0.0892	1.3125	0.0122	1.3298	0.0037
8	0.9845	0.1190	1.1435	0.0135	1.1466	0.0042
9	0.8948	0.1193	0.9924	0.0147	0.9814	0.0045
10	0.7337	0.1011	0.8526	0.0154	0.8582	0.0049

The search for a critical configuration begins with a bare sphere of radius 8.74 cm for which MC simulation (3000 histories for 50 cycles with 10 skip cycles) gives $k_{eff} = 0.98513\ 0.00180$ listed in Table 12.3; this is not critical due to the slightly decreased density compared with that for the Godiva sphere in Section 10.1. The mass of the HEU core is 52.0159 kg. To reduce the core size, a surrounding graphite layer of thickness 5.0 cm gives $k_{eff} = 01.10789\ 0.00170$ which is encouraging as it offers the possibility of reducing the core. In the next simulation, the core radius is decreased to 8.50 cm keeping the reflector thickness the same. The result shows that the system is supercritical with a reduction of about 4.2 kg in the core. A further reduction to a core radius of 8 cm is found to be slightly supercritical and another simulation is carried out with a core radius of 7.5 cm (HEU mass reduced by about 19 kg which is a ~37% mass reduction) with a 5 cm thick graphite reflector which gives $k_{eff} = 1.00058\ 0.00160$. Decreasing the core radius to 7.20 cm with the same reflector thickness results in a sub-critical system.

Based on the above results, it is possible to conclude on an approximate critical design that has a core radius in the range 7.20–7.50 cm with a 5.0 cm graphite reflector. Compare this with the experimental result of 7.39840 cm with a 5.1 cm graphite reflector (Whalen et al., 1991) for which $k_{eff} = 0.9981\ 0.0010$ for a sphere of radius 7.39840 cm.

Another design exercise is the selection of a "best" reflector. Table 12.4 shows MC simulation results for the critical HEU sphere of radius 7.39840 cm (31.5511 kg) described above with three reflectors-graphite, beryllium and light water showing beryllium to be the best.

Thus maximum core savings are possible with beryllium and the least effective of the three is graphite.

The effectiveness of a reflector is dependent on its atomic mass number and scattering and absorption cross-sections. Hydrogen has the lowest atomic number but it also has a high absorption cross-section and thus beryllium emerges as *the* most effective out of the three materials considered.

The critical masses for plutonium (density 19.5 g/cc α and 15.9 g/cc δ phase) are 11 and 15 kg respectively; for uranium (U^{233}, U^{235} density 18.9 g/cc) 16g and 48 kg respectively. For the Godiva and Jezebel criticality experiments, see

TABLE 12.3 Random search for criticality of a graphite-reflected spherical core.

Core		Reflector		Total mass	k_{eff}
Radius (cm)	Mass (kg)	Thickness (cm)	Mass (kg)	(kg)	
8.74	52.0159	Nil	nil	52.0159	0.98513 0.00180
8.74	52.0159	5.0	9.79459	61.8105	1.10789 0.00172
8.50	47.8474	5.0	12.9150	60.7624	1.10315 0.00168
8.00	39.8907	5.0	11.787	51.6777	1.04749 0.00166
7.50	32.8689	5.0	10.7115	43.5804	1.00058 0.00171
7.20	29.0803	5.0	10.0914	39.1717	0.96719 0.00195
7.20	29.0803	6.0	13.4779	42.5583	0.97962 0.00203
7.20	29.0803	7.0	17.4185	46.4988	0.99462 0.00197

TABLE 12.4 System multiplication k_{eff} for a reflected core.

Reflector				k_{eff}
Material	Density (g/cm)	Thickness (cm)	Mass (kg)	
Graphite	1.67	5.0	10.8246	0.9920 0.00229
Beryllium	1.85	5.0	11.9913	1.09337 0.00221
Light Water	1.0	5.0	6.48179	1.01864 0.00219

TABLE 12.5 Critical spherical systems reflected with 10 cm U.

Material	Density (g/cc)	Radius (cm)	Core mass (kg)	Total Mass (kg)
U^{233}	18.9	4.2	5.7	227
U^{235}	18.9	5.8	15.7	316
Pu^{239} α	15.9	3.8	4.5	209
Pu^{239} δ	19.5	4.7	7.0	250

problem no. 4 at the end of this chapter. The reduction in critical masses with a 10 cm natural uranium reflector is shown in Table 12.5 (DeVolpi, 1982).

The search for criticality is an exercise in which the critical dimensions are calculated for a specified core composition or the critical core composition is determined for a specified size.

One drawback in MC simulations has been the requirement of performing simulations for each separate design; this has been addressed to some extent by the perturbation algorithms which give the design sensitivities due to independent parametric changes. With this feature, the search for a critical design can be simplified.

Exercise 12.9: With the one-group effective multiplication given by

$$k_{\text{eff}} = \frac{\nu \Sigma_f}{DB^2 + \Sigma_a} \tag{12.14}$$

1. How would a decrease in the core radius affect k_{eff}?
2. How would the change in k_{eff} be estimated due to a 10% increase in the atomic number density of U-238 in Godiva?

12.4 Reactor core modeling

In this section, MCNP5B is used to model a Pressurized Water Reactor (PWR) fuel assembly starting from an input file by Hideki Matsumoto. Changes were made to obtain the assembly which this neutronic analysis is meant for.

12.4.1 Input file

Details of the input file can be found in the MCNP Documentation (Werner, 2017) (Volume I, Chapters 3 and 4). Here, parts of the input with reference to the PWR fuel assembly only will be described. The same geometry can be described (MCNP Vol II, Example 8, pp. 4–37) using LIKE m BUT and TRCL cards. The procedure is to define universes, fill a lattice cell, and then define the bounding window in which as many lattices will fit in as the bounding region would permit.

12.4.1.1 Defining the bounding window

The window, in which the lattice cell (universe 1) is to be filled, is defined as cell 1 with a void bounded by the surfaces 21 and 22 in the x direction (−10.7525, 10.7525, length 21.5050 cm), 23 and 24 in the y direction (same as in the x direction, i.e., 21.5050 cm), and surfaces 5 and 6 in the z direction (length 1 cm). The input line is:

1 0 21 −22 23 −24 5 −6 fill = 1

12.4.1.2 Defining the universe(s)

The universes are defined which will fill the lattice

Here, we have 3 universes, u = 1, u = 2, and u = 4

Universe 1 will be defined in the following section.

Universe 2 has cells 3, 4, and 5 with materials 1 (UO_2), 2 (zirconium), and 3 (water)

```
3 1 6.752111e-2    -1          u=2
4 2 4.310700e-2     1 -2       u=2
5 3 6.622400e-2     2          u=2
6 3 6.622400e-2    -3          u=4
7 2 4.310700e-2     3 -4       u=4
8 3 6.622400e-2     4          u=4
```

Exercise 12.10: Universes 2 and 4 are described radially. Where in the input would they be bounded axially?

12.4.1.3 Defining the lattice

Lattice 1 is defined, as cell 2 filled with universe 1 containing material 3 (water) and filled by itself with 19 × 19 × 1 elements in the x-y-z directions. In the x direction, it is bounded by the surfaces 11 and 12 and is of length 1.2650 cm; in the y direction by surfaces 13 and 14 also of length 1.2650 cm, and in the z direction by surfaces 5 and 6 and is of length 1 cm.

The input lines, describing cell 2 as the lattice 1 and filled with universe 1 defined itself with universes 1, 2, and 4, (with 19 × 19 elements) are:

```
2 3 6.622400e-2  11 -12   13 -14 5 -6 u=1 lat=1  fill=-9:9 -9:9 0:0
        1 1 1 1 1 1 1 1 1 1 1 1 1 1 1 1 1 1 1
        1 2 2 2 2 2 2 2 2 2 2 2 2 2 2 2 2 2 1
        1 2 2 2 2 2 2 2 2 2 2 2 2 2 2 2 2 2 1
        1 2 2 2 2 2 4 2 2 4 2 2 4 2 2 2 2 2 1
        1 2 2 2 4 2 2 2 2 2 2 2 2 2 4 2 2 2 1
        1 2 2 2 2 2 2 2 2 2 2 2 2 2 2 2 2 2 1
        1 2 2 4 2 2 4 2 2 4 2 2 4 2 2 4 2 2 1
        1 2 2 2 2 2 2 2 2 2 2 2 2 2 2 2 2 2 1
        1 2 2 2 2 2 2 2 2 2 2 2 2 2 2 2 2 2 1
        1 2 2 4 2 2 4 2 2 4 2 2 4 2 2 4 2 2 1
        1 2 2 2 2 2 2 2 2 2 2 2 2 2 2 2 2 2 1
        1 2 2 2 2 2 2 2 2 2 2 2 2 2 2 2 2 2 1
        1 2 2 4 2 2 4 2 2 4 2 2 4 2 2 4 2 2 1
        1 2 2 2 2 2 2 2 2 2 2 2 2 2 2 2 2 2 1
        1 2 2 2 4 2 2 2 2 2 2 2 2 2 4 2 2 2 1
        1 2 2 2 2 2 4 2 2 4 2 2 4 2 2 2 2 2 1
        1 2 2 2 2 2 2 2 2 2 2 2 2 2 2 2 2 2 1
        1 2 2 2 2 2 2 2 2 2 2 2 2 2 2 2 2 2 1
        1 1 1 1 1 1 1 1 1 1 1 1 1 1 1 1 1 1 1
```

12.4.2 Surrounding cells

Cell 10 surrounds cell 1 (PWR fuel assembly) and is filled with material 3 (water), while cell 9 is the outside universe surrounding the entire assembly and water; defined as everything in the universe that excludes cells 1 and 10.

The input lines are:

```
9 0          #1 #10
10 3 6.622400e-2 #1(25 -26 27 -28 5 -6)
```

12.4.2.1 Surface cards

```
Surface Cards
1    cz   0.412
2    cz   0.476
3    cz   0.570
4    cz   0.610
*5   pz   -0.5
*6   pz    0.5
11   px   -0.6325
12   px    0.6325
13   py   -0.6325
14   py    0.6325
21   px  -10.7525
22   px   10.7525
23   py  -10.7525
24   py   10.7525
*25  px  -10.8000
*26  px   10.8000
*27  py  -10.8000
*28  py   10.8000
```

12.4.3 Source description

The source description in MCNP can be done in one of three ways (MCNP Vol II, pp. 3–52) viz and explicit value, a distribution, or a distribution of a dependent variable. Here, the SDEF card is used to describe source location in a cell whose spatial distribution is represented by d1, and energy (erg) represented by Distribution 2.

For Distribution 1, the source information card SI1 is used with the l option indicating discrete source variable values. The source cell path is described as the location of the cell from the highest level to the lowest level. Here the lattice has 8 cells which have material 1 (fuel) so we need to locate cell 3 wherever it needs to be specified. So, since cell 3 is inside cell 2 which is inside cell 1, the "address" of this cell is through cell 1, and since cell 2 is a lattice, each position needs a location identifier such as (−1 −1 0) indicating the cell in the lowest row and most left column of the 3 × 3 lattice.

The energy Distribution 3 in MCNP represents the Watt fission spectrum described in Section 12.1.

The input parameters for source definition cards (SDEF), source information (SI) and source probability (SP) define the phase space parameters of the source. In the following lines, SDEF defines a source located in CEL = D1 with a space location defined by a probability Distribution 1, and with an energy distribution given by Distribution 2; SI is the source information card which gives the location of the source as Distribution 1 in the repeated geometry structure. There are eight cells in which the source appears and it will have equal probability as given in the SP1 card. Similarly, SP2 samples from the Watt fission spectrum with default parameters as stated above.

```
sdef  cel=d1  erg=d2
si1   l
          1:2( -1 -1  0):3  1:2(  0 -1  0):3  1:2(  1 -1  0):3
          1:2( -1  0  0):3  1:2(  1  0  0):3  1:2( -1  1  0):3
          1:2(  0  1  0):3  1:2(  1  1  0):3
sp1     3 7r
sp2    -3
```

Exercise 12.11: Is the cell sampling of the source adequate or can it be improved? What will be the effect of the cel = d1 specification on the quality of tallies?

12.4.4 Plotting the geometry

With the above surfaces, let's first calculate the length of the lattice cell which is bounded by surfaces 11 and 12 in the x direction, of length 0.6325 − (−0.6325) = 1.2650 cm. Cell 2 has 19 such-sized units in the x direction since it goes from −9:9 in the x direction, and 19 units in the y direction. They are numbered as

```
-9 9 0  -8 9 0  -7 9 0  ....  ....  7 9 0  8 9 0  9 9 0
-9 8 0  -8 8 0  -7 8 0  ....  ....  7 8 0  8 8 0  9 8 0




-9 -8 0  -8 -8 0  -7 -8 0  ....  ....  7 -8 0  8 -8 0  9 -8 0
-9 -9 0  -8 -9 0  -7 -9 0  ....  ....  7 -9 0  8 -9 0  9 -9 0
```

Thus, all 19 × 19 cells are of length 24.035 cm and extend from −12.0175 to +12.0175, but the bounding surfaces on cell 1 are −10.7525 to +10.7525. Similarly 18 × 18 cells are of length 22.77 cm extending from −11.385 to +11.385, and 17 × 17 cells are of length 24.035 cm extending from −12.0175 to +12.0175. Thus, the above input should show only 17 × 17 of the 19 × 19 defined cells.

The geometry is obtained by running mcnp5 in the ip (input and plot) mode; Fig. 12.6 below.

To obtain 3 × 3 cells from the above, the bounding window surfaces (21, 22, 23, 24) would be changed and the total length would now be: −1.8975 to +1.8975 = 3.795 = 3 × 2 × 0.6325.

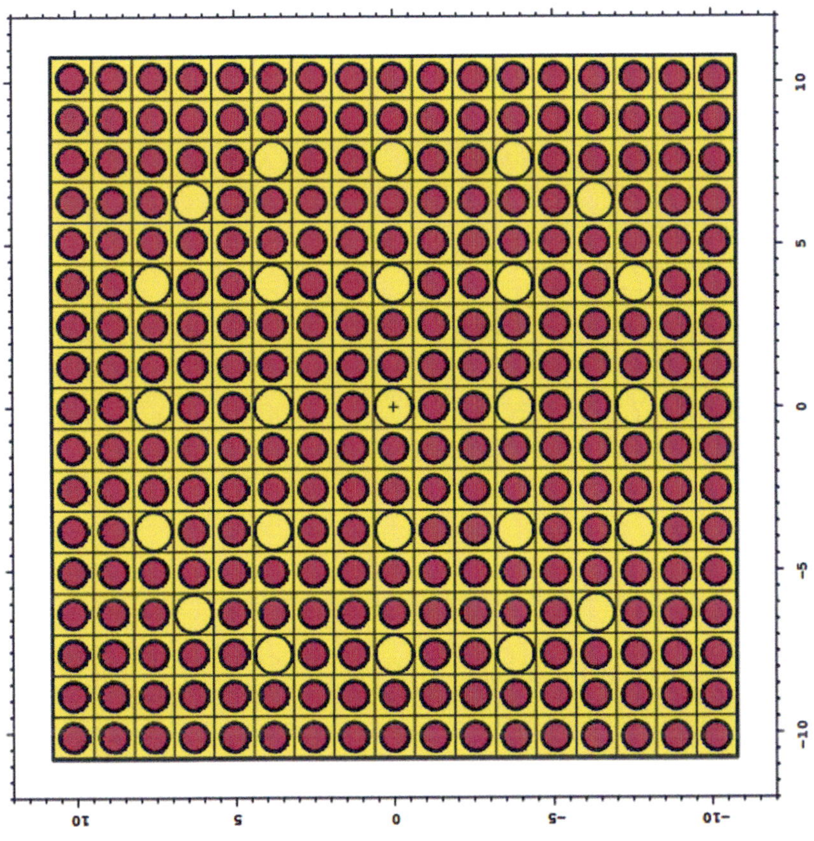

FIGURE 12.6 A typical PWR fuel assembly. (ZK1pwrA).

The 3 × 3 assembly is shown in Fig. 12.7 plotted with the commands

```
Input file ZKpwr3x3
Pz 0 extent 3 3 label 0 0 scales 1
```

with the input file ZKpwr3 × 3 listed below the figure.

```
14   py    0.6325
21   px   -1.8975
22   px    1.8975
23   py   -1.8975
24   py    1.8975
*25   px  -2.8000
*26   px   2.8000
*27   py  -2.8000
*28   py   2.8000

c geometry cards
mode  n
imp:n   1 7r  0 1
m1  92235.66c 9.3411e-4          $ u-235
    92238.66c 2.1573e-2          $ u-238
    8016.66c 4.5014e-2           $ o
m2  40000.66c 4.3107e-2          $ Zr
m3   1001.66c 4.4148e-2          $ h for 19 doc
    8016.66c 2.2076e-2           $ o
mt3  lwtr.60t
m4  92235.66c 1.0
m5  92238.66c 1.0
m6   8016.66c 1.0
c
e0 6.25e-7 10.0
f4:n   (3<2[-1,-1,0]<1) (3<2[0,-1,0]<1)  (3<2[1,-1,0]<1)
       (3<2[-1, 0,0]<1)               (3<2[1, 0,0]<1)
       (3<2[-1, 1,0]<1) (3<2[0, 1,0]<1)  (3<2[1, 1,0]<1)
sd4   1 7r
fmesh14:n GEOM=xyz ORIGIN = -2.8 -2.8 -0.5
        IMESH=2.8 IINTS=40 JMESH=2.8 JINTS=40 KMESH=0.5 KINTS=1
c source cards
kcode  1000 1.0  10  500
c
sdef  cel=d1  erg=d2
si1  l
       1:2( -1 -1  0):3  1:2(  0 -1  0):3  1:2(  1 -1  0):3
       1:2( -1  0  0):3  1:2(  1  0  0):3  1:2( -1  1  0):3
       1:2(  0  1  0):3  1:2(  1  1  0):3
sp1    3 7r
sp2  -3
prdmp  j -100 1 3
print
```

```
ZKpwr3X3 PWR fuel assembly benchmark: UO2
C 13 june 2021
c cell cards
 1 0           21 -22    23 -24 5 -6  fill=1
 2 3 6.622400e-2  11 -12    13 -14 5 -6 u=1 lat=1
            fill=-9:9 -9:9  0:0
          1 1 1 1 1 1 1 1 1 1 1 1 1 1 1 1 1 1 1
          1 2 2 2 2 2 2 2 2 2 2 2 2 2 2 2 2 2 1
          1 2 2 2 2 2 2 2 2 2 2 2 2 2 2 2 2 2 1
          1 2 2 2 2 2 4 2 2 4 2 2 4 2 2 2 2 2 1
          1 2 2 2 4 2 2 2 2 2 2 2 2 2 4 2 2 2 1
          1 2 2 2 2 2 2 2 2 2 2 2 2 2 2 2 2 2 1
          1 2 2 4 2 2 4 2 2 4 2 2 4 2 2 4 2 2 1
          1 2 2 2 2 2 2 2 2 2 2 2 2 2 2 2 2 2 1
          1 2 2 2 2 2 2 2 2 2 2 2 2 2 2 2 2 2 1
          1 2 2 4 2 2 4 2 2 4 2 2 4 2 2 4 2 2 1
          1 2 2 2 2 2 2 2 2 2 2 2 2 2 2 2 2 2 1
          1 2 2 2 2 2 2 2 2 2 2 2 2 2 2 2 2 2 1
          1 2 2 4 2 2 4 2 2 4 2 2 4 2 2 4 2 2 1
          1 2 2 2 2 2 2 2 2 2 2 2 2 2 2 2 2 2 1
          1 2 2 2 4 2 2 2 2 2 2 2 2 2 4 2 2 2 1
          1 2 2 2 2 2 4 2 2 4 2 2 4 2 2 2 2 2 1
          1 2 2 2 2 2 2 2 2 2 2 2 2 2 2 2 2 2 1
          1 2 2 2 2 2 2 2 2 2 2 2 2 2 2 2 2 2 1
          1 1 1 1 1 1 1 1 1 1 1 1 1 1 1 1 1 1 1
 3 1 6.752111e-2     -1              u=2
 4 2 4.310700e-2  1  -2              u=2
 5 3 6.622400e-2  2                  u=2
 6 3 6.622400e-2     -3              u=4
 7 2 4.310700e-2  3  -4              u=4
 8 3 6.622400e-2  4                  u=4
 9 0          #1 #10
 10 3 6.622400e-2 #1(25 -26 27 -28 5 -6)

c surface cards
 1   cz   0.412
 2   cz   0.476
 3   cz   0.570
 4   cz   0.610
*5   pz   -0.5
*6   pz    0.5
 11  px   -0.6325
 12  px    0.6325
 13  py   -0.6325
```

Exercise 12.12: How would you modify the water density to account for the temperature changes in the reactor from cold to hot states?

12.4.5 Tally cards

MCNP gives information "demanded" by the user; thus tallies need to be specified. In most cases, the tallies required are volume averaged flux, current, surface flux, energy deposition, and neutron heating. These are obtained from standard MCNP tallies F1, F2, F4, F5, F6 and associate energy values *F1, *F2, *F4, *F4, *F5, and *F6, as described below.

The flux tally is TALLY F14:N in the cells defined below:

```
f14:n (3<2[-8,-8,0]<1) ...    (3<2[ 8,-8,0]<1)  [17 entries]
      (3<2[-8,-7,0]<1) ...    (3<2[ 8,-7,0]<1)  [17 entries]
      (3<2[-8,-6,0]<1) ...    (3<2[ 8,-6,0]<1)  [14 entries]
      (3<2[-8,-5,0]<1) ...    (3<2[ 8,-5,0]<1)  [15 entries]
      (3<2[-8,-4,0]<1) ...    (3<2[ 8,-4,0]<1)  [17 entries]
      (3<2[-8,-3,0]<1) ...    (3<2[ 8,-3,0]<1)  [12 entries]
      (3<2[-8,-2,0]<1) ...    (3<2[ 8,-2,0]<1)  [17 entries]
      (3<2[-8,-1,0]<1) ...    (3<2[ 8,-1,0]<1)  [17 entries]
      (3<2[-8, 0,0]<1) ...    (3<2[ 8, 0,0]<1)  [12 entries]
      (3<2[-8, 1,0]<1) ...    (3<2[ 8, 1,0]<1)  [17 entries]
      (3<2[-8, 2,0]<1) ...    (3<2[ 8, 2,0]<1)  [17 entries]
      (3<2[-8, 3,0]<1) ...    (3<2[ 8, 3,0]<1)  [12 entries]
      (3<2[-8, 4,0]<1) ...    (3<2[ 8, 4,0]<1)  [17 entries]
      (3<2[-8, 5,0]<1)        (3<2[ 8, 5,0]<1)  [15 entries]
      (3<2[-8, 6,0]<1)  .     (3<2[ 8, 6,0]<1)  [14 entries]
      (3<2[-8, 7,0]<1) ...    (3<2[ 8, 7,0]<1)  [17 entries]
      (3<2[-8, 8,0]<1) ...    (3<2[ 8, 8,0]<1)  [17 entries]
```

Tallies available (for neutrons) include F1:N (particle current integrated over a surface, that is, no. of particles), F2: N (surface flux particles cm^{-2}), F4:N (flux at point or ring detector particles cm^{-2}), F6:n (energy deposition averaged over a cell MeV/g), F7:N (fission energy deposition averaged over a cell MeV/g). Together with these tallies, *F1 gives energy (MeV) integrated over a surface, *F2, *F4 and *F5 are energy quantities (MeV cm^{-2}), while *F6 and *F7 give jerks/g (1 jerk $\equiv 10^{-8}$s). In MCNP, adding 10 to a tally gives the same tally quantity; for example, F4, F14, F24, F34, ... are all flux quantities. Each tally can be obtained in bins of time, energy and direction.

12.4.6 Reaction rates

Any quantity of the form $C\int\phi(E)R_m(E)dE$, where $\phi(E)$ is the energy-dependent fluence (units particles cm^{-2}) and the (microscopic) reaction of interest R_m for material m is taken from the cross-section file. The material m needs to be specified on a separate *Mm* card and is not required to be present in any cell defined in the geometry. The reaction numbers R are taken from the Evaluated Nuclear Data File (ENDF/B) manual (where it is called the material type (MT) number), some commonly used MT numbers are

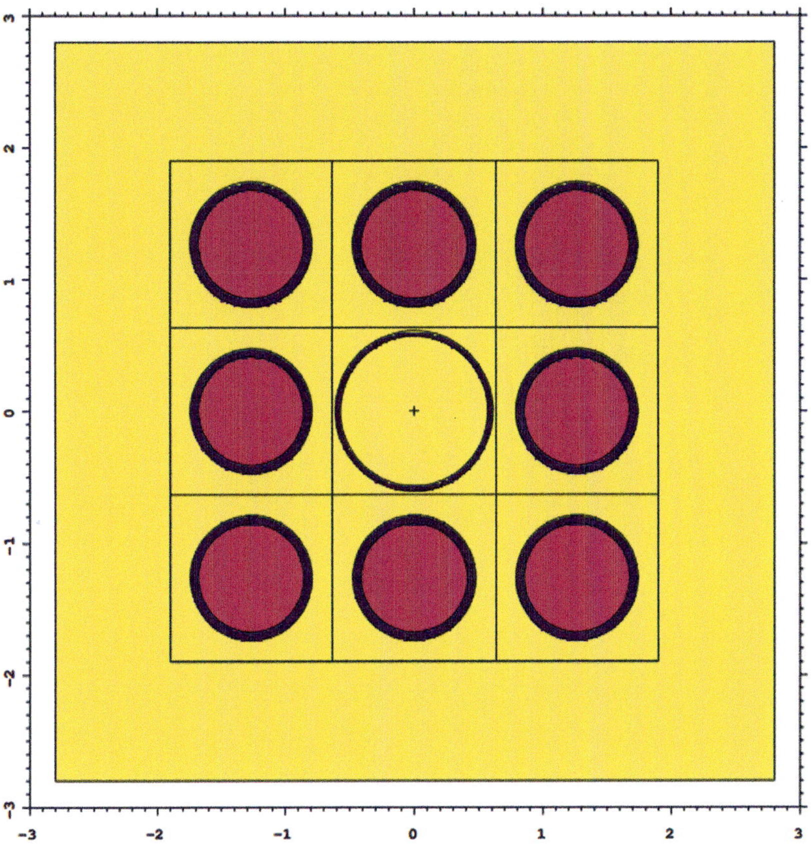

FIGURE 12.7 A 3×3 assembly (scales shown are centimeters).

MT	FM	σ
1	-1	Total
2	-3	Elastic
16		(n,2n)
17		(n,3n)
18		(n,fx)
18	-6	(n,f)
19		(n,f)
20		(n,n′f)
21		(n,2nf)
.		.
102		(n,γ)

For MT18 the condition is: if and only if MT = 18 is used for fission in the original evaluation

Total fission cross-section; equal to MT = 18 if MT = 18 exists (otherwise sum of MTs 19,20,21 and 38).

12.4.7 Plotting tallies

For plotting tallies and cross-sections, see MCNP5 Vol II Appendix B.III. The command `mcnp5 z options` invokes MCPLOT. Fig. 12.8 was plotted with the commands `kcode 1` and file which gave a postscript file plotm.ps which was then converted to the pdf file shown below. The average of the collision, absorption, and track-length estimates is $k_{eff} = 1.44256(0.00062)$. The standard deviation 0.00062, is *good*; it can be further improved as described in Chapter 4.

Fig. 12.9 shows the Figure of Merit for the average CE/absorption estimator (AE)/TLE of k_{eff}, which appears to have reasonably converged beyond 400 cycles.

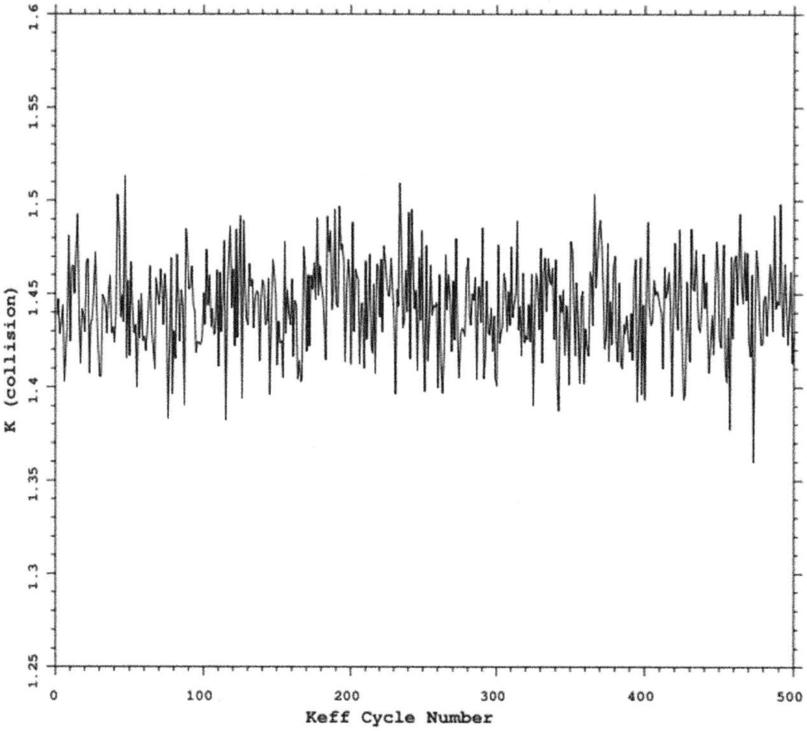

FIGURE 12.8 Collision estimate of k_{eff} versus Cycle number.

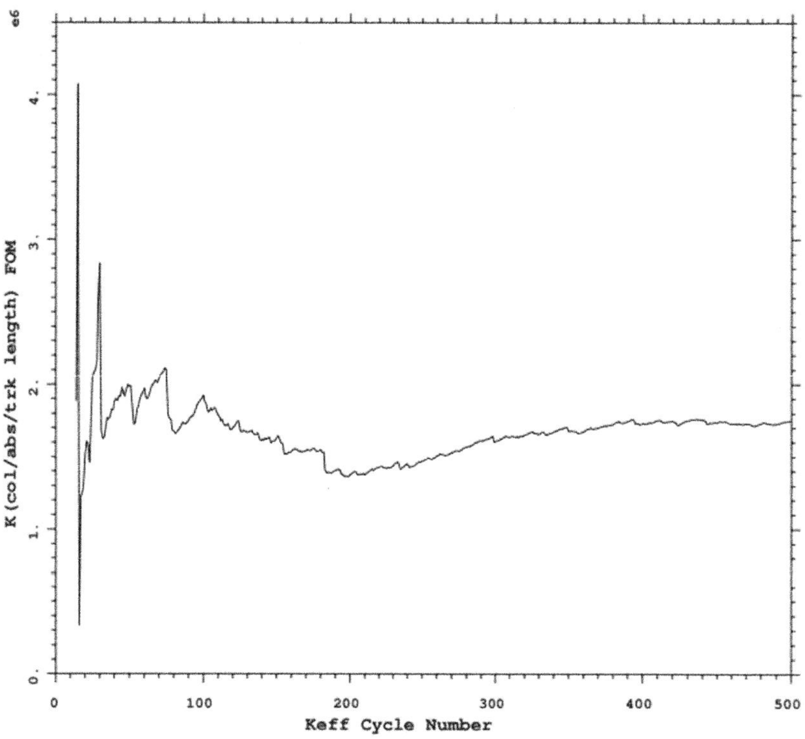

FIGURE 12.9 FOM of average k_{eff} versus Cycle number.

The energy-integrated neutron flux, plotted in Fig. 12.10 as a surface and in Fig. 12.11 on a xy plane, show a dip at the center of the 3 × 3 assembly.

These were obtained by using the fine mesh (FMESH) input "card" (Vol II, pp. 3–112) which gives an output in file MESHTAL processed in MATLAB.

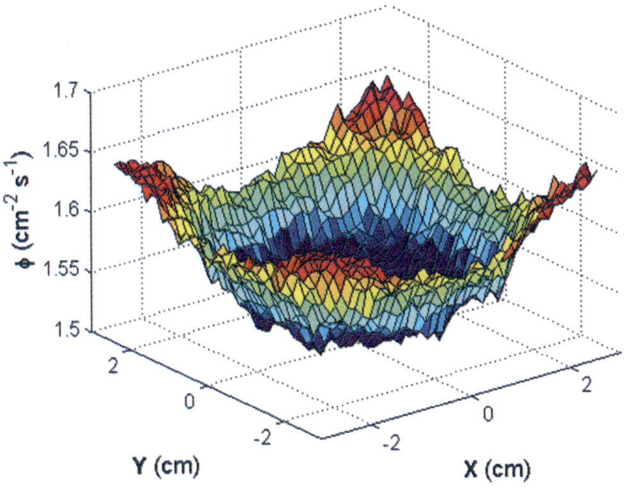

FIGURE 12.10 Neutron flux in a 3 × 3 PWR cell.

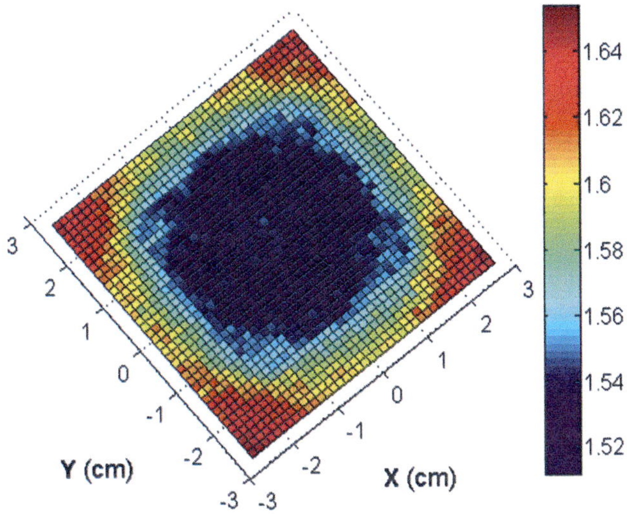

FIGURE 12.11 Neutron flux in a 3 × 3 PWR cell (xy plane).

12.5 Radiation safety and shielding

The basics of radiation and radiation shielding to protect people working in radiation environments, were highlighted in Chapter 1. The design of such shields is readily carried out by MC codes since detailed interaction models are used by sampling histories in a considered sample.

Nuclear criticality safety analysis, described in Chapter 10, is required at various stages in the fuel cycle from processing of uranium ore to the handling of process gases such as uranium hexafluoride (UF_6) as it is enriched, for example in a centrifuge plant Section 10.3.2, to its final stages when it takes the shape of enriched fissile fuel with reactor grade.

For analysis of nuclear criticality safety, simple cylinders containing solutions of uranium and plutonium to repeated structures in hexahedral (square) and hexagonal (triangular) lattices are easily modeled in MCNP.

In this section, MCNP simulations are carried out for a 20 cm thick slab of water in the x-direction, and 10 cm thick in the y, z-directions and with front, back, top, and bottom surfaces taken as reflecting surfaces. A point anisotropic neutron source of unit strength and energy 1 MeV is incident on the left face of the slab.

material	component nuclide, atom fraction	
3	1001, 6.66667E-01	8016, 3.33333E-01
90	1001, 1.00000E+00	
92	1001, 6.66667E-01	8016, 3.33333E-01

Exercise 12.13: A person is standing in front of a Cf^{252} source which has an activity of 532 Curies/g with a neutron dose $\sim$22 Sv m^2/g/h and a gamma dose 1.6 Sv m^2/g/h.

If the maximum permissible dose allowed is 5 rem/y (0.05 Sv/y), how much would have been accumulated more than the annual limit in 1 hour from 1 g of material at a distance of 1 m? [*Answer:* $\sim$400 times]

Consider a 1 MeV neutron source for which a water and an iron (Fe) slab are compared, each of thickness 20 cm. Results from the MCNP simulation, listed below, show that the total neutron dose across the iron shield 7.68320×10^{-13} Sv/(n/s) is about 100 times higher while the total photon dose 4.74296×10^{-16} Sv/(n/s) is about 3% less than that from a comparable water shield. MCNP mixed mode (N/P) simulations for 20 cm × 10 cm × 10 cm slabs of water (2 kg) and iron (15.720 kg) are given in Table 12.6.

So, it can be concluded that water is a much more effective neutron shield while iron is a much better gamma shield. The total dose for both is given below

Material	Neutron (Sv/(n-s))	Gamma (Sv/(n-s))	Total (Sv/(n-s))
Water	7.82087E−15 0.0477	1.74176E−14 0.0124	2.5239E−14
Iron	7.68320E−13 0.0100	4.74296E−16 0.0920	7.6879E−13

The total dose across the iron shield is thus about 30 times more. To estimate the effect of radiation, neutron and gamma radiation must be considered separately due to the different nature of both radiations.

For an annual limit of 0.05 Sv/y ($\sim$1.6 nSv/s), the maximum permissible neutron source would thus be $\sim$2000 n/s. Comparing this with the neutron emission of $\sim 10^{12}$ n/g from Cf^{252}, even a microgram would be far too dangerous for short durations

Exercise 12.14: Compare the half value layer thickness (HVL), at 1 MeV for iron and water to estimate the attenuation from 20 cm for both shields.

A detailed simulation shows (F1 current tally) gives a leakage of 5.85004E−03 gammas/(n, g) gamma compared to 0.1721 from water. The (n, γ) reactions are about ten times greater in water than in iron, thus requiring an effective iron shield to stop secondary gammas.

Relaxation length (width of material for intensity reduction to 1/e $\sim$37%) and HVL for 1 MeV photons is given in Table 12.7.

12.6 Perturbation calculations

One of the weaknesses of MC methods has been considered to be their inability to estimate the effect of small changes in independent parameters such as enrichment, material density or material composition (Section 6.6). This is due to the magnitude of the perturbation in the independent parameter which may be of the order of uncertainty in a stochastic estimate and hence the estimated change may be masked in the error itself.

As discussed in a previous chapter, perturbation computations can be carried out in schemes such as derivative sampling, correlated sampling or adjoint weighting in a Lagrangian variational formulation. This section considers the perturbation capability of MC simulation to estimate such design changes.

In Fig. 12.12 MCNP Godiva simulations (1000 × 130 with 30 skip cycles) show the increase in k_{eff} and decrease in J (particles crossing surface) as the relative density $\rho_r \equiv \rho/\rho_o$ increases (reference density $\rho_o = 18.75 \mathrm{g/cm^3}$). *Each run is an independent estimate.*

Consider now an elementary "one-group" diffusion estimate of the change in k_{eff} due to a material density change. The one-group diffusion equation gives

TABLE 12.6 Current, flux and dose for water and iron slabs.

Tally*/Mat/ reaction	Surface/ cell	Description	0–5 eV	5 eV–14 MeV	Total
F1:N	2	Current (n)	2.31323E−02 0.0146	3.26898E−03 0.0552	2.64013E−02 0.0144
			4.18566E−05 0.2921	2.34209E−01 0.0056	2.34251E−01 0.0056
F4:N	1	Cell flux (n)	1.14378E−02 0.0033	5.81316E−03 0.0020	1.72509E−02 0.0026
			6.73521E−06 0.1129	1.67608E−02 0.0037	1.67676E−02 0.0037
F2:N	2	Dose (n) Sv	3.57166E−15 0.0190	4.24921E−15 0.0864	7.82087E−15 0.0477
			6.64665E−18 0.3187	7.68314E−13 0.0100	7.68320E−13 0.0100
FM14:N	1	−1 90 102	3.15975E−04 0.0034	1.12125E−06 0.0034	3.17097E−04 0.0034
		−1 190 102	1.68707E−07 0.1265	9.18275E−06 0.0094	9.35145E−06 0.0101
FM24	1	1 90 102	3.15003E−03 0.0034	1.11780E−05 0.0034	3.16121E−03 0.0034
		1 190 102	1.99052E−06 0.1265	1.08344E−04 0.0094	1.10334E−04 0.0101
FM34	1	1 92 102	2.10062E−03 0.0034	7.45418E−06 0.0034	2.10808E−03 0.0034
		—	—	—	—
F11:P	2	Current (p)	0	1.72100E−01 0.0084	1.72100E−01 0.0084
			0	5.85004E−03 0.0420	5.85004E−03 0.0420
F12:P	2	Dose (p), Sv	0	1.74176E−14 0.0124	1.74176E−14 0.0124
			0	4.74296E−16 0.0920	4.74296E−16 0.0920

Upper and lower rows for each tally are for water and iron slabs respectively.

TABLE 12.7 Half value layer thickness for some shielding materials.

Material	Density (g/cm^3)	μ/ρ (cm^2/g)	$1/\mu\rho$ (cm)	HVT (cm)
Air	1.225×10^{-3}	0.0636	1.2835×10^4	1.2835×10^4
Water	1	0.0707	14.1443	9.8020
Concrete	2.4	0.0637	6.5411	4.5330
Iron	7.874	0.0599	2.1202	1.4693
Lead	11.34	0.0680	1.2968	0.8987

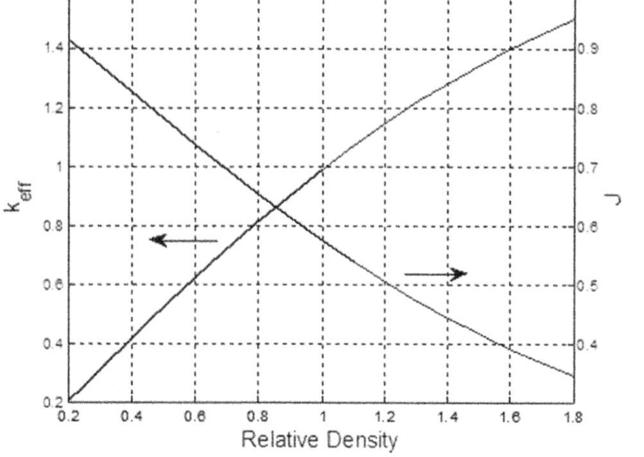

FIGURE 12.12 Effect of perturbation in material density on k_{eff}.

$$k_{\text{eff}} = \frac{\nu\Sigma_f}{DB^2 + \Sigma_a} \tag{12.15}$$

which can be used to estimate the perturbations, with respect to material density ρ for example, from the derivatives in a Taylor series of the form

$$k_{\text{eff}}^* = k_{\text{eff}} + \frac{\partial k_{\text{eff}}}{\partial \rho}\delta\rho + \frac{1}{2!}\frac{\partial^2 k_{\text{eff}}}{\partial \rho^2}(\delta\rho)^2 + \ldots \tag{12.16}$$

The first and second derivatives are obtained as:

$$\frac{1}{k_{\text{eff}}}\frac{\partial k_{\text{eff}}}{\partial \rho} = \frac{2\beta}{\rho}\frac{1}{\beta + \gamma\rho^2} \tag{12.17}$$

and

$$\frac{1}{k_{\text{eff}}}\frac{\partial^2 k_{\text{eff}}}{\partial \rho^2} = \frac{2\beta}{\rho^2}\frac{\beta - 3\gamma\rho^2}{(\beta+\gamma\rho^2)^2} \tag{12.18}$$

where $\beta \equiv \frac{B^2}{3\sigma_{tr}}$ and $\gamma \equiv \left(\frac{N_{av}}{M}\right)^2\sigma_a$.For Godiva, consider a pure U^{235} sphere of radius 8.37 cm of density $\rho = 18.75$ g/cm^3 ($m = 46.054$ kg) and use the fast spectrum data from Wirtz (1982): $\nu\sigma_f = 5.297$ atom b^{-1}, $\sigma_a = 2.844$ atom b^{-1} and $\sigma_a = 8.246$ atom b^{-1}. The first and second derivatives are: 0.0494 and -0.0030 respectively.

The above values are used to estimate k_{eff} for full re-runs (solid curve) compared with second order perturbation estimate from derivative sampling with MCNP, and with one-group exact analysis. Relative errors are shown for both sets of (MCNP) estimates.

Fig. 12.13 shows estimates for k_{eff} for full re-runs (solid curve) compared with second order perturbation estimate from derivative sampling and the elementary model. Relative errors are shown for both sets of estimates. It can be seen that the predicted derivative sampling estimates are good for up to a 60% decrease, and 40% increase, in the relative density. The lack of symmetry is due to the neutron histories simulated; as the density increases the mean free path decreases and more collisions take place, thus the sampling size has to be appropriately modified for statistical accuracy of an estimate. The elementary diffusion estimates have a larger error but indicate the trend very well.

Similar estimates, with similar conclusions, are shown for the surface current in Fig. 12.14.

12.7 MCNP geometry plotting in core neutronics

One of the strengths of MC simulations, particularly of MCNP, is the ability to represent realistic geometry of the configurations considered for systems such as the 4S reactor, the micronuclear reactor and the International Thermonuclear Reactor (ITER) tokamak.

In Fig. 12.15 a 4S-type configuration is shown; this has hexagonal cells with fuel elements and a central control rod.

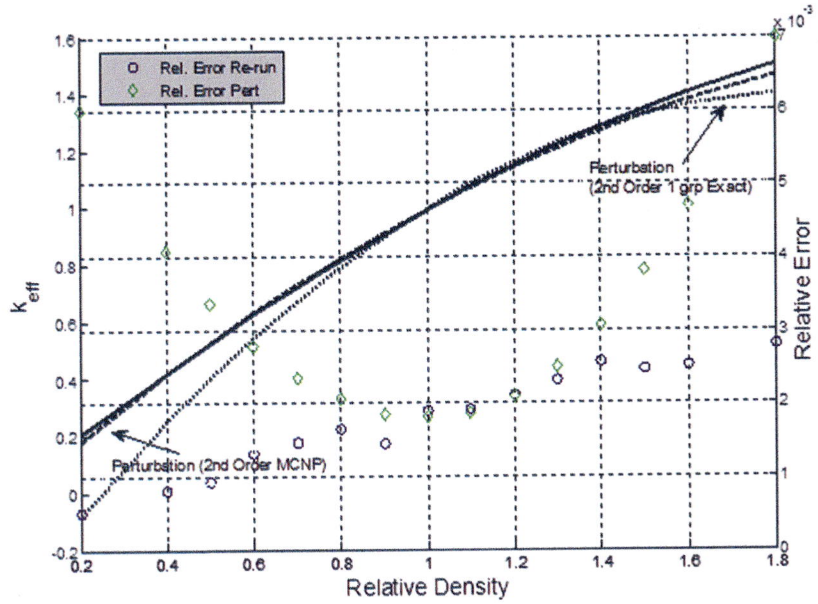

FIGURE 12.13 Perturbation estimates using one-group derivative sampling.

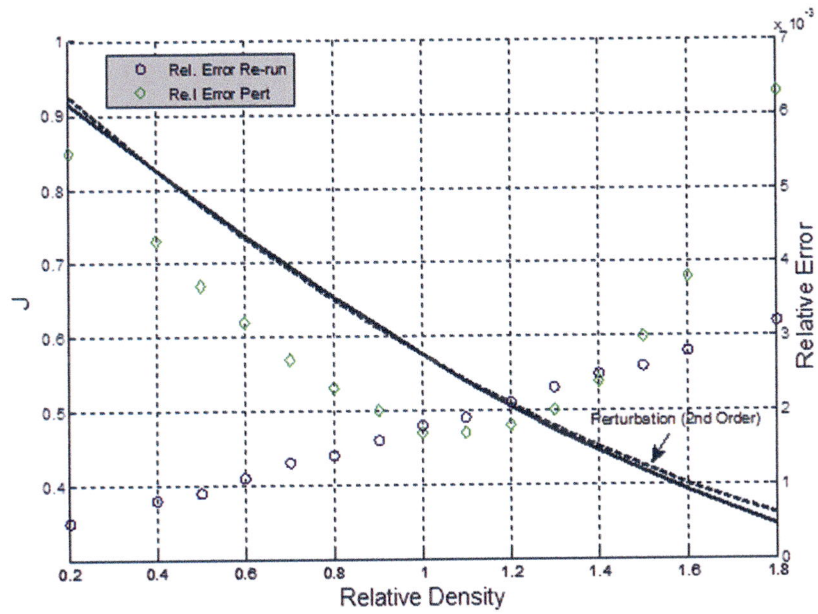

FIGURE 12.14 Effect of perturbation in material density on current J.

Exercise 12.15: The configuration given in Fig. 12.15 is of the 4S Gen-IV type reactor with metallic uranium fuel, sodium coolant and a beryllium reflector. Model the geometry with MCNP and carry out a simulation for the k_{eff} and flux tallies.

Consider a reactor core and reflector as shown below red being the fuel rods, blue the heat pipes, yellow the matrix, green the reflector and the dark strips in the six surrounding control drums representing the absorber material.

Exercise 12.16: Model the absorber regions in the control rod to face toward or away from the core.

This Exercise is thus to model the control drums with the absorbers spanning an angle 45 degrees to face the core or to be on the "opposite" side.

The first step is to generate the equations of the planes for the absorber regions. These are shown below; equations are $mx - y + c = 0$ in the MCNP format $Ax + By + Cz - D = 0$. The drums are labeled P to U with centers at (x, y). The

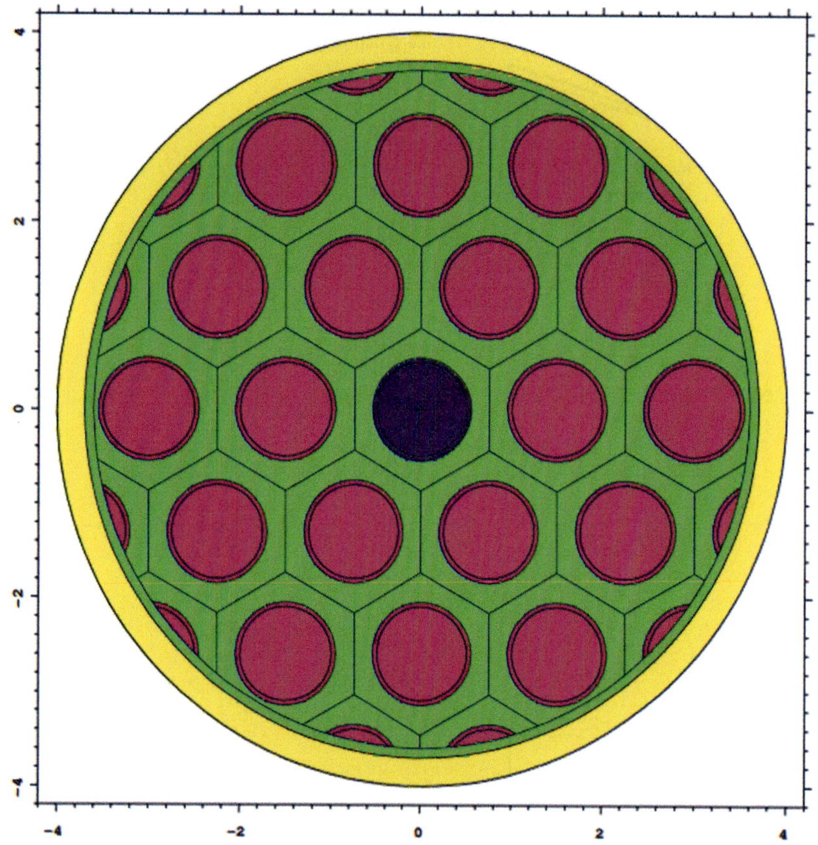

FIGURE 12.15 Reactor 4S-type core arrangement. (MCNP inp = ZK4Sfil).

TABLE 12.8 Surface coefficients.

		x	y	φ	m	c	surface
P	Top Right	a	b	$30-\theta$	0.131652497587396	10.881235923901318	40
				$30+\theta$	1.303225372841206	−17.721544252545193	41
Q	Top	0	28.1908	$90-\theta$	2.414213562373095	28.190799999999999	48
				$90+\theta$	−2.414213562373095	28.190799999999999	49
R	Top Left	-a	b	$150-\theta$	−1.303225372841206	−17.721544252545193	45
				$150+\theta$	−0.131652497587396	10.881235923901318	44
S	Bottom Left	− a	− b	$30+\theta$	1.303225372841206	17.721544252545193	47
				$30-\theta$	0.131652497587396	−10.881235923901318	46
T	Bottom	0	− 28.1908	$90+\theta$	−2.414213562373095	−28.190799999999999	50
				$90-\theta$	2.414213562373095	−28.190799999999999	51
U	Bottom Right	a	− b	$150+\theta$	−0.131652497587396	−10.881235923901318	43
				$150-\theta$	−1.303225372841206	17.721544252545193	42

$a = 24.4140$, $b = 28.1908$, $\theta = (45/2)$ degrees, numbers obtained with format "long" in MATLAB.

angles φ are relative to the horizontal axis, and the slope m and intercept c are calculated for each plane. The surface numbers are given in the "right-most column" below (Table 12.8).

The second step is to write the above in MCNP input format. Here, material 8 (density 3.01 g/cc) and material 9 (density 2.52 g/cc) are the reflector and absorber material, respectively. Cells are numbered 404–436.

```
c CONTROL DRUMS
404  8 -3.01 -25    15 -3  imp:n=1 $ Top control drum
405  8 -3.01  25 -26 15 -3  48 49 imp:n=1 $ upper strip
406  9 -2.52  25 -26 15 -3 -49 -48 imp:n=1 $  9 -2.52  B4C
407  8 -3.01  25 -26 15 -3  48 -49 imp:n=1 $ left strip
431  8 -3.01  25 -26 15 -3 -48  49 imp:n=1 $ right strip
c
408  8 -3.01 -27    15 -3 imp:n=1  $ TOP RIGHT
409  8 -3.01  27 -28 15 -3 -40 -41 imp:n=1
410  9 -2.52 27 -28 15 -3  40 -41 imp:n=1 $ B4C
411  8 -3.01  27 -28 15 -3  40 41 imp:n=1
432  8 -3.01  27 -28 15 -3 -40 41 imp:n=1
c
412  8 -3.01 -29    15 -3 imp:n=1  $ Bottom Right
413  8 -3.01  29 -300 15 -3 -43 -42 imp:n=1  $ lower strip
414  9 -2.52  29 -300 15 -3  43 -42 imp:n=1 $ B4C 9 -2.52
415  8 -3.01  29 -300 15 -3  43 42 imp:n=1 $ upper strip
433  8 -3.01  29 -300 15 -3 -43 42 imp:n=1 $ back strip
c
416  8 -3.01 -31  15 -3 imp:n=1  $ Bottom
417  8 -3.01 31 -32 15 -3 50 -51 imp:n=1 $ left strip
418  9 -2.52 31 -32 15 -3 -50 -51 imp:n=1 $ B4C
419  8 -3.01  31 -32 15 -3 51 -50 imp:n=1 $ right strip
434  8 -3.01  31 -32 15 -3 51 50 imp:n=1 $ bottom strip
c
420  8 -3.01 -33    15 -3 imp:n=1  $ Bottom Left
421  8 -3.01  33 -34 15 -3 -46 47 imp:n=1 $ upper strip
422  9 -2.52  33 -34 15 -3 -46 -47 imp:n=1 $ B4C
423  8 -3.01  33 -34 15 -3  46 -47 imp:n=1 $ lower strip
435  8 -3.01  33 -34 15 -3  46 47 imp:n=1 $ back strip
424  8 -3.01 -35    15 -3 imp:n=1     $ Top Left
425  8 -3.01  35 -36 15 -3   44 -45 imp:n=1 $ back strip
426  9 -2.52  35 -36 15 -3  -44 45 imp:n=1 $ B4C
427  8 -3.01  35 -36 15 -3  -44 -45 imp:n=1 $ lower strip
436  8 -3.01  35 -36 15 -3   44  45 imp:n=1 $ upper strip

C  surfaces (Top Right to Bottom)
40   p 0.131652 -1   0 -10.881236  $ TR
41   p 1.303225 -1   0 17.721544
42   p 1.303225  1   0 17.721544  $ BR
43   p 0.131652  1   0 -10.881236
44   p 0.131652  1   0  10.881236  $ TL
45   p 1.303225  1   0 -17.721544
46   p 0.131652 -1   0 10.881236  $ BL
47   p -1.303225 1   0 17.721544
48   p -2.414214  1   0 28.1908      $ T
49   p 2.414214  1   0 28.1908
50   p -2.414214 -1   0 28.1908     $ B
51   p 2.414214 -1   0 28.1908
```

The third step is to run MCNP in the ip mode with the commands:

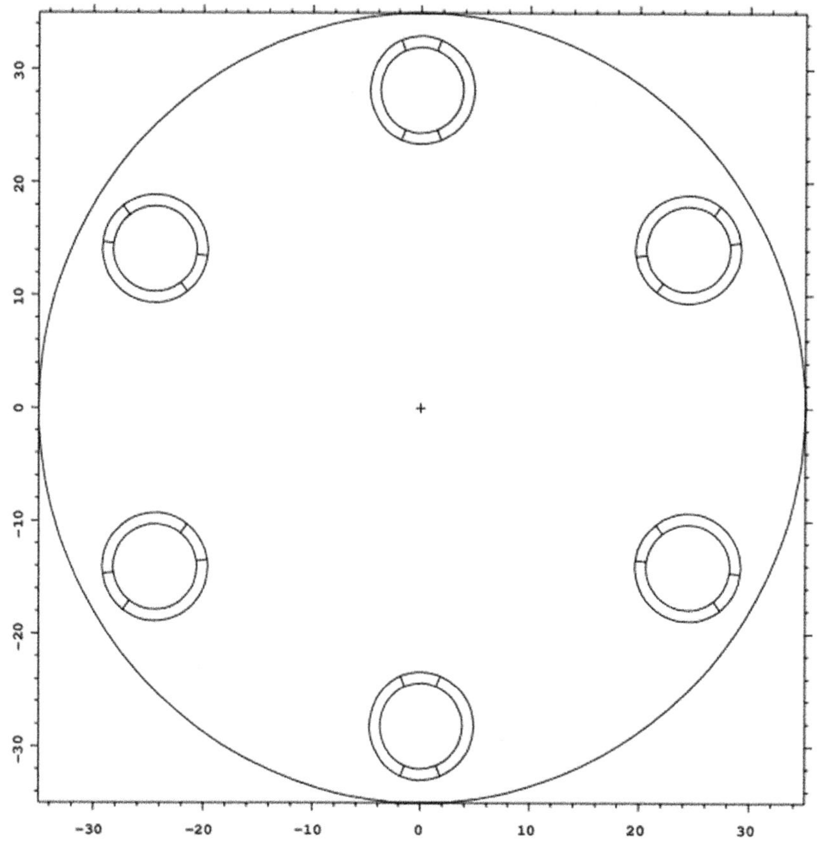

FIGURE 12.16 Absorbers on control drums in core barrel.

mcnp5 ip inp = inpfile plot> pz 0 plot > extent 35 35 plot> label 0 0 plot> scales 1plot> color on plot> color by mat plot> shade 1 red plot> shade 4 blue plot> shade 7 yellow plot> shade 8 green plot> shade 9 black plot > file

This gives a plotm.ps (or appropriately named file) which must be converted to pdf format.
Figs. 12.16–12.18 show the geometry with absorbers as required for the simulation.

Problems

1. Carry out a MC simulation for a graphite sphere of specified radius to determine the escape probability and compare with the diffusion theory estimate.
2. It was shown that the neutron flux in a sphere of radius R with a point isotropic source at $r = 0$ emitting S neutrons s^{-1} is given by

$$\phi(r) = \frac{S}{4\pi D \sinh\left(\frac{R+d}{L}\right)} \frac{\sinh\kappa(R+d-r)}{r}$$

   Carry out a MC simulation and determine the shape of the neutron flux and the leakage probability from the surface of a beryllium sphere of radius 2 m.f.p. with a 1 Ci ^{226}Ra-Be point source at its center emitting 10^7 neutrons s^{-1}.
3. Carry out a simulation to determine the albedo of a graphite slab of thickness 3 m.f.p and compare with previous estimates from diffusion theory and transport theory.
4. Compare MC simulation values for the k_{eff} values of some fast critical assemblies (Bell and Glasstone, 1979) given below (Table 12.9).

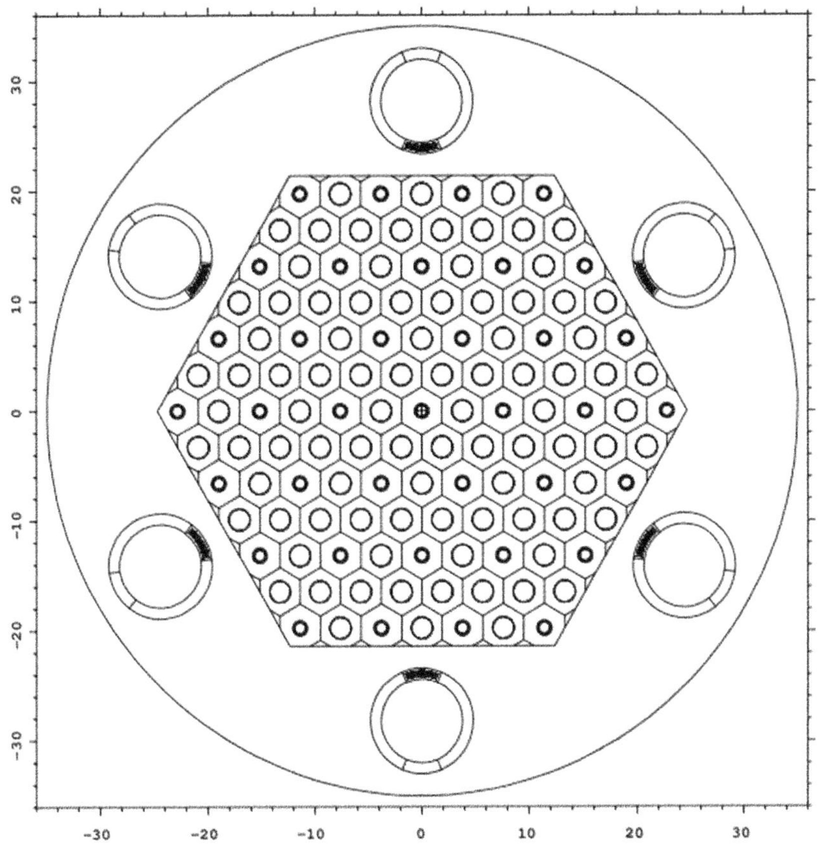

FIGURE 12.17 Absorbers on control drums facing toward the core.

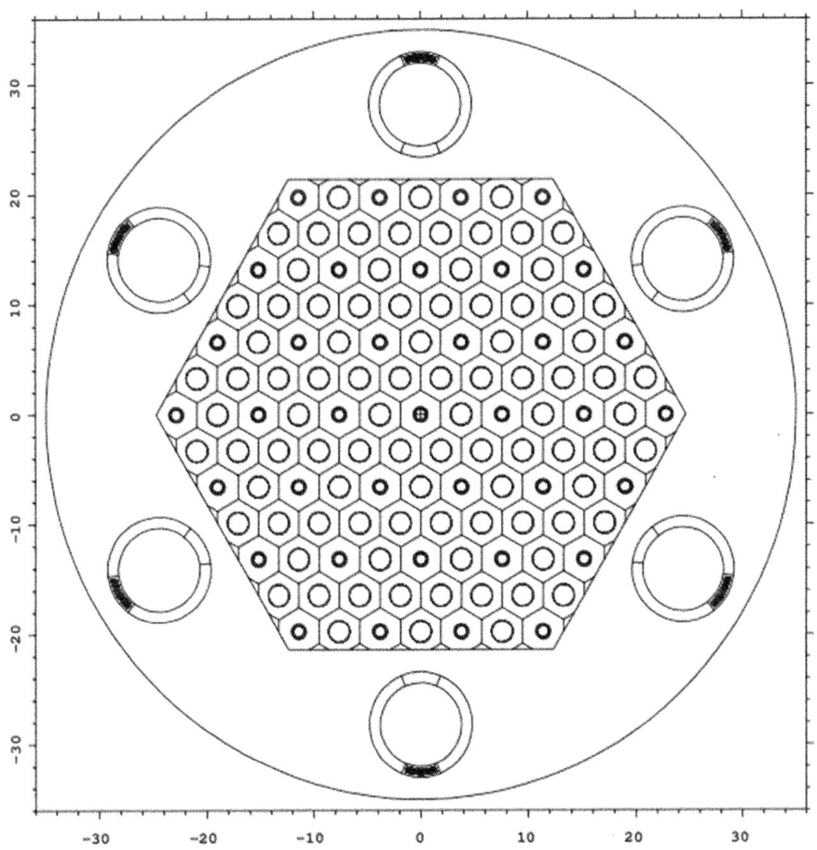

FIGURE 12.18 Absorbers on control drums facing away from core.

TABLE 12.9 Criticality estimates.

Assembly		Core radius (cm)	Calculated k
Core	Reflector		
Uranium-233	None	5.965	1.0115
Uranium-235 (Godiva)	None	8.710	0.9912
Plutonium-239 (Jezebel)	None	6.285	1.0039
Uranium-235 (16.7%)	7.6 cm uranium	20.32	0.9893
Uranium-235	8.9 cm uranium	6.391	0.9939
Uranium-235	1.6 cm thorium	7.80	0.9905
Uranium-235 (Topsy)	Thick uranium	6.045	0.9907

Conclusions

This chapter gives the necessary skills to apply the concepts learnt in the fundamentals covered in the earlier chapters of this book. Each section forms part of the underlying foundations of a Monte Carlo simulation. Though random sampling is done in a "black box" mode within a code, and quantities such as the average source energy and sampled energies of the first 50 particles appears in the output file, it gives confidence to the user to appreciate how these values were generated inside the code. The section on a fixed-source MC simulation was intended to show the easy steps of the simulation process; in eigenvalue problems, a source distribution is written at the end of each generation simulated and subsequently used in the next generation. In many cases, a large number of generations and source particles are specified in the input file assuming that convergence will take place. The statistics to determine stationarity are generally not well understood by a beginner. Thus, convergence is illustrated in these exercises. Other areas include geometry modeling, which typically takes a substantial part of the effort in setting up an input file, and the perturbation feature which is a very useful capability of MC codes such as MCNP5 and can help in carrying out design sensitivity studies leading to optimization. There is much more versatility in present MC codes to handle a far wider range of problems in terms of the number of particles that can be simulated, the detailed modeling and interfaces than earlier versions. The basics covered here are at the core foundation level.

Nomenclature

English lower case

d	extrapolation distance
f	probability distribution function (PDF)
k_{eff}	effective multiplication
k_∞	infinite multiplication
s	distance
w	statistical weight

English upper case

B	buckling
D	diffusion coefficient
F	cumulative distributive function (CDF)
J	neutron current
L	diffusion length
N_i	atomic density of the ith nuclide
R_c	critical radius
$\tilde{R}$	extrapolated radius
R	relative standard deviation

S source

Greek lower case

λ_{tr} transport mean free path
μ cosine of angle of scattering
μ mean value
μ_0 average cosine of scattering angle
ν_0 eigenvalue
ν number of neutrons produced per fission
ξ random number
ρ density
σ standard deviation
σ_a microscopic absorption cross-section
σ_f microscopic fission cross-section
σ_r microscopic removal cross-section
σ_s microscopic scattering cross-section
σ_{tr} microscopic transport cross-section
σ_t microscopic total cross-section
τ neutron age
φ azimuthal angle
ϕ neutron (photon) flux
χ fission spectrum

Greek upper case

$\sum_a$ macroscopic absorption cross-section
$\sum_f$ macroscopic fission cross-section
$\sum_r$ macroscopic removal cross-section
$\sum_s$ macroscopic scattering cross-section
$\sum_{tr}$ macroscopic transport cross-section
$\sum_t$ macroscopic total cross-section

Abbreviations

HVL half value layer
MC Monte Carlo
m.f.p mean free path
Sv sievert
TLE track-length estimator

References

Bell, G. I., & Glasstone, S. (1979). *Nuclear reactor theory*. New York: Robert E. Krieger Publishing Co., Inc.

Bowen, D. G., & Busch, R. D. (2005). *Hand calculational methods for criticality safety—A primer. 2005 NCSD Topical Meeting (American Nuclear Society Nuclear Criticality Safety Division).*

Brown, F. B. *Fundamentals of Monte Carlo particle transport.*

Werner, C. ed. (2017). *MCNP (R) user's manual code version 6.2.*

Cullen, D. E. (2004). Sampling ENDL Watt fission spectra. *Energy.*

DeVolpi, A. (1982). Denaturing fissile materials. *Progress in Nuclear Energy.*

Everett, C., Turner, G., & Cashwell, E. (1973). *A method of sampling certain probability densities without inversion of their distribution functions.*

Froehner, F. H., & Spencer, R. R. (1980). *Method for sampling from fission neutron energy spectra.*

Monte Carlo Team. (2005). *MCNP—A general Monte Carlo N-particle transport code, Version 5.* Los Alamos National Laboratory.

Sood, A., Forster, R. A., & Parsons, D. K. (2003). Analytical benchmark test set for criticality code verification. *Progress in Nuclear Energy.*

Wagner, J. C., Sisolak, J. E., & McKinney, G. (1992). *MCNP: Criticality safety benchmark problems.* LA-12415. New Mexico, United States.

Whalen, D. J., Cardon, D. A., Uhle, J. L., & Hendricks, J. S. (1991). *MCNP: Neutron benchmark problems.* LA-12212 DE92 004710.

Wirtz, K. (1982). *Lectures on fast reactors.* American Nuclear Society.

Annex A: MATLAB Program CH12_NormalSampling.m

```
% CH12_NormalSampling.m
% sample from a normal distribution
%
fid = fopen('Out','w');
mu = 5; sigma= 2;
C = 1.0/sqrt(2*pi)*sigma;
EL=0.0; EH=10;
Nw=1000;
step=(EH-EL)/Nw;
E=EL;
for i = 1:Nw
W = C * exp(-0.5*((E-mu)/sigma)^2);
xW(i)=E; yW(i)=W;
E=E+step;
end

sum1=0;sum2=0;
for i=1:Nw
    sum1 = sum1+ xW(i)*yW(i);
    sum2 = sum2+ yW(i);
end
% this is the value of mu from Nw points
Mean= sum1/sum2

Npts=10000;

%%%%%%%%%%%%%%%%%%%%%%%%%%%% sampling 1
for i=1:Npts
yy1(i)=sigma*randn +mu;
end
% this is the sampled mu from Npts points using the Matlab function randn
SampledMeanYY1= sum(yy1)/Npts
%%%%%%%%%%%% short command for the above to generate 10000 values and
compute the mean; for N=10000, Mean=4.9913, exact=5.0
%  N=10000;
%  y=2*randn(N,1)+5;
%  Mean=sum(y)/N

%%%%%%%%%%%%%%%%%%%%%%%%%%%% sampling 2
for i=1:Npts
x1=rand;
x2=rand;
yy2(i)=mu+sigma*sqrt(-2*log(x1)) * cos(2*pi*x2);
end
% this is the sampled mu from the above sampling scheme
SampleMeanYY2= sum(yy2)/Npts

% Make bins
Nbins=16; % Nbins+1 energy values
xx=[1 1.5 2.0 2.5 3.0 3.5 4.0 4.5 5.0 5.5 6.0 6.5 7.0 7.5 8.0 8.5 10];
```

```
%%%%%%%%%%%% how many samples fall into each bin?
N1 = histc(yy1,xx); % this gives the numbers in each bin
P1=N1/Npts;

N2 = histc(yy2,xx); % this gives the numbers in each bin
P2=N2/Npts;
for i=1:Nbins
    xmid(i)=(xx(i)+xx(i+1))/2;
end

xxmin=min(xx); xxmax=max(xx);
    figure(1)
    subplot(3,1,1)
    plot(xW,yW,'k-','LineWidth',2)
    set(gca,'FontSize',12)
grid off
%xlabel('{\bf E} (MeV)','fontsize',14)
ylabel('{\bf f(E)} ','fontsize',14)
    xlim([xxmin xxmax])
    %ylim([1e-6 1])
    subplot(3,1,2)
    bar(xx,P1,'FaceColor','w','EdgeColor','k')
    set(gca,'FontSize',12)
    ylabel('{\bf f_1(E)} ','fontsize',14)
    xlim([xxmin,xxmax])
    subplot(3,1,3)
    set(gca,'FontSize',12)
    bar(xx,P2,'FaceColor','w','EdgeColor','k')
    xlabel('{\bf E} (MeV)','fontsize',14)
    ylabel('{\bf f_2(E)} ','fontsize',14)
    xlim([xxmin xxmax])
    %ylim([1e-6 0.2])
fclose(fid)
```

Annex B MATLAB Program CH12_Watt Sampling.m

```
% CH12_WattSampling.m
% sample from a Watt fission distribution
%
fid = fopen('Out','w');

a=0.988; b=2.249; % low energy
C1 = 1.0/sqrt(pi*b*a^3);
C2 = 2*exp(-a*b/4);
C= C1*C2;
EL=0.000001; %MeV
EH=20;
Nw=1000;
step=(EH-EL)/Nw;
E=EL;

sumCDF=0.0;
for i = 1:Nw
 %   x(i)=0; y(i)=0;
W = C * exp(-E/a) *sinh( sqrt(b*E));
xW(i)=E; yW(i)=W;
E=E+step;

if(i==1)
    PDF(1)=W;
    CDF(1)=W;
end


if(i>1)
    PDF(i)=W;
CDF(i) = W + CDF(i-1);
end

end

TotPDF=sum(PDF); CDF=CDF/TotPDF;


sum1=0;sum2=0;
for i=1:Nw
    sum1 = sum1+ xW(i)*yW(i);
    sum2 = sum2+ yW(i);
end
% this is the value of mu from Nw points
Mean= sum1/sum2

%%%%%%%%%%%%%%%%%%%%%%%%%%%% sampling 1
Npts=10000;
for i=1:Npts
w=log(rand);
x=log(rand);
y=rand;
z=rand;

ww= a*(-w-x*(cos((pi*y/2))^2));
E= ww + a^2 *b/4+(2*z-1)*sqrt(a^2 *b* ww);
```

```
yy(i)=E;
end
% this is the sampled mu from Npts points using the Matlab funtion randn
SampledMeanYY= sum(yy)/Npts
%%%%%%%%%%%%%%%%%%%%%%%%%%% sampling 2
K=1+a*b/8; L=a*(K+sqrt(K^2-1));M=L/a-1;
Naccept=0;
for i=1:Npts
x=-log(rand);
y=-log(rand);
chk1=(y-M*(x+1))^2; chk2=b*L*x;
if (chk1<=chk2)
    Naccept=Naccept+1;
    zz(i)=L*x; % energy zz is accepted
end
end
SamplingEfficiency =100*Naccept/Npts

% this is the sampled mu from Npts points using the Matlab funtion randn
SampledMeanZZ= sum(zz)/Npts

figure(1)
loglog(xW,yW,'k-','LineWidth',2)
set(gca,'FontSize',12)
grid on
xlabel('{\bf E} (MeV)','fontsize',14)
ylabel('{\bf W(a,b,E)} ','fontsize',14)
xlim([EL EH])
%ylim([1e-6 1])
%set(gca,'XTickLabel','10^{-6}','10^{-5}','10^{-4}','10^{-3}','10^{-
2}','10^{-1}','10^{0}','10^{1}')


% Make bins
Nbins=16; % Nbins+1 energy values
xx=[1e-7 1e-6 1e-5 1e-4 1e-3 1e-2 1e-1  0.2  0.4  0.6  0.8  1.0 1.2  1.4
1.8 2 20];

%%%%%%%%%%%%%% how many samples fall into each bin?
% for Figure 2
Nyy = histc(yy,xx); % this gives the numbers in each bin
Pyy=Nyy/Npts;
Nzz = histc(zz,xx); % this gives the numbers in each bin
Pzz=Nzz/Npts;
xxmin=min(xx); xxmax=max(xx);
figure(2)
subplot(2,1,1)
bar(xx,Pyy,'FaceColor','w','EdgeColor','k')
set(gca,'FontSize',12)
set(gca,'XScale','log')
grid off
%xlabel('{\bf E} (MeV)','fontsize',14)
ylabel('{\bf f_3(E)} ','fontsize',14)
xlim([1e-2 20])
ylim([0 0.4])
```

```
    subplot(2,1,2)
    bar(xx,Pzz,'FaceColor','w','EdgeColor','k')
    set(gca,'FontSize',12)
    set(gca,'XScale','log')
    ylabel('{\bf f_4(E)} ','fontsize',14)
    xlim([1e-2 20])
    xlabel('{\bf E} (MeV)','fontsize',14)
    ylabel('{\bf f_2(E)} ','fontsize',14)
    ylim([0 0.4])
 %%%%%%%%%%%%%%%%%%%%%%%%%%%%%%%%%%%%%%%%%%%%%%%%%%%%%%%%%%
% PROBABILITIES VS MIDPOINTS IN EACH BIN
Myy=histc(yy,xx);Mzz=histc(zz,xx);
for i=1:Nbins
    xmid(i)=(xx(i)+xx(i+1))/2; % for Fig 2
    PROByy(i)=Myy(i)/Npts; % these will be Nbins values to plot in Fig 3
    PROBzz(i)=Mzz(i)/Npts; % these will be Nbins values to plot in Fig 3
end

figure(3)
    loglog(xmid,PROByy,'k-','LineWidth',2)
    hold on
    loglog(xmid,PROBzz,'k--','LineWidth',1.5)
    set(gca,'FontSize',12)
    xlim([1e-2 20])
    xlabel('{\bf E} (MeV)','fontsize',14)
    ylabel('{\bf f_s(E)} ','fontsize',14)
    %ylim([1e-6 1])
legend('\bf f_{s,3}(E)','\bf f_{s,4}','fontsize',12,'Location','SouthEast')
%%%%%%%%%%%%%%% PDF and CDF %%%%%%%%%%
    figure(4)
subplot(2,1,1)
xxmin2=min(xx);xxmax2=max(xx);
loglog(xW,yW,'k-','LineWidth',2)
set(gca,'FontSize',12)
grid on
%xlabel('{\bf E} (MeV)','fontsize',14)
ylabel('{\bf f(E)} ','fontsize',14)
xlim([1e-6 20])
ylim([1e-6 1])

subplot(2,1,2)
set(gca,'FontSize',12)
loglog(xW,CDF,'k-','LineWidth',2)
xlabel('{\bf E} (MeV)','fontsize',14)
ylabel('{\bf F(E)} ','fontsize',14)
xlim([1e-6 20])
ylim([1e-6 2])
grid on
fclose(fid)
```

Chapter 13

Optimization in nuclear systems

13.1 Introduction

The canvas of optimization in nuclear engineering is vast, diverse, and interconnected. It is by no means possible for an author to convey more than a glimpse to a reader in a book chapter, but it is hoped that the few descriptions and examples given here will give some insight into the work over the last few years and the emerging methods.

In nuclear fission and fusion system design optimization, the overall objective is to design the best systems which meet the requirements. This is more aptly described as multi-objective, multi-module, multi-constraints optimization in which the modules are inter-dependent (Rozin, Rubin, & Soboĺ, 1990; Stewart, Palmer, & DuPont, 2021). Thus, it is a fairly big challenge from the perspective of mathematical modeling and computing (Sections 9.1 and 9.6). The domains covered in it include fission and fusion neutronics, fusion reactor blanket design, thermal hydraulics (TH), control systems, and radiation shielding and since the constraints cannot be expressed in algebraic forms, the deterministic optimization techniques of Section 9.2.2 (differentiable functions in the Lagrangian) and Section 9.2.3 (Euler-Lagrange equation) cannot be applied to obtain analytical derivative expressions. The optimal formulation in the form of Pontryagin's Maximum (or Minimum) Principle (Section 9.2.5) can be applied as shown in this chapter. The suitable models are Dynamic Programming (Section 9.4) but more suitable are the stochastic *meta*-heuristic methods (Section 9.5) such as Genetic Algorithms (Section 9.5.1) and Particle Swarm Optimization (Section 9.5.2).

The analysis and particularly the engineering design results will be the focus for optimization calculations in reactor core design, fusion neutronics, radiation shielding, fuel loading pattern (LP), a thermal neutron activation analysis (TNAA)-based explosives detection system.

13.2 Reactor core design optimization

Important parameters related to core design are materials, dimensions and operating parameters. In a preliminary analysis, the overall design is achieved by carrying out a neutronic simulation coupled with thermal hydraulics (TH); this gives the basic initial design.

In a neutronic simulation depicted in Fig. 13.1, the materials and dimensions are input on the basis of some past experience and knowledge of a similar system to obtain the overall system multiplication k_{eff}. A number of runs are typically required to adjust the materials and dimensions to get $k_{\mathrm{eff}} = 1$. This would be a starting design for a coupled neutronic-TH simulation leading to more detailed and elaborate design parameters in a full optimization calculation.

In a preliminary model, the materials and geometry of the fuel rods, fuel assemblies and the whole core are specified as shown in Fig. 13.1 on the basis of a comparison with other similar designs and a neutronics solver, such as Monte Carlo N-Particle (MCNP), can be used to estimate k_{eff} and vary the parameters until a feasible design is obtained.

An example of a first-step neutronic calculation of a homogeneous model of the Korean KORI-1 nuclear reactor (Kim, 2019; Lee, 1973) can be carried out with a basic two-group model with cylindrical geometry in a neutronic code such as ANISN or DOT (Section 8.1).

A simulation carried out with MCNP5, with the input listed below, considers a homogeneous water-fuel mixture in a cylinder of radius 122.6 cm and height 365.8 cm.

MCNP input file: KORI-1 homogeneous reactor model

```
1 1 0.079394 −1 3 −4
2 1 0.079393 1 −2 3 −4
3 0 2:−3:4
1 cz 86.69
2 cz 122.6
```

Nuclear Engineering. DOI: https://doi.org/10.1016/B978-0-323-90618-0.00013-2

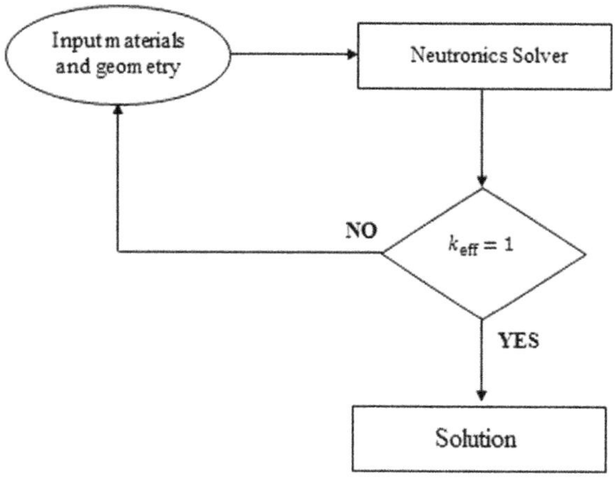

FIGURE 13.1 Basic geometry and materials feasibility for a neutronic calculation.

TABLE 13.1 KORI-1 reactor homogeneous model with MCNP5.

Enrichment (%)	k_{eff}
1.200	1.03204 (0.00109)
1.100	1.00168 (0.00109)
1.098	1.00446 (0.00105)
1.090	0.99965 (0.00106)
1.050	0.98414 (0.00110)
1.000	0.96907 (0.00107)
0.900	0.92754 (0.00098)

```
3 pz −182.9
4 pz 182.9
mode n
imp:n 1 1 0
c ------- enrich = 1.1% total 0.079394
m1 92235.66c 7.6934e-05 92238.66c 6.9171e-03 8016.66c 3.3460e-02 1001.66c 3.894e-2
kcode 500 1.0 10 500
ksrc 0 0 0
```

A number of such simulations have been carried out for fuel enrichment in the range 0.9%−1.2% and given the results shown in Table 13.1. The bare system with atomic densities given is critical at an enrichment of 1.09%. This matches the one-zone criticality result from the determinant (Section 5.3.2).

The atomic densities and masses of uranium are listed in Table 13.2

For criticality, the atomic densities and mass ratio are: $N_H = 3.8940 \times 10^{22}$, $N_O = 3.236 \times 10^{22}$ atoms cm^{-3}, $N_5/N_U = 0.01099$, $M_5/N_U = 0.01086$. The H/U^{235} ratio for criticality is 506.15 with 1.1% enrichment, $k_{eff} = 0.99974(0.00052)$ with a total uranium mass ~47.741 t out of which the fissile fuel U-235 is ~519 kg.

Exercise 13.1: Two-group criticality for a bare cylindrical reactor.

For a bare homogeneous nuclear reactor core (Lee, 1973) with the two-group data in Table 13.3, solve the criticality determinant to find the critical dimensions.

The atomic densities are:

$N^U = N^{235} + N^{238} = 6.994 \times 10^{21}$ atoms cm^{-3}

N^{U}_{Oxygen} = number of oxygen atoms in uranium = 1.399×10^{22} atoms cm^{-3}

$N_w = 1.947 \times 10^{22}$ molecules cm^{-3}

TABLE 13.2 Atomic density and mass of uranium in KORI-1 homogenous model.

Nuclide	N (atoms/cm^3)	N (atoms)	Mass (kg)
U-235	7.6934×10^{19}	1.3289×10^{27}	518.5958
U-238	6.9171×10^{21}	1.1948×10^{29}	47223.19
U	6.9940×10^{21}	1.2081×10^{29}	47741.79

TABLE 13.3 Two-group diffusion data.

Data	Group 1	Group 2
D (cm)	1.367	0.2294
σ_a^{235} (10^{-24} cm^2)	—	600.62
Σ_s (cm^{-1})	0.03922	—
Σ_a^{others} (cm^{-1})	—	0.028058

At the next level, a coupled neutronic-TH analysis, as depicted in Fig. 13.2, is carried out to obtain key parameters of thermal and flow characteristics. Such analysis typically gives the following set of design parameters:

1. Core dimensions (e.g., cylindrical radius and height)
2. Fuel road diameter (D) and length (L)
3. Fuel pitch to diameter ratio (P/D)
4. Size and pitch of fuel assemblies
5. Total thermal power
6. Channel flow characteristics such as coolant velocity (V_c)
7. Convective heat transfer
8. Pressure drop in coolant flow
9. Temperature differences (radial and axial).

The neutronic and TH modules are solved until the constraints are satisfied and a feasible solution is obtained. In case the initial parameters obtained from a preliminary neutronic analysis are not favorable for a feasible solution then a new set of values is obtained, if possible.

Once the neutronic-TH coupled simulations are performed, an optimization is performed to obtain the optimal solution. In case of genetic algorithms (GA), the method described in Section 9.5.1 is followed iteratively. This implies that a set of chromosomes, in a population, is used for each of the modules. Such a scheme can be prohibitively inefficient due to the large computation effort for each of the modules. For example, a GA code calling MCNP 100 times and then calling each other code the same number of times would be far too much to be realistic. Thus a more efficient scheme, shown in Fig. 13.3, utilizes regression and machine learning. In such a scheme, several runs are made initially with trial data to obtain output which trains a regression model which subsequently is able to produce predictive output.

The GA chromosomes are tested for fitness in the regression models rather than directly into the modules thus reducing the effort. Such schemes have been successfully implemented to give reliable optimized parameters.

Consider two optimizations where GA has been used for multi-objective multi-module multi-constraint optimization for a Gen-IV lead-cooled fast reactor (LFR) (Section 3.5.3) by Luo, Zhang, Wang, Jiang, and Chen (2021), a gas-cooled fast breeder reactor core (Kumar & Tsvetkov, 2015).

The optimization for a 1000 MWt medium-power modular lead-cooled fast reactor M^{2}LFR the optimization is:

$$\min \quad J(\mathbf{y}), \quad s.t. \quad \mathbf{x} \le \mathbf{x}_0, \quad k_{\text{eff}} \le k_{\text{eff},0}, \quad \text{PFF} \le \text{PFF}_0 \tag{13.1}$$

where

$$J = \sum_{i=1}^{4} a_i y_i(\mathbf{x}), \tag{13.2}$$

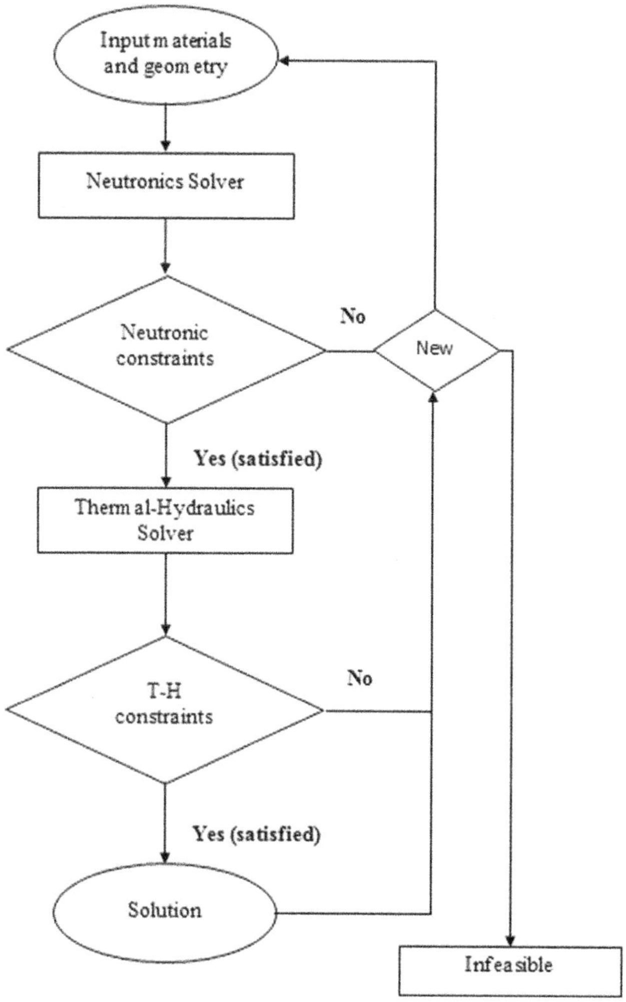

FIGURE 13.2 Coupled neutronics and thermal hydraulics calculation.

$x = (x_1, x_2, x_3)$ and the objective functions y_i are the system multiplication k_{eff}, the peak power factor (PPF), the cladding temperature T_c and the fuel pellet temperature T_p with weights a_i, and $x_0 = (x_{1,0}, x_{2,0}, x_{3,0})$ are given maximum values.

The fixed quantities are the number of fuel assemblies, control rods (CR) and safety rods (SR), fuel rod layers in an assembly, the fuel enrichment, the linear power density and the materials; these are specified at the beginning of the optimization.

In the first part of the neutronic optimization, the geometric design parameters are obtained for the fuel rods and overall core design. Then, in a coupled neutronic-TH optimization, the core is modeled as consisting of six layers namely the inner fuel zone, the SR zone with 12 safety rod assemblies, the outer fuel zone, the CR zone with 12 CR assemblies, the reflector zone and the pure lead (Pb) zone. The layers are homogenized, and cross sections are prepared for MCNP. The feasibility of the scheme is verified by comparing the maximum coolant velocity and fuel, cladding and coolant temperatures with the design constraints, For optimization variables: D, P/D and V_c, the search space for each independent variable is specified to obtain the optimum solution $D^* = 10\text{mm}$, $(P/D)^* = 1.3$, $V_c^* = 1.4\text{ms}^{-1}$. Thus, the optimized geometric design and overall thermal and flow parameters are found.

For a 100 MWt gas-cooled fast breeder reactor core optimization is carried out (Kumar & Tsvetkov, 2015) with the GA method in a scheme with a strategy similar to that depicted in Fig. 13.3. The four modules used in the optimization are for the fuel pin, the whole core, thermofluids and heat transfer, and Brayton cycle energy conversion. The multi-objective function comprises ten objectives namely high breeding of U-233 and Pu-239, desired radial and axial PPF, multiplication constants k_{eff} and k_∞ within prescribed limits, high fast fission factor, high energy conversion, limited pressure drop and high fuel burnup. The constraints and constants are prescribed similarly to the previous case.

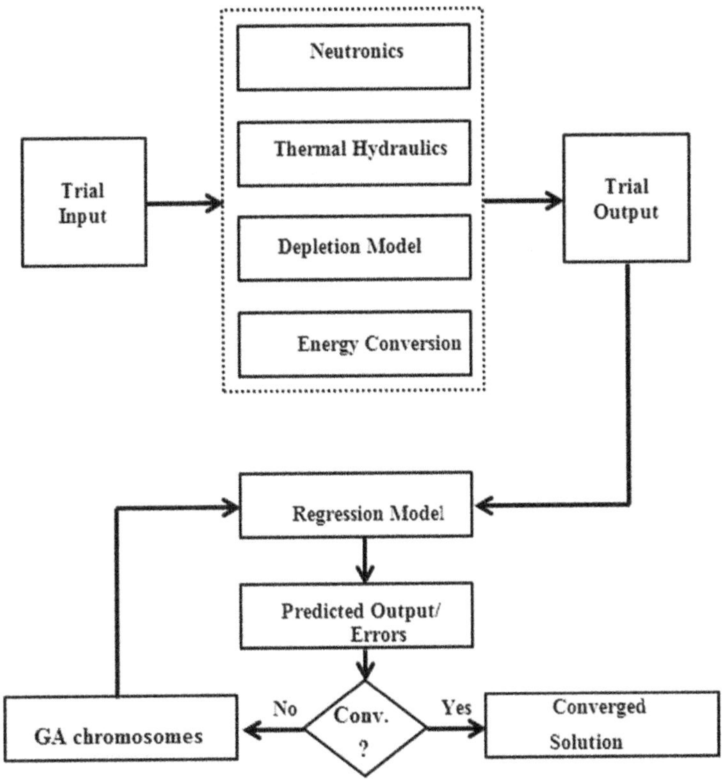

FIGURE 13.3 Coupled genetic algorithm-regression with multi-module optimization.

The independent variables are the fuel radius (r_F), the enrichment (ε), the coolant inlet temperature (T_{in}) and the mass flow rate $\dot{m}$.

The simulation parameters are: number of chromosomes in every generation = 40, number of bits per gene = 16, number of generations = 100. Crossover probability = 0.4, mutation probability = 0.05 and 2 elite elements preserved. The solutions seem to have converged within ~70 iterations, giving the optimized solution: $r_F^* = 0.217cm$, $\varepsilon^* = 18.467\%$, $T_{in}^* = 63.476^\circ$ and $\dot{m}^* = 35.937$ kg/s.

The two simulations just described are representative of current research in reactor core design optimization by GA; the second uses multivariate regression for improved computational efficiency.

To optimize fuel pin burnable poison in UO_2/Gd_2O_3 fuel placed in pressurized water reactor (PWR) fresh assemblies, Yilmaz, Ivanov, Levine, and Mahgerefteh (2006) have used GA to maximize the end of cycle (EOC) k_{eff}.

In several other works, core optimization has been successfully carried out with the particle swarm optimization (PSO) method (Domingos, Schirru, & Pereira, 2006), for multi-objective sodium-cooled fast test reactors (Zeng, Stauff, Hou, & Kim, 2020), for Small Modular Reactors (Akbari, Rezaei Ochbelagh, Gharib, Maiorino, & D'Auria, 2020), for design optimization for a High Flux Isotope Reactor conversion project (Bae, Betzler, Chandler, & Ilas, 2021; Betzler, Chandler, Cook, Davidson, & Ilas, 2019) and for a data-driven predictive model for predicting moisture carryover in a boiling water reactor using Artificial Intelligence (AI) technique (Wang, Gruenwald, Tusar, & Vilim, 2021). Considerable research effort is focused on the development of AI-based predictive models using data, look-up tables and regression for addressing non-linearities and uncertainties.

13.3 Fusion neutronics design optimization

In a fusion reactor such as the International Thermonuclear Experumental Reactor (ITER, Section 10.6), the neutron radiation from the plasma in the central cell produces secondary gamma rays during its interaction with structural and blanket materials. Two important factors in fusion, distinct from fission reactors, are activation and the requirement of a shield to ensure protection to the copper material in the surrounding superconducting coils which operate at cryogenic temperatures (to be superconducting).

In a fusion reactor, the inboard (IB) and outboard (OB) blanket modules serve to produce tritium to sustain the fusion reaction and to shield the superconducting coils. The blanket is exposed to intense radiation heating which is removed by the flow of coolant. Some blanket designs under study for ITER and the Demonstration Fusion Power Plant (DEMO) and the intermediate China Fusion Engineering Test Reactor are the helium-cooled solid breeding blanket, the water-cooled lithium lead (WCLL) breeding blanket and the water-cooled ceramic breeding blanket. The objective of blanket optimization studies is to maximize the tritium production, with a tritium breeding ratio (TBR) exceeding one, that is, more tritium produced than consumed and provide shielding for the blankets.

Some blanket/shield design optimization studies are:

1. optimization of the first wall armor, to maximize the TBR (Dai et al., 2021),
2. full 3D neutronic-TH coupled optimizations for TBR in IB and OB equatorial blankets, and the effect of beryllium and tungsten armor on the shielding performance of the blanket (Cui et al., 2017),
3. tritium self-sufficiency and shielding effectiveness in the WCLL breeder blanket of the DEMO design (Moro et al., 2020),
4. the design of an optimized WCLL mock-up experimental layout based on MCNP calculations (Flammini et al., 2020), and
5. the development of a coupled neutronics-TH code with multi-objective optimization, Neutronics/Thermal-hydraulic Coupling Optimization Code (NTCOC), for a radial build of the water-cooled ceramic breeder blanket (Li et al., 2016).

13.4 Radiation shielding design optimization

The design of radiation shielding in large nuclear power reactors consists mostly of structural steel surrounding the core and a large volume of concrete as a biological shield around the reactor. In swimming-pool type reactors, the water in the pool acts as a shield.

For small compact and portable micronuclear reactors, particularly for space and underwater applications, the design of a shield requires optimization of size, weight and cost. It is thus both desirable and necessary to find the optimal materials and configurations.

Shielding is thus a multi-objective, multi-constraint optimization for which heuristic, *meta*-heuristic and AI-based optimization methods are being applied and further developed for computational efficiency.

In the design of neutron and gamma radiation shielding, GA has also been used for design optimization with GA-MCNP calculations (Cai, Hu, Pan, Hu, & Zhang, 2018) for multi-region and composite shields as illustrated by Fig. 13.4.

In Fig. 13.4, the first material (labeled A) is typically iron (Fe) or tungsten (W) as a structural material and capable of attenuating fast neutrons and gammas. This is followed by a hydrogenous material such as water (labeled A) which further slows down the neutrons while producing gammas of ~ 2 MeV which need a third layer of a material such as lead (Pb) which is effective in attenuating energetic gammas. Optimization studies would thus typically give the dimensions of each of the iron, water, and lead layers obtained by a calculation which searches for the optimal configuration. Several other candidate materials and impurities include stainless steel SS316, tungsten, polyethylene, borated polyethylene, and B_4C.

A detailed calculation would use energy-dependent cross sections for radiative capture (n, γ) and alpha production (n, α) reactions. For example, for removal through the capture (n, γ) reaction, iron has an order of magnitude higher at low energies with prominent resonances at intermediate (>10 keV) energies extending well toward the high (~ 1 MeV) energies.

Optimization can also be carried out to determine the optimal compositions of a composite homogenized shield as shown in Fig. 13.4. A number of innovative combinations, such as mixing nanomaterials in concrete, have been investigated (Norhasri, Hamidah, & Fadzil, 2017) to make new and better radiation shielding materials. A lightweight and compact sixteen-layer gradient composite shield (Hu et al., 2020) has been designed by a GA-MCNP coupled simulation based on epoxy resin, B_4C, lead and a small amount of graphene oxide. The densities of three materials for the composite are 11.34 g/cm^3 for Pb, 1.20 g/cm^3 for epoxy resin and 2.52 g/cm^3 for B_4C. In the 30 cm composite shield the optimal density is found to be a parabolic shape starting from ~ 10 g/cm^3 in the beginning falling gradually to

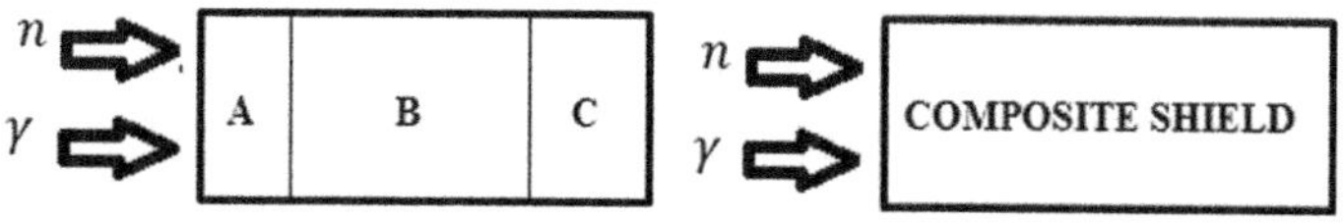

FIGURE 13.4 Radiation shield design optimization.

slightly above ~1 g/cm^3 by about 20 cm and then gradually increasing to ~10 g/cm^3. This shield was developed and tested and showed better performance than a uniform composite shield. In another GA-MCNP optimization for a shield subject to a U-235 spectrum with mixed neutrons and gamma rays (Cai, Hu, Pan, Sun, & Yan, 2018), Cai et al. have also shown that a composite multilayer shield optimized for volume and weight has the best performance. Extending the study to volume, weight as well as minimum activation for a vehicle carrying a D-T neutron source, a model using MCNP for neutronic calculations and a predictive model for mechanical and thermal properties a *synergistic* optimization model based on evolutionary algorithms has been developed using two multi-objective methods (priori and posteriori) for the design of composite shields (Cai, Hao, Yu, Wang, & Hu, 2020).

For small and portable micronuclear reactors, the emphasis is on both safety and lightweight effective shields (Ahmad, Chang, Li, Yang, & Liu, 2021; Kim & Moon, 2010). Coupling GA and neural network models (Song et al., 2020) with improved efficiency over Monte Carlo (MC) simulations for multi-objective and multi-constrained radiation shielding design optimization problem have been demonstrated it by the non-dominated sorting genetic algorithm (NSGA-II) and mini-batch stochastic gradient descent. An adaptive mutation rate operator has been proposed to improve the global searching capability. The optimization results show that the adaptive scheme can find the Pareto-optimal front of reactor shielding designs precisely and its calculation speed is two orders of magnitude faster than the pure MC method and similar computational accuracy.

Such an optimization based on GA has been used (Kim & Moon, 2010) for the search of an optimal shield subject to a set of constraints. A number of innovative combinations, such as mixing nanomaterials in concrete, have been investigated (Norhasri et al., 2017) to make new and better radiation shielding materials.

13.5 Fuel loading pattern optimization

A nuclear reactor begins with a fresh core with fuel that burns unevenly. Tin an ideal case, flux would have a more-or-less cosine distribution in the radial and axial directions. The variation in fluxes gives rise to a peaking factor described in the previous section in the context of optimization. With time, the fuel has burned unevenly and requires removal, shuffling, or replacement. This is the area of fuel loading where the best strategy is desired for both technical and economic reasons.

A utility that owns a nuclear reactor would typically have a planning horizon of a few years as shown in Fig. 13.5. As an example, consider a 7-year horizon where it is desired to produce, in the first 5 years, a certain amount of (electrical) energy E_i, within cycles of length τ_i $(i = 1, 2, \ldots 5)$ months, with each cycle having a beginning of cycle (BOC) and an EOC as shown in Fig. 13.6.

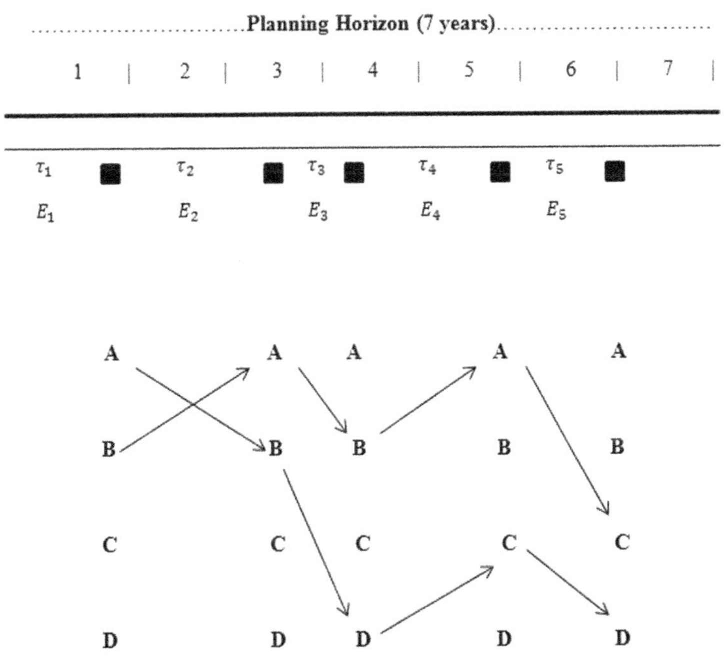

FIGURE 13.5 Possible paths include A-B-D-C-D and B-A-B-A-C.

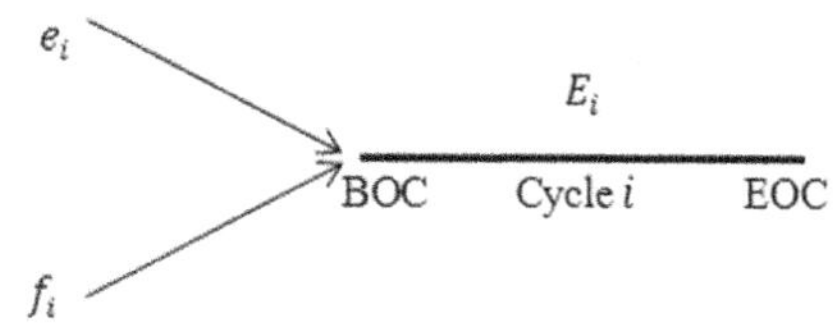

FIGURE 13.6 A fuel cycle with enrichment e_i and fuel loading fraction f_i at beginning of cycle.

At every BOC, the utility has a choice of adding fresh fuel of enrichment e_i for a fraction f_i of the total fuel in the core. As an example, a utility might decide to add 3.2% enriched fuel for one-third of the core.

A fuel cycle has a front-end (uranium ore to fuel fabrication), a middle part (reactor operation and refueling), and a back-end (open-cycle long-term storage or closed-cycle reprocessing).

During fuel reloading, typically one-third to one-half of the fuel may be removed and new fuel loaded in a LP while the remaining fuel is shuffled. A plant is shut down during the reloading period. It is desirable to have as long as possible operation between refueling. An outage may last several days or possibly up to a month.

The question then arises is how to best place it, how does this decision affect the burnup and cycle length, and how much electricity is planned in the coming cycle.

The reactor core would thus operate in cycles of different time durations during the planning horizon. Any decision would then require a consideration of the costs and benefits; this would be formulated as an optimization problem with the following parameters:

1. the number of fuel assemblies to remove
2. the location of fuel assemblies to remove
3. the next-cycle length
4. the reload fuel enrichment and batch fraction
5. the poison distribution to manage criticality, and
6. the pattern of remaining fuel for shuffling.

The constraints in this optimization problem would be the core power density ratios and the fuel burnup.

In the five cycles shown in Fig. 13.5, the nodes A, B, C, and D represent fuel types which are classified by their enrichment and burnup; finding the optimal path from the first stage to the terminal stage would thus constitute a Dynamic Programming (shortest path) problem solved by a recursive algorithm in Section 9.4.

Fig. 13.6 shows a 3-batch fuel LP in a 1/4th core with fresh fuel, once- and twice-burned fuel loaded in the assemblies.

Alternatively, a path such as. A-B-D-C-D or B-A-B-A-C could be considered a chromosome for a GA optimization scheme and its fitness could be evaluated. Both Dynamic Programming (DP) are GA and variants and advancements of these methods are applicable for performing optimization analyses to find the optimal path in a fuel loading strategy (Fig. 13.7).

A DP application of fuel loading (Kearney, 1973) is formulated in terms of minimizing the total cost T_C which is the sum over all cycles N of the revenue requirement in a refueling strategy

$$\min T_c = \min \sum_{i=1}^{N} C[x_i, \quad e_i, f_i] = \min \sum_{i=1}^{N} C_i \tag{13.3}$$

subject to the constraints that the energy produced in cycle i is E_i, the power peaking in the ith cycle $PP_i \le PP_{\max}$ and that the burnup at discharge $B_i \le B_{\max}$.

In Eq. (13.3), C_i is the total discounted revenue requirement for producing energy E_i in the ith cycle, and x_i, e_i and f_i denote the state, reload enrichment and reload batch fraction of the reactor at the beginning of the ith cycle.

The energy production in the reactor is dependent on the mass of the fissile nuclides uranium-235, plutonium-239 and plutonium-241 in each cycle.

In a preliminary model, a single decision variable can be taken to be the batch fraction, and with possible values [A] 0.37, [B] 0.33, [C] 0.29 and [D] 0.25 and the fitness of a path selected as a GA chromosome can be evaluate. This would of course require a coupled neutronic and depletion analysis for the burnup at the end of each cycle given a BOC composition.

A sample optimization problem carried out for the Zion-I 1065 Mwe reactor (Kearney, 1973) for a 3-zone refueling with a cycle energy production of 7363.2 GWHe and a discharge constraint of 50 GWD/T, gave C-A-D-D-C as the optimal path for reload batch fraction.

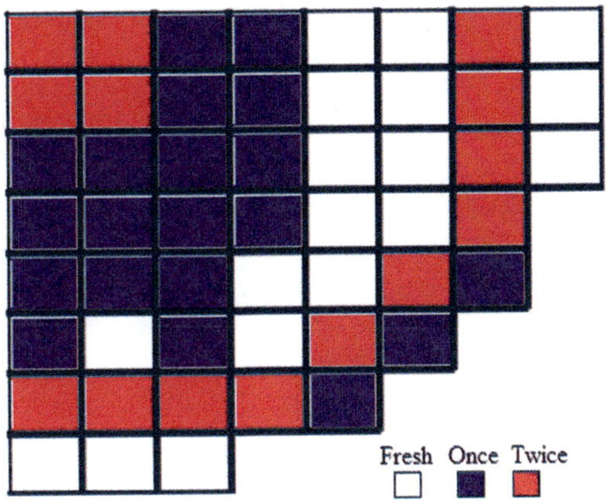

FIGURE 13.7 A nuclear reactor core with a 3-batch fuel loading pattern.

TABLE 13.4 Optimal refueling policy.

Cycle	E_i (GWHe/cycle)	f_i	$e_i(w/o)$	$C_i(10^6\$)$	$\sum_{i=1}^{N} C_i(\$10^6)$
1	7831.2	0.293	4.2145	12.420355	12.420355
2	9784.8	0.373	3.3637	14.446574	26.866923
3	7180.8	0.253	4.3932	9.472094	36.339008
4	6420.0	0.253	2.8467	7.428905	43.767904
5	6859.2	0.293	3.2650	7.537723	51.305616

In a more detailed model, for each cycle of the optimal path, the cycle energy production E_i, reload batch fraction f_i, reload enrichment e_i and the discounted cycle cost C_i were calculated using the two-group code and the DP algorithm for each stage beginning from the final stage.

Some of the results for the optimal path from stage 1 to 5 reproduced from Kearney are listed in Table 13.4. This is a simple illustration from *legacy* calculations when computing power was modest, but the design was based on elegant mathematics.

Maximum fuel utilization within the power demand requirement can be achieved by removing the fuel when it has achieved a maximum burnup, that is, there is minimum amount of fuel left in it. De Klerk et al. (1997) have demonstrated a nonlinear mixed integer formulation for calculating the reload optimization pattern with minimum discharge burnup. In this model, the two parameters used are the power density vector and the k_∞ values for fuel elements from n batches. Within a cycle, a linear decreasing function for the burnup equations is assumed for k_∞ obtaining values at each BOC ($t=0$) and EOC ($t=T$) such that the EOC value of k_∞ at a cycle becomes the BOC value of the fuel at the next cycle.

The GA method has been used with MCNP coupled with the PARET code (a point kinetics and TH to predict transients and Departure from Nucleate Boiling (DNB)) to replace five most burned fuel elements in a Training Research Reactor Idaho General Atomics research reactor (Chham et al., 2021) with reloaded fresh fuel on the basis of maximizing k_∞, minimizing the central fuel temperature (CFT) and maximizing the Departure from Nucleate Boiling Ratio (DNBR) as a safety margin. In this scheme, the chromosomes are input into MCNP where the neutronics calculation gives the excess reactivity and PPF. The PPF is input in to the PARET model to give the CFT; the fitness of the chromosomes is subsequently evaluated and updated for obtaining the converged solution. A similar fuel LP optimization for a materials test reactor with 29 fuel assemblies has been carried out using PSO (Section 9.5.2). The group constants were generated from Winfrith Improved Multigroup Scheme/D4S coupled with a diffusion theory code with objectives of maximizing k_{eff} while incorporating a penalty function to keep the PPF low (Ahmad & Ahmad, 2018).

In the present scenario, optimal strategies for fuel loading are obtained from stochastic heuristic and *meta*-heuristic methods with the help of expert-based systems. AI and Artificial Neural Networks (ANN) are designed on rule-based expert systems on the workings of the human brain. A strong feature in computer architecture favoring both AI and

ANN is the speedup based on vector processing and parallel processing. In ANN, for example, the structure of the brain is modeled in the form of layers of neutrons (decision points) connected by axons in which chemical ions carry signals to connect neurons. The complexity of the brain is so vast that of the order of ten billion neurons are interconnected for processing knowledge and information by which humans recognize, make decisions, and pass the command for action through signals across the spinal cord. Automated LP search tools may be attractive to solve the huge combinatorial LP optimization problem. Rules are developed based on the experience of people operating systems and are formulated in the form of *if this* then do *that*. This is the way humans operate by separating knowledge from action. Thus, modules such as neutronics, depletion models and TH are combined with heuristic rulesets on the placement of a particular type of fuel in terms of enrichment and burnup at a certain location in the core. As an example, consider two rules (Nissan, 2019): an elimination rule which says *Do not load a fresh assembly in such a position that it is adjacent to another position where there is another assembly of the same kind, except when one of these two positions is in a corner position*, and a preference rule which says *If it is a once-burned assembly that is currently being considered, then choose for it—from among those positions that were not forbidden by Rules 1–6 (the elimination rules)—that position whose distance from the center of the core is minimal.*

Rules are coded into a program for neural processing to make a rule-based expert system or an intelligent system based on GA.

Surrogate models encoding a set of heuristics have been proposed for use in quantum annealers, which can find global minima by tunneling through a barrier (as in quantum physics), applied to PWR fuel loading (Whyte & Parks, 2021).

13.5.1 Optimal distribution: Pontryagin's maximum principle

Pontryagin's Maximum Principle (Section 9.2.5) was described in Chapter 9 to obtain an optimal control when the admissible controls are discrete. Such a model can be used for the optimal placement of different enrichment fuels. A simple and elegant application is given by Lee (1973) for optimal placement of fuel to minimize the mass in a reactor.

In a two-group diffusion formulation, fluxes and currents are the state variables $x_1 = \phi_1$, $x_2 = r\dot{\phi}_1$, $x_3 = \phi_2$ and $x_4 = r\phi_2$ expressed as $\vec{x} = (x_1, x_2, x_3, x_3)$

$$\frac{dx_i}{dr} = f_i(\vec{x}, u) \qquad (i = 1, 2, 3, 4). \tag{13.4}$$

The independent control parameter u is the fuel enrichment. In this model, there are fuels of two enrichments u_{min}, u_{max} and the objective is to place them in such a way that the functional

$$J = 2\pi \int_0^R u(r) r dr \tag{13.5}$$

is minimized; J represents the mass of fissile material in a cylindrical reactor of radius R and unit height. Another state variable x_0 is defined as

$$x_0 = 2\pi \int_0^R u(r) r dr \tag{13.6}$$

A Hamiltonian is constructed as:

$$H\left(\vec{\Psi}, \vec{x}, u\right) = \sum_{i=0}^{n=4} \Psi_i f_i(\vec{x}, u) \tag{13.7}$$

where Ψ_i are the auxiliary functions defined as

$$\frac{d\Psi_i}{dr} = -\sum_{i=0}^{n=4} \frac{\partial f_i(\vec{x}, u)}{\partial x_i} \Psi_i \tag{13.8}$$

for $(i = 0, 1, 2, 3, 4)$. The above is equivalent to the system:

$$\frac{dx_i}{dr} = \frac{\partial H}{\partial \Psi_i}, \tag{13.9}$$

and

$$\frac{d\Psi_i}{dr} = -\frac{\partial H}{\partial x_i} \quad (13.10)$$

$$(i = 0, 1, 2, 3, 4)$$

The Hamiltonian is then arranged as $H = u\varphi + \vartheta$ so that the extremum on H can be determined by the sign change and the appropriate control can be implemented. With the normalization condition $2\pi\Psi_0 = -1$, gives

$$\varphi(r) = -r[1 + \alpha x_3(r)\Psi_2(r) - \beta x_3(r)\Psi_4(r)], \quad (13.11)$$

$$\vartheta(r) = \frac{1}{r}x_2\Psi_1 + \frac{1}{\tau_1}r^2 x_1\Psi_2 + \frac{1}{r}x_4\Psi_3 - \frac{r}{\tau_2}x_1\Psi_4 \quad (13.12)$$

where $\alpha = k_\infty/pD_1$, $\beta = 1/D_2$, $\tau_1 = D_1/\Sigma_1$ and $\tau_2 = D_2/p\Sigma_1$.

The objective is to ensure that H remains maximum for all $r \in (0, R)$; thus $u(r) = u_{\min}$ for $\varphi < 0$ and $u(r) = u_{\max}$ for $\varphi > 0$.

Further, it is evident that $\phi(0) = 0$ and $\phi(R) = -R^2 < 0$, since x_3, the thermal flux, vanishes at the extrapolated boundary $r = R$. The fluxes (Section 5.3.2 for cylindrical geometry) are

$$x_1 = AS_1\frac{\sin\mu r}{r} + BS_2\frac{\cos\mu r}{r} + CS_3\frac{\sinh\lambda r}{r} + DS_4\frac{\cosh\lambda r}{r} \quad (13.13)$$

and

$$x_3 = A\frac{\sin\mu r}{r} + B\frac{\cos\mu r}{r} + C\frac{\sinh\lambda r}{r} + D\frac{\cosh\lambda r}{r} \quad (13.14)$$

and the auxiliary functions are

$$\Psi_2 = a\frac{\sin\mu r}{r} + b\frac{\cos\mu r}{r} + c\frac{\sinh\lambda r}{r} + d\frac{\cosh\lambda r}{r} \quad (13.15)$$

and

$$\Psi_4 = aT_1\frac{\sin\mu r}{r} + bT_2\frac{\cos\mu r}{r} + cT_3\frac{\sinh\lambda r}{r} + dT_4\frac{\cosh\lambda r}{r} \quad (13.16)$$

The above are sufficient to determine the minimum critical mass by using appropriate values for the control variable, u depending on the sign of φ. This is a powerful tool, to determine the minimum critical mass but only after specifying the number of zones in the core.

Based on the numerical results and the intuitive fuel shuffling of the previous section, now consider an optimal control method which will take us to the desired configuration. The state variables x_1 (fast flux) and x_3 (thermal flux) can be used, with the boundary conditions (zero flux at the outer boundary $r = R$ and zero current at $r = 0$) to yield the critical determinant. The transversality conditions are used to obtain the boundary conditions of the auxiliary functions ($\Psi_1(0) = \Psi_3(0) = \Psi_2(R) = \Psi_4(R) = 0$).

The flux with appropriate constants A, B, C, D, and coupling coefficients $S_i^{(j)}$, for a homogeneous system, given by

$$\phi_1^{(j)} = A^{(j)}S_1^{(j)}X_1 + B^{(j)}S_2^{(j)}X_2 + C^{(j)}S_3^{(j)}Y_1 + D^{(j)}S_4^{(j)}Y_2 \quad (13.17)$$

and

$$\phi_2^{(j)} = A^{(j)}X_1 + B^{(j)}X_2 + C^{(j)}Y_1 + D^{(j)}Y_2 \quad (13.18)$$

where

$$X_1 = \frac{sin\mu r}{r}, \quad X_2 = \frac{cos\mu r}{r}, \quad Y_1 = \frac{sinh\lambda r}{r}, Y_2 = \frac{cosh\lambda r}{r}$$

$$S_1 = S_2 = \tau_2[\mu^2 + \beta\Sigma_{2Mc}(1 + u)]$$

$$S_3 = S_4 = -\tau_2[\lambda^2 - \beta\Sigma_{2Mc}(1 + u)]$$

and

$$\mu^2 = \frac{1}{2}(a+b), \qquad \text{and} \quad \lambda^2 = \frac{1}{\mu^2}\left[\frac{\alpha u}{\tau_2} - \frac{\beta\Sigma_{2Mc}(1+u)}{\tau_1}\right],$$

$$a \equiv \left\{\left[\beta\Sigma_{2Mc}(1+u) + \frac{1}{\tau_1}\right]^2 + 4\left(\frac{\alpha u}{\tau_2} - \frac{\beta\Sigma_{2Mc}(1+u)}{\tau_1}\right)\right\}^{\frac{1}{2}}, b \equiv -\left(\beta\Sigma_{2Mc}[1+u] + \frac{1}{\tau_1}\right)$$

Thus $\mu = \mu(u)$, $\lambda = \lambda(u)$, that is, both the variables are functions of the control variable, and the coupling coefficients have to be written separately for each physical zone in which an optimal control is sought.

In the one-zone arrangement and two-zone arrangements, the boundary conditions are insufficient to determine an optimal condition. The three-zone arrangement has 24 unknown constants and 24 boundary conditions and can therefore be considered an optimal arrangement. This gives a $[u_{min}, u_{max}, u_{min}]$ optimal fuel placement arrangement shown in Fig. 13.8.

The critical determinant, set equal to zero for the optimal condition (minimum critical mass) can have only one variable for which the criticality search can be carried out. Thus, for a nuclear reactor, when one value is given for the fuel enrichment, the other value can be determined from this analysis.

First consider that the enrichment is specified in Zone 2, and it is required to determine the enrichment in Zones 1 and 3. The control problem is then: "given $u = u_M$, $r_1 < r < r_2$, find $u = u_m$ for $0 \le r \le r_1$ and $r_2 \le r \le R$." Alternately, we can have the control variables all given and carry out a search for the critical radius of the core.

The control variable u is defined as

$$u \equiv \frac{\Sigma_{2Fc}}{\Sigma_{2Mc}} = \frac{N_5\sigma_{2,a5}}{N_w\sigma_{2,aw}} = u^*\gamma$$

The optimization goal is to determine a Hamiltonian for which

$$H(y^*, u^*, x) \ge H(y^*, u, x)$$

where $u_{min} \le u \le u_{max}$. As earlier stated, this discrete form readily yields "optimal" discrete values of u constant in subdomains of the problem. Thus where g is minimum (either sign), u_{max} is applied for an extremum. The zeros of the switching function will thus determine the controls applied.

The criticality determinant for the three-zone problem has ten simultaneous algebraic equations. As in Chapter 5, since these equations are homogenous, they are solved by Cramer's rule. The determinant is set equal to zero, the dimensions are entered and for a maximum control u_{max} in the central zone, the minimum control u_{min} is found from the determinant. These give pairs $[u_{min}, u_{max}]$ of feasible solutions for the Kori-1 data considered by Lee (1973).

u_{max}	0.9	1.0	1.1	1.2	1.3	1.4	1.5
u_{min}	0.89	0.80	0.71	0.65	0.61	0.58	0.56

The combination $[u_{min} = 0.65, u_{max} = 1.2]$ gives a minimum-mass combination.

The minimum-mass optimal placement PMP formulation was reviewed for its elegance and its capability to solve optimal LPs also. A variational formulation based on the adjoint formulation for higher-order perturbation theory has also been used for fusion design optimization (Graca et al., 1988). The preferred optimization schemes at present are AI-based *meta*-heuristic schemes with comparatively less use of deterministic or stochastic perturbation for optimization.

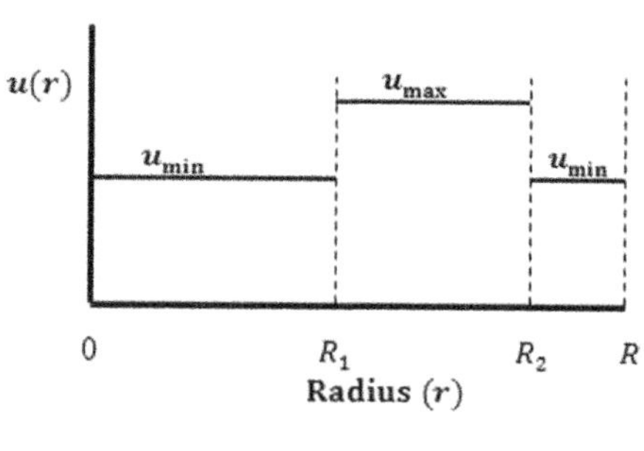

FIGURE 13.8 Optimal discrete (bang-bang) control.

13.6 Radiation detection or optimization

Radiation detection (Section 1.4.6) is used extensively for radiation monitoring of people, vehicles, cargo, and at airports to ensure security (Kouzes, Ely, Lintereur, Siciliano, & Woodring, 2009). Gamma radiation is monitored using scintillators such as NaI(TI) and for better resolution, HPGe solid state detectors, while neutron radiation is measured using gas proportional counters such as the He^3 and BF_3 detectors. Both gamma and neutron counting has to be done efficiently to be able to accurately classify materials and to raise alarms only when necessary.

The efficiency of radiation counters depend on a number of factors such as the moderation needed for obtaining a good signal from a gas proportional counter.

Consider, for example, an explosives detection system based on TNAA counting gammas with a NaI(TI) detector and neutrons with a BF_3 detector (Khan, Koreshi, & Yaqub, 2017).

Explosives such as TNT (trinitrotoluene) have carbon, oxygen, nitrogen, and hydrogen in varying fractions which allows their identification by neutron bombardment and subsequent analysis of the scattered and emitted signal.

Fast (14 MeV) neutron bombardment of hydrogen, carbon, oxygen, and nitrogen atoms produce the following reactions which produce neutrons and gammas:

$$612C + 01n(14\text{MeV}) \rightarrow 612C + 01n + \gamma \quad (4.4\text{MeV})$$
$$714N + 01n(14\text{MeV}) \rightarrow 714N + 01n + \gamma \quad (5.1\text{MeV})$$
$$816O + 01n(14\text{MeV}) \rightarrow 816O + 01n + \gamma \quad (6.1\text{MeV})$$

while thermal neutron bombardment reactions are significant for H and N in comparison with reactions for C and O producing characteristic gammas from hydrogen and nitrogen

$$11H + 01n(\text{thermal}) \rightarrow 12H + \gamma \quad (2.2\text{MeV})$$
$$714N + 01n(\text{thermal}) \rightarrow 715N + \gamma \quad (10.8\text{MeV})$$

For the (n,γ) reactions, the *disappearance* cross sections shown in Fig. 13.9 illustrate the extent to which thermal neutrons are beneficial in comparison with fast neutrons (KAERI Nuclear Data Center, 2019). Thus, the H and N thermal neutron (n,γ) reactions present an important method for the detection and characterization of explosives. Further, as there are no portable thermal neutron sources, the fast neutrons from a source such as Am-Be or Californium-252 will be required to be thermalized before activating a sample under investigation. This non-destructive method of characterization is called thermal neutron activation (TNA).

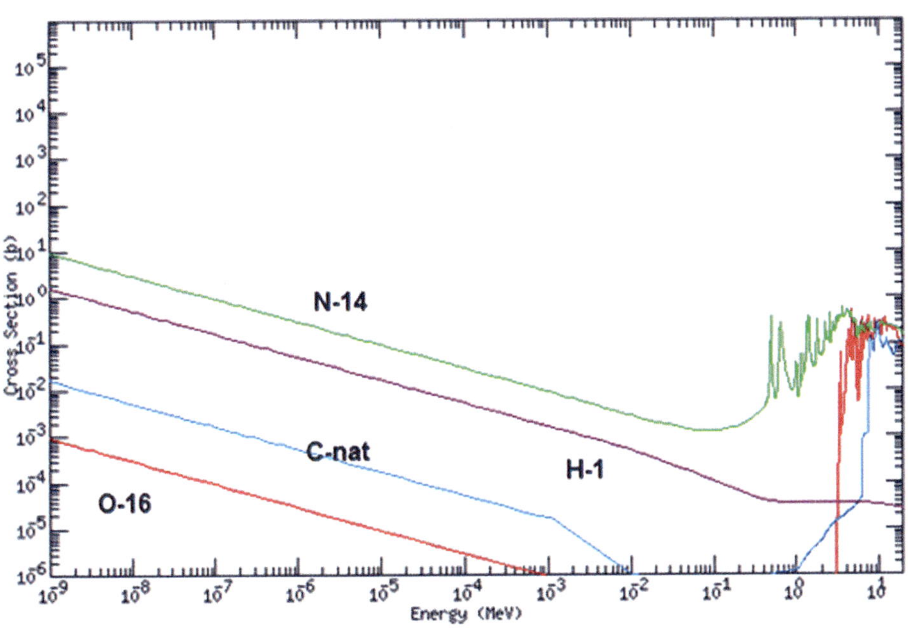

FIGURE 13.9 Disappearance cross sections.

For a sufficient signal to be produced from TNA, a detector needs to be optimized with a moderator. The design of the optimal moderator is obtained by optimization methods on GA, PSO, coupled GA-MCNP and AI methods described in the previous sections.

The functionals to maximize are the radiative capture reaction rates in the explosive for both hydrogen and nitrogen are $J_{H(n,\gamma)}$ and $J_{N(n,\gamma)}$ defined as

$$J_{H(n,\gamma)} = \iiint \Sigma^{H}_{n,\gamma}\phi(r, E, \Omega)dVdEd\Omega$$

$$J_{N(n,\gamma)} = \iiint \Sigma^{N}_{n,\gamma}\phi(r, E, \Omega)dVdEd\Omega.$$

The independent variables are the moderator material, its size and placement, the source energy spectrum, boron enrichment and gas pressure in the detector.

The optimization effort for moderator optimization is considerably less than that for a shielding optimization since one material is sufficient. Typical candidate moderators are low-Z materials such as wax, paraffin wax and polyethylene

The $B(n,\alpha)L_i$ reaction rate in a single BF_3 (cylindrical) detector with a Cf-252 source is shown in Table 13.5 for a wax moderator of thickness in the range 1–30 cm, showing an optimum thickness ~4 cm. The first five MCNP results have relative standard error within 2% while the last two are within 5% and 19%, respectively.

The energy spectrum of the neutron flux in the detector tube in shown in Fig. 13.10 for moderators of thickness 2, 4 and 6 cm together with the $B^{10}(n,\alpha)$ cross section. With an increase from a thickness of 2 to 4 cm, there is a slight increase in the low energy flux due to which there is an increase in the reaction rate shown in Table 13.5 while an

TABLE 13.5 $B(n,\alpha)Li$ reactions for a wax-moderated BF_3 detector.

ΔR (cm)	1	2	4	6	8	15	30
$B(n,\alpha)Li$	3.20627	8.96149	12.4770	9.72714	8.23782	1.31617	0.07004
10^{-6}/cm^3/s	0.0305	0.0204	0.0172	0.0200	0.0213	0.0489	0.1930

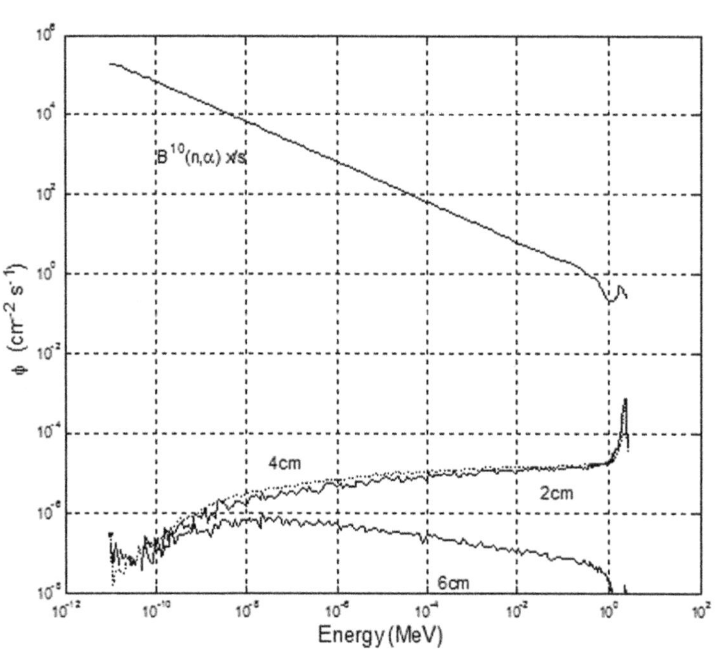

FIGURE 13.10 Neutron flux energy spectrum in the wax-moderated BF_3 detector.

TABLE 13.6 B(n, α) reaction rate as a function of pressure and ^{10}B weight fraction.

^{10}B in B	Gas density ρ_0 10^{-3} g/cm^3	10^{-5} (n, α) reactions cm^{-3} per source neutron s^{-1}		
		P = 1 atm $\rho=\rho_0$	P = 2 atm $\rho=2\rho_0$	P = 3 atm $\rho=3\rho_0$
0.2	2.7527×10^{-3}	1.15580	1.89182	2.4132
0.5	2.7406×10^{-3}	2.16983	3.07464	3.5792
0.9	2.7243×10^{-3}	3.89159	5.51797	6.4283

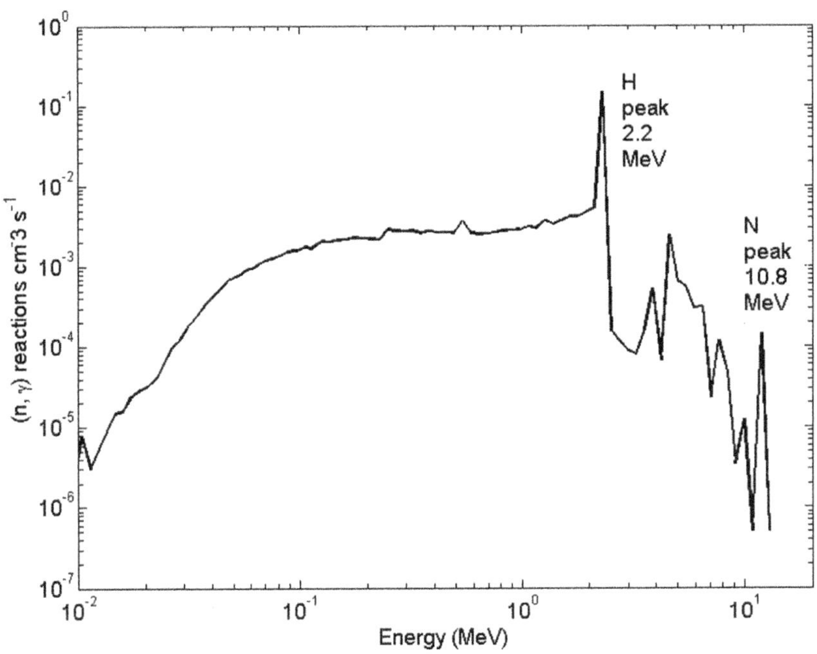

FIGURE 13.11 Characteristic H, N gamma peaks from trinitrotoluene.

increase in moderator thickness to 6 cm there is a reduction in the flux due to over-moderation which results in increased attenuation.

Simulations for polyethylene moderator shows an optimized thickness of 6 cm which is also reported as an experimental result in the literature (Waheed et al., 2017).

MCNP5 simulations were carried out to estimate the (n, α) reaction rate, at T = 300K as a function of pressure over the range 0.5–3 atm in a BF_3 tube and the isotopic enrichment of ^{10}B. The results of Table 13.6 show that the reaction rate is best for highest ^{10}B enrichment as well as for highest gas pressure giving a fourfold increase. While high enrichment is possible, there is limit on increasing gas pressure due to the high operating voltage that would be required at increased pressure.

In an explosives detection system, a source such as Californium-252 can be used to scan an area for mines by firing fast neutrons on the ground. Due to their penetrating power, the neutrons would undergo interactions with any subsurface material. In the case of concealed TNT, the back-scattered neutron and gamma radiation would be detected in the BF_3 and NaI (TI) detectors; from the resulting spectrum such as the pulse height of the gamma radiation shown in Fig. 13.11, it would be possible to estimate characterize an explosive.

13.7 Controller design optimization

Nuclear reactors have traditionally operated as baseload power plants with some maneuvering capability. Looking ahead, it is possible that a load-following capability might become necessary as renewables are added to electricity grids. Thus, a load-following mode would likely improve the operational costs. The technical issues with a strong bearing on load-following would be the speed by which temperatures and pressures in the moderator and coolant and the subsequent reactivity changes would be adequately controlled by the available mechanisms. This would mean that CR and poison management would require a careful re-appraisal. As an example, the ratio of gray to black CR would require re-adjustment. It is likely that with new modular reactors, load-following might not be as difficult as with a single large unit.

At present, countries such as France, with ~70% nuclear, and a few others have a significant nuclear share of electricity. For such countries, a load-following capability would become increasingly important.

One technical challenge for load-following would be an optimized controller which could respond in a *good* way. That translates into an optimization problem that requires minimization of a cost function that incorporates the overshoot, settling time, and stabilization time. Several optimization methods, such as GA, PSO, AI and fuzzy logic, have been designed and optimized. The Proportional Integral Derivative (PID) controller (Section 9.3) has been optimized for a load-following PWR using PSO (Mousakazemi, 2021) to minimize the overshoot, settling time and stabilization time. The agreement between the desired signal and the PID controller signal has been within a good range as confirmed by changes in the control rod movements, changes in precursor density, xenon concentration and coolant temperature.

In another load-following mode simulation of a PWR (Abdulraheem & Korolev, 2021), the nonlinear Integral sliding mode control has been used in combination with an optimal control Linear Quadratic Gaussian to show the robustness of the controller against disturbances incorporating the point kinetics equations with xenon reactivity. In simulation experiments the core is modeled with point kinetics with xenon reactivity feedbacks. The effectiveness of the hybrid controller is demonstrated by successful load-following under parametric variations. The issue of chattering disturbance, due to frequent switching of the controller, has been addressed by using a sliding mode controller for robust load-following in the presence of disturbances and uncertainties (Hui & Yuan, 2021). The proposed controller follows the load with a maximum power error of less than $\sim 10^4$W, compared with $\sim 5 \times 10^5$ W for a PID controller and 4×10^5 W for a conventional sliding mode controller.

There are several areas in nuclear engineering where optimization, particularly *meta*-heuristics are being applied and developed. Some of these areas have earlier been mentioned in chapter none. As stated in the introduction, optimization is indeed a vast area full of interesting ideas yet to be explored and developed.

Problems

1. Describe briefly how the micronuclear reactor described in Section 10.5.3 can be optimized in terms of weight and volume.
2. Describe a possible application of MC perturbation theory (Section 12.6) for calculating sensitivity coefficients of candidate materials in a shielding design optimization.
3. Considering the various breeder blankets for DEMO and the China Fusion Engineering Test Reactor (Section 13.3) describe how a blanket could be optimized for maximizing the TBR with a GA-MCNP coupled scheme.
4. Consider the composite gradient shield (Section 13.4) and write down a two-group diffusion model with variable atomic density (spatial dependence) for two candidate materials; iron and water to minimize the radiation dose across the shield.
5. With the two-group diffusion equation criticality equation (Section 5.3.2), show that a one-zone model does not have an optimal arrangement. Then consider the two-zone bare cylinder to find critical pairs for the fuel enrichment in each zone with the criticality determinant

$$\begin{vmatrix} S_{11}J_0(\mu R) & S_{13}I_0(\lambda R) \\ J_0(\mu R) & I_0(\lambda R) \end{vmatrix} = 0$$

$$\begin{vmatrix} -\mu_1 S_{11}J_1(\mu_1 r_1) & \mu_2 S_{21}J_1(\mu_2 r_1) & \mu_2 S_{22}Y_1(\mu_2 r_1) & \lambda_1 S_{13}I_0(\lambda_1 r_1) & -\lambda_2 S_{23}I_1(\lambda_2 r_1) & \lambda_2 S_{24}K_1(\lambda_2 r_1) \\ -\mu_1 J_1(\mu_1 r_1) & \mu_2 J_1(\mu_2 r_1) & \mu_2 Y_1(\mu_2 r_1) & \lambda_1 I_1(\lambda_1 r_1) & -\lambda_2 I_1(\lambda_2 r_1) & \lambda_2 K_0(\lambda_2 r_1) \\ S_{11}J_0(\mu_1 r_1) & -S_{21}J_0(\mu_2 r_1) & -S_{22}Y_0(\mu_1 r_1) & S_{13}I_0(\lambda_1 r_1) & -S_{23}I_0(\lambda_2 r_1) & -S_{24}K_0(\lambda_2 r_1) \\ J_0(\mu_1 r_1) & -J_0(\mu_2 r_1) & -Y_0(\mu_1 r_1) & I_0(\lambda_1 r_1) & -I_0(\lambda_2 r_1) & -K_0(\lambda_2 r_1) \\ 0 & S_{21}J_0(\mu_2 R) & S_{22}Y_0(\mu_2 R) & 0 & S_{23}I_0(\lambda_2 R) & S_{24}K_0(\lambda_2 R) \\ 0 & J_0(\mu_2 R) & Y_0(\mu_2 R) & 0 & I_0(\lambda_2 R) & K_0(\lambda_2 R) \end{vmatrix}$$

where, S_{ij} are the coupling coefficients for zone i and index j with

$$S_{11} = \tau_2(\mu^2 + \beta \mathrm{u}) = S_{21} = S_{22}, S_{13} = -\tau_2(\lambda^2 - \beta \mathrm{u}) = S_{23} = S_{24}$$

$$\mu^2 = \frac{1}{2\tau_1 L^2}\left[-(\tau_1 + L^2) + \sqrt{(\tau_1 + L^2)^2 + 4(k_\infty - 1)\tau_1 L^2}\right]$$

$$\lambda^2 = \frac{1}{2\tau_1 L^2}\left[(\tau_1 + L^2) + \sqrt{(\tau_1 + L^2)^2 + 4(k_\infty - 1)\tau_1 L^2}.\right.$$

Nomenclature

English Lower Case

e	enrichment
f	fraction
k_{eff}	effective multiplication factor
k_∞	infinite multiplication factor
n	neutron density
r	radius
u	control parameter

English Upper Case

C	precursor concentration
D	diffusion coefficient
D	diameter
E	energy
J	integral performance index
P	pitch
P	pressure
S	coupling coefficient
T	temperature
T_c	total cost

Greek Lower Case

β	delay fraction
ε	enrichment
λ	decay constant
ϕ	neutron flux
ρ	reactivity
σ	microscopic cross section
τ	cycle time

Greek Upper Case

$\mathscr{F}$	objective function
Σ_x	macroscopic cross section (reaction x)
$\mathbf{H}$	Hamiltonian
Ψ	auxiliary function

Abbreviations and acronyms

AI	Artificial Intelligence
ANN	Artificial Neural Networks
BOC	beginning of cycle
CR	control rod
EOC	end of cycle
GA	genetic algorithm

LFR Lead-cooled fast reactor
LP loading pattern
PPF peak power factor
PSO particle swarm optimization
PWR pressurized water reactor
SR safety rod
TH thermal hydraulics
TNAA thermal neutron activation analysis
WCLL water-cooled lithium lead

References

Abdulraheem, K. K., & Korolev, S. A. (2021). Robust optimal-integral sliding mode control for a pressurized water nuclear reactor in load following mode of operation. *Annals of Nuclear Energy, 158*, 108288.

Ahmad, A., & Ahmad, S. U. L. I. (2018). Optimization of fuel loading pattern for a material test reactor using swarm intelligence. *Progress in Nuclear Energy*.

Ahmad, S., Chang, B., Li, B., Yang, Q., & Liu, C. (2021). Mass optimization of the radiation shadow shield for space nuclear power system. *Progress in Nuclear Energy*.

Akbari, R., Rezaei Ochbelagh, D., Gharib, A., Maiorino, J. R., & D'Auria, F. (2020). Small modular reactor full scope core optimization using Cuckoo Optimization Algorithm. *Progress in Nuclear Energy*.

Bae, J. W., Betzler, B. R., Chandler, D., & Ilas, G. (2021). Automated fuel design optimization for high flux isotope reactor low enriched uranium core design. *EPJ Web Conference*.

Betzler, B. R., Chandler, D., Cook, D. H., Davidson, E. E., & Ilas, G. (2019). Design optimization methods for high-performance research reactor core design. *Nuclear Engineering and Design*.

Cai, Y., Hao, R., Yu, S., Wang, C., & Hu, G. (2020). Comparison of two multi-objective optimization methods for composite radiation shielding materials. *Applied Radiation and Isotopes: Including Data, Instrumentation and Methods for Use in Agriculture, Industry and Medicine*.

Cai, Y., Hu, H., Pan, Z., Hu, G., & Zhang, T. (2018). A method to optimize the shield compact and lightweight combining the structure with components together by genetic algorithm and MCNP code. *Applied Radiation and Isotopes: Including Data, Instrumentation and Methods for Use in Agriculture, Industry and Medicine*.

Cai, Y., Hu, H., Pan, Z., Sun, W., & Yan, M. (2018). Metaheuristic optimization in shielding design for neutrons and gamma rays reducing dose equivalent as much as possible. *Annals of Nuclear Energy*.

Chham, E., et al. (2021). Fuel reloads optimization for TRIGA research reactor using Genetic Algorithm coupled with neutronic and thermal-hydraulic codes. *Progress in Nuclear Energy*.

Cui, S., et al. (2017). Evaluation and optimization of tritium breeding, shielding and nuclear heating performances of the helium cooled solid breeder blanket for CFETR. *International Journal of Hydrogen Energy*.

Dai, T., Cao, L., He, Q., Wu, H., Feng, K., & Cao, Q. (2021). Neutronics optimization study on the first wall design of CFETR for TBR enhancement. *Fusion Engineering and Design, 172*.

De Klerk, E., et al. (1997). Optimization of nuclear reactor reloading patterns. *Annals of Operations Research*.

Domingos, R. P., Schirru, R., & Pereira, C. M. N. A. (2006). Particle swarm optimization in reactor core design. *Nuclear Science and Engineering: The Journal of the American Nuclear Society*.

Flammini, D., et al. (2020). Pre-analysis of the WCLL breeding blanket mock-up neutronics experiment at the frascati neutron generator. *Fusion Engineering and Design*.

Graca, G. T., Lewins, C. O., Parks, J. D., Graca, G. T., Lewins, C. O., & Parks, J. D. (1988). Optimization of blanket design for fusion reactors by higher order perturbation theory. *Nuclear Energy, 27*(121), 121–130.

Hu, G., Shi, G., Hu, H., Yang, Q., Yu, B., & Sun, W. (2020). Development of gradient composite shielding material for shielding neutrons and gamma rays. *Nuclear Engineering and Technology*.

Hui, J., & Yuan, J. (2021). Chattering-free higher order sliding mode controller with a high-gain observer for the load following of a pressurized water reactor. *Energy, 223*, 120066.

Kearney, J. P. M. E. A. (1973). Simulation and optimization techniques of nuclear in-core fuel management decisions. *Massachusetts Institute of Technology*.

Khan, H., Koreshi, Z. U., & Yaqub, M. (2017). The sensitivity studies of a landmine explosive detection system based on neutron back scattering using Monte Carlo simulation. *Nuclear Technology & Radiation Protection*.

Kim, D. W. (2019). Site-specific PSR process for Kori-1 permanent shut down NPP. In PBNC 2018—Pacific Basin nuclear conference, San Francisco, California, USA, 30 September - 4 October 2018. ISBN: 978-1-5108-7472-5. http://toc.proceedings.com/42142webtoc.pdf

Kim, B. S., & Moon, J. H. (2010). Use of a genetic algorithm in the search for a near-optimal shielding design. *Annals of Nuclear Energy, 37*(2), 120–129.

Kouzes R. T., Ely J. H., Lintereur A. T., Siciliano E. R. & Woodring M. L. BF_3 neutron detector tests. Pacific Northwest National Laboratory http://www.pnl.gov/main/publications/external/technical_reports/PNNL-19050.pdf (Date of access: 03/12/2016) (2009).

KAERI Nuclear Data Center. (2019). https://atom.kaeri.re.kr/nuchart/ last accessed 20 June 2021.

Kumar, A., & Tsvetkov, P. V. (2015). A new approach to nuclear reactor design optimization using genetic algorithms and regression analysis. *Annals of Nuclear Energy*.

Lee, C. K. (1973). Critical mass minimization of a cylindrical geometry reactor by two-group diffusion equation. *Journal of the Korean Nuclear Society*, *5*(2), 115–131.

Li, J., Jiang, K., Zhang, X., Nie, X., Zhu, Q., & Liu, S. (2016). Nuclear-thermal-coupled optimization code for the fusion breeding blanket conceptual design. *Fusion Engineering and Design*.

Luo, X., Zhang, X., Wang, S., Jiang, R., & Chen, H. (2021). Automated core design code development for a lead-cooled fast reactor and its core optimization. *International Journal of Energy Research*.

Moro, F., et al. (2020). Nuclear analysis of the water cooled lithium lead DEMO reactor. *Fusion Engineering and Design*.

Mousakazemi, S. M. H. (2021). Control of a pressurized light-water nuclear reactor two-point kinetics model with the performance index-oriented PSO. *Nuclear Engineering and Technology*.

Nissan, E. (2019). An overview of AI methods for in-core fuel management: Tools for the automatic design of nuclear reactor core configurations for fuel reload, (re)arranging new and partly spent fuel. *Designs*.

Norhasri, M. S. M., Hamidah, M. S., & Fadzil, A. M. (2017). Applications of using nano material in concrete: A review. *Construction and Building Materials*.

Rozin, S. G., Rubin, I. E., & Soboĺ, I. M. (1990). On multicriterial optimization in reactor problems using Monte Carlo calculations. *Progress in Nuclear Energy*.

Song, Y., et al. (2020). Research on fast intelligence multi-objective optimization method of nuclear reactor radiation shielding. *Annals of Nuclear Energy*.

Stewart, R. H., Palmer, T. S., & DuPont, Bryony (2021). A survey of multi-objective optimization methods and their applications for nuclear scientists and engineers. *Progress in Nuclear Energy*, *138*.

Waheed, A., et al. (2017). Optimization of moderator assembly for neutron flux measurement: Experimental and theoretical approaches. *Nuclear Science and Techniques*.

Wang, H., Gruenwald, J. T., Tusar, J., & Vilim, R. (2021). Moisture-carryover performance optimization using physics-constrained machine learning. *Progress in Nuclear Energy*.

Whyte, A., & Parks, G. (2021). Quantum annealing optimization of a heuristic suurogate model for PWR fuel loading. *EPJ Web Conference*.

Yilmaz, S., Ivanov, K., Levine, S., & Mahgerefteh, M. (2006). Genetic algorithm to optimize the UO2/Gd2O3 fuel pin designs in a pressurized water reactor. *Nuclear Technology*.

Zeng, K., Stauff, N. E., Hou, J., & Kim, T. K. (2020). Development of multi-objective core optimization framework and application to sodium-cooled fast test reactors. *Progress in Nuclear Energy*.

Chapter 14

Monte Carlo simulation in medical physics

14.1 Introduction

In medical physics (MP), radiation plays a very important role for diagnostics, pathological investigations and therapy, with Monte Carlo (MC) simulation appearing as a high-preference methodology just as it has become in nuclear engineering (Section 7.2). One measure of the number of research journals, the Web of Science Master Journal List (2021) lists 5,121 journals relevant to a search on "nuclear medicine", 2,278 for "medical physics" and 1,404 for "biomedical physics" out a total of 24,930 results. The growth in the number of papers in *Physics in Medicine and Biology* in the initial years of computing was more than 20-fold in the years from 1970 to 2005 (Rogers, 2006) with electron-photon transport as the foundation of the versatile Electron Gamma Shower (EGS) system of codes (Bielajew & Rogers, 1989; Hirayama, 2005) just as the Monte Carlo N-Particle (MCNP) code was in the nuclear community for criticality and nuclear reactor calculations. A search on "Monte Carlo Simulation" in the ScienceDirect database of 2021 lists 1,013 results for the number of research articles published in Medicine and Dentistry. In MC simulation, the MCNP code is extensively used in medical physics as given in the primer on *Medical Physics Calculations with MCNP* (Lazarine & Goorley, 2005; Reed, 2007) covering the modeling of tumors in tissue, phototherapy, and brachytherapy as well as an elaborate description on the use of the Zubal phantom (a Computerized Tomography(CT)-based torso and a CT-based head for use in MC simulations of realistic nuclear medicine (NM) imaging geometries).

For imaging, liquids containing radio-isotopes are injected in a patient and used as radiotracers giving images as they collect in the vicinity of an organ or move in an available channel. The images are produced from the emitted radiations and captured in a camera such as the Single-Photon Emission Computed Tomography (SPECT/CT) 3D scan. Similarly, a positron emitter such as fluorine-18 is used as a radiotracer in a Positron Emission Tomograhy (PET)/CT for imaging by the production of gamma rays formed by positron-electron annihilation.

Cancer cases globally are expected to grow from 14.1 million in 2012 to 24 million by 2035 with the top three: lung (Yu, Lewis, Luisa Trejos, Patel, & Malthaner, 2011), prostate and colorectal cases accounting for over 47% of all cancers in men (Blanchard et al., 2018; Rice et al., 2019). The treatments for cancer, in order of general preference, are chemotherapy, surgery, and radiation therapy. However, in the case of recurrent cancer, brachytherapy is the preferred treatment due to the localized and non-invasive effects.

Another use of radiation is for the treatment of cancer by applying radiation from an external source placed some distance away from a person or by placing seeds of a radio-isotope inside the body of a person close to the organ where a tumor is to be destroyed. An example of the former, called *tele-therapy* is the cobalt knife where gamma rays from a cobalt-60 source focus on an organ which is dangerous to be operated upon by surgery such as a tumor inside the brain. The treatment by placing a source within the body is called *brachytherapy* from the Greek work *brachus* meaning short.

In NM, radiation nuclide therapy is used for the imaging of the thyroid, bones, heart, liver and for the treatment of cancer in organs such as the lungs, breast, colon and rectum, prostate, stomach, and liver *etc*. Nuclear radiation, in the form of X-rays, gamma rays and charged particles, can damage the DNA (deoxyribonucleic acid) molecules of cancer cells that carry the genetic information and pass it from one generation to the next, to stop their further division.

More than two hundred radio-isotopes are used in NM common radio-isotopes include technetium ^{99m}Tc, iodine ^{125}I, palladium ^{103}Pd, iridium ^{192}Ir, cesium ^{137}Cs, and cobalt ^{60}Co. Important radionuclides for brachytherapy include the conventional encapsulated (Perez-Calatayud et al., 2019) low dose rate (LDR) sources (<200 cGy per hour) Ir-192, and high dose rate (HDR) sources (> 1200 cGy per hour). Permanent brachytherapy sources (Lechtman et al., 2013)

Nuclear Engineering. DOI: https://doi.org/10.1016/B978-0-323-90618-0.00014-4

with energy <50 keV, such as ^{125}I and ^{103}Pd are commonly used LDR sources, with typical implants of 50–80 metallic seeds encasing isotopes, used for the treatment of cancer.

There are several radiations which are used for radiotherapy according to the energy emitted (Section 1.2) and the ways in which they interact with matter (Section 1.3), for example α, have a short range and high energy with a high linear energy transfer (LET) which makes them favorable for short-range treatments. The β rays are, in contrast, longer range radiation so they would not be expected to kill tumor cells at a short distance; the same would hold true for gamma radiation.

14.1.1 The production of radio-isotopes

In nuclear power reactors, as fission takes place, there are several fission fragments such as in the case of fission from Uranium-235, molybdenum is produced in about 6.1% of the nuclear fission reactions. Molydenum (Mo-99) is unstable and decays with a half-life of 66 years which means that in 33 years its activity gets reduced to half of what it was; so it is fairly stable. During its decay, it produces technetium-99m, a metastable state, with a half-life of six hours which compared with Mo-99 is very short. If a patient had to undergo a procedure with technetium in the evening and the radio-isotope was procured the previous day then by the time of the procedure it would have only a quarter of its strength left. These are the typical problems associated with radio-isotopes; they are unstable.

The other places where radio-isotopes are produced are nuclear research reactors, cyclotrons and generators where a particle is bombarded and after transmutation, it becomes the desirable radio-isotope.

Coming back to technetium-99m, a radio-isotope used in over 85% of all 30 million patient examinations every year worldwide, is so useful because of the gamma radiation it gives off during its decay

$$Tc_{43}^{99m} \rightarrow Tc_{43}^{99} + \gamma \quad (140.5\text{keV}).$$

Technetium-99m (Tc-99m) is used in NM for the diagnostic scans of the organs. In the United States alone, with about 10 million procedures/year over half are cardiac-related. Together with Canada, Germany, France, Japan, Italy, Spain, Belgium, and UK, these countries account for about 9% of the procedures for diagnostic scans of a broad range of body parts and for the diagnoses of cancer, heart disease and neurological disorders including dementia and movement disorders (OECD/NEA, 2019).

The gamma radiation coming out of a technetium radio-isotope as it stays placed near or in an organ, is captured in a gamma camera such as the NM technologies SPECT/CT and PET/CT with little slits to differentiate parts of space where the intensity is greater than that in other parts. This spatial distribution in slices is put together into a 3D image.

Some other imaging techniques are based on strong magnetic fields and radio waves in the Magnetic Resonance Imaging (MRI) machine, and the X-ray based CT scans; these are fundamentally different from NM imaging technologies. The NM techniques are functional as they give information on an organ as well as on its activity while the non-NM technologies give anatomical information without organ activity.

The objective in this chapter is to understand how MC simulations play such an important role in estimating the effect of radiations so that a doctor may be able to plan a surgery for a patient according to the size and condition of the tumor.

The calculation schemes for MC simulations of α, β and γ radiation transport (Section 1.6) are suitable for medical physics because they can model a great number of relevant mechanisms. At the same time, it would be true to say that while MC simulations for nuclear engineering calculations may be excellent for performing dosimetry calculations and give very accurate estimates of the dose (Section 1.4) from a given amount of a radio-isotope, they may be insufficient to model the interaction of radiation with biology particularly the DNA which has a biological structure that is still not completely understood.

In spite of a remaining biology interface, MC simulation has been successfully applied for modeling ionizing radiation (IR) useful for dosimetry and therapy (Chatzipapas et al., 2020) by codes such as MCNP (Pelowitz et al., 2013), EGS, FLUKA (Fluctuating Kaskade), and PENELOPE (Penetration and ENErgy LOss of Positrons and Electrons (Archambault & Mainegra-Hing, 2015) at the tissue and organ level.

Efforts in the modeling of IR-induced DNA Damage, DNA Damage Response (DDR) and the DNA Repair mechanisms referred to as *computational radiobiology* is still in the developing and emerging stage. For simulations in biological matter at small scales (nm-μm), Monte Carlo Track Structure (MCTS) codes are used while the GEometry ANd Tracking code (Geant4)-DNA incorporates radiobiological functionalities (Chatzipapas et al., 2019; Engels et al., 2020) capable of modeling interactions of gold nanoparticles (GNPs) to quantify their dose enhancement. Inter-comparisons, using seven well-known MC codes have been carried out to estimate dose enhancement effects (Li et al., 2020) by irradiation of single GNPs of 50 and 100 nm diameter by X-rays generated from 50 to 100 kilovolts. The results have showed large variations and uncertainties in the electron energy spectra in the range 100–500 eV indicating that modeling effort is still required.

In this chapter, the discussion and use of MC will be largely connected with the calculation of the angular flux $\phi(\bar{r}, \hat{\Omega}, E, t)$ in an organ or a tissue of given composition. The appropriate reaction rate yielding the dose D in medically relevant quantities as the gray, rad, or rem will be used to make the connection with the dosage for a medical treatment. In MCNP, this is the F6 tally defined as

$$D = \frac{\rho}{m}\int dE \int dt \int dV \int d\Omega \quad \sigma_t(E)H(E)\phi(\bar{r}, \hat{\Omega}, E, t)\text{MeV/g}$$

where all quantities are defined in the preceding chapters of neutron diffusion, transport and MC simulation. Recall that a specific energy in MeV/g is converted to yield the dose rate Gy/h with 1 Gy (1 gray) = 1 J/kg = 100 rad. Also recall that the background radiation dose is of the order 2–3 mSv/year, or 0.2–0.3 rem/year (1 rem = 0.01 Sv = 10 mSv). A typical NM treatment could expose a person to a few millisievert while to destroy a tumor more than a few hundred Gy could be required. Regulators use the linear non-threshold dose model (LNT) according to which there is no safe level of radiation as cancers and heredity affects are understood to be stochastic, rather than deterministic, events. Therefore, there is no safe limit as regards the amount of radiation that can be considered safe. Technically, it is preferred that the amount of radiation should satisfy the ALARA (As Low As Reasonably Achievable) approach which amounts to saying that no amount of radiation is safe. Thus, a radiation of about 6 millisievert, though small, is still undesirable when accumulated in a hospital treatment.

The following sections consider MC applications for alpha and gamma radiations in some MP treatments.

14.1.2 Alpha radiation therapy

As mentioned above, the use of α particles in MP is preferable for short-range (70–100 μm) target tumor cells due to their high LET. The relative biological effectiveness (RBE) of α radiation is about 20 times higher than for photons, electrons and protons; thus, a higher dose is delivered. Ra-223, for example, with the reaction

$$Ra_{88}^{223} \rightarrow Rn_{86}^{219} + He_2^4$$

emits a 5.77 MeV α and a 0.141 MeV gamma ray, has a half-life of 11.4 days and is therefore a suitable radio-isotope for cancer treatment such as in boron neutron capture $B(n,\alpha)$ therapy (Section 1.2.4). Some high energy alpha emitters good for targeted radiotherapy are astatine At-211 ($E = 6$ MeV, $t_{1/2} = 7.2$ hours), bismuth Bi-213 ($E = 6$ MeV, $t_{1/2} = 46$ minutes) and actinium Ac-225 ($E = 5.9$ MeV, $t_{1/2} = 10$ days) (Zalutsky & Pruszynski, 2012).

A boron neutron capture therapy (BNCT) would function as depicted in Fig. 14.1. Consider a group of tumor cells in the vicinity of normal cells in the human brain with blood flowing and an intricate mechanism operating in the brain; so intricate that no physical surgery would be considered free of a big risk.

With BNCT, an amount of boron, in whatever form, would be injected in the brain and neutrons would be fired from outside initiating the nuclear reaction

$$B_5^{10} + n_0^1 \rightarrow He_2^4 + Li_3^7$$

which would give off α (He_2^4) radiation which would travel to the desired site and destroy the tumor. Of course, there are still many mechanisms by which the IR could find its way to the tumor cells rather than to the normal cells; that is

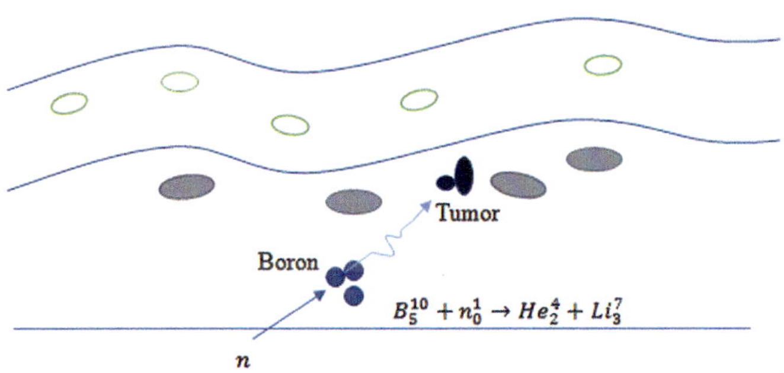

FIGURE 14.1 Boron neutron capture therapy.

another technology beyond the scope of this chapter but as a mention, it uses one of several schemes such as polymer coated nanoparticles on which a magnetic field is applied to recognize some characteristic of a tumor cell and a drug is released to penetrate the surface of the tumor cell.

BNCT has been assessed (Dymova, Taskaev, Richter, & Kuligina, 2020) in view of the development of new accelerators and has been found to be useful if performed alone or with chemotherapy or radiotherapy provided more selective boron delivery agents are developed to transport boron compounds into the brain, and an epithermal neutron beam with definite characteristics can be developed. MC simulations have been used with the Particle and Heavy Ion Transport Code System (Matsuya, Fukunaga, Omura, & Date, 2020; Sato et al., 2015) to show good agreement with in vitro experimental data for BNCT with Boron-10 concentrations of 10 ppm. Several MC simulations have been carried out for modeling the epithermal flux for BNCT treatment (Darda, Soliman, Aljohani, & Xoubi, 2020; Hassanein, Hassan, Mohamed, & Abou Manoury, 2018; Kasesaz, Khalafi, & Rahmani, 2014; Monshizadeh, Kasesaz, Khalafi, & Hamidi, 2015; Shaaban & Albarhoum, 2015) including the development of a head phantom for accurate calculations of the dose requirement (Bavarnegin, Khalafi, Sadremomtaz, & Kasesaz, 2016). To determine the place for a patient to be treated by BNCT, a MC simulation of the BAEC TRIGA reactor (Darda et al., 2020) using the OpenMC concludes that the thermal column with an epithermal flux of $\sim 10^9$ near the thermal column graphite is a suitable location. In another study, with a miniature source neutron reactor (MNSR), using MCNP4C (Monshizadeh et al., 2015), the thermal and epithermal neutron fluxes are reported as 1.39×10^9 n/cm^2/s and 0.635×10^9 n/cm^2/s, respectively, with a treatment time estimated to be about 70 minutes.

14.2 Brachytherapy

Consider now a brachytherapy procedure in which MC simulations are used to calculate the dose absorbed by a tissue in an organ.

An MC simulation could give useful information in the size, composition and configuration of GNPs for example, to maximize the dose since that is the ultimate goal of brachytherapy. The constraint is that there should be minimum collateral damage to the surrounding normal tissue and hence this could be formulated as an optimization problem amenable to the solutions described in Chapter 9, particularly the *meta*-heuristic applications discussed Chapter 13. In brachytherapy, there is still no consensus on what the optimal size, shape and distribution of GNPs should be. There is some experimental and pre-clinical evidence for mouse tumors showing a 1-year survival rate of 86% mice given a dose of 26 Gy with 1.9 nm intravenously administered GNPs versus 20% for tumors not laden with GNPs (Jain, Hirst, & O'Sullivan, 2012). This is the motivation for performing detailed MC simulations for brachytherapy to reduce the human exposure with maximum benefit of whatever radiation is administered.

14.2.1 Monte Carlo simulation in brachytherapy

MC simulation has been used in several studies for brachytherapy (Jangjoo, Ghiasi, & Mesbahi, 2019; Khan, Aziz, & Koreshi, 2019; Yu et al., 2017). For brachytherapy, radio-isotopes with energy <50 keV, such as ^{125}I and ^{103}Pd sources with typical implants of 50–80 metallic seeds encasing isotopes, are used as LDR therapies for the treatment of prostate cancer, uveal melanomas and brain tumors.

The dose is estimated by a number of computer codes such as EGSnrc, GEANT, PENELOPE, and MCNP based on MC methods.

In order to enhance the effectiveness of brachytherapy the injection of GNPs through fenestrations of cancer cells is being considered. The requirements of particle size necessitate the range down to nanoscales ($\sim 10^{-9}$ m) which compares with the diameter of an atom ($\sim 10^{-10}$ m). For MC simulations in brachytherapy, it has been demonstrated that the dose enhancement factor (DEF) depends on the source energy and concentration of GNP solution, while the size of GNPs, according to some research results, is not significant above the K-edge energy (Chatterjee et al., 2013; Jain et al., 2012; Jangjoo et al., 2019; Lechtman et al., 2011; Mesbahi, Jamali, & Gharehaghaji, 2013; Sharabiani, Vaezzadeh, & Asadi, 2016).

Several studies have thus used a homogenous model with considerable savings on the computational effort required for a full heterogeneous model, is used to extract crucial information on the DEF. Calculations are performed for the dose resulting from the presence of gold in small concentration, MC codes such as MCNP come with a capability of modeling very detailed heterogeneous configurations so this assumption is not a limitation of MC simulations.

The dose delivered to a prostate tumor. For example, by radio-isotopes ^{125}I, ^{103}Pd, and ^{131}Cs are assessed particularly in the context of dose enhancement in the presence of GNPs.

The radio-isotope cesium-131 (^{131}Cs) with ($E = 29 - 30.4$keV, $t_{1/2} = 9.7$d) is found to yield the required dose in the least less time compared with ^{125}I ($E = 35.49$keV, $t_{1/2} = 59.4$d) and ^{103}Pd ($E = 20.8$keV, $t_{1/2} = 17$d) due to its shorter half-life and higher energy.

Typically, iodine ^{125}I is used with a radiation dose of 145 Gy or more in accordance with the prescribed dose of 145 Gy suggested by the American Association of Physicists in Medicine Task Group 64 (Pons-Llanas et al., 2018; Yu et al., 1999).

One of the main advantages of ^{131}Cs is that it offers an initial dose rate of ~32 cGy/h at the periphery which is 1.5 and 4 times higher than that from ^{103}Pd and ^{125}I, respectively. This initial dose rate advantage is a vital radiobiological parameter for a tumor that is growing at a high speed in comparison with slow-growing tumors such as prostate adenocarcinomas.

The MC code MCNP5 has been used (Khan et al., 2019) to estimate the rose distribution in a coupled photon-electron simulation in the range 1 keV–100 MeV for prostate tumor brachytherapy. The radio-isotope sources considered are ^{125}I, ^{103}Pd, and ^{131}Cs in the form of "seeds" modeled as point sources.

As shown in Fig. 14.2, 98 ^{125}I, seeds each of activity 0.31 mCi and 115 ^{103}Pd seeds each of activity1.4 mCi were considered spatially distributed in tumor tissue. The placement of needles considers a number of factors including the location of vital organs surrounding the prostate, such as the bladder, urethra and rectum. For simulating the effect of ^{131}Cs, the ^{103}Pd seeds were replaced by ^{131}Cs seeds of the same initial activity and spatial distribution. The energy deposition track length and the pulse height tallies are both used for estimating energy deposition to get reliable estimates in case of a few interactions in a region of interest. The simulation is repeated with gold-tissue solution. The photon and electron data for both tissue and gold are based on Evaluated Nuclear Data File ENDF/B-VI (Release 8).

Several MC codes have been used and compared (Šídlová & Trojek, 2010) on the basis of calculating bremsstrahlung, energy deposition in matter, electron ranges and production of secondary electrons by gamma radiation.

The activity of a radio-isotope is $A(t) = A_o e^{-\lambda t}$ and the number of transformations in an interval $\tau = t_2 - t_1$ gives an estimate for the absorbed dose

$$\text{Abs Dose} = 1.6 \times 10^{-10} \frac{A_o}{\lambda} \frac{NE}{m} \text{Gy}$$

where A_o is the initial activity of a radio-isotope, N is the number of radio-isotope seeds, and E is the energy pulse height tally in MeV for a tumor of mass m grams.

Thus, the absorbed dose varies directly with the energy deposition and source, which in turn depends on the number of transformations, that is, the product of initial activity and half-life. The longer decay time for ^{125}I while faster decay rates of ^{103}Pd and ^{131}Cs give the number of transformations (integrated over time) of 2.74×10^{14}, 7.84×10^{13}, and 4.47×10^{13} for ^{125}I, ^{103}Pd, and ^{131}Cs. Thus, iodine has a slower build-up, but a higher dose from ^{125}I after ~6 months compared with the much faster effects of ^{103}Pd and ^{131}Cs. The absorbed dose calculated from a MC simulation is shown in Fig. 14.3 showing the extent of dose enhancement resulting from GNPs. This makes it possible to achieve an absorbed dose as high as about 300 Gy with a GNP assisted brachytherapy.

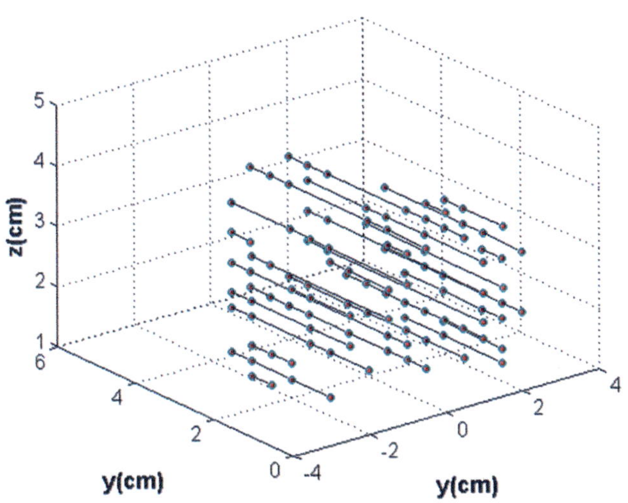

FIGURE 14.2 Placement of 29 needles of 115 seeds of ^{103}Pd/^{131}Cs in the x-y plane.

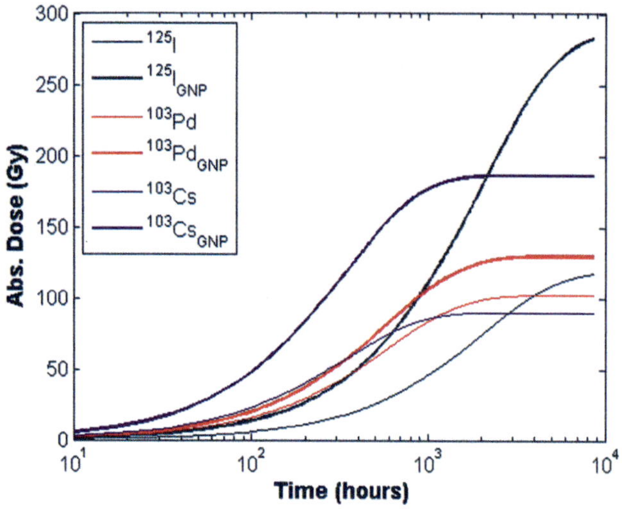

FIGURE 14.3 Absorbed dose (Gy) for ^{125}I, ^{103}Pd and ^{131}Cs.

Iodine ^{125}I starts with an activity 0.31 mCi, with a distribution shown in Fig. 14.2 while ^{103}Pd and ^{131}Cs both have initial activity 1.4 mCi. The results indicate that ^{131}Cs is clearly the best in terms of delivering the highest and fastest dose to the tumor, reaching 50 Gy in the first 10 days and 100 Gy when GNPs in a solution of 25 mg/g tissue are injected in the tumor while on a longer timescale, ^{125}I gives the highest dose.

The initial dose ^{131}Cs over the first ~300 hours is the highest exceeding 5.7 times and 1.5 times that of ^{125}I and ^{103}Pd respectively in the initial period and falling gradually about 300 hours for ^{103}Pd and about 400 hours for ^{125}I.

In MCNP the "detailed physics" simulation incorporates coherent (Thomson) scattering and fluorescent photons produced from photoelectric absorption. Electrons produced from photon collisions are transported in a "condensed history" method that accumulates the effects of many individual collisions into single steps sampled probabilistically.

The effects of such artifacts for electron transport have been investigated by inter-comparisons (Section 14.2) with other MC codes, such as EGSnrc, GEANT and PENELOPE codes.

In some cases "large discrepancies" (> 3%) have been found between MCNP5 dose distributions and the "reference codes" concluding that MCNP5 electron transport calculations are not accurate at all energies and in every medium by general clinical standards (Almansa, Guerrero, Al-Dweri, Anguiano, & Lallena, 2007; Archambault & Mainegra-Hing, 2015; Koivunoro et al., 2012). It can thus be anticipated that MCNP5 may differ due to its inadequate low-energy treatment of electron transport. The differences with are reported to have been reduced with improved electron transport in MCNP6.

These results are in line with the decay rates of the three isotopes which favor ^{125}I in terms of energy but result in a slow dose delivery. The dose enhancement with gold GNP-tissue found in these simulations is the effect of both source energy of each radio-isotope and the solution concentration. The photoelectric effect plays a dominant for high-Z materials, such as gold, for which the cross-section varies as $\mu_{PE} \sim \rho Z^3/E^3$ so that low energy and high-Z are desirable for dose enhancement which is localized to the tumor due to the short range of photoelectrons and Auger electrons in the surrounding medium which for electrons of energy 0.1 MeV is ~100 microns in water and ~15 microns in gold. Thus GNPs are used in thin layers of < 100 nm thickness to utilize the energy of photoelectrons in water.

14.2.2 Monte Carlo simulation to calculate energy deposition and dose distribution for brachytherapy

In continuation of the previous section, the dose distribution and the subsequent DEF by the use of gold, a high-Z biocompatible element, in solution are both estimated by MC codes as a function of source energy typical of brachytherapy sources (40 keV–1 MeV), solution concentration (5–25 mg Au/g H_2O) and solution placement (1–2 cm concentric shells).

Results from the MC simulation code MCNP5 are compared with other widely used MC codes such as PENELOPE and GEANT, to validate the dose estimates which may vary considerably due to artifacts and data libraries. MC perturbation estimates for radiation oncology, are yet to be used extensively to carry out sensitivity studies which can be used to obtain optimal experimental parameters such as radiation energy, concentration of gold in solution and the optimal distribution of material for maximizing an objective function of interest. MC simulation, using MCNP5, has been carried out for simulating coupled photon-electron radiation transport from X-rays emanating from a radiation source implanted in a cancer cell, modeled by a spherical water phantom, to estimate the energy deposited and the subsequent DEF using a water sphere of radius 15 cm to represent tissue.

Since the dose enhancement is mainly due to photoelectron production from gold, the use of water for soft tissue, as defined by the International Committee for Radiological Protection (ICRP) is justifiable due to similar density and photon interaction cross-sections. For carrying out sensitivity studies, the MC perturbation feature, with material perturbations, was used to sample derivatives in a single run which were used in a Taylor series to estimate subsequent dose and enhancement.

From the validation studies, good agreement is found between MCNP5 yields dose estimates, distributions and enhancements which values reported for the MC codes PENELOPE and GEANT with lower energy cutoffs for electron transport as compared with MCNP5. While the dose increases with source energy, the DEF was found to increase inversely with the source energy, for a given concentration, achieving a value as high as 1.8–2.5 for a source of 40 keV and concentration 5–10 mg Au/g H_2O.

This enhancement was found to occur at energies near the *K*- and *L*-shell electrons of gold and no significant enhancement was found at higher (MeV) energies. The validity of first- and second-order perturbation theory was confirmed for small changes ($\sim 2\%$) in the material density of gold-water solution enabling a single run to estimate the DEF from concentrations in the range 5–25 mg Au/g H_2O with a considerable computational speedup.

For calculations for dose distribution MC methods and general purpose codes, such as EGSnrc, GEANT, PENELOPE, and MCNP have been extensively used for simulating the transport of radiation from radiation seeds in the medium consisting of tissue, and material in the vicinity of the cancer. These codes have produced fast and accurate results which have been experimentally validated and benchmarked.

At the small scales of nanotechnology, comparable with the atomic scale, the behavior of materials and subsequently their electrical and thermal properties also depend on size and shape. Nanotechnology, driven largely by the opportunities in electronics and semiconductors, has emerged as one of the frontiers of science capable of revolutionizing technology in areas including communications and computing, materials and medicine.

The source energy of interest is 40 keV–1 MeV typical of brachytherapy sources, while the gold particles in solution form with water were assumed to comprise a homogeneous mixture with concentrations ranging form 5–25 mg Au/g H_2O. While MCNP has the capability of modeling very detailed heterogeneous configurations, this paper considers a homogeneous model solely for the purpose of demonstrating (1) results from benchmarked problems and (2) for obtaining MC sensitivity estimates for a "bulk" material to demonstrate validity of MC perturbation and quantifying increased computational efficiency.

MC perturbation analysis has applications in brachytherapy for estimating dose perturbations when implements are present in the vicinity of an organ receiving radiation from an implanted source (see e.g., Yu et al., 1999). However, it is yet to be used to simulate brachytherapy studies and can provide great computational efficiency leading to optimal designs based on "best" experimental parameters such as radiation energy, concentration of gold in solution and material placement for maximizing an objective function of interest.

The MC code MCNP5 is used to carry out a coupled photon-electron simulation of radiation transport in the range 1 keV–100 MeV to estimate the dose distribution, from a point isotropic photon source typical of brachytherapy sources (30 keV for ^{125}I –1.25 MeV for ^{60}Co) located at the origin of a sphere, in concentric shells of water of thickness 0.5 mm up to a radius of 15 cm. A solution of gold and water is then considered in the shells located 1–2 cm from the center and the F6 and *F8 tallies are used for phantom dosimetry. Photon and electron data for air, water and gold are based on ENDF/B-VI (Release 8). In the MCNP "detailed physics" simulation, coherent (Thomson) scattering is included and fluorescent photons produced from photoelectric absorption are included and electrons produced from photon collisions are transported in a "condensed history" method that accumulates the effects of many individual collisions into single steps sampled probabilistically.

Perturbation algorithms in MC simulation, developed extended the capability of MC methods to sensitivity studies and optimization. The change in a response function such as dose *D*, due to a variation in an independent parameter, such as material density ρ, such as the dose, expressed as a Taylor series

$$D(\rho) = D(\rho_o) + D'(\rho_o)\delta\rho + \frac{1}{2!}D''(\delta\rho)^2 + \cdots$$

can be used with first- and second-order derivatives D' and D'' from a single run. Thus, when the change in a parameter is small enough to be of the order of the statistical uncertainty of a MC estimate, then the difference from two independent runs may mask the actual difference. The energy deposition for a "perturbed" design (D^{new}) can thus be estimated from a Taylor series in terms of a "reference" design (D^{ref}).

The material used in this work is water which has density $\rho = 1.04$ g/cm^3 very close to that of soft tissue $\rho = 1.04$ g/cm^3 (with a four-component simplified composition hydrogen (H), carbon (C), nitrogen (N) and oxygen (O) with weight percentages: 10.454%, 22.663%, 2.49%, 63.525%, respectively) and the photon electric cross-sections for both are almost indistinguishable so that conclusions drawn for energy deposition would hold for both materials to a reasonably acceptable order. While water phantoms are used for representing configurations similar to soft tissue, full anthropomorphic phantoms are also used to carry out detailed simulation for the dose.

To compare MCNP5 results for energy deposition from a mono-energetic point isotropic source at the center of a sphere of radius 15 cm, the F6 tally in 0.5 mm thick concentric shells gives the energy distribution. When normalized to a reference dose from a 1 MeV source, it is seen that the lowest energy, 15 keV, has an energy deposition about 65% of the reference which falls off rapidly to less than 10% in 1.5 cm of water. With a further increase in energy, the relative intensities drop but the deposition sustains to greater depths which is less important in brachytherapy than to give a localized deposition from a low-energy source.

The DEF shown in Fig. 14.4 gives the extent and magnitude as a function of source energy and concentration for E = 40 keV, 50 keV, 90 keV, 1 MeV for varying concentrations 5, 10, 25 mg Au/g H_2O confirming the trend of direct variation of DEF with concentration and inverse variation with source energy. The spatial effect of high energy photons extends further in the medium as compared with low-energy photons as observed from results in Fig. 14.5 when the 1 MeV photons are able to persist to longer distances with a rising slope for GNP size.

MC simulations have also been used for ophthalmic brachytherapy with ^{125}I with mean energy 35.49 keV with a full three-dimensional heterogeneous model simulating 10^7-10^9 histories for 50 nm GNPs (Asadi et al., 2015) estimating the DEF for 7, 10, 18 and 30 mg GNP concentration/g as varying from 1.9, 2.2, 3.2 to 4.6 for the tumor phantom compared with 1.9, 2.3, 3.3 and 4.8 in the water phantom, respectively, which indicates that the tumor and water phantoms are very similar.

Preliminary MC simulations with a homogenous mixture model for a water phantom with 10^6 simulations gives, for 40 keV and concentrations of 5, 10, 25 mg/g maximum DEFs of 1.8, 2.5. 4.6.

Mesbahi et al. (2013) report DEF from 1.4 to 3.7 with the highest DEF for 90 keV; they also concluded, as did Lechtman et al. (2011), that the effect of GNP size was not considerable while concentration and energy were important. Of their simulations carried out for 7 and 18 mg Au/g H_2O for 30, 50, 100 nm GNPs and energies between 50 to 120 keV , they reported highest DEF of 3.5 at 90 keV followed by 3.0 at 50 keV both for 30 nm GNPs and 18 mg/g concentration.

For high energy sources (Banoqitah & Djouider, 2016) a maximum DEF of 1.45 is found within the tumor when implanted with 70 mg/g Gd for the lowest energy (^{192}Ir). Thus MC simulations find that the DEF varies directly with concentration and inversely with source energy, for example, for a concentration of 30 mg/g, it decreases from 1.24 to 1.09 when the photon energy increases from 0.380 MeV (^{192}Ir) to 1.20 MeV (^{60}Co).

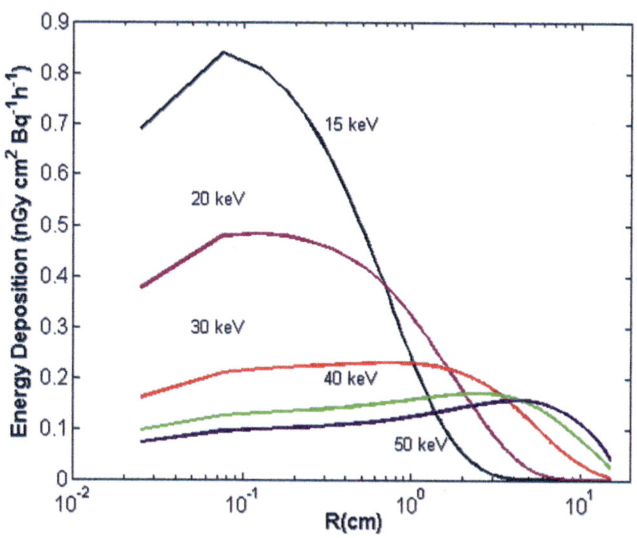

FIGURE 14.4 Energy deposition for mono-energetic photons of energy E: 15 keV–1 MeV.

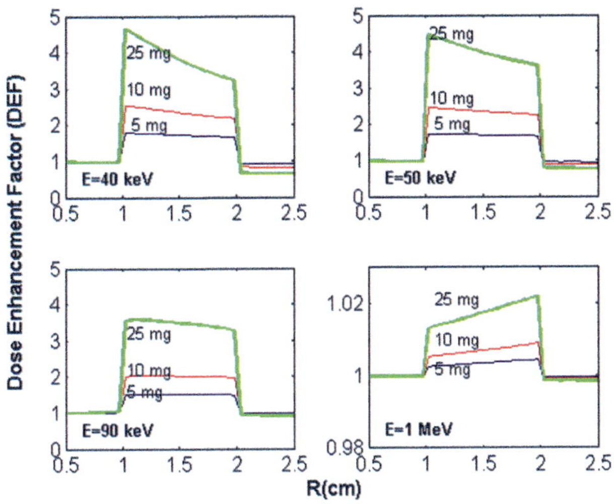

FIGURE 14.5 Dose enhancement factor (DEF) versus distance.

When full heterogeneities are modeled, MC sensitivity analysis can be useful for calculating sensitivity coefficients for the design optimization of GNP size with varying concentration.

Nomenclature

English

m	mass
t	time
D	dose
E	energy
H	heating number (MeV/collision)
N	number of particles simulated
V	volume
Z	atomic number

Greek

μ_{PE}	attenuation coefficient (photoelectric)
ρ	density
σ_t	microscopic total cross-section
ϕ	scalar flux scored over volume (track length/volume) cm^{-2}
Ω	solid angle

Abbreviations

BAEC	Bangladesh Atomic Energy Commission
DEF	dose enhancement factor
GNP	gold nanoparticle
Gy	gray
HDR	high dose rate
LDR	low dose rate
TRIGA	Training Research Isotopes General Atomics

References

Almansa, J. F., Guerrero, R., Al-Dweri, F. M. O., Anguiano, M., & Lallena, A. M. (2007). Dose distribution in water for monoenergetic photon point sources in the energy range of interest in brachytherapy: Monte Carlo simulations with PENELOPE and GEANT4. *Radiation Physics and Chemistry*.

Archambault, J. P., & Mainegra-Hing, E. (2015). Comparison between EGSnrc, Geant4, MCNP5 and Penelope for mono-energetic electron beams. *Physics in Medicine and Biology*.

Asadi, S., Vaez-zadeh, M., Masoudi, S. F., Rahmani, F., Knaup, C., & Meigooni, A. S. (2015). Gold nanoparticle-based brachytherapy enhancement in choroidal melanoma using a full monte carlo model of the human eye. *Journal of Applied Clinical Medical Physics/American College of Medical Physics.*

Banoqitah, E., & Djouider, F. (2016). Dose distribution and dose enhancement by using gadolinium nanoparticles implant in brain tumor in stereotactic brachytherapy. *Radiation Physics and Chemistry.*

Bavarnegin, E., Khalafi, H., Sadremomtaz, A., & Kasesaz, Y. (2016). Construction of a head phantom for mixed neutron and gamma field dosimetry in TRR. *Measurement: Journal of the International Measurement Confederation.*

Bielajew, A. F., & Rogers, D. W. O. (1989). *Lecture notes: Electron Monte Carlo simulation.* Canada: National Research Council of Canada Ottawa.

Blanchard, P., et al. (2018). Patient-reported health-related quality of life for men treated with low-dose-rate prostate brachytherapy as monotherapy with 125-iodine, 103-palladium, or 131-cesium: Results of a prospective phase II study. *Brachytherapy.*

Chatterjee, D. K., et al. (2013). Convergence of nanotechnology with radiation therapy-insights and implications for clinical translation. *Translational Cancer Research.*

Chatzipapas, K. P., et al. (2019). Quantification of DNA double-strand breaks using Geant4-DNA. *Medical Physics.*

Chatzipapas, K. P., et al. (2020). Ionizing radiation and complex dna damage: Quantifying the radiobiological damage using monte carlo simulations. *Cancers.*

Darda, S. A., Soliman, A. Y., Aljohani, M. S., & Xoubi, N. (2020). Technical feasibility study of BAEC TRIGA reactor (BTRR) as a neutron source for BNCT using OpenMC Monte Carlo code. *Progress in Nuclear Energy.*

Dymova, M. A., Taskaev, S. Y., Richter, V. A., & Kuligina, E. V. (2020). Boron neutron capture therapy: Current status and future perspectives. *Cancer Communications.*

Engels, E., et al. (2020). Advances in modelling gold nanoparticle radiosensitization using new Geant4-DNA physics models. *Physics in Medicine and Biology.*

Hassanein, A. M., Hassan, M. H., Mohamed, N. M. A., & Abou Mandour, M. A. (2018). An optimized epithermal BNCT beam design for research reactors. *Progress in Nuclear Energy.*

Hirayama, H., et al. (2005). *The EGS5 code system.*

Jain, S., Hirst, D. G., & O'Sullivan, J. M. (2012). Gold nanoparticles as novel agents for cancer therapy. *British Journal of Radiology.*

Jangjoo, A. G., Ghiasi, H., & Mesbahi, A. (2019). A Monte Carlo study on the radio-sensitization effect of gold nanoparticles in brachytherapy of prostate by 103Pd seeds. *Polish Journal of Medical Physics and Engineering.*

Kasesaz, Y., Khalafi, H., & Rahmani, F. (2014). Design of an epithermal neutron beam for BNCT in thermal column of Tehran research reactor. *Annals of Nuclear Energy.*

Khan, H., Aziz, U., & Koreshi, Z. U. (2019). Radiation dose delivered by 125I, 103PD and 131CS and dose enhancement by gold nanoparticle (gnp) solution in prostate brachytherapy: A comparative analysis by monte carlo simulation. *IIUM Engineering Journal.*

Koivunoro, H., Siiskonen, T., Kotiluoto, P., Auterinen, I., Hippeläinen, E., & Savolainen, S. (2012). Accuracy of the electron transport in mcnp5 and its suitability for ionization chamber response simulations: A comparison with the egsnrc and penelope codes. *Medical Physics.*

Lazarine, A. D., & Goorley, T. (2005). Medical physics calculations with MCNP^{TM}: A primer. *Transactions of the American Nuclear Society.*

Lechtman, E., Chattopadhyay, N., Cai, Z., Mashouf, S., Reilly, R., & Pignol, J. P. (2011). Implications on clinical scenario of gold nanoparticle radiosensitization in regards to photon energy, nanoparticle size, concentration and location. *Physics in Medicine and Biology.*

Lechtman, E., et al. (2013). A Monte Carlo-based model of gold nanoparticle radiosensitization accounting for increased radiobiological effectiveness. *Physics in Medicine and Biology.*

Li, W. B., et al. (2020). Intercomparison of dose enhancement ratio and secondary electron spectra for gold nanoparticles irradiated by X-rays calculated using multiple Monte Carlo simulation codes. *Physica Medica.*

Matsuya, Y., Fukunaga, H., Omura, M., & Date, H. (2020). A model for estimating dose-rate effects on cell-killing of human melanoma after boron neutron capture therapy. *Cells.*

Mesbahi, A., Jamali, F., & Gharehaghaji, N. (2013). Effect of photon beam energy, gold nanoparticle size and concentration on the dose enhancement in radiation therapy. *BioImpacts.*

Monshizadeh, M., Kasesaz, Y., Khalafi, H., & Hamidi, S. (2015). MCNP design of thermal and epithermal neutron beam for BNCT at the Isfahan MNSR. *Progress in Nuclear Energy.*

OECD/NEA. (2019). *The supply of medical isotopes: An economic diagnosis and possible solutions.*

Pelowitz, D. B., et al. (2013). MCNP6 user's manual. *La-Cp-13–00634.*

Perez-Calatayud, J., et al. (2019). GEC-ESTRO ACROP recommendations on calibration and traceability of LE-LDR photon-emitting brachytherapy sources at the hospital level. *Radiotherapy and Oncology: Journal of the European Society for Therapeutic Radiology and Oncology.*

Pons-Llanas, O., et al. (2018). Permanent seed implant brachytherapy in low-risk prostate cancer: Preoperative planning with 145 Gy vs real-time intraoperative planning with 160 Gy. *Reports of Practical Oncology and Radiotherapy.*

Reed, A. (2007). *Medical physics calculations with MCNP: A primer.* Boston, MA: Los Alamos Natl. Lab. X-3 MCC.

Rice, S. R., et al. (2019). A phase II study evaluating bone marrow-sparing, image-guided pelvic intensity-modulated radiotherapy (IMRT) with Cesium-131 brachytherapy boost, adjuvant chemotherapy, and long-term hormonal ablation in patients with high risk, nonmetastatic prostate can. *American Journal of Clinical Oncology: Cancer Clinical Trials.*

Rogers, D. W. O. (2006). Fifty years of Monte Carlo simulations for medical physics. *Physics in Medicine and Biology.*

Sato, T., et al. (2015). Overview of particle and heavy ion transport code system PHITS. *Annals of Nuclear Energy.*

Shaaban, I., & Albarhoum, M. (2015). Design calculation of an epithermal neutronic beam for BNCT at the Syrian MNSR using the MCNP4C code. *Progress in Nuclear Energy*.

Sharabiani, M., Vaez-zadeh, M., & Asadi, S. (2016). Size dependence of GNPs dose enhancement effects in cancer treatment – Geant4 and MCNP code. *Radiotherapy and Oncology: Journal of the European Society for Therapeutic Radiology and Oncology*.

Šídlová, V., & Trojek, T. (2010). Testing Monte Carlo computer codes for simulations of electron transport in matter. *Applied Radiation and Isotopes: Including Data, Instrumentation and Methods for use in Agriculture, Industry and Medicine*.

Yu, E., Lewis, C., Luisa Trejos, A., Patel, R. V., & Malthaner, R. A. (2011). Lung cancer brachytherapy: Robotics-assisted minimally invasive approach. *Current Respiratory Medicine Reviews*.

Yu, P. C., et al. (2017). CT-based MCNPX dose calculations for gynecology brachytherapy employing a Henschke applicator. *Radiation Physics and Chemistry*, *140*, 392–397.

Yu, Y., et al. (1999). Permanent prostate seed implant brachytherapy: Report of the American Association of Physicists in Medicine Task Group No. 64. *Medical Physics*.

Zalutsky, M. R., & Pruszynski, M. (2012). Astatine-211: Production and availability. *Current Radiopharmaceuticals*.

Index

Note: Page numbers followed by "*f*," "*t*," and "*b*" refer to figures, tables, and boxes, respectively.

O

S

T

CPI Antony Rowe
Eastbourne, UK
April 20, 2022